ALDO CASSOL STAMM   WOLFGANG DRAF (Eds.)

Micro-endoscopic Surgery of the Paranasal Sinuses and the Skull Base

Springer

*Berlin
Heidelberg
New York
Barcelona
Hong Kong
London
Milan
Paris
Singapore
Tokyo*

Aldo Cassol Stamm   Wolfgang Draf (Eds.)

Micro-endoscopic Surgery of the Paranasal Sinuses and the Skull Base

With 711 Figures and 101 Tables

Aldo Cassol Stamm, M.D.
Director of ENT São Paulo Center
Rua Afonso Braz 525 – Cj 13
04511-010 São Paulo
Brazil

Wolfgang Draf, M.D. FRCS
Direktor der Klinik für Hals-Nasen-Ohren-Krankheiten,
Kopf-, Hals- und Plastische Gesichtschirurgie
Klinikum Fulda
Pacelli-Allee 4
36043 Fulda
Germany

ISBN 3-540-66629-X   Springer Verlag Berlin Heidelberg New York

Cataloging-in-Publication Data applied for.
Die Deutsche Bibliothek – CIP-Einheitsaufnahme
Micro-endoscopic surgery of the paranasal sinuses and the skull base / ed.: Aldo Stamm ; Wolfgang Draf. – Berlin ; Heidelberg ; New York ; Barcelona ; Hongkong ; London ; Mailand ; Paris ; Singapur ; Tokio ; Springer, 2000
   ISBN 3-540-66629-X

This work is subject to copyright. All rights are reserved, whether the whole or part of the material is concerned, specifically the rights of translation, reprinting, reuse of illustrations, recitation, broadcasting, reproduction on microfilm or in any other way, and storage in data banks. Duplication of this publication or parts thereof is permitted only under the provisions of the German copyright Law of September 9, 1965, in its current version, and permission for use must always be obtained from Springer-Verlag. Violations are liable for prosecution under the German Copyright Law.

Springer-Verlag is a company in the BertelsmannSpringer publishing group
© Springer-Verlag Berlin Heidelberg 2000
Printed in Germany

The use of general descriptive names, registered names, trademarks, etc. in this publication does not imply, even in the absence of a specific statement, that such names are exempt from the relevant protective laws and regulations and therefore free for general use.

Cover Design: d&p, 69121 Heidelberg, Germany
Production: ProEdit GmbH, 69126 Heidelberg, Germany
Printed on acid free paper   SPIN 10746412   24/3135 Re – 5 4 3 2 1 0

Preface

The diagnosis and treatment of the diseases of the nose, paranasal sinuses and skull base have changed dramatically in the past few years, transforming rhinology in one of most exciting and attractive medical fields. The increasing advance in this area has addressed the ENT physicians to keep up with the expanding information in the high tech area of sinus surgery and its expansion into skull base surgery.

The intention of this book is to present in a unique way basic information on Anatomy, Endoscopy, Rhinomanometry, Imaging, Allergy, Nasosinus Infection, and Polyposis followed by by clinical and surgical chapters written by some of the most experienced rhinosurgeons around the world. We belive that this book will be of value for all levels of Otolaryngology, from house officers to experienced surgeons and, although principally for Otolaryngologists, Radiologists, Pathologists, Maxillofacial surgeons, Ophthalmologists, Neurosurgeons, and Infectious Diseases Specialists may find the book of value because of its overlap with their interests.

We would like to show that, both from the technical and conceptual points of view, success can be achieved using different techniques and philosophies. Certainly, both endoscopic instrumentation and the surgical microscope have proved to be of great assistance in nasal and sinus surgery, and now image systems are providing further progress.

There are several textbooks dealing with nasal sinus surgery, but we believe this to be most comprehensive and up-to-date. Many monographies represent the efforts and beliefs of one or two individuals, whereas we present viewpoints and techniques of many individuals around the world with sometimes quite divergent opinions. We belive that physicians in Europe, Asia, Africa, Australia and South America want to have the latest information about these different techniques and philosophies, even if they currently use one particular method. We believe that this is also true for many physicians in the United States. The book may also stimulate more international scientific cooperation.

Most important is, that this publication is not only dealing with inflammatory diseases of the nose and sinuses as the majority of the existing textbooks, but also with surgery of malformations, of lacrimal drainage system, orbital and optic nerve decompression, duraplasty, tumor surgery and management of complications.

We hope, this real global reflection of Rhinology can serve as guide and stimulus for daily practise and building interdisciplinary, international scientific links.

February 2000

ALDO CASSOL STAMM
WOLFGANG DRAF

Acknowledgments

There are no adequate words to express our gratitude to everyone who helped with the creation of this book. Although it's not possible to list all the wonderful individuals who contributed, we want to take this opportunity to give special recognition and thanks to a few who made unique or particularly generous contributuions.

Without the encouragement and direction of Ms. Gabriele M. Schroeder and Mr. Antonio Tendero from Springer-Verlag Heidelberg, the project would never have become a reality.

To Dr. Lee A. Harker, our eternal gratitude for his generous time and effort with the orthographic corrections of the text, and for his valuable suggestions. It is a privilege to share his friendship.

A very special thanks to Dr. Shirley Pignatari for her generous help in every step of this project.

We also wish to thank our colleagues at the São Paulo ENT Center, Dr. André Bordasch, for his valuable contribution, and also Dr. Cleonice H. Watashi, Dr. Glaura Ferreira, Dr. José Figueiredo, Dr. Levon Mekhitarian, Dr. Luis A.S. Freire, Dr. Moacir Pozzobon, Dr. Monica M. Miyake and Dr. Paulo F. Malheiros. Our thanks also to the doctors and staff of the ENT-Department Klinikum Fulda.

We are indebted to the expertise of Suely Knoll for preparing the majority of the drawings.

Thanks to Dr. Alfredo Herrera and Dr. Ricardo Cohen for their generosity in helping with revision work.

Our final and special thanks to Professor João C. Navarro, for his collaboration in providing continuous knowledge of the anatomic basis upon the development of the sinus surgery is based.

February 2000

ALDO CASSOL STAMM
WOLFGANG DRAF

Contents

Part I
Anatomy, Pathophysiology, Diagnosis, Staying and Medical Management

Chapter 1
**Development and Use of Microscopic and Endoscopic Surgery
of the Nose and Sinuses** 1
JOACHIM HEERMANN and RALF HEERMANN

Chapter 2
**Surgical Anatomy of the Nose, Paranasal Sinuses
and Pterygopalatine Fossa** 17
JOÃO A. CALDAS NAVARRO

Chapter 3
Endoscopic Diagnosis: Nasosinus Disorders 35
EULALIA SAKANO

Chapter 4
Rhinomanometry 43
JUAN EUGENIO SALAS GALICIA

Chapter 5
Imaging of the Nose and Paranasal Sinuses 53
RAINER G. HAETINGER

Chapter 6
Immunology and Allergy of the Nose and Sinuses 83
MONICA AIDAR MENON MIYAKE and PEDRO CAVALCANTI FILHO

Chapter 7
Bacterial, Viral and Fungal Infections of the Nose and Paranasal Sinuses 97
PAUL VAN CAUWENBERGE and DE-YUN WANG

Chapter 8
Nasal Polyps 103
MIRKO TOS, PER L. LARSEN, KNUD LARSEN, and PER CAYÉ-THOMASEN

Chapter 9
Sinusitis: Medical Management 127
LUC L.M. WECKX and PAULO A.L. PONTES

Chapter 10
Biomechanical Effects of the Ethmoidal Isthmi on the Pathogenesis of Chronic Paranasal Sinusitis 135
Malte Erik Wigand

Chapter 11
Pre-operative Staging, Grading and Post-Operative Follow-Up of Chronic Inflammatory Sinus Disease 141
Valerie J. Lund

Chapter 12
Surgical "Grading" System for Inverting Papilloma 147
Aldo Cassol Stamm

Chapter 13
Teaching and Learning in Endonasal Sinus Surgery 153
Rainer Keerl, Rainer Weber, and Wolfgang Draf

Part II
Surgica Treatment of the Nose and Paranasal Sinus Diseases

Chapter 14
Micro-endoscopic Surgery of the Turbinates and Nasal Septum 161
Aldo Cassol Stamm, Elisabeth Araujo,
Antonio Douglas Menon, and Karen Borne Teufert

Chapter 15
Surgical Approaches to Sinusitis: Overview of Techniques 179
Renato Roithmann, Ian Witterick, Michael Hawke,
and Aldo Cassol Stamm

Chapter 16
High-Risk Areas in Endoscopic Sinus Surgery 195
Toshio Ohnishi

Chapter 17
Micro-endoscopic Surgery of the Paranasal Sinuses 201
Aldo Cassol Stamm

Chapter 18
Combined Microscopic and Endoscopic Surgery of the Ethmoid Sinus 237
Heinrich H. Rudert and Christoph G. Mahnke

Chapter 19
Micro-endoscopic Surgery of the Maxillary Sinus 249
Rodolfo Arias, Hector Ariza, Ivan Correa,
and Aldo Cassol Stamm

Chapter 20
Endonasal and External Micro-endoscopic Surgery of the Frontal Sinus 257
Wolfgang Draf, Rainer Weber, Rainer Keerl,
Jannis Constantinidis, Bernhard Schick, and Anjali Saha

Chapter 21
Mini-anterior and Combined Frontal Sinusotomy and Drilling of Nasofrontal Beak 279
Kanit Muntarbhorn and Sanguansak Thanaviratananich

Chapter 22
Recurrence of Polyposis: Risk Factors, Prevention, Treatment and Follow-Up 287
Pierre Rouvier and Roger Peynegre

Chapter 23
What Is the Place of Endonasal Surgery in Fungal Sinusitis? 309
Valerie J. Lund

Chapter 24
Combined Fiberoptic Headlight and Endoscopic Sinus Surgery with Adjunct Use of Middle-Meatal Stenting 315
Paul H. Toffel

Chapter 25
Sinus Mucosal Wound Healing Following Endoscopic Sinus Surgery: Mucosal Preservation 323
Hiroshi Moriyama

Chapter 26
Wound Healing after Endonasal-Sinus Surgery in Time-Lapse Video: a New Way of Continuous In Vivo Observation and Documentation in Rhinology 329
Rainer Weber, Rainer Keerl, Andreas Huppmann, Wolfgang Draf, and Anjali Saha

Chapter 27
Pediatric Endoscopic Endonasal Sinus Surgery 347
Gerald Wolf

Chapter 28
Micro-endoscopic Sinus Surgery in Children 357
Shirley Shizue Nagata Pignatari and Aldo Cassol Stamm

Chapter 29
Revision Endoscopic Sinus Surgery 371
Jean-Michel Klossek

Chapter 30
Power Instrumentation in Rhinologic and Endoscopic Sinus Surgery 377
Daniel G. Becker and David W. Kennedy

Chapter 31
Image-Guided Surgical Navigation in Functional Endoscopic Sinus Surgery 387
Winston C. Vaughan and Frederick A. Kuhn

Part III
Transnasal Micro-Endoscopic Advanced Surgery

Chapter 32
Severe Epistaxis: Micro-endoscopic Surgical Techniques 393
Aldo Cassol Stamm, Glaura Ferreira,
João A. Caldas Navarro and Luiz A. Silva Freire

Chapter 33
Choanal Atresia: Transnasal Micro-endoscopic Surgery 405
Aldo Cassol Stamm, Levon Mekhitarian,
and Shirley Shizue Nagata Pignatari

Chapter 34
Endoscopic Transnasal Dacryocystorhinostomy 415
Gustavo A. Riveros-Castillo and Alfredo Campos

Chapter 35
Endonasal Surgery of the Lacrimal System 425
Joachim Heermann and Ralf Heermann

Chapter 36
Endoscopic Transnasal Orbital Decompression 433
Peter H. Hwang and David W. Kennedy

Chapter 37
Endoscopic Optic-Nerve Decompression in Traumatic Optic Neuropathy 441
Luis Alfonso Parra Duque

Chapter 38
Cerebrospinal Fluid Rhinorrhea – Transnasal Micro-endoscopic Surgery 451
Aldo Cassol Stamm and Luiz A. Silva Freire

Chapter 39
Endoscopic Repair of Cerebrospinal-Fluid Rhinorrhea 465
Alfredo Herrera, Emiro Caicedo

Chapter 40
**Endonasal Micro-endoscopic Surgery of Nasal
and Paranasal-Sinuses Tumors** 481
Wolfgang Draf, Bernhard Schick, Rainer Weber,
Rainer Keerl, and Anjali Saha

Chapter 41
Micro-endoscopic Surgery of Benign Sino-Nasal Tumors 489
Aldo Cassol Stamm, Cleonice Hirata Watashi,
Paulo Fernando Malheiros, and Lee Alan Harker

Chapter 42
Juvenile Nasopharyngeal Angiofibroma – Transantral Microsurgical Approach 515
Alejandro E. Terzian

Chapter 43
Endoscopic Approach to Lesions of the Anterior Skull Base and Orbit 529
Dharambir S. Sethi

Chapter 44
**Endonasal, Microscopic, Trans-Septal, Sphenoidal Approach
to the Sella and Parasellar Lesions** 543
RICARDO SERGIO COHEN, ALDO CASSOL STAMM, and ANDRÉ BORDASCH

Chapter 45
Transnasal Endoscopic Surgery of Sellar and Parasellar Regions 555
ALDO CASSOL STAMM, ANDRÉ BORDASCH, EDUARDO VELLUTINI,
and FELIX PAHL

Chapter 46
Midfacial Degloving – Microsurgical Approach 569
ALDO CASSOL STAMM, MOACIR POZZOBON,
and OSWALDO L. MENDONÇA CRUZ

Chapter 47
Complications of Micro-endoscopic Sinus Surgery 581
ALDO CASSOL STAMM

Subject Index 595

Contributors

ELISABETH ARAUJO, M.D.
Associate Professor,
Departament of Medicine, Federal University of Rio Grande do Sul,
Porto Alegre, Brasil

RODOLFO ARIAS A, M.D.
Associated Professor,
Division of Otorhinolaryngology, Nueva Granada Military Hospital,
Hospital Militar Central,
Santafé de Bogotá, Colombia

HECTOR ARIZA, M.D.
Division of Otorhinolaryngology,
National University of Colombia,
Santafé de Bogotá, Colombia

DANIEL G. BECKER, M.D.
Assistant Professor,
Division of Facial Plastic & Reconstrutive Surgery,
Department of Otorhinolaryngology, Head and Neck Surgery,
University of Pennsylvania
Philadelphia, USA

ANDRÉ BORDASCH, M.D.
Center of Otorhinolaryngology and Audiology of São Paulo,
Hospital Prof. Edmundo Vasconcelos,
São Paulo, Brasil

PAUL VAN CAUWENBERGE, M.D.,PhD.
Professor and Chairman,
Department of Otorhinolaryngology,
University Hospital Ghent,
Ghent, Belgium

PEDRO CAVALCANTI FILHO, M.D.
Assistant Professor,
Division of Otorhinolaryngology,
University of Rio Grande do Norte,
Natal, Brasil

Emiro Caicedo, M.D.
Department of Otolaryngology, Hospital and Clinics San Rafael,
Santafé de Bogotá, Colombia

Alfredo Campos, M.D.
Department of Otolaryngology, Hospital and Clinics San Rafael,
Nueva Granada Military University,
Santafé de Bogotá, Colombia

Jannis Constantinidis, M.D.
Department of Otorhinolaryngology Diseases
Head-Neck and Facial Plastic Surgery,
Communication Disorders-Klinikum Fulda,
Department of ENT, University of the Saarland,
Homburg, Germany

Ricardo S. Cohen, M.D.
Director,
Center of Otorhinolayngology of Tucumán,
San Miguel de Tucumán, Argentina.

Ivan Correa, M.D.
Department of Otorhinolaryngology,
Javeriana's University,
Santafé de Bogotá, Colombia

Oswaldo Laercio Mendonca Cruz, M.D., PhD.
Inviting Professor,
Department of Otolaryngology and Comunication Disorders,
Federal University of São Paulo,
São Paulo, Brasil

Wolfgang Draf, M.D., Med.Sci., F.R.C.S.
Director,
Department of Otorhinolaryngology, Head-Neck and Facial Plastic Surgery,
Communication Disorders-Kinikum Fulda,
Fulda, Germany

Glaura Pimentel Ferreira, M.D.
Center of Otorhinolaryngology and Audiology of São Paulo,
Hospital Prof. Edmundo Vasconcelos,
São Paulo, Brasil

Luiz A Silva Freire, M.D.
Center of Otorhinolaryngology and Audiology of São Paulo,
Hospital Prof. Edmundo Vasconcelos,
São Paulo, Brasil

Rainer Guilherme Haetinger, M.D.
Chief,
Division of Head and Neck Radiology,
MED, IMAGEM, Hospital Beneficência Portuguesa,
São Paulo, Brasil

LEE ALAN HARKER, M.D.
Deputy Director,
Boys Town National Research Hospital,
Vice Chairman,
Department of Otolaryngology and Human Communication,
Creighton University School of Medicine,
Omaha, USA

MICHAEL HAWKE, M.D., F.R.C.S.
Professor,
Department of Otolaryngology and Pathology,
Saint Joseph´s Health Centre,
University of Toronto,
Toronto, Ontario, Canada

JOACHIM HEERMANN, M.D.
Professor and Chairmam,
Department of Otorhinolaryngology, Alfried Krupp Hospital,
Essen, Germany

RALF HEERMANN, M.D.
Deparment of Otorhinolaryngology,
University of Hannover,
Hannover, Germany

ALFREDO HERRERA, M.D.
Assistant Professor,
Department of Otorhinolaryngology, Hospital and Clinics San Rafael,
University Militar Nueva Granada,
Santafé de Bogotá, Colombia

ANDREAS HUPPMANN, M.D.
Department of Internal Medicine, Klinikum Fulda,
Fulda, Germany

PETER H. HWANG, M.D.
Assistant Professor,
Department of Otolaryngology, Head and Neck Surgery,
Oregon Health Sciences, University of Portland,
Portland, USA

RAINER KEERL, M.D.
Senior Staff,
Department of Otolaryngology, Head-Neck and Facial Plastic Surgery,
Communication Disorders, Klinikum Fulda,
Fulda, Germany

DAVID W. KENNEDY, M.D.
Professor and Chair,
Department of Otorhinolaryngology, Head and Neck Surgery,
University of Pennsylvania Health System,
Philadelphia, USA

JEAN MICHEL KLOSSEK, M.D.
Professor,
Department of Otorhinolaryngology, University of Poitiers,
Service ORL et Chirurgie Cervico,
Faciale, Hopital Jean Bernard,
Poitiers, France

FREDERICK A. KUHN, M.D., F.A.C.S.
Director
Georgia Rhinology & Sinus Center,
Georgia Ear Institute at Memorial Medical Center,
Savannah, USA

PER L. LARSEN, M.D.
ENT Department, Gentofte Hospital,
University of Copenhagen,
Copenhagen, Denmark

KNUD LARSEN, M.D.
ENT Clinic,
Esbjerg Hospital,
Esbjerg, Denmark

VALERIE J. LUND, M.D., F.R.C.S.
Professor of Rhinology,
Institute of Laryngology and Otology University College,
London, United Kingdom

PAULO F. MALHEIROS, M.D.
Center of Otorhinolaryngology and Audiology of São Paulo,
Hospital Prof. Edmundo Vasconcelos,
São Paulo, Brasil

CHRISTOPH G. MAHNKE, M.D.
Department of Otorhinolaryngology, Head and Neck Surgery,
University of Kiel,
Kiel, Germany

LEVON MEKHITARIAN NETO, M.D.
Center of Otorhinolaryngology and Audiology of São Paulo,
Hospital Prof. Edmundo Vasconcelos,
São Paulo, Brasil

ANTONIO DOUGLAS MENON, M.D.
Center of Otorhinolaryngology and Audiology of São Paulo,
Hospital Prof. Edmundo Vasconcelos,
Hospital Sírio Libanês,
São Paulo, Brasil

MONICA MENON MIYAKE, M.D.
Division of Allergy
Center of Otorhinolaryngology and Audiology of São Paulo,
Hospital Prof. Edmundo Vasconcelos,
Hospital Sírio Libanês,
São Paulo, Brasil

HIROSHI MORIYAMA, M.D.
Professor and Chairman,
Department of Otorhinolaryngology,
The Jikei University School of Medicine,
Tokio, Japan

KANIT MUNTARBHORN, M.D., F.R.C.S.
Associate Professor,
Department of Otolaryngology, Faculty of Medicine, Ramathibodi Hospital,
Mahidol University,
Bangkok, Thailand

JOÃO A. CALDAS NAVARRO, D.D.S.
Professor and Chairman,
Department of Anatomy, Dental School,
University of São Paulo,
Bauru, Brasil

TOSHIO OHNISHI, M.D.
Department of Otolaryngology,
St. Luke's International Hospital,
Tokio, Japan

FELIX PAHL, M.D., PhD.
Assistant Professor,
Department of Neurosurgery,
University of São Paulo,
São Paulo, Brasil

LUIS ALFONSO PARRA DUQUE, M.D.
Director,
Otorhinolaryngological Medical Group, Clínica de Occidente,
Cali, Colombia

ROGER PEYNEGRE, M.D.
Chairman,
Department of Otorhinolaryngology, Head and Neck Surgery,
CHIC Hospital,
Creteil, France

SHIRLEY S. NAGATA PIGNATARI, M.D., PhD.
Visiting Professor,
Division of Pediatric Otolaryngology, Federal University of São Paulo,
Center of Otorhinolaryngology and Audiology of São Paulo,
Hospital Prof. Edmundo Vasconcelos,
São Paulo, Brasil

PAULO AUGUSTO DE LIMA PONTES, M.D., PhD.
Full Professor,
Department of Otolaryngology and Comunication Disorders,
Federal University of São Paulo,
São Paulo, Brasil

MOACIR POZZOBON, M.D.
Center of Otorhinolaryngology and Audiology of São Paulo,
Hospital Prof. Edmundo Vasconcelos,
São Paulo, Brasil

GUSTAVO A. RIVEIROS C., M.D.
Associate Professor,
Department of Otorhinolaryngology, Nueva Granada Military University,
Chairman,
Department of Otolaryngology, Hospital and Clinics San Rafael,
Santafé de Bogotá, Colombia

RENATO ROITHMANN, M.D., PhD.
Professor,
Division of Otorhinolaryngology, Brazilian Luteran University,
Canoas, Brasil
Associate Scientific Staff, Mount Sinai Hospital,
Toronto, Canada

PIERRE ROUVIER, M.D.
Chairman,
Department of Otolaryngology, Head and Neck Surgery,
J.Imbert Hospital,
Arles, France

HEINRICH H. RUDERT, M.D.
Professor and Chairman,
Department of Otorhinolaryngology, Head and Neck Surgery,
University of Kiel,
Kiel, Germany

ANJALI SAHA, M.D.
Visiting Doctor,
Department of Otolaryngology, Head, Neck and Facial Plastic Surgery,
Communication Disorders, Klinikum Fulda,
Fulda, Germany
Hospital & Research Centre,
Calcutta, Índia

EULALIA SAKANO, M.D.
Director,
Rhinology Division,
Department of Ophtalmology-Otorhinolaryngology,
State University of Campinas (UNICAMP),
Campinas, Brasil

JUAN EUGENIO SALAS GALICIA, M.D.
Professor,
Department of Otorhinolaryngology,
University of Veracruz, "Adolfo Ruiz Cortinez",
National Medical Center,
Vera Cruz, México

BERNHARD SCHICK, M.D.
Department of Otolaryngology,
Head, Neck, Facial Plastic Surgery and Communication-Disorders,
Klinikum Fulda
Fulda, Germany

DHARAMBIR S. SETHI, M.D., F.R.C.S., F.A.M.S.
Consultant & Acting Head,
Department of Otolaryngology,
Singapore General Hospital,
Republic of Singapore

ALDO CASSOL STAMM, M.D., PhD.
Director,
Center of Otorhinolaryngology and Audiology of São Paulo,
Hospital Prof. Edmundo Vasconcelos,
Hospital Sirio Libanes,
São Paulo, Brasil

KAREN BORNE TEUFERT, M.D.
Center of Otorhinolaryngology and Audiology of São Paulo,
Hospital Prof. Edmundo Vasconcelos,
São Paulo, Brasil

ALEJANDRO ERNESTO TERZIAN, M.D.
Associate Professor, Head,
Rhinology Division, Department of Otorhinolaryngology,
School of Medicine, University of Buenos Aires,
Buenos Aires, Argentina

SANGUANSAK THANAVIRATANANICH, M.D.
Assistant Professor,
Department of Otorhinolaryngology, School of Medicine,
Khon Kaen University,
Khon Kaen, Thailand

PER CAYÉ THOMASEN, M.D.
Department of Otorhinolaryngology,
Gentofte Hospital,
University of Copenhagen,
Copenhagen, Denmark

MIRKO TOS, M.D., PhD.
Professor and Chairman,
Department of Otorhinolaryngology,
Gentofte Hospital,
University of Copenhagen,
Copenhagen, Denmark

PAUL H. TOFFEL, M.D.
Clinical Professor,
Department of Otolaryngology Head & Neck Surgery,
University of Sourthern California, School of Medicine,
Los Angeles, USA

Winston C. Vaughan, M.D.
Georgia Rhinology & Sinus Center,
Georgia Ear Institute,
Savannah, USA

Eduardo Vellutini, M.D., PhD.
Assistant Professor,
Department of Neurosurgery,
University of São Paulo,
São Paulo, Brasil

De-Yun Wang, M.D., PhD.
Director,
Rhinology and Clinical Immunology Laboratory Division,
Department of Otorhinolaryngology, University Hospital of Ghent,
Ghent, Belgium

Cleonice H. Watashi, M.D., PhD.
Center of Otorhinolaryngology and Audiology of São Paulo,
Hospital Prof. Edmundo Vasconcelos,
São Paulo, Brasil

Luc Louis Maurice Weckx, M.D., PhD.
Associate Professor & Head,
Division of Pediatric Otolaryngology,
Department of Otolaryngology and Comunication Disorders,
Federal University of São Paulo,
São Paulo, Brasil

Malte Erik Wigand, M.D, PhD
Professor and Chairman,
Department of Otorhinolaryngology,
University of Erlangen, Nürnberg,
Erlangen, Germany

Ian Witterick, M.D., F.R.C.S.
Professor,
Department of Otolaryngology,
University of Toronto,
Department of Otolaryngology, Mount Sinai Hospital, Toronto,
Toronto, Ontario, Canada

Rainer Weber, M.D.
Professor,
Department of Otolaryngology,
Otto-von-Guericke University,
Magdeburg, Germany,
Staff Member, Department of Internal Medicine, Klinikum Fulda,
Fulda, Germany

Gerald Wolf, M.D.
University Professor,
Department of ENT, Head & Neck Surgery,
University Hospital Graz,
Graz, Austria

Part I

Anatomy, Pathophysiology, Diagnosis, Staying and Medical Management

Chapter 1
Development and Use of Microscopic and Endoscopic Surgery of the Nose and Sinuses 1
Joachim Heermann and Ralf Heermann

Chapter 2
Surgical Anatomy of the Nose, Paranasal Sinuses and Pterygopalatine Fossa 17
João A. Caldas Navarro

Chapter 3
Endoscopic Diagnosis: Nasosinus Disorders 35
Eulalia Sakano

Chapter 4
Rhinomanometry 43
Juan Eugenio Salas Galicia

Chapter 5
Imaging of the Nose and Paranasal Sinuses 53
Rainer G. Haetinger

Chapter 6
Immunology and Allergy of the Nose and Sinuses 83
Monica Aidar Menon Miyake and Pedro Cavalcanti Filho

Chapter 7
Bacterial, Viral and Fungal Infections of the Nose and Paranasal Sinuses 97
Paul Van Cauwenberge and De-Yun Wang

Chapter 8
Nasal Polyps 103
Mirko Tos, Per L. Larsen, Knud Larsen, and Per Cayé-Thomasen

Chapter 9
Sinusitis: Medical Management 127
Luc L.M. Weckx and Paulo A.L. Pontes

Chapter 10
Biomechanical Effects of the Ethmoidal Isthmi on the Pathogenesis of Chronic Paranasal Sinusitis 135
Malte Erik Wigand

Chapter 11
**Pre-operative Staging, Grading and Post-Operative Follow-Up
of Chronic Inflammatory Sinus Disease** 141
Valerie J. Lund

Chapter 12
Surgical "Grading" System for Inverting Papilloma 147
Aldo Cassol Stamm

Chapter 13
Teaching and Learning in Endonasal Sinus Surgery 153
Rainer Keerl, Rainer Weber, and Wolfgang Draf

Development and Use of Microscopic and Endoscopic Surgery of the Nose and Sinuses

Joachim Heermann and Ralf Heermann

Introduction

Endonasal surgery parallels the historical development of stapes surgery. The first era of stapes extraction began with Kessel's report in 1876 [46] and ended with its condemnation by Siebenmann in 1900 [65]. Endonasal ethmoidal surgery was developed approximately 25 years later [6, 7, 20, 21, 43, 45, 47, 51, 52, 57, 66]. Because the risk was judged to be too high (as in stapes surgery), endonasal surgery was temporarily abandoned by most surgeons within 25 years and was continued by only a few [5, 22, 23, 25, 60], including four generations of physicians in the Heermann family at our hospital.

Endoscopes were introduced in sinus surgery by Hirschmann in 1903 [39]. They were tried in our hospital in the 1950s and 1960s but, because of the poor light from the bulb, we also used mirrors for control of the maxillary sinus and nasofrontal ducts.

Binocular microsurgery of the ear without coaxial light was introduced by Holmgren (in Sweden) in 1928. Within 55 years after its condemnation, stapes surgery had achieved worldwide acceptance. This occurred because of newly developed antibiotics and the binocular Zeiss operating microscope, which featured coaxial illumination.

In 1957, Hans Heermann [24] demonstrated the use of the binocular Zeiss operating microscope with coaxial illumination for endonasal removal of the uncinate process and the ethmoidal cells. In the subsequent years, all endonasal surgery (Fig. 1.1) – including septal operations [28, 30, 36] and 14,000 ethmoid, maxillary (Figs. 1.2–1.4), sphenoid, and frontal sinus operations – have been performed with the operating microscope [30]. In 1974, we advocated [30] discontinuing Caldwell-Luc surgery. Extensive large upper (and, if necessary, lower) fenestrations gave enough exposure for surgery. The larger the fenestrations, the fewer the recurrences. Our tech-

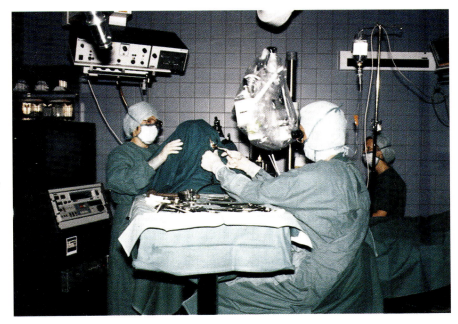

Fig. 1.1. Patient in semi-Fowler's position. An adjustable headrest is necessary for changes of the patient's position by a nurse. To facilitate focusing, a microscope placed in a balanced, movable position is directed by head motions of the surgeon. The surgeon can relax both elbows and has both hands free for surgery sitting in a convenient position. The view of the anatomy is the same in the office and the operating room. Less bleeding occurs and less suction results in less damage to the mucosa, shorter surgical procedure, less complications and better results.

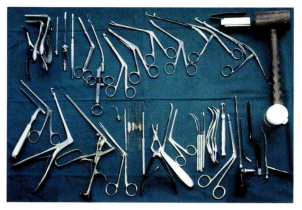

Fig. 1.2. Top row – three nasal speculas, needles for local anesthesia, septal knife, elevator, turbinate scissors and snare, Bruening forceps, Craig forceps, Struyken forceps, straight ethmoid punch, angulated Heermann punchs, bone compress, hammer. Bottom row – hollow chisel with handle, double spoon forceps, Hajek punch, choanal atresia chisel punch, lacrimal sac cannula, Bowman probes 1-4 with drill, neurosurgical scalpel, Luer power forceps, Watson-Williams rasp, double spoon forceps, angulated ear curettes, Ritter frontal sinus probes 1-4, hollow chisel for osteoplastic frontal sinus surgery, toothed forceps, endoscope, monopolar and bipolar cautery, suction.

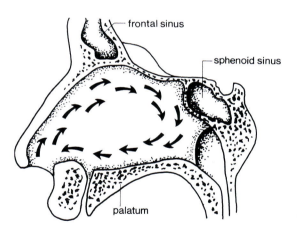

Fig. 1.3. Septum surgery starts (*on the more deviated side*) with an incision on the anterior caudal cartilaginous septum; the incision is made downwards to avoid damaging the nostrils. Mobilizing first the upper and posterior mucosa results in fewer iatrogenic perforations

nique of leaving intact mucosa in the sinuses was controversial at that time. The general opinion changed during the subsequent years.

Endoscopes with improved cold-light illumination (Hopkins optics) stimulated increased usage of endonasal surgery by Messerklinger [54, 55], Draf [15–17], Steiner and Wigand [75–78], and Stammberger [68, 69]. Messerklinger [56] stressed enlargement of the outflow of the paranasal sinuses in the anterior ethmoid area. Endoscopes are excellent for photodocumentation and for control of areas invisible to the operating microscope in the maxillary sinus and the nasofrontal ducts. However, whenever possible, we prefer to use the binocular operating microscope and to have both hands free for surgery. Continuing the comparison with ear surgery, we expect to see a similar worldwide acceptance of endonasal ethmoid surgery after further development of hypotensive anesthesia [30, 34, 36, 37].

General Preoperative Preparation

Premedication

A sleeping pill is given to the patient the night before surgery. The patient receives midazolam or flunitrazepam 1 h prior to surgery. About 80% of patients have no recollection of being moved into the operating room.

Hypotensive Anesthesia

Since 1959, we have used intravenous chlorpromazine to maintain a systolic blood pressure of 55–90 mmHg and, for the last 9 years, intravenous propofol has been given prior to intubation. Since 1983, we have also used enflurane or isoflurane gas insufflation with intubation and fentanyl or alfentanyl hydrochloride–nitroglycerine in about 25% of elderly patients and have used α-blockers (clonidine) and β-blockers (esmolol) in approximately 20% of younger patients. Remifentanil perfusors are also in use.

Semi-Fowler's Position of the Patient

The patient is placed in a semi-Fowler's (half-sitting) position (Fig. 1.1). An adjustable, specially designed headrest is necessary to enable the nurse to change the position of the patient's head. This position has several advantages [36, 37].
1. The surgeon can relax his elbows and can sit in a convenient position
2. The view of the anatomy is the same in the office and in the operating room
3. The operating microscope, which is in a balanced, movable position, can be focused by head motions of the surgeon, improving the depth of focus immensely
4. Less bleeding occurs

Fig. 1.4. a Position of extensive upper and (if necessary) lower fenestration of the maxillary sinus, to the lacrimal sac and ethmoidal arteries below the ethmoid roof [30].
b Endonasal microsurgical removal of choanal atresia and the floor and anterior wall of the sphenoid sinus [26]

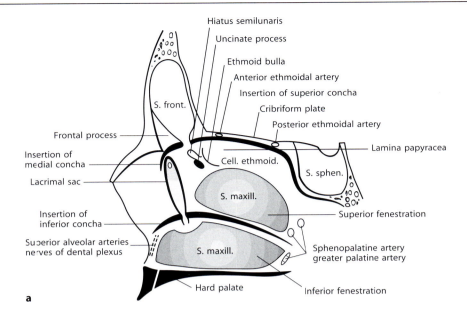

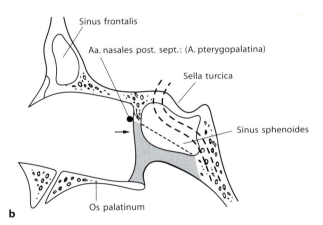

5. Fluids drain from the skull base without suction
6. Further correction of blood pressure is possible by moving the patient's legs up or down
7. A dry operating field and less suction results in less damage to the mucosa, shorter surgical procedure, less complications and better results meke navigation [80] not necessary in normal cases and advanced surgeons

Microsurgery of the Septum

Using the operating binocular microscope [24], we determine the side of the more severe deviation; the incision is made downwards (to avoid hurting the nostrils) along the caudal end of the anterior cartilaginous septum [28, 33]. After first mobilizing the mucosa upwards to the cribriform plate (Figs. 1.2–1.4, 1.6, 1.7), we proceed inferiorly with the elevator to the choana; from there, we proceed anteriorly. This procedure is easier to perform and has less chance of iatrogenic perforation. After incising the mucosa on both sides of the vomer in the choanae in order to prevent postoperative hematomas, we reinsert [60] all corrected deviated pieces into the septum. The lower part is filled with cartilage, and the upper part is filled with bone because of better fixation to the mucosa. If necessary, the posterior ends [43] of the lower turbinates can also be resected.

Maxillary Sinus Surgery

Since 1974 [30], we have not used the Caldwell-Luc [8, 52] approach except in cases with malignancies, oroantral fistulas [27] or malformations.

Procedure

Indication [1, 12, 13, 44, 49, 50] for upper and lower fenestration surgery (Fig. 1.4) of the maxillary sinus is confirmed by X-ray, computed tomography (CT), magnetic resonance, or sinoscopy. A No. 3 Ritter probe is inserted [36] into the base of the uncinate process, just above the inferior turbinate and posterior to the lower portion of the lacrimal sac. In 37 years of using this technique, we have observed no permanent damage to the orbit or the lacrimal sac. Probing superiorly and inferiorly and medializing the uncinate process into the upper nasal meatus allows part of the maxillary sinus to be visualized in a binocular view (using the operating microscope) and avoids opening into the orbit. The uncinate process [24] is then in the correct position [36] for removal with straight and retrograde cutting forceps (Fig. 1.2). We do remove the bone in the fenestration field with the attached mucosa, but without pulling intact mucosa out of the maxillary sinus. The opening of the punch is directed to the nasal cavity, so no pieces of bone can disappear into the sinus cavity [36]. As we advocated in 1974 [30], an extensive, enlarged upper fenestration (Fig. 1.4) is created. Establishing a smooth, convex continuation (Fig. 1.5) from the nasal cavity to the sinus is important to prevent postoperative adhesions and closure of the window.

If necessary, a secondary large window is created in the inferior [9] nasal meatus beneath the inferior turbinate (Figs. 1.4, 1.5). If required, these two large windows allow complete removal of the residual sinus contents under microscopic [24, 30] and endoscopic [16, 17, 56, 68, 69, 75, 78] vision, with preservation of any intact mucosa [1, 13, 14, 16, 30, 41, 49, 56, 58, 61, 62, 67, 72].

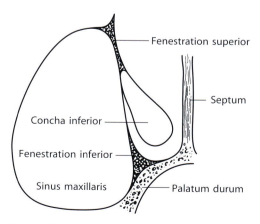

Fig. 1.5. Smooth edges in upper and (if necessary) lower fenestrations of the maxillary sinus [37]

Packing in Fungal Infections

The sinus cavity is packed with gelfoam saturated with antibiotic and antifungal [29, 30, 36, 37] solutions. This packing is left in place for 3 weeks and is removed with irrigation.

Allergy

When there are eosinophilic polyps or asthma and aspirin idiosyncrasy, recurrence is possible [18], and this is explained to the patient. Medical antiallergic therapy is advised postoperatively.

Surgery of Ethmoid and Sphenoid Sinuses

Indications for 942 intranasal ethmoidectomies performed in 1993 were:
1. Acute infections, either with severe exophthalmus or chronic with or without empyema or mucocele
2. Polyposis
3. Tumors
4. As a initial surgical approach for:
   - Orbital decompression or maxillary, sphenoid, or frontal sinus surgery
   - Osteoplastic frontal-sinus surgery
   - Lacrimal-sac surgery
   - CSF leakage
   - Pituitary-gland surgery
   - Epistaxis

Procedure

After mobilization of the upper posterior septum, the inferior turbinate is displaced laterally and the middle turbinate is displaced medially. A straight punch forceps (Fig. 1.2) is used to enter the ethmoid bulla and to remove the large cells. Using the middle turbinate as a guide, the surgical procedure is continued posteriorly until the anterior wall of the sphenoid sinus is exposed or removed [4, 20, 26]. The resection follows the ethmoidal roof from the posterior end to the anterior end [33, 37], identifying the bony canals of the posterior and anterior ethmoidal arteries, which are situated a few millimeters below the ethmoidal roof (Figs. 1.4, 1.7). An angulated cutting punch forceps with a 2-cm opening (as used by Heermann in 1908) enters the ethmoidal cells, permitting direct visualization with the surgical microscope [24]. This punch forceps can also be used as a curette for identifying the ethmoidal roof. It preserves the arc between the anterior middle turbinate and the agger nasi, preventing adhesions.

Preservation of the Middle Turbinate

It is important to preserve the middle turbinate as a landmark [30, 36, 37] for possible secondary procedures. If the middle turbinate contains cells [37] (concha bullosa) or is doubled, the lateral part may be resected [37]. The cribriform plate (Figs. 1.4, 1.6–1.8) is directly medial to the base of the middle turbinate and extends lower than the skull base of the ethmoid sinus. Therefore, it is relatively safe only if the surgeon operates lateral to the middle turbinate.

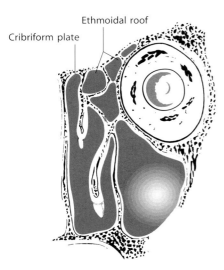

Fig. 1.6. Endonasal approach to the ethmoid sinus and cribriform plate [33]

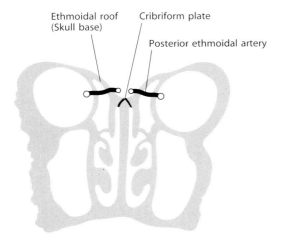

Fig. 1.7. The ethmoid artery below the ethmoid roof but superior to the cribriform plate [23]

Results in Nasal Polyposis

Our long-term results after 5 years were reported by Hohenhorst and Wielgosz [40], who evaluated 302 cases (Fig. 1.9). Bronchial asthma was present in 26.5% of patients, anosmia in 15.7%, and hyposmia in 31.9%. Recurrences occurred in 17.3%, and reoperations were necessary in 6.5%; there were no cases of CFS leakage, blood transfusion, or disturbance of the vision due to surgery, and post-operative synechia (no self-retaining retractor, short ointment packing) was seen. Improvement of anosmia was seen in 62.1% of patients, hyposmia in 83.1%, and improvement of nasal obstruction in 91.4%. Overall, 35.6% of patients felt they were without any symptoms, 48.6% thought their symptoms improved, 11.9% felt unchanged, and 3.9% felt worse.

Procedure in Tumors

To prevent recurrences in cases involving inverted papillomas or small esthesioneuroblastomas, laser or electrocoagulation is used, and packs saturated with metacresolsulfonic acid (Albothyl) are placed for 2 days. During a 37-year period, all cases of inverted papillomas were treated by the intranasal approach. We observed esthesioneuroblastomas that did not recur even after more than 12 years. Carcinomas and melanomas require more extensive laser surgery and radiation.

Cauterization of the Ethmoid Artery Through the Bony Canal during Epistaxis

The main blood supply to the ethmoid (Fig. 1.8b) arteries stems from the internal carotid artery. The larger anterior ethmoid artery is responsible for epistaxis more often than the posterior artery. The distance to the optic nerve is 2 cm for the anterior and 1 cm for the posterior ethmoid artery. We use a fine monopolar needle (insulated to the tip) insinuated through the vessel's bony canal for cauterization in the ethmoid sinus [35, 37]. Cauterization of epistaxis in the cribriform plate is dangerous [23] because of the risk of CSF leakage. After cauteriza-

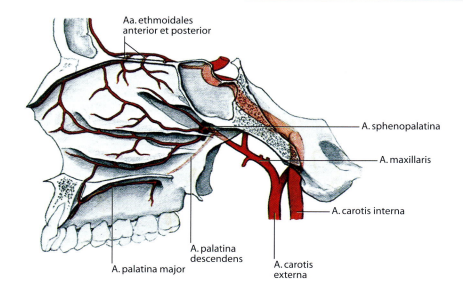

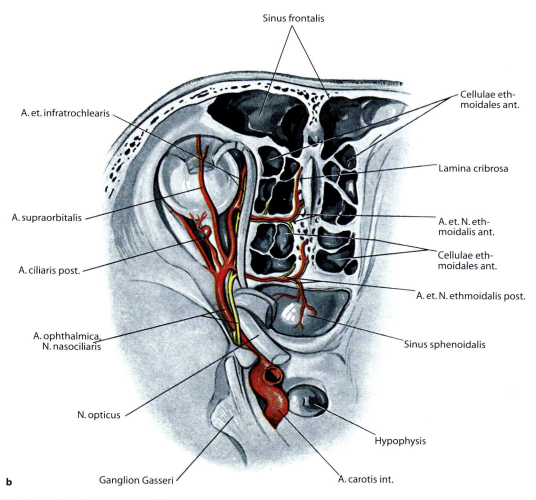

Fig. 1.8 a,b. Vessels of the nose [48]

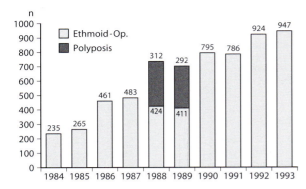

Fig. 1.9. After 1983, using general anesthesia, the number of ethmoid surgeries increased. Results in polyposis nasi after 5 years: improvement of obstruction in 91.4% of patients, recurrent polyps in 17.2%, and reoperations in 6.5% [40]

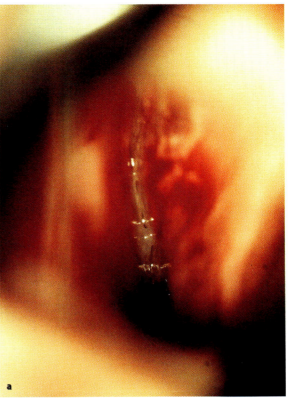

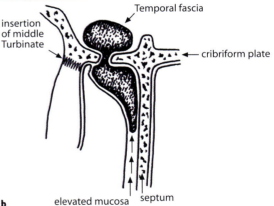

Fig. 1.10. a,b. We found spontaneous CSF fistulas in the posterior cribriform plate in 70% of patients. **b** Our new hourglass technique, with insertion of temporal fascia through the septum into the fistula without glue and with optimal fixation, is more effective than a direct approach.

tion of the ethmoid arteries on both sides, epistaxis from the cribriform plate stops without complications. For the past 35 years, we have been able to control severe epistaxis in all our patients by an intranasal procedure without embolization. Bipolar cauterization is used for more inferior epistaxis. An external approach for epistaxis or embolization is more dangerous [2, 11].

Spontaneous CSF Leakage

We have found the sources of *spontaneous* CSF leakage (Fig. 1.10A) in the posterior cribriform plate in 70% of cases [36]. We developed a new technique, placing the temporalis fascia through the septum [36] into the fistula in the shape of an hourglass for better fixation (Fig. 1.10b). Glue is not necessary for this procedure. This technique is easier than a direct approach to the cribriform plate, and we have had excellent results in all cases.

We observed two cases where the CSF spaces in the head were replaced by air [36] as seen on CT scans (Fig. 1.10c). These patients had only minor headaches and were ambulatory without other complaints. One patient had a cleft palate. Large defects in the posterior wall of the sphenoid sinus were closed with temporalis fascia and cartilage from the auricle.

Some of the fistulas caused repeated bouts of meningitis. However, we also saw cases where CSF leakages were present for 7 months, 1.5 years and 3 years without causing meningitis.

Extensive traumatic defects can require an open frontal-craniotomy approach. Rib cartilage is trimmed into the shape of a bottle cork [31, 36, 37] and is wedged from above into the present defect (Fig. 1.10d). This procedure has successfully closed large defects for more than 20 years.

Fig. 1.10. c Computed tomography scans (without CSF) of two patients with spontaneous fistulas in the sphenoid sinus.
d Conic rib cartilage is inserted using a craniotomy osteoplastic approach for closure of large traumatic defects of the ethmoid roof [35]

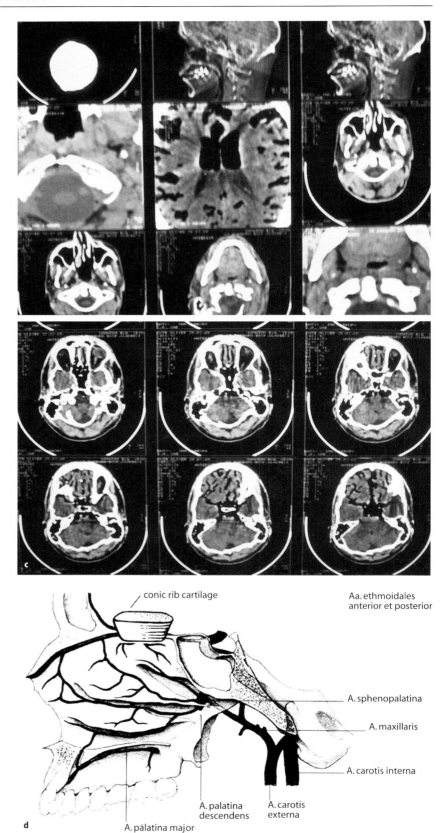

Fig. 1.11. a Computed tomography revealing an air bubble in the brain after use of Beck's burr-hole technique. **b,c** Revision of the left frontal sinus by removing a conically chiseled bone plate through an incision below the eyebrows [67]

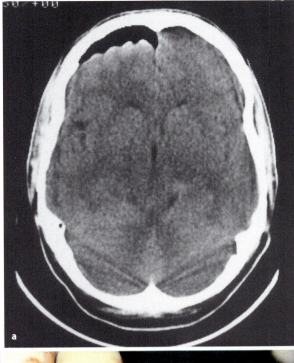

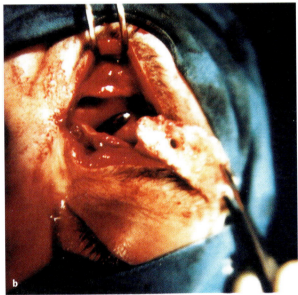

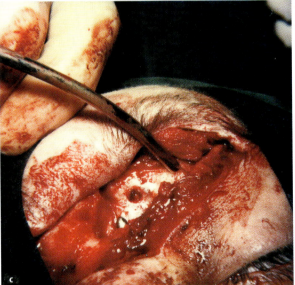

Endonasal and Osteoplastic Surgery of Frontal Sinus

We never use the Beck's burr-hole technique [3], which is not less dangerous than intranasal surgery. We saw one patient who had a burr hole inadvertently "alio loco" in the posterior wall of the frontal sinus, resulting in a large air bubble in the brain (Fig. 1.11a), necessitating osteoplastic surgery (Figs. 1.11b,c) [10].

The intranasal approach has been used in 96% of our patients. Contraindications are tumors of the frontal sinus (including osteomas and meningoceles), bony obliterations of the frontal ducts, and reoperations with defects of the posterior frontal-sinus wall.

Procedure

After ethmoidectomy and removal of the frontal process, direct visualization of the infundibulum of the frontal sinus is possible with the operating microscope. Gentle placement of a Ritter probe No. 4 or No. 3 (never use No. 1 or No. 2 first! [36]) will allow free drainage of pus into the nose in acute infections, with reduction of exophthalmus, if present (Fig. 1.4a). Excellent results in managing mucoceles can be achieved (Fig. 1.12).

In osteoplastic procedures [10], we use Gussenbauer's incision [19] below the eyebrows (Figs. 1.11B, C) and over the glabella, and remove the bony plate of the frontal bone with a chisel in a conical fashion so it can be replaced easily without the need for glue, wires, or screws. The interfrontal septum is resected, and (if necessary) both nasofrontal ducts are united (in some cases, with removal of the upper septum). In cases of bony defects, a conically shaped bone plate from the mastoid is transplanted (Fig. 1.13; Table 1.1) [36]. In large defects, bone from both mastoids is used. We have never obliterated the frontal sinus with fat [16].

Embolization of Juvenile Angiofibromas for Endonasal Surgery

New developments in radiology allow successful embolization (Fig. 1.14a) of nasopharyngeal tumors. Hours later, intranasal removal [38] using hypotensive anesthesia is possible. The consistency of the tumor is still firm after embolization. In difficult cases with deep invasion into the vertebral and sphenoid bones, a second embolization is performed; this makes the tumor softer [36], and removal is easier (Fig. 14b).

Microsurgery of Aesthetic Rhinoplasty

Nasal-tip surgery including hump reduction and trimming of the upper and lower cartilage (with control of open roof) can be performed with direct visualization using the operating microscope.

Operation Certificate

After surgery, the patient receives a certificate (Table 1.2) describing the performed operation and containing further instructions concerning postoperative treatment.

Complications with the Microscope and Endoscope

For residents and novices in intranasal surgery, thorough training on 50 cadavers [33, 36, 37] is essential (Fig. 1.15), as is frequent observation of surgery, as seen through the observing tubes of the surgical microscope or on a video monitor. As with ear surgery, nasal surgery should start with binocular visualization using the operating microscope (including septum surgery). When the above admonitions are observed, we have never seen a persistent postoperative disturbance of vision or ocular mobility due to surgery. We observed temporary blindness in one case [32] after injecting local anesthesia into the ethmoid. However, the condition was reversed with a stellate ganglion block. Since 1983, we have used general anesthesia (without local anesthesia) [34, 36, 37] in the ethmoid.

During the past 37 years, only five iatrogenic CSF leakages (0.03%) have occurred in more than 14,000 cases. All the leakages were discovered during the operation and were treated with temporalis fascia without the need for a reoperation. In all cases, epistaxis was controlled by intranasal cauterization [35, 37] of the ethmoid arteries. Transfusions of blood were not necessary, (hypotensive anesthesia). Self-retaining retractors [59, 64] are obsolete (pressure to the mucosa causes synechia) if surgery is done with hypotensive anesthesia.

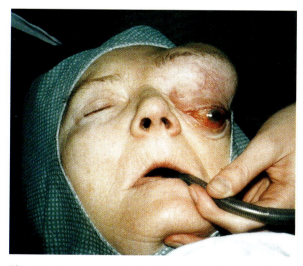

Fig. 1.12. Large mucocele that developed during a 12-year period, with extreme protrusion of the eye; the mucocele was removed using an endonasal approach [34, 67]

Using a transnasal monocular-endoscopic approach, Wigand [76] reported CSF leakages in 1.6% of cases and, in 1991, Wigand and Hosemann [42] observed leakages in 0.5% of 2100 operations. Draf et al. [16, 17] reported CSF leakages in 2.3% of 678 operations, and von Illberg et al. [72] had leakages in 1.02% of 295 cases. Wigand and Hosemann [77] and Weber and Draf [73, 74] each reported a case with fatal bleeding from the internal carotid artery in the sphenoid sinus. Wigand and Hosemann [77] observed blindness in one eye after bleeding, as did Halle [22] in three cases and Wirth [79] in one case. Further cases of blindness have been reported by Stankiewicz [70, 71], and complications were reported by Maniglia [53], Clemens, Pelausa, Stammberger [41] and Rosenbaum [63]. At the 1992 Sisson Congress in Vail (Colorado, USA), a case of bilateral blindness after ethmoid surgery (Fig. 1.15) by a less experienced surgeon was discussed.

According to our experience, primary binocular microscopic surgery is safer than controlled endoscopic surgery alone [78]. We use mirrors or endoscopes only if the view into the maxillary sinus or

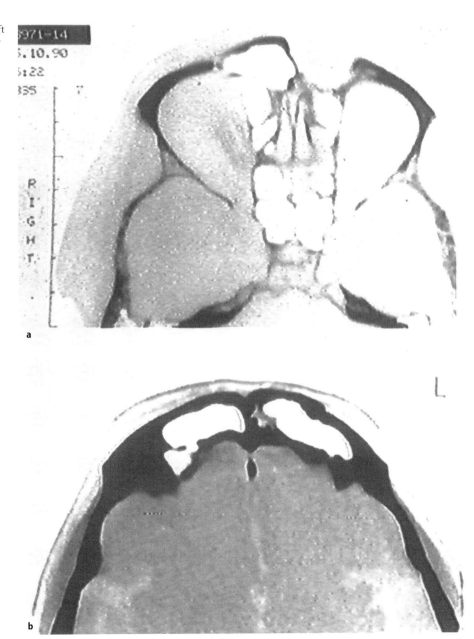

Fig. 1.13. a Defect of the left anterior frontal-sinus bony wall after four operations and insertion of several prosthesis.
b Two years after transplantation of conically removed bone plate from the mastoid

Table 1.1. The Heermann concept of endonasal microendoscopic surgery [40]

General preoperative preparation	
Surgeon	Intensive studies of the anatomy, and training on at least 50 cadavers
General anesthesia with controlled hypotension	Premedication: 1–2 mg flunitrazepam or 7.5 mg midazolam 1 h preoperation Anesthesia: N_2O/O_2 (50:50–70:30) °Propofol °Fentanyl or alfentanyl or remifentanil perfusors Hypotension (70–90 mmHg is desirable): clonidine (at the beginning of the operation) and/or nitrate (elderly patients)
Semi-sitting (Fowler's) position of the patient	Advantages: Blood flows off the skull base and does not accumulate in the sinuses Less bleeding in the operation area as a result of lower blood pressure Correction of blood pressure is possible by moving the patient's legs up or down
Adjustable headrest	The headrest can be adjusted in every desired position by a nurse
Binocular/microscope (300-mm lens)	Three dimensional stereoscopic view, even in the depth of the nose The microscope should be in a balanced, movable position and can be directed by head movements of the surgeon
Endoscope (30–70°)	To microscopically control hidden parts of the sinuses
Operative procedure	
Septum	Submucous septoplasty (the upper, posterior part is very important)
Turbinates	Resection of the posterior ends of the inferior turbinates but preservation of the middle turbinate (or only lateral resection) if the turbinate contains cells or if it is doubled
Maxillary sinus	The indication for superior or superior and inferior fenestration is confirmed by sinuscopy The large superior fenestration and (only if necessary) the second, extensively enlarged inferior fenestration allow the complete removal of the sinus contents under microscopic and endoscopic control [30]
Ethmoid sinuses and sphenoid sinus	Septoplasty (mainly of the upper, posterior part) Lateral displacement of the inferior turbinate and medial displacement of the medial turbinate Ethmoidectomy (posterior to anterior) following the lateral base of the middle turbinate Use of the angulated punch (developed by Heermann in 1908), which allows good control under the microscope The middle turbinate is preserved (important as a landmark) Opening of the sphenoid sinus in its anterior wall, which may be resected medially and downwards Smooth surfaces from one cell system to the next are important to prevent circular stenosing scars
Frontal sinus	Preliminary septoplasty, anterior ethmoidectomy Insertion of Ritter probe No. 4 (or at least probe No. 3) If necessary. the frontal process of the maxilla is resected with the help of a chisel Endoscopic control of the lumen
Epistaxis	In cases of arterial bleeding [from the arteria ethmoidalis (anterior or posterior) or the arteria sphenopalatina], cauterization of the arteries within their bony canal after ethmoidectomy
CSF leakage (traumatic, spontaneous or iatrogenic)	Rate of iatrogenic CSF leakage: 0.03% in more than 13,000 cases treated according to the Heermann concept Obturation with fascia temporalis and in cases of large defects with septum or auricular cartilage (to prevent brain prolapse and obliteration of the nasofrontal duct)
Lacrimal sac	Septoplasty (not necessary in children) Anterior ethmoidectomy Removal of the medial bone over the lacrimal sac; exposing of the membranous sac with a chisel Inserting of Bowman probes and incision and resection of the medial part of the membranous sac If necessary, excision of presaccal stenoses Dilatation of the horizontal duct with Bowman probes (tear-duct prostheses are necessary in less than 3% of patients)

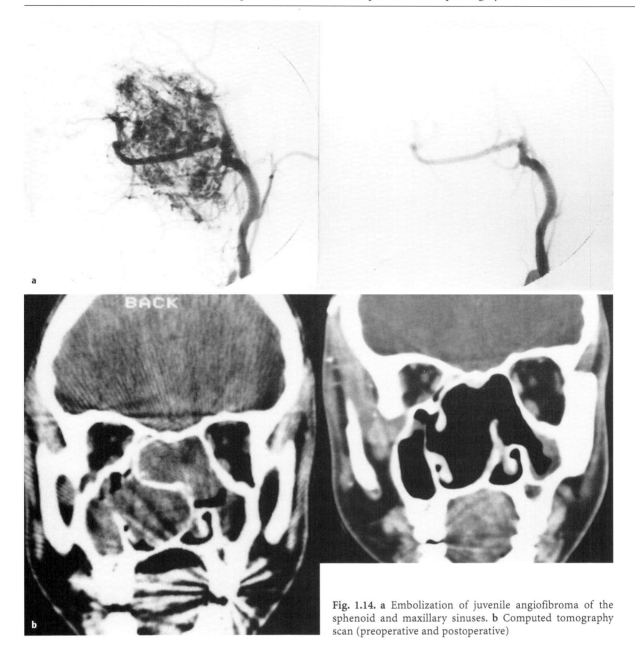

Fig. 1.14. a Embolization of juvenile angiofibroma of the sphenoid and maxillary sinuses. b Computed tomography scan (preoperative and postoperative)

into the infundibulum of the frontal sinus is not possible with the operating microscope.

We observed mortality only in one case. In 1960, a reoperation for a purulent ethmoiditis resulted in metastatic brain abscesses 2 weeks later. In 1963, a purulent lacrimal sac operation (as an outpatient surgical procedure) was followed by meningitis, which healed well with antibiotics. By routinely using antibiotics, we have not seen postoperative meningitis in more than 30 years.

References

1. Arnes E, Anke IM, Mair WS (1985) A comparison between middle and inferior meatal antrostomy in the treatment of chronic maxillary sinus infection. Rhinology 23:65–69
2. Beall J, Scholl P, Jafek B (1985) Total ophthalmoplegia after internal maxillary artery ligation. Arch Otolaryngol Head Neck Surg 111:696
3. Beck K (1937) Weitere Erfahrungen mit der Stirnhöhlenpunktion und der Drainage von aussen bei Stirnhöhleneiterungen. Arch Otolaryngol 142:205–209

Table 1.2. After surgery, the patient receives a certificate of the performed surgery and further instructions concerning postoperative treatment [36]

Please show this information to your doctor at the first follow-up examination. Do not discard this paper! After a change of residence, show this paper to your new doctor!

Intranasal microsurgical procedures (operations) performed:
- ❏ Septum
- ❏ Augmenting of the nasal dorsum
- ❏ Rhinoplasty
- ❏ Swab (bacteriology)
- ❏ Dorsal ends of lower turbinate
- ❏ Narrowing of the columella with suture for 3 weeks
- ❏ Biopsy

Time of intranasal packing:

Right maxillary sinus	Left maxillary sinus
❏ Endoscopy	❏ Endoscopy
❏ Upper fenestration	❏ Upper fenestration
❏ Lower fenestration	❏ Lower fenestration
❏ Radical operation	❏ Radical operation
❏ Caldwell-Luc procedure	❏ Caldwell-Luc procedure

Polyposis (ethmoid, sphenoid, maxillary sinus, frontal sinus, olfactory region):

❏ Right ethmoid	❏ Left ethmoid
❏ Right sphenoid	❏ Left sphenoid
❏ Right lacrimal sac	❏ Left lacrimal sac
❏ Right frontal sinus	❏ Left frontal sinus

- ❏ External approach (osteoplastic)
- ❏ Atresia of the choana
- ❏ Hypophysis
- ❏ Mucocele
- ❏ Tumor
- ❏ Osteoma
- ❏ Decompression of the optic nerve
- ❏ Decompression of the bulbus oculi
- ❏ Nasopharyngeal fibroma

❏ Widening ostium of the right maxillary sinus	❏ Widening ostium of the left maxillary sinus

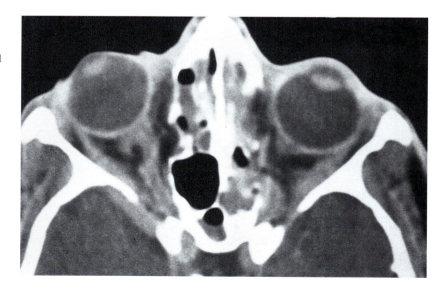

Fig. 1.15. Computed tomography scan after bilateral cutting of the optic nerves on both sides of the orbit by an inexperienced ethmoid surgeon, resulting in blindness in both eyes (Sisson Congress Vail, Colorado, USA, 1992)

4. Berger E (1890) La chirurgie du sinus sphenoidale. Octave Doin, Paris
5. Boenninghaus G (1923) Die Operationen an den Nebenhöhlen der Nase. In: Katz L, Blumenfeld F (eds) Handbuch der speziellen Chirurgie des Ohres und der oberen Luftwege. Kabitzsch, Leipzig
6. Bothworth FH, Bryan JH, Mackenzie JN, Ino O (1894) The surgery of the accessory sinuses of the nose. Zentralbl Laryngol 95:3402
7. Brieger O (1894) Über chron. Eiterungen der Nebenhöhlen der Nase und Demonstration zur Behandlung chron. Mittelohreiterungen. 72. Jahresbericht schles. Ges. vaterl. Kultur 91–94
8. Caldwell GW (1893) Diseases of the accessory sinuses of the nose, and an improved method of treatment for suppuration of the maxillary antrum. N Y Med J 58:526–528
9. Claoué E (1904) Traitement des suppurations chroniques du sinus maxillaire, par la resection large de la partie inferieure de la paroi nasale du sinus, Resultats. Societé francaise de Laryngologie. Ann Mal Oreil Larynx 30:221–230
10. Czerny V (1895) Osteoplastische Eröffnung der Stirnhöhle. Arch Klin Chir 544–550
11. Denecke H (1968) Carotis interna Verletzung mit unstillbarem Nasenbluten, geheilt durch intraarterielle Thrombininjektion. Arch Klin Exp Ohren Nasen Kehlkopfheilkd 191:217–404
12. Dixon HS (1976) Microscopic antrostomy in children a review of the literature in chronic sinusitis and a plan of medical and surgical treatment. Laryngoscope 86:1796–1814
13. Dixon HS (1983) Microscopic sinus surgery, transnasal ethmoidectomy and sphenoidectomy. Laryngoscope 93:440–444
14. Dixon HS (1985) The use of the operating microscope in ethmoid surgery. Otolaryngol Clin N Am 18:75–86
15. Draf W (1978) Endoskopie der Nasennebenhöhlen, Springer, Berlin Heidelberg New York
16. Draf W (1982) Die chirurgische Behandlung entzündlicher Erkrankungen der Nasennebenhöhlen. Arch Klin Exp Ohren Nasen Kehlkopfheilkd 235:133–305
17. Draf W (1992) Endonasale Mikro-endoskopische Pansinusoperation bei chronischer Sinusitis. III. Endonasale mikroendoskopische Stirnhöhlenchirurgie: eine Standortbestimmung. Otorhinolaryngol Nova 2:118–125
18. English GM (1986) Nasal polypectomy and sinus surgery in patients with asthma and aspirin idiosyncrasy. Laryngoscope 96:374–380
19. Gussenbauer K (1895) Die temporäre Resektion des Nasengerüstes zur Freilegung des Sinus frontalis, ethmoidalis, sphenoidalis und der Orbitahöhlen. Klin Wochenschr 21:377–380
20. Hajek M (1903) Pathologie und Therapie der entzündlichen Erkrankungen der Nebenhöhlen der Nase. Bd 1. Deutike Verlag, Leipzig, 1–361
21. Halle M (1906) Externe oder interne Operation der Nebenhöhleneiterungen. Berl Klin Wochenschr 43:1369–1372, 1407
22. Halle M (1923) Nebenhöhlenoperationen. Z HNO 4:489–492
23. Heermann H (1954) Endonasale Unterbindung der A. ethmoidalis anterior und posterior bei unstillbarem Nasenbluten aus der Riechspalte. Arch Klin Exp Ohren Nasen Kehlkopfheilkd 165:507–510
24. Heermann H (1958) Endonasal surgery with the use of the binocular Zeiss operating microscope. Arch Klin Exp Ohren Nasen Kehlkopfheilkd 171:295–297
25. Heermann J (1922) Nasal surgery of the lacrimal ducts. Z Laryngol Rhinol 11:67–69
26. Heermann J (1962) Microsurgical resection of the floor and anterior wall of the sphenoid sinus in choanal atresia. Laryngo-Rhino-Otol. 41:390–393
27. Heermann J (1962) Closure of oro-antral fistulas with periosteum-connective tissue through the maxillary sinus. Laryngo-Rhino-Otol. 41:213–216
28. Heermann J (1965) Microsurgery of the septum. HNO 14:352
29. Heermann J (1974) Application of antibiotic ointment into the cavum nasi in children with maxillary sinusitis. Laryngo-Rhino-Otol. 53:836–837
30. Heermann J (1974) Endonasal microsurgery of the maxillary sinus. Laryngo-Rhino-Otol. 53:938–942
31. Heermann J (1979) CSF leakage. Arch Otolaryngol Head Neck Surg 223:457–458
32. Heermann J (1980) Temporäre Amaurose bei mikrochirurgischer Ethmoid- und Saccus-lacrimalis-Operation in Lokalanästhesie. Laryngo-Rhino-Otol. 59:433–437
33. Heermann J (1982) Endonasal microsurgery of the ethmoid sinus on a patient in a semi-sitting-Fowler position under hypotensive anesthesia. HNO 30:180–185
34. Heermann J (1985) Intranasale Mikrochirurgie aller Nasennebenhöhlen und des Tränensackes am halbsitzenden Patienten in Hypotension – 25 Jahre Erfahrung. In: Majer und Zrunek E (ed) Aktuelles in der Otorhinolaryngologie 1984. Thieme, Stuttgart, pp 66–69
35. Heermann J (1986) Intranasal microsurgery of epistaxis in the cribriform plate and further surgical procedures under hypotensive anesthesia. HNO 34:208–215
36. Heermann J (1995) Micro-endoscopic surgery of the septum, all paranasal sinuses, and the lacrimal sac with removal of a presaccal stenosis under hypotensive anesthesia. In: Tos M (ed) Rhinology: a state of the art (15th Congress, Copenhagen, 1994). Kugler, Amsterdam, pp 51–57
37. Heermann J, Neues D (1986) Intranasal microsurgery of all paranasal sinuses, the septum, and the lacrimal sac under hypotensive anesthesia. Ann Otol Rhinol Laryngol 95:631–638
38. Heermann J, Kühne D, Strasser K (1988) Endonasal removal of juvenile neurofibroma under hypotensive anesthesia after embolization. Arch Otolaryngol Head Neck Surg 114:261–263
39. Hirschmann A (1903) Über Endoskopie der Nase und deren Nebenhöhlen. Eine neue Untersuchungsmethode. Arch Laryngol Rhinol 14:195–202
40. Hohenhorst W, Wielgosz R (1995) Long-term results of endonasal microsurgery in polyposis nasi. In: Tos M (ed) Rhinology: a state of the art (15th Congress, Copenhagen, 1994). Kugler, Amsterdam, pp 71–77
41. Hosemann W (1996) Die endonasale Chirurgie der Nasennebenhöhlen – Konzepte, Techniken Ergebnisse, Komplikationen, Revisionseingriffe. Eur Arch Otorhinolaryngol Suppl 1:155–169
42. Hosemann W, Nitsche N, Rettinger G, Wigand ME (1991) Die endonasale, endoskopisch kontrollierte Versorgung von Defekten der Rhinobasis. Laryngo-Rhino-Otol. 70:115–119
43. Jarvis WC (1882) Removal of hypertrophied turbinated tissue by ecrasement with the cold wire. Arch Laryngol 3:105–111
44. Jorgensen RA (1991) Endoscopic and computed tomographic findings in ostiomeatal sinus disease. Arch Otolaryngol Head Neck Surg 117:279–287

45. Kaspariant Z (1900) 13e Congress intern de med sect de Rhinol, P s. 90 Paris, zit. Nach Boenninghaus, Paris, pp 1923
46. Kessel J (1876) Über die Durchschneidung des Steigbügelmuskels beim Menschen und über die Extraktion des Steigbügels, rep. Columella bei Tieren. Arch Ohr hk 11:199–217
47. Killian G (1900) Die Krankheiten der Kieferhöhle in Heymann P. Handbuch der Laryngologie und Rhinologie, III. Bd, 2. Hälfte Hölder, Wien, pp 1004–1096
48. Krmpotic-Nemanic J, Draf W, Helms J (1985) Chirurgische Anatomie des Kopf-Hals-Bereiches. Springer, Berlin Heidelberg New York
49. Lavelle R, Harrison MS (1971) Infection of the maxillary sinus: the case for the middle meatal antrostomy. Laryngoscope 81:90–106
50. Litton WB (1973) Surgery of the paranasal sinuses. In: Paparella MM, Shumrick DA (eds) Otolaryngology. Saunders, Philadelphia
51. Lothrop HA (1897) Empyema of the antrum of Highmore. A new operation for the care of obstinate cases. Boston Med Surg J. 136:455–462
52. Luc J (1897) Une nouvelle methode operatoire pour la cure radicale et rapide de l'empy'ème chronique du sinus maxillaire. Arch Laryngol Otol Rhinol Bronchesophag 1:273–285
53. Maniglia AJ (1989) Fatal and major complications secondary to nasal and sinus surgery. Laryngoscope 99:276–283
54. Messerklinger W (1972) Technik und Möglichkeiten der Nasenendoskopie. HNO 20:133–135
55. Messerklinger W (1978) Endoscopy of the nose. Urban and Schwarzenberg, München
56. Messerklinger W (1987) Die Rolle der lateralen Nasenwand in der Pathogenese, Diagnose und Therapie der rezidivierenden und chronischen Rhinosinusitis. Laryngo-Rhino-Otol. 66:293–299
57. Mikulicz H (1887) Zur operativen Behandlung des Empyems der Highmorhöhle. Arch Klin Chir 34:626–634
58. Neto AF (1979) Microcirurgia endonasal. Rev Bras Otorhinolaryngol 45:215–223
59. Park IY (1988) Improved endonasal sinus surgery by use of an operating microscope and a self-retaining retractor speculum. Acta Otolaryngol Suppl (Stockh) 458:27–33
60. Passow A (1926) Die Erkrankungen der Nasenscheidewand. Denker und Kahler, vol.2. Springer, Berlin Heidelberg New York, pp 444–517
61. Paulsen K (1995) Endonasale Mikrochirurgie. Thieme, Stuttgart
62. Prades J, Bosch J, Tolosa A (1977) Microcirurgia endonasal. Garsi, Madrid, pp 120–170
63. Rosenbaum AL, Astle WF (1985) Superior oblique and inferior rectus muscle injury following frontal and intranasal sinus surgery. J Pediatr Ophthalmol Strabismus 22:194–202
64. Rudert H (1988) Mikroskop und Endoskop gestützte Chirurgie der entzündlichen Nasennebenhöhlenerkrankungen. HNO 36:475–482
65. Siebenmann F (1900) Traitement chirurgical de la sclerose otique. Ann Mal Oreille 26:467
66. Siebenmann G (1900) Die Behandlung der chronischen Eiterung der Highmorhöhle durch Resektion der oberen Hälfte (pars supra-turbinalis) ihrer nasalen Wand. Münch Med Wochenschr 47:31–33
67. Stamm A (1995) Microcirurgia Naso-Sinusal. Revinter, Rio de Janeiro, pp 1–436
68. Stammberger H (1985) Unsere endoskopische Operationstechnik der lateralen Nasenwand – ein endoskopisches chirurgisches Konzept zur Behandlung entzündlicher Nasennebenhöhlenerkrankungen. Laryngo-Rhino-Otol. 64:559–566
69. Stammberger H (1994) The evolution of functional endoscopic sinus surgery. Ear Nose Throat J 73:451–455
70. Stankiewicz JA (1989) Blindness in intranasal ethmoidectomy: prevention and management. Otolaryngol Head Neck Surg 101:320–329
71. Stankiewicz JA (1989) Complications of endoscopic intranasal ethmoidectomy: an update. Laryngoscope 99:686–690
72. von Illberg C, May A (1997) Bemerkungen zur endonasalen Chirurgie der Nebenhöhlen, zur Indikation und operativen Technik. Eur Arch Otorhinolaryngol Suppl 2:204–208
73. Weber R, Draf W (1992) Endonasale mikro-endoskopische Pansinusoperation bei chronischer Sinusitis. Otorhinolaryngol Nova 2:63–69
74. Weber R, Draf W (1992) Komplikationen der endonasalen mikro-endoskopischen Siebbeinoperation. HNO 40:170–175
75. Wigand ME (1981) Transnasale, endoskopische Chirurgie der Nasennebenhöhlen bei chronischer Sinusitis. I. Ein biomechanisches Konzept der Schleimhautchirurgie. HNO 29:215–221
76. Wigand ME (1981) Transnasale endoskopische Chirurgie der Nasennebenhöhlen bei chronischer Sinusitis. II. Die endonasale Siebbeinausräumung. HNO 29:287–293
77. Wigand ME (1989) Die endoskopische Chirurgie der Nasennebenhöhlen und der vorderen Schädelbasis. Thieme, Stuttgart
78. Wigand ME, Steiner W (1977) Endonasale Kieferhöhlenoperation mit endoskopischer Kontrolle. Laryngo-Rhino-Otol. 56:421–425
79. Wirth G (1963) Entstehung einseitiger Amaurose bei Nasennebenhöhlenerkrankungen, insbesondere nach operativen Eingriffen. HNO 11:21–25
80. Heermann R, Lenarz Th (1998) Navigationssysteme in der Orbitachirurgie in W. Steiner (Ed.) Referateband der Jahrestagung der Deutschen Gesellschaft für Hals-Nasen-Ohrenheilkunde, Kopf-, Halschirurgie. Springer Verlag, Berlin, 111–123

Surgical Anatomy of the Nose, Paranasal Sinuses, and Pterygopalatine Fossa

João A. Caldas Navarro

Nose

The nose is part of the upper portion of the upper respiratory tract and is divided into the external nose and the nasal cavity, from which the paranasal sinuses originate and extend. The external nose is that portion of the nose that emerges from the face, and it is formed by an osteocartilaginous skeleton based on the frontal processes, maxillary palatines and nasal bones. The cartilaginous portion is formed by two pairs of large cartilage pieces (the superior and inferior lateral cartilages) and a variable number of smaller or accessory cartilages. These pieces of cartilage are attached to the nasal septal cartilage at each side of the median sagittal plane (Fig. 2.1). The arrangement of these anatomical elements keeps the air pathway open at the aditus or nasal vestibules, which are externally seen as the nostrils. Differences in nasal shapes depend on the nasal contours, which are determined externally by the root, spine and nasal apex. The entrance of the nostril usually has an oval shape, and in this region, there are several vibrissae, described as small hairs with tactile and protective functions for the nose vestibule.

The term nasal pyramid is used because of the triangular pyramidal shape of the external nose. The base of the pyramid is formed by the nostrils, and the corners are formed by the nasal spine; the rims are implanted in the frontal and maxillary palatine processes, and the root originates from the junction of the corners at the upper portion of the nasal bones.

The internal or nasal surface of the lateral nasal wall forms a triangle with its base inferior. The concave nostril, covered by vibrissae, is found in this inferior part (the vestibular region). Immediately above the nostril is what looks like an open arch to the nasal limen, a localized accumulation of fibrous tissue in the mucosa; the limen is the superior limit of the inferior alar cartilage. Just above the limen is an open regular area, the nasal aditus, which is anterior to the insertion of the middle and inferior nasal turbinates (Fig. 2.2). This region is formed by a very hard bone corresponding to the frontal process of the maxilla. The nostrils, formed by the inferior alar cartilages, are extremely motile and work as a valve system that is bordered internally and superiorly by the nasal limens [8, 11, 12, 15, 25, 29, 34].

The nasal cavity is described as a frustum pyramid with a larger inferior base (floor) and a smaller superior base (roof). It has a medial (septal) and a lateral (orbital) wall. It is divided into right and left nasal cavities by the nasal septum. Thus, the septum is the common medial wall of both cavities.

The pyriform recess of the nasal cavity is bordered by the maxilla through its frontal and palatine processes. These processes meet in the median sagittal plane to form the anterior nasal spine; the frontal processes articulate medially with the two nasal bones, superiorly with the frontal bone and maxillary processes, and laterally with the lacrimal bones.

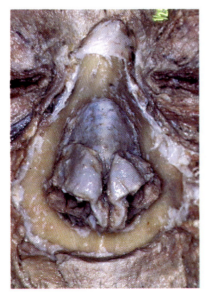

Fig. 2.1. The external nose. *1*, Nasal bone; *2*, superior lateral cartilage; *3*, septal cartilage; *4*, inferior lateral cartilage; *5*, nostril

The projections of the inferior and middle turbinates inserted in the lateral nasal wall can be seen through this pyramidal opening. The bony nasal septum, often deflected from the median plane, is formed superiorly by the perpendicular plate of the ethmoid bone, inferiorly by the vomer, and is completed by the septal cartilage anteriorly [11, 15, 25]. The lateral surfaces of the osseous pyramid are always irregular, especially in the upper, nasomaxillary segment of the pyramid (Fig. 2.3).

The choana is the posterior aperture of the nasal cavity, at the opposite side of the piriform aperture.

It is a broad, oval area with small, variably sized bony borders. The choanal borders are formed by the inferior face of the sphenoid bone (the most posterior region of the roof of the nasal cavity), beneath which are the vomer and the nasal septum. The area of the posterior nasal spine (in the nasal septum) consists of the junction of the vomer with two maxillary or horizontal processes of the palatine bones. This is an important area for fibromuscular insertions of the soft palate [15, 25, 30, 34].

The elongated vertical pterygoid process, whose plate and medial face form the lateral borders of the choanae, is the posterior part of the nasal cavity and is projected downward from the junction formed by the sphenoid bone and its wings (Fig. 2.4). The bony skeleton of the nasal cavity is complex, formed by the articulation among several (mostly facial) bones. The majority of the bones of the nasal lateral wall and nasal cavity floor are part of the maxillae; however, the roof of the nasal cavity is mostly formed by the ethmoid bone.

In the earliest embryological phase, the nasal cartilaginous capsule, which is anteriorly placed in the sphenoid region, is like a cover lying over the lateral parts of the future nasal cavity, the future nasal septum, and an extension of the skull that will eventually become the *crista galli*. From this framework, both the ethmoid bone and the maxillae will develop as the main components of the nasal and paranasal skeleton.

Fig. 2.2. The external nose and lateral nasal wall: left side. *1*, Nostril; *2*, nasal limen; *3*, inferior nasal meatus; *4*, inferior turbinate; *5*, middle meatus; *6*, middle turbinate; *7*, superior meatus; *8*, superior turbinate; *9*, sphenoid sinus; *10*, agger nasi; *11*, atrium of the middle meatus

Fig. 2.3. The nasal cavity, anterior opening. *1*, Nasal bone; *2*, frontal process of the maxilla; *3*, maxilla; *4*, inferior turbinate; *5*, middle turbinate; *6*, nasal septum; *7*, anterior nasal spine

Fig. 2.4. The nasal cavity, posterior opening (choanae). *1*, Pterygoid process of the sphenoid; *2*, posterior ethmoidal cell; *3*, middle turbinate; *4*, middle meatus; *5*, inferior turbinate; *6*, palate; *7*, nasal septum; *8*, occipital bone; *9*, foramem magnum

The nasal, frontal, ethmoid, and sphenoid bones participate in the formation of the nasal roof. The ethmoid bone is the major contributor, with its horizontal plate; the cribriform plate, resembling a horseshoe, continues cranially as the *crista galli*. The nasal portion of the cribriform plate is composed of a thin plate occupied throughout by foramina and terminal branches of the olfactory nerve. At the distal part of this plate, where the ethmoid bone joins the lesser wing of the sphenoid, there is an common area occupied by expansions of the ethmoid or sphenoid sinuses. Divided by the bony septum, the nasal roof is always narrow, giving the impression of two anteroposterior ducts that begin and end on each side of the septum (Fig. 2.5).

The bony part of the floor of the nasal cavity is formed by both the maxillary palatine process and the palatine process, which also forms the osseous palate on the buccal side. Each side has articulations on the medial sagittal surface, where they form the crest of the nasal floor, upon which the vomer is positioned. The roof of the nasal cavity is almost planar, inclining downward from the anterior side to the posterior side, ending in an open, wide canal that extends laterally as a marked concavity (the inferior nasal meatus) under the inferior turbinate. These structures are the anatomical elements of the lateral wall of the nasal cavity (Fig. 2.5) [8, 11, 12, 25, 28]. The lateral wall of the nasal cavity is certainly the most complex and important anatomic structure from both clinical and surgical standpoints. The frontal process of the maxillary bone borders the whole nasal cavity aditus, giving it a vertical anatomical shape with compact hard bone. Superiorly, this process is found between the nasal and lacrimal bones. Immediately distal to the frontal maxillary process, the lacrimal bone has a well-marked visible wrinkle, the lacrimal fossa, where the lacrimal sac is lodged. This bone continues in an osseous plate that will form the nasolacrimal duct along with the frontal maxillary process and the anterior edge of the uncinate process. The area anterior to the duct is the middle nasal meatus aditus, the posterior continuation of the aditus of the nasal cavity. It corresponds to the space between the insertion of the inferior and middle turbinates.

A prominence that is the projection of one or two anterior ethmoidal cells in the nasal wall is frequently superior to the lacrimal bone, close to the nasal roof. This bulge corresponds to the *agger nasi* region, coinciding with the anterior end of the uncinate process, to which these cells may extend. The most superior anterior portion of the head of the middle turbinate can be found slightly above this level, close to the nasal roof (Fig. 2.6) [1, 8, 11, 15, 20, 25, 28, 29].

A vertical line tangential to the posterior nasal aditus border would pass through the nasal projections of the middle and inferior turbinates, which are inserted in the lateral nasal cavity; these projections reach the nasal septum. These osseous plates have different origins; the inferior turbinate is an individual bone articulated at the wall in a longitudinal, almost horizontal plane with a posterior conver-

Fig. 2.5. The nasal cavity: lateral nasal wall: right side. *1*, Palate; *2*, anterior nasal spine; *3*, inferior meatus; *4*, inferior turbinate; *5*, middle meatus; *6*, middle turbinate; *7*, superior meatus; *8*, superior turbinate; *9*, sphenoid sinus; *10*, pterygoid process of the sphenoid; *11*, frontal process of the maxilla; *12*, nasal bone; *13*, frontal sinus; *14*, anterior cranial fossa; *15*, middle cranial fossa; *16*, pituitary fossa

Fig. 2.6. The lateral nasal wall, middle meatus: right side. *1*, Palate; *2*, inferior meatus; *3*, inferior turbinate; *4*, middle meatus; *5*, uncinate process; *6*, anterior nasal fontanelle; *7*, posterior nasal fontanelle; *8*, ethmoidal process of the inferior turbinate; *9*, hiatus semilunaris; *10*, ethmoid bulla; *11*, maxillary sinus; *12*, agger nasi; *13*, frontal sinus

Fig. 2.7. The lateral nasal wall: right side. *1*, Nostril; *2*, nasal limen; *3*, palate; *4*, inferior meatus; *5*, inferior turbinate; *6*, aditus of the middle meatus; *7*, middle meatus; *8*, accessory ostium of the maxillary sinus; *9*, middle turbinate; *10*, superior meatus; *11*, superior turbinate; *12*, agger nasi; *13*, sphenoid sinus; *14*, choanal region; *15*, pharyngeal opening of the eustachian tuba; *16*, septum of the frontal sinus

gence. Its anterior third is thicker and articulates with the maxilla; further along its articulation with the palatine vertical plate, it becomes narrower. The medial portion of the inferior turbinate is related to the lacrimal process of the ethmoid bone and the uncinate process. The superior edge of the inferior turbinate is convex and borders the most complex anatomical space of the nasal wall, the middle nasal meatus. Due to its origin and the anatomical relationship of the inferior turbinate with the maxilla, it is also called the nasomaxillary concha. The lateral wall region under the inferior turbinate is called the inferior nasal meatus (Figs. 2.5, 2.7). In this meatus, close to the point where the turbinate articulates into the nasal wall between its anterior and middle thirds, the nasal ostium of the nasolacrimal duct is seen as a bell-shaped, osseous aperture.

The middle turbinate and the other osseous projections from the lateral wall or from the middle nasal meatus toward the nasal cavity are generally parts of the ethmoid bone and are responsible for the anatomical structure of this complex and important region. The bony part of the middle turbinate is smaller than that of the inferior turbinate, but it has the same basic anatomical shape. The head of the middle turbinate is very close to the nasal roof and is very near the cribriform-plate area, and sometimes it is pneumatized in areas surrounding the floor of the anterior cranial fossa, increasing the risk for cranial complications when it must be approached surgically. The posterior aspect of the middle turbinate is lopsided anteroposteriorly and articulates in the vertical plate of the palatine bone, under the inferior edge of the sphenopalatine foramen.

The embryological development of the middle turbinate is fundamental for the paranasal sinus arrangement. The origin of this turbinate precedes the origin of the paranasal cavities. Therefore, when the nasal epithelium invaginates in the nasal walls to create the respective sinuses and cells, the middle turbinate insertion has already defined the supra- and infraconchal spaces, namely the middle and superior meati, respectively. This line of insertion is the basal lamella of the middle turbinate. Consequently, paranasal cavities presenting their ostia in the middle nasal meatus are classified as *anterior*; likewise, if the ostia are in the superior nasal meatus, the paranasal cavities are regarded as *posterior* [7, 15, 29, 32].

The nasal middle meatus presents several important elements for the air flow. The individual and bilateral complexity of these formations is probably linked to genetic factors and to an adaptive capacity whose details are still unknown (Fig. 2.7).

Posterior to the nasal middle meatus aditus, the uncinate process is seen as an elongated osseous structure oblique in its superoinferior and anteroposterior aspects. The uncinate process constitutes the posterior border of the nasolacrimal duct and functions as a guide for paranasal-cavity drainage, being related to the recesses, wrinkles, and openings. In the skull, the uncinate process is located between two openings: the anterior and posterior fontanelles of the middle nasal meatus. These fontanelles are closed only by conjunctive-epithelial soft tissues, which help to define the nasal wall of the maxillary sinus (Figs. 2.8, 2.9).

The ethmoid bulla is an ethmoid anatomic element and is somewhat above and posterior to the uncinate process. It may present one or more anterior ethmoidal cells in its interior. Similar to the ethmoid bulla, the uncinate process constitutes an ethmoidal element that adjoins to other non-ethmoidal formations (Fig. 2.8, 2.9; Table 2.1) [1, 8, 11, 15, 20, 25, 28–31, 32].

The junction of the maxillary tuberosity with the anterior edge of the vertical plate of the palatine bone can be seen as a thin and planar surface followed by a vertical osseous thickening in the posterior portion of the middle meatus. Another very thin area, usually translucent and presenting some dehiscences, is located adjacent to this region; this area is important, because the sphenopalatine foramen is present in its superior border. The bony part of the lateral wall of the nasal cavity is completed by the medial plate of the pterygoid process of the sphenoid bone; the pterygoid process is as hard as the frontal process of the maxillary bone (Fig. 2.5) [1, 8, 11, 15, 20, 25, 28–32].

All of these bony anatomical structures can be seen in illustrations of the formalized anatomical

Fig. 2.8. The lateral nasal cavity, middle meatus: right side. *1*, Inferior meatus; *2*, inferior turbinate; *3*, aditus of the middle meatus; *4*, uncinate process; *5*, hiatus semilunaris; *6*, ethmoid bulla; *7*, infundibulum; *8*, accessory ostium of the maxillary sinus; *9*, middle meatus; *10*, middle turbinate; *11*, sphenoid sinus; *12*, choanal region; *13*, pharyngeal opening of the eustachian tuba; *14*, agger nasi

Fig. 2.9. The lateral nasal cavity, middle meatus: left side. *1*, Inferior turbinate; *2*, nasal limen; *3*, atrium of the middle meatus; *4*, middle meatus; *5*, accessory ostium of the maxillary sinus; *6*, uncinate process (reflected); *7*, infundibulum; *8*, maxillary ostium; *9*, ethmoid bulla; *10*, agger nasi; *11*, middle turbinate (reflected)

Table 2.1. Ethmoid cells distribution according to various groups and authors

Authors	Anterior	Middle	Posterior	Total
Figún and Garino [11]	5–10	–	2–4	8–10
Bouche and Cuilleret [8]	–	–	–	8–10
Hollinshead [12]	–	3	–	2–18
Mattox [16]	–	–	–	4–17
Roviere [25]	5	–	2–4	8–10
Terracol and Ardouin [29]	2–5	–	–	2–5
Testut and Latarjet [30]	–	7–9	–	5–16
Van Alyea [32]	–	–	–	4–17

preparations. The periosseous, submucous and respiratory mucous lining can be turgid in some areas and can add considerable thickness to the basic osseous structures. The inferior turbinate is usually not pneumatized and presents a special respiratory lining rich in venous and arterial blood vessels. Moreover, it is highly innervated, appearing hypertrophic or atrophic, with or without wrinkles and folds. These anatomical variations determine the volume of the concha and the size of the inferior nasal meatus and the common nasal meatus (defined as the space between the nasal septum and the septal edges of the turbinates; Figs. 2.8, 2.9) [6, 15, 23, 26, 28, 29, 32].

The middle turbinate, also called nasoethmoidal bone I, presents an anatomical structure similar to that of the inferior turbinate. As part of the ethmoid bone, the middle turbinate may be pneumatized by an anterior and/or posterior ethmoid cell or cells, thereby converting it into the concha bullosa [15, 28, 29]. Thus, its volume increases, reducing the meatal spaces. Any other middle meatus anatomic formation originating from the ethmoid cells can also be pneumatized. The hiatus semilunaris is the space between the uncinate process and the ethmoid bulla. The hiatus has an arc-like shape, with a regular superior concavity corresponding to the characteristics of the uncinate process and the anteroinferior edge of the bulla. The space, limited medially by the uncinate process and laterally by the lamina papyracea, is called the infundibulum. In some cases, extensive pneumatization of the ethmoid bulla is the result of hypertrophy of the ethmoid cells. The volume of the uncinate process may be increased by pneumatization, reducing the space between all these structures and impairing paranasal sinus drainage (Figs. 2.7–2.9) [7, 9, 15, 23, 28].

The superior turbinate, also named nasoethmoidal bone II, is smaller and is located superior and posterior to the middle turbinate, near the sphenoethmoidal recess. It is frequently pneumatized by a large posterior ethmoid cell or by several smaller cells. The superior nasal meatus, under this turbinate, is very small and has few anatomical elements. It is the place where the posterior ethmoid cells' ostia and recesses are located, and it follows a posterior course in the sphenoethmoid recess, where the sphenoid sinus ostium is located at approximately the same level as the caudal end of the superior turbinate (Figs. 2.5, 2.7).

A third ethmoidal structure, named the supreme turbinate, may be present; however, it is always atrophied or modified by the presence of posterior ethmoid cells. The supreme turbinate is frequently described in the fetus and the newborn but is no much frequent in adults [15, 28, 29].

It has become important to determine the exact position of the nasolacrimal duct because of the increased use of the endonasal surgical approach to this important structure. Direct access to the nasolacrimal duct may be hampered by the frontal process of the maxilla which, in some cases, grows over the ductal nasal wall. In order to circumvent this problem, a wide osteotomy may be necessary. Nevertheless, if the (usually very thin) lacrimal bone is prominent in the nasal wall, the displacement of the periosteum is sufficient to expose it. An anatomical relationship with the anterior ethmoid cells (*agger nasi* or perilacrimals) occurs when the lacrimal sac is localized more superiorly; however, the duct itself begins lower and remains free. A pneumatized uncinate process may partially cover the back side of the duct. The nasal duct opens at the inferior nasal meatus, at the anterior third of the nasal wall. Several classic publications describe the variable anatomy of the ostium (Fig. 2.10) [11, 12, 15, 25, 26, 28–30].

The nasal septum has a relatively simple anatomy. The perpendicular plate of the ethmoid forms its upper and posterior parts; the superior, posterior, inferior, and anterior sides of the vomer complete the bony septum like a posteriorly based triangle. The anterior edges of the osseous plates of the ethmoid bone and vomer are attached to the four-sided septal cartilage forming the medial walls of the nostrils. The interosseous and osteocartilaginous sutures of these elements are very important for nasal, facial, and craniofacial growth. The commonly observed septum configuration, deflected from the median sagittal plane, is a consequence of genetic and environmental factors. The occurrence of septal deviations and unilateral or bilateral exostoses are very frequent and have been noted to be present before birth. Indications for nasal septal surgery in children under 11 years old are still very controversial. The bilateral septal lining is made of periosteum, perichondrium, mucosa, and submucosa; it is very rich in vessels and nerves, especially in the anteroinferior region near the incisive foramens. Those foramens are related to the premaxillae, which are still seen in fetuses and newborns. In the superior part of the nasal septum, portions of the lateral wall, and the nasal roof, there is a special mucosa called the olfactory or pituitary mucosa. It contains the terminal branches of the olfactory nerves that pass through the ethmoid cribriform foramens to reach the olfactory bulbs (Fig. 2.11) [11, 25, 28–30].

The nasal cavity has a very rich blood supply that comes bilaterally from branches of the internal and external carotid arteries. The maxillary artery begins at the external carotid artery bifurcation into the parotid gland and, after a extensive and tortuous path, enters the pterygopalatine fossa, where several branches extend to the craniofacial region and into the nasal cavity and paranasal complexes. Two terminal branches of the maxillary artery (the septal and posterior lateral arteries) originate very close to the medial wall of the pterygopalatine fossa and

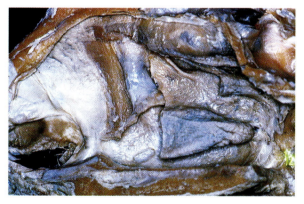

Fig. 2.10. The lateral nasal wall, nasolacrimal duct: right side. *1*, Palate; *2*, nostril; *3*, nasal limen; *4*, inferior meatus; *5*, inferior turbinate (sectioned); *6*, nasolacrimal duct, nasal opening; *7*, atrium of the middle meatus; *8*, nasolacrimal duct; *9*, middle meatus; *10*, middle turbinate (sectioned); *11*, posterior ethmoidal cells; *12*, sphenoid sinus

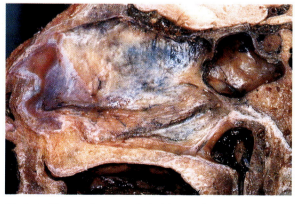

Fig. 2.11. The nasal septum: left side. *1*, Palate; *2*, septal cartilage; *3*, perpendicular plate of the ethmoid bone; *4*, vomer; *5*, sphenoid sinus; *6*, choanal region

course through the sphenopalatine foramen bordering the superior and inferior edges of the pterygopalatine fossa, respectively. The septal artery runs under the periosteum, close to the anterior wall of the sphenoid body; approximately 15 mm below the sphenoid sinus ostium, the artery splits into a nasal branch and a sinusal branch, which continue toward the upper and posterior portions of the nasal septum, respectively. This artery is encompassed by the septal periosteum and delivers branches throughout the septum; its ascending branches anastomose with the terminal septal branches of the posterior and anterior ethmoidal arteries.

The posterior and anterior ethmoidal arteries are branches of the ophthalmic artery, which originates from the internal carotid artery beneath the optic canal. (Fig. 2.12). They are very fine where they anastomose, and their submucosa is quite difficult to dissect. All anterior and septal branches of these arteries contribute to a large vertical and trans-septal anastomotic network. The nasopalatine artery is bilateral and runs diagonally along the septum bordering the upper edge of the vomer, reaching the homolateral incisive canal. These canals join in a "Y" shape where the two arteries anastomose. The nerves from each side communicate to reach the anterior palate, where they are called the incisive vasculonervous pedicle; the nerves then continue to the periodontium of the anterior palatine region. Although it presents a lining less turgescent than that of the nasal turbinates, the nasal septum has arteriovenous concentrations in its anteroinferior regions to deal with the impact of air and foreign material [1, 7, 8, 11, 12, 15, 23, 25, 28–30].

The posterior lateral nasal artery arises from the inferior edge of the sphenopalatine foramen and descends beneath the periosteum in a vertical or oblique direction. It is usually single but may be double or triple if the sphenopalatine foramen is multiple. In such cases, the size is proportional to the foramenal diameter. Where the posterior lateral nasal artery surrounds the foramen edge, it emits a large branch to the caudal end of the middle turbinate; this branch is the middle turbinate artery, which enters the middle turbinate at its head portion, ramifying to its turgescent lining, where the arteriovenous terminal plexuses are located. Ascending and descending branches proceed toward the middle and superior meatus; in the middle meatus, branches extend to the eustachian tube and choana regions. In the aditus of the middle meatus, terminal branches of the middle turbinate artery and branches of the facial and infraorbital arteries extending through the external nose wall form the external nasal vasculonervous plexus of the anterior ethmoidal complex. The posterior lateral nasal artery reaches the caudal end of the inferior turbinate.

The inferior turbinate is innervated and vascularized in a similar fashion. The vasculonervous concentration in this turbinate is even higher because of its volume and turgescence. Volume and turgescence patterns are probably related to the level of contact with the air flow. The terminal branches of the posterior lateral nasal artery extend to the inferior meatus and the nasal floor, where they meet the transpalatine and anterior superior alveolar arteries. Usually, vascular branches accompany their nervous homologs (Fig. 2.12, 2.13).

The main branches of the ophthalmic artery (the anterior and posterior ethmoidal arteries) begin in the orbit, and an intermediate branch (the middle

Fig. 2.12. Arteries of the lateral nasal cavity: left side. *1*, Palate; *2*, sphenoid sinus; *3*, posterior lateral nasal artery and branches to inferior meatus (*4*); *5*, inferior nasal concha; *6*, choanal region; *7*, middle meatus; *8*, middle turbinate; *9*, superior meatus; *10*, superior turbinate; *11*, nasal roof; *12*, atrium of the middle meatus; *13*, nasal limen; *14*, nostril

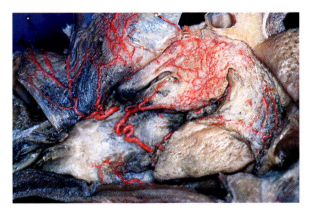

Fig. 2.13. Arteries of the nasal cavity: left side. *1*, Septal artery and branches to the nasal septum (*2*); *3*, posterior lateral nasal artery and branches to inferior meatus (*4*); *5*, inferior turbinate (sectioned); *6*, middle meatus; *7*, middle turbinate (sectioned); *8*, superior meatus; *9*, superior turbinate; *10*, nasal roof; *11*, atrium of the middle meatus

ethmoid artery) may exist. These branches and their corresponding nerves leave the orbit through foramens and analogous canals that extend toward the lateral edge of the ethmoid cribriform plate. The blood flow in such canals may occur at the roof of the ethmoid sinus, between and under ethmoid cells. Blood flow occurs this way because the origin of these vasculonervous components precedes the origin of the respective canals. In embryonic development, while the arteries extend from the orbit to the endocranium, the ethmoid cells remain protected by a periorbital wrapper around which the osseous canals and ethmoid cells will be formed. After growing and expanding, these ethmoid cells will play different roles in their anatomical relationships to each other and to the ethmoid canals (Fig. 2.14).

Venous drainage of the nasal cavity is via the tributary veins of the pterygoid plexus (due to an extensive ramification throughout the nasal and paranasal mucosa), while the external nose drains to the facial tributaries. Lymphatic ducts of the nasal cavity converge at the nasopharynx, whereas the external ducts extend to the submandibullary and genian regions [7, 8, 12, 15, 20, 23, 25, 26, 29, 34].

The primary nerves supplying the nasal cavity are the maxillary nerve, its pterygopalatine ganglion extensions, and the pericarotid sympathetic plexus, which is distributed along the internal and external carotid-artery branches. The maxillary nerve emerges from the foramen rotundum at the pterygopalatine fossa. From the inferior edge of this foramen, a canal containing this nerve leads laterally toward the zygomatic fossa. This canal descends and serves as a guide to the descending palatine branch. The maxillary nerve follows in a double lateral arch in the zygomatic fossa where, together with the infraorbital artery, it enters the infraorbital canal as an infraorbital branch. Very close to the foramen rotundum inside the pterygopalatine fossa, the maxillary nerve extends branches to the pterygopalatine ganglion. At this point, or inside the canal, the zygomatic orbital nerve emerges laterally; this nerve is the parasympathetic supply for the lacrimal gland. Due to its involvement with the maxillary artery and branches and with the pterygopalatine ganglion itself, there is enormous complexity in the nerve distribution in the interior of the pterygopalatine fossa. Several communications between the ophthalmic and maxillary nerves have been observed at the superior orbital fissure.

Immediately after leaving its foramen, the pterygoid canal nerve posteriorly reach the pterygopalatine ganglion. The pterygopalatine-canal nerve leaves the ganglion and enters its respective foramens and canals. While the pterygoid canal nerve has pre-ganglionic sympathetic and parasympathetic fibers, the pterygopalatine canal nerve has postganglionic sympathetic and parasympathetic fibers that supply the pharyngeal aperture of the eustachian tube. In conclusion, the nervous structures described above carry afferent and secretomotor fibers to the bucconasal, sinusal, nasopharyngeal, and orbital regions.

Sufficient knowledge of the topographic anatomy of this particular region is of great importance, especially when surgical access to the endonasal area is planned. The anatomical complexity of the paranasal cavities and their multiple characteristics are still considered a challenge for rhinosurgeons [1, 7, 8, 11, 12, 15, 20, 25, 26, 29, 30, 34].

Paranasal Sinuses

The paranasal sinuses are extensions of the nasal cavity originating from the cartilaginous nasal capsule after invagination of the nasal epithelium into the craniofacial bones. However, the original relationship to the nasal cavity is preserved through their original ostia. These cavities (or paranasal sinuses) are named according to the bones inside which they develop and grow; they are called the frontal sinuses, maxillary sinuses, ethmoid sinuses, and sphenoid sinuses.

Fig. 2.14. Anterior and posterior ethmoid arteries, cranial view. *1*, Orbit; *2*, optic nerve; *3*, supraorbital nerve; *4*, ophthalmic artery; *5*, posterior ethmoidal cells; *6*, sphenoid sinus; *7*, posterior ethmoid artery; *8*, anterior ethmoidal cells; *9*, anterior ethmoid artery; *10*, frontal sinus; *11*, superior oblique muscle

All paranasal sinuses are bilateral (even the sphenoid, which is located in a midline bone). These bones originally exhibit bilateral ossification, later providing space for sinusal development. Consequently, an osseous septum remains between each side of the sinus cavities; this septum is usually paramedian, rendering the paranasal cavities asymmetrical in shape, with irregular growth on one side or the other. This leads to a lack of symmetry between analogous sinuses.

The paranasal sinuses start to develop at approximately 2 months of intrauterine life, in the period of transition between embryo and fetus; development begins with the anterior ethmoidal cells and the maxillary sinus. Development of the sphenoid and frontal sinuses begins at approximately the fourth month.

The sphenoid and frontal sinuses originate from the sphenoethmoidal and frontoethmoidal recesses, respectively. They are formed by the pneumatization of the epithelial canals in a single stage.

The growth of the paranasal sinuses is slow during the fetal period but, after birth, it accelerates in spurts of growth, mainly in the middle and superior faces and middle and anterior cranial fossae. These growth spurts occur during childhood, puberty, and adolescence. The paranasal sinuses may expand and occupy significant portions of the respective bones; the sinuses grow in different directions and form sinusal expansions and recesses. Sinusal expansions may create intrinsic weaknesses in the respective bones by making the walls of the bones thinner, changing their architecture, and rendering them prone to fractures and invasion by diseases.

Moreover, the sinuses may affect surrounding structures by creating dehiscences into the sinusal cavity; this represents a great risk during surgical procedures. All paranasal sinuses attached to the nasal cavity are originally linked to the ethmoid bone and should be described together with it, i.e., as ethmoidofrontal, ethmoidomaxillary, and ethmoidosphenoidal sinuses.

As an essentially ethmoidal anatomical element, the middle meatus is probably the center of origin of most of the other paranasal sinuses, which maintain communication with the ethmoid sinus through the middle meatus or a structure close to it. As a result of the expansion of the sinus cavities and their contiguity with other regions of the head, one could give them additional names; for example, the frontal sinus could be categorized as a "brain sinus", the maxillary as a "dental sinus", and the sphenoid as a "skull sinus" [1, 3, 7, 11, 12, 15, 20, 22, 24, 28, 32]. The paranasal sinuses are actually pneumatized nasal expansions from the most primitive olfactory formation; they are inferiorly and laterally directed, defining wide cavities in different bones. The sinuses remain covered by altered respiratory epithelium that retains continuity with the nasal epithelium.

There are several examples of sinusal expansions that can affect surrounding structures, such as posterior ethmoidal cells near the optical canal or anterior ethmoid cells in the middle turbinate. The sphenoid sinus may present recesses to the anterior clinoid process, pituitary fossa, posterior clinoid process, optical and round canals, the pterygoid process, maxillary sinus, pterygopalatine fossa, superior orbital fissure, internal carotid artery, and the pterygopalatine ganglion and foramen ovale [1, 3, 7, 15, 22, 25, 26, 28, 29].

Frontal Sinus

The frontal sinus is formed in each hemifrontal bone, from the frontal recess. A parasagittal septum separates the two sinuses, completely, individualizing them anatomically, physiologically, and pathologically. Incomplete frontal septae are found in the interior of each sinus, dividing it into recesses. However, a complete septum resembling an intersinusal septum may occur (Fig. 2.15).

The frontal sinuses exhibit a wide variety of morphologies and volumes and may reach enormous dimensions with zygomatic, supraorbital, and parietal expansions. The frontal sinuses are absent in 16% of cases [14, 15, 24, 28, 29].

Since anterior ethmoid cells originate simultaneously around the ethmoidofrontal recess, there is competition for space; frequently, the frontal sinus is

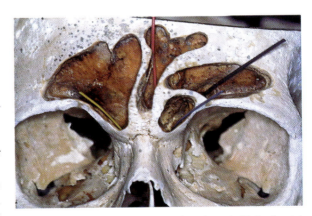

Fig. 2.15. The frontal sinus, anterior view. *1*, Right frontal sinus; *2*, septum of the frontal sinus; *3*, left frontal sinus; *4*, accessory septae of the frontal sinus; *5,6*, probes into the frontonasal ducts and anterior ethmoidal cells; *7*, orbit; *8*, nasal cavity

Fig. 2.16. The nasofrontal duct: left side. *1*, Probe into the nasofrontal duct; *2*, nasofrontal duct opening into the infundibulum; *3*, uncinate process; *4*, ethmoid bulla; *5*, middle meatus; *6*, maxillary sinus accessories ostia; *7*, middle turbinate (reflected); *8*, inferior turbinate (sectioned); *9*, inferior meatus; *10*, nasal septum (reflected); *11*, frontal sinus

duct that expands in the frontal sinus as it reaches the frontal bone (Fig. 2.15).

The anatomical relationship between the frontal sinus and the supraorbital cells is highly variable. Wide supraorbital expansions may weaken the superior wall of the orbit and may culminate in dehiscences.

Some studies have tried to relate the growth pattern of the frontal sinuses to the development of the sense of smell. In human beings, the frontal sinuses and olfaction are relatively underdeveloped. The arterial supply to the frontal sinus comes from the maxillary and ophthalmic arteries, which originate at the external and internal carotid arteries, respectively.

Tributaries of both ophthalmic and intracranial veins participate in the venous drainage. Lymphatic drainage also follows the orbital, nasal, and anterior cranial avenues.

Sensory innervation comes from anterior ethmoidal, supratrochlear, and supraorbital branches of the ophthalmic division of the trigeminal nerve. The secretomotor and vasoconstrictor innervation is mediated by parasympathetic branches of the nervus intermedius and by sympathetic carotid plexuses, which are incorporated by the trigeminal sensory branches at the pterygopalatine ganglion [1, 3, 7, 8, 11, 15, 22, 24, 26–30].

Maxillary Sinus

The maxillary sinus is the first to originate in the middle meatus and remains linked to it through the main maxillary sinus ostium and through one or more non-functional accessory ostia (Giraldes' ostia). The maxillary sinus can occupy the entire maxilla by expanding within it in every possible direction. At the second month of fetal life, the maxillary sinus is merely an epithelial bud of the nasal lining invaginating into the ossifying maxilla (as a first pneumatization).

At birth, the maxillary sinus is a small oval cavity a few millimeters in size overlying the dental germ of the first deciduous molar, far from the orbital floor and close to the nasal wall. The infraorbital canal is approximately the same distance laterally. Before age 2 years, the sinus reaches the dental bud of the first permanent molar.

During the first years of life, the maxillary sinus floor is a little higher than the nasal floor. At approximately 8 years of age, they are at the same level and, after 12 years of age, the floor of the sinus is about 4 mm lower than the nasal floor, allowing access to the maxillary sinus through the inferior meatus.

compressed beginning at the start of development, becoming superiorly elongated and reaching supraorbital spaces. Consequently, its ostium remains connected to the floor of the sinus cavity that leads to the frontonasal duct; this cavity has a variable location and extent (Fig. 2.16).

In general, the frontal sinus develops from the frontal recess, close to the nasal surface of the lateral wall; it can also (though not often) develop from the infundibular recess together with other anterior ethmoid cells. Occasionally, the frontal sinus can develop from a prebullar ethmoidal cell. Although one cell usually forms the frontal sinus, others may exhibit delayed expansion toward the sinus floor, establishing intrasinusal prominences known as frontal bullae (Fig. 2.15) [14, 15, 22, 24, 26, 28, 29]. In some cases, more than a single cell invades the frontal bone, thus giving rise to multiple sinuses. An important fact is that each cavity or sinus has its own ostium. The location of the frontonasal duct facilitating frontal-sinus drainage will depend on the placement of the sinus cell of origin. If the cell is close to the bone basis, the duct is practically unnecessary, because the opening will be wide and direct and will probably open into the frontonasal recess. Usually, the more distant the original cell, the longer its pathway to the frontal bone. When the cell reaches the bone, it can be compressed and narrowed by other cells, resulting in a generally straight (but narrow)

Maxillary sinus growth accelerates following maxillary dental eruptions (Fig. 2.17) [2, 3, 7, 11, 15, 20, 22–25, 28–30, 32, 34].

As the sinus expands superiorly, its roof becomes more intimately related to the bone of the ethmoidal cells and the infraorbital canal, which can become dehiscent over wide areas, exposing vasculonervous structures (Fig. 2.18). In children, the maxillary sinus has the shape of a solid geometric pyramid or cube. In reality, the cuboidal shape is a better representation of the sinus anatomy, since it has one roof, one floor, and four walls.

The sinus roof is related to the inferior orbital wall, and the floor is related to the alveolar process of the maxilla (Fig. 2.18). The anterior wall of the maxillary sinus corresponds to the analogous face of the maxilla; the lateral wall corresponds to the maxillary pyramidal process, where the zygomatic process is found. The posterior wall extends from the lateral wall and corresponds to the maxillary tuberosity. Finally, the medial wall of the maxillary sinus corresponds to the lateral wall of the nasal cavity, where the maxillary sinus ostia are located. Usually, the thicknesses of the lateral and posterior walls are symmetric, even in well-pneumatized sinuses, where the walls are quite thin. The maxillary artery (or some of its branches) can be found close to the posterior wall, which can have areas of dehiscence. Terminal branches of the maxillary and trigeminal nerves (in the zygomatic region at the entrance of the pterygopalatine fossa) can also be present.

Near its facial opening, the infraorbital canal enlarges and becomes the anterior superior canal. As the roof of the sinus involves the region of the infraorbital canal, there may be dehiscences with lateral recesses. The lateral and posterior walls and the roof may be very thin, and the superior alveolar arterial and nervous branches may be visible through the thin, bony wall (Fig. 2.18).

The maxillary sinus medial wall is quite irregular and concave in its superior part, where it joins the sinus roof. Its main ostium is placed slight anteriorly in the medial wall. Just adjacent to it is a fibrous septum binding the medial wall and the roof of the sinus, protecting the ostium. Anterior to the ostium, near the anterior sinus wall, a large vertical convexity that corresponds to the nasolacrimal duct can be seen. Just below it, a vertical convexity extends horizontally and posteriorly, corresponding to the inferior nasal meatus concavity. The maxillary sinus accessory ostium is usually a smooth surface between these two convexities in the medial wall. Superior and anterior to the ostium, irregular prominences and recesses from the ethmoidal cells impinging on the middle meatus sometimes occur [2, 9, 15, 22, 26, 28].

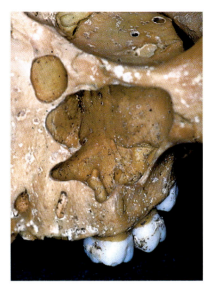

Fig. 2.17. The maxillary sinus, anterolateral view: left side. *1*, Orbital floor; *2*, maxillary sinus; *3*, dental roots extending into the maxillary sinus floor

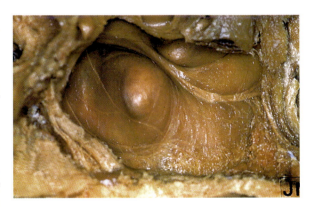

Fig. 2.18. The maxillary sinus, nasal view: left side. *1*, Maxillary sinus, posterolateral wall; *2*, roof of the maxillary sinus; *3*, infraorbital canal; *4*, anterior superior alveolar nerve; *5*, vasculonervous structures of the pterygopalatine fossa; *6*, palate; *7*, superior dental nervous plexus

The main ostium of the maxillary sinus is not a simple opening but is a superoinferiorly and posteroanteriorly oblique fibromucous canal a few millimeters in diameter and length. Its sinusal and nasal openings vary depending on the turgescence of the nasosinusal mucosa.

In the roof of the maxillary sinus are the infraorbital vasculonervous branches of the maxillary artery and nerve, which extend to the anterior wall and the alveolar process, where they are distributed to the periodontium and superior teeth. This is the reason they are frequently damaged during the transmaxillary sinusal approaches.

The maxillary sinus floor has a varied anatomical relationship to the osseous alveolar process (and particularly to the superior premolar and molar teeth); this can affect later eruption of the permanent dentition. The osseous recesses and prominences in the sinus floor are generally related to the molar roots, whose tips may have no bony covering and are then exposed to the sinus lining. Such a condition also increases surgical risks in the maxillary sinus and in exodontic and endodontic procedures. Prominent transverse osseous septae may also develop and create recesses, which can add to surgical difficulties (Fig. 2.17) [2, 20, 22, 23, 28–30].

Sinus endostoses may be observed; these sometimes reduce the size of the sinus cavity (agenesis of the maxillary sinus). The maxillary-sinus blood flow comes from orbital, nasal, tuberal, and palatine sources via the maxillary-artery branches. Venous drainage occurs via the nasal tributaries and the pterygopalatine plexus, and the lymphatics drain into the nasal, oral, and submandibular regions. Sensory innervation is via the superior alveolar, infraorbital, nasal, palatine, and gingival branches. The secretomotor activity and vasoconstriction are controlled by terminal parasympathetic branches of the nervus intermedius via the trigeminal nerve, and sympathetic fibers come from the carotid plexus [2, 15, 22, 26, 28, 29, 33].

Ethmoid Sinus

The ethmoid sinus is a complex series of multicellular expansions. Because of its morphological characteristics and spatial distribution, this sinus is considered the most important of the paranasal sinuses. Consequently, surgical approaches are often risky due to the close relationship of these cells with noble structures, such as the orbit, optic nerve, and anterior cranial fossa. In the transition from embryo to fetus during the second or the third month, the folds and thickenings of the middle meatus epithelium may already be seen as a precursor to the anterior ethmoid cells, which will give rise to the future paranasal sinuses.

The middle meatus starts to develop by the end of the third month of fetal life. The nasal elements, such as the ethmoid bulla, frontal recess, infundibulum, and uncinate process are already designated along a prolonged and irregular wall. Some supra- and infrabullar ostia are already present, marking the beginning of the epithelial expansions that will form the anterior ethmoid cells [3, 7, 26, 28, 29, 32].

The majority of the expansions of these cells occur in the first years of life. Their growth toward the cribriform plate develops the ethmoid fovea; as the cells expand into the supra- and infraorbital, maxillary, middle turbinate, sphenoid, and lacrimal areas, they assume the names of these structures (although their morphology depends on the adjoining osseous architecture) [3, 7, 26, 28, 29, 32]. A growth spurt of the ethmoid cells is seen in puberty, and some developmental expansion may last until approximately 25 years of age.

The ethmoid cells maintain anatomical relationships with the osseous plates (such as the middle, superior, and supreme turbinates) that project from the lateral wall into the nasal cavity and with the ethmoid bulla and the uncinate process. The anterior group formed by the bullar, infundibular and frontal cells is more numerous but less voluminous and develops drainage into the middle meatus, while the posterior group (which forms larger but less numerous cells) drains into the superior and supreme meati. The suprainfundibular and suprabullar cells are found in the superior recess of the middle meatus. They drain into the bullar–uncinate junction at the terminal region of the hiatus semilunaris (Fig. 2.19) [4–6, 13, 16, 17, 24].

The most frequent infundibular cell is the *agger nasi*, which results from an infundibular cell expansion toward this region. The supreme posterior cells can be compressed between the ethmoid bulla and the sphenoid sinus when excessive growth of the bulla toward the skull occurs.

Expansions of the ethmoid cells toward the maxillary infraorbital plate, maxillary sinus, and pterygopalatine fossa may also occur. Great expansions may border the anterior wall of the sphenoid sinus and the pituitary fossa, and sometimes these expansions cross the midline to reach the opposite side.

Fig. 2.19. Ethmoid cells, orbital view: right side. *1*, Anterior ethmoidal cells; *2*, posterior ethmoidal cells; *3*, maxillary sinus; *4*, infraorbital nerve; *5*, optic nerve (sectioned); *6*, ophthalmic artery and branches; *7*, ophthalmic nerve and branches

Projections of the ethmoid cells into the maxillary sinus are generally covered by the sinusal lining, maintaining regular, concave surfaces of the superior, posterior, and medial walls of the cavity. The removal of the sinus mucoperiosteum exposes the orbitomaxillary cells prominences in these regions [1, 4, 5, 7, 13, 15–17, 28, 29].

The anterior ethmoid cells are related to the anterior half of the medial orbital wall, whereas the posterior cells relate to the posterior half (Fig. 2.20) [15, 26, 28, 29, 32]. The anterior ethmoid foramen is the anatomical border of these two groups of cells. The ethmoid foramens in the medial orbital wall are important anatomical landmarks for orbitoethmoid surgical approaches [4, 13, 16, 17, 28, 29, 32].

The most anterior ethmoid cells, the *agger nasi*, are deeply positioned with regard to the lacrimal fossa. There are usually only one to four of these cells (which are poorly developed in human beings), but they may invade the lacrimal bone, reaching the frontal process of the maxillae (Figs. 2.21, 2.22).

The cells of the frontal recess drain into the recess itself above the infundibulum; sometimes, one of these cells gives rise to the frontal sinus. In such a case, the frontal duct appears to be compressed by other cells of the recess, impairing frontal-sinus drainage.

One to four bullar cells may open into the anterior wall of the ethmoid bulla in the hiatus semilunaris or into a medial fold superior to the bulla called the suprabullar fold or lateral sinus. The posterior ethmoid cells number between two and six and tend to be larger than the anterior ethmoid cells (Figs. 2.20–2.22).

The distance between the lacrimal crest and the anterior ethmoid foramen is approximately 24 mm; there is approximately 12 mm between the anterior and posterior foramens and 6 mm between the posterior and optical foramens. However, the posterior ethmoid foramen is multiple in 30% of cases [1, 4, 5, 13, 15–17, 22, 24, 26, 28, 29].

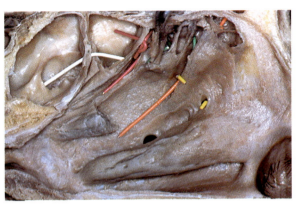

Fig. 2.21. The nasal lateral wall and paranasal sinuses: left side. *1*, Nostril; *2*, nasal limen; *3*, palate; *4*, inferior meatus; *5*, inferior turbinate; *6*, middle meatus; *7*, maxillary sinus, accessories ostia; *8*, uncinate process; *9*, frontal sinus drainage into the infundibulum (*orange probe*); *10*, ethmoid bulla; *11*, lateral sinus; *12*, nasal ostia of anterior ethmoidal cells (*green probe*); *13*, middle turbinate (sectioned); *14*, nasal ostium of a large posterior ethmoidal cell (*red probe*); *15*, nasal ostium of the sphenoid sinus (*white probe*); *16*, sphenoid sinus; *17*, agger-nasi cell

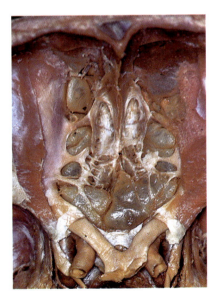

Fig. 2.20. Ethmoid cells, cranial view. *1*, Optic chiasma; *2*, optic nerve; *3*, internal carotid artery; *4*, cavernous sinus; *5*, sphenoid sinus; *6*, posterior ethmoidal cells; *7*, posterior ethmoidal foramen; *8*, middle ethmoidal foramen; *9*, anterior ethmoidal foramen; *10*, anterior ethmoidal cells; *11*, cribriform plate; *12*, frontal sinus; *13*, orbit

Fig. 2.22. The nasal lateral wall, sagittal section: right side. *1*, Inferior turbinate; *2*, atrium of the middle meatus; *3*, middle meatus; *4*, accessory ostium of the maxillary sinus; *5*, middle turbinate (sectioned); *6*, ethmoid bulla; *7*, anterior ethmoidal cells; *8*, uncinate process, pneumatized; *9*, large posterior ethmoidal cell; *10*, sphenoid sinus; *11*, frontal sinus

The ethmoid complex receives its blood supply from the ethmoid arteries and the maxillary artery's nasal branches. The venous drainage is to the cavernous sinus, facial veins, and pterygoid plexus tributaries. The lymphatic vessels converge to the nasal cavity and meninges.

Sensory innervation is supplied by the trigeminal nerve via the maxillary nerve, the superior, posterior, and nasal branches of the trigeminal ophthalmic division, and the anteroposterior ethmoid branches. Secretomotor fibers reach the ethmoid cell mucosa with both parasympathetic supply fibers (intermediate nerve, superficial petrosal nerve, pterygopalatine ganglion, maxillary nerve branches) and sympathetic fibers from the superior cervical plexus [1, 7, 11, 16, 19, 24–26, 28–30, 34].

Sphenoid Sinus

At approximately the fourth month of fetal life, the sphenoethmoid recess is joined by the sphenoid concha, marking the beginning of the sphenoid sinus. Between the fifth intrauterine month and the fifth month after birth, the growth process proceeds slowly. At age 3 years, the sinus cavity is still limited to the soft parts of the nasal surface of the sphenoid body, but osseous reabsorption and pneumatization then begin. Until age 7 years, the recess remains between the concha and the sphenoid body, but the sinus then grows to its pre-sphenoid configuration and later reaches the base of the sphenoid bone. As with the frontal sinus, sphenoid pneumatization is bilateral, presenting a paramedian sagittal osseous septum (usually placed slightly to the left), which completely separates the two sides into compartments that vary in shape and volume.

Other accessory or secondary vertical or oblique septations may be present, but they are always incomplete. A partial septum, visible in an axial computed-tomography scan, is important because of its vertical trajectory toward the internal carotid artery region close to the posterolateral wall of the sphenoid sinus (Fig. 2.24) [3, 5–7, 10, 11, 17, 22, 24, 26, 27, 29, 32, 33].

The basic sphenoid sinus shape is a cube, with four walls, one roof and one floor, all of which are important because of their anatomosurgical relationships. The anterior wall of the sinus is concave, lodging the homolateral sphenoid sinus ostium in its superior and most medial portion. This wall is related to the posterior wall of the last ethmoid cell. The lateral wall of the sphenoid sinus is not bony in its upper portion but borders the cavernous sinus, which originates from the dura mater, is shaped like an oval purse, contains a net-like fibrous parenchyma filled with venous blood, is crossed from posterior to anterior by the internal carotid artery, the VI cranial nerve, and is laterally and superiorly bordered by the III, IV, and VI cranial nerves. The lateral wall of the sphenoid sinus may expand as far as the foramen rotundum, making the sinus dehiscent and exposing its vasculonervous structures. The inferior boundary of the cavernous sinus coincides with the superior edge of the foramen rotundum; hence, expansions of the sphenoid sinus to the greater wing of the sphenoid bone occur under the cavernous sinus.

The posterior or basilar wall of the sphenoid sinus is related to the basilar process of the sphenoid body. The occipital venous sinus, which communicates with both cavernous sinuses, is located under the dura mater of this region. A rupture of the sinus wall exposes this venous sinus, which may allow the development of infection. Just outside the area of the junction of the lateral and posterior sinus walls is the vertical segment of the internal carotid artery in its ascending course as it enters the cavernous sinus. A posterior expansion of the sinus brings it into contact with the carotid canal, creating a dehiscence and exposing the internal carotid artery itself (Figs. 2.23, 2.24) [27, 28, 33].

The medial wall of the sphenoid sinus consists of its septum which, in most cases, is positioned slightly laterally. This septum may be either extremely thin or very thick and may be parasagittal, vertical, or oblique.

In the nasal cavity, the anterior wall of the sphenoid sinus articulates with the vomer, forming the superior choanal region, which is the transition to the nasopharynx. On each side, the sphenoid sinus ostia are lateral to the septal vomer at the most medial and superior aspects of the anterior sinus wall.

The roof of the sphenoid sinus abuts the pituitary fossa. Its superior expansion usually occurs anterolaterally, toward the lesser wing and the anterior clinoid process, approaching the optic canal and sometimes resulting in dehiscence of the optic nerve. This extremely close anatomical relationship occurs in 8% of cases (Fig. 2.24).

Expansion of the sphenoid sinus into its floor very frequently brings it into contact with the pterygoid-process root. The pterygoid canal is frequently involved in this expansion and may become dehiscent, increasing the risk of damage during surgical approaches to the sphenoid sinus (Figs. 2.23, 2.24). Due to the multiple expansions of the sphenoid sinus, partial incompletely formed septae (sagittal or oblique) are always present, forming several recesses that make surgical access more difficult [1, 5–7, 10, 15, 17, 27–29, 32, 33].

Fig. 2.23. The basilar sphenoid sinus: left side. *1*, Sphenoid sinus; *2*, pterygoid canal; *3*, sinusal prominence of the internal carotid artery; *4*, carotid-optic recess (*arrow*) and intrasinusal prominence of the optic canal; *5*, pituitary gland

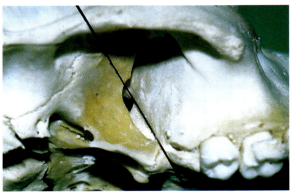

Fig. 2.25. Zygomatic and pterygopalatine fossae, lateral view: left side. *1*, Pterygoid process; *2*, maxillary tuberosity; *3*, zygomatic arc; *4*, sectional plane (*black*) through the pterygopalatine fossa

Fig. 2.24. The basilar sphenoid sinus: left side. *1*, Accessory sinus septum; *2*, sinusal prominence of the internal carotid artery; *3*, sinusal prominence of the optic canal; *4*, sinusal prominence of the round canal; *5*, sphenoid ostium; *6*, pituitary gland; superior (*7*), middle (*8*) and inferior (*9*) nasal turbinates; carotid-optic recess (*arrow*)

Pterygopalatine Fossa

The pterygopalatine fossa is a bilateral, interosseus space at the craniofacial junction. Due to its location and anatomical features, it is considered together with the structures of the paranasal sinuses. It has great anatomic and surgical importance, because it frequently harbors tumors and vascular pathological processes and because it is frequently approached surgically in cases of trauma and orthognathic operations.

The anatomical shape of the pterygopalatine fossa resembles a four-sided pyramid with an imaginary base, anterior, medial, and posterior walls, and an imaginary wall inferiorly converging to the vertex (Fig. 2.25). Its base corresponds to the region of the orbital vertex and the inferior and superior orbital fissures. Its anterior wall is bordered by a small vertical portion of the maxillary tuberosity close to its junction with the palatine vertical plate. The medial wall is formed by the vertical plate of the palatine bone and is crossed by the sphenopalatine foramen (which, in turn, is a result of the orbital and sphenoid processes articulating this plate with the posterior ethmoid and sphenoid body, respectively). The posterior wall corresponds to the anterior face of the pterygoid process of the sphenoid bone. The lateral wall of the fossa lies against the skull and is sealed by fibrous tissue that only allows the passage of vascular and nervous structures. The pyramid's vertex or inferior aspect is the junction of the walls, where the palatine osseous canals connect the pterygopalatine fossa with the oral cavity through the hard palate. Canals and foramens open in the walls of the pterygopalatine fossa, covering vascular and nervous structures of vital importance to the craniofacial regions.

Interosseous spaces, such the superior and inferior orbital fissures, are sealed both in life and in death, avoiding direct communication and imbalance of pressure between the orbit and the pterygopalatine fossa. Adipose tissue is present in the fossa to maintain and support its contents (Figs. 2.26–2.28) [11, 18–20, 35].

During the embryonic period, the dura mater spreads exocranially to the orbit, becoming the periorbita; this also occurs with the optic and pterygoid canals, the foramen rotundum, and the pterygopalatine fossa itself (with enclosed contents) in a dura mater sheath that accompanies every vascular and nervous element leaving the fossa (Fig. 2.27).

Superiorly and laterally, the foramen rotundum is seen in the posterior wall of the fossa (as is the

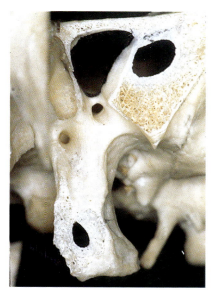

Fig. 2.26. The pterygopalatine fossa, posterior wall: left side. *1*, Superior orbital fissure; *2*, foramen rotundum; *3*, pterygoid foramen; *4*, sphenoid sinus; *5*, pterygoid process; *6*, nasal cavity; *7*, zygomatic fossa

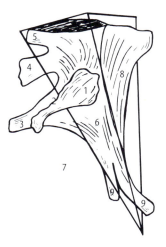

Fig. 2.28. Schematic drawing of the pterygopalatine fossa: left side. *1*, Sphenopalatine foramen region; *2*, pterygopalatine canal; *3*, pterygoid canal; *4*, foramen rotundum; *5*, superior orbital fissure; *6*, medial or nasal wall of the fossa; *7*, posterior or pterygoid wall of the fossa; *8*, anterior or sinus wall of the fossa; *9*, palatine canals

Fig. 2.27. A pterygopalatine fossa enveloped in the periorbital extension. *1*, Sphenopalatine foramen region; *2*, pterygopalatine canal; *3*, pterygoid canal; *4*, canal rotundum; *5*, superior orbital fissure; *6*, medial or nasal wall of the fossa; *7*, posterior or pterygoid wall of the fossa (*arrow*) and anterior or sinusal wall of the fossa; *8*, maxillary sinus; *9*, pharyngeal opening of the eustachian tube; *10*, sphenoid sinus

pterygoid foramen, the medial, inferior opening between the pterygoid canal and the fossa). The pterygopalatine fossa is linked to the foramen rotundum, bringing to it the artery and nerve and to the pterygoid canal, with its pre-ganglionic parasympathetic and post-ganglionic sympathetic fibers, the branches of maxillary artery, and of the internal carotid artery (which anastomoses, inside the canal). One or two smaller pterygopalatine foramens are located medially and inferiorly; these continue in canals between the sphenoid process of the palatine bone and the sphenoid body (posteroinferior to the pharyngeal opening of the eustachian tube). The foramens carry vascular and nervous parasympathetic elements to the mucosa of the tube's ostium (Figs. 2.26–2.28). In the medial wall of the fossa, beneath the anterior wall of the sphenoid sinus, the sphenopalatine foramen is covered by the fibrous sac of the pterygopalatine fossa. Most of the vascular and nervous structures of the nose and paranasal sinuses reach the nasal cavity through the sphenopalatine foramen (Figs. 2.27, 2.28) [11, 18–21, 35].

The vascular and nervous structures of the pterygopalatine fossa are complex but important for the neurovisceral structures of the head. In the maxilla tuberosity at the zygomatic fossa, the maxillary artery presents its tuberal portion before entering the pterygopalatine fossa. This artery is very sinuous and splits into the superior, posterior, gingival, oral, descending palatine, and infraorbital alveolar branches, all of which are followed by their respective nerves. The maxillary artery and its anterior and superior posterior alveolar nerves are fixed firmly to the tuberal wall, thus maintaining their positions. The infraorbital artery is the most sinuous branch of the maxillary artery and, in approximately 40% of cases, the maxillary artery and the superior posterior alveolar artery originate from the same trunk. After splitting into these branches, the maxillary artery continues toward the pterygomaxillary fissure, crosses its fibrous wall, and enters the pterygopalatine fossa (generally near the superior edge).

In the interior of the fossa, the maxillary artery extends from the lateral end to the medial end and from the anterior ganglion to the pterygopalatine ganglion, which is a motor neural structure for the parasympathetic pathways of the nervus intermedius. This nerve is localized close to the aperture of the pterygoid canal in the posterior wall of the fossa (the pterygoid foramen, a funnel-shaped opening located very close to the sphenopalatine foramen and, consequently, to the medial wall of the fossa and the nasal cavity). The branches of the maxillary artery in this short course are tiny and accompany the maxillary nerve in the round canal, the pterygoid and pterygopalatine canals, and the orbital vertex. They extend downward into the small palatine canals. In front of the ganglion, the maxillary artery is surrounded by several nervous branches and bifurcates into the septal and posterior lateral nasal branches, which reach the sphenopalatine foramen to be distributed into the nasal cavity. The septal artery extends along the superior edge of the foramen, whereas the posterior lateral nasal artery extends along the inferior edge. The locations and branchings of these arteries were described in the section concerning the nasal cavity (Figs. 2.29, 2.30).

The maxillary nerve has a short length in the pterygopalatine fossa; leaving the foramen rotundum, it splits into the zygomatic orbital branch, whose parasympathetic components join the lacrimal nerve and proceed to the lacrimal gland. Near the posterior wall of the pterygopalatine fossa, the maxillary nerve divides into communicating branches that extend toward the pterygopalatine ganglion. After this segment, the trigeminal branches contain parasympathetic elements. Later, the maxillary nerve leaves the pterygopalatine fossa laterally and courses anteriorly through the zygomatic fossa to reach the infraorbital fissure, where it enters the infraorbital canal together with its analogous artery, becoming the infraorbital vasculonervous pedicle (Figs. 2.29, 2.30). The venous drainage inside the fossa connects the nasal veins to veins of the pterygoid plexus; the lymphatic drainage follows the drainage of the zygomatic fossa toward the cervical and submandibular regions and may partially follow the nasopharyngeal drainage [11, 18, 19, 20, 21, 35].

Fig. 2.29. The pterygopalatine fossa, posterolateral dissection: left side. *1*, Maxillary artery; *2*, palatine artery; *3*, posterior superior alveolar artery and nerves; *4*, pterygopalatine ganglion (sectioned); *5*, septal artery; *6*, posterior lateral nasal artery; *7*, nasal cavity, lateral wall; *8*, palatine nerves; *9*, infraorbital artery; *10*, maxillary nerve; *11*, maxillary tuberosity

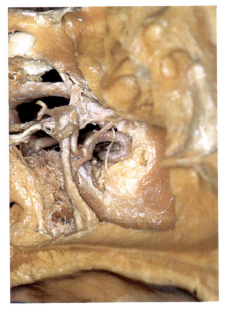

Fig. 2.30. The pterygopalatine fossa, nasosinus dissection: left side. *1*, Maxilary sinus; *2*, nasal cavity, lateral wall; *3*, sphenoid sinus; *4*, maxillary artery; *5*, maxillary nerve; *6*, infraorbital artery; *7*, infraorbital nerve; *8*, septal artery; *9*, posterior lateral nasal artery; *10*, foramen rotundum; *11*, pterygoid canal; *12*, pterygopalatine canal; *13*, palatine artery; *14*, palatine nerve; *15*, palate; *16*, ethmoid cells

References

1.Ademà JM, Massegur H, Sprekelsen BM (1994) Cirugia Endoscópica nasosinusal. Editoral Garsi, Madrid
2. Alberti PW (1976) Applied surgical anatomy of the maxillary sinus. Otolaryngol Clin North Am 9:3–20
3. Anon JB, Rontal M, Zinreich SJ (1996) Anatomy of the paranasal sinuses. Thieme, New York
4. Bagatella F, Guiraldo CR (1983) The ethmoid labyrinth. An anatomical and radiological study. Acta Otolaryngol Suppl (Stockh) 403:3–19
5. Bansberg FS, Harner, SG, Fordes G (1987) Relationship of the optic nerve to the paranasal sinuses as shown by computed tomography. Otolaryngol Head Neck Surg 96:331–335
6. Becker SP (1994) Applied anatomy of the paranasal sinuses with emphasis on endoscopic surgery. Ann Otol Rhinol Laryngol 103:3–32
7. Blitzer A (1985) Surgery of the paranasal sinuses. Saunders, New York
8. Bouchet A, Cuilleret J (1979) Anatomia descriptiva, topográfica y funcional. Panamericana, Buenos Aires
9. Caliot P, Midy D, Pressis JL (1990) The surgical anatomy of the middle nasal meatus. Surg Radiol Anat 12:97–101
10. Canuyt J, Terracol G (1925) Le sinus sphenoidal. Libraires de L'Academie de Medicine, Paris
11. Figún ME, Garino RG (1989) Anatomia odontológica funcional e aplicada. Panamericana, São Paulo
12. Hollinshead WH (1980) Anatomia humana. Harper and Row, São Paulo
13. Kainz J, Stammberger H (1989) The roof of the anterior ethmoid: a place of least resistance in the skull base. Am J Rhinol 3:191–199
14. Kasper KA (1936) Nasofrontal connections. Arch Otolaryngol 23:322–343
15. Lang J (1989) Clinical anatomy of the nose, nasal cavity, and paranasal sinuses. Thieme, New York
16. Mattox DE, Delaney RG (1985) Anatomy of the ethmoid sinus. Otolaryngol Clin North Am 18:3–14
17. Meloni F, Mini R, Rovasio R (1992) Anatomic variations of surgical importance in ethmoid labyrinth and sphenoid sinus. A study of radiological anatomy. Surg Radiol Anat 14:65–70
18. Morgenstein KM (1987) Anatomy and surgery of the pterygopalatine fossa. In: Goldman JL (ed). The principles practice of rhinology. John Wiley and Sons, New York, pp 651–668
19. Navarro JAC, Toledo Filho JL, Zorzetto NL (1982) Anatomy of maxillary artery into the pterigomaxillopalatine fossa. Anat Anz 152:413–433
20. Paturet L (1950) Traité d'anatomie humaine, vol 1. Masson, Paris
21. Rabischong P (1981) Anatomical bases for the surgical access to the pterygopalatine fossa. Anat Clin 2:209–222
22. Rice DH, Schaefer SD (1993) Endoscopic paranasal sinus surgery. Raven, New York
23. Ritter FN (1982) The middle turbinate and its relationship to the ethmoidal labyrinth and the orbit. Laryngoscope 92:479–482
24. Rontal M, Rontal E (1991) Studying whole-mounted sections of the paranasal sinuses to understand the complications of endoscopic sinus surgery. Laryngoscope 101:361–366
25. Rouviére H (1961) Anatomia humana descriptiva y topográfica Tome I. Bailly-Bailiere, Madrid
26. Shankar L, Evans K, Hawke M, Stammberger H (1994) An atlas of imaging of the paranasal sinuses. Martin Dunitz, Singapore
27. Siebert DR (1994) Anatomia dos seios esfenoidais. Rev Bras ORL 60:28–34
28. Sieur C, Jacob O (1901) Recherches anatomiques, cliniques et opératoires sur les fosses nasales et leurs sinus. Rueff, Paris
29. Terracol J, Ardouin P (1965) Anatomie des fosses nasales et des cavités annexes. Maloine, Paris
30. Testut L, Laterjet A (1959) Tratado de anatomia humana Tomo I, IV. Salvat, Barcelona
31. Tondüry G (1958) Anatomia topográfica y aplicada. Científico-Médica, Barcelona
32. Van Alyea OE (1942) Nasal sinuses. Williams and Wilkins, Baltimore
33. Vídic SB (1969) Extreme development of the paranasal sinuses. Ann Otol Rhinol Laryngol 78:1291–1298
34. Vinelli BB (1943) Anatomia humana v.1. Scientífica, Rio de Janeiro
35. Wentges RT (1975) Surgical anatomy of the pterygopalatine fossa J Laryngol Otol 89:35–45

Endoscopic Diagnosis: Nasosinus Disorders

Eulalia Sakano

Introduction

The endoscopic examination of the nasal cavity is now a standard part of the diagnostic evaluation for all rhinosinus pathologies [1, 4, 7, 12]. Excellent images and direct visualization of structures (even those that were previously difficult to visualize, e.g., the lateral nasal wall, with its clefts and recesses [5]) have allowed diagnosis and better evaluation of pathologic processes previously thought to be uncommon or difficult to diagnose [2]. A further benefit of endoscopic examination is better understanding of the physio-pathological mechanisms of chronic sinus infections. The evaluation of the changes in the mucous membrane or the structures of the lateral wall of the nose and computed-tomography scans of the paranasal sinuses have become powerful tools to locate the most probable sites of the pathology in sinus disease.

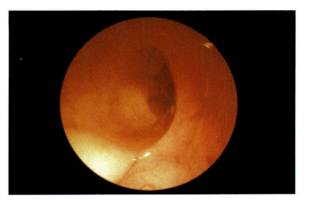

Fig. 3.1. Nasolacrimal duct ostium in an inferior meatus

Routine Examination

To carry out the examination, a mixture of topical anesthetics and vasoconstrictors is applied to the nasal mucosa in adults or older children. In small children or infants, general anesthetic allows greater, complete, easy examination.

A 0°- or 30°-angle endoscope is routinely used to examine the nasal cavity to assess the character of the mucous and the presence of secretions, structural alterations (such as septal deviations) and pathological alterations. Taking care to avoid damaging the nasal mucosa (which would cause edema or bleeding), the endoscope is first introduced along the floor of the nasal cavity, accessing the inferior meatus and the nasolacrimal duct (Figs. 3.1, 3.2) and reaching the region of the nasopharynx.

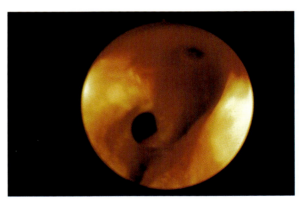

Fig. 3.2. Opening of the nasolacrimal duct in a left inferior meatus (*arrow*). Note the puncture site for inferior antroscopy. *it* Inferior turbinate

Nasopharynx

The structures of the orifices of the Eustachian tubes, their mucous membranes and their motions during deglutition are always assessed and compared. Depending on the clinical situation, even small unilateral alterations can be important (Fig. 3.3).

Using a 30°-angle endoscope, evaluation of the adenoidal tissue of the posterior wall of the nasopharynx is performed, especially in children. After peak adenoidal development, regression proceeding from inferior to superior occurs in older children. In post-adenoidectomy cases with persistence of nasal respiratory difficulty and in some children under age 2 years with persistent nasal obstruction, the endoscope has frequently shown the cause to be adenoidal tissue in the region of the choana (Figs. 3.4, 3.5).

Lateral Nasal Wall

With the posterior margin of the septum and the posterior end of the inferior turbinate as reference points, the 30°-angle endoscope is passed along the septum, passing the posterior end of the middle turbinate and arriving at the free lower margin of the upper turbinate, the reference point for location of the sphenoethmoidal recess and the sphenoid-sinus ostium. This ostium usually has an oval or rounded shape (Fig. 3.6). The endoscope is withdrawn and redirected above the posterior end of the inferior turbinate, initiating examination of the middle meatus. If insertion via this path is difficult, the endoscope is withdrawn and introduced through the anterior portion of the middle meatus. Here, evaluation of the lateral wall of the nose initially reveals the uncinate process, part of the ethmoidal bulla; posterior to this location, examination of the lateral wall of the nose reveals the accessory ostium, through which (if the ostium is large enough) visualization of the maxillary sinus is possible (Figs. 3.7, 3.8). The ethmoidal bulla can be seen directly if it is sizeable enough or if the free margin of the uncinate process is laterally positioned. Directing the endoscope superiorly across the semilunar hiatus, various ostia and openings of the ethmoidal cells (and, more superiorly, the frontal recess) can be seen. Evaluation in this

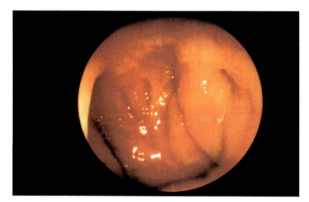

Fig. 3.3. Left Eustachian-tube ostium. The discrete edema of its lymphoreticular tissue was revealed at biopsy to be a carcinoma

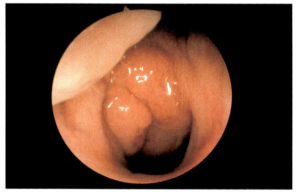

Fig. 3.5. The adenoidal mass is obstructing the choana in a 5-year-old boy. The previous radiography showed a normal nasopharynx

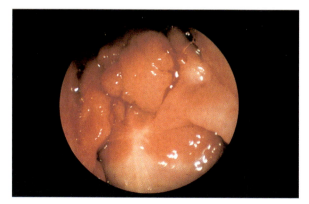

Fig. 3.4. Recurrence of adenoidal tissue (right nasal choana) 8 months after adenoidectomy

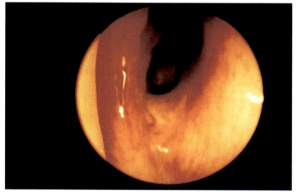

Fig. 3.6. The sphenoidal ostium (*arrow*) can be seen in the sphenoethmoidal recess. *S* Septum

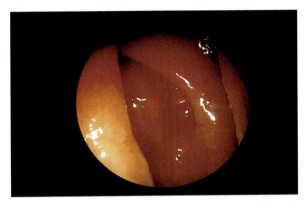

Fig. 3.7. Left middle meatus. The uncinate process and the ethmoidal bulla can be seen. *mt*, Middle turbinate; *up*, uncinate process; *eb*, ethmoidal bulla

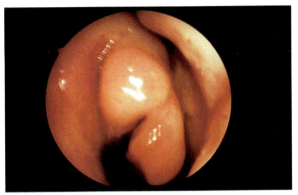

Fig. 3.9. Frontal sinusitis. Anatomical variation of the right middle turbinate (*mt*). The superior portion of the head of the middle turbinate is swollen, causing a narrowing of the entrance of frontal recess

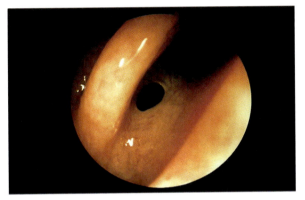

Fig. 3.8. Normal middle meatus. Accessory maxillary ostium

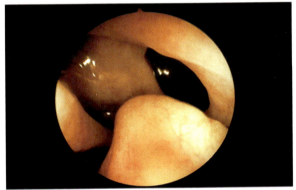

Fig. 3.10. Septum deviation. Left nasal cavity

region is difficult because it is narrow; it is often necessary to use a 2.7-mm telescope. The configuration of the frontal recess can also be affected by variations in the surrounding structures, such as an enlarged ethmoidal bulla, a large concha bullosa, pneumatization of the agger nasi cells or structural variations of the middle turbinate (Fig. 3.9).

In chronic or recurrent sinus infections, the anatomic variants of the structures of the middle meatus that encourage persistent blockage of the ostia have been widely studied and described radiologically [10]. Endoscopically, analysis of these variations is more subjective, but careful assessment of the mucosa is essential because it cannot be judged radiographically and because the mucosa may be a critical factor in the persistence of disease (Fig. 3.10). In narrower areas, there is a greater chance of contact between the mucous surfaces (as happens in the frontal recess, between the uncinate process and the middle turbinate or between the ethmoidal bulla and the middle turbinate), giving rise to local mucosal edema or hyperplasia with the formation of small polyps (Figs. 3.11, 3.12).

Nasal Polyps

Nasal polyps may be present to a variable degree and generally originate in the middle meatus (Fig. 3.13) [6]. After clinical treatment to reduce the inflammatory process, endoscopic examination is carried out, allowing better evaluation of the affected areas and more accurate assessment of the origin of the polyps. In children, polyposis is uncommon and, if it is seen in small children, cystic fibrosis must be suspected. In cases of recurrent sinusitis in children, we can usually endoscopically detect abnormalities in the middle meatus mucosa, occasionally with small polyps in regions that block the ostia.

Antrochoanal polyps can be present in both children and adults; these polyps generally present uni-

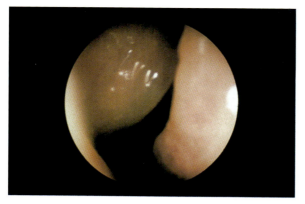

Fig. 3.11. Right nasal cavity. Note the mucous membrane edema (*) at the contact between middle turbinate (*mt*) and the ethmoidal bulla

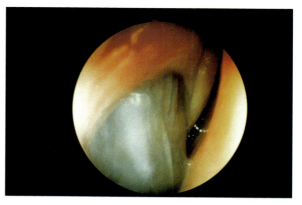

Fig. 3.14. Antrochoanal polyp (*p*) that extends through a large accessory maxillary ostium

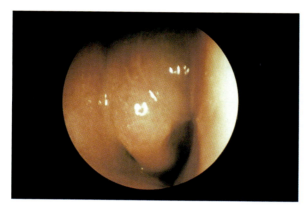

Fig. 3.12. Left nasal cavity. At the site of contact between the free margin of the middle turbinate (*mt*) and the lateral wall, mucous membrane hyperplasia (*) is visualized

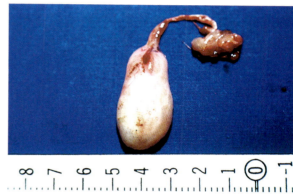

Fig. 3.15. Antrochoanal polyp. Note the solid part of the polyp and the cystic component (*c*) after resection

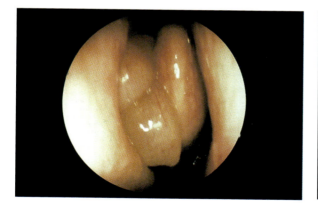

Fig. 3.13. Polyps arising from the right middle meatus. *mt*, Middle meatus; *p*, polyp

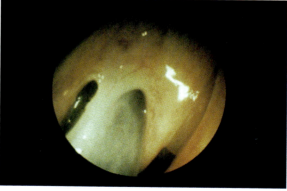

Fig. 3.16. Antrochoanal polyp (*p*). Note the implantation site around the accessory ostium

laterally, without an obvious cause. They fill the maxillary sinus and have a cystic component (Figs. 3.14–3.16). These cysts spread to the nasal cavity and, depending on their size, can reach the nasopharynx. This extension occurs through an enlarged accessory ostium that can be visualized endoscopically. Sphenochoanal polyps present in the same way (Fig. 3.17).

Maxillary Sinus

The maxillary sinus can be evaluated endoscopically only through the inferior meatal approach or through the canine fossa [8]. Maxillary sinuscopy is recommended in cases of chronic or recurrent inflammatory processes, in symptomatic cysts, to remove foreign bodies, and when neoplastic processes are suspected (Fig. 3.18).

Secretions that are thick and caseous suggest fungal sinusitis or a chronic inflammatory process with retained secretions. The endoscopic suggestion of a fungal ball is a whitish or dark color that is sometimes seen in the region of the accessory ostium (Fig. 3.19) [9]. The patient usually complains of considerable pain and sometimes experiences bloody nasal drainage.

Depending on their size, foreign bodies can be removed endoscopically [11, 13]. Most foreign bodies are dental roots, dental filling material or pins for dental implants and are typically surrounded by infected, inflamed, thickened sinus mucosa (Figs. 3.20, 3.21).

Removal of cysts depends not on size but on location and on the symptoms they cause (if any). They may vary from thin-walled, translucent cysts containing serous liquid to opaque, mucous-containing

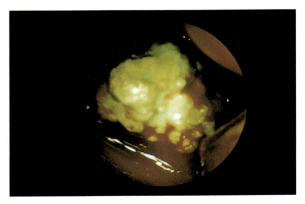

Fig. 3.19. Noninvasive fungal sinusitis. *F*, fungal ball

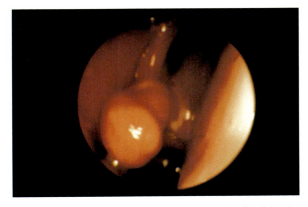

Fig. 3.17. Sphenochoanal polyp. Note the small polyp (*p*) originating from the sphenoethmoidal recess

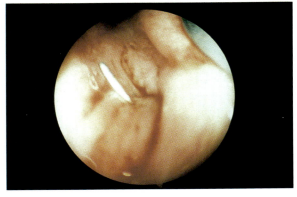

Fig. 3.20. Foreign body of maxillary sinus. The *arrow* shows a dental-implant pin

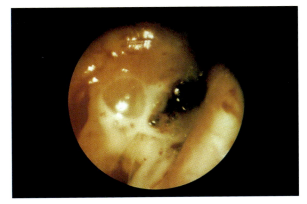

Fig. 3.18. Purulent sinusitis. Antroscopy shows purulent secretion and hyperplasia of the mucosa

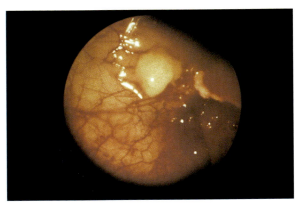

Fig. 3.21. Dental root (*) in the floor of the maxillary sinus

cysts with a thicker capsule. Often, the mucous membranes of residual cysts in the sinus and the sinus ostium are normal (Fig. 3.22).

X-ray images with suspicious findings call for endoscopy and possible biopsy, as illustrated by a patient who had a small tumor in the region of the infraorbital nerve, with normal mucosa in the rest of the maxillary sinus. Biopsy showed it to be a carcinoma.

Post-Operative Management

In post-operative management of the sinus cavities, endoscopic examination is helpful for removal of eschars and clots, for lysing synechias, for removing recurrent polyps (thus facilitating healing) and in the long-term control of recurrent polyposis (Figs. 3.23, 3.24) [3].

Lacrimal Semiology

In nasolacrimal diseases, endoscopic evaluation of the nasal cavity after dacryocystorhinostomy has greatly improved surgical results. Removal of granulation in the region of the surgical ostium and evaluation of the length and position of the Lester-Jones tube is helpful to the ophthalmologist (Figs. 3.25, 3.26). In congenital dacryocystitis not amenable to lacrimal sac massage and probing of the nasolacrimal duct, systematic evaluation of the inferior meatus has frequently detected the presence of a cyst filled with thick secretions in the terminal portion of the nasolacrimal duct. Opening of the cyst with the removal of the capsule is fundamental to the control of dacryocystitis (Fig. 3.27).

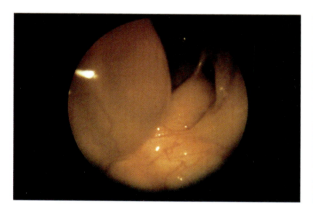

Fig. 3.22. Maxillary-sinus cyst near the ostium

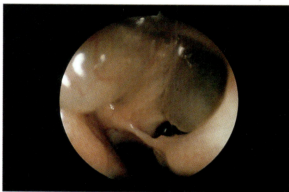

Fig. 3.24. Three months after surgery, we can observe two small persistent or recurrent polyps

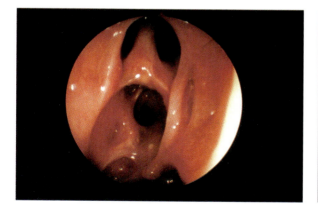

Fig. 3.23. Endoscopic appearance 8 months after ethmoidal surgery. The mucosa of the ethmoid is normal

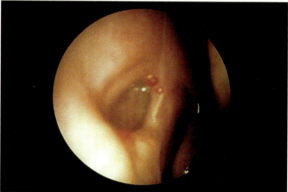

Fig. 3.25. External dacryocystorhinostomy. On the right lateral wall, the osteotomy (*o*) aspect 8 months after surgery is visible. *mt*, Middle turbinate

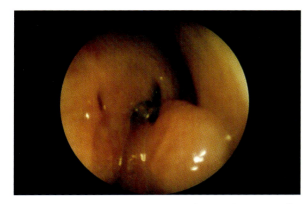

Fig. 3.26. Recurrent dacryocystitis after implantation of a Lester-Jones tube (*t*). The distal end was blocked by the middle turbinate (*mt*)

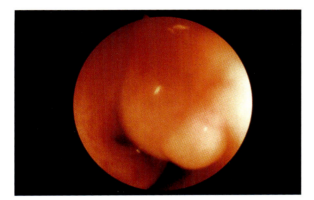

Fig. 3.27. A frequent endoscopic finding in the inferior meatus of a congenital dacryocystitis is a nasolacrimal duct cyst (*c*). Left nasal cavity of a 1-year-old boy

References

1. Draf W (1983) Endoscopy of the paranasal sinuses. Springer, Berlin Heidelberg New York
2. Edelstein DR, Arlis HR, Bushkin S, Han JC (1991) Posterior sinus anatomy: clinical correlation and pitfalls. Op Tech Otolaryngol Head Neck Surg 2:222–225
3. Josephson JS, Linden BE (1990) The importance of postoperative care in the adult and pediatric patient treated with functional endoscopic sinus surgery. Op Tech Otolaryngol Head Neck Surg 2:112–116
4. Kennedy DW, Zinreich SJ, Rosenbaum AE, Johns ME (1985) Functional endoscopic sinus surgery: theory and diagnostic evaluation. Arch Otolaryngol 111:576
5. Lang J (1989) Clinical anatomy of the nose, nasal cavity and paranasal sinuses. Thieme, Stuttgart
6. Larsen PPL, Tos M (1996) Anatomic site of origin of nasal polyps. Am J Rhinol 10:211–216
7. Messerklinger W (1978) Endoscopy of the nose. Urban and Schwarzenberg, Baltimore
8. Rice DH, Schaefer SD (1988) Endoscopic paranasal sinus surgery. Raven, New York
9. Roithmann R, Shankar L, Hawke M, Chapnik J, Kassel E, Noyek A (1995) Diagnostic imaging of fungal sinusitis: eleven new cases and literature review. Rhinology 33:104–110
10. Stackpole SA, Edelstein DR (1996) Anatomic variants of the paranasal sinuses and their implications for sinusitis. Curr Opin Otolaryngol Head Neck Surg 4:1–6
11. Stammberger H (1991) Functional endoscopic sinus surgery. Decker, Philadelphia
12. Terrier G (1978) L'endoscopie rhinosinusale moderne. Inpharzan, Lugano
13. Wigand ME (1990) Endoscopic surgery of the paranasal sinuses and anterior skull base. Thieme, New York

Rhinomanometry

JUAN EUGENIO SALAS GALICIA

Introduction

The purpose of this chapter is to demonstrate the clinical utility of rhinomanometry in the evaluation as an integral part of preoperative studies of nasosinusal microendoscopic surgery in patients with nasal obstructions associated with rhinosinusal pathology. Nasal obstruction is the most frequent symptom of patients with nasosinusal pathology, and rhinomanometry is the only objective and dynamic method by which air pressure and air flow through the nasal cavities can be measured [4, 8, 10–14, 17, 18]. Acoustic rhinometry is a good, objective and static method to evaluate the morphology of the nasal cavity and rhinopharynx; acoustic rhinometry can complement rhinomanometry, but the indications and usefulness of each are different [1, 6, 14].

The basic principles of nasosinusal endoscopic and microendoscopic surgery are to promote and allow optimum nasal air flow and drainage of the middle meatus and its cavities. In order to achieve this, there must be a preoperative evaluation of the area of the nasal valve because, under normal conditions, it directs 90% of the air flow to the superior aspect of the middle turbinate and its meatus. Therefore, stenosis or valve insufficiency results in inadequate aeration of the osteomeatal complex and a resulting accumulation of mucus and inflammation there. Rhinomanometry is fundamental in the dynamic appraisal of this area Fig. 4.1 [8, 10, 14, 19, 20].

For many years, rhinomanometry has been disdained by otorhinolaryngologists, including rhinologists. In 1968, Williams suggested that, in North America, rhinomanometry was regarded similar to the way in which audiometry was regarded in 1920 and electronystagmography in 1958 [21].

One of the main reasons that rhinomanometry did not gain wider acceptance was the lack of standardization of the procedure and the lack of results that could contribute clear and concise data with obvious clinical application. Thus, it was considered to be a complicated and slow procedure with little clinical correlation; it was thought that only the masters knew about rhinomanometry and handled it properly [2, 3, 5, 9, 17]

This chapter will attempt to demonstrate the simplicity and utility of the procedure using currently available computerized rhinomanometers. Kern maintains that rhinomanometry is as essential to the rhinologist as the audiometer is to the otologist [14].

History

Zwaardemaker and Kayser (1894, 1895) were the first to use hygrometric studies and to apply aerodynamics principles to the measurement of nasal air flow; Kayser was the first to use the term "nasal resistance" and to apply his theories and principles to assess his surgical results on patients with a nasal obstruction. Unfortunately, his surgical results and techniques

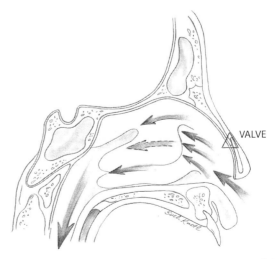

Fig. 4.1. Through the nasal valve, the major portion of the air stream is directed toward the head of the middle turbinate and the entrance of the middle meatus

were no better than those of the other surgeons of the time [14, 21].

Rhinomanometry experienced renewed interest with the work of Reynolds (1901) and Lillie (1923), and the latter was the first to mention the nasal cycle [21]. This was followed by a period of relative disinterest until 1958, when Ogura published his studies on nasopulmonary reflexes and rhinomanometry and the importance of nasosinusal pathology in nasopulmonary physiology. He reported that, under normal conditions, resistance to air passage at the nasal level represents between 40% and 60% of the total resistance of the air passages, with an additional 7% at the level of the pharynx, larynx and trachea and the remaining 45% at the bronchus and lungs. However, with nasal or nasosinusal pathologies, there are repercussions in the lower aerial passages (rhinosinobronchial syndrome) [15, 16].

During the 1970s and 1980s, Stoksted, Guillen, Monserrat, Cottle, Masing, Kern, Clement, et al. contributed valuable studies about the clinical use of rhinomanometry [2, 3, 4, 7, 8, 9, 11, 12, 13]. On February 24, 1983, in Brussels, Belgium, a report was given by the International Committee on Standardization of Rhinomanometry. This committee was started by Kern (USA) and other experienced rhinologists, including Bachman (Germany), Clement (Belgium), Eccles (UK), Klaassen (Netherlands), Malm (Sweden), McCaffrey (USA), with the main objective of standardizing the procedure and its results to make them useful worldwide. In this chapter, we will make frequent references to that report [2].

Terminology and Definitions

Rhinomanometry is an objective procedure that measures the difference of air pressure and the speed of air flow in the nasal air passage during breathing; with these data, the nasal resistance to air flow can be calculated. In order to understand rhinomanometry, it is necessary to understand the following terms:
- **Active rhinomanometry.** A study in which the flow of air through the nasal cavities corresponds to the breathing movements of the person being studied
- **Passive rhinomanometry.** A study in which a fixed flow of air coming from an external source of positive pressure is sent through the nasal cavities of the patient while he maintains apnea
- **Uninasal rhinomanometry.** A study in which each nostril is examined separately
- **Binasal rhinomanometry.** A study in which both nostrils are examined simultaneously

- **Anterior rhinomanometry.** A study in which the difference in pressure is measured contralateral to the nostril in study by the placement of a pressure tube in the vestibule
- **Posterior rhinomanometry.** A study in which the difference of pressure is measured by placing a pressure tube in the oropharynx
- **Rhinomanometry with olives.** A study in which olives (metal or plastic) are used and placed in the nasal vestibule in order to measure both flow and pressure
- **Rhinomanometry with a mask.** A study in which a panoramic or oronasal mask is used in order to avoid distortion of the nasal vestibule being studied

For a better understanding of the units and basic terminology used in rhinomanometry, see Table 4.1.

Techniques

At present, there are several ways to perform rhinomanometry.
- **Active anterior rhinomanometry with olives.** In this study, one of the olives reads the pressure and the other the air flow.
- **Active anterior rhinomanometry with an oronasal or panoramic mask.** In this study, one of the nostrils is sealed tightly with transparent adhesive tape, sealing in place a plastic tube used for measuring the variation of pressure; the flow is

Table 4.1. Basic terminology and units used in rhinomanometry

Term	Definition
Pressure (P)	The action of a force (air) against a force of opposition (nasal septum and nasal lateral wall) measured in cm H_2O or Pa
Flow (F)	The speed of the change of volume, expressed in l/s or cm^3/s
Resistance (R)	Represents the factors that act to impede the flow of air through an air route; it requires measures of pressure and flow R=P/F, and it is measured in cm H_2O/l/s, Pa/l/s or Pa/cm^3/s
Difference in pressure (DP)	The difference between the pressures measured at two points in a P1–P2 system (generally at both ends of a system of air conduction). It is measured in cm H_2O or Pa
Pa	An international standard unit of pressure corresponding to a force of 1 N/m^2. One pascal is equal to 0.0102 cm H_2O

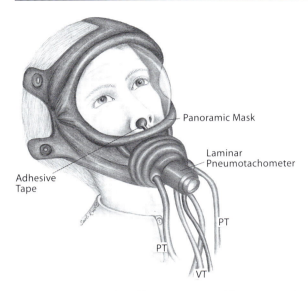

Fig. 4.2. Active anterior rhinomanometry with a panoramic mask. The adhesive tape occludes the patient's contralateral nasal vestibule. *PT*, tubes to measure pressure; *VT*, tubes to measure volume (flow)

measured at open field via a connection with the mask (Fig. 4.2).

- **Active posterior rhinomanometry.** In this study, the patient breathes through both nostrils inside a mask connected to a pneumotachometer in order to measure the flow; a tube located in the mouth and connected to a transductor of pressure measures the pressure of the nasopharynx. This method requires good patient cooperation and relaxation of the oropharyngeal muscles, which many patients cannot maintain (Fig. 4.3).
- **Passive anterior rhinomanometry.** In this study, air is insufflated through an olive placed in one nostril at a constant and pre-established rate of flow (usually 250 cm^3/s), with the patient in apnea. The pressure induced by the nasal resistance at a level given by the olive is measured.

Each of these methods has its specific techniques, advantages and disadvantages (Tables 4.2–4.4).

The International Committee on Standardization of Rhinomanometry (1984) recommended active anterior rhinomanometry using a transparent (panoramic) mask as the preferred physiological technique. As a method of occlusion, they recommended an adhesive tape that will not modify the

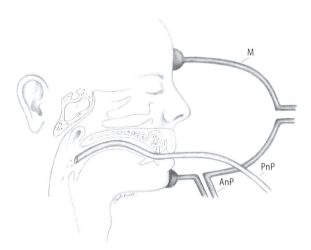

Fig. 4.3. Active posterior rhinomanometry. Cannula in the oropharynx. *AnP*, anterior nasal pressure; *M*, mask; *PnP*, posterior nasal pressure

Table 4.3. Passive anterior rhinomanometry

Advantages	Disadvantages
Easy and fast	High price
Possible in children	Expensive
Possible to measure one nostril while the other is blocked	Can only measure one nostril at a time
Objective graphic registration	Some subjects have difficulty maintaining the apnea
	Cannot obtain written record
	Dynamic studies are not possible
	Necessary to calibrate the equipment at each determination
	A little sensitive

Table 4.2. Active anterior rhinomanometry

Advantages	Disadvantages
Easy and relatively fast	Possible deformation of the valve by nasal pieces (not with a mask)
Always possible in children	Not possible with septal perforation
Pressure and flow can be measured	Only possible to record one nostril at a time Graphic recording
Objective graphic registration	Equipment calibration necessary at each determination
	Problems where the mask meets the face

Table 4.4. Posterior rhinomanometry

Advantages	Disadvantages
No valve deformation	Very painstaking and time consuming
Measures total nasal resistance	The mask causes face problems
Simultaneously measures pressure and flow	It is not possible to study the nostrils separately
Can use with septal perforation	30–50% of patients cannot complete the test
Objective graphic Registration	It is necessary to calibrate the equipment at each recording

nasal vestibule [2]. Passive posterior rhinomanometry is reserved for nasal breathing-research studies and for cases of septal perforation (where it is desirable to know the total nasal resistance). Passive anterior rhinomanometry is indicated for screening studies performed on large groups or students.

In the cases discussed in this chapter, we used active anterior rhinomanometry with a panoramic mask applied using adhesive tape to effect nasal occlusion. This was connected to a plastic tube to allow the pressure to be varied; the air flow was measured in an open field via a connection to the mask (Fig. 4.4). We used a Rhinospir 164 Sibelmed computerized rhinomanometer (Fig. 4.5) connected to a computer and a printer. Air flow was measured in cubic centimeters per second in both nostrils (separately) at 75, 100, 150 and 300 Pa of pressure

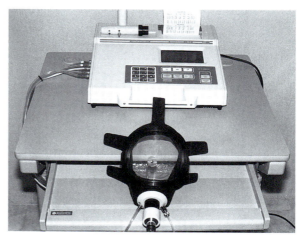

Fig. 4.5. Computerized rhinomanometry using a Rhinospir 164 with registration over the x–y axis and a sinusoidal oscillograph with flow–time and pressure–time curves

during inhalation and exhalation, recording the most significant 30 s of the test (approximately seven cycles of inhalation and exhalation).

Methods

We follow the recommendations of the International Committee for the Standardization of Rhinomanometry [2]:

1. The patient to be examined is instructed to avoid the use of any substance or the performance of any activity that could modify the nasal mucosa (and, therefore, the nasal resistance) beginning 24–48 h prior to the study. Such substances and activities include: local or systemic antihistamines, decongestants of any kind, steroids or irritators, tobacco, alcohol, cigarettes, plentiful meals, steam baths, exercise or physical effort.
2. The patient should be at rest, wearing loose clothes for at least 30 min before the test, and the room should be at a relatively constant humidity and temperature.

Fig. 4.4. Active anterior rhinomanometry. Note the adhesive tape occluding the patient's left nasal vestibule

3. With the patient sitting comfortable and erect, the skin surrounding the nostrils should be cleaned, removing any cutaneous secretions and occluding the opposite nostril with a transparent adhesive tape that encloses a small metallic catheter that is fixed to a flexible plastic cannula that leads to a transductor, an amplifier and an electronic recorder to record the pressures Fig. 4.6. Deformation of the nasal vestibule, which could affect in the contralateral valve area, should be avoided.
4. The mask is connected to a pneumotachometer, which is placed firmly on the patient's face. To obtain the flow velocity, the recorded pressure is transmitted to a differential transductor, an amplifier and a registration recorder system (Fig. 4.6). The pneumotachometer is in the lower part of the mask, so there is no distortion of the valve area or nostril under study with rhinomanometers that use olives.
5. The patient is asked to breathe normally and unhurriedly until stable breathing is evident. Both nostrils are studied, obtaining a minimum of five cycles of inhalation and exhalation for each side. This takes approximately 30 s for each side.
6. For comparison recording after application of vasoconstrictors or exercise for 10 min, we use either of the two recommended vasoconstrictors: 1% oxymetazoline and phenylephrine or hydrochloride of metazoline with adrenaline. Two sprays are applied to each nostril. After a 10-min wait, the rhinomanometry is repeated, recording the use and type of vasoconstrictor on the record. Thus, we obtain two traces for each side: before and after vasoconstriction.

Calibration

The International Committee for the Standardization of Rhinomanometry [2] indicates that the equipment must be calibrated at least once each day, with a yearly maintenance check of the electrical and mechanical elements of the systems. For the daily calibration, the use of a resistance pattern with a known value of 150 Pa connected to the pneumotachometer together with breathing through the mouth is effective. This is known as the "artificial nose" (Fig. 4.7). When the equipment is calibrated, an alarm sounds.

Hygiene

The mask should be sterilized by wiping with benzine or alcohol after each patient. The small tube and the adhesive tape that are fastened to the nostrils are disposable. Both the mask and the tubes of the pneumotachometer must be sterilized periodically by gas or a sterilizing solution.

Fig. 4.6. Diagram of a computerized rhinomanometer with a multiplexer and an analog–digital converter

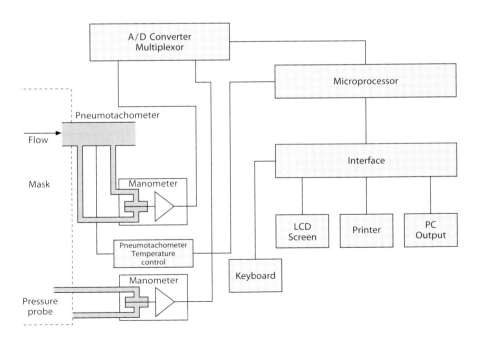

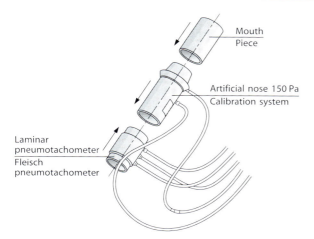

Fig. 4.7. Calibration of the rhinomanometer using a 150-Pa pattern calibrator known as an "artificial nose". Calibration must be done once per day

Recording

The data obtained in the rhinomanometry are recorded in a rhinogram, a graph depicting the relationship between flow (ordinate scale) and pressure (abscissa scale).

The graph corresponds to a mirror image with four quadrants: quadrants II and IV represent the left side nostril, and quadrants I and III represent the right nostril. Inhalation (negative pressure) is recorded at the right side of the graph, and exhalation (positive pressure) is recorded on the left side of the graph (Fig. 4.8) [2, 3, 9, 14, 17].

Results

The International Committee on Standardization indicated that the results should be expressed in units featured in the international unit system (SI units), i.e., pascals (Pa) for pressure and cubic centimeters per second (cm^3/s) for flow. Similarly, resistance should be expressed at a fixed pressure instead of a fixed flow. The reference pressure was 150 Pa.

The currently available rhinomanometers obtain the values of flow rates expressed in cubic centimeters per second at 75, 100, 150 and 300 Pa for both inhalation and exhalation at each nostril at the different and partial resistances.

Normal test values have been obtained from studies at pressures of 150 Pa:
1. The total flow (added from both nostrils at 150 Pa) should be equal to or greater than 700 cm^3/s.

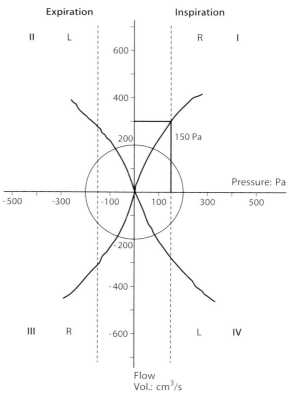

Fig. 4.8. Registration rhinograph with a mirror image of the x–y coordinates with quadrants *I–IV*; *I* and *III* are for the right nostril, and *II* and *IV* are for the left nostril. The ordinate represents the flow (in cm^3/s), and the abscissa is the pressure gradient (in Pa). Inhalation (negative pressure) is on the right (*R*), and exhalation (positive pressure) is on the left (*L*)

2. The difference in flow between the better and the poorer nostril should not exceed 1.5 cm^3/s.
3. The total respiratory resistance of each nostril should be less than 0.30 $Pa/cm^3/s$ (3.0 cm $H_2O/l/s$).

Presentation of Cases

Two cases are presented to demonstrate the utility of rhinomanometry in the diagnosis and treatment of patients with a nasal obstruction.

Case 1

A 46-year-old male executive with seasonal allergic rhinitis presented with left nasal unilateral obstruction of 2 years duration. He had undergone septoplasty and cauterization of turbinates 15 years ago (with a good results, until the beginning of his current complaint). Physical examination revealed a straight septum with hypertrophy of turbinates, especially the left inferior turbinate, with polypoid degeneration and slightly congested edematous mucosa.

The rhinogram at rest (basal) on the x–y coordinate axis showed that the resistance of the right nostril was 0.28 Pa/cm^3/s (2.8 cm H$_2$O/l/s) in response to 150 Pa of air pressure; the resistance of the left nostril was 0.49 Pa/cm^3/s (4.9 cm H$_2$O/l/s) in response to 150 Pa of air pressure. Thus, there is an elevated resistance in the right nostril; this is compensated by the resistance in the contralateral nostril and the increased flow (823 cm^3/s; normal>700 cm^3/s; Fig. 4.9). After the application of a vasoconstrictor, the resistance of the right nostril showed a slight decrease from 0.28 cm H$_2$O/l/s to 0.26 Pa/cm^3/s (2.8 cm H$_2$O/l/s to 2.6 cm H$_2$O/l/s), and the resistance of the left nostril decreased from 0.49 Pa/cm^3/s to 0.20 Pa/cm^3/s (4.9 cm H$_2$O/l/s to 2.0 cm H$_2$O/l/s; Fig. 4.10).

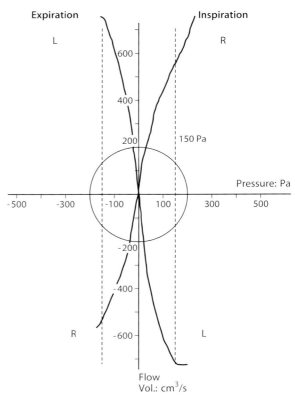

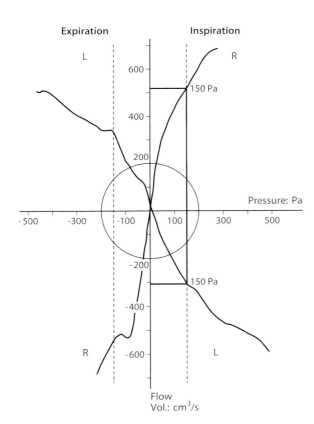

Fig. 4.9. Basal rhinogram from case 1. The flows were marked at 150 Pa in both nostrils on inhalation. At the bottom is a computerized report of flows and resistances for each nostril and the totals at pressures of 75, 100, 150 and 300 Pa. *R*, right; *L*, left

Inspiration

Pressure (Pa)	75	100	150	300
Flow right (cm^3/s)	377	441	557	0
Flow left (cm^3/s)	521	608	721	0
Flow total (cm^3/s)	898	1049	1278	0
Res. right (Pa·s/cm^3)	0.19	0.22	0.26	0.00
Res. left (Pa·s/cm^3)	0.14	0.16	0.20	0.00

Expiration

Pressure (Pa)	75	100	150	300
Flow right (cm^3/s)	363	422	540	0
Flow left (cm^3/s)	518	606	762	0
Flow total (cm^3/s)	881	1028	1302	0
Res. right (Pa·s/cm^3)	0.20	0.23	0.27	0.00
Res. left (Pa·s/cm^3)	0.14	0.16	0.19	0.00

Fig. 4.10. Rhinogram from case 1 after vasoconstriction, with a computerized report of the flows and resistances for each nostril and the totals at pressures of 75, 100, 150 and 300 Pa. *R*, right; *L*, left

Rhinomanometric Diagnosis

The rhinomanometric diagnosis was nasal obstruction of mucous type (functional) in left nostril. The left nostril obstruction was reversible with a vasoconstrictor. This patient underwent nasal endoscopy in the doctor's office; the posterior portion of the left turbinate had a raspberry appearance (polypoid degeneration associated with hyperplasic rhinitis). Cauterization of the lower left turbinate with a partial turbinectomy was indicated. The right nostril did not merit any treatment due to the risk of causing atrophic rhinitis. The right nostril was clinically, rhinomanometrically and endoscopically normal. Postoperative basal rhinomanometry 6 months later showed a right-side inspiratory airway resistance of 0.26 Pa/cm^3/s in response to 150 Pa air pressure and showed a left-side inspiratory airway resistance of 0.18 Pa/cm^3/s in response to 150 Pa air pressure. There was an increased flow of 927 cm^3/s.

Postoperative Rhinomanometric Diagnosis

The postoperative rhinomanometric diagnosis was normal.

Case 2

A 43-year-old male merchant reported bilateral nasal obstruction (worse on the left side); for more than 20 years, he was a chronic snorer and usually pulled on his left cheek in order to sleep (Cottle's sign). Upon physical examination, there was a left obstructive septal deviation in area II, with valvular collapse and a contacting right deviation in area II–IV, hypertrophied inferior turbinates, pale mucosa and a positive "U" sign ("nasal tension").

On the x–y coordinate axis, the basal rhinogram showed that the resistance of the right nostril was 0.41 Pa/cm^3/s (4.1 cm H$_2$O/l/s) in response to 150 Pa air pressure; the resistance of the left nostril was 2.67 Pa/cm^3/s (26.70 cm H$_2$O/l/s). The increased flow was at 420 cm^3/s at 150 Pa air pressure (Fig. 4.11). The resistance in the left nostril was abnormally high. During the study, the patient was apprehensive due to the almost total inability to ventilate the left nostril.

After the application of the vasoconstrictor, the resistance in the right nostril essentially did not change; it decreased from 0.41 Pa/cm^3/s to 0.39 Pa/cm^3/s. In the left nostril, the resistance significantly improved from 2.6 Pa/cm^3/s to 0.89 Pa/cm^3/s, but the persistence was still above normal (0.30 Pa/cm^3/s). Unfortunately, the improvement was because the patient took two breaths through his mouth; he would not allow the mask to be tightly placed. Nevertheless, his added flow was less than 700 cm^3/s (548 cm^3/s). The collapsed zone did not disappear with vasoconstriction (Fig. 4.12).

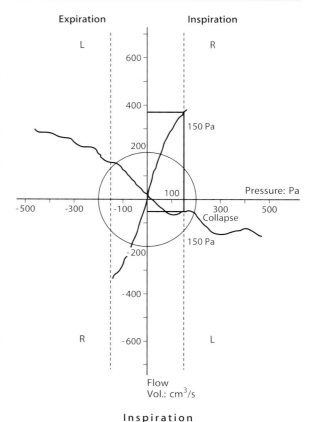

Inspiration				
Pressure (Pa)	75	100	150	300
Flow right (cm^3/s)	252	298	364	0
Flow left (cm^3/s)	55	67	56	151
Flow total (cm^3/s)	307	365	420	151
Res. right (Pa·s/cm^3)	0.29	0.33	0.41	0.00
Res. left (Pa·s/cm^3)	1.36	1.49	2.67	1.98

Expiration				
Pressure (Pa)	75	100	150	300
Flow right (cm^3/s)	227	276	0	0
Flow left (cm^3/s)	94	124	154	242
Flow total (cm^3/s)	321	400	154	242
Res. right (Pa·s/cm^3)	0.33	0.36	0.00	0.00
Res. left (Pa·s/cm^3)	0.79	0.80	0.97	1.23

Fig. 4.11. Basal rhinogram from case 2. Observe the left nostril on inhalation at 150 Pa. Valvular collapse occurs with a resistance of 2.67 Pa/cm^3/s. The patient had a severe nasal obstruction. *R*, right; *L*, left

Rhinomanometric Diagnosis

The nasal obstruction was of the structural type; it was anatomical and bilateral, with left valvular collapse. The right nostril obstruction was irreversible with vasoconstriction. The left nostril obstruction was irreversible with vasoconstriction and featured

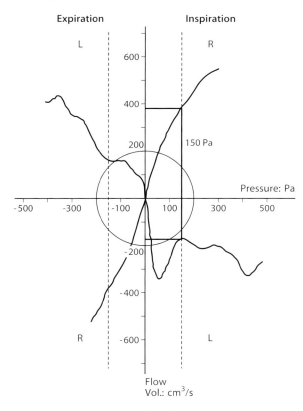

Inspiration

Pressure (Pa)	75	100	150	300
Flow right (cm³/s)	251	300	380	0
Flow left (cm³/s)	310	250	168	205
Flow total (cm³/s)	561	550	548	205
Res. right (Pa · cm³/s)	0.29	0.33	0.39	0.00
Res. left (Pa · cm³/s)	0.24	0.40	0.89	1.46

Expiration

Pressure (Pa)	75	100	150	300
Flow right (cm³/s)	237	289	373	0
Flow left (cm³/s)	147	159	163	345
Flow total (cm³/s)	384	448	536	345
Res. right (Pa · cm³/s)	0.31	0.34	0.40	0.00
Res. left (Pa · cm³/s)	0.51	0.62	0.92	0.86

Fig. 4.12. Rhinogram from case 2 after vasoconstriction. Observe the persistence of the image of the valvular collapse. *R*, right; *L*, left

visible valvular collapse. The patient was a candidate for septorhinoplasty due to nasal tension, septal deviation and left valvular collapse. This abnormality was treated with a Lopez Infante flap [8].

Postoperative basal rhinomanometry 7 months later showed a right-sided inspiratory airway resistance of 0.18 Pa/cm³/s in response to 150 Pa air pressure and showed a left-sided inspiratory airway resistance of 0.30 Pa/cm³/s in response to 150 Pa air pressure. An increased flow of 823 cm³/s was observed.

Postoperative Rhinomanometric Diagnosis

The postoperative rhinomanometric diagnosis was normal.

Conclusions

Rhinomanometry is an objective study of the nasal airway; it provides information enabling one to differentiate three basic types of obstruction:
1. Mucous or functional nasal obstruction (reversible with a vasoconstrictor)
2. Structural or anatomical nasal obstruction (irreversible with a vasoconstrictor)
3. Mixed-type nasal obstruction (mucous/structural)

Rhinomanometry is also useful in the diagnosis of valvular collapse, which causes a characteristic curve in the x–y coordinate axis (case 2). It can also support the diagnosis of atrophic rhinitis when there is absence of nasal resistance without vasoconstriction.

Rhinomanometry assists the rhinologist in establishing and choosing the most beneficial therapy for each individual patient, in evaluating pre- and postoperative patients with nasal obstruction and in analyzing results more objectively in order to facilitate future changes or modifications in technique. Rhinomanometry should be a routine study for the rhinologist, as audiometry is for the otologist.

References

1. Buenting JE, Dalston RM, Drake AF (1994) Nasal cavity area in term infants determined by acoustic rhinometry. Laryngoscope 104:1439–1445
2. Clement PAR (1980) The use of mathematical model in rhinomanometry. Rhinology 18:197–207
3. Clement PAR (1984) Committee report on standardization of rhinomanometry. Rhinology 22:151–155
4. Cottle MH (1968): Rhino-sphygno-manometry. An aid in physical diagnosis. Rhinology 6:7–26
5. Eichler J, Lenz H (1985) Comparison of different coefficients and units in rhinomanometry. Rhinology 23:149–157
6. Fisher EW, Lund VI, Scadding GK (1994) Acoustic rhinometry in rhinological practice: discussion paper. J R Soc Med 87:411–413

7. Gordon A, McCaffrey TV, Kern EB (1989) Rhinomanometry for preoperative and postoperative assessment of nasal obstruction. Otolaryngol Head Neck Surg 101:20–26
8. Kasperbauer JL, Kern EB (1987) Nasal valve physiology. Otolaryngol Clin North Am 20:699–719
9. Kern EB (1977) Standardization of rhinomanometry. Rhinology. 25:115–119
10. Knops JL, McCaffrey TV, Kern EB (1983) Inflammatory diseases of the sinuses:physiology. Clinical applications. Otolaringol Clin North Am 4:515–531
11. McCaffrey TV, Kern EB (1979) Clinical evaluation of nasal obstruction: A study of 1000 patients. Arch Otolaryngol Head Neck Surg 105:543–545
12. McCaffrey TV, Kern EB (1986) Rhinomanometry. Facial Plast Surg 3:217–223
13. Monserrat V (1974) Rhinomanometría clínica. Exploracion funcional de las resistencias nasales. Ann Otorhinolaryngol Ibero Am 2:13–63
14. Nienhuis DM, McCaffrey TV, Kern EB (1993) Rhinomanometry: clinical applications in the evaluation of nasal obstruction. Practical endoscopic sinus surgery. McGraw-Hill, New York, pp 31–43
15. Ogura JH (1966) Nasal obstructions and mechanics of breathing physiologic relationship and effects of nasal surgery. Arch Otolaryngol Head Neck Surg 83:135
16. Ogura JH, Stokstead P (1958) Rhinomanometry in some rhinologic diseases. Laryngoscope 68:2001–2014
17. Salas JE (1996) Nasal function evaluation. Rhinology: science and art. Mexican Society of Otorhinolaringology, Head and Neck Surgery. Masson-Salvat Med 1:84–94
18. Salas JE, Chavez M (1996) Anterior active rhinomanometry: method, interpretation and utility in office. 46th annual meeting, Mexican Society of Otorhinolaryngology, Head and Neck Surgery. May 1–5, 1996. Mexican Society of Otorhinolaryngology, Head and Neck Surgery, Mexico City
19. Santiago Diez de Bonilla J, McCaffrey TV, Kern EB (1986) The nasal valve: a rhinomanometric evaluation of maximum nasal inspiratory flow and pressure curves. Ann Otol Rhynol Laryngol 229–232
20. Stammberger H (1991) Secretion transportation. Functional endoscopic sinus surgery. Decker, Philadelphia, pp 17–47
21. Williams HL (1968) The history of rhinomanometry in North America. Int Rhinol 22:151

Imaging of the Nose and Paranasal Sinuses

Rainer G. Haetinger

"To solve a problem, we need to make clear:
(a) what we know, (b) what we don't know,
(c) what we are looking for."

E. Hodnett

Introduction

The planning of endoscopic and/or endonasal microendoscopic surgery requires a close association between the clinical data and the methods and results of image investigation. Knowledge of anatomy and of anatomical variations through direct vision, as in computed tomography (CT) or magnetic resonance imaging (MRI), constitutes the first steps in this investigation. The purpose of this chapter is to demonstrate the anatomical aspects most relevant for surgical procedure and to exemplify situations in which an anatomical variation can be either a predisposing factor for inflammatory disease or a cause of nasal obstruction [5, 15, 23, 27].

Examination Technique

Computed Tomography

Image investigation of the ostiomeatal unit and of the other drainage channels of the sinuses is primarily made by CT. Whenever possible, coronal images should be made with the patient in a prone position with overextension of the head, as this is the angle of endoscopic vision and surgical access. Sometimes it is not possible to obtain direct coronal images (for example, in young children, patients who are intubated or have tracheostomies, patients with severe cervical arthropathy or those who are otherwise debilitated) [26]. In these cases, it is best to obtain helical contiguous images in the axial plane and to perform coronal reformations. There are no fixed rules for the technical programming of a CT examination, since this depends on the purpose of the examination. The best option is to adapt the study to the condition and requirements of each patient. In general, the slices should be fine (between 1.5 mm and 3 mm wide, although some authors prefer 5 mm) and should have a maximum dislocation of 5 mm.

With helical equipment, we use a width and dislocation of 3 mm (pitch 1:1). The helical acquisition of the images allows for the reconstruction of "intermediate" images (for example, 3 mm width with 1.5 mm of dislocation) by the computer, substantially improving the quality of sagittal, coronal or oblique reformations and making high quality three-dimensional reconstructions possible (when necessary).

In the axial plane, the slices should examine the area from the alveolar dental edge to the cranial limit of the frontal sinuses; on the coronal plane, the slices should reach from the anterior cartilaginous portion of the nasal septum to the posterior limit of the sphenoid sinuses. The hard palate is helpful as a reference for the angling of the axial plane in the "scout view" (digital radiography using previously programmed slices). The coronal slices should be as perpendicular as possible to the axial plane. The reconstruction algorithm can be "standard" or "detail" (we use the "bone" algorithm only in the case of bony lesions after reconstruction based on the raw data). At the ethmoid labyrinth, the window level should always be used to demonstrate bone tissue (when i.v. contrast media is used, the ethmoid should be documented with a window for the bone and for soft tissues). It is very important to change the window level during the examination in order to detect soft tissue or bony lesions.

Usually, patients should not have eaten for at least 3 h before an i.v. contrast media injection is required. Prior to the tomographic examination, a nasal vasoconstrictor is applied, and nasal hygiene is requested in order to eliminate the maximum amount of nasal secretions.

Use of i.v. Contrast Media in CT

The use of i.v. contrast media is variable. We consider it to be obligatory in epistaxis cases (because there is a need to look for vascular lesions), when neoplasia is suspected (for the purpose of staging, it is essential to establish the limits of the lesions, which usually enhance), in complications of acute sinusitis (subperiosteal abscess, epidural or subdural abscess, osteomyelitis, orbital cellulitis or abscess, thrombosis of the cavernous sinus) [7] and, sometimes, in nasal obstructions caused by polypous lesions (to clearly differentiate between the polyps and the adjacent mucosa and to detect angiomatous polyps or neoplasms). When there is doubt, images are usually obtained without contrast media up to the upper limit of the maxillary sinus. A decision concerning whether or not to inject is then made, depending on information from the first images.

Naturally, the risk/benefit of the use of i.v. contrast media has to be taken into consideration. Common sense should prevail in patients with history of allergy or other relative or absolute contraindications to iodine contrast. In some cases, a combination of CT without i.v. contrast and MRI can also produce good results.

Magnetic Resonance Imaging

MRI has advantages and disadvantages relative to CT. The advantages include a clear differentiation among the different types of soft tissue, the absence of ionizing radiation, the possibility of obtaining multiple image planes, the resources of water saturation and fat suppression, the fact that it is only slightly invasive (in cases of i.v. injection of the paramagnetic agent) and the possibility of performing magnetic resonance angiography. MRI demonstrates intracranial complications of sinusitis, such as septic thrombosis and empyema, and differentiates inflammatory disease from neoplastic disease. Midline congenital malformations are best evaluated with both CT and MRI. The disadvantages of MRI include reduced sensitivity for the detection of calcification and for the demonstration of the cortical bone (despite the fact that MRI is greatly superior to CT for the visualization of bone marrow) and the possible degradation of the image caused by movement artifacts or by the presence of obturations or metallic dental prosthesis when the oral cavity or teeth are being studied. Another disadvantage is the similar (and sometimes identical) signal intensities (in both T1- and T2-weighted images) of air, mycetomas, desiccated secretions, acute hemorrhage, calcium, bone and enamel, which appear as low-intensity signals or signal voids (an acute hemorrhage or a mycetoma may be misdiagnosed as a well-aerated sinus, for example) [6]. The disadvantages are being progressively reduced with the rapid development of MRI equipment, software and coils.

The planning of a MRI study may vary in accordance with the purpose of the examination. A relatively general evaluation of the face includes sagittal T1-weighted, axial pre-gadolinium T1-weighted, axial T2-weighted, axial post-gadolinium T1-weighted (with or without fat suppression) and coronal post-gadolinium T1-weighted sequences. Lesions that involve the midline region [for example, the nasopharynx, nasal cavities, sphenoid sinuses and the cribriform plate (*lamina cribrosa*) of the ethmoid] should also be studied in a sagittal T1-weighted post-gadolinium or T2-weighted sequence (in this case, the axial T1-weighted pre-gadolinium sequence can be dispensed with). Some sequences can be eliminated, substituted by others or added to those mentioned.

For simplicity, we will use the terms T1-weighted imaging, proton density and T2-weighted imaging, regardless of the sequences used (spin echo, fast spin echo, inversion recovery, etc.). Explanation of each sequence mode and its variations would be very lengthy and is not the main purpose of this chapter.

In a simplified way, the overall picture (Table 5.1) shows the principal behaviors of the different tissues and substances in the T1-weighted and T2-weighted sequences. The behavior of other structures, such as the muscles (whose signal intensity is intermediate on T1-weighted images and hyperintense on T2-weighted images) and cartilages (whose signal intensity may vary in accordance with the quantity of calcium, depending on the age of the patient), could be added to Table 5.1.

MRI allows the different planes (sagittal, axial and coronal) to be formatted by the computer without the need to change the position of the patient (who continues to lie in a supine position, with the head in a neutral and comfortable position). MRI investigation techniques have many variables, principally as a result of the constant development of new sequences. Compared with CT studies, there is a greater need for MRI studies to be adapted to each patient and disease.

The Objectives of Image Investigation

The principal indications for CT and MRI in the study of the paranasal sinuses and the nose include:
- Benign lesions
  - Inflammatory disease (recurrent or persistent sinusitis)
  - Complications of infection [14]

Table 5.1. The principal behaviors of the different tissues and substances in T1-weighted and T2-weighted sequences

Hyperintense	Hypointense
T1-weighted images	
Fat	Cortical bone calcification
Bone marrow	Flow (blood vessels); "flow-void"
Cartilage	Water (the majority of lesions)
Hyperprotein fluids	Fluids in general with low protein concentration
Enhancement by paramagnetic agent (gadolinium)	Inflammation/edema
Paramagnetic effects	Tumor
Cysts with high protein content	Hemosiderin
	Acute hemorrhage (deoxyhemoglobin)
	Cyst (with low internal protein concentration)
	Fibrosis
	Iron
	Air
	Mycetoma
T2-weighted images	
Water and fluids in general	Cortical bone calcification
Inflammation/edema	Flow (blood vessels); "flow-void"
Inflammatory polyps	Hemosiderin
Tumor	Acute hematoma (deoxyhemoglobin)
Mucosa	Fibrosis
Lymphatic tissue	Iron
	Teeth; obturations
	Air
	Mycetoma

- Single or multiple paranasal sinus disease [14]
- Spread to adjacent tissues and anatomical compartments [14]
- Tissue characterization [14]
- Differentiation between solid and cystic lesions [14]
- Differentiation between neoplastic and inflammatory disease (by MRI)
- Trauma evaluation (by CT) [14]
- Search for hidden disease [14]
- Malformation identification
- Malignant lesions (staging)
  - Pterygopalatine fossa and/or infraorbital fissure involvement [14]
  - Pterygoid fossa extension [14]
  - Invasion of the mastigatory space
  - Invasion of the anterior cranial fossa through the cribriform plate [14]
  - Extension to the middle cranial fossa by the sphenoid sinus or cavernous sinus [14]
  - Orbital extension [14]
  - Nasopharyngeal involvement [14]
  - Skin lymphatic involvement [14]

With both plain radiographs and planigraphs, a totally opacified paranasal sinus always presents a diagnostic dilemma. The cause could be fluid, tumor, fibrosis, chronic infection or a combination of these factors. On CT, debris-free fluids exhibit low density, while lesions associated with thickening of the mucosa and fluids usually show variable densities within the paranasal sinus. Solid tumors are normally enhanced with the i.v. contrast. Unfortunately, in a totally opacified sinus, the differentiation between tumor and chronic inflammatory disease is not always possible by CT alone. MRI can make this distinction in most cases, but surgical exploration or biopsy continue to be the definitive procedures for diagnosis and treatment.

Through MRI, it is generally possible to differentiate between the inflammatory conditions (with or without accumulated secretions) and solid neoplastic tumors, because the large amount of water in the inflammatory conditions results in a marked high signal on T2-weighted images and is reduced on T1-weighted images. However, there are limitations to this evaluation. More benign tumors and those of the glandular type, such as polyps, papillomas, minor salivary gland tumors and schwannomas, usually have sufficient water content to produce high signal intensity on T2-weighted images [12]. However, subacute or chronic inflammatory states show progressive loss of hydration and an accumulation of protein, salts and minerals, resulting in variable signal intensities in both T1- and T2-weighted images [18]. These aspects will be dealt with in detail later when mucoceles are discussed [19].

In the acute inflammatory state, the behavior of secretion and edematous mucosa is similar to that of water, i.e., hypointense on T1-weighted and hyperintense on T2-weighted images. Mucous retention cysts and inflammatory polyps behave in the same way. Intrasinus hemorrhage initially shows a reduced signal intensity in all the MRI sequences but, as the blood undergoes more oxidation to methemoglobin, its signal intensity becomes high in T1-weighted and proton-density-weighted images and intermediate in T2-weighted images.

Plain Radiographs

Plain radiographs are still widely used (despite the innovations resultant from other imaging methods), principally because of the low cost and ease of acquisition; they are a good method for screening patients who may have sinusitis without complications, foreign bodies or fractures. It is well known that the conventional radiographs underestimate inflammatory sinus disease (especially chronic disease), detect ethmoidal disease poorly and cannot evaluate the ostiomeatal unit adequately. Consequently, they are not sufficient for a preoperative evaluation.

Plain radiography can serve as a guide for maxillary puncture and endonasal or radical antrostomy (for example, detecting hypoplastic sinuses, intranasal bone septa or thickening of the bone walls). Other indications include a Caldwell view for constructing a mold for osteoplastic surgery of the frontal sinus or for trephination of the frontal sinus [2].

Planning of Endoscopic or Microendoscopic Surgery

CT is the examination of choice for the detailed demonstration of the anatomical information necessary for planning different stages of endoscopic or microendoscopic nose and paranasal-sinus surgery, as tabulated below and commented on subsequently [13, 16]:
1. Evaluation of the ostiomeatal unit
2. Pneumatization of the orbital roofs
3. The distance between the fovea ethmoidalis and the cribriform plate
4. Dehiscence of the lamina papyracea
5. Pneumatization of a nasal turbinate (concha bullosa)
6. Anatomic variations which narrow the middle meatus
7. Anatomic variations which narrow the infundibulum
8. Continuity or interruption of the ethmomaxillary plate.
9. Relationship between the maxillary and sphenoid sinuses
10. Distance between the floor of the sphenoid sinus and the sphenopalatine foramen
11. Relationship between the posterior ethmoidal cells and the sphenoid sinuses
12. Relationship between the posterior ethmoidal cells and the maxillary sinus
13. Relationship among the posterior paranasal sinuses and the neurovascular structures, principally in cases of bone dehiscence
14. Basilar pneumatization of the sphenoid sinus, with thinning of the clivus
15. Sphenoid sinus extensions (pneumatization of the pterygoid process, the anterior clinoid process, the greater wing of the sphenoid and the posterior portion of the nasal septum)
16. Maxillary sinus extensions (palatine recess, infraorbital recess, alveolar recess, zygomatic recess)
17. Frontal sinus extensions (frontal bulla, pneumatization of the crista galli, orbital plates of the frontal bones)
18. Prominent agger nasi cells

Evaluation of the Ostiomeatal Unit

CT provides very clear and extensive anatomic and pathologic information, is an essential tool for preoperative evaluation and is an excellent guide for surgical approach. The evaluation of the ostiomeatal unit (or complex) is the principal area of study in non-neoplastic disease.

The ostiomeatal unit has been referred to as the maxillary sinus ostium, the ethmoid infundibulum, the hiatus semilunaris, the anterior and middle ethmoidal cells' ostia, the frontal recess (frontonasal duct), the middle meatus, the sphenoethmoidal recess and the superior meatus and is the object of evaluation for functional endoscopic sinus surgery. These anatomical details are shown in Figs. 5.1–5.7.

Imaging of the Nose and Paranasal Sinuses 57

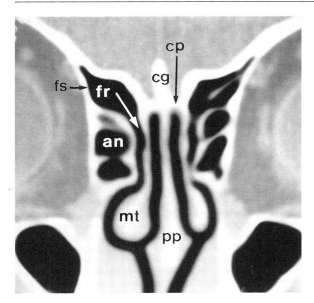

Fig. 5.1. Computed tomography of the coronal plane: drainage of the frontal sinus. *an*, agger nasi cell; *cp*, cribriform plate (lamina cribrosa); *cg*, crista galli, *fr*, frontonasal recess; *fs*, frontal sinus; *mt*, middle turbinate; *pp*, perpendicular plate

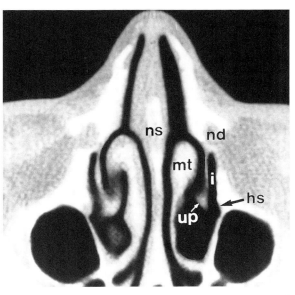

Fig. 5.3. Computed tomography of the axial plane: ostiomeatal unit. *hs*, hiatus semilunaris; *i*, infundibulum; *mt*, middle turbinate; *nd*, nasolacrimal duct; *ns*, nasal septum; *up* unciform process

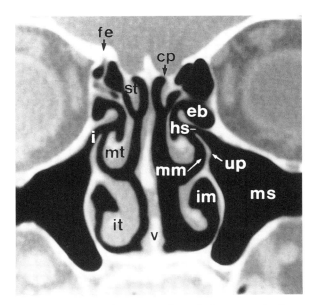

Fig. 5.2. Computed tomography of the coronal plane: ostiomeatal unit. *cp*, cribriform plate (lamina cribrosa); *eb*, ethmoidal bulla; *fe*, fovea ethmoidalis; *hs*, hiatus semilunaris; *i*, infundibulum; *im*, inferior meatus; *mm*, middle meatus; *ms*, maxillary sinus; *mt*, middle turbinate; *st* superior turbinate; *v* vomer

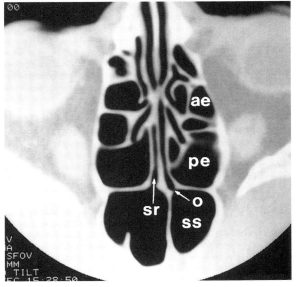

Fig. 5.4. Computed tomography of the axial plane: ethmoidal labyrinth and sphenoid sinus drainage. *ae*, anterior ethmoidal cell; *o*, ostium; *pe*, posterior ethmoidal cell; *sr*, sphenoethmoidal recess; *ss*, sphenoidal sinus

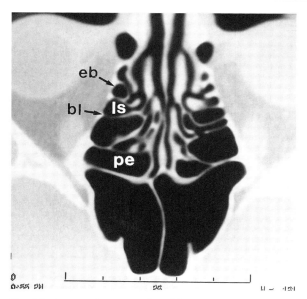

Fig. 5.5. Computed tomography of the axial plane: ethmoidal labyrinth; showing the basal lamella and the lateral sinus. *bl*, basal lamella; *eb*, ethmoidal bulla; *ls*, lateral sinus; *pe*, posterior ethmoidal cell

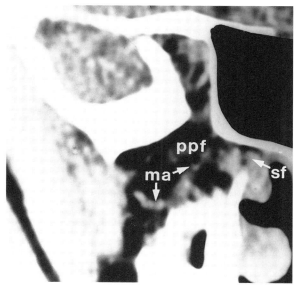

Fig. 5.6. Computed tomography of the coronal plane: pterigopalatine fossa; with the sphenopalatine foramen and the maxillary artery. *ma*, maxillary artery; *ppf*, pterygopalatine fossa; *sf*, sphenopalatine foramen

Pneumatization of the Orbital Roofs (Anatomic Variation)

The orbital roof may be pneumatized by the frontal sinus, ethmoidal cells or both. The ethmoidal cell is usually located posteriorly to the frontal sinus in the ceiling of the orbit and is termed the supraorbital recess or supernumerary frontal sinus (Fig. 5.8). It is very important to know whether this condition is present, as it may influence the choice of surgical approach (endonasal versus external access).

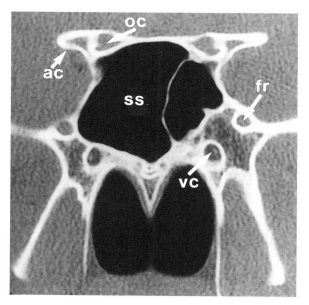

Fig. 5.7. Computed tomography of the coronal plane at the level of the foramen rotundum and vidian canal. *ac*, Anterior clinoid process; *fr*, frontonasal recess; *oc*, optic canal; *ss*, sphenoidal sinus; *vc*, vidian canal (pterygoid canal)

The Distance Between the Fovea Ethmoidalis and the Cribriform Plate

Coronal sections are the best way to evaluate the distance between the fovea ethmoidalis (ethmoid roof) and the cribriform plate (where the olfactory bulb is lodged). According to Keros [8], who studied 450 skulls (1962), the distance from the ethmoid roof to the cribriform plate measured between 4 mm and 7 mm in 70.16% of the cases, between 8 mm and 16 mm in 18.25% and between 1 mm and 3 mm in 11.59%. Preoperative knowledge of this distance is very important in the case of intranasal surgery, because the greater the difference, the stronger the relationship between the ethmoidal cells and the endocranium (both the roof and the medial wall of the upper ethmoid abut the latter; Fig. 5.9).

Imaging of the Nose and Paranasal Sinuses

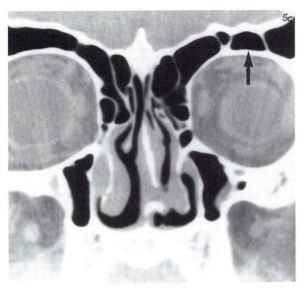

Fig. 5.8. Pneumatization of the orbitary roofs (anatomic variation). The *arrow* shows an ethmoidal cell at the orbitary roof

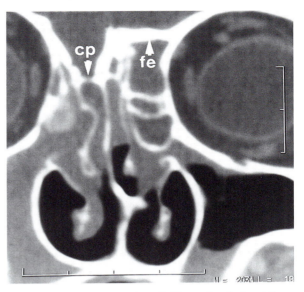

Fig. 5.9. Distance between the cribriform plate (*cp*) and fovea ethmoidalis (*fe*)

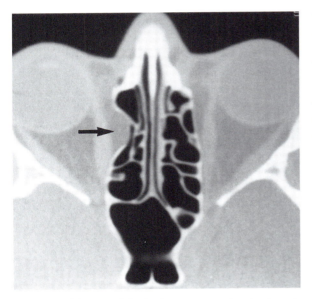

Fig. 5.10. Dehiscence of the lamina papyracea (anatomic variation). The *arrow* shows the bulging of the orbitary fat

Dehiscence of the Lamina Papyracea

Spontaneous dehiscence of the ethmo-orbital wall is a relatively common situation. Previous surgery or trauma have to be investigated in the differential diagnosis. Nasosinusal polyposis is also a cause for dehiscence. The lamina papyracea is a very thin bone structure, and the medial rectus and the superior oblique muscles lie close behind it when exentera-tion of the most lateral ethmoidal cells is necessary. A dehiscence may increase the risk of orbital complications. The usual finding is a medial bulging of the extraconal orbital fat (Fig. 5.10).

Pneumatization of a Nasal Turbinate (Concha Bullosa; Anatomic Variation)

The pneumatization of the middle turbinate is one of the most common findings on CT examination, with a great variation in size. A pneumatized superior turbinate is not uncommon, but pneumatization of the inferior turbinate (derived from the maxillary sinus) is very rare (Figs. 5.11–5.13). Usually, the indications to operate on the concha bullosa are severe nasal obstruction caused by an enlarged or bulged concha bullosa (narrowing the meatus) or an infected cell (which may be a cause of recurrent sinusitis), sometimes containing a mucocele (Figs. 5.14, 5.15). There is a predisposition for infection inside a concha bullosa, because its drainage is sometimes less efficient then that of the rest of the ethmoidal complex.

Anatomic Variations that Narrow the Middle Meatus

- Enlarged concha bullosa
- Enlarged middle concha

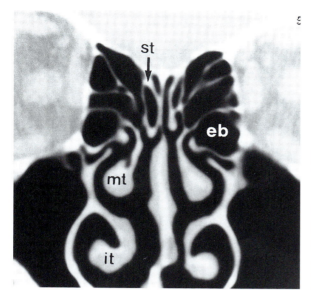

Fig. 5.11. Superior concha bullosa (anatomic variation). *eb*, ethmoidal bulla; *it*, inferior turbinate; *mt*, middle turbinate; *st*, superior turbinate

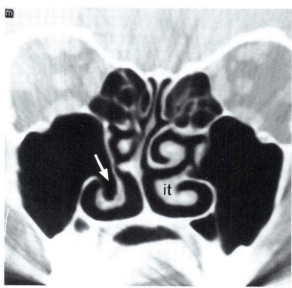

Fig. 5.13. Inferior concha bullosa (anatomic variation). *it*, inferior turbinate

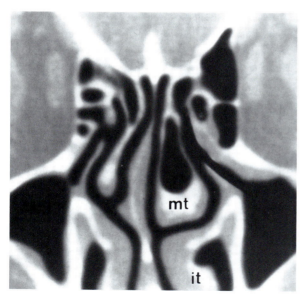

Fig. 5.12. Middle concha bullosa (anatomic variation). *it*, inferior turbinate; *mt*, middle turbinate

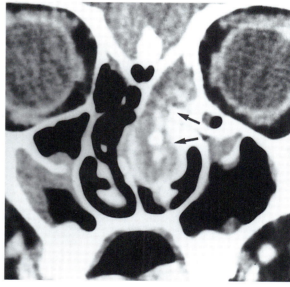

Fig. 5.14. Mucocele within concha bullosa. Enhanced computed tomography of the coronal plane showing a mucocele within a concha bullosa (courtesy of Dr. Senair Ambros; Hospital S. Vicente de Paulo; Passo Fundo, RS, Brazil)

- Enlarged ethmoidal bulla
- Paradoxical middle turbinate
- Deviation of the nasal septum
- Bulla within the uncinate process
- Hooked unciform process

The middle meatus can sometimes be narrowed by anatomic variations, such as a enlarged concha bullosa, enlarged ethmoidal bulla, paradoxical middle turbinate, deviation of the nasal septum or a bulla within the uncinate process. Earwaker [3], studying

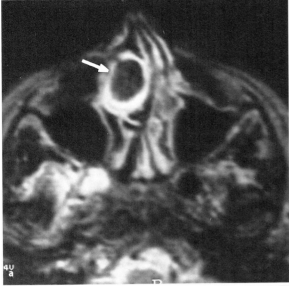

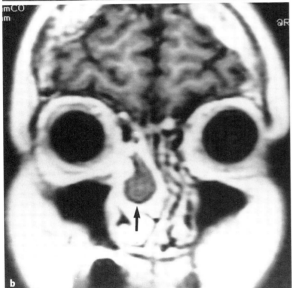

Fig. 5.15. Mucocele within concha bullosa. a Magnetic resonance (MR); T2-weighted image; axial plane. b MR; T1-weighted image; post gadolinium; coronal plane

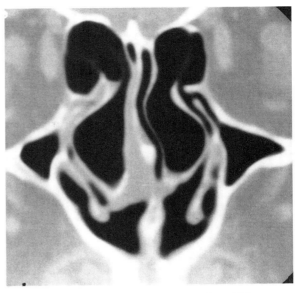

Fig. 5.16. Large concha bullosa narrowing the middle meatus bilaterally

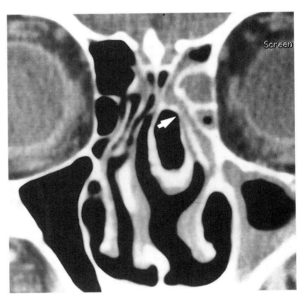

Fig. 5.17. Large concha bullosa narrowing the middle meatus and the hiatus semilunaris (*arrow*), causing obstructive maxillary sinusitis; and frontal and anterior ethmoidal sinusitis

800 patients in 1993, found pneumatization of the middle turbinate (concha bullosa) in 55% of the cases; pneumatization involved the vertical lamella, only the bulb or the whole turbinate, demonstrating that more than one half of the patients have this variation. However, only 37% of the patients had pneumatization that produced significant distortion of the turbinate (Figs. 5.16, 5.17).

The middle turbinate may also be enlarged without being pneumatized. When the bony component of the turbinate is large, the whole turbinate is enlarged, including the soft-tissue component. Therefore, the size of the middle meatus is dependent on both the bone and soft-tissue components of the middle turbinate (Fig. 5.18).

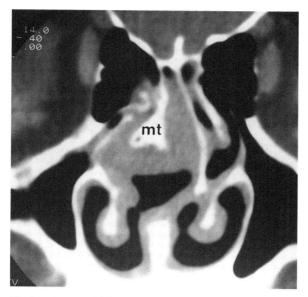

Fig. 5.18. Large middle turbinate narrowing the middle meatus (*mm*)

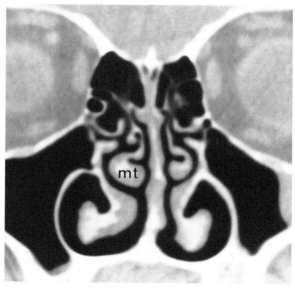

Fig. 5.20. Paradoxical middle turbinate (*mt*, anatomic variation)

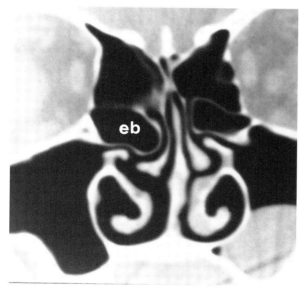

Fig. 5.19. Large ethmoidal bulla (*eb*) narrowing the middle meatus

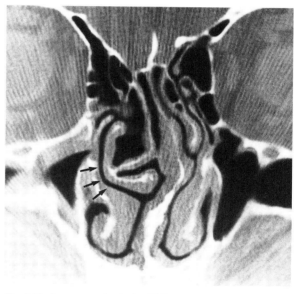

Fig. 5.21. Large paradoxical middle turbinate narrowing the middle meatus (*arrows*)

Large ethmoid bullae may prolapse into the middle meatus, displacing the vertical lamella of the middle turbinate medially (Fig. 5.19). A paradoxical middle turbinate has a laterally faced convexity (instead of the usual medially convex face similar to the inferior turbinate) and may have a bulged configuration, sometimes narrowing the middle meatus (Figs. 5.20, 5.21).

Septal deviation is considered when the nasal septum diverges from the midline with a single or double curve (S shaped) associated with significant asymmetry of the nasal conchae and/or nasal-wall structures. A septal spur at the junction between the lamina perpendicularis and the vomer, with great variation in size, is also a relatively common finding (Fig. 5.22). When the spur is large, the adjacent mid-

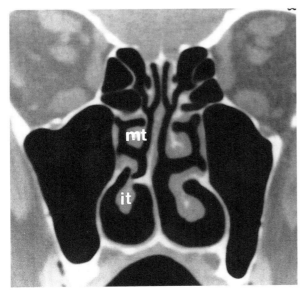

Fig. 5.22. Septal spur at the junction of the vomer and the perpendicular lamina of the ethmoid touching the inferior turbinate. *it*, inferior turbinate; *mt*, middle turbinate;

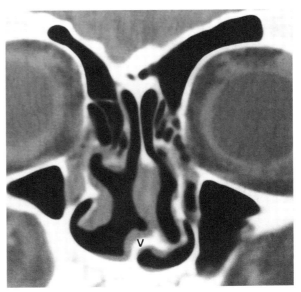

Fig. 5.24. Deformity of the condrovomeral junction

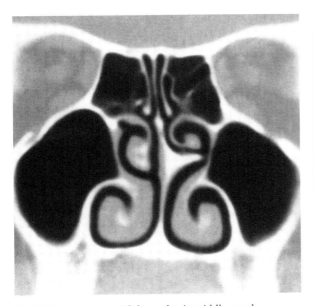

Fig. 5.23. Septal spur with hypoplastic middle concha

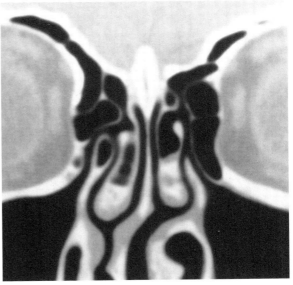

Fig. 5.25. Pneumatization of the unciform process (anatomic variation)

dle concha usually appears hypoplastic (Fig. 5.23). Another finding can be a ridge at this location. A deformity of the condrovomeral junction is commonly associated with deviation of the anterior curvature (Fig. 5.24).

The pneumatization of the unciform process (also called bulla within the unciform process) is a less common cause of narrowing of the middle meatus.

The pneumatization commonly occurs as an extension of the agger nasi cells, which lie anteriorly (Figs. 5.25, 5.26).

The unciform process may have a hook configuration, deviated medially and inferiorly (Fig. 5.27). Depending on its size, the hooked unciform process may narrow the middle meatus (Fig. 5.28).

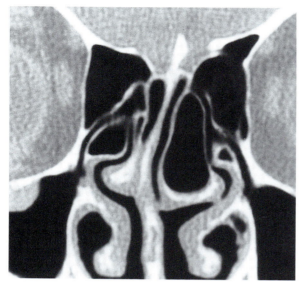

Fig. 5.26. Pneumatization of the unciform process narrowing the middle meatus

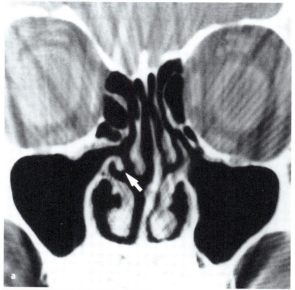

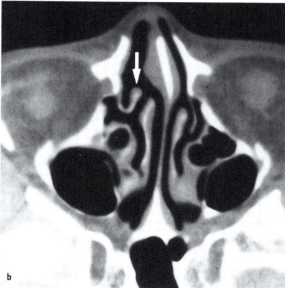

Fig. 5.27. Hooked unciform process (*arrows*). a Coronal plane (anatomic variation). b Axial plane

Anatomic Variations which Narrow the Infundibulum

- Enlarged ethmoidal bulla
- Bulla within the uncinate process
- Haller's cell
- Lateralization of the unciform process

An enlarged ethmoidal bulla and a bulla within the uncinate process are causes of narrowing of both the middle meatus and the infundibulum. Other causes of the narrowing of the infundibulum are the presence of Haller's cells or a lateralization of the unciform process.

When the ethmoidal bulla is enlarged (usually with a downward development), it is accompanied by a horizontalization of the unciform process and is a potential cause of obstruction of maxillary drainage (Fig. 5.29).

Air cells extending into the roof of the maxillary sinus were described by A. von Haller in the 18th century. They commonly originate from the anterior ethmoid cells and are directly related to the limits of the infundibulum (Fig. 5.30). Therefore, in case of large Haller's cells, the infundibulum may be narrowed (Fig. 5.31). This anatomic variant is also a relatively common incidental finding. Like any other ethmoidal cell, it can also be the site of a mucocele (Fig. 5.32). Any lateralization of the unciform process diminishes the infundibulum (because of the proximity to the limit of the orbit) and is also a potential cause of obstructive sinusitis (Fig. 5.33).

Continuity or Interruption of the Ethmomaxillary Plate

The ethmomaxillary plate, which separates the ethmoid sinus from the maxillary sinus, is frequently interrupted in pathologic conditions, such as advanced nasosinusal polyposis, inverted papilloma or neoplastic disease. When the interruption results from non-neoplastic causes, a transantral ethmoidectomy provides good access (Fig. 5.34).

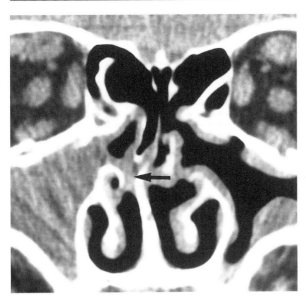

Fig. 5.28. Hooked unciform process (*arrows*) narrowing the middle meatus

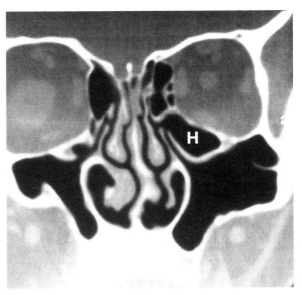

Fig. 5.30. Haller's cell (*H*). Anatomic variation

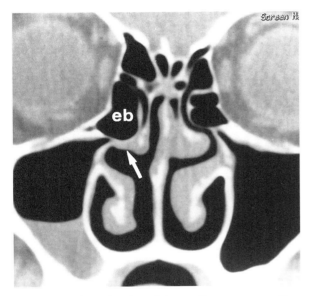

Fig. 5.29. Horizontalization of the unciform process; with a large ethmoidal bulla (*eb*) and narrowing of the infundibulum (*arrow*)

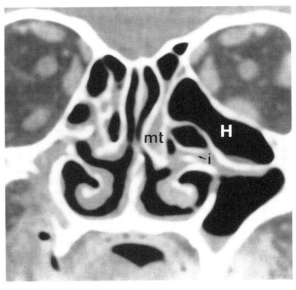

Fig. 5.31. Large Haller's cell (*H*) narrowing the infundibulum (*i*). *mt*, middle turbinate;

Relationship Between the Maxillary and Sphenoid Sinuses (Anatomic Variation)

When there is a direct relationship between the maxillary and sphenoid sinuses, the posterior part of the ethmomaxillary plate becomes the sphenomaxillary plate (Fig. 5.35). Knowledge of its presence is important for transantral surgery and should be appreciated to avoid mistaking the sphenoid sinus for the most posterior ethmoid cell.

Distance Between the Floor of the Sphenoid Sinus and the Sphenopalatine Foramen

The sphenopalatine foramen is localized very close to the anterior and inferior edges of the sphenoid

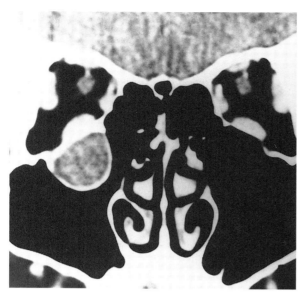

Fig. 5.32. Mucocele within a Haller's cell

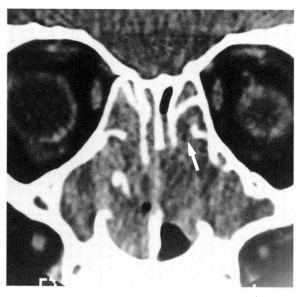

Fig. 5.34. Interruption of the ethmomaxillary plate in advanced sinonasal polyposis

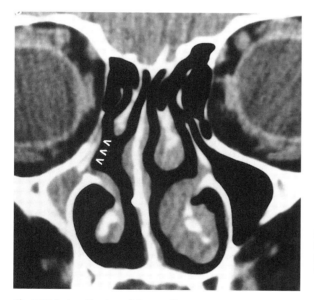

Fig. 5.33. Lateralization of the unciform process narrowing the infundibulum (*white arrowheads*) and maxillary sinusitis

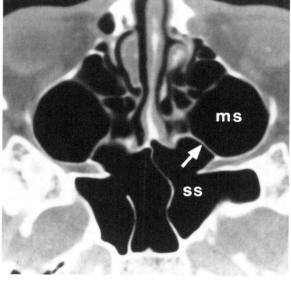

Fig. 5.35. Sphenomaxillary plate (*white arrow*). Contiguity between the sphenoid (*ss*) and maxillary sinuses (*ms*). Anatomic variation

sinus. The distance between these landmarks is important for surgery involving the sphenoethmoidal recess and the sphenoid sinus in order to avoid bleeding from the maxillary artery (Fig. 5.36).

Relationship Between the Posterior Ethmoidal Cells and the Sphenoid Sinuses (Anatomy or Anatomic Variation)

The sphenoid sinus may be approached by surgery through the posterior ethmoid cell. It is important to remember that the posterior wall of the ethmoid cell is not always flat or entirely in contact with the sphe-

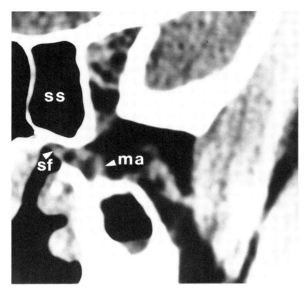

Fig. 5.36. Distance between the floor of the sphenoid sinus (*ss*) and the sphenopalatine foramen (*sf*). Anatomic landmark. *ma*, maxillary artery

noid sinus. This limit is commonly irregular and, when there is an extension of the posterior ethmoid cell above, beside or below the sphenoid sinus, it is called Onodi's cell (Fig. 5.37). The most common extension is above the sphenoid sinus, and it is not uncommon to have a direct relationship between the Onodi's cell and the optic canal or the intracavernous internal carotid artery.

Relationship Between the Posterior Ethmoidal Cells and the Maxillary Sinus (Anatomic Variation)

Sometimes the posterior ethmoid cells extend laterally into the maxilla and into the floor of the orbit [10, 11, 17]. It is necessary to identify the drainage into the superior meatus (as with any other posterior ethmoid cell; Fig. 5.38). Usually, this condition does not narrow the ethmoidal infundibulum. It is important to differentiate this condition from the second-

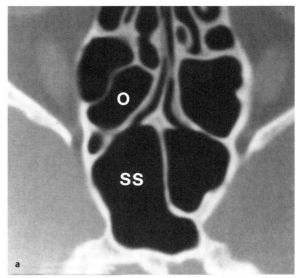
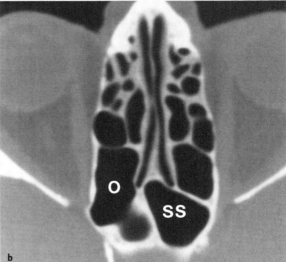
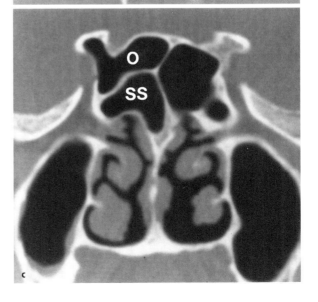

Fig. 5.37. Onodi's cell. **a** Computed tomography (CT) in the axial plane at the level of the ostia of the sphenoid sinuses showing the relation between the Onodi's cell (*O* on the right side) and the sphenoid sinus (*ss*). Anatomic variation. **b** CT in the axial plane (more cranial than the image in **a**) showing the most posterior ethmoid cell on the right side over the right sphenoid sinus. **c** CT in the coronal plane demonstrating the relation between the Onodi's cell (*O*) and the optic canal.

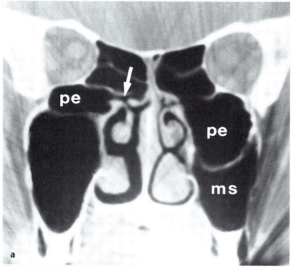

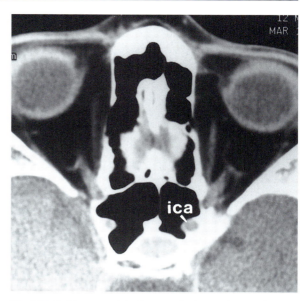

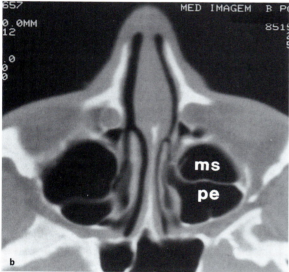

Fig. 5.38. Relationship between large posterior ethmoidal cells (*pe*) and the maxillary sinus (*ms*). **a** Computed tomography (CT) in the coronal plane. **b** CT in the axial plane (anatomic variation)

Fig. 5.39. Dehiscence of the bone at the internal carotid artery (*ica*) (anatomic variation)

ary maxillary sinus [10, 17], a rare anomaly in which two independent cavities in the same maxilla drain into the middle meatus through independent ostia.

Relationship among the Posterior Paranasal Sinuses and the Neurovascular Structures (Anatomy or Anatomic Variation)

- Optic nerves
- Internal carotid arteries
- Maxillary and vidian nerves

Usually, the optic nerves, the internal carotid artery and the maxillary and vidian nerves are in a close relationship with the sphenoid sinuses, protected by bony walls. As anatomical variations, dehiscenses of the bone may exist, increasing the potential risk of lesion during a surgical procedure (Fig. 5.39). In addition, there may be a direct relationship among an Onodi's cell and these structures (with or without dehiscence; Fig. 5.40). The vidian and maxillary nerves may also pass inside the sphenoid sinuses in cases of pneumatization of the pterygoid process (see "Sphenoid-Sinus Extensions") or may be directly related to Onodi's cells, with extension below or beside the sphenoid sinuses.

Basilar Pneumatization of the Sphenoid Sinus with Thinning of the Clivus (Anatomic Variation)

When inflammatory disease is present and the sphenoid sinus is studied endoscopically (in the same way that other paranasal sinuses are studied), the bone wall adjacent to the pathological process is frequently not well defined. For this reason, if there is an extreme thinning (or dehiscence) of the bone, there is a higher risk of complication by perforation. One of the critical areas is the posterior wall of the sphenoid sinus, as behind it is the posterior fossa and, more importantly, the basilar artery and the brain stem (Fig. 5.41).

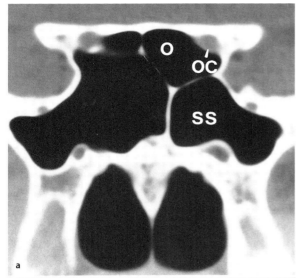

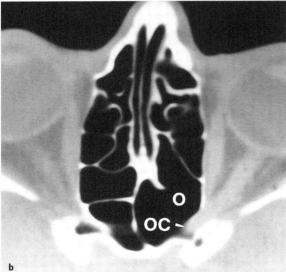

Fig. 5.40. Onodi's cell (*O*) and optic canal (*OC*). **a** Computed tomography (CT) in the coronal plane. **b** CT in the axial plane. *ss*, sphenoid sinus (anatomic variation)

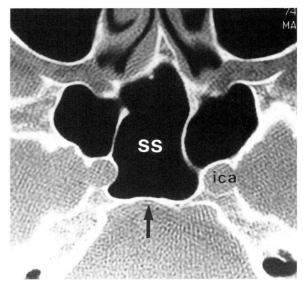

Fig. 5.41. Basilar pneumatization of the sphenoid sinus (*ss*). There is a very thin bony wall between the sphenoid sinus and the posterior fossa (anatomic variation). *ica*, internal carotid artery

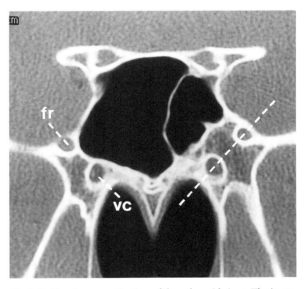

Fig. 5.42. Usual pneumatization of the sphenoid sinus. The imaginary line between the foramen rotundum (*fr*) and the vidian canal (*vc*) delimits the usual extension of the sphenoid sinus

Sphenoid-Sinus Extensions
(Pneumatizations; Anatomic Variations)

- Pterygoid process
- Anterior clinoid process
- Greater wing of the sphenoid
- Posterior portion of the nasal septum
- Basilar Pneumatization

The contour of the sphenoid cavity is sometimes very irregular, with multiple recesses due to a great variety of pneumatizations, including the pterygoid process, the anterior clinoid process, the greater wing and the posterior portion of the nasal septum. These variations can be associated with dehiscenses of the bone, and it is very important to know the relationship of these variations to the optic nerves, the vidian nerves and the internal carotid arteries.

Usually, the sphenoid sinus does not extend laterally beyond a line joining the foramen rotundum and the vidian (pterygoid) canal (Fig. 5.42). An extension beyond this line characterizes the pneumatization of the pterygoid process, which may be uni- or bilateral, in different degrees (Fig. 5.43).

An extension into the lesser wing of the sphenoid determines uni- or bilateral pneumatization of the anterior clinoid process (Fig. 5.44) and may be associated with bone dehiscence related to the optic canal (Fig. 5.45). An inferolateral pneumatization of the greater wing of the sphenoid frequently occurs downward and laterally and may extend into the posterior aspect of the lateral orbital wall (Fig. 5.46).

The posterior portion of the nasal septum pneumatizes from the anterior sphenoid sinus. This is a relatively common finding and, usually, it does not play an important role in endoscopy or endoscopic surgery (Fig. 5.47). It may eventually be important in cases of septoplasty if this part of the septum is very extensive anteriorly.

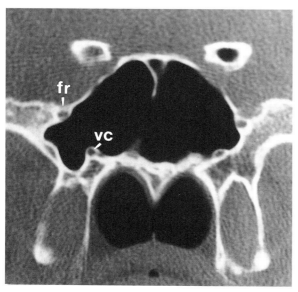

Fig. 5.43. Pneumatization of the pterygoid process (anatomic variation). *fr*, foramen rotundum; *vc*, vidian canal

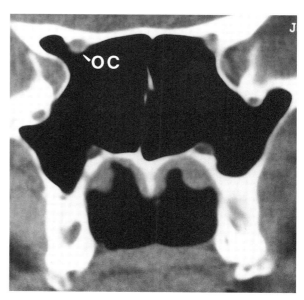

Fig. 5.45. Bone dehiscence related to the optic canal (*oc*) (anatomic variation)

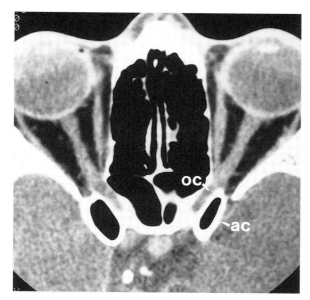

Fig. 5.44. Pneumatization of the anterior clinoid process (*ac*), *oc*, optic canal

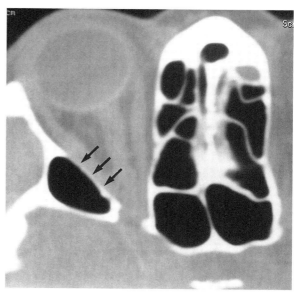

Fig. 5.46. Inferolateral pneumatization of the greater wing of the sphenoid (anatomic variation)

Maxillary-Sinus Extensions (Anatomy and Anatomic Variations)

- Palatine recess (*recessus palatinus*)
- Infraorbital recess (*recessus infraorbitalis*)
- Alveolar recess (*recessus alveolaris*)
- Zygomatic recess (*recessus zygomaticus*)

The palatine recesses extend inferomedially into the hard palate and are usually bilateral and symmetric. The distance between the two recesses has to be less than half the width of the nose at the level of the inferior meatus for this to be considered an anatomic variation (Fig. 5.48) [3, 10].

The infraorbital recess extends anteriorly and often medially to the infraorbital canal along the roof of the maxillary sinus. Sometimes, it extends anteriorly to the nasolacrimal duct (the so-called prelacrimal recess). This type of recess is usually unilateral and, when bilateral, is asymmetric (Fig. 5.49).

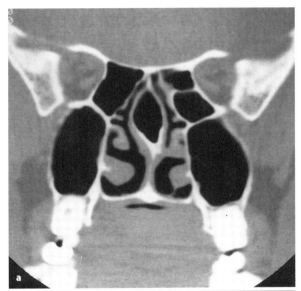

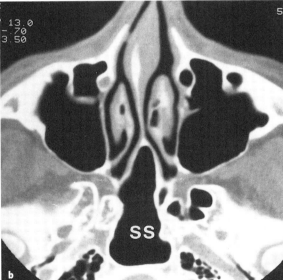

Fig. 5.47. a Pneumatization of the posterior portion of the nasal septum (anatomic variation). Computed tomography (CT) in the coronal plane. **b** Pneumatization of the posterior portion of the nasal septum. CT in the axial plane. *ss*, sphenoid sinus

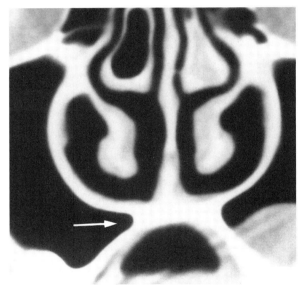

Fig. 5.48. Palatine recess (anatomic variation)

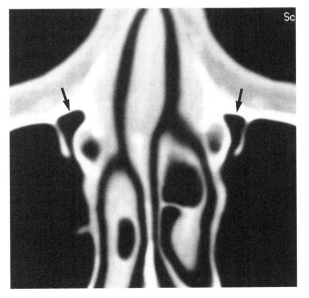

Fig. 5.49. Prelacrimal recess

The alveolar recesses extend into the alveolar margins close to the roots of the molar and premolar teeth, at a lower level than the floor of the nasal cavity. They are usually bilateral and symmetric (Fig. 5.50). The zygomatic recesses (or lateral recesses) extend into the malar bones and are commonly bilateral and symmetric (Fig. 5.51) [3, 24].

Frontal-Sinus Extensions (Anatomic Variations)

- Frontal bulla
- Pneumatization of the crista galli
- Orbital plates of the frontal bones (orbitary roof)

The frontal bulla develops from the anterior ethmoid into the frontal sinus, may be uni- or bilateral and is usually asymmetric (Fig. 5.52) [21]. It may also be the site of a mucocele (Fig. 5.53).

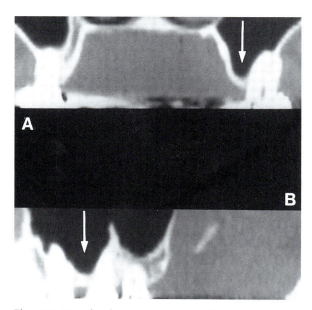

Fig. 5.50 A,B. Alveolar recess. Coronal (A) and sagittal (B) reformatations

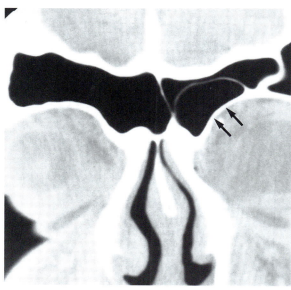

Fig. 5.52. Frontal bulla (anatomic variation)

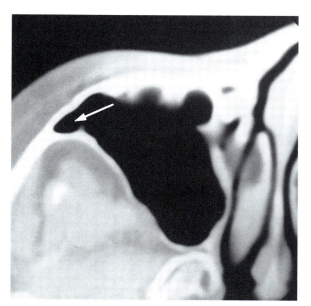

Fig. 5.51. Zygomatic recess (anatomic variation)

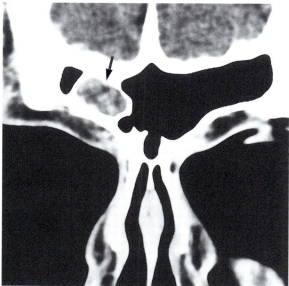

Fig. 5.53. Frontal bulla with mucocele

Extension into the crista galli originates from the frontal sinus. This recess is a relatively common incidental finding on CT examination (Fig. 5.54).

Pneumatization of the orbital roofs (orbital plates) may originate from the frontal sinuses or from the ethmoid cells and usually occurs bilaterally. This recess is significant when it reaches the level of the posterior ethmoid cells, and it may also extend to the planum sphenoidale (see "Pneumatization of the Orbital Roofs").

Enlarged Agger Nasi Cells

Extensive pneumatization of the agger nasi cells may result in obstruction of the frontal recess or the infundibulum (through a pneumatized uncinate process).

Clinical Significance of Anatomic Variations

Although anatomic variations may frequently impair drainage and ventilation of the paranasal sinuses may or cause nasal obstruction (especially when they involve the ostiomeatal unit), they may also be asymptomatic and may only predispose the patient to blockage with mucosal swelling during infection. Frequently, they constitute only incidental findings of paranasal sinus examinations [9, 23].

When anatomic variations occur in the posterior ethmoid or in the sphenoid sinus, they may increase the risk of damage to optic nerves, internal carotid arteries [3], maxillary and vidian nerves and the anterior structures within the posterior cranial fossa (behind a dehiscent clivus). Nasal-septum deviation (especially with a septal spur) and large middle turbinates may hinder endoscopic access to distal areas. Variations of the fovea ethmoidalis may increase the risk of anterior cranial-fossa damage during endoscopic surgical procedures. The relevance of these findings depends on a very careful evaluation of clinical aspects, endoscopy and imaging.

Non-neoplastic Disease

Air-fluid levels occur most commonly in patients with acute bacterial sinusitis. However, a sinus lavage in the treatment of acute bacterial sinusitis can form an air-fluid level that takes at least 2–4 days to empty from the sinus. Therefore, studies should not be obtained until at least 6 days or 7 days after an antral lavage. Air-fluid levels can also be caused by hemorrhage but, in this case, the intrasinus blood is denser than mucosal edema and inflammatory secretions on CT scans.

The thickened, inflamed mucosa exhibit low signal intensity on T1-weighted images and high signal intensity on T2-weighted images, reflecting the high water content of the inflamed tissue. This appearance is different from that of fibrosis, which appears as a low to intermediate signal intensity tissue in both T1- and T2-weighted images. Neoplastic disease also

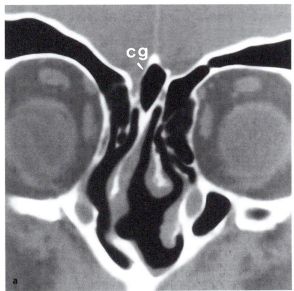

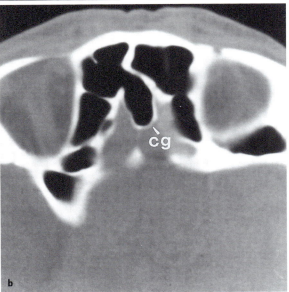

Fig. 5.54. a Pneumatization of the crista galli (*cg*, anatomic variation). Computed tomography (CT) in the coronal plane. b Pneumatization of the crista galli. CT in the axial plane

exhibits intermediate signal intensity on T2-weighted images.

In CT scans, there are two criteria that allow for differentiation between chronic inflammatory disease (eventually associated with polyposis) and neoplastic disease within a paranasal sinus. The first is the chronic inflammation of the ethmoidal complex, which obliterates the cells in a symmetric or asymmetric way, sometimes causing a bulge, but without destroying the delicate bone trabeculae among them (which are merely full of thickened mucosa and secretions). A solid mass, on the contrary, tends to destroy these trabeculae. Therefore, the preservation of the ethmoid trabeculae indicates a benign process (Fig. 5.55). The second criteria of benignity is characterized by occupation of the sinus by soft tissue that is curved or in a cascade form surrounding the inner limits of the walls and is intermixed with hypoattenuating material that corresponds to secretion. To this criteria can be added the fact that, occasionally, the material can be hyperattenuating, which is common in chronic inflammatory disease and fungal disease and, less frequently, corresponds to hemorrhage (Fig. 5.56).

Polyps and Cysts

Multiple polyps, especially in the nasal cavities, can sometimes form conglomerations that can be difficult or even impossible to differentiate from tumors, such as papillomas and lymphomas. At times, i.v. contrast injection can help, because some tumors enhance (in contrast to polyps, which remain hypodense or show only a slight peripheral enhancement; Fig. 5.57). Polyps may result from allergy, infection or vasomotor impairment and can be found within the nasal cavity or within the paranasal sinuses.

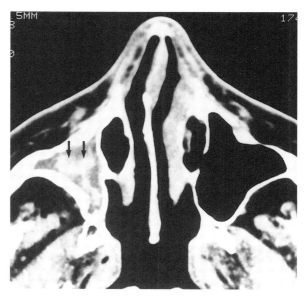

Fig. 5.56. Chronic inflammatory disease within the maxillary sinus. Thickened wall and mucosa with dense soft tissue inside

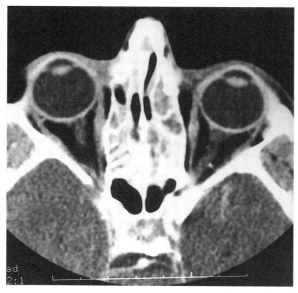

Fig. 5.55. Inflammatory disease within the ethmoidal complex. Post-contrast computed tomography demonstrating soft-tissue peripheric enhancement and thickened bone trabeculae

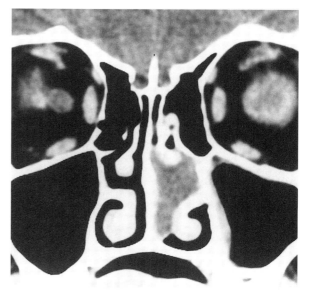

Fig. 5.57. Nasal polyp. Computed tomography after i.v. contrast; showing enhanced nasal turbinates and a nasal polyp without enhancement

The mucous-retention cyst, which results from obstruction of submucosal mucinous glands, is very common and can occur in any paranasal sinus (most commonly within the maxillary sinus), usually as an incidental finding. The serous retention cyst, which results from the accumulation of serous fluid in the submucosal layer of the sinus mucosal lining, is usually found at the base of the maxillary sinus (Fig. 5.58).

Cysts and intrasinus polyps cannot be differentiated by imaging methods. Both are homogeneous soft-tissue masses with outwardly convex and smooth borders on plain radiographs and CT and intermediate signals on T1-weighted and high signal on T2-weighted images on MRI (reflecting the high water and low protein content) [18].

Antrochoanal (Killian's Polyp) and Sphenochoanal Polyps

The antrochoanal and sphenochoanal polyps have a similar behavior on CT and MRI, have the same range of density on CT and have an intermediate signal on T1-weighted images and a high signal on T2-weighted images. Both types are similar to other nasal polyps, except for the presence of a pedicle attached to the sinus [25].

The antrochoanal polyp was first described by Killian in 1906, is characteristically attached to the wall of the maxillary sinus and protrudes through the ostium into the nasal cavity, extending into the choana. It can protrude through any ostium, but usually extends through the secondary one (Fig. 5.59). This lesion is usually unilateral and solitary but can be bilateral and is sometimes associated with other polyps. It occurs more often in teenagers and young adults, affecting males more than females. The sphenochoanal polyp is rare and originates inside the sphenoid sinus, protruding through its ostium into the nasopharynx (Fig. 5.60).

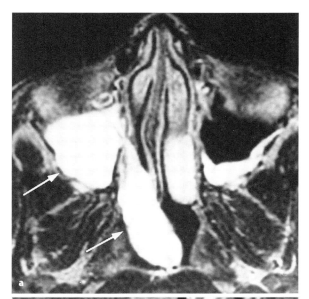

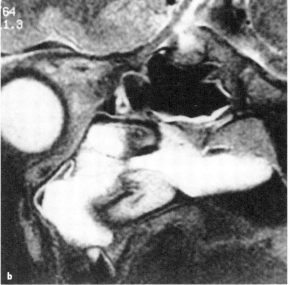

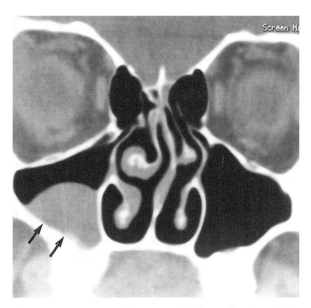

Fig. 5.58. Mucous retention cyst; typically on the floor of the maxillary sinus

Fig. 5.59 a,b. Antrochoanal polyp (Killian's polyp). Magnetic resonance; T2-weighted image. **a** Axial plane. **b** Oblique sagittal plane

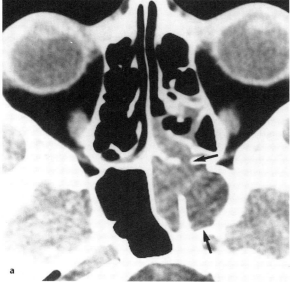
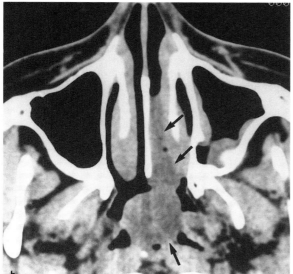
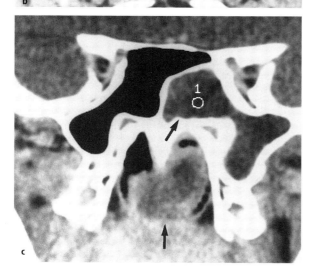

Turbinate Hypertrophy

An enlargement of the turbinates doesn't necessarily correspond to hypertrophy, based only on CT or MR. It is clinically easy to differentiate between hypertrophy and physiological volume increase due to the nasal cycle, but this is not always possible based only on CT or MR. The enlargement of the tails is more characteristic of hypertrophy (Fig. 5.61), while a unilateral, uniform increase suggests momentary change of size within the nasal cycle. It is essential to correlate this finding with the endoscopic evaluation. Sagittal reformations are very helpful for the demonstration of the turbinates and their relation to the ethmoidal labyrinth (Fig. 5.62).

Mucocele

Within a sinus or a part of a sinus, the usual presentation of a mucocele on CT is characterized by the presence of sinus contents whose density is similar to water (about 0–15 HU) or slightly higher (between 20 HU and 40 HU), depending on the protein content. In some cases, there may be slight enhancement of the periphery after i.v. contrast-media injection, and there is almost always some degree of bone remodeling (Fig. 5.63). Sclerosis of adjacent bone is very common, but erosion and thinning of the sinus wall (or part of it) may also occur.

Although some authors consider that peripheral enhancement of the mucosa by i.v. contrast is indicative of a mucopyocele, experience at our institution has shown poor correlation between this finding and the surgical data. Naturally, the chance that a lesion such as this is a mucopyocele is higher, but peripheral enhancement should not always be expected.

Inflammatory lesions show a variety of behaviors on MRI. At the least chronic stage, the content of the mucous secretion shows a high amount of water (~95%) and, as time passes (usually some months), the protein content and the accumulation of salts and minerals increase, while the lesions dehydrate.

Initially, the signal intensity is low on T1-weighted images, intermediary on proton-density images

Fig. 5.60 a–c. Sphenochoanal polyp. **a** Computed tomography (CT) in the axial plane. **b** CT in the axial plane; more cranial than the image in **a**; at the level of the ostium and the sphenoethmoidal recess. **c** CT in the coronal plane

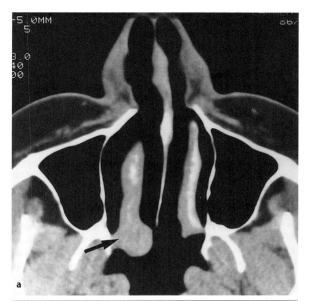

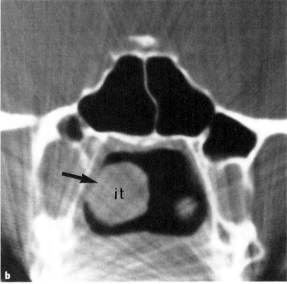

Fig. 5.61 a,b. Turbinate hypertrophy. a Computed tomography (CT) in the axial plane. b CT in the coronal plane. *it,* inferior turbinate

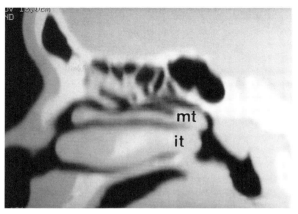

Fig. 5.62. Sagittal reformatation based on helical computed tomography, demonstrating the nasal turbinates. *it,* inferior turbinate; *mt,* middle turbinate

and high on T2-weighted images. The signal intensity progressively increases on T1-weighted and proton-density-weighted images. Signal intensity remains high on T2-weighted images during almost the whole period, declining only in the much more chronic phases (Fig. 5.63). In the more chronic stage, the mucocele is characterized by the presence of hypointensity in both T1- and T2-weighted images (Fig. 5.15).

Fungal Disease

The most frequent fungal agent is the *Aspergillus* (90% of type *A. fumigatus,* the remainder of types *A. flavus* and *A. niger*). Other etiologies include mucormycosis, candidiasis, histoplasmosis, cryptococcosis, blastomycosis, rhinosporidiosis and myospherulosis [18, 19].

Although pathognomonic findings have not been definitely identified, fungal disease can exhibit very characteristic features in both CT and MRI. In the very early stages of the infection, an unspecified reaction of the mucosa may be present in the nasal cavity or in a paranasal sinus. At a more chronic stage, calcification can exist inside the compromised sinuses, as can be more easily seen in CT scans (Fig. 5.64). On MRI, the most typical finding is the presence of hypointensity inside the sinus on T1- and T2-weighted images. The most frequent sites are the maxilla and the ethmoid. Frontal-sinus involvement is very rare. Bone erosion may occur, eventually simulating a carcinoma. When aspergillosis involves immunocompromised patients, especially those with neutropenia and hematologic malignancies (leukemia, granulocytopenia), the degree of aggressivity is so great that the paranasal sinuses can be destroyed in a few days; this is called invasive aspergillosis. This kind of aspergillosis is acute and has a high mortality rate. CT and/or MRI demonstrate tissue infiltration associated with extensive necrosis (Fig. 5.65). Other complications include vascular thrombosis and hemorrhagic infarct.

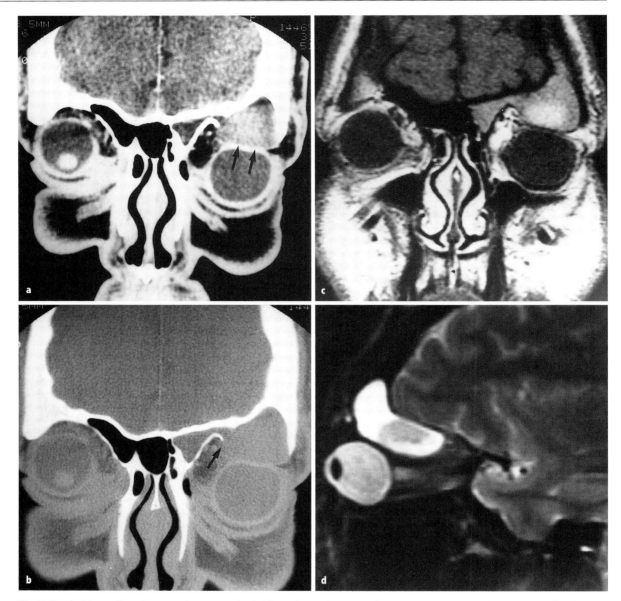

Fig. 5.63 a–b. Mucocele. **a** Computed tomography (CT) on the coronal plane. **b** CT on the coronal plane; demonstration of the bone atrophy at the orbitary roof.

Fig. 5.63 c–d. c Magnetic resonance (MR); T1-weighted image; coronal plane. **d** MR; T2-weighted image; sagittal plane

CSF Leak (CSF Fistula)

CSF leak may be secondary to trauma, surgery (it is one of the most common complications of endonasal surgery), tumor (especially in the pituitary gland) or congenital anomalies and may also be spontaneous. The most common sites of fistula are the cribriform plate and the fovea ethmoidalis.

The investigation is best performed by CT associated with intrathecal injection of non-ionic contrast (computed cisternotomography). The examination technique includes [20]:

1. Subarachnoid (intrathecal) injection of approximately 10 ml of non-ionic contrast media (200 mg iodine/ml). We prefer lumbar-spine-level (L3) access.
2. Trendelenburg's positioning of the patient for a short period (<1 min).
3. Application of cotton plugs in both nasal cavities.
4. Prone positioning of the patient.

5. Acquisition of CT slices in the coronal plane (the first slice is at the level of the cotton plugs; subsequent slices are located from the anterior wall of the frontal sinuses up to the sella turcica and are 1 mm or 1.5 mm wide for every 3 mm of dislocation).
6. Acquisition of slices in the axial plane in a supine position, if necessary.

The first aspect to analyze is the cotton plug in each nasal cavity. A cotton plug drenched by the contrast media indicates the presence of the CSF fistula (unilateral or bilateral). The next point is to look for the passage of the contrast media into the nasal cavity or into a paranasal sinus, scanning from the anterior wall of the frontal sinus up to the sella turcica (in the coronal plane; Fig. 5.66). Sometimes, the axial plane is also helpful, especially in cases of a fistula into a paranasal sinus. Other examples and a detailed approach are discussed elsewhere in this volume.

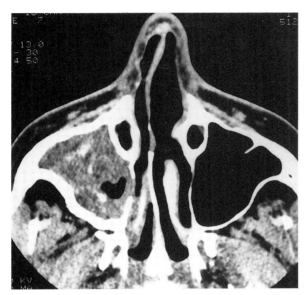

Fig. 5.64. Aspergillosis. Computed tomography on the axial plane. Multiple calcifications within the maxillary sinus

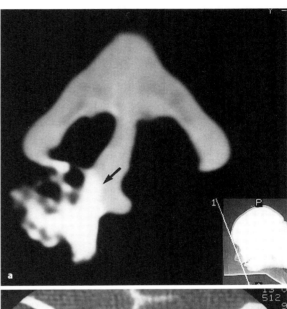

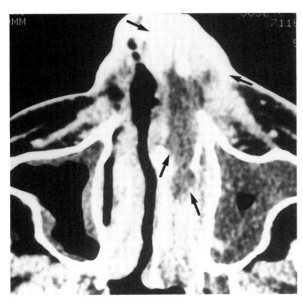

Fig. 5.65. Aspergillosis in an immunocompromised patient. Enhanced computed tomography on the axial plane. Extensive infiltration and necrosis in the nasal cavity

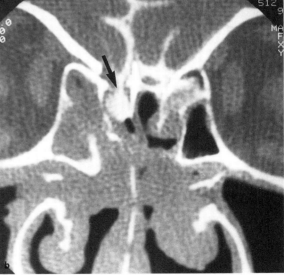

Fig. 5.66 a,b. CSF fistula. **a** Computed tomography (CT) on the coronal plane; showing the drenched cotton plug. **b** CT on the coronal plane; demonstrating the exact point of the fistula at the cribriform plate (*arrow*)

Virtual Endoscopy

The use of post-processing software for virtual reality is another modality for documentation based on helical CT. We have used the program "Navigator" (GE Medical Systems) and a Sun Ultra-Sparc workstation. This technique allows views not normally accessible by conventional endoscopy and permits structural visualization with unconventional perspectives and locations (Fig. 5.67) [1, 4, 22].

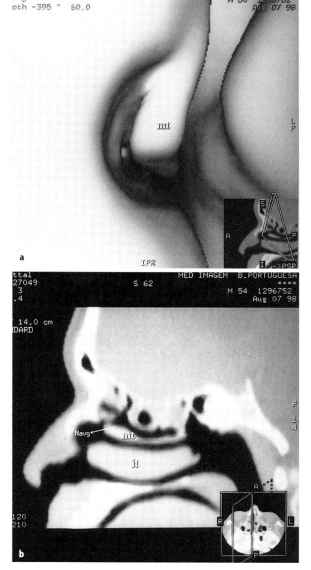

Fig. 5.67. a Virtual endoscopy. **b** Sagittal reference image showing the angle of view (*arrow*)

Acknowledgements. Special thanks to Dr. Sergio Santos Lima, founder and director of Med Imagem, Hospital Beneficência Portuguesa São Paulo, and to team colleagues who directly or indirectly contributed to this work.

References

1. Dessl A, Giacomuzzi SM, Springer P, Stoeger A, Pototschnig C, Völklein C, Schreder SG, Jaschke W (1997) Virtuelle Endoskopie mittels Postprocessing helikaler CT-Datensätze. Aktuelle Radiol 7:216–221
2. Draf W, Weber R, Keerl R (1995) Endonasal micro-endoscopic frontal sinus surgery versus frontal sinus surgery by external access. In: Stamm A (ed) Micro-cirurgia naso-sinusal. Revinter, Rio de Janeiro, pp 223–239
3. Earwaker J (1993) Anatomic variants in sinonasal CT. Radiographics 13:381–415
4. Gilani S, Norbash AM, Ringl H, Rubin GD, Napel S, Terris DJ (1997) Virtual endoscopy of the paranasal sinuses using perspective volume-rendered helical sinus computed tomography. Laryngoscope 107:25–29
5. Gomes ACP, Mendonça RA, Haetinger RG (1995) CT of the nose, paranasal sinuses and correlated structures. In: Stamm A (ed) Micro-cirurgia naso-sinusal. Revinter, Rio de Janeiro, pp 79–100
6. Hasso AN, LeBeau D (1993) MRI atlas of the head and neck. Martin Dunitz, London, pp 55–79
7. Hudgins PA (1996) Sinonasal cavities and ostiomeatal complex II. Paranasal sinus imaging: inflammatory and nonneoplastic lesions. RSNA special course in head and neck imaging. Radiological Society of North America, New York, pp 43–48
8. Keros P (1962) Über die praktische Bedeutung der Niveauunterschiede der Lamina cribrosa des Ethmoids. Z Laryngol Rhinol Otol Ihre Grenzgeb 11:808–813
9. Kopp W, Stammberger H, Potter R (1988) Special radiologic imaging of paranasal sinuses: a prerequisite for functional endoscopic sinus surgery. Eur J Radiol 8:153–156
10. Lang J (1988) Klinische Anatomie der Nase, Nasenhöle und Nebenhölen. Georg Thieme Verlag, Stuttgart, pp 70–71
11. Maffee MF (1993) Preoperative imaging anatomy of nasal-ethmoid complex for functional endoscopic sinus surgery. Radiol Clin North Am 31:1–20
12. Maffee MF (1993) Nonepithelial tumors of the paranasal sinuses and nasal cavity. Radiol Clin North Am 31:75–90
13. Maffee MF, Carter BL (1995) Nasal cavity and paranasal sinuses. In: Valvassori GE, Maffee MF, Carter BL (eds) Imaging of the head and neck. Georg Thieme Verlag, Stuttgart, pp 248–292
14. Mancuso AA, Hanafee WN (1985) Computed tomography and magnetic resonance imaging of the head and neck. Williams and Wilkins, Baltimore
15. Mendonça RA, Haetinger RG, Gomes ACP (1995) MRI of the nose, paranasal sinuses and correlated structures. In: Stamm A (ed) Micro-cirurgia naso-sinusal. Revinter, Rio de Janeiro, pp 101–107
16. Oliverio PJ, Benson ML, Zinreich SJ (1995) Update on imaging for functional endoscopic sinus surgery. Otolaryngol Clin North Am 28:585–608

17. Shankar L, Evans K, Hawke M, Stammberger H (1994) An atlas of imaging of the paranasal sinuses. Lippincott, Philadelphia, pp 41-81
18. Som PM, Curtin HD (1993) Chronic inflammatory sinonasal disease including fungal infections. Radiol Clin North Am 31:33-44
19. Som PM, Curtin HD (1996) Head and neck imaging, 3rd edn. Mosby, St. Louis, pp 97-315
20. Stamm AC, Freire LAS, Braga FM (1995) CSF leak: transnasal microsurgery. In: Stamm A (ed) Micro-cirurgia naso-sinusal. Revinter, Rio de Janeiro, pp 265-277
21. Takahashi R (1983) The formation of the human paranasal sinuses. Acta Otolaryngol Suppl (Tokyo) 408:4-28
22. Tarjan Z, Pozzi Mucelli F, Pozzi Mucelli R (1995) Ottimizzazione dei parametri di scansione e di elaborazione nella riconstruzione tridimensionale in tomografia computerizzata del massiccio facciale. Radiol Med (Torino) 89:578-585
23. Teatini G, Masala W, Meloni F, Simonetti G, Rovasio F, Dedola GL, Salvolini U (1987) Computed tomography of the ethmoid labyrinth and adjacent structures. Ann Otol Rhinol Laryngol 96:239-250
24. Terrier F, Weber W, Ruefenacht D, Porcellini B (1985) Anatomy of the ethmoid: CT, endoscopic, and macroscopic. AJNR Am J Neuroradiol 6:77-84
25. Wenig BM (1993) Atlas of head and neck pathology. Saunders, Philadelphia, pp 7-28
26. Zinreich SJ (1996) Sinonasal cavities and ostiomeatal complex I. Imaging of the ostiomeatal complex for functional endoscopic sinus surgery. RSNA special course in head and neck imaging. Radiological Society of North America, New York, pp 33-41
27. Zinreich SJ, Kennedy DW, Rosenbaum AE, et al. (1987) Paranasal sinuses: CT imaging requirements for endoscopic surgery. Radiology 163:769-775

Immunology and Allergy of the Nose and Sinuses

Monica Aidar Menon Miyake and Pedro Cavalcanti Filho

Introduction

Immunology and allergy are sometimes felt to be relatively unimportant for surgical specialists. However, misdiagnosed nasal allergies and immune diseases in patients with nasal symptoms may be important causes of unsuccessful nasal and paranasal sinus surgery and persistence of complaints during the postoperative period. Allergy is one of the most common causes of nasal and paranasal sinus diseases but is still poorly understood by many otolaryngologists and patients. Also, allergy is frequently invoked as a cause of numerous nasal and sinus symptoms when there is no other apparent explanation [6].

Today, the rhinologist must be knowledgeable in other fields (such as immunology, allergy and microbiology) in order to bring important developments from those disciplines (and others) to the diagnosis and treatment of patients with nasal complaints. Formerly, knowledge of immunoglobulin E (IgE) and the role of histamine in allergic mechanisms was sufficient, but much more knowledge is now necessary, even when referral to an allergist/immunologist is appropriate.

In this chapter, we intend to update the otolaryngologist concerning immunologic principles and nasal physiopathology based on allergy, one of the most thoroughly investigated immune disorders. We present what we hope is clinically useful information and refer readers to the vast literature for further information, because new findings come quite quickly.

Principles of Immunopathology of the Nose

The specific defense of normal respiratory mucosa depends primarily on secretory immunity. B cells (B lymphocytes) are initially stimulated in lymphoid tissue, including the tonsils, adenoid and other parts of the Waldeyer ring. The immunoglobulins they produce, IgA and IgM, are important in mucosa because of their secretory components. The secretory antibodies inhibit uptake of soluble antigens and block epithelial colonization of microorganisms. Physical mechanisms, such as the presence of mucin and normal ciliary function (natural immunologic tools or barriers), must be present for the mucosal defense to be intact [1].

Many immunopathological disorders may be responsible for nasal symptoms, but allergy is by far the most common and well studied. Local production of IgE is rarely seen in the respiratory mucosa, but mast cells "armed" with IgE are common in allergic patients. IgE is synthesized and released into the bloodstream by B lymphocytes and can be produced in abnormal amounts either in response to an abnormally large stimulus or by an abnormally large response to a normal stimulus in atopic subjects [6].

Atopy

This term was first used in 1923 by Coca and Cooke, approximately 45 years before IgE was described. Atopy refers to the clinical presentations of type-I hypersensitivity: rhinitis, asthma, eczema and urticaria [21]. It usually occurs in subjects with a positive family history, in whom there is a genetic predisposition to produce large amounts of IgE, even in response to small levels of allergenic challenge. A locus linked to the atopy phenotype has been shown to be present on chromosome 11q12–13 [22], although it is probably under polygenic control.

Allergic rhinitis is traditionally thought to be the result of an immediate (type-I) hypersensitivity response mediated by IgE and linked to a mast cell that releases histamine and other soluble mediators in the nasal mucosa after contacting an antigen. However, the immediate response is frequently followed by a delayed response that involves many cell types and mediators and is associated with inflammatory changes in the nose. There are several substances and mediators that influence these reactions, such as cytokines, adhesion molecules, etc. [19, 21].

Immunoglobulin E

IgE is a large molecule that weights approximately 190,000 Da. It has a fairly complex chemical structure (for an immunoglobulin) and can be well represented graphically by the letter Y, as in Fig. 6.1a [19]. There are five types of immunoglobulin, each with a different structure: IgG, IgM, IgA, IgD and IgE. Although IgG_4 (an IgG subtype) may have a similar role, IgE is the most important antibody involved in allergic processes, because its Fc fragment can become fixed to mast cells or basophils, which are rich in mediators of the allergic reaction (Fig. 6.1b).

Immediate Allergic Reaction

The allergic state requires two phases before an immediate reaction is possible: a sensitizing step and a subsequent challenge.

Sensitization Phase

This occurs during the initial contact with a foreign antigen or allergen. An interaction with other cells of the immune system must occur for the sensitization to be completed.

The antigen is processed by a macrophage or another antigen processing cell and is presented to lymphocytes. In either normal or allergic nasal mucosa, two thirds of T lymphocytes are CD4+ (helper) cells and one third are CD8+ (cytotoxic/suppressor) cells. When an antigen is presented to a Th_0 lymphocyte in an allergic patient, the lymphocyte differentiates into a Th_2 type cell, a response that probably is genetically determined. Next, the Th_2 lymphocyte begins to produce interleukin 2 (IL-2), IL-3, IL-10, IL-13 and granulocyte–macrophage colony-stimulating factor (GM-CSF). IL-2 is a potent stimulator capable of promoting clonal expansion of helper T cells (only specific to certain antigens) and stimulating differentiation of B cells into antibody-secreting plasma cells. The lymphocytes also produce IL-4, IL-5 and IL-9. IL-4 plays a crucial role in stimulating the immunoglobulin isotype switch to IgE, which leads to the production of allergen-specific IgE (Fig. 6.2) [6,13].

The IgE molecules are specific and combine (through their Fab fragment) only with the antigen that stimulated their synthesis [6]. After IgE synthesis is complete, IgE diffuses to adjacent lymphoid tissues in order to encounter tissue mast cells; it also circulates (via the blood) to distant lymphoid tissues, where it binds to other target cells. The high affinity of IgE for receptors on mast cells and basophils explains why minimal doses of IgE can sensitize the cells and why such sensitization is persistent.

When the lymphocyte differentiates into Th_1, the production of interferon γ (IFN-γ) and tumor necrosis factor α (TNF-α) begins. This usually happens in situations involving bacterial or non-allergic antigens. IFN-γ is also found in allergic nasal mucosa, and helps to induce the expression of adhesion molecules and to increase inflammation [13, 15].

- Th_0 cells are the activated lymphocytes that produce IL-2, IL-4, IL-5 and IFN-γ. They are the precursors of subpopulations Th_1 and Th_2.
- Th_1 cells mainly produce IL-2 and IFN-γ, which are associated with inflammatory processes. Th_1 cells are induced by bacterial antigens.
- Th_2 cells mainly produce IL-4 and IL-5, which are associated with allergic reactions. Th_2 cells are

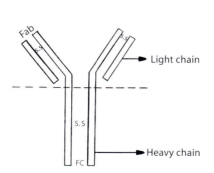
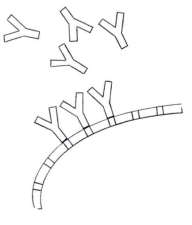

Fig. 6.1a,b. Immunoglobulin E (IgE) structure, including light and heavy chains linked by disulfide bridges (*s.s.*). *Fab*, site of the antigen link. A bridge formation (made by the antigen) between Fab sites of two different IgE molecules is necessary for mast cell degranulation. *Fc*, links to the mast cell or basophil membrane site

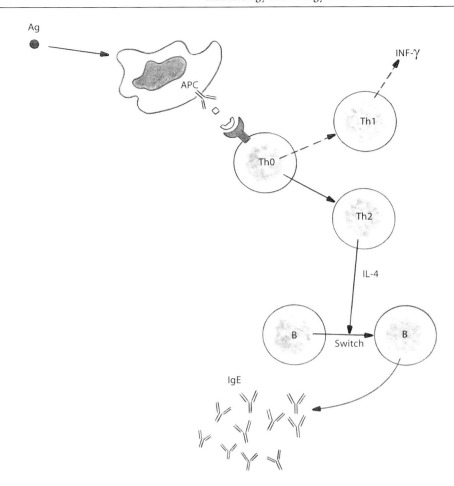

Fig. 6.2. Sensitization phase

stimulated by allergens and parasitic antigens and are highly present in atopic individuals.

Challenge Phase

Once sensitized to an allergen, the clinical effects of nasal allergy will be produced after a second or subsequent exposure. The antigen in the nasal mucosa cross-links surface-bound IgE (which involves the bridging of adjacent IgE Fab by divalent antigens), allowing an increase of intracellular Ca^{2+} and exocytosis of granules of the mast cell (Fig. 6.3).

This triggers the release of histamine and other pre-formed mediators, including heparin, triptase, kinins, TNF-α, IL-3, IL-4, IL-5, IL-13 and the chemokine known as "regulated on activation, normal T-cell expressed and secreted" (RANTES). IL-5 and RANTES are chemotactic and are activating factors for eosinophils.

Pharmacologically active substances (called newly synthesized substances) derived from the mast cell's phospholipid layer are formed there. These substances include platelet activating factor and arachidonic acid, whose metabolites via cyclo-oxygenase include prostaglandins and thromboxanes and whose metabolites via lipoxygenase include the leukotrienes (LTB_4, LTC_4, LTD_4 and LTE_4). Each of these metabolites has important clinical roles (Fig. 6.4) [6, 13, 21]. The tissue responses include vasodilatation, smooth-muscle contraction, increased postcapillary edema, mucus-gland secretion, platelet activation, eosinophilic and neutrophilic accumulation and other post-inflammatory responses [6].

Cross-linking of cell-bound IgE also results in a decline of intracellular cyclic adenosine monophosphate (cAMP), facilitating degranulation. Pharmacological agents can block the release of these mediators.

Histamine is the most important mediator in nasal allergy, directly activating the H_1 histamine receptors and causing edema and nasal congestion. Its effects on nerve terminals trigger parasympathetic reflexes that lead to hypersecretion. It also increases mucosal permeability, facilitating allergen pene-

Fig. 6.3. Challenge phase

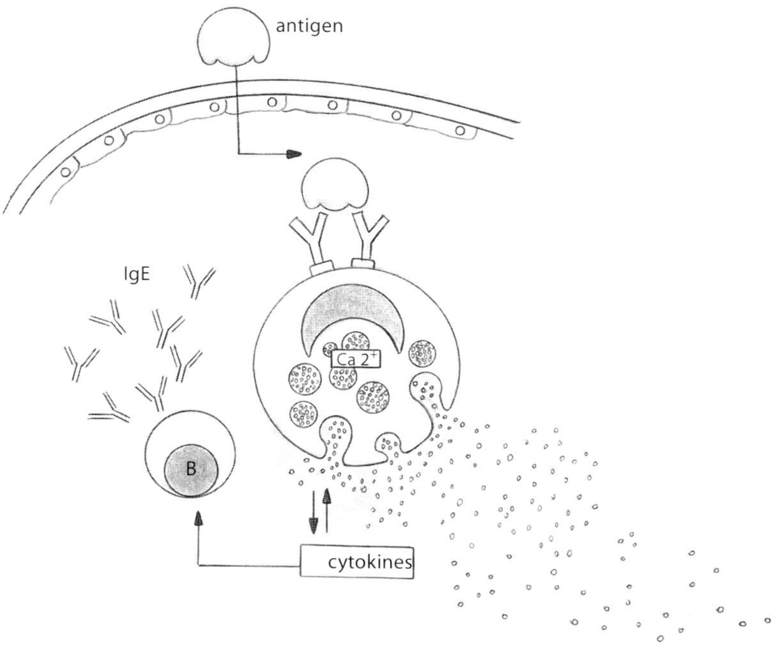

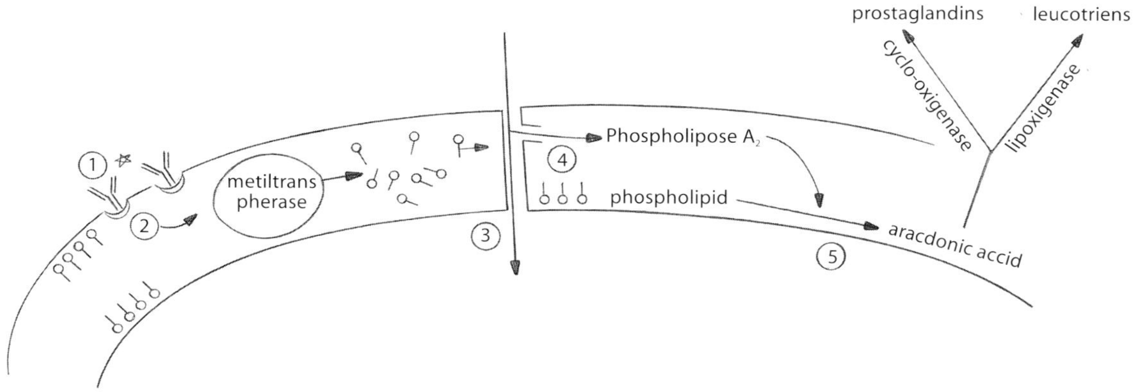

Fig. 6.4. Arachidonic-acid metabolism. *1*, Ag–Ab reaction on mast cell membrane. *2*, Disorganization of membrane phospholipds by the enzyme methyltransferase. *3*, Calcium influx. *4*, Activation of phospholipase A2. *5*, Metabolism of phospholipids to arachidonic acid [16]

tration. There are different kinds of histamine receptors: H_1, H_2 and probably H_3. All of them are present in nasal tissue, but only the H_1-receptor antagonists appear to be of therapeutic benefit in allergic rhinitis. Usually, antihistamine therapy (with H_1-receptor antagonists) is effective against the immediate nasal effects of allergens [10, 16, 19].

The secretory response of the mast cell depends on:
1. The quantity of mediators that can be released present in the cells
2. The amount of IgE bound to the cell's surface
3. The concentration of intracellular Ca^{2+}
4. The ratio of intracellular cAMP to cyclic guanosine monophosphate (cGMP) [15]

Late Phase

In many cases (~50%), the allergic response includes both an *immediate reaction* that starts 15–30 min after challenge and a *late reaction* that does not begin until 2–8 h later and may persist for 24–48 h or longer. Both primary and secondary mediators may be involved, as may cytokines. A sequence of recurrent delayed reactions superimposed on one another can lead to chronic disease and tissue destruction.

The late response is more likely to occur if the patient is highly sensitive to the allergen or has an exposure that is larger or of longer duration. Late-phase reactions are characterized by inflammation and tissue infiltration by eosinophils, basophils and sometimes neutrophils, with fibrin deposition. Eosinophils are by far the most important; they are virtually absent from normal mucosa, and the extent of eosinophilic infiltration is closely related to the severity of the chronic manifestations of the disorder. T lymphocytes play an important role in the survival of mucosal eosinophils through their release of pro-inflammatory cytokines. The penetration of eosinophils and basophils into the nasal mucosa depends on some adhesion molecules, such as E-selectin, P-selectin, ICAM-1 (intercellular cell-adhesion molecule 1) and VCAM (vascular cell-adhesion molecule) [1, 2, 13, 15].

Nasal obstruction is a common feature of this phase, as is anosmia. In seasonal rhinitis, symptoms may persist for days or weeks after pollen counts have fallen. In chronic rhinitis, symptoms are not as directly related to allergen exposure, because the mucosal hyper-reactivity is not a reaction to a specific antigen, so non-allergenic environmental irritants, such as smoke and cold air, may trigger the reaction. Simple antihistamines have little or no effect against nasal blockage and anosmia, but newer antihistamines with additional anti-inflammatory action may be of some help. Corticosteroids are beneficial therapeutically; cromolyn and immunotherapy are prophylactic and prevent the symptoms of this phase [16].

Nonimmune Mast Cell Degranulation

Various agents and mechanisms may be responsible for release of pharmacological mediators. These non-immune, non-atopic stimuli can cause reactions that simulate allergy and may be clinically indistinguishable, especially when there are simultaneous, IgE-dependent mechanisms. Some of these triggers may be endogenous, such as acetylcholine, complement, prostaglandins and hormones. Other are extrinsic chemicals, such as anesthetics, contrast media, salicylates, antibiotics, opiates, lectins and certain food constituents, etc. Weather changes, respiratory infections, stress, etc. may also play a role [6]. Some hydrocarbon products derived from diesel burning increase IgE and some interleukins important in allergic reaction (for example, IL-4). This may explain why the incidence of allergic rhinitis is increased at industrial sites [13].

Cyclic Nucleotides and the Autonomic Nervous System

Activation of mast cells is also modulated by their intracellular cAMP concentration. A normal balance exists between intracellular levels of cAMP and cGMP, with a slight predominance of cAMP. An increase in the predominance of cAMP leads to decreased secretion of inflammatory mediators, whereas an increase of the role of cGMP leads to increased secretion of mediators [6].

An imbalance in the autonomic nervous system (ANS), with the predominance of the parasympathetic system, may explain watery rhinorrhea, nasal congestion and sneezing attacks. In addition to acetylcholine and norepinephrine released by the ANS, other neurotransmitters are also found in nasal mucosa, leading to vasodilatation, increased vascular permeability, glandular secretion and ciliary beating. These neurotransmitters are named substance P, calcitonin-gene-related peptide, gastrin-releasing peptide, vasoactive intestinal peptide, neurokinin A and nitric oxide; the only neurotransmitter with sympathetic actions is neuropeptide Y [13]. Nasal hypersecretion in allergic and non-allergic rhinitis is predominantly reflex induced and, since it is mediated by glandular cholinergic receptors, anticholinergic agents should be effective (at least in theory) [15].

Cytokines

In the past, modulating substances were thought to be derived only from lymphocytes or monocytes and, thus, were called lymphokines and monokines. The current term cytokines was adopted because such substances are now known to be derived from virtually every cell. They are currently considered as proteic hormones with important roles in the effector mechanisms of allergic diseases and general inflammation processes [13].

In fact, there are many cytokines, including some that have only recently been described. The otolaryngologist needs to know that some therapies for allergic rhinitis target the action of these mediators [7]. For example, topical steroids decrease the expression of IL-4, and immunotherapy increases the expression of IL-2 and IFN-γ, which antagonizes IL-4. The best-studied cytokines are interferons (such as IFN-γ and a large subfamily), interleukins (IL-1 to IL-13 have been identified so far), tumor necrosis factors (TNF-α and TNF-β) and GM-CSF.

Cell-Adhesion Molecules

Cell-adhesion molecules (CAMs) are a new group of substances that have only recently been studied; they have some functional activities, some of which are very important in the immune response. They are involved in intercellular adhesion, the adhesion of leukocytes to epithelium and endothelium, recruitment, selective migration, homing and tissue-architecture maintenance.

The main groups of CAMs are:
1. The immunoglobulin superfamily (glycoproteins, of which the most import members are the immunoglobulins themselves). Also, ICAM-1 plays a pivotal role in allergic inflammation.
2. Integrins, which include β1 integrins, (very late antigens), β2 integrins (or leukocyte integrins) and β3 integrins (or cytoadhesins).
3. Selectins (P-selectin, E-selectin and L-selectin are the main members).

Another type of CAM (cadherins) and integrins participate in epithelial integrity. Some H_1 blockers and corticosteroids have the pharmacological effect of reducing ICAM-1 expression [2].

Eosinophils and Allergic Inflammation

Eosinophils usually are not present in normal mucosa, but they have the potential to damage tissue and play a pivotal role in the nasal mucosa in many of the most important inflammatory nasal diseases, such as allergic rhinitis, non-allergic rhinitis with eosinophilia syndrome and nasal polyposis. Eosinophils release important mediators, including eosinophilic cationic protein, main basic protein, eosinophil neurotoxin and eosinophilic peroxidase (EPO). EPO leads to cell disaggregation and epithelial desquamation, which is called "allergic inflammation" [13].

Eosinophils also produce LTC_4, IL-3, IL-4, IL-5, IL-8, IL-10, RANTES and GM-CSF. Their survival in the tissue is promoted by IL-3, IL-5 and GM-CSF and potentiates the maintenance of the inflammation.

In fact, inflammatory processes in the nasal mucosa lead to malfunction of the mucociliary barrier, facilitating the penetration of antigen and non-specific irritants. This stimulates nervous terminations, producing neuropeptides and parasympathetic symptoms that continue the inflammation in a vicious immunopathologic circle. This is what happens to a patient with constant clinical features of rhinitis; such a patient never becomes better, even on treatment with antihistamines. In such a patient, we cannot identify what triggers the symptoms.

Nasal Allergy

It is not uncommon today for an otolaryngologist to see a patient seeking help for allergy symptoms, commonly due to allergic rhinitis, and occasionally infectious complications are also present.

The diagnosis of respiratory tract allergy results from following a systematic protocol that allows the doctor to confirm what is clinically presumed (Fig. 6.5). Knowledge of the diseases that mimic allergic rhinitis is necessary in order to avoid confusion and erroneous diagnosis. The otolaryngologist also needs basic medical information and an adequate knowledge of immunology and allergy.

Symptoms such as sneezing, rhinorrhea, and nasal obstruction may be present as part of the clinical manifestations of several types of rhinitis and cannot be considered pathognomonic of allergic rhinitis. The family history is always important. There is a 30% chance that a child will be atopic when one of the parents is atopic, and over a 50% chance if both parents have atopy. Information about primary etiologic factors (allergens), predisposition and aggravating factors, and associated diseases are essential for the correct diagnosis and staging of the allergic disease.

The patient should also be questioned about factors that aggravate symptoms, such as the time of the day, and places and season, and the doctor should inquire about the patient's housing conditions. This detailed clinical history suggesting respiratory allergy is added to general and otorhinolaryngological examination, carefully observing the structural aspects of the nasal cavity and the characteristics of the mucosa to elucidate the diagnosis. History of otitis media (acute, recurrent or even chronic), and Eustachian tube dysfunction may be present.

Some clinical signs of chronic rhinitis can be seen in the general examination, such as edema and infra-orbital venous congestion, Denny's marks (eyeshiners) in the skin of the inferior eyelid, and a well-established nasal dorsal wrinkle.

Adequate examination of the nasal cavity consists of evaluation of the nasal vestibule and valves, complete examination of the nasal cavity, (nasal septum, lateral nasal wall, inferior and middle turbinates, and ostiomeatal complex), preferably using good optics with magnification. In addition, careful observation of the choana and nasopharynx is important.

In allergic patients, the nasal mucosa exhibits a pearly white color, especially the inferior turbinates,

Fig. 6.5. Diagnostic algorithm

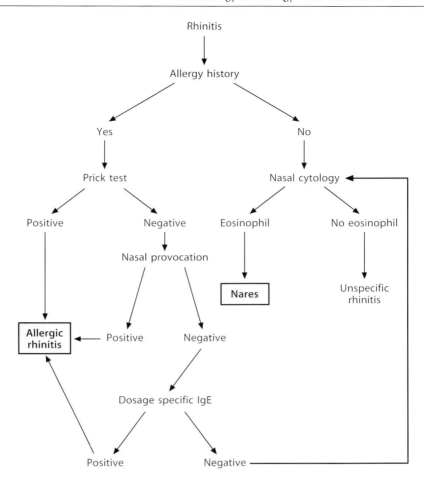

due to the edema of the mucosa. Secretions may or may not be present, varying in character, according to the stage of the disease and the presence of concomitant infection.

Patients with allergic rhinitis usually exhibit non-specific reactions to external stimuli; thus, it is possible sometimes to observe these reactions during the clinical examination, characterizing a state of nasal hyperactivity (NHR). It is important to remember that the NHR may be present in normal people or in patients with allergic and non-allergic rhinitis.

Because allergy is a chronic process, we can often observe an "adenoid facies," which is often associated with structural deformities of the hard palate, and dental arch. Many patients present with postnasal drainage, cough, and dysphonia, which may be directly related to the allergic process, resulting in larynx edema, especially of the vocal cords. The chest examination is imperative, because of the frequent association of upper respiratory tract diseases such as sinusitis, with bronchopulmonary disturbances.

Complementary Diagnostic Protocol

If a clinical history and physical examination are highly suggestive of nasal allergy, it is necessary to confirm the diagnosis and identify the specific etiology, in order to direct the therapy, which may be different for each patient. Negative test results do not necessarily indicate that the clinical diagnosis is wrong since false-negative results are not uncommon and may reflect the technique, the materials used, allergen standardization or even on the clinical condition of the patient at that moment (Table 6.1).

"In Vivo" Tests

Although "in vivo" allergic tests are relevant to confirm the diagnosis of allergy and to identify specific triggering agents, their results may be influenced by local and systemic factors such as psychological disturbances; age; local responsiveness; physical, chem-

Table 6.1. Diagnostic protocol

"In Vivo" Tests
Nasal cytology
Skin tests
Nasal challenge tests
"In vitro" laboratory tests
Total serum IgE level
Specific IgE level – allergologic profile

ical and biologic irritative agents; drugs; and habits. Laboratory tests, radiological exams, computerized rhinomanometry, pulmonary function tests, and esophageal pHmetry provide important data to understand the extent and consequences of the allergic process.

Nasal Cytology

Nasal cytology may play an important role in the diagnosis of allergic rhinitis. The most help is given by the results of smears taken from the middle meatus and inferior turbinate areas and specimens taken before and after nasal challenge with specific allergens. However, eosinophils can also be present in the nasal secretion in non-specific, non-IgE-related stimulation, as well as in cases of nonallergic eosinophylic rhinitis syndrom (NARES).

Skin Tests

For a long time, skin tests have been used as a "screening" test for qualitative and quantitative diagnosis in allergic patients. Currently, the puncture technique (PRICK TEST) performed with disposable instruments is highly effective and has shown an excellent reproducibility compared to the simple puncture tests, the scratch technique, or even to intradermal testing, which, although highly specific, has not been well accepted by patients.

Skin tests are specific for detection of local IgE at site of application. The results are evaluated after 20 min by an experienced professional, able to accurately evaluate the results [8]. Interpretation of results may vary: erythema and papule would correspond to positive "3" of Histamine result, when the negative control (excipient) is "0." There is a lack of standardized positive and negative controls, as well as a pattern for allergenic extracts. This test should obtain results with maximum specificity, especially for inhalants. Several factors influence the interpretation of a response: texture and skin thickness, color and reactivity. Despite having a subjective character, in our experience, the magnitude of the reactions is measured from 0 to +5, always remembering that a small percentage of patients may exhibit false-positive or negative responses.

Nasal Challenge Tests

The use of nasal challenge has increased in the last few years, particularly if a false-positive or negative PRICK TEST is suspected. It has also been used to evaluate variations in nasal flow and respiratory tract resistance observed by computerized rhinomanometry. The results are highly specific. Accurate positioning of the nebulizer is important; it should direct the spray towards the inferior turbinate and lateral wall of the nasal cavity.

This test is useful to evaluate both quantitative and qualitative effects of nasal allergy and also the response to medication, with pre- and postnasal cytology and rhinomanometry challenge tests.

"In Vitro" Laboratory Tests

"In vitro" laboratory tests of allergy evaluation may be necessary for the diagnosis of IgE- and IgG-mediated allergy, especially that which is induced by food or ingestants, pollens, and fungal allergens. Results may provide quantitative and qualitative data which suggest introducing a specific therapy.

Total Serum IgE Level

Some patients present with a clinical history and physical examination highly suggestive of allergy, however, react negatively to skin tests and to nasal challenge tests. In these cases, determination of the serum IgE (specific and unspecific) should be done as "screening" tests, but because of its limitations as a predictive test, it is not routinely recommended in allergic patients.

Specific IgE Level: Allergologic Profile

Currently, many tests can be used for the level of the specific IgE, but the radio allergosorbent test

(RAST) is the method most utilized worldwide [9]. It was the first to be standardized from the paper-radioimmunosorbent test (PRIST) method. Newer, less expensive methods such as the enzyme-linked immunosorbent assay (ELISA) and the multiallergen specific test- chemiluminescent assay (MAST-CLA) are very precise, with increased sensitivity and specificity for IgE, IgG, and subclasses of IgG(IgG2 e IgG3), especially in allergy induced by food allergens. These tests can be conducted in an office setting, since very little equipment is required.

Interpretation of the results varies from undetectable to highly positive, similar to the results observed from skin tests and by the puncture method.

"In vivo" and "in vitro" tests are important to establish a therapeutic approach when the etiologic diagnosis is confirmed; however, they are not used to monitor responses to treatment, until after 1 year, except in very special cases.

Treatment of Nasal Allergy

Therapeutic assessment in patients with respiratory allergy, includes:
1. Indoor and environmental control and prophylaxis
2. Drug therapy
   - Antihistamines
   - Decongestants
   - Mast cell stabilizers
   - Anticholinergics
   - Antileukotrienes
   - Topical corticosteroids
   - Systemic Corticosteroids
3. Immunotherapy
4. Nasosinus surgery

Environmental Control and Prophylaxis

The avoidance of nonspecific irritants such as atmospheric pollution, inhaled substances, and specific allergen is the first step to control and treat allergy. In everyday practice, the instructions given by the doctor are carefully followed, especially if reinforced by written material [11]. Patients also show considerable clinical improvement after control of the bedroom and the elimination of indoor tobacco smoke. However, such changes are often followed for only a short time, and the patient returns to the doctor with recurrence of the symptoms. Complete avoidance of inhalants is impossible, especially house dust mite allergens. On the other hand, danders are easier to avoid, but this is only effective if the animal is permanently removed from the environment. Molds and spores cannot be avoided, particularly in cities where the humidity and the temperature are always high. Dust cannot be avoided either, but indoor control may be accomplished by adequate ventilation and permitting sunshine into the house.

Drug Therapy

Antihistamines

Many topical and systemic H1 and H2 receptor antihistamine antagonists are now available. The anti H1 agents are used to treat nasal allergy, and the anti H2 antagonists are not effective for that (although the nasal mucosa has H2 receptors), but are useful in gastroesophageal diseases.

Anti H1 anti-histamines act against the histamine effects in the nerves, vessels, and glands of the nasal mucosa [3] by competitive inhibition of the H1 receptors, and they also have anti-cholinergic effects. This is the best group of medicines for nasal symptons such as sneezing, pruritus and rhinorrhea, but not for nasal obstruction.

The very first antihistamines became available in the 1940s, and some of them are still sold as "over the counter" medicines, alone or associated with decongestants, because of their efficacy and low cost. They are called "first generation" or "classic" antihistamines. Their principal side effect is sedation because they cross the blood–brain barrier and also interact with alcohol and other drugs.

The "second generation" or nonsedating antihistamines have much less CNS action and inhibit H1 receptors more effectively [17]. Another advantage is easier dosing. They act better on the "late phase" of the allergic reaction, in which symptoms occur 5–8 h after the first contact with the antigen. In other words, they are more effective in alleviating the allergic inflammation and antagonize the adhesion molecules. They also can help patients with mild to moderate asthma associated with rhinitis.

Although these new antihistamines are safer with respect to CNS side effects, they can cause problems in large doses or in cases of cardiopathy, liver insufficiency, or when the patient takes other medicines that use the P-450 cytochrome liver pathway, such as cetoconazol, and macrolides. In these cases, antihistamines may cause some cardioarrhythmia and may be contraindicated.

The first newer and widely used systemic antihistamine was the very potent terfenadine, which was recently removed from the market because of side effects. Its active metabolite fexofenadine is already available. Astemizole, also one of the first newer antihistamines, can cause the same side effects as well as increased appetite; after 6 weeks of treatment it inhibits the wheal-and-flare reaction.

Loratadine is a very safe antihistaminic at recommended doses that has very low concentrations in breast milk and appears to have no effects on the QT interval. Azelastin and Levocabastine are available in topical metered dosage form in an aqueous solution and are very safe because of their low absorption and toxicity. Cetirizine, Ebastine, Epinastine, Mequitazine, Noberastine are also effective and secure newer antihistaminics available in different countries.

A third generation of antihistamines has been described and includes some drugs synthetized from first or second generation drugs with the hope of minimizing side effects and increasing potency. Cetirizine (from hidroxizine) and fexofenadine (from terfenadine) are classified in this group; other newer medicines are still under development.

Decongestants

Current use of anti-H1 agents and nasal vasoconstrictors is very frequent today; the anti-H1 decreases sneezing and rhinorrhea, and the α-adrenergic receptor agonists are effective against nasal obstruction and stimulate the CNS. Pseudo-ephedrine is used most, but should be used with caution in children (because it sometimes causes behavioral changes), in the elderly (in whom it may provoke prostatic disturbances), and in patients with cardiomyopathies or hypertension (because of its vasoconstrictor effects). These agents can be used systemically or topically. Topical decongestants work quickly and effectively, but with prolonged use, they predispose to rhinitis medicamentosa.

Mast Cell Stablizers

Cromolyn sodium is an antiallergic compound that inhibits immunological degranulation of mast cells. It results in blockage of chloride/calcium transport and leads to reduced release of histamine, prostaglandin D2 and leukotriene C4, and also reduced eosinophils in upper airways. Its activity and effects are different from antihistamines and corticosteroids and it is used primarily for prophylaxis of the perennial allergic rhinitis.

Cromolyn can be used safely in children (6 years and up) because absorbance is low, they are well tolerated, and they cause very few side effects. It is recommended that they be taken 4–6 times a day, which can be a potential disadvantage because of compliance problems.

Anticholinergics

Ipatropium bromide, a topical anti-cholinergic, may be helpful in patients with allergic rhinitis, because it antagonizes the parasympathetic stimulation that leads to vasodilatation and increased glandular secretion and watery rhinorrhea. [18]. It does not have significant side effects [5] and can be an effective help in these patients.

Antileukotrienes

Leukotriene-receptor antagonists are a new class of antiasthma drugs. They have a unique action in that they are a hybrid of anti-inflammatory and bronchodilator effects and can be taken as a tablet once or twice daily. Montelukast or zafirlukast have good antiasthmatic activity. They also appear to be effective in treating allergic rhinitis, which commonly coexists in patients with asthma. Clinical trials with a combination of anti-leukotrienes and antihistamines are promising, and this combination may be available in the near future [12]. The development of eosinophilic conditions such as Churg-Strauss syndrome has been described.

Corticosteroids

Corticosteroids inhibit multiple steps of the inflammatory processes that are clinically important in allergic rhinitis [14]. They cause vasoconstriction, decrease capillary permeability, decrease glandular response to cholinergic stimulation, and interfere with arachdonic acid metabolism, leading to decreased mediator production, reduced mediator release, and decreased production of cytokines from TH-2-type lymphocytes, and inhibit influx of eosinophils and basophilic cells to the nasal epithelium. Although very effective, they are only palliative.

Glucocorticosteroids are rapidly absorbed either by oral or intramuscular injection. They cross the placenta and may be distributed into breast milk in very low amounts. Newer intranasal corticosteroids are very active and reduce the risk of systemic known glucocorticoid reations.

Systemic Steroids

For allergic rhinitis treatment, systemic corticosteroids should be reserved for acute, severe exacerbations to reduce nasal and sinus inflammation. Short courses are recommended in order to minimize the risks of side effects, but sometimes longer-term administration is required. Attention to specific contraindications is needed, such as pregnancy, diabetes, hypertension, gastrointestinal diseases, and glaucoma.

Topical Steroids

Intranasal turbinate injections of corticosteroid were reported to be useful for allergic rhinitis in the past, but no controlled studies have been conducted and there were some reports of loss of vision and other side effects. Oral corticosteroids should be better recommended in these cases.

The development of topically active intranasal corticosteroids has reduced the need for the use of systemic steroids. They have low absorption by the nasal mucosa and are especially indicated in patients in whom other therapeutic choices have already failed because the intranasal steroid potency exceeds that of antihistamines and cromolyn sodium.

Beclomethasone dipropionate was introduced in 1973 as the first modern and safe topical steroid. Fluticasone, triamcinolone, flunisolide, budesonide, and mometasone furoate are now available for reducing pruritus, rhinorrhea, sneezing, and also nasal blockage and for controlling eosinophylic rhinitis and chronic polypoid rhinosinusitis. They are very useful also in postoperative control of nasal diseases. Adverse effects do not include ocular changes or suppression of the hypothalamic–pituitary–adrenal axis. Long-term intranasal beclometasone has been associated with a slower growth rate in children.

The commercial products are available either as freon-driven aerosols, dry powder, or pump sprays with aqueous or glycol solutions. The aqueous vehicle seems to be better tolerated by the patients, although it may cause some dryness and crusting, epistaxis, and local irritation. Rarely, septal perforation or *Candida* colonization has been reported. All patients should be instructed about optimal positioning and methods to instill the drug.

Immunotherapy

Many otolaryngologists consider immunotherapy a controversial issue, because it is sometimes used inappropriately or even abused. In general, the indication for hyposensitization treatment should fulfill the following criteria:
- Correct diagnosis
- Perennial allergy
- Permanent use and/or unsatisfactory results from symptomatic medication
- Moderate to severe allergy
- Allergenic exposure strongly triggering the symptoms

The objective of the immunotherapy is to increase the production of IgG circulating antibodies, IgG subclasses, and secretory IgA [4].

Blocking antibodies leads to a reduction of the serum specific IgE levels, diminishing release of histamine and vasoactive substances from mast cells degranulation and, consequently, improving the symptoms and making it possible to reduce symptomatic medications. It is well established that basophils also release histamine; thus, in some patients, immunotherapy directed to the basophils can decrease allergic sensitivity as well as the lymphocytic response to the allergenic stimulation. When hyposensitization therapy is employed, it is important to be sure that the extracts contain active allergens with the capacity to evoke a memory of the immunological response. Allergenic suppliers should not alter the power of the allergens from the established values, and the allergen should not contain irritative low-molecular-weight substances.

Aqueous and slow release extracts are now available, but these should only be used by physicians with considerable experience in immunotherapy management. In the future, modifid allergen extracts (alergoids) which induce complete and permanent tolerance will be safer because of high antigenicity and low allergenicity and will be convienent to use.

When considering immunotherapy, the doctor's decision should be also based on the patient's age, severity of the symptoms, probable adverse effects, and projected cost/benefit ratio in order to avoid noncompliance and discontinuation of therapy. It is usually scheduled in two phases: (1) initial and (2) maintenance.

During the initial phase, the dosage is periodically and progressively increased, and, during the maintenance phase, the highest tolerated dose is given periodically for a long time. Hyposensitization therapy does not exclude medical treatment, which in combination with environmental prophylaxis may be considered coadjuvants to the immunotherapy.

The duration of immunotherapy is variable; however, the doctor can schedule an annual follow-up, measuring specific IgE levels, especially in cases with minimal clinical improvement, in order to determine whether to continue treatment.

The immunotherapy injections should only be given under medical supervision in a clinical setting where immediate assistance can be given if necessary. The patient must know about delayed reactions, which should be immediately reported. If treatment is interrupted for any reason, it should be reinitiated only after medical evaluation. Patients who become pregnant before or during the treatment must be informed about the benefits and risks of the immunotherapy.

Sinonasal Surgery

When all the previous measures to control the symptoms fail, surgical procedures may be useful. Surgical treatment attempts to minimize the symptoms of the mucosal hyperactivity and many times is essential to the success of the treatment of an allergic rhinosisusitis and its repercussions.

The partial removal of the turbinates (turbinoplasty) reduces population of glands, venous pools, and sensitive branches, which all are responsible for the nasal symptomatology. An obstructing process of the middle meatus due to structural anomalies such as concha bullosa, ethmoid bulla hypertrophy, hypertrophic uncinate process, and mucosal must all be surgically corrected because of their relation and interaction with the paranasal sinuses and inferior airway [23].

If these conditions are noticed at the physical examination, the patients should be informed of the possible need for a surgical procedure, after evaluation of the efficacy of the medical approach and before the allergic medical treatment is begun.

Allergic Rhinitis and Food Allergy

Allergic rhinitis is frequently related to inhalant allergens and its severity, based on IgE levels. The advances and routine use of "in vitro" tests have identified some cases of food allergy mediated by IgE and several mediated by IgG. It is estimated that around 40% of food allergy cases are mediated by IgG and exhibit nasal manifestations, but no etiologic diagnosis has been made.

In childhood, before age 10, IgE mediates most cases of food allergy. After this age they can be mediated by IgE and specific IgG, initiating a chronic inflammatory process of the upper respiratory tract. The diagnosis is usually confirmed clinically, through special restrictive diets and subsequent introduction of the suspected food allergen.

The treatment consists of a rotating diet, excluding and slowly reintroducing the food allergen over several months, until an alimentary tolerance has developed, keeping the patient partial or completely assymptomatic [20].

References

1. Brandtzaef P, Jahnsen FL, Farstad IN (1996) Immune functions and immunopathology of the mucosa of the upper respiratory pathways. Acta Otolaryngol (Stockh) 116: 149–159
2. Canonica GW, Ciprandi G, Pesce GP, Buscaglia S, Paolieri F, Bagnasco M (1995) ICAM-1 on epithelial cells in allergic subjects: a hallmark of allergic inflammation. Int Arch Allergy Immunol 107:99–102
3. Cavalcanti Filho PO (1995) Bases imunológicas da alergia nasal. In: Stamm AC (ed) Microcirurgia naso-sinusal. Revinter, Rio de Janeiro, pp 123-128
4. Creticos PS, Norman PS (1987) Immunotherapy in inhalant allergy. In: Lessof MH, Lee TH, Kemeny DM (eds) Allergy, an international textbook. The Bath Press, Bath, p 617
5. Dolovich J, Kennedy L, Vickerson E (1987) Control of the hypersecretion of vasomotor rhinitis by tropical ipratropiumm bromide. J Allergy Clin Immunol 80: 274
6. Fadal RG (1993) IgE-mediated hypersensitivity reactions. Otolaryngol Head Neck Surg 109:565–578
7. Fireman P (1997) Citoquinas en la rinitis alergica. Allergy Asthma Proc 11:1–4
8. Imber WE (1977) Allergic skin testing: a clinical investigation. J Allergy Clin Immunol 60:47–55
9. Ishizaka T, Ishizaka K (1983) Immunology of IgE-mediated hypersensitivity. In: Middleton E, Reed CE, Ellis EF (eds) Allergy: principles and practice, 2nd edn. Mosby, St. Louis, pp 43–74
10. Jackson W (1995) Management update in rhinitis and nasal polyposis: a symposium report. Clinical Vision, Astra Draco
11. King HC (1990) An otolaryngologist's guide to allergy. Thieme, New York
12. Lipworth BJ (1999) Leukotriene-receptor antagonists. Lancet 353: 57–62
13. Mello JF, Mello JF Jr (1997) Imunofisiopatologia. In: Castro FFM (ed) Rinite alergica: modernas abordagens para uma clássica questão. Lemos, São Paulo
14. Meltzer EO (1995) An overview of current pharmacotherapy in perennial rhinitis. J Allergy Clin Immunol 95:1097-1110
15. Mygind N (1982) Pathophysiology of rhinitis and the role of anticholinergic therapy. In: Wilson JD (ed) New dimensions in the management of airways disease and rhinitis. Adis, Aukland, pp 55–57
16. Mygind N (1993) Alergia: um texto ilustrado. Revinter, Rio de Janeiro
17. Naclerio RM, Togias AS (1990) The role of antihistamines in allergic rhinitis therapy. J Resp Dis 11:11
18. Nalebuff DJ, Fadal RG (1979) In vitro determination of immunotherapy dose – a new application of the radioallergosorbent test. In: Johnson F (ed) Allergy: including IgE in diagnosis and treatment. Symposia Specialists, Miami
19. Rihoux JP (1993) Allergic reaction, 2nd edn. Imprimerie Lielens, Brussels

20. Rinkel HJ, Randolph TG, Zeller M (1951) Food allergy. Thomas, Springfield, IL, p 5
21. Roitt I, Brastoff J, Male D (1993) Immunology, 3rd edn. Mosby, London
22. Sandford AJ, Moffatt MF, Daniels SE, Nakamura Y, Lathrop GM, Hopkin JM, Cookson WO (1995) A genetic map of chromosome 11q, including the atopy locus. Eur J Hum Genet 3:188–194
23. Stammberger H (1986) Endoscopic endonasal surgery – concepts in treatment of recurring rhinosinusitis. Part I. Anatomic and pathophysiologic considerations. Otolaryngol Head Neck Surg 94:143-146

Bacterial, Viral and Fungal Infections of the Nose and Paranasal Sinuses

Paul Van Cauwenberge and De-Yun Wang

Introduction

Rhinitis and sinusitis are the most common diseases that occur at any age of life. It is now well understood that mucosal inflammation caused by many different pathogenic mechanisms is the principal symptom of these diseases.

There is also a significant relationship between the pathogenic mechanisms of rhinitis and those of sinusitis. Rhinitis may lead to (or may be associated with) sinusitis, because the mucous membranes of the nose and sinuses are contiguous, and sinusitis does not usually develop without prior rhinitis. Therefore, the term "rhinosinusitis" is more appropriate than either rhinitis or sinusitis.

During the past decades, there has been extensive research investigating the basic biology and immunological functions of the components that constitute the framework of the mucosal immune system in the nose and sinuses. The research was focused on microbiological identification and was extended to the understanding of the local immune defense system. Different impairments of individual local immunity and the influences of bacteria and respiratory viruses on immune responses can interact, thus perpetuating the pathophysiological events of rhinosinusitis. This has greatly enhanced our understanding of the pathophysiology of common nasal diseases and their treatment.

Etiology of Rhinosinusitis

The most common etiologic factors in rhinosinusitis are viral upper-respiratory infections (URIs) and allergic reactions. Since children have an average of six to eight URIs per year (compared with two or three in adults), sinusitis is more common in the pediatric age group. Allergic rhinitis is one of the most common diseases and affects approximately 10–20% of the world's population.

There is still some debate about the pathogenesis of sinusitis. There is universal acceptance of the clinical observation that URIs are associated with episodes of acute and recurrent acute sinusitis and flare-ups of chronic sinusitis. Approximately 0.5% of adult URIs [4] and 5–10% of childhood URIs [26] are complicated by acute sinusitis. While flare-ups of chronic disease may also be related to infection, the basic pathologic process is usually the result of abnormalities of the ostiomeatal complex (with consequent ventilation and drainage problems) rather than of the infection itself. Pathogenic microorganisms associated with sinusitis include viruses, bacteria (aerobic and anaerobic) and fungi.

Physiology and Immunologic Functions of the Nose and Sinuses

The most has several important physiological functions, such as (1) conditioning and filtration of the inspired air and (2) allowing smell. It is known that the nose also fulfills an important defense function as the nasal mucosa is the first site of interaction between the host tissue and foreign invaders, i.e., bacteria, allergens, chemicals and other stimuli. The function of the paranasal sinuses is open to debate [25]. They decrease the weight of the skull and protect the intracranial structures. In addition, the sinuses may contribute to the unconditioning of the inspired air, to olfaction, to protection against barotrauma, and to resonance of the voice.

The mucosal lining of the nasal cavity covers an area of 100–200 cm^2, extends into the sinuses and is coated by a mucus layer 10–15 μm thick. Mucus is supplied by goblet cells in the epithelium and submucousal seromucous glands. Rhinosinusal secretions are a mixture of glycoproteins, other glandular products and plasma proteins. Secretions are rich in lysozyme, lactoferrin, albumin, secretory leukoprotease inhibitors and mucoproteins [17]. Lysozyme, a 14-kDa secretory product of submucosal glands, is found in all body secretions. It represents 15–30% of nasal proteins. Lysozyme is bactericidal against many airborne bacteria and some microorganisms normally resident on respiratory mucosa. Lactoferrin is also produced by serous cells of the submucosal glandular acini. It constitutes 2–4% of nasal proteins and exerts its antibacterial action by chelating iron required for microbial growth. Albumin represents approximately 15% of total nasal proteins. It is transudated from mucosal blood vessels and may play a role in binding particulate materials.

Immunoglobulin G (IgG) and IgA are also major components of respiratory secretions. IgG, derived from mucosal plasma cells (25%) and plasma (75%), is found diffusely throughout the mucosa, with its highest concentrations near the basement membrane [13]. Although it comprises only 2–4% of the total secretory product, IgG is found in higher concentrations in the interstitial fluid. Increased vascular permeability during an inflammatory process can elevate the concentration of IgG in respiratory secretions more than 100-fold.

Dimeric IgA molecules are produced by periglandular plasmocytes and are transported by serous epithelial cells. During this process, IgA molecules acquire a glycoprotein secretory piece that facilitates transport of IgA into the secretions and inhibits proteolysis. Secretory IgA inhibits bacterial invasion by binding microorganisms in the lumen and blocking attachment of pathogens to the mucosa.

Lymphocytes are the major cellular elements in the most superficial 200 µm of the lamina propria of the nasal epithelium [12]. Most of these cells are T lymphocytes with a helper (CD4+) immunophenotype. Cytotoxic/suppressor T cells (CD8+) and B lymphocytes are in the minority. The CD4:CD8 ratio is approximately 2.5:1. In the absence of active inflammation, most of these cells are in a quiescent stage and do not express the low-affinity interleukin-2 receptor (CD25).

Microbiology of Rhinosinusitis

Viral Infection

Viral rhinitis (or common cold) is, by definition, an acute rhinitis induced by respiratory viruses. Viral rhinitis can be caused by any of more than 200 different strains of various families of viruses, such as rhinovirus, coronavirus, respiratory syncytial viruses, influenza, parainfluenza and adenoviruses [22].

Rhinoviruses are the most common viruses affecting adults and are considered to be the etiological organisms in approximately 50% of cases. Coronaviruses are the second most common viral pathogens (~15%). In approximately 30% of cases, no virus or other microorganism can be identified [9]. In children, there is a wider variety of responsible viruses and, in addition to rhinoviruses and coronaviruses, respiratory syncytial viruses, parainfluenza viruses and adenoviruses also occur [9].

People living in crowded places and debilitated persons are more prone to viral infections. Especially for rhinoviruses, there is evidence that rhinosinusitis spreads by direct contact and not by air [21]. Self inoculation with the virus via the eye or nose results in infection of nasal epithelial cells, including ciliated cells [27]. Rhinovirus infection of an epithelial cell may trigger an inflammatory cascade thought to be responsible for cold symptoms.

In some instances, both viruses and bacteria are cultured from the same specimen. An infection can start from a small viral inoculum. Following exposure, the viral particle binds to specific surface antigens of the target cells. In 90% of rhinovirus infections, binding is to intercellular adhesion molecule-1 molecules on the surface of the cell [8]. Binding is very target- and host-specific. For example, rhinoviruses are difficult to study experimentally because they infect only humans and higher primates. Following cellular invasion and replication, viremia may occur, or the infection may remain localized at the target cells and, the regional lymphoid tissue. Target specificity and regionalization can be documented by viral culture studies. The organism can be recovered from the nose in 90% of rhinovirus-infected individuals, from the throat in 70% and from saliva in 50%, but not from sputum [19]. In an experimental design where healthy volunteers were inoculated with rhinoviruses, a 1:16 titer of specific neutralizing antibodies was protective against infection with the same species. However, immunity was not long lasting; 9–12 months after infection, the antibodies had disappeared from the serum, and re-infection with the same species again became possible [22].

While computed tomography (CT) scans of patients with the common cold can document evidence of sinus disease, the changes are transient and do not seem to influence the symptoms. Involvement is most frequent in the maxillary sinus (80%), followed by the ethmoidal infundibulum (70%) and the frontal sinus (30%) [10]. In a series of experiments, we inoculated 1100 young, healthy volunteers without a history of sinusitis, otitis media or allergies with rhinovirus types 2 or 39 or Hank's virus. Irrespective of the size of the viral inoculum, none of the subjects developed clinical acute sinusitis.

However, viruses are clearly involved in the pathogenesis in patients with co-morbidities. This results from obstruction of sinus ostia by edematous mucosa, impairment of mucociliary clearance and destruction of epithelial integrity. Rhinovirus, influenza-A virus and parainfluenza virus have been recovered from aspirates in patients with acute purulent maxillary sinusitis [11]. Saito et al. [18] documented high titers of neutralizing antibodies to parainfluenza virus 3 in 18 of 31 patients following acute exacerbation of sinusitis.

At the present time there is no specific treatment for viral rhinitis. In the future, treatment of colds should focus on combinations of antiviral agents and anti-inflammatory medications. Applied topically, α-interferon can diminish rhinovirus shedding, but its effects are limited, because it cannot reverse the progression of the inflammatory cascade. Partial symptomatic relief can be achieved by different anti-inflammatory compounds but, because several of them increase viral shedding, there is a need for simultaneous use of antivirals in such future treatments.

Bacterial Infection

The normal bacterial flora of the nose includes corynebacteria, staphylococci and α-hemolytic streptococci. This normal flora is an important part of the defense mechanisms of the nasal mucosa against pathogenic microorganisms. Acute bacterial rhinitis is nearly always caused by aerobic bacteria like pneumococci, *Staphylococcus aureus* or *Haemophilus influenzae* [20]. The nasal mucosa is smooth, red, swollen and tender and may be coated by purulent secretions or, more rarely, a gray membrane of fibrin.

The microbiology of acute maxillary sinusitis is best known, since this sinus is frequently involved in acute pathology and because representative specimens are relatively easy to obtain. There is strong agreement among the authors who have studied this subject [24]. In a multicenter European study, puncture of the medial wall of the antrum for intrasinusal aspiration was performed [16]. No growth was found in 32.3%, while the "infernal trio" (*H. influenzae, S. pneumoniae* and *Moraxella catarrhalis*) was cultured in 44.1% of the patients.

The infernal trio is also present in chronic sinusitis, but to a much lesser degree, and a variety of other bacteria is found [24]. Very often, more than one bacterial species can be found, in contrast to what is seen in acute sinusitis. It is, however, difficult to compare the results of various authors. In general, the samples used to investigate the microbiology of sinusitis are aspirates obtained by sinus puncture through the inferior nasal meatus or are biopsies taken during a surgical procedure. The method of sampling will certainly have an influence on the bacteriological result, but other factors probably also play a role. These factors include patient selection (age, duration and extent of disease, presurgical treatment with antibiotics, etc.), site of culture (maxillary sinus versus ethmoidal region), transport method and media, and culturing techniques (including the amount of time between sample gathering and media inoculation) [24].

An interesting topic is the presence and importance of anaerobes. The frequency of anaerobes reported in chronic sinusitis samples ranges from 0% [5] to 100% [1]. According to the literature, the most prevalent anaerobic bacteria are *Propionibacterium*, *Bacteroides* and *Peptococcus* species. Results are difficult to compare, because different inclusion criteria, sampling and culturing methods are used. Doyle and Woodham [5] cultured biopsy specimens taken from anterior ethmoidal cells and found no anaerobic organisms, which contrasts with findings reported by Brook [2]. Muntz and Lusk [15] have suggested that these conflicting results may be a result of differences in culture site and the culturing technique used for these fastidious organisms. Doyle and Woodham [5] have suggested that the ethmoid sinus may be less susceptible to anaerobes than the other sinuses, because it is less likely to be obstructed and is more exposed to inspired oxygen.

Obstruction of the ostium of the maxillary or frontal sinus leads to an increase in pCO_2, a decrease in pO_2, reduced pH, decreased mucosal blood flow, decreased ciliary activity and subsequent bacterial infection, with chemotaxis of polymorphonuclear leukocytes. Consequently, a purulent secretion is produced, with a low pO_2 and a low oxidation–reduction potential, which is a good medium for the growth of anaerobes. In normal sinuses, the pO_2 is about 17%, decreases to about 12% in acute sinusitis and becomes nearly 0% in chronic maxillary sinusitis [3]. However, since anaerobic organisms are also present in normal paranasal cavities, they could also be considered "normal colonizers".

We investigated 48 ethmoidal biopsies and 31 maxillary biopsies from 25 patients with chronic sinusitis [23]. Only three samples showed no growth. The average number of strains was 2.9 per specimen. In general, cultures of the samples revealed a slight to moderate growth. Anaerobes (mostly *Propionibacterium* sp.) were cultured in 72% of cases. "Classical" pathogens like *H. influenzae* and *S. aureus* were seen in 6% and 18% of cases, respectively. Enterobacteriaceae were present in 22% of cases. As far as bacteria with low intrinsic pathogenic potential are concerned, *Propionibacterium* species were found in 70% of the samples, while *Corynebacterium* and/or *S. epidermidis* were cultured in 77%. The bacteriology of chronic ethmoidal sinusitis was very similar to that of maxillary sinusitis.

From these bacteriological findings in chronic sinusitis, it seems that bacteria do not really play a major role. In addition, chronic sinusitis is seldom controlled or cured by antibiotics, even after many weeks of treatment. Consequently, it can be postulated that chronic sinusitis is not primarily an infectious process but instead is caused by an ostial obstruction, stasis of mucus and impaired ciliary activity, resulting in an overgrowth of colonizing bacteria. Thus, conservative therapy is often insufficient in the management of chronic sinusitis, making surgical intervention necessary.

Fungal Infection

As a result of better recovery and identification of the organisms and increased numbers of patients with predisposing factors, the incidence of fungal sinusitis has increased in recent years. Four fungal syndromes can be differentiated by histologic study:
1. *Allergic fungal sinusitis* [23]. This is identified by the presence of noninvasive fungal elements in an eosinophil-rich mucin.
2. *Mycetoma*. These structures are fungus balls lying within the sinus cavity. They are not associated with tissue invasion or an inflammatory reaction, and antifungal therapy is not usually required after endoscopic removal.
3. *Indolent fungal sinusitis* [7]. Patients with this disorder may have a moldy smell, nasal crusting or a peanut butter-like exudate in the sinus cavity. Although biopsy shows a granulomatous reaction, there is no deep invasion.
4. *Invasive fungal sinusitis* [6, 14]. This disorder occurs most often in immunocompromised hosts and is associated with prolonged granulocytopenia in, e.g., bone-marrow transplant recipients and acute leukemia patients, patients undergoing chronic steroid therapy and acidosis (diabetic or renal) patients. Invasion of the sinus wall can be followed by orbital, meningeal or other central nervous system complications. The most common invasive fungi in patients with immune compromise are *Aspergillus* sp., *Rhizopus* sp., *Candida albicans*, *Fusarium* and *Alternaria* sp. The high morbidity and mortality of invasive fungal infections should prompt emergency CT or magnetic resonance imaging scans when fungal sinusitis or rhinitis is suspected. Radical surgery and antimycotic treatment are needed.

Microbiological Pathophysiology of Sinusitis

Most cases of acute sinusitis are believed to represent bacterial complications of viral URIs. As documented by CT studies of otherwise normal patients with the common cold [10], URIs are accompanied by transient occlusion of the ethmoid infundibulum, secretions in the sinus cavity and thickened mucosa of the turbinates and sinus lining. These changes are bilateral in 85% of subjects. While the normal paranasal sinuses are sterile, small volumes of inspired air may carry normal nasal flora into the sinus. Normally, these bacteria are rapidly cleared by the mucociliary transport system.

In viral infections, however, the mucosal lining of both the nose and sinuses becomes inflamed and edematous. Contrary to infections with rhinoviruses, infections resulting from influenza and parainfluenza viruses may produce focal ulceration or extensive mucosal necrosis and, in addition, may affect the lower airway. Edema with or without tissue necrosis compromises ostiomeatal drainage and produces stasis of sinus contents. An additive impairment is infection-induced ciliary dysfunction. Impairment of the normal clearance mechanisms allows sinus secretions to accumulate and traps bacteria within the antrum. As the intrasinus hydrostatic pressure increases, local blood flow decreases, mean oxygen tension falls, and sinus pCO_2 rises. These changes favor the growth of organisms, such as *S. pneumoniae* and *H influenzae*. Bacterial colony-forming units in aspirates of acutely infected sinuses are greater than 10^5/ml and may exceed 10^8/ml. In response to bacterial metabolites and mediators released by damaged tissue, an exudate containing polymorphonuclear leukocytes (usually in excess of 5000 cells/ml) develops within the cavity. As the normal defense mechanisms come into action, sometimes aided by judicious antibiotic therapy, the inflammatory process subsides, edema lessens, coordinated mucociliary flow returns and stagnation disappears.

In patients with chronic sinusitis, it is the lack of self cleansing (rather than infection) that produces the basic disorder. As a result, chronic sinusitis is resistant to measures used to treat uncomplicated acute or recurrent acute sinusitis. Anatomic abnormalities, allergic disorders, foreign bodies, mucoviscidosis, and ciliary dyskinesia are accompanied by abnormalities of the ostiomeatal complex, impaired mucociliary flow or both. Under these conditions, repeated episodes of infection result in epithelial hyperplasia, squamous metaplasia, in-growth of well-vascularized connective tissue, polyp formation, periosteal reaction and accumulation of eosinophils, lymphocytes and plasma cells. Together with the primary lesion, these changes further impair drainage. A vicious cycle ensues. With chronic stasis and inflammation, oxygen tension can be reduced to anaerobic levels. The low redox potential favors the growth of upper-respiratory-tract anaerobes.

Conclusions

Rhinitis and sinusitis are the result of complex interactions between environmental, infectious, anatomic and physiologic factors. Altered ventilation and impaired drainage of the sinus cavities lead to abnormal secretion in the cavities, stagnation of fluid and release of inflammatory mediators. A secondary bacterial infection adds to the chronicity of the inflammatory and infectious process. Bacteria found in chronic sinusitis include microorganisms already present in acute sinusitis, additional anaerobic pathogens and, sometimes, yeasts and molds. It is doubtful that bacteria play a major causative role in chronic sinusitis; instead, they complicate the process as opportunistic organisms or contaminants. Surgery is often necessary in cases resistant to conservative therapy.

References

1. Brook I (1981) Bacteriologic features of chronic sinusitis in children. JAMA 246:967–969
2. Brook I (1989) Bacteriology of chronic maxillary sinusitis in adults. Ann Otol Rhinol Laryngol 98:426–428
3. Carenfelt C, Lundberg C, Nord C-E, Wretlind B (1978) Bacteriology of maxillary sinusitis in relation to quality of the retained secretion. Acta Otolaryngol (Stockh) 86:298–302
4. Diaz I, Bamberger DM (1995) Acute sinusitis. Semin Respir Infect 10:14–20
5. Doyle PW, Woodham JD (1991) Evaluation of the microbiology of chronic ethmoid sinusitis. J Clin Microbiol 29:2396–2400
6. Drakos PE, Nagler A, Or R, et al. (1993) Invasive fungal sinusitis in patients undergoing bone marrow transplantation. Bone Marrow Transplant 12:203–208
7. Evans KL (1994) Diagnosis and management of sinusitis. BMJ 309:1415–1422
8. Greve JM, Davis E, Meyer AM, et al. (1989) The major human rhinovirus receptors is ICAM-1. Cell 56:839–847
9. Gwaltney JM, Hayden FG (1982) The nose and infection. In: Proctor DF, Andersen I (eds) The nose. Elsevier, Amsterdam
10. Gwaltney JM, Phillips CD, Miller RD, et al. (1994) Computed tomographic study of the common cold. N Engl J Med 330:25–30
11. Hamory BH, Sande MA, Sydnor A Jr, et al. (1979) Etiology and antimicrobial therapy of acute maxillary sinusitis. J Infect Dis 139:197–202
12. Igarashi Y, Kaliner MA, Hausfeld JN, et al. (1993) Quantification of resident inflammatory cells in the human nasal mucosa. J Allergy Clin Immunol 91:1082–1093
13. Meredith SD, Raphael GD, Baraniuk JN, et al. (1989) The pathogenesis of rhinitis. III. The control of IgG secretion. J Allergy Clin Immunol 84:920–930
14. Morrison VA, Pomeroy C (1995) Upper respiratory tract infections in the immunocompromised host. Semin Respir Infect 10:37–50
15. Muntz RH, Lusk RP (1991) Bacteriology of the ethmoid bullae in children with chronic sinusitis. Arch Otolaryngol Head Neck Surg 117:179–181
16. Nord CE (1988) Efficacy of penicillin treatment in purulent maxillary sinusitis. A European multicenter trial. Infection 16:209–214
17. Raphael GD, Baraniuk JN, Kaliner MA (1991) How the nose runs and why. J Allergy Clin Immunol 87:457–467
18. Saito H, Takenaka H, Hoshino A, et al. (1981) Antiviral defense mechanisms of the nasal mucosa. Rhinology 19[suppl]:19–27
19. Van Cauwenberge P (1979) Diagnosis of infectious rhinopathy. Acta Otorhinolaryngol Belg 33:607–614
20. Van Cauwenberge P (1981) Micro-organisms involved in nasal and sinusal infections. Rhinology 19[suppl]:29–40
21. Van Cauwenberge P (1985) Epidemiology of common cold. Rhinology 23:273–282
22. Van Cauwenberge P, Ectors L (1994) Antibodies against rhinoviruses. In: Mogi G, Veldman JE, Kawauchi H (eds) Immunobiology in otorhinolaryngology. Progress of a decade. Kugler, Amsterdam, pp 539–542
23. Van Cauwenberge P, Ingels K (1996) Effects of viral and bacterial infection on nasal and sinus mucosa. Acta Otolaryngol (Stockh) 116:316–321
24. Van Cauwenberge P, Vander Mijnsbrugge A-M, Ingels KJAO (1993) The microbiology of acute and chronic sinusitis and otitis media. Eur Arch Otorhinolaryngol 250:S3–S6
25. Wagenmann M, Naclerio RM (1992) Anatomic and physiologic considerations in sinusitis. J Allergy Clin Immunol 90:419–423
26. Wald ER (1992) Sinusitis in children. N Engl J Med 326:319–323
27. Winther B (1994) Effect on the basal mucosa of upper respiratory viruses (common cold). Dan Med Bull 41:119–242

Nasal Polyps

Mirko Tos, Per L. Larsen, Knud Larsen, and Per Cayé-Thomasen

Definition

It is very difficult to exactly define "nasal polyps". They have been described as edematous nasal mucosa and benign neoplasms that have a polyp body and a stalk. Edematous nasal mucosa with small polypoid protrusions in the middle meatus contrast greatly with large single polyps; edematous nasal mucosa resembles a benign neoplasm that obstructs the entire nasal cavity. The commonly used definition (that a nasal polyp is a non-neoplastic mucosal swelling that usually arises from the middle meatus and prolapses into the nasal cavity) relates to its pathogenesis, but there is no proof that a polyp is a prolapse of the mucosa. Polyps are also called "pale bags" of edematous tissue, but a morphologist will not designate the polyp epithelium as a bag and the polyp stroma as the content of the bag. Any benign neoplasm covered with an epithelium, e.g., inverted papilloma, would then be called a bag. A "neutral", acceptable definition might be that a polyp is a pedunculated, soft, elongated structure attached to the nasal mucosa by a slim stalk or a broad base. Polyps are usually soft, mobile, have a smooth, shiny surface and are of bluish-gray or pink translucent appearance. Polyp size varies considerably but is usually approximately 1 cm in diameter (Fig. 8.1).

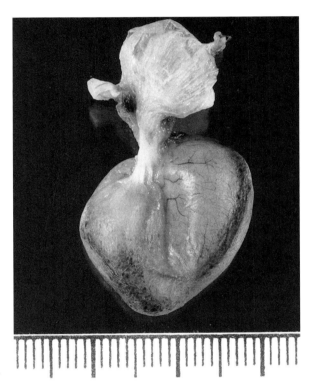

Fig. 8.1. Nasal polyp removed by a snare. Divisions are in millimeters

Incidence

Nasal polyposis is a relatively common condition found in 1–4% of the general population and in high percentages of some selected groups of patients. The polyps are found in 36% of patients with aspirin intolerance, in 7% of those with asthma (in 13% with non-atopic asthma and in 5% with atopic asthma), in 20% of those with cystic fibrosis [66] and in 2% of those with chronic rhinosinusitis (in 5% with non-allergic rhinosinusitis and in 1% with allergic rhinosinusitis). Other conditions are also associated with nasal polyps; these conditions include Young's syndrome, Churg-Strauss syndrome, ciliary dysfunction syndrome (Kartagener's) and allergic fungal sinusitis (with polyps in 66–100% of cases).

In cadaver studies, polyps are found in surprisingly high percentages of investigated specimens, extending from 2% when investigating by anterior rhinoscopy [40] to 26% or even 42% when investigated by endoscopic sinus surgery [42, 43]. Most of the polyps from the latter two materials are, however, small and without clinical symptoms. Finding such a high incidence of asymptomatic nasal polyps in a randomized group of elderly hospitalized patients may indicate a high reversibility of nasal polyps, with only a certain percentage of polyps becoming symptomatic and clinically recognizable. However, larger cadaver studies are needed to resolve these issues.

In several clinical series of patients with nasal polyps, the frequency of asthma is high, but it depends on the character and grade of selection of the patients. In our random sample of 180 patients from a well-defined geographic area, treated for the first time for nasal polyps during an 8-year period (January 1984 to December 1991), 21% had asthma [35]. Among 103 patients from the same area and period but with previous polypectomy, 31% had asthma [36]; this indicates that, in more severe and long-lasting polyposis, the frequency of asthma is high. In some other series with patients having polyps, the frequency of asthma is 70–100% [5, 19, 65]. There are no epidemiological studies on the prevalence of nasal polyps in large, randomized, normal population groups; most studies are based on clinical populations.

Morphology

Morphologically, the polyp consists of the epithelium and a stroma. It has a stalk and a body that can be divided into a proximal part near the stalk, a middle part and a distal part. The polyp surface is either anterior (facing the entrance of the nose), posterior (facing the choana), medial or lateral.

Epithelium

The nasal-polyp epithelium is a vulnerable barrier continuously influenced by the air stream, which contains dead or living material, such as viruses, bacteria and air-pollution irritants. The air stream per se has a great influence on the epithelium and can lead to transformation of the polyp epithelium (an event called metaplasia by some authors).

Types and Distribution of Epithelia

Different types of epithelia have been found, most often a typical respiratory, pseudostratified epithelium with ciliary cells and goblet cells [1, 21] but also low cubic or cylindric epithelium, stratified squamous, non-keratinized epithelium [2] and transitional epithelium [39]. In our study of the distribution of epithelia [39], the major parts of the nasal polyps were covered by a pseudostratified, cylindric epithelium of varying height (Fig. 8.2) found in 62% of anterior polyps and 78% of posterior polyps, respectively. Transitional epithelium (Fig. 8.3) was found in 33% of anterior polyps and 19% of posterior polyps. Stratified squamous epithelium (Fig. 8.4) was found in 5% of anterior and 3% of posterior polyps (Table 8.1).

In areas covered with pseudostratified epithelium, a high or hyperplastic epithelium (Fig. 8.2a) dominated (Table 8.1). The hyperplastic and high transitional epithelia were more often found in the anterior polyps, replacing pseudostratified epithelium. A transition (most often gradual) between different epithelial types (Fig. 8.5) could be seen [39]. Other researchers have reported finding transitional [5], stratified, squamous, non-keratinized epithelium in anterior polyps [1, 2, 24, 70].

Goblet Cell Density

The presence of goblet cells in nasal polyps has been mentioned by several authors. By counting the goblet cells in whole-mount specimens, the distribution of goblet cells was found to be extremely irregular (Fig. 8.6), and there were great variations in goblet cell density, not only among different

Table 8.1. Number and percentage of localities with various epithelium types of various heights

Epithelium type	Height	Anterior polyps (105 localities) (%)	Posterior polyps (98 localities) (%)
Pseudostratified	Hyperplastic	18	39
	High	42	35
	Low	2	4
Transitional	Hyperplastic	22	10
	High	9	9
	Low	2	–
Squamous	Hyperplastic	5	2
	High	–	1

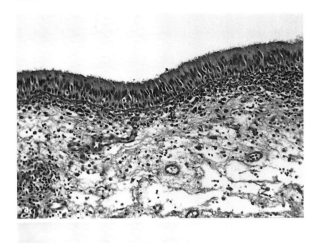

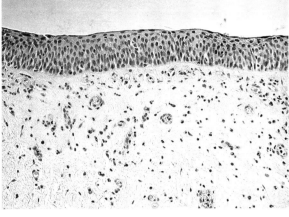

Fig. 8.4. Stratified squamous epithelium

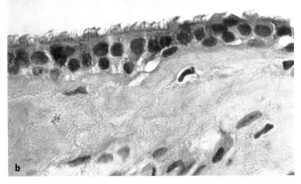

Fig. 8.2. a Pseudostratified cylindric epithelium with ciliary cells and goblet cells. Normal height. PAS–Alcian blue and hematoxylin and eosin (HE) staining. b Low cylindric epithelium with ciliary cells (HE staining)

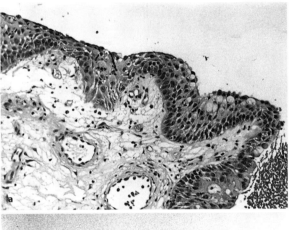

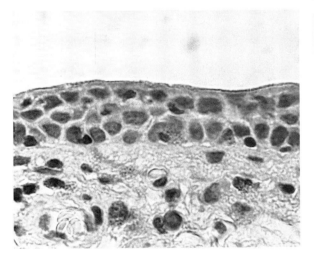

Fig. 8.3. Transitional epithelium

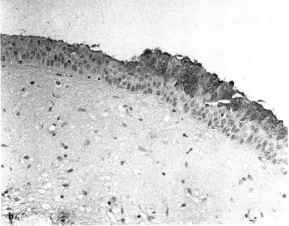

Fig. 8.5. a Gradual transition between squamous and pseudostratified transitional epithelium. b Sudden transition between squamous and pseudostratified epithelium with goblet cells. Hematoxylin and eosin and PAS-Alcian blue staining

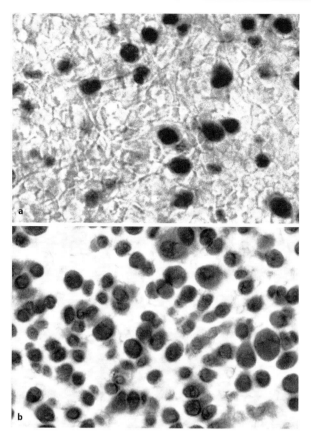

Fig. 8.6 a,b. Goblet cells in PAS–Alcian blue-stained. Whole mount. Magnification ×500. **a** Low density. **b** High density

polyps but also among different areas within the same polyp, ranging from areas completely devoid of goblet cells to areas with a relatively high density. The median density of the individual polyps varied from 280 cells/mm^2 to 10000 cells/mm^2. The median density for all 15 polyps situated anteriorly in the nose was 3450 cells/mm^2, compared with 6050 cells/mm^2 in 14 polyps situated posteriorly (in the middle meatus, but behind the anterior polyps) [83].

The study of goblet cell density in whole-mount specimens revealed four conspicuous findings:
1. The distribution was extremely irregular
2. The density was very low compared with the densities of the nasal septum, nasal turbinates, paranasal sinuses [55, 56, 71]
3. There are large differences in the median density among individual polyps
4. There is no general regularity in the distribution of goblet cells

The density is, however, lower in the anterior polyps than in the posteriorly positioned polyps.

Relationship Between Goblet Cell Density and Type of the Epithelium

The goblet cell density was highest in *pseudostratified cylindric epithelium* (Table 8.2), whereas it was four times lower in transitional epithelium and even lower in the squamous, non-keratinized epithelium, although great variation could be found. The height of the cylindric epithelium has some influence on the goblet cell density. A high epithelium with pronounced basal cell hyperplasia had the highest density whereas, in lower pseudostratified cylindric epithelium, the density was generally lower. One- two- and three-layered stratified cuboidal epithelia sometimes had a relatively low goblet cell density. However, there were several exceptions to this rule.

Transitional epithelium, with basal cylindric cells and superficial round cells (similar to the structure of the bladder), was encountered in quite a few places. In areas with a high transitional epithelium, the goblet cell density was higher than in areas with a low transitional epithelium.

Stratified squamous non-keratinized epithelium with a low density of goblet cells was rarely found. Transitional and squamous epithelia contained areas completely devoid of goblet cells in both anterior and posterior polyps (Tables 8.1, 8.2).

Intraepithelial Glands

In some polyps, intraepithelial glands are found; in one polyp, they were found in large numbers (Fig. 8.7). These pathological structures consist of 30–50 mucous cells arranged radially around a lumen; they are found in various numbers and of various distribution in the trachea [18] and Eustachian-tube mucosa [73]. Messerklinger [52] considers them to be unstable formations, and he has experimentally induced intraepithelial glands in the guinea pig trachea 1 h after injection of pilocarpine; this had a parasympaticomimetic effect, thus illustrating the dynamics of the histopathology.

Dynamics of Epithelial Covering

The presence of intraepithelial glands in some polyps and their complete absence in other polyps and the histopathological picture of the epithelium (with large variations in goblet cell density and epithelium type in different locations in individual

Table 8.2. Goblet cell densities in different epithelium types in 203 localities of all 29 polyps (15 anterior and 14 posterior)

Epithelium type	Goblet cells/field Median	Range	95% Confidence limits	Cells/mm² Median
Pseudostratified	115	0–227	97–124	6500
Transitional	33	0–184	16–56	1870
Squamous	14	2–93	2–93	790
All localities	93	0–227	79–110	5260

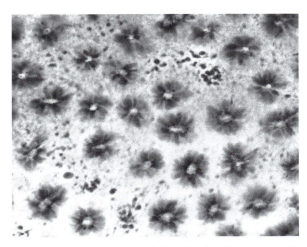

Fig. 8.7. Star-shaped intraepithelial glands with mucous cells arranged around a lumen. PAS–Alcian blue staining. Whole mount

polyps) illustrate that a range of dynamics takes place within a single polyp. Many factors are involved in these processes.

Air flow is considered a constant trauma to the surface epithelium, bombarding the polyp with particles and damaging the cilia, as demonstrated experimentally in rabbits by surgical closure of one nostril [57, 78–80]. The air-flow trauma of the polyp epithelium increases its regeneration rate and its thickness, because the damaged cells have to be replaced by new cells, and these processes require hyperplasia of the epithelium and increased activity of division of basal cells. An increased goblet cell density in the nasal mucosa has been described in *nasal infections* [71] and in *nasal allergy* [26, 64]. However, quantitative analysis using the wholemount method indicated a lower-than-normal goblet cell density in patients with allergy [54]. Polyps from patients with clinical signs of allergy did not show a higher density than polyps from patients without allergy.

Other factors, such as *contact between the surfaces* of the epithelium, the regeneration rate of the epithelium and the *growth rates* of the polyps, may influence the goblet cell density in the nasal-polyp epithelium. The presence of the thymus gland was shown to influence goblet cell density [44].

It is reasonable to assume that the great variation in the epithelial covering of the polyp and the variations in the thickness of the epithelium and the goblet cell density are reflections of the constant external and internal influences on the polyp. Therefore, any explanation of the pathogenesis of nasal polyps that is based on the polyp epithelium [3] is questionable.

Stroma

The polyp stroma includes loose fibrous tissue with vessels and a varying density of different antigen-containing, immunocompetent cell types, such as neutrophils, eosinophils, basophils, mast cells, plasma cells, lymphocytes and others. However, for some cell types, a great variation in density at different locations in the same polyp was found, suggesting that a dynamic process within the polyp takes place. The ready availability of nasal-polyp tissue and modern histochemical and molecular biological techniques for identifying chemical mediators of inflammation have resulted in important research in nasal mediator-producing cells (such as mast cells, basophils, eosinophils, macrophages and neutrophils) in polyp stroma [23].

Mucous Glands

The presence and character of mucous glands in nasal polyps have been discussed by several famous surgeons and rhinologists since the last century. In 1885, Billroth [4], in his thesis *Structure of Nasal Polyps*, described nasal glands as long, tubular, new formations; others [89] have considered them as nasal glands. We [76] have thoroughly studied mucous glands in nasal polyps and demonstrated that they play an important role in understanding the pathogenesis and growth of nasal polyps [72, 77].

Distribution and Density

The glandular orifices are of very irregular distribution. There is no particular concentration of glands in the stalk or in the most distal end of the polyp. In some polyps, only a few glands can be found (less than ten) and, in others, more than 100 glands can be found. In most polyps, the density is between 0.1 glands/mm^2 and 0.5 glands/mm^2 of polyp surface. The density of glands in nasal polyps is considerably lower than in the nasal mucosa [71]. In nasal mucosa of the lateral and medial walls of the middle turbinate, the density is 7 glands/mm^2. The glands are regularly distributed throughout the mucosa; thus, they are completely different from the glands of polyps.

Shape and Structure

The polyp glands are tubular, of different shapes and sizes and differ widely from those in the nose [71]. They are of various types (Fig. 8.8). The most striking glands are long, tubular glands that may be 1–8 mm in length; they most often arise from the middle or distal part of the polyp and grow towards the stalk. They are parallel to each other and to the longitudinal axis of the polyp. Some are very simple, narrow tubes (Fig. 8.9); others have prominences of small, round, alveolar bulges on their sides (Figs. 8.10, 8.11). Some glands are small, simple tubuli without dichotomous division (Fig. 8.12); other small tubulous glands undergo dichotomous division.

Architecture

The epithelial lining of the tubules is extremely polymorphous. Some long glands are lined with pseudostratified, columnar, ciliated epithelium with goblet cells (Fig. 8.13); distally, these glands become thinner (two layered or one layered). Others are lined with tall, simple columnar epithelium in which all the cells are mucosal.

Formation and Growth

The glands most often have their orifices in their lower halves, and the long tubules run up towards the stalk. The shape and architecture of the glands differ a great deal from those in the nasal mucosa and indicate that all the glands are formed during the growth of the polyps and that none grow into the polyp from the original nasal mucosa.

When the first glands form, the polyps have already attained a certain size. This is the only explanation for the shape and orientation of the long glands. Presumably, the long glands are the first to

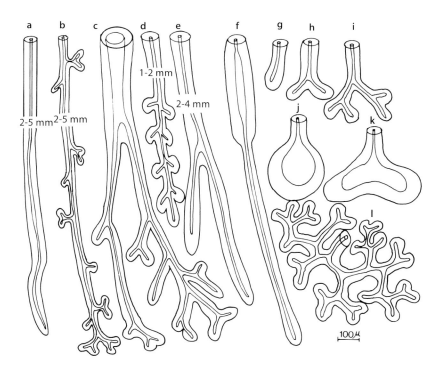

Fig. 8.8. Various types of mucous glands in nasal polyps. Long, simple tubulous glands (several millimeters in length, *A, F*). Long tubulous glands with some branching (*B–E*). Short, simple tubulous glands (*G*). Short, branched tubulous glands (*H, I*). Tubulous glands with flask-shaped dilatation (*J, K*). Tubuloalveolar glands found only in two polyps in a limited area (*L*)

Fig. 8.9. Long simple tubulous glands in a polyp. PAS–Alcian blue staining. Whole mount

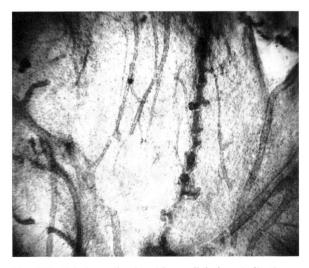

Fig. 8.10. Tubulous glands with small bulges indicating a dichotomous division. PAS–Alcian blue staining. Whole mount

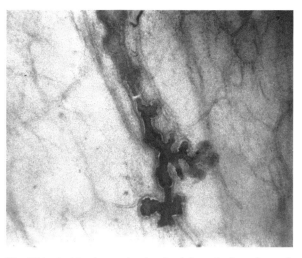

Fig. 8.11. A thin, long, simple gland branched at the end. PAS–Alcian blue staining. Whole mount

Fig. 8.12. A relatively high density of small, simple tubulous glands formed in the polyp after its formation

form in the polyp, growing from the basal layer of the surface epithelium down towards the depth of the polyp and then becoming canalized. As the polyp continues its growth and elongates, the glands become long and stretched (Fig. 8.14). Passive stretching of the glandular ducts indicates that growth of the polyp is also passive, i.e., there is an increase in length.

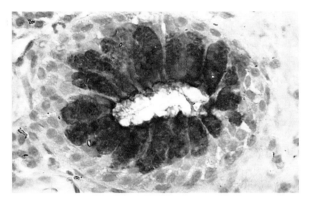

Fig. 8.13. Cross-section of a tubular gland covered with active, pseudostratified epithelium. PAS–Alcian blue–hematoxylin–eosin-stained section

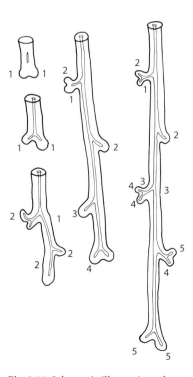

Fig. 8.14. Schematic illustration of growth and passive stretching of long tubulous glands in nasal polyps. The dichotomous divisions are numbered *1–5*; only one side of each division stretches and grows, making the glands asymmetrical

Degeneration

All the types of glands described above have been observed as active and as completely degenerated types. The most striking glands are the degenerated, long glands in which the entire long duct and the small lateral ducts are distended twofold or threefold and are filled with mucus (Fig. 8.15). The degeneration of glands starts with the stagnation of mucus within the tubulus and the duct, which then becomes distended. The secretory epithelium stretches, the cells become cuboidal and flat, lose their secretory ability and gradually become entirely inactive. Such degenerated ducts in whole-mount preparations are seen as dilatated; in sections, they are seen as small cysts. Loss of secretory activity in the glands of nasal polyps has also been demonstrated using immunofluorescent techniques [61].

Histological Classification

Based on histopathological findings, some histopathologists [22] subdivide nasal polyps into four types:
1. The *edematous*, *eosinophilic* or *allergic* polyp, which is the most common, comprises 86% of polyps [10] and is characterized by edematous stroma, goblet cell hyperplasia, respiratory epithelium and numerous eosinophils and mast cells.
2. The *chronic inflammatory polyp* or fibrous inflammatory polyp comprises 10%. The epithelium is devoid of goblet cell hyperplasia and frequently shows squamous and cuboid metaplasia. The inflammatory infiltrate is often intense, but lymphocytes predominate; the stroma contains numerous fibroblasts and fibrosis. They may have slight gland hyperplasia.

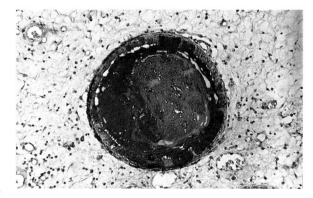

Fig. 8.15. Degeneration of glands. Lumen of a gland tubulus completely filled with stagnated mucus. PAS–Alcian blue staining. Whole mount

3. Polyps with *hyperplasia of glands*.
4. Polyps with *stroma atypia*, which can easily be mistaken for neoplasms.

This subdivision is rather academic, but it may be of some help in the differential diagnosis of neoplasms. We have shown that the epithelium may change and that various epithelia (such as squamous epithelia) may be seen in the same polyp. Furthermore, the goblet cell density varies very much, making a meaningful subclassification based on the epithelia difficult.

Etiology

Etiology is supposed to establish why the nasal polyps form. Nasal polyps are the end result of an inflammatory process associated with various disease states, which implies a multiplicity of etiologies. These diseases include: various forms of adult asthma (non-allergic or non-atopic, intrinsic and allergic or atopic), aspirin intolerance, chronic rhinosinusitis with non-allergic rhinitis or allergic rhinitis, allergic fungal sinusitis and some congenital, genetically transmitted diseases or syndromes, such as cystic fibrosis, Kartagener's syndrome (bronchiectasis, sinusitis, situs inversus and ciliary dyskinesia), Churg-Strauss syndrome (asthma, fever, eosinophilia, vasculitis and granuloma) and Young's syndrome (sinopulmonary disease and azoospermia). In very rare cases, childhood asthma and rhinitis may be associated with nasal polyps, and the association is common in cases of cystic fibrosis in children [66]. For the basic diseases, we propose the term "nasal-polyp-causing diseases". They differ from each other, but they have some abnormalities in the upper or lower respiratory tract mucosa in common, particularly inflammation of the nasal mucosa, dysfunction of cilia (as in Kartagener's syndrome) and abnormal composition of the mucus (as in cystic fibrosis). The latter diseases cause stagnation of mucociliary transport and secondary infection of the nasal mucosa. Thus, the etiology of nasal polyps is very complex and difficult to understand; it probably involves several different mechanisms.

Since only some of the patients with nasal-polyp-causing diseases form polyps, the severity and duration of the basic polyp disease state must have an influence on polyp formation. Since epidemiology of nasal polyps is closely related to etiology, further studies of epidemiology may give some etiological answers.

Pathogenesis

The precise pathogenesis of nasal polyps is not known, although there are several factors that are important for the development of polyps:
1. Chronic and recurrent inflammation and infection of the nasal or nasal-sinus mucosa, producing edema. However, it is well known that chronic or recurrent infection or inflammation does not always produce polyp formation.
2. Abnormal vasomotor responses, hypersensitivity of the vascular bed (as in aspirin intolerance and allergy), producing mucosal edema. However, here again, an abnormal vasomotor state occurs in rhinitis medicamentosa without producing nasal polyps, and nasal mucosal edema occurs in many local or systemic illnesses without the association of polyps.
3. Air-flow blockade at the place of origin of nasal polyps, such as the middle meatus, the middle and superior turbinates and outlets of the paranasal sinuses. However, polypoid and papillary hyperplasias of the nasal mucosa also occur at the posterior end of the inferior turbinate, which is well aerated.

Several pathogenetic theories based upon the mucosa or edema of the mucosa (and several that also involve the presence of glands and cysts of the polyp) and upon the glands of the nasal mucosa have been adduced [72].

The Pathogenetic Theories

Adenoma and Fibroma Theories

The famous surgeon Theodor Billroth [4] found increased numbers of long, tubulous glands in polyps. He interpreted them as new formations in the nasal mucosa. The nasal polyps were interpreted as adenomas that began by growing under the nasal mucosa, pushing the epithelium and the original nasal glands outward (Fig. 8.16).

In our quantitative studies, we found relatively few glands. In the 84 polyps we examined [76], the median number of glands was less than 20, and the gland density was less than 0.5 glands/mm^2, which is quite low compared with the density in nasal mucosa (7 glands/mm^2) [74, 75]. The distribution and arrangement of glands in the nasal polyp does not resemble those of an adenoma.

Hopmann [25] did not find glands at all in nasal polyps, which he interpreted as soft fibromas. Both

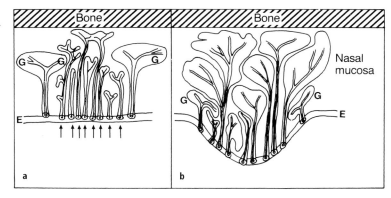

Fig. 8.16 a,b. Schematic illustration of the adenoma theory on polyp formation. **a** New formation of glands in nasal mucosa (*arrows*). **b** Newly formed glands pushing the original epithelium outward as a polyp; *E*, epithelium of the nasal mucosa; *G*, original tubuloalveolar glands

the adenoma and fibroma theories have been refuted adequately in the past century.

Necrotizing Ethmoiditis Theory

Ethmoiditis leads to periostitis and osteitis of the bone of the ethmoid and middle turbinate and causes necrosis (Fig. 8.17) [87]. At the sites of bone necrosis, myxomatous tissue projects towards the nasal mucosa. The nasal mucosa is pressed caudally as a polyp. Hayek [21] argued strongly against this theory; in many patients with nasal polyps, he could not find bone necrosis in the ethmoid sinus, which was without exudate or infection.

Glandular-Cyst Theory

This theory is based upon the presence of cystic glands and mucus-filled cysts in the polyps [20]. This is the oldest of several theories involving the mucous glands directly. The author believes that edema of the nasal mucosa causes obstruction of the ducts of basal glands, leading to the formation of cysts in the nasal mucosa (Fig. 8.18). The cysts expand and pull and push the nasal mucosa downward, forming a polyp. Billroth suggested that the gland duct may be blocked by the duct epithelium. However, Taylor [70] and we [76] have shown that cystic dilatation occurs, particularly among newly formed glands after the polyp has been formed.

Mucosal Exudate Theory

Hayek believed that the formation of nasal polyps starts via an exudate localized deep in the mucosa, which is pressed caudally [21]. A vascular stalk then forms, and vascular congestion increases the volume of the polyp (Fig. 8.19).

According to this theory, the distal part of the polyp should contain seromucous nasal glands displaced there by exudate. We found none [76]. Furthermore, these glands are supposed to be of a special orientation and shape, compressed subepithelially in the polyp. This was not found. All the glands of the nasal polyp are developed after the polyp has been formed and has attained a certain size. Presumably, none of the glands is derived from the nasal or ethmoidal sinus mucosa.

Cystic Dilatation of the Excretory Duct and Vessel Obstruction

In chronic inflammation of the nasal mucosa, excretory ducts of nasal tubuloalveolar glands are obstructed, distended, and dilatated into cystic structures. The capillaries and veins (which are arranged around the excretory ducts and the gland mass) become stretched and obstructed, resulting in increased permeability, transudation and edema. Folds or projections are formed by the dilatated glandular ducts on the infiltrated and edematous mucous membrane. This theory is based on the presence of excessive activity of the nasal glands and a periglandular plexus, which also surrounds the excretory ducts [88].

Again, cystic dilatation of glands occurs among the newly formed glands only after the polyp is formed. Only in cystic fibrosis of the pancreas is the dilatation of tubuloalveolar glands seen as a result of the hereditary gene defect of mucus transport through the duct system. This theory has been used to explain polyp formation in cystic fibrosis. We found exactly the same newly formed, long, tubulous glands in patients with and without cystic fibrosis [81].

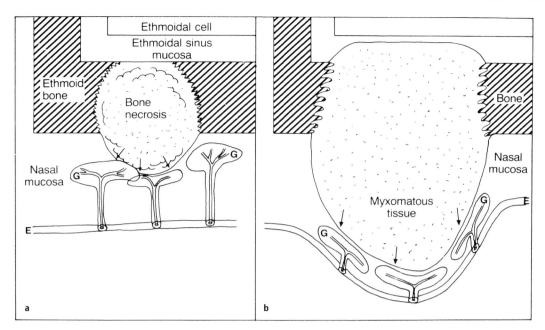

Fig. 8.17 a,b. Schematic illustration of necrosing-ethmoiditis theory. **a** Bone necrosis within the ethmoid bone, with myxomatous tissue protrusion toward the nasal mucosa (*arrows*). **b** Growth of the myxomatous tissue pressing the nasal tubuloalveolar glands (*G*) and the nasal epithelium (*E*) downward as a polyp

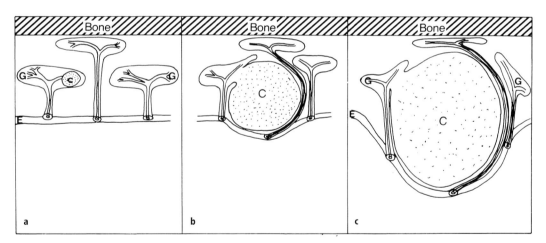

Fig. 8.18 a–c. Schematic illustration of glandular-cyst theory. **a** A cyst is formed within the nasal gland (*G*). **b, c** The cyst is expanding, pressing the nasal epithelium outward, and forming a polyp; *C*, glandular cyst; *E*, nasal epithelium; *G*, tubuloalveolar nasal glands

Blockade Theory

This theory is based on the fact that polyp formation is always preceded by the same degree of chronic inflammation [28]. This inflammation may be either infectious or allergic. The polyp itself is an accumulation of intercellular fluid dammed up in a localized tissue. The dam is usually caused by an infiltration of round cells, producing blockade of intercellular spaces and local lymph edema. If the blockade persists, a typical polyp forms and, if the blockade covers a large area, multiple polyps may form. If the blockade of the same round cells is lifted, accumulated fluid in the polyps will be absorbed, and the polyps will disappear. This is one of the many theories of polyp formation based on edema but, in our opinion, explains nothing about why and how polyps can arise in one particular place but not in another.

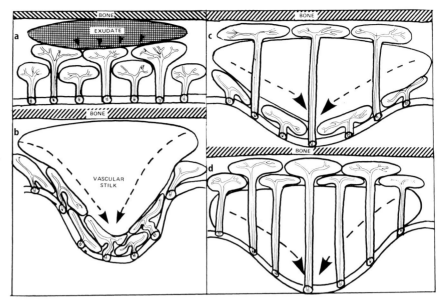

Fig. 8.19 a–d. Schematic illustration of the mucosal-exudate theory. **a** Nasal mucosa with tubuloalveolar glands and epithelium with an exudate localized between the deep gland layer and the bone, pressing the glands outward (*arrows*). **b** The nasal glands are displaced by the exudate and are compressed subepithelially in the polyp. **c** Edema predominantly between the deep and superficial glandular layers. **d** Predominantly subepithelial edema

Periphlebitis and Perilymphangitis Theory

Recurrent infections lead to periphlebitis and perilymphangitis, which lead to blocking of intercellular fluid transport in the mucosa and edema of the lamina propria. If the edema involves major areas, the result is prolapse of the mucosa and formation of polyps. This theory is based upon the demonstration of chronic vascular changes in the nasal mucosa, but these changes are diffuse, and the theory cannot explain how the polyp is formed in a particular place [17]. The theory is an explanation for edema formation rather than polyp formation. If the edema forms predominantly deep in the lamina propria (beneath the deepest glandular layer), this would lead to a situation already discussed by Hayek's theory (Figs. 8.19a, b) [21]. If it arises predominantly between the deeper and the superficial layers of glands, the superficial glands will be pressed into the polyps and will be found there, but we did not find the tubuloalveolar glands distally in the polyps [76]. The duct from the deeper layer of nasal glands would be stretched and would gradually become quite long, coursing through the stalk of the polyp, while the gland mass would be pressed towards the bone (Fig. 8.19c). We found long, tubulous glands in the polyp [76], but they did not reach the stalk, and they were developed after the polyp had been formed (Fig. 8.9).

If the edema is localized predominantly subepithelially, i.e., between the epithelium and the superficial layer of glands, the epithelium would bulge out in the form of a polyp and pull out the ducts of the nasal glands (Fig. 8.19d). In this case, the stalk would contain many long ducts, which we did not find.

Glandular-Hyperplasia Theory

In cases of chronic infection or allergy, localized infiltrates in the nasal mucosa and localized hyperplasia of nasal glands will occur. These will increase in size and cause bulging of the mucosa (Fig. 8.20) [32]. Apart from the gland hyperplasia, the change of blood vessels and the edema in the nasal mucosa in the region of the middle nasal meatus will lead to mucosal prolapses in the form of polyps.

We did not find tubuloalveolar glands in the polyps. In chronic hypertrophic rhinitis, there is very little gland hyperplasia in nasal mucosa, and we did not observe localized bulging of the nasal mucosa on the inferior turbinate caused by hyperplasia of nasal glands [71, 74, 75]. The number and density of glands was the same in patients with chronic hypertrophic rhinitis and in normal subjects.

Epithelial-Rupture Theory

We studied the shape, distribution, density and histologic profile of the glands by staining the glands in nasal polyps using the whole-mount method. Based on these studies, a new theory for the pathogenesis of nasal polyps was described by us [77]. It was postulated that, in the initial stage of polyp formation, an epithelial rupture or epithelial necrosis caused by inflammation and tissue pressure from the edematous and infiltrated lamina propria takes place (Fig. 8.21a). Lamina propria protrudes through the epithelial defect, and the adjacent epithelium tends to cover the defect by migration of epithelial cells

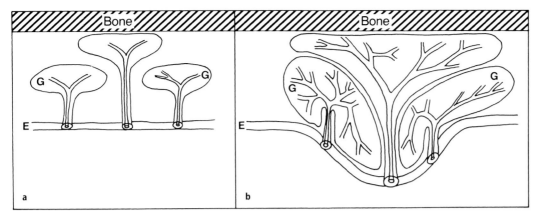

Fig. 8.20 a,b. Schematic illustration of glandular-hyperplasia theory. **a** Normal nasal tubuloalveolar glands (*G*). **b** Hyperplasia of nasal glands, causing protrusion of nasal mucosa; *E*, nasal epithelium

from the surroundings (Figs. 8.21 b,c). If the epithelial defect is not covered soon enough, or if it is insufficiently covered, the prolapsed lamina propria continues to grow, and the polyp (with its vascular stalk) is established. After epithelialization of the polyp, the characteristic new, long, tubulous glands are formed (Figs. 8.21d, e). Whole-mount studies elucidated the structure and density of glands in nasal polyps and showed that their shape and distribution were completely different from those of normal nasal-mucosa seromucous glands (Figs. 8.9–8.14). Our studies strongly indicate that the glands are newly formed structures and that the polyp is not simply a prolapse of the original nasal mucosa. We have been able to confirm the epithelial-rupture theory in experimental otitis media in rats [8]. We illustrated these early stages of polyp formation:
1. Localized rupture of the epithelium (Fig. 8.22)
2. Luminal protrusion of the lamina propria through the epithelial defect
3. Re-epithelialization of the protruded tissue and formation of a polyp
4. Growth of the polyp

During these processes, the glands are formed and, with further growth and stretching of the polyp, the glands became elongated and stretched.

Polyp formation (including initiation by rupture of the epithelium, by prolapse of the lamina propria and by re-epithelialization of the protruded tissue) was also demonstrated in chronic tubal occlusion in rats (Figs. 8.23, 8.24) [41, 45]. Polyp formation initiated by epithelial defects was also documented in experimental sinusitis in rabbits [62].

Chemical Mediators

The polyp tissue is readily available for study and, during the last several years, several chemical mediators have been identified; some of them may influence the pathogenesis of nasal polyps. A chemical mediator is defined as a molecule (formed by one cell) that has the ability to influence or affect changes in other, responding cells [23].

Several mediators, including histamine, platelet-activating factor (PAF) prostaglandins, leukotrienes, substance P and vasoactive intestinal peptide, affect vasoactive or spasmogenic actions. Other mediators, such as transforming growth factor and tumor necrosis factor, can change the composition of the extracellular matrix in a tissue.

All inflammatory processes involve control of cell migration towards the site of inflammation, and several mediators, such as high molecular weight-neutrophil chemotactic factor, histamine, prostaglandins, leukotrienes, PAF and interleukins, directly attract leukocytes. The discovery of adhesion molecules on the cell surface has revealed another mechanism influencing the migration of leukocytes to specific areas.

The production or release of chemical mediators is initiated by the activation of the cells that produce them, and this activation can be initiated by other chemical mediators, especially several interleukins and adhesion molecules. For histamine, the mediator-producing cells in nasal polyps are basophils and mast cells. The histamine content in nasal-polyp fluid varies widely, and the increase of histamine, which mediates the formation of polyps, is thought to occur by the following mechanisms: first, insufficient blood supply to the ethmoid sinus inhibits degenerative

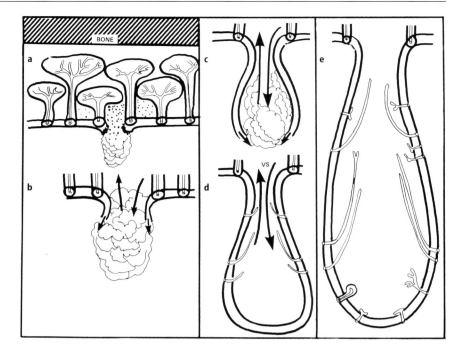

Fig. 8.21 a–e. Epithelial-rupture theory or glandular new-formation theory. An epithelial defect (**a**) with prolapse of the underlying lamina propria (**b**), epithelialization of the lamina propria prolapse (**c**), formation of a vascular stalk and formation of glands from the newly formed epithelium (**d**). There is further growth of the polyp and of the glands, with passive stretching of the gland ducts. A fully developed epithelialized polyp with long tubular glands is the final result (**e**)

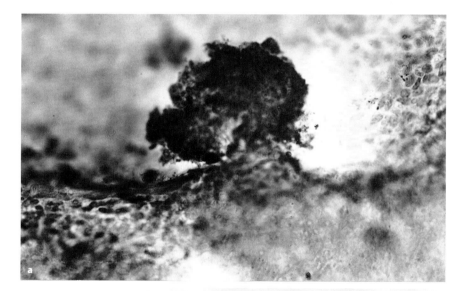

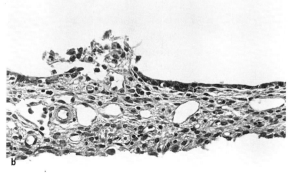

Fig. 8.22. a Small polypoid prominence seen in a whole mount 16 days after inoculation of rat middle ear with pneumococci.
b A section of the same polyp, illustrating epithelial rupture, incipient prolapse of the fibrous tissue from the lamina propria and re-epithelialization (magnification ×400)

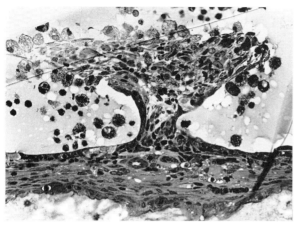

Fig. 8.23. Initial, partially epithelialized polyp in the rat middle ear after long-term tubal occlusion

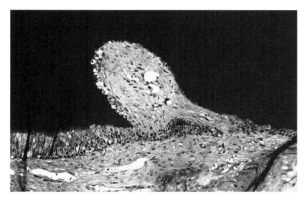

Fig. 8.24. A fully epithelialized polyp in the rat middle ear after chronic experimental tubal occlusion

enzymes; this causes increased histamine and results in edema and polyp formation [14]. Prostaglandins, produced by mast cells, cause vasodilatation and increased vascular permeability and may also be important chemotactic mediators in polyp formation.

Several other chemical mediators, such as interleukins, cytokines, growth factors and neurotransmitters, have been implicated in polyp formation, growth and perpetuation. However, even if mediators maintain and regulate inflammation and edema, this does not explain why the polyps form in some places in the nose but not in others.

Origin of Nasal Polyps

There are very few systematic studies about the origin of nasal polyps. In 29 out of 39 cases studied, Zuckerkandl [89] found that the origin of nasal polyps was the nasal mucosa of the lateral wall of the nose, near the orifices of the ethmoid. In patients treated by endoscopic sinus surgery, Stammberger [68] usually found polyps in the middle meatus, originating from the mucosa of narrow places and outlets of the paranasal sinuses. Polyps were found at the infundibulum between the ethmoidal bulla and the hiatus semilunaris in 65% of cases and at the frontal recess in 48% of the cases.

Autopsy Materials

In an attempt to determine the site of origination of nasal polyps, we studied three separate autopsy series; these included screening of the paranasal sinuses for polyps or polypoid hypertrophic mucosa. In the first series [40], nasal polyps were found by anterior rhinoscopy in six out of 300 cadavers (2%). The naso-ethmoidal blocks were removed for further examination to determine the origin of the polyps (Fig. 8.25). Eight out of 44 polyps originated in sinus ostia, clefts or recesses.

In the second series [42], 19 naso-ethmoidal blocks were removed without a preceding rhinoscopy. Polyps were found in five blocks (26%). Nine polyps were present, and all except one (89%) were found in the sinus clefts, ostia and recesses (Table 8.3).

The third series [43] was comprised of 31 autopsies in which endoscopic examination and sinus surgery of the nose and paranasal sinuses were performed. Polyps were found in 13 cases (42%).

In total, out of the 80 polyps from all three series, 35 polyps (44%) originated in sinus ostia, clefts or recesses (Table 8.3). Polyps were found more often in

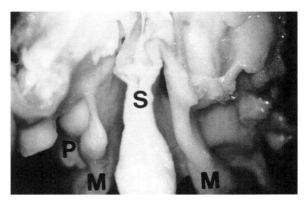

Fig. 8.25. Naso-ethmoidal block from material I. A polyp is seen in the anterior part of the middle meatus. *M*, middle turbinate; *P*, polyp; *S*, septum

Table 8.3. Origin of nasal polyps in 24 autopsy cases from three studies found by anterior rhinoscopy (study I, n=6), by transcranial-block removal (study II, n=5) and by endoscopic nasal and paranasal surgery (study III, n=13). Study I includes 44 polyps, study II includes nine polyps, and study III includes 27 polyps

Origin	Relationship with sinus ostium/cleft/recess					
	No Study I	No Study II	No Study III	Yes Study I	Yes Study II	Yes Study III
Middle meatus, med. (middle turbinate)	–	–	4	–	–	–
Middle meatus, lat. (uncinate process)	1	–	1	–	–	–
Middle meatus, lat. (bulla)	–	–	1	–	–	–
Middle meatus, lat./top (frontal recess)	–	–	–	–	–	4
Middle meatus, med.	–	1	–	–	–	–
Middle meatus lat. (anterior ethmoidal cleft)	27	–	–	8	4	6
Middle meatus, lat. (supra-retrobullar recess/sinus lat.)	–	–	–	–	1	–
Middle meatus, lat. (maxillary sinus ostium)	–	–	–	–	–	4
Superior meatus, lat./top (posterior ethmoidal cleft)	–	–	–	–	3	4
Middle turbinate, med.	–	–	1	–	–	–
Agger nasi	1	–	1	–	–	–
Sphenoethmoidal recess	1	–	–	–	–	1
Septum	6	–	–	–	–	–
Total	36	1	8	8	8	19
Total (all three studies)	45 (56%)			35 (44%)		

lat., lateral; *med.*, medial

the outlet of the paranasal sinuses in the last two series (transcranial blocks and endonasal surgery) than in the first series.

Most of the polyps (56%) were associated with the anterior ethmoidal cleft (Fig. 8.26). Four (15%) were associated with the frontal recess (Fig. 8.27) and maxillary sinus ostia and one (4%) was associated with the sphenoid sinus ostium (Table 8.3).

Nine cases (38%) had bilateral polyps. Seven (29%) of these patients had more than two polyps. In series I, where only the anterior part of the nasal cavity could be visualized, 50% of the cases revealed several nasal polyps. In the last two series, the polyps were unilateral in most of the autopsies, and only one polyp in each nasal cavity was found in 13 (54%) of the cases.

Although the nasal polyps are associated with and are caused by a systemic disease, they start in areas of the nasal mucosa related to outlets from the sinuses (the ethmoidal clefts, frontal recess and maxillary sinus ostia) and may progress to a chronic polypoid sinusitis with nasal polyps. One or several local factors (probably ventilation or blockage of the mucociliary clearance in the pathways of the ostiomeatal complex) contribute to persistence and growth of the polyps. The small polyps found in cadavers are probably reversible. Maintenance of the polyp in a polyp-causing disease can be explained by a vicious cycle. Infection or inflammation results in formation of one polyp, which leads to blockage of

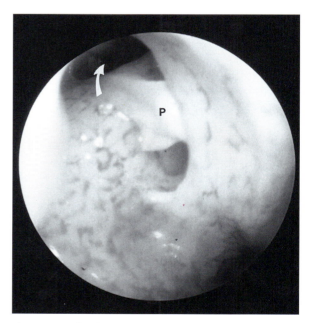

Fig. 8.26. Small polyp originating from the mucosa in the outlet (*arrow*) from the anterior ethmoid. *P*, polyp

ventilation in the meatus, leading to formation of more polyps in the meatus, which leads to further blockage of the sinus ostia, recesses and clefts, which leads to infection and stagnation of secretion in the ethmoid and other paranasal sinuses, which finally

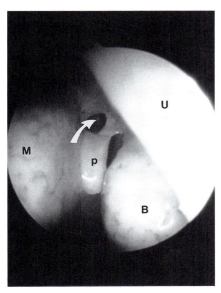

Fig. 8.27. Small polyp originating in the anterior part of the outlet from the frontal recess (*arrow*). Material III. *B*, ethmoidal bulla; *M*, middle turbinate; *P*, polyp; *U*, uncinate process

leads to hypertrophic mucosa/polyp formation in the paranasal sinuses.

Diagnosis and Differential Diagnosis

Clinical Symptoms

The predominant complaint in nasal polyps is constant nasal obstruction, that varies in severity with the size of the polyps. The quality of the voice is changed, and snoring is often present. Many patients complain of anterior rhinorrhea or postnasal discharge of mucopurulent secretion (or both), but facial pain and headaches are rare, despite obstructed paranasal sinuses. Partial or complete loss of smell and/or taste is a common associated effect of nasal polyps.

Clinical Examination

Clinical examination can be performed by anterior rhinoscopy and endoscopy with a flexible endoscope or (ideally) with a rigid endoscope. Even in the absence of a nasal speculum or endoscope, the nasal cavity can be adequately examined with an aural speculum or an otoscope. Clinically, polyps are pale bags of edematous tissue; they arise most commonly from the middle meatus and are relatively insensitive. The pale color is due to the poor blood supply but, in the presence of repeated trauma and inflammation, they may become reddened. They are commonly bilateral and, when unilateral, they require histological examination to exclude inverted papilloma or malignancy. Nasal polyps may be classified (according to Johansen et al. [29]) as:

- **Grade 1.** Mild polyps (including small polyps not reaching the lower edge of the middle turbinate) causing only slight obstruction
- **Grade 2.** Moderate polyps (including medium-sized polyps reaching between the upper edge of the inferior and the lower edge of the middle turbinate) causing troublesome obstruction
- **Grade 3.** Severe polyps (including large polyps reaching below the lower end of the inferior turbinate) causing total or almost-total obstruction

Furthermore, the number of polyps can be assessed, but the exact number can only be determined after polypectomy. Sometimes, it can be difficult to distinguish between hypertrophic inferior or middle turbinates and nasal polyps; gentle palpation will show that the polyps are insensitive (in contrast to the turbinate, which has an excellent sensory supply).

Several objective examinations, such as nasal inspiratory peak flow, anterior rhinomanometry and acoustic rhinometry, plain X-rays or (preferably) computed tomography (CT) scans, are available to quantify the restriction in the nasal airway. The CT scan will show the anatomy of the nose and sinuses, including any alterations of the mucosa and various degrees of sinus clouding, up to a complete opacification of the sinuses and nose. Sometimes, expansion of the ethmoid or some degree of decalcification or resorption of bony septa can be seen. When polyps begin in young adults or at an earlier age, there is often some degree of ethmoidal expansion, which can sometimes reach significant levels.

Differential Diagnosis of Nasal Polyps

In the presence of an unilateral nasal polypoid mass, several benign epithelial and mesenchymal tumors and some sinonasal anatomical abnormalities have to be considered in differential diagnoses. *Concha bullosa* can produce a large mass in the nose that might, at the first examination, simulate a polypoid tumor.

Inverted papilloma is a rare tumor that often exhibits rapid growth with bone destruction and has a high recurrence rate. The incidence is 0.6 patients per 100,000 inhabitants per year [6, 7]. It is usually impossible to distinguish it from a nasal polyp. There is a potential for malignant transformation; thus, it is important to establish the histological diagnosis of nasal polyps.

Juvenile *angiofibroma* occurs almost exclusively in male children or young adolescents who present with nasal obstruction and epistaxis.

The differential diagnosis should also include *epithelial tumors* (papilloma, pleomorphic adenoma), *benign mesenchymal tumors* (such as fibroma, ossified fibroma, hemangioma, Schwannoma, meningeoma, leiomyoma, angioleiomyoma), *inflammatory granulomatous conditions* (such as Wegener granulomatosis, sarcoidosis), *malignant lesions in adults* (such as adenocarcinoma, squamous cell carcinoma). In children, the differential diagnosis of a unilateral mass in the nose includes congenital lesions (such as encephalocele, glioma, dermoid cysts, nasal lacrimal duct cysts) and neoplasms (such as craniopharyngioma, hemangioma, neurofibroma, malignant rhabdomyosarcoma) [3].

Treatment of Nasal Polyps

There is a general agreement that rational treatment should be based on understanding the disease process involved and tailoring the treatment accordingly. During the last 10 years, there has been great deal of research on the etiology, pathogenesis and origination of nasal polyps, but the results of these studies are not always taken into account when deciding on treatment of nasal polyps. Today, disagreements about the various treatment modalities for nasal polyps are even greater than before [82]. Endonasal sinus surgery is becoming more and more important in the treatment of nasal polyps.

Goals of Treatment

The goal of management of nasal polyps, as defined by Mygind and Lildholdt [60], are:
1. Disappearance or considerable reduction in the size of nasal polyps. This can be achieved, at least temporarily, by local or systemic steroid treatment and by surgery.
2. Improvement or re-establishment of nasal breathing and open nasal airways. This goal is temporarily achieved by all treatment modalities.
3. Freedom from rhinitis symptoms is (at least in most patients) achieved to a higher degree with medical than with surgical treatment.
4. Normal sense of smell and improvement of olfaction.

Requirements of ideal treatment are:
1. High patient compliance (best achieved when the patient experiences no pain or discomfort, low cost, short duration of treatment and long-lasting effects)
2. No risk of serious adverse effects
3. No changes in normal nasal structure and function

None of the many treatment modalities reaches the ideal objectives of treatment, completely eliminates the symptoms for the long term and completely prevents recurrences. Several modalities can reduce the subjective and objective symptoms considerably and can approach the goals of treatment.

Treatment Modalities of Nasal Polyps

The various treatment modalities (Table 8.4) are combinations of the four basic treatments: (1) local steroids, (2) systemic steroids, (3) simple polypectomy and (4) endonasal sinus surgery. Some modalities are applied often, usually as a routine procedure, e.g., as combination of local steroids followed by simple polypectomy or vise versa (first simple polypectomy, followed by local steroids). Some modalities, e.g., systemic steroids, are applied relatively rarely. Today, many rhinosurgeons include endonasal sinus sur-

Table 8.4. Treatment modalities of nasal polyps

1. Long-term local steroids
2. Long-term local steroids combined with short-term systemic steroids
3. Long-term local steroids followed by simple polypectomy
4. Short-term steroids followed by simple polypectomy
5. Short-term steroids followed by ESS
6. Long-term local steroids followed by ESS
7. Systemic steroids
8. Simple polypectomy
9. Simple polypectomy and, eventually, local steroids
10. Simple polypectomy followed by long-term local steroids
11. Simple polypectomy followed by local and systemic steroids
12. Simple polypectomy, local and systemic steroids followed by ESS
13. Simple polypectomy followed by ESS in recurrence
14. ESS as primary surgery

ESS, endonasal sinus surgery

gery, often combined with preoperative local steroids, in their treatment of nasal polyps. In general, the four basic treatments are seldom applied alone in the long-term (i.e., during the entire course of the polyp disease). Even if local treatment with steroids alone or simple polypectomy alone considerably improves or eliminates the symptoms in easy cases, more protracted cases will require a combination of treatment with local steroids and simple polypectomy, and the most protracted cases will require endonasal sinus surgery. In our opinion, this should be necessary in less than 5% of patients with nasal polyps.

Local Steroids

Use of intranasal local steroids is the most well-documented type of treatment for nasal polyps. It is well documented that the glucocorticoid steroids have a multifactorial effect initiated by their binding to a specific cytoplasmic glucocorticoid receptor. This binding induces a change of protein synthesis in the cells and is responsible for the clinical effect, which only becomes manifest after several days of treatment. Glucocorticoid administration reduces the number of glucocorticoid receptors [58]. Long-term beneficial effects of treating nasal polyps with beclomethasone dipropionate aerosol have been demonstrated by several groups [30, 48–50, 58–60, 63]. Even though the local application of steroids results in a significant effect compared with placebo in all controlled studies, it is evident that nasal steroids do not solve all the problems for a number of patients with nasal polyposis. Some patients do not respond to topical glucocorticoid steroids (for example, patients with cystic fibrosis or primary ciliary dyskinesias). Lack of response to topical steroids is also seen when there is intense local tissue infiltration with neutrophils, in patients with purulent infection and in patients who have large polyps blocking the nose and preventing access of the steroid spray to the nasal mucosa. There was obvious reduction of polyp size in each of the three series, but the polyps did not totally disappear [9, 16, 29, 49, 50, 84]. Therefore, after long-term treatment with local glucocorticoid steroids, a simple polypectomy is often performed to remove the residual polyp tissue. This combination of local corticosteroid treatment followed by simple polypectomy often gives satisfactory results. Differences between nasal drops and sprays have not been investigated.

Prospective randomized studies with local corticosteroids usually report only short term or 1-year results, without long-term follow-up. Often, patients who were treated successfully with local steroids need polypectomy months later. Long-term follow-up of these clinical trials is necessary to find out which of the many treatment modalities (Table 8.4) is beneficial in the long term.

Systemic Steroids

Systemic steroid treatment of nasal polyps has gained increasing interest, but is still insufficiently investigated. No placebo-controlled studies have been performed. In two studies, short-course systemic steroid treatment (depot injection of 14 mg β-methasone, corresponding to approximately 100 mg prednisolone) has achieved results as good as those resulting from simple polypectomy [49, 50]. Oral prednisolone (starting with a burst dose of 60 mg for 4 days and a successive reduction of 5 mg daily, for a total dose of 500 mg) has also been used. Systemic steroids have a marked effect on polyp size and, in most cases, improve the sense of smell. In many cases, preoperative systemic steroid therapy can facilitate surgical treatment but, when systemic steroids are the sole treatment, there is a strong tendency for polyp recurrence within 5 months. The obvious disadvantage of systemic steroids is the risk of adverse systemic effects, but two or three treatments per year are generally regarded as safe.

Simple Polypectomy

Simple polypectomy has been used for at least 2000 years. The *sponge method* of Hippocrates delivered the polyps by pulling a rough sponge through the nose. *Caustic agents* and *poisonous fluids* have since been used, as have *ligatures* around the polyps. A type of *nasal snare* was also used by Hippocrates, but the first major advance in the principle of the snare was made by Fallopius (1523–1562 AD) [85]. The instruments have improved since but are still relatively simple.

In the term "simple polypectomy", we include both the *classical snare polypectomy* by anterior rhinoscopy and *polypectomy with endoscopic control* (using both Weil and cutting forceps or rotating suction/cutting instruments used in endonasal sinus surgery) but not invasive surgery of the sinuses (expect the cleaning of spontaneously created ethmoid cavities). Also included is the use of powered instrumentation for polyp removal (a refined variant of endoscope- or microscope-guided simple polypectomy) when its use is restricted to soft-polyp-tissue removal from the nasal cavity. *Laser polypectomy* constitutes a similar refined technique.

In our comprehensive long-term follow-up study [35–38] after simple polypectomy with snare (including all 283 patients treated for nasal polyps between January 1984 and December 1991 in a geographically well-defined ear, nose and throat practice), patients were followed for 1–8 years after their initial surgery. In the group of 180 patients entering the study for *first-time polypectomy*, 65% of patients needed only one polypectomy during the entire observation period, and 4% needed five or more polypectomies (Table 8.5). In the more severe group of 103 patients having had polypectomies *before* entrance into the study, 45% needed only one polypectomy during the observation period; 8% needed five or more polypectomies.

At the follow-up of this material in 1996 (with a median observation time of 8 years and a range of 3–12 years) [35], the symptoms of blocked nose and nasal secretion were non-existent or slight in 70–80% of cases, but the sense of smell was impaired or lost in most cases. Only 3% had major polyps on inspection. Most patients also received topical steroids in addition to simple surgical treatment. In a comparable follow-up study [13–15], severe recurrences were found in 5% of the patients. Based on our data, only minor surgery (a simple polypectomy and/or medical treatment) is needed in the vast majority of patients. Long-term studies on simple polypectomies from other centers are needed, however.

The problem in the decision between simple polypectomy and more extensive surgery is the difficulty in determining which of the patients will be the more recalcitrant cases. However, performing extensive surgery on all patients with polyps constitutes over-treatment, from both the patient's point of view and an economical point of view.

Endonasal Sinus Surgery

All series reported essentially good results after endonasal sinus surgery [31, 46, 47, 51, 69, 86] but, in most series, sinusitis and polyp patients were reported together, and it is not always possible to ascertain the results from patients with polyposis only. It is often stated that patients are included as surgical candidates only if medical treatment has failed, but the criteria for failure are not stated, nor is the length of treatment or whether it was medical or featured simple polypectomy. It is necessary to answer these questions before judging the success of endonasal sinus surgery.

Obstructing polyps would lead to failure of medical topical steroid treatment unless the major polyps were first removed with simple polypectomy to facilitate the local application of the active drug. We believe a proper approach is to consider extensive surgery only when combinations of medical treatment and minor polyp surgery have failed. Although such a decision should be made on an individual basis, some guidelines can be drawn from our own studies (Table 8.5). If the number of polypectomies is used as a parameter, three or more polypectomies within 5 years might indicate a success rate below 80% and may eventually call for more radical surgery. It is mandatory that there be optimal patient compliance with topical steroids in combination with simple polypectomy. Most studies of functional endoscopic sinus surgery show success rates above 80%, though with somewhat shorter observation times than indicated above [35]. Lower rates are reported in asthmatics [46, 47]. Others have found asthma to have no significant effect and found that the outcome was more related to the extent of disease (as seen on CT before surgery) [31]. In our series, asthmatics had the poorest outcome after simple polypectomy. Based on a meta-analysis of the relationship of polyps to asthma, there is no evidence

Table 8.5. Number of simple polypectomies and median observation time in two groups of patients with nasal polyps. The patients were followed from January 1984 to December 1991

Polypectomies	First simple polypectomy			Previous simple polypectomy		
	Patients (n=180) (%)	Observation time months		Patients (n=180) (%)	Observation time months	
		Median	Range		Median	Range
One	65	52	12–95	45	58	13–95
Two	18	66	18–95	28	72	16–95
Three	10	53	37–93	10	82	14–94
Four	3	72	49–87	9	84	22–95
Five or more	4	85	60–93	8	79	53–96

that endonasal sinus surgery is superior to simple polypectomy combined with topical steroid in patients with asthma and polyps [34].

Four features could argue for endonasal sinus surgery as the primary procedure:
1. Improvement of sense of smell
2. Prolongation of intervals between intranasal procedures
3. Elimination of sinusitis problems
4. Improvement of lower-airway symptoms

Improvement of the Sense of Smell

Improvement of the sense of smell after endonasal sinus surgery has been found in 71–81% of patients [11, 51]. There is no evidence that these results persist for a longer observation time. In comparison, 75% of patients with nasal polyps treated with budesonide powder showed improvement of the sense of smell [48, 49, 84]. Nevertheless, the sense of smell is a critical aspect of the treatment of nasal polyps.

Prolonged Intervals Between Simple Polypectomies

It is often argued that endonasal sinus surgery increases the intervals needed for subsequent intervention in patients with recurrent polyposis. However, it is well known that numerous intranasal cleansing procedures are needed after sinus surgery to secure good long-term results. These long-term, postoperative, simple surgical treatments can be equated to simple polypectomies; therefore, the cost–benefit analysis, even in the most protracted cases needing several polypectomies (Table 8.5), greatly favors simple polypectomy.

Sinusitis

There is no doubt that endonasal sinus surgery is beneficial for the treatment of recurrent acute or chronic sinusitis. In polyp patients, there is a good rationale for using endonasal sinus surgery to remove any obstruction to the sinuses.

However, in patients treated with topical steroids combined with simple polypectomy, we found that only 3% of patients had three or more treatments for sinusitis in the 3-year period before follow-up [35]. No comparative study exists but, obviously, much can be achieved by simple treatment modalities.

Improvement of Lower-Airway Symptoms

It was long ago stated that simple polypectomy could accelerate or aggravate asthma. This statement has been rejected several times in the literature but, occasionally, it finds its way into current literature [27], where it is used to contrast with the situation after endoscopic surgery. From a survey of the literature, it is obvious that most patients are improved or at least unchanged regarding the lower-airway after upper-airway improvement by surgery for polyps. In that respect, no difference between ethmoidectomy and simple polypectomy could be extracted from the literature, which describes clinical parameters (such as pulmonary function, medical need and global setting of the control of asthma) [37].

Reduction of non-specific bronchial hyper-reactivity has been reported to be reduced in 30% of patients undergoing radical sphenoethmoidectomy [27, 67]. In fact, similar results were observed in studies after simple polypectomy [12, 53], although only half as many patients were studied. Recently, it was found that patients selected for intranasal ethmoidectomy who did not respond to topical steroids showed greater deterioration of bronchial hyper-reactivity after the surgery compared with patients receiving topical steroids [33].

From these observations, there is no evidence that ethmoidectomy is superior (in terms of its effects on the lower airway) to other simple treatment modalities for polyps. Furthermore, there is nothing to suggest that either early or late ethmoidectomy or polypectomy can delay or prevent the development of lower-airway disease. There is no doubt that endonasal sinus surgery has a place in the treatment of nasal polyposis, but it is mandatory to find the right indications, i.e., the type and duration of medical treatment and the number of previous simple polypectomies in a certain time interval.

References

1. Andersen HC (1943) Nasal polyps and hyperplastic sinusitis. Thesis. Thanning and Appel, Copenhagen
2. Berdal D (1954) Histology of polyps, particularly cellular infiltration. Acta Otolaryngol Suppl (Stockh) 115:60–80
3. Bernstein J (1997) The immuno-histopathology and pathophysiology of nasal polyps. In: Settipane GA, Lund VJ, Bernstein JM, Tos M (eds) Nasal polyps: epidemiology, pathogenesis and treatment. Oceanside, Providence, pp 85–95
4. Billroth T (1855) Über den Bau der Schleimpolyppen. Georg Reimer, Berlin, pp 1–90

5. Blumstein GI, Tuft L (1957) Allergy treatment in recurrent nasal polyps. Am J Med Sci 234:269–280
6. Buchwald C, Holme-Nielsen L, Nielsen PL, Ahlgren P, Tos M (1989) Inverted papilloma. A follow-up study including primarily unacknowledged cases. Am J Otolaryngol 10:273–281
7. Buchwald C, Holme-Nielsen L, Ahlgren P, Nielsen PL, Tos M (1990) Radiologic aspects of inverted papilloma. Eur J Radiol 10:134–139
8. Cayé-Thomasen P, Hermansson A, Tos M, Prellner K (1995) Polyp pathogenesis. A histopathological study in experimental otitis media. Acta Otolaryngol (Stockh) 115:76–82
9. Chalton R, Mackay I, Wilson R, Cole P (1985) Double-blind placebo controlled trial of β-methasone nasal drops for nasal polyposis. BMJ 291:788
10. Davidsson Å, Hellquist HB (1993) The so-called allergic nasal polyps. ORL J Otorhinolaryngol Relat Spec 55:30–35
11. Delank KW, Stoll W (1994) Sense of smell before and after endonasal sinus surgery in chronic sinusitis with polyps. HNO 42:619–623
12. Downing ET, Braman S, Settipane GA (1982) Bronchial reactivity in patients with nasal polyposis before and after polypectomy. J Allergy Clin Immunol 69:102
13. Drake-Lee AB (1996) Magical numbers and the treatment of nasal polyps. Clin Otolaryngol 21:193–197
14. Drake-Lee AB (1997) The pathogenesis of nasal polyps. In: Settipane GA, Lund VJ, Bernstein JM, Tos M (eds) Nasal polyps: epidemiology, pathogenesis and treatment. Oceanside, Providence, pp 57–64
15. Drake-Lee AB, Lowe D, Swanston A, Grace A (1984) Clinical profile and recurrence of nasal polyps. J Laryngol Otol 98:783–793
16. Drettner B, Ebbesen A, Nilsson M (1982) Prophylactic treatment with flunisolide after polypectomy. Rhinology 20:149–158
17. Eggston AA, Wolff D (1947) Histopathology of the ear, nose and throat. Williams and Wilkins, Baltimore, pp 613–676
18. Ellefsen P, Tos M (1972) Pathological intraepithelial glands in human trachea. Acta Otolaryngol (Stockh) 73:443–453
19. English GM (1986) Nasal polypectomy and sinus surgery in patients with asthma and aspirin idiosyncrasy. Laryngoscope 96:374–380
20. Frerichs (1843) Über den Bau der Schleimpolyppen. Georg Reimer, Berlin, pp 1–90
21. Hajek M (1986) Über die Patologischen Veränderungen der Siebbeinknochen im Gefolge der Entzündlichen Schleimhauthypertrophie und der Nasenpolypen. Arch Laryngol Rhinol 4:277–301
22. Hellquist HB (1997) Histopathology of nasal polyps. In: Settipane GA, Lund VJ, Bernstein JM, Tos M (eds) Nasal polyps: epidemiology, pathogenesis and treatment. Oceanside, Providence, pp 31–39
23. Hoffman HB, Wasserman SI (1997) Chemical mediators in polyps. In: Settipane GA, Lund VJ, Bernstein JM, Tos M (eds) Nasal polyps: epidemiology, pathogenesis and treatment. Oceanside, Providence, pp 85–95
24. Holopainen E, Mäkinen J, Paavolainen M, Palva T, Salo OP (1979) Nasal polyposis. Relation to allergy and acetylsalicylic acid intolerance. Acta Otolaryngol (Stockh) 87:330–334
25. Hopmann A (1885) Über Nasenpolyppen. Monatsschr Ohrenheilkunde 19:161–167
26. Jahnke V (1972) Ultrastructure of the normal nasal mucosa in man. Z Laryngol Rhinol Otol 51:152–162
27. Jankowsky R, Moneret-Vautrin DA, Goetz R, Wayoff M (1992) Incidence of medico-surgical treatment for nasal polyps on the development of associated asthma. Rhinology 30:249–258
28. Jenkins J (1932) Blockade theory of polyp formation. Laryngoscope 42:703–704
29. Johansen LV, Illum P, Kristensen S, Winther L, Petersen SV, Synnersted B (1993) The effect of budesonide (Rhinocort) in the treatment of small and medium-sized nasal polyps. Clin Otolaryngol 18:524–527
30. Karlsson G, Rundcrantz H (1982) A randomized trial of intranasal beclomethasone diproprionate after polypectomy. Rhinology 20:144–148
31. Kennedy DW (1992) Prognostic factors, outcomes and staging in ethmoid sinus surgery. Laryngoscope 57[suppl]:1–18
32. Krajina Z (1963) A contribution to the aetiopathogenesis of the nasal polyps. Pract Otorhinolaryngol 25:241–246
33. Lamblin C, Tittlie-Leblond J, Darras F, Dubrulle D, Chevalier E, Cardot T, Perz B, Valleart JJ, Piguet J, Tonnel AB (1997) Sequential evaluation of pulmonary function and bronchial hyper-responsiveness in patients with nasal polyposis. A prospective study. Am J Respir Crit Care Med 155:99–103
34. Larsen K (1997) Nasal polyps: relationships to asthma. In: Settipane GA, Lund VJ, Bernstein JM, Tos M (eds) Nasal polyps: epidemiology, pathogenesis and treatment. Oceanside, Providence, pp 65–84
35. Larsen K, Tos M (1994) Clinical courses of patient with primary nasal polyps. Acta Otolaryngol (Stockh) 114:556–559
36. Larsen K, Tos M (1994) Recurrence of nasal polyps. In: Passali D (ed) Rhinology up to date. Industrial Grafica Romania, Rome, pp 225–228
37. Larsen K, Tos M (1995) Polypectomy with snare. Is it still relevant? In: Tos M, Thomsen J and Balle V (eds) Rhinology: a state of the art. Kugler, Amsterdam, pp 525–528
38. Larsen K, Tos M (1997) A long-term follow-up study of nasal polyp patients after simple polypectomies. Eur Arch Otorhinolaryngol Suppl 254:S85–S88
39. Larsen PL, Tos M (1990) Nasal polyps. Epithelium and goblet cell density. Laryngoscope 99:1274–1280
40. Larsen PL, Tos M (1991) Origin of nasal polyps. Laryngoscope 101:305–312
41. Larsen PL, Tos M (1991) Polyp formation by experimental tubal occlusion in the rat. Acta Otolaryngol (Stockh) 11:926–933
42. Larsen PL, Tos M (1995) Origin of nasal polyps. Transcranially removed naso-ethmoidal blocks as a screening method for nasal polyps in an autopsy material. Rhinology 3:185–88
43. Larsen PL, Tos M (1996) Origin of nasal polyps. Endoscopic nasal- and paranasal sinus surgery as a screening method for nasal polyps in an autopsy material. Am J Rhinol 10:211–216
44. Larsen PL, Sørensen CH, Rygaard J, Tos M (1987) Mucous elements of the respiratory tract in athymic rats. A quantitative histologic study. In: Lim DJ (ed) Fourth International Symposium on Recent Advances in Otitis Media. Decker, Toronto
45. Larsen PL, Tos M, Kuijpers W, Van der Beek JMH (1992) The early stages of polyp formation. Laryngoscope 102:670–677
46. Lawson W (1994) The intranasal ethmoidectomy: evolution and an assessment of the procedure. Laryngoscope 64[suppl]:1–49

47. Levine HL (1990) Functional endoscopic sinus surgery: evaluation, surgery and follow-up of 250 patients. Laryngoscope 100:79–84
48. Lildholdt T, Fogstrup J, Gammelgaard N, Kortholm B, Ulsoe C (1988) Surgical versus medical treatment of nasal polyps. Acta Otolaryngol (Stockh) 105:140–143
49. Lildholdt T, Rundcrantz H, Lindquist N (1995) Efficacy of topical corticosteroid powder for nasal polyps: a double-blind, placebo-controlled study of budesonide. Clin Otolaryngol 20:26–30
50. Lildholdt T, Rundcrantz H, Bende M, Larsen K (1997) Glucocorticoid treatment for nasal polyps. A study of budesonide powder i.m. deposit β-methasone and surgical treatment. Arch Otolaryngol Head Neck Surg 123:595–600
51. May M, Levine HL, Schaitkin B, Mester SJ (1993) Results of surgery. In: Levine HL, May M (eds) Endoscopic sinus surgery. Thieme, New York, pp 176–192
52. Messerklinger W (1958) Die Schleimhaut der oberen Luftwege im Blickfeld neurer Forschung. Arch Ohren Nasen Kehlkopfheilk 193:1–150
53. Miles-Lawrence R, Kaplan M, Chang K (1982) Methacholine sensitivity in nasal polyps and the effects of polypectomy. J Allergy Clin Immunol 69:102
54. Mogensen C, Tos M (1976) Bechercellen bei der Nasenallergie. Z Laryngol Rhinol Otol 55:955–961
55. Mogensen C, Tos M (1977) Density of goblet cells in the normal adult human nasal septum. Anat Anz 141:237–247
56. Mogensen C, Tos M (1977) Density of goblet cells on normal adult nasal turbinates. Anat Anz 142:322–330
57. Mogensen C, Tos M (1978) Experimental surgery of the nose. I. Air flow and goblet cell density. Acta Otolaryngol (Stockh) 86:289–297
58. Mygind N (1993) Glucorticosteroids and rhinitis. Allergy 48:476–490
59. Mygind N, Brahe Pedersen C, Prytz S, Sørensen H (1975) Treatment of nasal polyps with intranasal beclomethasone dipropionate aerosol. Clin Allergy 5:159–164
60. Mygind N, Lildholdt T (1997) Treatment: medical management. In: Settipane GA, Lund VJ, Bernstein JM, Tos M (eds) Nasal polyps: epidemiology, pathogenesis and treatment. Oceanside, Providence, pp 40–56
61. Nakashima T, Hamashima Y (1979) Loss of secretory activity in the glands of nasal polyps. Ann Otol Rhinol Laryngol 88:210–216
62. Nordlander T, Fukami M, Westrin KM, Stierna P, Carlsöö B (1993) Formation of mucosal polyps in the nasal and maxillary sinus cavities by infection. Otolaryngol Head Neck Surg 109:522–529
63. Pedersen CB, Mygind N, Sørensen H, Prytz S (1976) Long-term treatment of nasal polyps with beclomethasone dipropionate aerosol. Acta Otolaryngol (Stockh) 82:256–259
64. Puskas F, Modis L, Jakabfi I, Kosa D (1969) Histologische und histochemische Untersuchungen bei allergischen Rhinitis. Z Laryngol Rhinol Otol 48:188–197
65. Settipane GA (1987) Nasal polyps: epidemiology, pathology, immunology and treatment. Am J Rhinol 1:119–126
66. Settipane GA (1997) Epidemiology of nasal polyps. In: Settipane GA, Lund VJ, Bernstein JM, Tos M (eds) Nasal polyps: epidemiology, pathogenesis and treatment. Oceanside, Providence, pp 17–24
67. Slavin RG (1982) Relationship of nasal disease and sinusitis to bronchial asthma. Ann Allergy 49:74–80
68. Stammberger H (1991) Functional endoscopic sinus surgery. The Messerklinger technique. Decker, Toronto
69. Stammberger H. Posawetz W (1990) Functional endoscopic sinus surgery. Eur Arch Otorhinolaryngol 247:63–76
70. Taylor M (1963) Histochemical studies on nasal polypi. J Laryngol Otol 77:326–341
71. Tos M (1982) Goblet cells and glands in the nose and paranasal sinuses. In: Proctor D, Andersen J (eds) The nose, upper air-way physiology and the atmospheric environments. Elsevier, Amsterdam, pp 99–140
72. Tos M (1990) The pathogenetic theories on formation of nasal polyps. Am J Rhinol 4:51–56
73. Tos M, Bak-Pedersen K (1972) Intraepithelial glands in the human Eustachian tube. Arch Otolaryngol (Stockh) 96:546–552
74. Tos M, Mogensen C (1977) Density of mucous glands in normal adult nasal turbinates. Arch Otorhinolaryngol 215:101–111
75. Tos M, Mogensen C (1977) Nasal glands in nasal allergy. Acta Otolaryngol (Stockh) 83:498–504
76. Tos M, Mogensen C (1977) Mucous glands in nasal polyps. Arch Otolaryngol 103:407–413
77. Tos M, Mogensen C (1977) Pathogenesis of nasal polyps. Rhinology 15:87–95
78. Tos M, Mogensen C (1978) Experimental surgery of the nose. II. Changes of the mucosa in altered air-flow. Illustrated by blind quantitative histology. J Laryngol Otol 92:667–680
79. Tos M, Mogensen C (1979) Experimental surgery of the nose. Changes of the epithelium in the vestibular region in altered air-flow. Acta Otolaryngol (Stockh) 87:317–323
80. Tos M, Mogensen C (1979) Experimental surgery of the nose. Anteroposterior changes of the mucosa on altering the air-flow. Rhinology 17:215–225
81. Tos M, Mogensen C, Thomsen J (1977) Nasal polyps in cystic fibrosis. J Laryngol Otol 91:827–835
82. Tos M, Drake-Lee AB, Lund VJ, Stammberger H (1989) Treatment of nasal polyps. Medication or surgery and which technique. Rhinology 8[suppl]:45–49
83. Tos M, Larsen PL, Møller K (1990) Goblet cell density in nasal polyps. Ann Otol Rhinol Laryngol 99:310–315
84. Tos M, Svendstrup F, Arndal H, Ørntoft S, Jakobsen J, et al. (1998) Efficacy of an aqueous and a powder formulation of nasal budesonide compared in patients with nasal polyps. Am J Rhinol 12:183–189
85. Vancil ME (1969) A historical survey of treatments for nasal polyposis. Laryngoscope 79:435–445
86. Vleming M, Stoop AE, Middleweerd RJ, De Vries N (1991) Results of endoscopic sinus surgery for nasal polyps. Am J Rhinol 5:173–176
87. Woakes E (1885) Über nekrotisierende Ethmoiditis und ihre Beziehung zur Entwicklung von Nasalpolyppen. BMJ 4:701–705
88. Yonge ES (1907) Observations on the determining cause of the formation of nasal polypi. BMJ 12:964–968
89. Zuckerkandl E (1882) Normale und Pathologische Anatomie der Nasenhöhle und Ihrer Pneumatischen Anhänge. Braumüller, Wien

Sinusitis: Medical Management

Luc L.M. Weckx and Paulo A.L. Pontes

Introduction

The paranasal sinuses originate during the second month of embryonic life; at birth, there is a cleft for the maxillary sinus, and ethmoid sinuses are present. The floor of the maxillary sinus progressively descends and reaches the level of the nasal cavity by the age of 8 years, with complete development by age 15 years, while the ethmoid sinus increases in size until puberty. The sphenoid sinus begins to develop at age 2–3 years, and adult size is attained between 12 years and 15 years of age. The frontal sinus begins its development between the ages of 6 years and 8 years, growing slowly until reaching adult size between ages 15 years and 20 years [14].

It is estimated that 0.5–5% of upper-airway infections are complicated by acute sinusitis. Since the average adult has two to three colds per year and children have from six to eight colds per year, sinusitis is a common disease [16].

Under normal conditions, mucus secreted by the cylindrical and ciliated respiratory mucosa of the sinus is swept through the sinus ostium towards the nose by the action of the cilia, like a carpet in constant motion (Figs. 9.1, 9.2). Three key elements are important in maintaining the physiology of the paranasal sinuses: patency of the ostium, normal function of the cilia and mucus with correct chemical/physical characteristics.

The small caliber of the sinus ostia (2.5–3 mm diameter for the maxillary sinus and from 1 mm to 3 mm for the anterior and posterior ethmoid sinuses) facilitates easy and frequent obstruction. Predisposing factors for ostium obstruction include anything that causes mucosal edema [viral upper respiratory infection (URI), allergic inflammation, swimming] or mechanical obstruction (nasal septal deviation, nasal polyps or foreign bodies). As a consequence of ostium obstruction, negative intrasinus pressure develops, facilitating the entrance of nasal bacteria via sneezing, sniffing or blowing the nose. Another consequence is a decrease in oxygen pres-

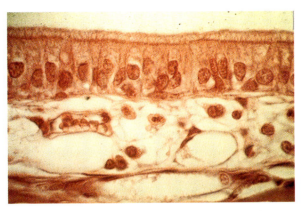

Fig. 9.1. Light photomicrograph of nasal mucosa, showing cylindrical and ciliated epithelium and capillary vessels of corium

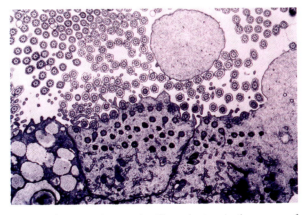

Fig. 9.2. Electron micrograph: cilia and a tear in the mucus of the respiratory mucosa

sure inside the sinus that favors multiplication of certain bacteria [6].

Factors that can impair *mucociliary transport* include cold, dry air, alterations in the composition of mucus, chemical products, systemic or topical medications and viral infections. Thus, a viral infection of the upper airways that causes a cytotoxic effect on the ciliated cells, thickens the mucus and impairs ciliary motion will decrease the protective function of the epithelium and facilitate a secondary infection of the nose and sinuses. In chronic sinusitis, there is a marked slowing of the mucociliary transport and a relationship between a higher degree of delay and surgical indication [18]. Advances in nasal endoscopy and computed tomography (CT) have expanded our current diagnostic and therapeutic possibilities and have changed some of the traditional concepts of sinus disease.

Etiology

It is currently accepted that, in most cases, maxillary sinusitis is secondary to impaired drainage and disease in the anterior ethmoid–middle meatus region (the ostiomeatal complex). Thus, the most common site of sinus pathology is the ethmoid sinus. However, drainage of the maxillary sinus is difficult, because the ostium is at a higher level than the floor. The principal causes of sinusitis are shown in Table 9.1 [32].

- *Acute viral rhinitis* (common cold) decreases ciliary activity and causes edema and obstruction of the ostium; organisms that normally inhabit the anterior portion of the nasal cavity rapidly proliferate and reach the sinus, where there is an accumulation of mucus. As a consequence, these organisms act as pathogens, producing secondary infections of the sinus.
- *Anatomic defects that affect the ostiomeatal complex* include nasal septal deviation, deformed uncinate process, concha bullosa, an abnormally large ethmoidal bulla and a lateralized middle turbinate.
- *Allergy* produces a chronic inflammatory process of the mucous membrane, with narrowing of the airways and ostia. Although many authors still regard this as a controversial cause, recent studies suggest a relationship between allergy and sinusitis. Hyperemia and an increased sinus metabolic activity after exposure to an allergen have been demonstrated and, in our series, children with allergic rhinitis exhibited radiological alteration of the sinuses in 87% of cases [12, 25, 29].
- *Adenoiditis*, adenoid hyperplasia and other conditions that produce infection or obstruction in the area (such as choanal atresia, foreign bodies and cleft palate) can predispose patients to the development of sinus infections, especially in children [16].
- *Swimming-pool-water contamination, diving and barotrauma* are common causes of sinusitis. The

Table 9.1. Sinusitis etiologies

- **Acute viral rhinitis**
  Anatomic defects in the ostiomeatal complex
  Deviation of the nasal septum
  Deformities in the uncinate process
  Concha bullosa
  Enlarged ethmoidal bulla
  Lateralized middle turbinate
- **Allergy**
- **Infections and neighboring obstructions**
  Adenoiditis
  Adenoid hyperplasia
  Choanal atresia
  Cleft palate
  Foreign body
- **Maxillofacial trauma**
- **Swimming pool water contamination, diving and barotrauma**
- **Inappropriate environmental conditions**
  Inadequate housing
  Climate changes
  Air conditioning
  Air pollution
  Smoking
- **Local irritants**
  Topical vasoconstrictor abuse
  Cocaine
- **Odontogenic factors**
- **More unusual causes of sinusitis associated with other infections**
  Cystic fibrosis
  Mucociliary disease
  Acetylsalicylic acid (aspirin) intolerance
  Immunodeficiency states

water that sometimes enters the sinus during swimming is generally not contaminated and is well tolerated by the cilia of the mucosa; however, chlorine and other impurities may be present in sufficiently high concentrations to cause chemical irritation. This is most common in children who stay in the water for many hours and jump in an upright position.
- *Barotrauma* leading to sinusitis is most common in individuals who fly in airplanes during an URI, but it also can occur after flying without an URI.
- *Maxillofacial trauma* often has an infectious component and can easily involve the sinuses secondarily.
- *Other conditions that locally or systematically favor the appearance of a cold* will increase the probability of secondary sinusitis. These include low resistance, crowded housing, exposure to epi-

demic viral infections, rapid changes in climate and air conditioning, pollution and smoking, and topical-vasoconstrictor abuse.
- *Odontogenic causes* include a periapical or periodontal abscess, commonly involving the first maxillary molar. Such infections may spread and destroy the bony lamina between the apex of the tooth and the floor of the maxillary sinus and may initiate maxillary sinusitis.
- The form of *cocaine* used for topical nasal use has an acid pH and is a local irritant. Its chronic use may cause ischemia of the mucoperichondrium, with necrosis and subsequent septal perforation, loss of sense of smell and taste, and sinusitis with osteitis and osteolysis. Additionally, the rebound vasodilatation and nasal obstruction that occur after prolonged cocaine abuse leads the patient to continue topical nasal-vasoconstrictor use and abuse, with increasing irritation and damage to the nasal mucous membrane [28].

When the sinus infection is recurrent (particularly when there are other infectious manifestations) other, more unusual causes should be considered.
- *Cystic fibrosis*, or mucoviscidosis, is a systemic, hereditary disease characterized by anomalies of the exocrine glands. Upper-airway involvement is manifested by chronic pansinusitis with accumulation of thick mucus, a high incidence of nasal polyps and cultures revealing *Pseudomonas aeruginosa*.
- *Congenital or acquired ciliary dysfunction* is a the result of ultrastructural abnormalities of the cilia, with loss of the coordination of ciliary beating. The cilia may be immotile, but this is not always the case; they may move, beating in an inefficient, uncoordinated manner. The condition is characterized by frequent episodes of sinusitis, recurrent otitis media, relapsing pneumonia and bronchiectasis [24].
- *Aspirin intolerance* may be part of a syndrome described more than 50 years ago, when it was noted that nasal polyps, sinusitis and asthma could be induced by the use of aspirin and nonhormonal anti-inflammatory drugs. Clinically, the nasal symptoms appear slowly and progressively [7].
- *Congenital and acquired immunodeficiencies* have increased in prevalence as causes of sinusitis because of acquired immunodeficiency syndrome (AIDS). Thirty percent of AIDS patients have some degree of sinusitis. The congenital immunologic deficiency states are less common. Many patients with serum immunoglobulin A (IgA) deficiencies have normal levels of secretory IgA and are asymptomatic. Currently, selective IgG deficiencies are more frequently diagnosed, though they may be manifested subclinically and are difficult to recognize. Undernourishment and deficient cellular immunity may also be causes of chronic sinusitis.

Clinical Picture

History

- **Acute sinusitis.** Most of the time, the patient has had a URI for 7–20 days, with yellowish or greenish, malodorous, unilateral or bilateral rhinorrhea accompanied by nasal obstruction and facial pain. Frequently, the pain becomes worse when the patient bends the head forward and down; the pain is also accentuated in the morning. There are some variations, depending on which sinus is affected.
    - **Frontal sinusitis.** Intense frontal headache with temporal or occipital radiation.
    - **Maxillary sinusitis.** Pain in the malar region, which radiates towards the frontal or maxillary-tooth region, where it is sometimes localized.
    - **Ethmoid sinusitis.** Localized pain between or behind the eyes, with radiation toward the temporal region worsened by visual effort.
    - **Sphenoid sinusitis.** Headache in the occipital region, (in the vertex or retro-orbital regions).
- **Chronic sinusitis.** The most frequent symptom is postnasal drainage of mucopurulent secretions; pain is less commonly found and, when present, is more commonly referred to mainly as a periorbital weight on the face. The possibility of chronic sinusitis should always be considered in patients with recurrent acute pharyngitis.
- **Sinusitis in children.** Sinusitis should be suspected when there is persistent cough, rhinorrhea more than 7–10 days fetid breath, facial tics or sniffing, recurrent otitis media or recurrent URIs. Usually, it is not the severity of the symptoms but the persistence that arouses suspicion [32].

Otorhinolaryngologic Examination

The presence of purulent secretion in the nasal fossae, mainly at the level of the middle meatus [15], suggests sinus infection; when the examination reveals polyps, it is usually chronic sinusitis. Most of the time, the mucous membrane is congested and edematous. Occasionally, there is a normal otorhinolaryngologic presentation in a patient with acute sinusitis and intense headache, an observation that can be explained by edema with obstruction of the sinus ostia.

The traditional inspection of the nasal cavity with frontal light and speculum provides a very limited view of the critical area (the ostiomeatal complex). With the appearance of endoscopes with rigid and flexible optical fibers, systematic nasal endoscopy is now a mandatory office procedure, with or without topical anesthesia, making possible the study of the middle meatus and abnormalities of neighboring structures that may narrow the meatus. Such procedures have changed both the criteria for surgical intervention and the surgical techniques. Nasal endoscopy is as dramatically superior to rhinoscopy as CT is to simple X-rays.

Diagnosis

In addition to history and otorhinolaryngologic examination, several tests may help to establish the diagnosis of sinusitis and its extent and nature.

1. *Nasal cytometry* is a microscopic study of the lining cells and white blood cells on the surface of the nasal mucous membrane (obtained by a smear of the surface of the membrane). The presence of a great number of neutrophils is indicative of an acute bacterial infectious process, while a higher number of lymphocytes suggests a chronic process. The finding of numerous goblet cells and eosinophils characterizes allergic processes [3, 12, 19].
2. *Diaphanoscopy* is the transillumination of the facial bones and cavities using an electric bulb at the infraorbital rim and observing the light transmitted through the hard palate into the oral cavity. It is only valid for maxillary and frontal sinuses and, due to its limitations and the need of an experienced examiner, it is now rarely used.
3. *Conventional radiography*, including frontal–nasal, submentovertex and lateral views, remains the most necessary test in the diagnosis of acute sinusitis. Generally, radiography discriminates quite well between healthy sinuses and sinusitis, but it is less effective in distinguishing the type of inflammatory process present inside the sinus. Complete opacification of one sinus may represent either an acute bacterial sinusitis with pus or a chronic sinusitis with polyps completely filling the sinus. This was illustrated by our study, in which the radiological aspects of 50 inflammatory sinuses were compared with the findings of an inferior-meatus maxillary puncture (Table 9.2).

This study showed that there was not always reliable correlation. For instance, the same patient had only slight thickening of the mucous membrane in one sinus on X-ray and had a moderate amount of pus on puncture, while the other sinus, which exhibited a more intense thickening, had no pus. However, all the sinuses with complete hypotransparency (cotton-like clouding of the whole sinus) or with total opacification had secretions on puncture [23]. When radiologic findings are compared with endoscopic findings, the agreement varies from 32% to 92%. It is important to interpret the X-ray carefully, remembering that a radiological shadow is not an anatomopathological diagnosis [10]. On sinusoscopy, sinuses with normal radiographs may exhibit slight to moderate alterations of the sinus mucous membrane, and an X-ray with slight alterations may be seen in a patient who has a normal presentation on sinusoscopy. At present, the greatest limitation of plain X-rays is the underestimation of disease in the ethmoid sinus (for example, in acute sinusitis, where the ethmoid sinus frequently shows minimal involvement on X-ray but is clinically involved). In children, there is need for even more caution in interpreting radiographs; 87% of allergic children have abnormal sinus X-rays. It is difficult to know if this represents purulent secretions that need antibiotic treatment or uninfected viscid mucus due only to the participation of the sinus mucosa in the allergic process. Radiological opacification of the sinuses during and after crying in children less than 1 year old is considered normal [1, 25].
4. *CT* is indicated in chronic sinusitis and when sinusitis recurs or persists after treatment with antibiotics, and should be performed after the period of acute infection. While simple conventional X-rays demonstrate maxillary and frontal disease, CT demonstrates the cells of the anterior ethmoid and the upper two thirds of nasal cavity and frontal recess, the areas where the inflammatory process begins. Inflammation extends to the maxillary and frontal sinuses only later. CT is also indicated when there are complications of sinusitis. The advantage of CT is its demonstration of the extent of the disease in the ostiomeatal complex, thus helping to determine the need for the route of surgery [11, 36]. CT findings must be correlated with the history and endoscopy findings, because Manning has shown that up to 30–40% of the population shows areas of mucous-membrane thickening on CT [20].
5. *Ultrasound* uses high-frequency sound waves inaudible to humans. The sound waves pass through the soft tissues and reflect 50–70% of

Table 9.2. Correlation between radiological aspects of maxillary sinusitis and the findings of maxillary puncture

Citrine	Clumps/catarrh	No secretions	X-ray appearance	Purulent (<1 cc)	Purulent (1–3 cc)	Purulent (>3 cc)
0	2	5	Slight thickening	0	1	1
0	1	6	Moderate thickening	2	2	0
2	1	2	Subtotal clouding	0	1	0
0	1	0	Global hypotransparency	0	2	1
0	1	0	Total opacification	1	5	7
0	1	1	Air–fluid	0	1	2
0	0	1	Cystic image	0	0	0

their intensity when reaching bone; 100% of the waves is transmitted through air. Its clinical utility is limited to the maxillary sinus, with its thin bony plate; in cases of radiological opacification, it may be possible to determine whether the opacification represents secretion, solid tumor or cyst. It is particularly useful in evaluating pregnant women with maxillary sinusitis [27].

6. *Rhinosinusal endoscopy* and *sinusoscopy* started at the beginning of the last century, when a cystoscope was introduced into the maxillary sinus through a buccosinusal fistula secondary to a dental excision. Sinusoscopy examines the nasal cavity and rhinopharynx under local anesthesia (general anesthesia in children). After careful examination of the areas that are difficult to see on rhinoscopy, examination of the maxillary sinus (antroscopy) is performed via an inferior meatus or canine fossa (in adults) puncture with cannula and trochar. Endoscopic exploration, in addition to providing the opportunity for direct visual examination to help determine if the cause of persistent radiological maxillary opacification represents pus, polyposis or mucous membrane fibrosis or thickening, also makes biopsy possible. It is thus possible to establish the diagnosis [4].

7. *Magnetic resonance imaging* does not employ ionizing radiation, and the images obtained are the result of the measurement of the different characteristics of each tissue. This technique is unnecessary for the diagnosis and management of most patients with sinus pathology, but it can be valuable in assessing the characteristics of tissue within the paranasal sinuses. For example, it is very useful in detecting fungal infections and can differentiate neoplastic processes from inflammatory diseases in 90% of cases [22].

Bacteriology

In acute sinusitis, the literature documents a predominance of *Streptococcus pneumoniae* and *Haemophilus influenzae*; there are greater numbers of anaerobic organisms in acute sinusitis of dental origin (especially in chronic sinusitis) [13, 15, 17, 35]. However, in 15–30% of patients, the cultures are negative, suggesting viral etiology.

Using cultures obtained by maxillary puncture, we studied 26 cases of *acute sinusitis* and *acute sinusitis superimposed on chronic sinusitis*. These results give a good picture of the bacteriology and are shown in Table 9.3 [23]. Twenty-seven percent of the cultures were negative. Of the positive cultures (73%), 61.5% featured anaerobic organisms (27% had only anaerobes and 34.5% were associated with aerobes), and 11.5% featured only aerobes. Thus, in more than 84% of the positive cultures, anaerobic organisms were found (frequently more than one organism per sinus).

In children with acute sinusitis, there is a predominance of *Moraxella catarrhalis* in addition to *S. pneumoniae* and *H. influenzae*. Most studies have not cultured anaerobic organisms in acute sinusitis. In one Brazilian study, maxillary puncture in children with sinusitis (clouding of the maxillary sinus) who underwent adenoidectomy or adenotonsillectomy revealed a predominance of *S. pneumoniae* and *Staphylococcus aureus* [1]. In diabetic and immunodepressed patients, it is important to consider the possibility of a mycotic etiology. In 1989, the classic concept that there is no correlation between cultures collected from the nose and cultures obtained from the maxillary sinuses was questioned by Clement et al. [5], who believe that the method of collection and transport may influence the results.

Table 9.3. Acute and chronic-turned-acute sinusitis: the frequency with which organisms are found on maxillary puncture in adults

Germs	Frequency
Anaerobic	
Fusobacterium sp.	8
Bacteroides sp.	5
Bacteroides fragilis	4
Bacteroides melaninogenicus	2
Peptostreptococcus sp.	2
Eubacterium sp.	1
Acinetobacter sp.	1
Peptococcus sp.	1
Actinomyces sp.	1
Aerobic	
Streptococcus pneumoniae	8
Staphylococcus aureus	4
Streptococcus pyogenes	2

Table 9.4. Antibiotic therapy in the treatment of sinusitis

Acute sinusitis
- **First choice**
  Amoxicillin
  Cephalexin
  Erythromycin + azithromycin
  Clarithromycin + roxithromycin
- **When there is therapeutic failure**
  Amoxicillin + clavulanate
  Cefaclor
  Axetil + cefuroxime
  Cefprozil
  Cefpodoxime

Chronic or recurrent sinusitis
  Doxycycline
  Amoxicillin + clavulanate
  Cefaclor
  Axetil + cefuroxime
  Cefprozil
  Cefpodoxime
  Associated (or not) with clindamycin, metronidazol

Ocular or intracranial complications
  Needing hospitalization
  Cefoxitin + ceftriaxone
  Associated (or not) with vancomycin + metronidazol

Management

In the treatment of sinusitis, it is important to treat the bacterial infection, to attempt to relieve the obstruction of the sinus ostia and to normalize the mucociliary transport function. In addition, it is important to eliminate any underlying factors, such as nasal septum deviation, adenoid hyperplasia, respiratory allergy and poor environmental conditions.

Medical Management

Antibiotic Therapy

In uncomplicated *acute* and *subacute sinusitis*, the drugs of first choice are oral, semi-synthetic penicillins, such as amoxicillin (50 mg/kg/day three times per day; Table 9.4). Cephalexin (50 mg/kg/day), erythromycin and the new macrolides (azithromycin, clarithromycin and roxithromycin) can also be administered. In children, sulfamethoxazole/trimethoprim (40/10 mg/kg/day) twice daily is also recommended. It is always important to remember that the rate of spontaneous resolution in acute sinusitis is approximately 60%. Treatment should be continued for 10–14 days.

In cases where there is *no adequate response* after first-choice antibiotics or when there is a clinical suspicion or laboratory confirmation of organism-produced penicillinase, the combination of amoxicillin and clavulanic acid (50 mg/kg/day, divided into three daily doses) or second- and third-generation cephalosporins – such as cefaclor (40/mg/kg, three doses per day), axetil–cefuroxime (10–25 mg/kg/day in two doses), cefprozil (15–30 mg/kg/day in two doses) and cefpodoxime – should be used for an additional 14 days.

In *chronic-turned-acute sinusitis* and *odontogenic sinusitis*, in which there are most likely both aerobes and anaerobes, the above antibiotics may be tried (amoxicillin–clavulanate or second- and third-generation cephalosporins), often in combination with drugs active against anaerobic organisms, such as clindamycin (10–20 mg/kg/day) or metronidazole (7 mg/kg/day). Doxycycline could also be administered, but only in adults and not in pregnant women. In terms of spectrum, the new fluoroquinolones, such as levofloxacin, moxifloxacin and gatifloxacin, seem to be very promising for chronic and recurrent sinusitis in adults, although there is still a lack of clinical studies. The duration of treatment should be at least 14–21 days. In patients with *ocular or intracranial complications of sinusitis*, hospitalization, CT and intravenous antibiotics, such as second-generation IV cephalosporins (cefoxitin) and third-generation cephalosporins (ceftriaxone, ceftazidime) with or without vancomycin and metronidazole, are indicated [8, 30, 31, 33].

Anti-Inflammatory Drugs

The use of anti-inflammatory drugs in sinusitis is recommended when there is evidence of obstruction of the sinus ostium (for instance, intense headache with little rhinorrhea on examination), in chronic-turned-acute sinusitis with polyps or evidence of a significant mucus-membrane thickening, or in patients with obvious allergic etiology. In these cases, the anti-inflammatory drugs that give the best response are oral corticosteroids (prednisone, β-methasone and dexamethasone) given for a period of 5–7 days.

Nasal Decongestants

Topical nasal vasoconstrictors can be used during the first four or five days of therapy; after this period, oral systemic decongestants should be used [9, 34]. However, it should be remembered that:
1. In addition to the beneficial effects of decongesting the nasal mucous membrane and promoting ostial drainage, topical vasoconstrictors also decrease cilary motility and nasociliary flow [2]
2. Simultaneous use of oral vasoconstrictors and antibiotics should be avoided, because the oral vasoconstrictors may impair adequate concentration of the antibiotics in the paranasal cavity
3. The decongestant effect of systemic vasoconstrictors on the sinus ostium is still controversial [21]
4. Long-term use of systemic decongestants tends to dry the secretions, because the antihistamines in these medicines increase the viscosity of the secretions due to of their anticholinergic action; this can delay drainage and impair sinus ventilation
5. The use of local nasal decongestants should not exceed four or five days, because extended use leads to rhinitis medicamentosa

Supportive Measures

The use of *mucolytic drugs* is indicated when nasal and sinus secretions are thick and difficult to eliminate. The use of dilution by maintaining adequate hydration and inhaling humidified air is preferred.

Puncture of the Maxillary Sinus

Transmeatal puncture of the maxillary sinus is indicated in cases resistant to clinical treatment, in cases of sinusitis in immunocompromised patients and in patients with intraorbital or intracranial complications or severe symptoms, such as headache. In addition to being a diagnostic method that can assess the local conditions of the sinus (ostium patency and character of the mucous membrane and its secretions), puncture can be therapeutic, permitting lavage of the maxillary sinus and instillation of medicines into the sinus cavity (Fig. 9.3). The combination of maxillary puncture and nasosinus endoscopy will achieve better evaluation and treatment of sinusitis [30].

Finally, the need to *investigate the possible underlying causes* of sinusitis, such as allergy, swimming-pool water, dental abscesses, the chronic use of topical nasal vasoconstrictors and others [26], can not be overstressed. Such investigation will decrease the need for frequent antibiotics and will prevent unnecessary surgical intervention.

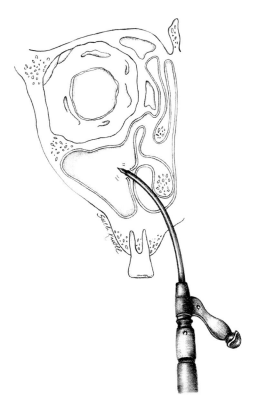

Fig. 9.3. Puncture of the maxillary sinus

References

1. Arruda LK, Mimiça IM, Solé D, Weckx LLM, Schoettler J, Heiner DCC, Naspitz CK (1990) Abnormal maxillary sinus radiographs in children: do they represent bacterial infection. Pediatrics 85:553–558
2. Bende M, Loth S (1986) Vascular effects of topical oxymetazoline on human nasal mucosa. J Laryngol Otol 100:285–302
3. Bryan WTK, Bryan MP (1959) Cytologic diagnosis in otolaryngology. Trans Am Acad Ophthalmol Otolaryngol 63:597
4. Castagno LA (1993) Nasal endoscopy in the ENT clinic. Rev Bras ORL 50:181–185
5. Clement PAR, Bijloos J, Kaufman L, Lauwers L, Maes JJ, Van der Veken P, Zisis G (1989) Incidence and etiology of rhinosinusitis in children. Acta Otorhinolaryngol Belg 43:523–543
6. Drettner B (1988) Therapeutical aspects of sinusitis in relation to pathogenesis. Acta Otolaryngol Suppl (Stockh) 458:13–16
7. English GM (1986) Nasal polypectomy and sinus surgery in patients with asthma aspirin idiosyncrasy. Laryngocope 96:374–380
8. Fairbanks DNF (1996) Pocket guide to antimicrobial therapy in otolaryngology: head and neck surgery, 8th edn. American Academy of Otolaryngology, Head and Neck Surgery, Alexandria
9. Falck B, Svanholm A, Aust R (1990) The effect of xylometazoline on the mucosa of human maxillary sinus. Rhinology 28:239–247
10. Gomes CC, Sakano E, Endo IH, Bilechi MMC, Lucchezi MC (1994) Chronic maxillary sinusitis. Comparative study between radiological and endoscopic exams. Evaluation of 104 patients. Rev Bras ORL 60:180–185
11. Grasel SS, Sanchez TG, Giardini L, Murano E, Almeida ER, Butugan O, Miniti A (1996) Comparison of plain radiographs and coronal CT scans in adults with recurrent chronic sinusitis. Rev Bras ORL 62:378–385
12. Gungor A, Corex JP (1997) Pediatric sinusitis: a literature review with emphasis on the rate of allergy. Otolaryngol Head Neck Surg 116:4–15
13. Gwaltney JM Jr, Sydnor A Jr, Sande MA (1981) Etiology and antimicrobial treatment of acute sinusitis. Ann Otol Rhinol Laryngol 90:68–71
14. Isaacson G (1996) Sinusitis in childhood. Pediatr Clin North Am 43:1297–1318
15. Karma P, Jokipii L, Sipila A, Luotonen J, Jokipii AMM (1979) Bacteria in chronic maxillary sinusitis. Arch Otolaryngol Head Neck Surg 105:386–390
16. Lee D, Rosenfeld RM (1997) Adenoid bacteriology and sinonasal symptoms in children. Otolaryngol Head Neck Surg 116:301–307
17. Lundberg C, Carenfelt C, Engquist S, Nord CE (1979) Anaerobic bacteria in maxillary sinusitis. Scand J Infect Dis Suppl 19:74–76
18. Lusk RP, Wolf G (1992) Pathophysiology of chronic sinusitis. In: Lusk RP (ed) Pediatric sinusitis. Raven, New York, pp 7–13
19. Mangabeira Albernaz PL, Ganança MM, Toseti GEA, Guerra CCC (1967) O citograma nasal: sua importância na orientação terapêutica das sinusites. Rev Paul Med 70:211–220
20. Manning SC, Biavati MJ, Philips DL (1996) Correlation of clinical sinusitis signs and symptons to imaging finding in pediatric patients. Int J Pediatr Otorhinolaryngol 37:65–74
21. Melen I, Friberg D, Andreasson L, Ivarsson A, Jannert M (1986) Effects of phenyl propanolamine on ostial and nasal patency in patients treated for chronic maxillary sinusitis. Acta Otolaryngol Stockh 101:494–500
22. Moore J, Potchen M, Siena A, Waldenmaier N, Potchen EJ (1986) High-field magnetic resonance imaging of paranasal sinus inflammatory disease. Laryngoscope 96:267–271
23. Moreira CA, Fernandez H, Nunes CTA, Trabulei LR, Weckx LLM (1983) Estudo bacteriológico em sinusites maxilares. Rev Assoc Med Bras 29:218–219
24. Mygind N, Pedersen M, Nielsen MH (1983) Primary and secondary ciliary dyskinesia. Acta Otolaryngol (Stockh) 95:681–694
25. Naspitz CK, Solé D, Hilário MOE, Weckx LLM, Iorio MCM, Azevedo MF, Ledermann H, Ajzen J (1986) Estudo clínico laboratorial de crianças portadoras de rinite alérgica e alterações radiológicas dos seios da face. Rev Bras Alergia Immunol 8:9–31
26. Parsons DS (1996) Chronic sinusitis. A medical or surgical disease? Otolaryngol Clin North Am 29:1–10
27. Revonta M (1980) Ultrasound in the diagnosis of maxillary and frontal sinusitis. Acta Otolaryngol (Stockh) 370:1–55
28. Schweitzer VG (1986) Osteolytic sinusitis and pneumomediastinum: deceptive otoryngologic complications of cocaine abuse. Laryngoscope 96:206–210
29. Shapiro GG (1985) Role of allergy in sinusitis. Pediatr Infect Dis 4:55–58
30. Stankievicz J, Osguthorpe JD (1994) Medical treatment of sinusitis. Otolaryngol Head Neck Surg 110:361–362
31. Sydow C, Axelsson A, Jensen C (1986) Acute maxillary sinusitis: a comparison between 27 different treatment modes. Rhinology 20:223–229
32. Wald ER (1986) Epidemiology, pathophysiology and etiology of sinusitis. Pediatr Infect Dis J 4[suppl 6]:51–53
33. Wald ER (1992) Medical management of sinusitis in pediatric patients. In: Lusk RP (ed) Pediatric sinusitis. Raven, New York, pp 71–75
34. Weckx LLM (1983) Vasoconstritores nasais tópicos. Rev Bras Med 40:294–296
35. Winther B, Vickery CL, Gross CW, Hendley JO (1996) Microbiology of the maxillary sinus in adults with chronic sinus disease. Am J Rhinol 10:347–350
36. Zinreich JS (1990) Paranasal sinus imaging. Otolaryngol Head Neck Surg 103:863–868

Biomechanical Effects of the Ethmoidal Isthmi on the Pathogenesis of Chronic Paranasal Sinusitis

Malte Erik Wigand

Introduction

Only a few decades ago, the causes of chronic respiratory mucositis – either chronic otitis media or sinusitis – were ascribed to constitutional defects of the affected individuals (Aschoff, Wittmaack). Contemporary literature has confirmed that some occur due to the uncommon disorders mucoviscidosis or immotile cilia syndromes; however, this does not adequately explain why so many people develop chronicity or polyposis from acute sinusitis, but the majority does not. Since an experimental model of human chronic, progressive, polypoidal sinusitis has not yet been developed (except for a few investigations of maxillary sinusitis in rabbits), our knowledge of its pathogenesis is based on clinical observations. These investigations suggest that, while a variety of factors may influence the resolution (or non-resolution) of acute sinusitis, biomechanical factors are among the most prominent. This hypothesis is strongly supported by the observations that: (1) paranasal surgery with mechanical remodeling of the related anatomy often stops the disease and restores physiological function to an abnormal mucosa and (2) nasal trauma, which also mechanically changes the anatomy, frequently induces chronic sinusitis.

Many observers since Zuckerkandl, Hajek and Killian have accepted the concept that the sites of predilection of ascending chronic inflammation are narrow outlets of the paranasal sinus system, especially the middle meatus and the ethmoidal infundibulum, which Flottes has called the ostiomeatal complex [2]. In addition, endonasal endoscopic surgery was developed to remove such narrow straits within the mucociliary transport system and was termed isthmus surgery [6]. It is also true, however, that individual anatomic variations and postoperative tissue reactions create isthmi in many regions of the nasal and paranasal structures beyond the ethmoidal infundibulum. Accordingly, endoscopic sinus surgery should follow the principles of isthmus surgery in practically all areas [7] (not just in the middle nasal meatus, which had been the focus of surgical measures by Messerklinger and his advocates [5]).

The following paragraphs will describe the typical isthmic configurations (Table 10.1) that may need to be altered during and after sinus surgery in order to cure chronic sinusitis. The extent of the pathology will dictate how many of them need to be removed. These recommendations are based on the personal experiences of the author.

Basic Features of Endonasal Isthmus Surgery

When reviewing the endoscopic anatomy of the nose and paranasal sinuses of a patient with chronic sinusitis, one can recognize many narrow passages that may contribute to obstruction of ventilation and mucociliary clearance. Though the ethmoid sinus is of primary interest for the origin of paranasal sinusitis, anatomical abnormalities of the large cavities and the nasal meatuses must also be considered (Table 10.1). It must be emphasized that the removal of bottleneck configurations is to be accomplished without damage to the lining mucosa. Bone removal is thus carried out by sharp, cutting instruments, avoiding unnecessary laceration of the adjacent mucosa. This principle of isthmus surgery can be adapted to the individual extent of pathology by various types of intervention (Table 10.2).

Flanking Nasal Surgery

The nasal pyramid and the nasal septum, if deformed by growth or trauma, may induce severe disturbances of the meatal air stream. Septal corrections, and sometimes rhinoplasty, are then prerequisites for success. Septoplasty in conjunction with ethmoidectomy is advantageous for three reasons:

Table 10.1. Typical isthmic configurations of the ethmoid

Middle nasal meatus below a thick middle turbinate
Semilunar hiatus and the ethmoidal infundibulum
Entrance of the anterior ethmoidal artery
Frontal recess with the naso-frontal duct
Primary maxillary ostium in the form of a tunnel
Postoperative laterofixation of the middle turbinate
Postoperative synechia between the ground lamella and naso-antrostomy

Table 10.2. Endoscopic operations of the paranasal sinuses

Uncovering of the ethmoidal infundibulum (infundibulectomy)	
Anterior partial ethmoidectomy	
Posterior partial ethmoidectomy	
Total ethmoidectomy, antero-posterior, postero-anterior dissection	
Supraturbinal naso-antrostomy	with or
Infraturbinal naso-antrostomy	without
Frontal sinusotomy	intracavity
Sphenoidal sinusotomy	surgery
Combinations (pan-sinus operation)	
Combinations with external approaches	

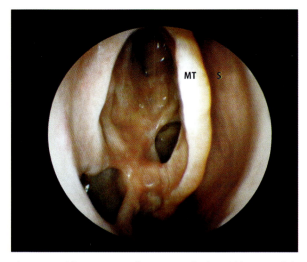

Fig. 10.1. Wide exposure of an operated ethmoid by remodeling of the middle turbinate with partial resection of the ground lamella. *MT*, middle turbinate; *S*, septum

enhanced ventilation, improved exposure of the surgical field and less painful postoperative treatment.

The middle turbinate is one of the important determinants of isthmic pathology in the middle nasal meatus. Its base, especially if bullous and curved laterally, can narrow the ethmoidal infundibulum considerably. For the past few years, I have systematically remodeled its shape and trimmed its ground lamella (Fig. 10.1). These measures help to medialize the middle turbinate and widen the narrow isthmus. This kind of turbinoplasty allows excellent exposure of both the posterior ethmoid and the sphenoid and makes resection of the posterior third of the middle turbinate unnecessary. If well executed, this technique of turbinal remodeling respects the full height of the turbinal body, thus preserving all of the olfactory rim while definitively opening the middle nasal meatus.

The Ethmoidal Labyrinth

The normal anatomy of the ethmoidal labyrinth has many narrow communications via tunnels and ostia, which, in cases of infectious mucosal swelling, can induce a vicious cycle of retention and inflammation. The semilunar hiatus and the infundibulum are especially delicate structures that affect the reversibility and chronicity of a process. The success of their surgical widening by infundibulotomy, with subsequent recovery of the mucosa, has proven that biomechanical conditions play an important role in chronic ethmoiditis. The same is true for two other ethmoidal isthmi. One is the physiological bottleneck at the entrance of the anterior ethmoidal artery. This vessel often pulls the medical orbital wall into the ethmoid sinus, thus narrowing the distance to the base of the middle turbinate (Fig. 10.2). This anterior ethmoidal isthmus is a site of predisposition for obstructive scar formation and for iatrogenic injuries of the orbital wall.

Another critical place, where initial central ethmoiditis may produce stenosis and subsequent frontal sinusitis, is the frontal recess of the infundibulum (Fig. 10.3). This variable cell, from which a naso-frontal communication often starts, is frequently an indicator of the extent of polypoidal proliferation. If it is clean, the more anterior compartments will be healthy. If it is obstructed, one may expect polyps in the fronto-ethmoidal transition, and an elective frontal sinusotomy may be indicated.

The Frontal Sinus

The naso-frontal duct forms an isthmus that can either promote or cause frontal-sinus mucositis. Its shape is quite variable, and can be identified as a slit into an ethmoidal cell, a hole or a tunnel; it is often

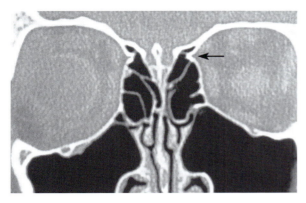

Fig. 10.2. Physiological isthmus of the anterior ethmoid. Note the entrance of the anterior ethmoidal artery in a computed tomography scan (*arrow*)

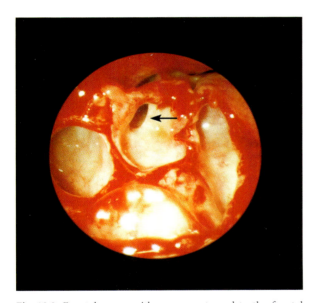

Fig. 10.3. Frontal recess with a narrow tunnel to the frontal sinus infundibulum (*arrow*)

particularly deformed by a protruding internal nasal spine [3]. The possibility that anatomic abnormalities are causing sinusitis is strongest when repeated recurrences occur. Surgical widening of the nasofrontal duct can be extremely difficult [1], but enlargement to a diameter of 4 mm will almost always cure the disease.

The Sphenoid Sinus

The isthmus of the sphenoid sinus is its natural ostium, which opens into the superior nasal meatus. It is unique, because it is not covered by overhanging cells, unlike at the frontal and maxillary sinuses. The anterior wall of the sphenoid sinus should be surgically removed during a pansinus operation so that one third of the opening communicates with the nasal cavity and two thirds communicate with the posterior ethmoid (the attachment of the posterior tip of the middle turbinate separates the two parts like a pillar). This broad communication avoids an isthmic configuration, which may account for the fact that recurrences of chronic sphenoiditis are rare. A second important step is to remove the intrasphenoid material so that intracavity stenoses cannot form.

The Maxillary Sinus

Chronic maxillary sinusitis is the most frequent indication for endoscopic sinus surgery. We know from the broad application of computed tomography imaging before surgery that chronic maxillary sinusitis almost always coincides with anterior or central ethmoiditis, and this supports the assumption that inflammatory stenosis of the ethmoidal infundibulum is the initial trigger of maxillary mucositis. The primary ostium, which provides the mucociliary clearance of the cavity, is often a channel several millimeters long and is of a typical isthmic configuration. If swelling is great enough to cause functional disturbances, intense decongestion of this area may help abort early maxillary sinusitis.

For no other "new window" do so many recommendations exist. For the supraturbinal naso-antrostomy (above the inferior turbinate), the establishment of an oversize window is neither possible nor necessary. The curved orbital floor and the attachment of the inferior turbinate are vertical limitations; the sphenopalatine groove and the nasolacrimal duct delineate the limits of the horizontal margins. In my opinion, a window of moderate size, e.g., 4–6 mm in diameter, is satisfactory. However, I always try to preserve the original natural ostium and to place the new window caudal to it. Two other features can help prevent the formation of a new critical isthmus. They are valid for both supraturbinal and infraturbinal windows:

1. The new opening should not have angles of less than 45°, because narrower angles have a tendency to allow reclosure of the window by shrinkage of the almost-parallel collagenous fibers.
2. Even more important is the need for immediate epithelization of the new window. The best way to accomplish this is via the eversion technique [6], which transposes maxillary mucosa into the middle nasal meatus. When this is performed, reclosure of the window is extremely unlikely.

Development and Prevention of Postoperative Isthmi

After any type of sinus surgery, there are opportunities for the formation of isthmic structures within the remodeled sinus system. Isthmic structures are formed by scar-tissue formation or new bone growth in vulnerable areas, initiated by granulation tissue or by the presence of blood clots or secretions. They have the same pathogenetic effect on the recurrence or persistence of chronic, hyperplastic sinusitis as the original anatomic isthmus. Therefore, postoperative care with removal of these masses from the tunnels and cavities during the exudative and proliferative phases of wound healing [4] is mandatory. Some sites of predilection for such postoperative structures are the anterior ethmoid sinus, the naso-frontal communication and the supraturbinal naso-maxillary window. The middle turbinate plays a prominent part in all three cases:

1. The anterior ethmoid sinus is the most frequent site of recurrent sinusitis and polyposis. The space between the middle turbinate and the lateral nasal wall (agger nasi, medial orbital wall) can be bridged easily by scar tissue if the opposing wound surfaces are denuded and the space is narrow, especially if the middle turbinate is lateralized (Fig. 10.4).
2. The narrow naso-frontal communication (duct or window) is particularly vulnerable to fibrous obstruction if its walls are denuded of their mucosal lining. A tiny opening may persist but, functionally, it behaves as an obstructed isthmus and does not allow free passage of air and secretions. This is the typical morphological substrate of postoperative frontal headache. A definitive cure can be obtained by surgical enlargement to a diameter of at least 6 mm.
3. A third frequent site of postoperative obstruction by fibrous adhesions is the narrow cleft between the posterior end of the ground lamella and the postero-inferior margin of the new naso-antrostomy (Fig. 10.5). Care must be taken to establish a 4- to 8-mm space between both structures during the initial surgery and to keep the isthmus open during the productive phase of wound healing. Other postoperative isthmi may develop but are less frequent. For the prevention of the recurrent obstructions from uncontrolled scar formation, an appropriate remodeling of turbinates and ethmoidal isthmi is very important.

Summary

Endoscopic sinus surgery following the principle of enlarging narrow, bony openings, windows and tunnels (with preservation of most of the chronically diseased mucosa lining them) has successfully replaced the older technique (radical exenteration of the paranasal cavities). It has now been established that isthmic structures can exert a pathogenetic influence on their mucosal lining. Co-factors (such as infection, localized edema of the mucosa and impaired drainage of secretions) lead to obstruction and a vicious cycle of chronicity. An intimate know-

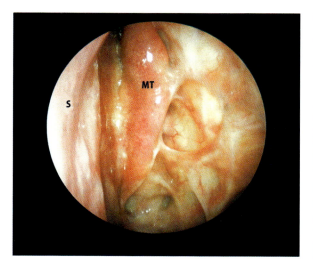

Fig. 10.4. Typical postoperative isthmus by laterofixation of the middle turbinate occluding the frontal sinus. *MT*, middle turbinate; *S*, septum

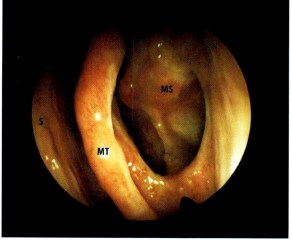

Fig. 10.5. Obliteration of the ethmoidal isthmus between the ground lamella and the postero-inferior edge of a naso-antrostomy. *MS*, maxillary sinus; *MT*, middle turbinate; *S*, septum

ledge of the common isthmic formations within the sinus system and of the common sites of postoperative development of secondary isthmi is, therefore, a prerequisite for successful surgical management of chronic sinusitis, including the severe forms of polyposis.

References

1. Draf W (1992) Endonasale mikro-endoskopische Pansinusoperation bei chronischer Sinusitis. III. Endonasale mikro-endoskopische Stirnhöhlenchirurgie. Otorhinolaryngol Nova 2:118–125
2. Flottes L, Clerc P, Riu R, Devilla F (1960) La physiologie des sinus. Librairie Arnette, Paris
3. Gross R, Hosemann W, Göde U (1995) Anatomische Studie zur endonasalen Stirnhöhlenchirurgie. HNO Informationen 20:38
4. Hosemann W, Wigand ME, Göde U, Länger F, Dunker I (1991) Normal wound healing of the paranasal sinuses: clinical and experimental investigations. Eur Arch Otorhinolaryngol 248:390–394
5. Messerklinger W (1978) Endoscopy of the nose. Urban and Schwarzenberg, München
6. Wigand ME (1990) Endoscopic surgery of the paranasal sinuses and anterior skull base. Thieme, Stuttgart
7. Wigand ME, Steiner W, Jaumann MP (1978) Endonasal sinus surgery with endoscopic control: from radical operation to rehabilitation of the mucosa. Endoscopy 10:225–260

Pre-operative Staging, Grading and Post-Operative Follow-Up of Chronic Inflammatory Sinus Disease

Valerie J. Lund

Introduction

The diagnosis and surgical management of chronic rhinosinusitis has been greatly enhanced by the combination of rigid endoscopy and computed tomography (CT) scanning. However, in rhinology, it has proved remarkably difficult to quantify and qualify our patients' diseases and their responses to treatment. Although a number of objective tests of nasal-airway and mucociliary clearance exist, they have not entered routine clinical practice for a variety of reasons. There is clearly a need for some universally acceptable method of assessing disease extent; this might allow comparison of results between one centre and another. The acceptance of such a method would largely depend upon its ease of application.

Whilst it is known that symptom severity and appearances on nasal endoscopy do not necessarily correlate with extent of disease on sinus CT, the latter investigation is performed in a comparable fashion in virtually all patients undergoing endoscopic sinus surgery. As a consequence, most staging, scoring or grading systems have largely been based on the CT appearances. These systems may be broadly divided into: those that use the distribution of opacification within the sinuses to define four stages (Tables 11.1–11.4) [2, 4, 6, 7], those using a numerical score for each sinus group (Tables 11.5, 11.6) [5, 10] and one that utilises actual measurement of mucosal thickness (Table 11.7) [13]. Another system uses five stages derived from an overall score based on scores for site, surgery, polyps, infection and immune status, which is by far the most complex (Table 11.8) [3]. A number of attempts to compare inter- and intra-observer agreement for some of these systems have been made. Gliklich and Metson [4] compared the systems of Kennedy [6], Friedman et al. [2], Levine and May [7] and their own Harvard method; perhaps not surprisingly, they found their own system to be superior to the others. Kennedy [6], in a monograph on prognostic factors and outcomes, concluded that factors like allergy, asthma and aspirin sensitivity were not significant when the radiographic extent of disease was taken into account. This view was supported when Friedman and Katsantonis [1] com-

Table 11.1. The staging system of Friedman et al. [2]

Stage 0	Normal
Stage I	Single-focus disease (involving a single focus or sinus unit)
Stage II	Multifocal disease (includes bilateral or multiple areas of disease that are not confluent or are diffuse throughout the ethmoid labyrinth, bilateral middle-meatal polyps)
Stage III	Diffuse disease (extensive bilateral involvement of multiple sinuses) without bony changes
Stage IV	Diffuse disease associated with bony changes

Table 11.2. The staging system of Kennedy [6]

Stage 0	Normal
Stage I	Anatomic abnormalities
	All unilateral sinus disease
	Bilateral disease limited to ethmoid sinuses
Stage II	Bilateral ethmoid disease with involvement of one dependent sinus
Stage III	Bilateral ethmoid disease with involvement of two or more dependent sinuses on each side
Stage IV	Diffuse sinonasal polyposis

Table 11.3. The staging system of Levine and May [7]

Stage 0	Normal
Stage I	Disease limited to ostiomeatal complex
Stage II	Incomplete opacification of one or more major sinuses (frontal, maxillary, sphenoid)
Stage III	Complete opacification of one or more major sinuses, but not all
Stage IV	Total opacification of all sinuses

Table 11.4. The staging system of Gliklich and Metson [4]

Stage	
Stage 0	Normal (<2 mm of mucosal thickening on any sinus wall)
Stage I	All unilateral disease or anatomic abnormality
Stage II	Bilateral disease limited to ethmoid or maxillary sinuses
Stage III	Bilateral disease with involvement of at least one sphenoid or frontal sinus
Stage IV	Pansinusitis

Table 11.5. Worksheet for the scoring system of Jorgensen [5]

Structure	Left	Right
Frontal-sinus opacification[a]		
Maxillary-antrum opacification[a]		
Anterior ethmoidal-labyrinth opacification[a]		
Posterior ethmoidal-labyrinth opacification[a]		
Sphenoid-sinus opacification[a]		
Maxillary-antrum polyp[b]		
Hiatus-semilunaris occlusion[c]		
Maxillary-sinus ostium occlusion[c]		
Frontal-recess occlusion[c]		
Ethmoidal-infundibulum occlusion[c]		

[a] Score for opacification: none=0, mild=1, moderate=2, marked=3, complete=4
[b] Score for size of maxillary sinus polyp: none=0, small=1, medium=2, large=3
[c] Score for occlusion: none=0, mild=1, moderate=2, complete=3

Table 11.6. Worksheet for the radiological grading of sinus systems [9]

Sinus system	Left	Right
Maxillary (0, 1, 2)		
Anterior ethmoid (0, 1, 2)		
Posterior ethmoid (0, 1, 2)		
Sphenoid (0, 1, 2)		
Frontal (0, 1, 2)		
Ostiomeatal complex (0 or 2 only)		
*Total points		

For all systems except the ostiomeatal complex, 0=no abnormalities, 1=partial opacification, 2=total opacification
*For the ostiomeatal complex, 0=not occluded, 2=occluded

pared the results of their own system with those of Kennedy [6] and Lund and Mackay [9]. A more recent study by Oluwole et al. [14] compared the systems of the Jorgensen [5], Levine and May [7], Lund and Mackay [9] and Newman et al. [13], finding that the Lund-Mackay system facilitated the highest level

Table 11.7. Worksheet for the grading system of Newman et al. [13]

Structure	Left	Right	Left and/or right
Maxillary sinus[a]			
Frontal sinus[a]			
Sphenoid sinus[a]			
Ethmoidal sinus[b]			
Ostiomeatal complex[c]			
Nasal passages[c]			

[a] Mucosal thickening scores: 0 (0–1 mm); 1 (2–5 mm); 2 (6–9 mm); 3 (>9 mm)
[b] Mucosal thickening scores: 0 (0 mm); 1 (0–1 mm); 2 (2–3 mm); 3 (>3 mm)
[c] Degree of obstruction scores: 0 (none); 1 (mild); 2 (partial); 3 (complete)

of both intra- and inter-observer agreement in the evaluation of chronic rhinosinusitis.

The Lund-Mackay Scoring System

The Lund-Mackay scoring system (Tables 11.5, 11.9–11.13), which is based on a simple numerical score derived from CT scans, was used for a number of years prior to its publication [9]. After some minor modifications, it was included in the supplement resulting from the International Conference on Sinus Disease [8] and, more recently, it was recommended by the task force of the American Academy of Otorhinolaryngology, Head and Neck Surgery to form the basis for further outcome research.

Outline of Lund-Mackay Scoring System

Demographic information is collected (Table 11.9), including nasal diagnosis, which is classified as follows:
1. Chronic rhinosinusitis
2. Acute recurrent rhinosinusitis
3. Nasal polyposis
4. Miscellaneous

The last group includes frontoethmoidal mucocoeles, repair of cerebrospinal fluid leaks, orbital decompressions, dacrocystorhinostomy and all other extended applications of endoscopic sinus surgery. Systemic diagnosis might include asthma (with or without aspirin sensitivity), cystic fibrosis, primary abnormalities of mucociliary clearance (primary ciliary dyskinesia, Young's syndrome), immune defi-

Table 11.8. Stages of surgical sinus disease [3]

Stage 0	Score=0; no surgical sinus disease
Stage I	Score<1.3
Site	Inflammation limited to the ostiomeatal complex area
Surgery	No prior sinus/nasal surgery except septoplasty and/or inferior meatal antrostomies
Polyps	No polyps or localised to <10% of the sinus space
Infection	Well-controlled infection with no active mucopurulent drainage
Immunology	No underlying immunologic disease except well-controlled allergy
Stage II	Score=1.3–2.3
Site	Inflammation confined to the maxillary/ethmoid/ostiomeatal areas
Surgery	Prior Caldwell-Luc or polypectomy
Polyps	Polyp disease, with involvement of 10–50% of the nasal/sinus cavities
Infection	Persistent, localised infection with some active purulent drainage
Immunology	Low-grade immune disorder or fair allergy control
Stage III	Score=2.3–4
Site	Pansinus involvement (unilateral or bilateral); isolated sphenoid disease
Surgery	Prior anterior ethmoidectomy/middle-turbinate surgery
Polyps	Nasal/sinus polyposis filling more than 50% of the nasal and sinus cavities
Infection	Poorly controlled multisinus infection with active mucopurulent drainage; active fungal sinusitis
Immunology	Poorly controlled allergic rhinitis or significant immune disorder; history of long-term steroid treatment
Stage IV	Score>4
Site	Sinus disease with extranasal/sinus extension, orbital or intracranial; frontal disease above the nasofrontal duct
Surgery	Prior complete ethmoidectomy or sphenoidectomy
Polyps	Inverting papilloma or other potentially malignant nasal/sinus neoplasms
Infection	Osteomyelitis or infection eroding into the orbit or cranium; mucormycosis
Immunology	End-stage immunologic disease; profoundly immunocompromised patient

Table 11.9. Demographic information collected for the Lund-Mackay scoring system [9]

Last name
First name
Sex
Date of birth
Age
Hospital no.
Operation
Operation date
Surgeon
Nasal diagnosis (0–4)
Systemic diagnosis
General or local anaesthetic
Duration (min)
Postoperative medication
Complications

Table 11.10. Worksheet for radiological grading of anatomic variants in the Lund-Mackay scoring system [9]

Anatomic variant	Left	Right
Absent frontal sinus		
Concha bullosa		
Paradoxical middle turbinate		
Everted uncinate process		
Haller cells		
Agger nasi cells		
Total points		

Scoring: 0=no variant, 1=variant present

Table 11.11. Worksheet for surgery scores in the Lund-Mackay scoring system [9]

Surgery	Left	Right
Uncinectomy		
Middle meatal antrostomy		
Anterior ethmoidectomy		
Posterior ethmoidectomy		
Sphenoidectomy		
Frontal recess surgery		
Reduction of the middle turbinate		
Total points each side		

Score: 0=no procedure done, 1=surgery done. The maximum score is 14 (7 each side)

ciency, bronchiectasis, sarcoidosis and other conditions such as diabetes mellitus or multiple myeloma, which might be relevant to the development of infection. It may be of interest to record additional information on smoking history, allergic status and previous and present medication and/or surgery.

The *scoring system* has been devised based on CT-scan appearances (Table 11.5), the scans generally having been performed after an adequate trial of medical treatment, though what this constitutes

Table 11.12. Worksheet for symptom scores in the Lund-Mackay scoring system [9]. Determine the scores by the visual-analogue method

Symptom	Pre-operation	After 3 months	After 6 months	After 1 year	After 2 years
Nasal blockage or congestion					
Headache					
Facial pain					
Problems of smell					
Nasal discharge					
Sneezing					
Overall					
Total points					

Score each category 0–10 according to the degree of symptom severity, with 0=symptoms not present and 10=great severity

Table 11.13. Worksheet for endoscopic appearance in the Lund-Mackay scoring system [9]

Characteristic	Baseline	After 3 months	After 6 months	After 1 year	After 2 years
Polyp, left[a]					
Polyp, right[a]					
Oedema, left[b]					
Oedema, right[c]					
Discharge, left[c]					
Discharge, right[c]					
Postoperative scores (to be used for outcome assessment only)					
Scarring, left[b]					
Scarring, right[b]					
Crusting, left[b]					
Crusting, right[b]					
Total points					

[a] Scoring: 0=absence of polyps; 1=polyps in the middle meatus only; 2=polyps beyond the middle meatus
[b] Scoring: 0=absent; 1=mild; 2=severe
[c] Scoring: 0=no discharge; 1=clear, thin discharge; 2=thick, purulent discharge

may be the subject of some debate. Each sinus group is graded between 0 and 2 (where 0=no abnormality, 1=partial opacification and 2=total opacification).

The sinus groups are comprised of the maxillary, frontal, sphenoid, anterior ethmoid and posterior ethmoid sinuses. As it is difficult to apply this gradation to the ostiomeatal complex, it is simply scored as 0 (not obstructed) or 2 (obstructed). A total score between 0 and 24 is possible, and each side can be considered separately (score range for each side = 0–12).

Various *anatomic variants* are also noted as being present (1) or absent (0), but they do not contribute to the sinus score; these variations include absent frontal sinuses, concha bullosa, paradoxical middle turbinates, Haller cells, everted uncinate process and agger nasi pneumatisation (Table 11.10). The estimate of disease extent relies on the CT appearances, but it may be of interest to quantify other aspects of the disease and its treatment. Thus, a *surgical score* (0 if not performed, 1 if undertaken) may be derived, which renders a maximum score between 0 and 14 (0–7 for each side) and allows a quantification of the operation (which may, if desired, be correlated with other parameters; Table 11.11).

Symptoms are assessed by the patient on a visual analogue score (VAS) of 0–10 (where 0 means no symptom is present and 10 is the most severe rating) for nasal obstruction or congestion, headache, facial pain, sense of smell, nasal discharge and sneezing. This is combined with an overall symptomatic assessment (Table 11.12). This method is well established in the evaluation of rhinological patients [12], but it is also of interest to ask patients to prioritise their three worst symptoms; this can distinguish the relative importance of symptoms given the same scores and does not always equate with the VAS. The symptom score is evaluated pre-operatively and at regular intervals post-operatively.

The endoscopic appearance of the nose is also quantified on a 2-point basis for the presence of polyps (0=none, 1=confined to the middle meatus, 2=beyond the middle meatus), discharge (0=none, 1=clear and thin, 2=thick and purulent), oedema, scarring, and adhesions and crusting. These appearances are assessed pre-operatively and at regular post-operative visits, but they are not included in the staging system. The results for both the symptoms and endoscopic appearances are recorded on a relational database (Table 1.13).

Discussion

All such attempts at grading disease are open to criticism, but it is the complexity of many of the other systems that have prevented them from entering routine clinical practice. No account of the degree of partial opacification is taken. This avoids interobserver variation, and the interpretation of the scan requires no formal radiological training. The scan represents a snapshot at one moment in time, and it is generally stated that the scan should be performed after *adequate* medical treatment and *reasonably* close to the time of the surgical procedure. Most authors have been careful to avoid defining these terms, though I would regard 2–3 months as a reasonable rule of thumb in each case.

The scoring of a scan where no frontal sinus is apparent is also an area of compromise. It actually occurs in less than 1% of the Caucasian population and could be overcome by determining a percentage (i.e. out of 11 rather than 12). This would significantly decrease the user-friendliness of the method and, as a consequence, an absent frontal sinus is simply scored as zero. Similarly, previous surgery (such as a Caldwell-Luc) may lead to permanently thickened mucosa, which may be fibrotic rather than inflammatory. Notwithstanding this, in the absence of pathological confirmation, the sinus should be scored as previously described.

Due to the problems of interpretation posed by previous surgery, the occasional natural absence of the frontal sinus and the difficulty of differentiating between opacification due to inspissated mucus and mucosal inflammation (which may have prognostic significance), the reliance on CT appearances for defining the extent of disease is by no means ideal. It is also well known that the correlation between symptoms and disease extent is generally poor. Notwithstanding these criticisms, CT does offer one of the few assessments of rhinosinusitis that can be objectively quantified and which the vast majority of patients undergo in routine clinical practice. As simplicity is the key to general utility, we have recommended a method that has proved easy to apply and reproduce and which seems highly suited to the validation of outcomes in large clinical studies.

Post-Operative Follow-Up

The major drawback with all such attempts at quantifying disease extent is that patients suffer from symptoms and, in most countries, it would be regarded as unethical to perform post-operative CT scans in the absence of significant symptoms. Consequently, most of the larger follow-up studies have relied on semi-quantitative assessment of symptomatology with or without endoscopic appearances. Objective tests of nasal function, whilst offering important validation of therapeutic response, nonetheless remain largely research tools [11]. A few studies have compared pre- and post-operative scans [12] but, unsurprisingly, the correlation with symptoms remains poor. However, it can be stated that if the surgical cavity appears to be macroscopically healed, the chances of relapse of chronic rhinosinusitis are small. When VASs after 1 year are compared with those after 3–4 years of follow-up, the results remain the same or are improved in 78% of cases [10].

Conclusion

Clearly, we need methods of estimating the severity of disease that are readily available to all clinicians and are easy to use. At present, these largely depend on imaging, which we know correlates poorly with endoscopic appearances and clinical symptoms; however, imaging at least provides some tangible evidence of disease.

References

1. Friedman WH, Katsantonis GP (1994) Staging systems for chronic sinus disease. Ear Nose Throat J 73:480–484
2. Friedman WH, Katsantonis GP, Sivore M, Kay S (1990) Computed tomography staging of the paranasal sinuses in chronic hyperplastic rhinosinusitis. Laryngoscope 100:111–1165
3. Gaskins RE (1992) A surgical staging system for chronic sinusitis. Am J Rhinol 6:5–12
4. Gliklich R, Metson R (1994) A comparison of sinus computed tomography (CT) staging systems for outcomes research. Am J Rhinology 8:291–297

5. Jorgensen RA (1991) Endoscopic and computed tomographic findings in ostiomeatal sinus disease. Arch Otolaryngol Head Neck Surg 117:279–287
6. Kennedy DW (1992) Prognostic factors, outcomes and staging in ethmoid sinus surgery. Laryngoscope 57[suppl]:1–18
7. Levine H, May M (eds) (1993) Rhinology and sinusology. Thieme, New York, p 261
8. Lund VJ, Kennedy DW (eds) (1995) Quantification for staging sinusitis. Ann Otol Rhinol Laryngol Suppl 167:17–21
9. Lund VJ, Mackay IS (1993) Staging in rhinosinusitis. Rhinology 107:183–184
10. Lund VJ, Mackay IS (1994) Outcome assessment of endoscopic sinus surgery. J R Soc Med 87:70–72
11. Lund VJ, Holmstrom M, Scadding GK (1991) Functional endoscopic sinus surgery in the management of chronic rhinosinusitis: an objective assessment. J Laryngol Otol 105:832–835
12. Mantoni M, Berner B, Larsen PL, Orntof S, Hansen H, Winther B, Tos M (1995) Pre and post-operative evaluation of functional endoscopic sinus surgery by coronal CT. In: Tos M, Thomsen J, Balle V (eds) Rhinology: a state of the art. Kugler, Amsterdam, pp 285–288
13. Newman LJ, Platts-Mills TAE, Phillips DC, Hazen KC, Gross CW (1994) Chronic sinusitis relationship of computed tomographic findings to allergy, asthma and eosinophilia. JAMA 271:363–367
14. Oluwole M, Russell N, Tan L, Gardiner Q, White P (1996) A comparison of computerized tomographic staging systems in chronic sinusitis. Clin Otolaryngol 21:91–95

Surgical "Grading" System for Inverting Papilloma

Aldo Cassol Stamm

Introduction

The sinonasal papilloma is a benign neoplasm of the mucosal epithelium [4]. Nasosinus papillomas were divided into three histopathological types by the World Health Organization in 1991 [11]: inverting papilloma (IP), exophytic papilloma and columnar papilloma.

The IP (also known as Schneiderian papilloma, Ewing's papilloma, transitional cell papilloma and papillary sinusitis) usually arises from the lateral nasal wall into the middle meatus, with local extension to the paranasal sinuses and neighboring structures [1]. The etiology of IP is uncertain but is thought to involve allergy, tobacco, environmental carcinogens and viruses, especially human papilloma virus [1, 2, 4].

IP has been reported to occur in all age groups, but it is most common during the fifth to seventh decades of life [3]. The male-to-female ratio ranges from 3:1 to 5:1 [5]. The most common symptoms are nasal obstruction, epistaxis and nasal discharge. Vrabec [14] has attributed three characteristics to the IP tumor: a tendency to recur, a destructive capacity and a propensity to be associated with malignancy.

Traditional treatment of IP is surgical removal, and various approaches have been recommended. Vrabec [14] recommends lateral rhinotomy in all cases of IP, independent of the size of the lesion. Rouvier [10] treats IP via different approaches according to the location of the lesion (lesions in the ethmoid and sphenoid sinuses are treated with an endonasal approach; lesions in the maxillary sinus are treated via sublabial access; lesions in the frontal sinus are treated via an external approach). When multiple sinuses are involved, combined procedures can be used. A local resection has been advocated by Lawson et al. [7] when the IP lesion is confined to the lateral nasal wall, with limited extension into the ethmoid sinus and maxillary antrum. Kennedy et al. [6] believe that an endonasal endoscopic approach is the treatment of choice both in cases in which the IP lesion is small and can be easily removed with a margin of normal tissue and for treatment of recurrent disease. In some areas that are difficult to visualize endoscopically (such as the maxillary sinus), combining an endoscopic and external approach is a reasonable consideration. Extension of the lesion to the frontal sinus is probably a contraindication for an endoscopic approach. In this situation, an external approach (osteoplastic approach or external fronto-ethmoidectomy) is preferable [6]. IP with anterior cranial fossa extension and dural penetration is very rare and represents an extremely aggressive form. Miller et al. [8] recommended a craniofacial resection, and radiotherapy should be considered in selected cases.

According to the location of the IP lesion, a surgical grading system has been proposed [12, 13]. This grading system helps the surgeon choose the appropriate surgical approach and will standardize the reporting of the end results according to the endoscopic, computed-tomography and magnetic-resonance findings.

Material and Methods

A retrospective review revealed that 36 patients with sinonasal papillomas were diagnosed and treated between 1982 and 1999 at the Ear, Nose and Throat Center of São Paulo. Three patients had exophytic papilloma, one had columnar cell papilloma and 32 patients had IP and are the subject of this report. The age at diagnosis ranged from 15 years to 72 years, and the average age overall was 52 years. There were 28 men and four women. The main symptoms of the IP cases were nasal obstruction (80%), epistaxis (15%) and nasal discharge (20%). Three patients also had nasal polyposis, and one patient had squamous cell carcinoma (Table 12.1). The treatment was surgical in all 32 cases, and the operation selected varied according to the surgical grading system.

Classification (Surgical Grading System)

Many surgical approaches to remove IPs (including lateral rhinotomy, midfacial degloving, external ethmoidectomy, Caldwell-Luc, transnasal endoscopic techniques and craniofacial resection) have been described. The surgical removal should be based on the location and extent of the neoplasm. Our surgical grading system for IP is based on endoscopic, computed-tomography and magnetic-resonance findings (Table 12.2, Fig. 12.1) [12, 13].

- **Grade I.** Involvement of the middle meatus (ethmoid sinus), the medial aspect of the maxillary sinus, the sphenoid sinus and the nasal septum. The surgical approach recommended in this grade is the transnasal micro-endoscopic technique.
- **Grade II.** Involvement of the same areas as in grade I, plus lateral extension of the IP into the maxillary sinus. A midfacial degloving approach with micro-endoscopic technique is recommended.
- **Grade III.** Involvement of the all paranasal sinuses unilaterally with or without extension to uncommon sites [such as the nasopharynx (Fig. 12.2), orbit, soft tissues of the cheek, dura attachment or other regions]. In this grade, we recommend combining transnasal and external approaches (such as lateral rhinotomy) using the operating microscope.
- **Grade IV.** This grade is very uncommon and indicates involvement of the intra-cranial space or intradural extension of the lesions. A cranio-facial approach with the operating microscope is recommended.

According to our surgical grading system, the 32 cases of IP were distributed as illustrated in Table 12.3.

Table 12.1. Sinonasal papillomas during the period 1982–1999 (*n*=36)

	n	%
Inverted papilloma	32	89
Exophytic papilloma	3	8
Columnar papilloma	1	3
Inverted papilloma/polyposis	3	8
Inverted papilloma/carcinoma	1	3

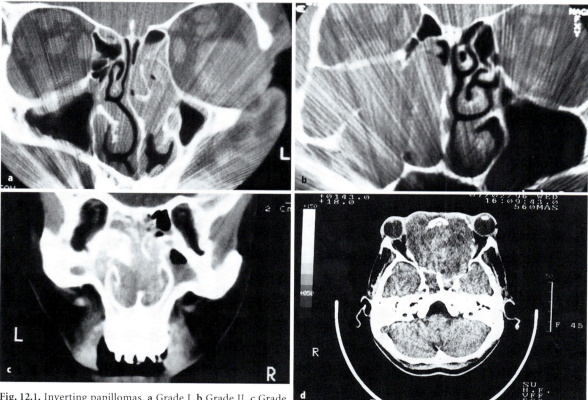

Fig. 12.1. Inverting papillomas. **a** Grade I. **b** Grade II. **c** Grade III. **d** Grade IV (courtesy of Unicamp, São Paulo)

Table 12.2. Surgical grading system for inverting papilloma [12, 13]

Grade	Location	Surgical approach
I	Middle meatus; ethmoid/maxillary (medial)/sphenoid sinuses; nasal cavity	Transnasal (microscope/endoscope)
II	Same as in grade I + maxillary sinuses (lateral)	"Degloving" (microscope/endoscope)
III	All paranasal sinuses; nasal cavity; uncommon sites[a]	Combined procedures (microscope)
IV	Intracranial extension	Cranio-facial approach (microscope)

[a]Nasopharynx, orbit, supra-orbit cells, soft tissues of the cheek, dura attachment and other locations

Table 12.3. Inverting papilloma: distribution according to the surgical grading system during the period 1982–1999 (n=32)

Grade	n	%
I	6	19
II	22	69
III	4	12
IV	–	–

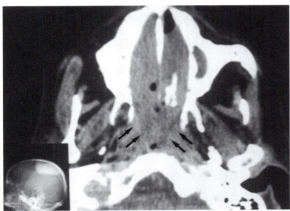

Fig. 12.2. Grade-III inverting papilloma with extension to the nasopharynx (*arrows*)

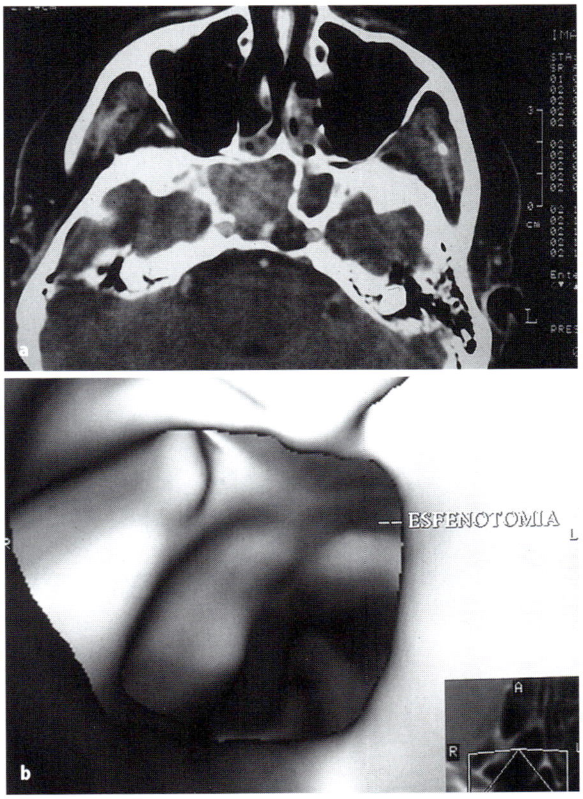

Fig. 12.3. a Grade-I inverting papilloma envolving the ethmoid and sphenoid sinuses. b Postoperative virtual endoscopic image showing no residual tumor after endoscopic surgical resection

Treatment and Results

According to Fu [4], the nasosinus papillomas have a propensity to involve the underlying seromucous glands and bone at multiple sites. The aims of IP surgery are to remove the lesions with the underlying bone attachment (subperiosteal dissection), to leave the orbital periosteum if possible, to sacrifice the naso-lacrimal duct if necessary and to create a wide-open cavity [6].

In our 32 IP cases, six (grade I) were operated on using a transnasal micro-endoscopic technique (Fig. 12.3). In two cases, recurrence was observed, and they were re-operated using the same technique. Waitz-Wigand [15] reported an incidence of 17% recurrence in their experience with endoscopic techniques.

Twenty-two patients (grade II) were treated using a midfacial degloving approach with an operating microscope and using an angulated endoscope to visualize areas around the corners (Figs. 12.4, 12.5). In four patients, recurrence of IP was identified in the new cavity. They were treated using an endonasal micro-endoscopic approach (one patient was treated three times and two patients were treat-

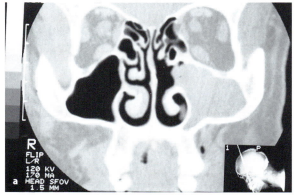

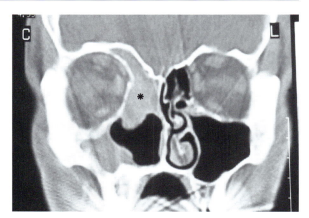

Fig. 12.6. Recurrence of an inverting papilloma at the ethmoid sinus and frontal recess after an initial midfacial degloving procedure (*asterisk*)

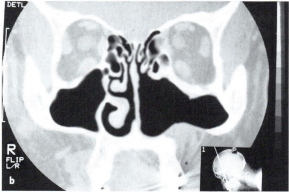

Fig. 12.4. a Grade-II inverting papilloma. **b** Post-operative aspect after a midfacial degloving approach

Table 12.4. Recurrence of inverting papilloma (range=5 years; $n=32$)

Grade	n	Number of recurrences	%
I	6	2	33
II	22	4	18
III	4	1	25
IV	-	-	-
Total	32	7	22

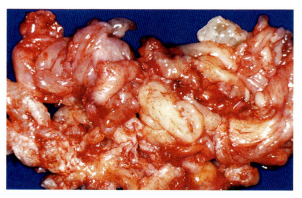

Fig. 12.5. Surgical specimen removed en bloc

ed two times; Fig. 12.6). Dolgin et al. [3] noted recurrence in two of nine patients (22%) treated with midface degloving.

Four patients of our series had grade-III IP. Two were treated using lateral rhinotomy and two were treated with external fronto-ethmoidectomy plus an endonasal approach. All these patients were operated on using an operating microscope. One patient with carcinoma was treated with cranio-facial surgery associated with radiotherapy. Myers et al. [9] noted 4% recurrence in 26 patients treated with lateral rhinotomy with en bloc excision of the lateral nasal wall, and Buchwald et al. [2] observed 50% recurrence after using a similar technique. No patient was classified as grade IV in our study. Miller et al. [8] reviewed the English literature from 1962 to 1992 for the presence of intracranial extension and dural invasion of IP, and only five cases demonstrated extension into the anterior cranial fossa. Our recurrence rate (including all grades) was 22% (Table 12.4).

Surgical Complications

Potential complications include cerebrospinal-fluid leak, diplopia, epiphora, dacryocystitis, lid edema, epistaxis, mucoceles, crusting, etc. In 32 patients with IP, our surgical complications were: epiphora [2], epistaxis [1], frontal mucocele [2], maxillary mucocele (Fig. 12.7) [1]; nasal crusting, infraorbital paresthesia, sinus-cavity infection and synechiae were observed in 15% of the patients.

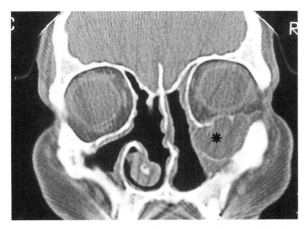

Fig. 12.7. Post-operative maxillary mucocele (*asterisk*) after midfacial degloving. This patient had nasosinus polyposis in the opposite side and was treated with a microendoscopic technique

Conclusions

1. We propose the above surgical grading system to help the surgeon choose the appropriate approach for each case. This may lead to a decreased recurrence rate and decreased surgical morbidity.
2. The use of surgical microscopes and telescopes for endonasal or external approaches and the use of a subperiosteal dissection increases the likelihood of a complete removal of the lesion and a lower recurrence rate.
3. Use of such a surgical grading system will allow better comparisons of surgical complications, recurrences and final results among different surgeons.
4. Follow-up examination with nasal endoscopy and CT scanning is imperative.

References

1. Batsakis JG (1979) Tumors of the head and neck, 2nd edn. Williams and Wilkins, Baltimore, pp 132–137
2. Buchwald C, Franzmann MB, Tos M (1995) Sinonasal papillomas: a report of 82 cases in Copenhagen county, including a longitudinal epidemiological and clinical study. Laryngoscope 105:72–79
3. Dolgin SR, Zaveri VD, Cassiano RR, Maniglia AJ (1992) Different options for treatment of inverting papilloma of the nose and paranasal sinuses: A report of 41 cases. Laryngoscope 102:231–236
4. Fu YF (1995) Histopathology of inverted papilloma and surgical implications. Am J Rhinology 9:75–76
5. Hyams VJ (1971) Papillomas of the nasal cavity and paranasal sinuses: a clinicopathological study of 315 cases. Ann Otol Rhinol Laryngol 80:192–206
6. Kennedy DW, Keogh B, Senior B, Lanza DC (1996) Endoscopic approach to tumors of anterior skull base and orbit. Oper Tech Otolaryngol Head and Neck Surg 7:257–263
7. Lawson W, Le Benger J, Som P, et al. (1989) Inverting papilloma: an analysis of 87 cases. Laryngoscope 99:1117–1124
8. Miller PJ, Jacobs J, Roland T Jr, et al. (1996) Intracranial inverting papilloma. Head Neck 18: 450–454
9. Myers EN, Fernau JL, Johnson JT, et al. (1990) Management of inverted papilloma. Laryngoscope 100:481–490
10. Rouvier P (1996) Personal communication
11. Shanmugaratnan K, Sobin LH (1991) Histological typing of tumors of the upper respiratory tract and ear. Springer, Berlin Heidelberg New York, pp 20–21
12. Stamm A (1996) Surgical grading system for inverting papilloma. 16th Congress of the European Rhinologic Society. Isian, Ghent, p 235
13. Stamm A (1996) Surgical grading system for inverting papilloma. In: McCafferty G, Coman W, Carrol R (eds) 16th World Congress of Otorhinolaryngology, Head and Neck Surgery. Monduzzi, Sydney, pp 1423–1427
14. Vrabec DP (1994) The inverted Schneiderian papilloma: a 25-year study. Laryngoscope 104:582–605
15. Waitz G, Wigand ME (1992) Results of endoscopic sinus surgery for the treatment of inverted papillomas. Laryngoscope 102:917–922

Teaching and Learning in Endonasal Sinus Surgery

Rainer Keerl, Rainer Weber, and Wolfgang Draf

Introduction

In medicine, educational problems arise while learning and using new or improved surgical techniques and during the repetition of technical procedures. Although endonasal sinus surgery is a standard procedure in managing most sinus problems, many ear, nose and throat (ENT) surgeons did not practice it during their education [12]. However, the challenging anatomy of the paranasal sinuses and the difficult problem of choosing the best operation require special and intensive teaching and practice [1, 2, 5, 9, 10, 13, 16, 17, 20, 21, 25, 26].

Learning Curve

Before we discuss teaching, we have to introduce the concept of the so-called learning curve. When learning a new technique, (1) the surgical goal may not be achieved (success rate) or (2) complication may be frequent during the first operations (complication rate).

After a certain amount of experience, the number of complications declines. Stankiewicz [22] demonstrated his learning curve by discussing complications occurring in endoscopic intranasal ethmoidectomies he performed. In his first 53 patients, he had 21 complications, but he had only seven in his next 127 patients. A more subtle example of a learning curve is presented by Hughes [7] concerning stapedectomy; this article shows his personal continuous development in terms of the success rate. He performed 50 stapedectomies before his success became reliable (criteria: an air–bone gap smaller than 10 dB) in 80% of the patients. The aims of teaching are to have surgical success beginning with the first patient and to have a low complication rate, because "accepting poorer results than the experts can achieve is a disservice for the patient, the surgeon and the institution" [24].

We examined the way doctors in training develop into experienced surgeons regarding sinus surgery by analyzing their individual learning curves. In a retrospective study, we evaluated 1000 endonasal pansinus operations [including 444 revisions (44%)] carried out by four different ENT surgeons training in the city hospital of Fulda, Germany. We found a dura lesion six times (0.6%), found an opening of the periosteum of the orbit in 30 cases (3%) and encountered bleeding from anterior ethmoidal artery 36 times (3.6%). No other complications occurred.

The complications were managed as follows. All dura lesions were closed successfully via the endonasal route without any further complications. When a lesion of the periosteum of the orbit occurred, further manipulation of this area was avoided. Except for lid ecchymosis, no other orbital sequelae arose. Ointment inside the nose (using paraffin as basic medium) should be avoided to prevent a long and unhappy history of paraffin granuloma on the face [8]. Bleeding of anterior ethmoidal arteries did not lead to further problems, because the surgeons coagulated the artery.

Surgeon 1 ($n=177$ procedures) encountered three lesions of dura during his first ten operations. The surgeon then changed the direction of surgery. Prior to the change, he searched his way along the skull base from the posterior wall of frontal sinus to the sphenoid sinus; after the change, he worked ventrally from the roof of sphenoid sinus. Regarding the orbit, no significant trend was evident. Assistance during surgery was rarely available.

Surgeon 2 performed 186 surgical procedures. During the first 30, he needed help in 20%; one dura lesion occurred during the first 30 procedures. From the 60th operation onward, the rate of periosteum lesions rose, until it dropped to a minimum after 100 operations. Injuries to the anterior ethmoid artery were rare.

Surgeon 3 ($n=333$ procedures) needed help with two early cases and with the 190th operation, because he feared that the internal carotid artery had been injured. Fortunately, it was only severe bleeding from the sphenopalatine artery. There was no dura lesion. Damage to the periosteum of the orbit occurred in waves and was less frequent with every hundred cases. The ethmoidal artery was coagulated quite often.

Surgeon 4 (304 procedures) had a very low rate of complications. One lesion of the periosteum and one dura injury occurred at the beginning. He had help or supervision during his first 100 procedures, and the percentage of cases requiring revision did not rise to the normal level until after the first 100 operations.

Complications were not distributed equally (Fig. 13.1), but the learning process proceeded in three stages, which we describe in terms of a traffic light. There is a "red" phase (procedures 1–30) with a high risk of injury to the dura, resulting in a CSF leakage. Next is a "yellow" phase (approximately procedures 30–100) with a high incidence of opening of the periorbital periosteum but less risk of dural damage. One possible explanation is that the surgeon in this stage of his career tends to be more "aggressive" toward the disease without having adequate anatomical orientation. A "green" phase (procedure 100 onward) characterizes the trained surgeon. This learning curve is correct for surgeons who monitor their actions intensively and try to avoid complications, even at the price of not removing the entire disease. As litigation shows, unpleasant and even deadly complications can occur to a very experienced surgeon.

How Can We Improve Education?

Important topics in understanding and learning endonasal sinus surgery includes knowledge of (or from):

- Theory (physiology and pathology, basic knowledge of procedure, different techniques)
- Anatomy
- Computed tomography (CT)
- Assistance during surgery
- Attending surgical courses
- Interactive multimedia teaching software

Theory

For the inexperienced surgeon, it is indispensable to acquire a broad theoretical knowledge by studying basic texts concerning sinus surgery, physiology and pathology [1, 2, 4, 10, 11, 16, 18, 20–23, 27]. In addition to gaining familiarity with complications and pitfalls, he will learn general rules, guidelines for the indications for surgery and the extent of the appropriate surgery.

Anatomy

The key to sinus surgery is to know the secrets of nasal and paranasal-sinus anatomy. Cadaver dissections are highly recommended [19]. In this way, the surgeon will become familiar with important anatomical landmarks. After this anatomical training, his intraoperative orientation will be much better, beginning with the first real operation. Hillen [6] invented a new technique of studying anatomy by developing an interactive compact disk (CD), "Paranasal Sinuses and Anterior Skull Base" (for technical requirements, see the appendix). To perform an operation, the surgeon needs a three-dimensional mental image of the surgical site. For this reason, he must know the positional relationship

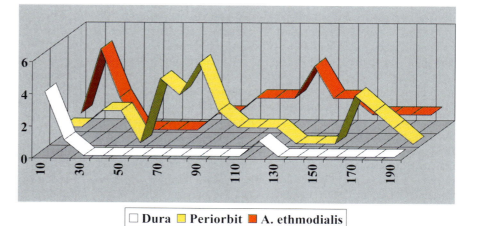

Fig. 13.1. Relative rate of complication depending on surgical experience ($n=1000$; the numbers *1–3* indicate when each surgeon participated in the operations)

between the surgical field and (for example) the optic nerve and ocular muscles. This may be achieved by preparing these structures in cadaver dissections but, after this, the surrounding structures are destroyed and are not accessible for further preparation.

Therefore, Hillen cut the head of a 57-year-old woman into 78-μm-thick slices in the coronal plane. These cuts were digitized with the help of specialized software. Sagittal and transverse planes were reconstructed. To this, CT and magnetic-resonance imaging (MRI) pictures of a 47-year-old volunteer were added, as were Hillen's measures of the cadaver data. The user can switch interactively between the different anatomical planes, the corresponding CT and MRI pictures and the original histological cut. By touching unknown structures with the mouse pointer, the corresponding anatomical descriptions appear. The most important part of the program is the so-called video animation. It allows the user to study three-dimensional anatomy at different magnifications and from different directions, which helps him to achieve a complex picture of the anatomy.

Computed Tomography

Just as knowledge of anatomy is crucial for sinus surgery in general, CT is the key to the analysis of the individual patient and his personal anatomy, anatomic variants and disease. Which questions should the radiologist answer for the surgeon? First, the extent of the disease should be explained. As our analysis of the learning curve shows, the shape of the skull base, the lamina papyracea and its possible variations and the position of anterior ethmoidal artery are all of great importance. However, the positions of the lacrimal drainage system, infraorbital nerve and Haller cells and the distance between the uncinate process and the lamina papyracea, roots of the teeth, ocular muscles, optic nerve, carotid artery, hypophysis, the natural opening of the sphenoid sinus and other important structures also have to be considered and studied in detail.

What kind of CT should we use? It is absolutely necessary to get an overview of the preoperative pathology, the anatomy and especially its variations. There may be dangerous traps for the surgeon. Our examination consists of the following points. First, transverse cuts of the paranasal sinuses are performed parallel to the infraorbitomeatal line, with a slice thickness of 2 mm high in the ethmoid sinuses and with 4-mm cuts in all other areas. Next, a coronal reconstruction is made. Eventually, the results are documented in the bony and soft-tissue structures. In computing the coronal plane from the axial cuts, problems of tooth artifacts are avoided. As experience shows, these are most often found in the area of the posterior ethmoid cells and the sphenoid sinus. Another argument for primary axial cuts is that important structures (such as the lamina papyracea, optic nerve and the internal carotid artery) are more easily seen in this plane [3].

Tomographic Imaging

A new concept for the visualization of two- and three-dimensional pictures has been developed by Zeppelin company (for technical details, see appendix). After loading the digital data from the CT or MRI computer into a personal computer (PC), tomographic imaging (TIM) allows the surgeon to analyze the two- or three-dimensional reconstructions in different planes in his office or operation theater. Furthermore, interesting structures, such as tumors, may be colored for better visualization. From this data, the program reconstructs a three-dimensional structure, which is presented in relationship to surrounding structures. A three-dimensional animation or stereo three-dimensional picture enhances this impression. In our opinion, these presentations of imaging data have several advantages. A user-friendly software package on a PC platform gives the surgeon a thorough evaluation of diagnostic CT and MRI scans without time loss. Brightness and magnification of the picture may be modified on the computer monitor. For special indications, complex anatomical and pathological presentations may be computed within minutes in order to outline an individual approach to the disease.

The CD-ROM "Endonasal Sinus Surgery"

Consistent with the above-mentioned requirements, the structure of our multimedia system for further surgical training in pansinus operations consists of the following components:
1. A title screen leading to five chapters
   I. Introduction
   II. A video surgery demonstration
   III. CT atlas
   IV. Anatomy
   V. General complications

2. Two structural trees (Figs. 13.2, 13.3) give a detailed overview of the chapters of interest.

Introduction

The introduction provides general information concerning indications, complications and techniques and provides preoperative explanations for the patient. A special section, "Program Service", explains the structure and the rules for using the compact disc. A subchapter, "Useful Requirements", contains instruments and pharmaceutical products for perioperative management.

Surgical Demonstration

The surgical demonstration begins with presentation of the patient with reference to the clinical and CT findings and consists of four corresponding parts:
1. The original surgical video. Demonstration of a complete pansinus operation using a microscope and endoscope. For didactic reasons, it is subdivided into 13 steps.
2. A revised surgical video (a surgical demonstration including graphic animation and detailed explanations).
3. Complications. With reference to each step of the operation, a complete description of possible complications (including their prevention and treatment) is provided (Fig. 13.4).
4. CT. For each of the 13 steps of the operation, the corresponding CT of the involved anatomical region (axial, coronal and sagittal) is displayed.

It is recommended that the CD be used in the following way. First, the video section should be viewed. One can stop at any time with the "stop" button and can move further with the "continue" button. It is possible to restart the sequence from the beginning with the "start" button. Furthermore, it is possible to view the CT scan relevant to the surgical step or to view the corresponding chapter concerning complications.

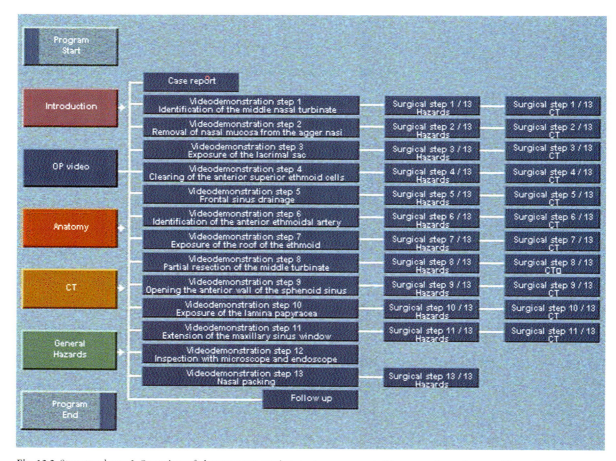

Fig. 13.2. Structural tree I. Overview of chapters concerning anatomy, computed tomography and general hazards

CT Atlas

One can obtain an orientation to the special anatomy of this region with the CT atlas (Fig. 13.5); all CT pictures (axial, coronal, sagittal) of the operated patient are available.

Anatomy

In the section concerning anatomy, the topographic anatomy of the nose and sinuses is demonstrated (Fig. 13.6). Its relevance to operations is highlighted with the help of graphics, anatomical preparations (kindly placed at our disposal by Prof. Lang, emeritus director of the Anatomical Institute, University of Wuerzburg) and CT images.

General Complications

In addition, a special chapter dealing with the anatomical structures at risk (optic nerve, internal carotid artery) is provided. The option "Index" contains an alphabetical list of all relevant terms; it is possible to access the corresponding chapter immediately. Furthermore, the option "Notebook" allows one to make personal notes for each chapter.

What Are the Advantages of this New Teaching Device?

The actual technique of endonasal surgery, the basic anatomy (with important variations), the individual's anatomy (as shown by the preoperative CT

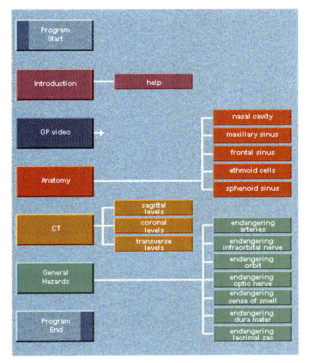

Fig. 13.3. Structural tree II. Overview of video sequences of the CD-ROM

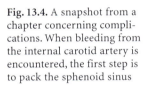

Fig. 13.4. A snapshot from a chapter concerning complications. When bleeding from the internal carotid artery is encountered, the first step is to pack the sphenoid sinus

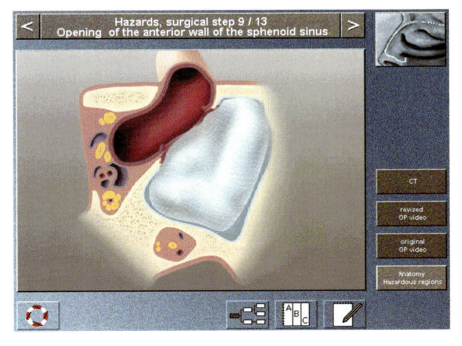

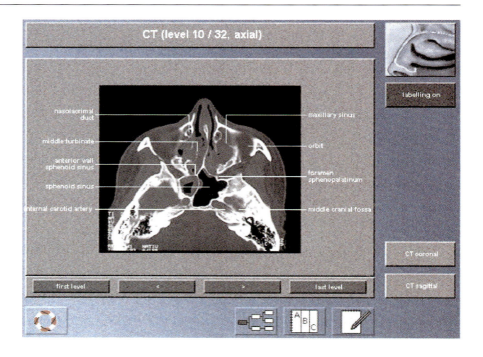

Fig. 13.5. One axial picture from a radiological paranasal-sinus atlas, with explanations

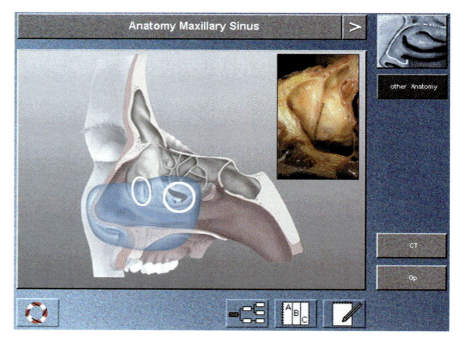

Fig. 13.6. Detail of an anatomical chapter concerning the maxillary sinus

examination) and the pathology must constantly be considered by the surgeon. The individual steps of the operation must be tailored individually so that the intraoperative situation and the CT findings correspond.

The experienced surgeon is subconsciously aware of the relevant CT anatomy and operative information. This is of the utmost importance in preventing complications and completing the task of removing pathological tissue. The inexperienced surgeon must learn to compare information from various sources and must learn to integrate, process and translate the information into a conclusive operative concept.

The multimedia system can be of definite help in the process of gaining experience because of the

CD's immediate and discrete interactive approach to the individual learning steps. The CD is programmed to train the user to perform the subconscious mental switching from one point of view to another.

The following advantages can be attained only with the multimedia system:
- Faster and targeted approach to details
- Combined work with more colleagues (group work)
- Repetition of demonstration at the user's preferred speed
- Dynamic demonstration of surgery (live video)

Furthermore, the interactive services of the system lead to active learning, thus heightening the learning effect.

What Are the Consequences for Education?

We recommend the following steps for education, to enhance the safety and success of sinus surgery:
1. Intensive theoretical studies are necessary.
2. Cadaver dissections are indispensable for anatomic orientation in the complex labyrinth of the paranasal sinuses.
3. Visiting surgical courses are highly recommended so that the student is able to see experts performing surgery. In addition, the knowledge of anatomy is deepened, as is the ability to interpret CT findings and to evaluate indications for surgery.
4. It is very helpful for students to assist more experienced surgeons. A training clinic should have video equipment in order to allow the trainee to follow the progress of the operation [14, 15].
5. Interactive study of multimedia teaching software results in a more sophisticated introduction to the surgical concepts and possible complications. Furthermore, it leads to a deeper knowledge of the network of surgical sites, the interpretation of CT findings and the repair of complications.
6. It is recommended that the novice surgeon perform the first 50 procedures with the help of an experienced surgeon. The next 50 operations should be performed with help readily available.

Appendix: Technical Requirements

CD "Paranasal Sinuses and Anterior Skull Base"

- **Software.** The program is sold by Elsevier (Amsterdam) in two versions: a CD-I (compact disk, interactive) and a CD-ROM for use with MS-Windows and Video for Windows (Microsoft). The latter will be installed from the CD-ROM if not present.
- **Hardware.** A CD-I player (Philips CDI 220) with a full-screen, full-motion option (digital video) or an IBM-compatible PC (33-MHz 486 or faster, 4 MB of RAM, 5 MB of free space on the hard disk), a CD-ROM device, a mouse and a 256-color VGA graphics board with a resolution of 640×480 pixels.

TIM

- **Software.** Installed by Zeppelin (Munich)
- **Hardware.** An IBM-compatible 486 or Pentium PC, a SVGA graphics board with 1 MB of video memory, 8 MB of RAM and 120 MB free space on the hard disk (allows storage for 25 patients)

Exchange of information between the computer and the diagnostic tool takes place via a computer network (online) or a magneto-optical disc (offline).

CD-ROM "Endonasal Sinus Surgery"

- **Software.** Compact disk (ISBN 3-933755-01-8)
- **Hardware.** Multimedia PC 166 MHz or faster, Windows 95/98
  www.GuebelVerlag.de

References

1. Draf W (1983) Endoscopy of the paranasal sinuses. Springer, Berlin Heidelberg New York
2. Draf W (1991) Endonasal micro-endoscopic frontal sinus surgery: the Fulda concept. Operative Tech Otolaryngol Head Neck Surg 2:234–240
3. Draf W, Weber R (1992) Endonasale Chirurgie der Nasennebenhöhlen. Das Fuldaer mikro-endoskopische Konzept. In: Ganz H, Schätzle W (ed) HNO Praxis Heute, vol 12. Springer, Berlin Heidelberg New York, pp 59–80
4. Draf W, Weber R (1993) Endonasal pansinus operation in chronic sinusitis. I. Indication and operation technique. Am J Otolaryngol 14:394–398
5. Heermann H (1958) Über endonasale Chirurgie unter Verwendung des binokularen Mikroskopes. Arch Ohr Nase Kehlkopfheilkd 171:295–297
6. Hillen BI (1993) Paranasal sinuses and anterior skull base. In: Hillen BI (ed) Atlas of continuous cross-sections on full-motion CD-I. (Elsevier's interactive anatomy) Elsevier, Amsterdam

7. Hughes GB (1991) The learning curve in stapes surgery. Laryngoscope 101:1280–1284
8. Keerl R, Weber R, Draf W (1996) Periorbital paraffin granuloma following paranasal sinus surgery. Am J Otolaryngol 17:264–268
9. Keerl R, Weber R, Draf W (1996) Multimedia in ENT surgery. The endonasal pansinus operation CD- ROM. Mosby, Wiesbaden
10. Kennedy DW (1985) Functional endoscopic sinus surgery. Technique. Arch Otolaryngol 111:643–649
11. Kennedy DW (1994) Sinus disease: guide to first-line management. Health Communications, Darien
12. Kennedy DW, Shaman P, Han W, Selman H, Deems D, Lanza D (1994) Complications of ethmoidectomy: a survey of fellows of the American Academy of Otolaryngology, Head and Neck Surgery. J Otolaryngol Head Neck Surg 111:589–599
13. Levine H, May M, Schaitkin B, Mester S (1996) Complications of endoscopic sinus surgery. In: Levine H, May M (eds) Endoscopic sinus surgery. Thieme, New York
14. May M, Schaitkin B (1993) Educational programs catalogue. The Shadyside Sinus Surgery Center, Pittsburgh
15. May M, Korzec KR, Mester SJ (1990) Video telescopic sinus surgery technique for teaching. Trans Pa Acad Opthalmol Otolaryngol 42:1037–1039
16. Messerklinger W (1970) Die Endoskopie der Nase. Monatsschr Ohrenheilkd Laryngorhinol 104:451–456
17. Messerklinger W (1972) Technik und Möglichkeiten der Nasenendoskopie. HNO 20:133–135
18. Messerklinger W (1978) Endoscopy of the nose. Urban and Schwarzenberg, München
19. Rivron RP, Maran AG (1991) The Edingburgh FESS Trainer: a cadaver-based bench-top practice system for endoscopic ethmoidal surgery. Clin Otolaryngol 16:426–429
20. Rudert H (1988) Der Stellenwert der Infundibulotomie nach Messerklinger. HNO 36:475–482
21. Stammberger H (1991) Functional endoscopic sinus surgery. Decker, Philadelphia
22. Stankiewicz JA (1991) Complications of endoscopic sinus surgery. Otolaryngol Clin North Am 22:749–758
23. Stankiewicz JA (1991) Cerebrospinal fluid fistula and endoscopic sinus surgery. Laryngoscope 101:250–256
24. Vernick DM (1986) Stapedectomy results in a residency training program. Ann Otol Rhinol Laryngol 95:477–479
25. Wigand ME (1981) Ein Spül-Saug-Endoskop für die transnasale Chirurgie der Nasennebenhöhlen und der Schädelbasis. HNO 29:102–103
26. Wigand ME (1981) Transnasale, endoskopische Chirurgie der Nasennebenhöhlen bei chronischer Sinusitis I: Ein bio mechanisches Konzept der Schleimhautchirurgie. HNO 29:215–221
27. Wigand ME (1989) Endoskopische Chirurgie der Nasennebenhöhlen und der vorderen Schädelbasis. Thieme, Stuttgart

Part II

Surgica Treatment of the Nose and Paranasal Sinus Diseases

II

Chapter 14
Micro-endoscopic Surgery of the Turbinates and Nasal Septum 161
Aldo Cassol Stamm, Elisabeth Araujo,
Antonio Douglas Menon, and Karen Borne Teufert

Chapter 15
Surgical Approaches to Sinusitis: Overview of Techniques 179
Renato Roithmann, Ian Witterick, Michael Hawke,
and Aldo Cassol Stamm

Chapter 16
High-Risk Areas in Endoscopic Sinus Surgery 195
Toshio Ohnishi

Chapter 17
Micro-endoscopic Surgery of the Paranasal Sinuses 201
Aldo Cassol Stamm

Chapter 18
Combined Microscopic and Endoscopic Surgery of the Ethmoid Sinus 237
Heinrich H. Rudert and Christoph G. Mahnke

Chapter 19
Micro-endoscopic Surgery of the Maxillary Sinus 249
Rodolfo Arias, Hector Ariza, Ivan Correa,
and Aldo Cassol Stamm

Chapter 20
Endonasal and External Micro-endoscopic Surgery of the Frontal Sinus 257
Wolfgang Draf, Rainer Weber, Rainer Keerl,
Jannis Constantinidis, Bernhard Schick, and Anjali Saha

Chapter 21
**Mini-anterior and Combined Frontal Sinusotomy and Drilling
of Nasofrontal Beak** 279
Kanit Muntarbhorn and Sanguansak Thanaviratananich

Chapter 22
**Recurrence of Polyposis: Risk Factors, Prevention,
Treatment and Follow-Up** 287
Pierre Rouvier and Roger Peynegre

Chapter 23
What Is the Place of Endonasal Surgery in Fungal Sinusitis? 309
Valerie J. Lund

Chapter 24
Combined Fiberoptic Headlight and Endoscopic Sinus Surgery with Adjunct Use of Middle-Meatal Stenting 315
Paul H. Toffel

Chapter 25
Sinus Mucosal Wound Healing Following Endoscopic Sinus Surgery: Mucosal Preservation 323
Hiroshi Moriyama

Chapter 26
Wound Healing after Endonasal-Sinus Surgery in Time-Lapse Video: a New Way of Continuous In Vivo Observation and Documentation in Rhinology 329
Rainer Weber, Rainer Keerl, Andreas Huppmann, Wolfgang Draf, and Anjali Saha

Chapter 27
Pediatric Endoscopic Endonasal Sinus Surgery 347
Gerald Wolf

Chapter 28
Micro-endoscopic Sinus Surgery in Children 357
Shirley Shizue Nagata Pignatari and Aldo Cassol Stamm

Chapter 29
Revision Endoscopic Sinus Surgery 371
Jean-Michel Klossek

Chapter 30
Power Instrumentation in Rhinologic and Endoscopic Sinus Surgery 377
Daniel G. Becker and David W. Kennedy

Chapter 31
Image-Guided Surgical Navigation in Functional Endoscopic Sinus Surgery 387
Winston C. Vaughan and Frederick A. Kuhn

Micro-endoscopic Surgery of the Turbinates and Nasal Septum

Aldo Cassol Stamm, Elisabeth Araujo,
Antonio Douglas Menon, and Karen Borne Teufert

Introduction

The principal structures of the nasal cavity are the nasal septum and the inferior and middle turbinates. Functional and anatomical alterations of these structures, such as nasal-septum deviations and septal spurs [isolated or associated with hypertrophy or hyperplasia of the turbinates (usually the inferior turbinate)], can result in a great number of symptoms, especially nasal obstruction. A large concha bullosa or degeneration of the middle turbinate can also cause impaired ventilation and drainage of the paranasal sinuses, leading to chronic and/or recurrent sinusitis. In these cases, the final diagnosis and treatment plan should reflect the specialist's goals of both correcting the problem and restoring a good airway for the patient.

Anatomy of the Nasal Turbinates and Septum

Anatomically, the lateral nasal wall is one of the most complex and important structures from both clinical and surgical standpoints. The nasal turbinates are found in the lateral wall of the nasal cavity. The ethmoid bone, the maxilla, the palatine bone, the lacrimal bone, the inferior turbinate and medial surface of the medial pterygoid plate and a small area of the nasal bone together form the lateral nasal wall (Fig. 14.1). Usually, two ethmoidal conchae project into the nasal cavity; often, a third one is found. Below them lies the inferior nasal concha. Medial to the concha, a single air space termed the common meatus is present. The spaces under the conchae are termed the superior, middle and inferior meati, respectively. They are connected to the common nasal meatus at the lower edge of each concha. The superior ethmoidal concha is also termed the supreme nasal concha and is present in 60–70% of subjects [36]. The inferior nasal concha develops as an independent bone with a complex area of attachment to the skeleton of the lateral wall of the nose.

The nasal septum is composed of cartilage and bone that is primarily covered by respiratory mucous membrane. It is divided in to anterior (dorsal), posterior, caudal, cephalic and superior and inferior portions. It is formed by the nasal spine of the frontal bone, the perpendicular plate of the ethmoid bone (upper and posterior parts), the vomer (superior, posterior, inferior and anterior sides, like a triangle

Fig. 14.1. **a** Osseous lateral nasal wall. **b** Anatomy of the lateral nasal wall. *IT*, inferior turbinate; *MT*, middle turbinate; *SS*, sphenoid sinus; *ST*, superior turbinate

Fig. 14.2. Anatomical configuration of the nasal septum. *FB*, frontal bone; *NB*, nasal bone; *PP*, perpendicular plate of the ethmoid bone; *SC*, septal cartilage; *SR*, sphenoid-sinus rostrum; *V*, vomer

based posteriorly), the rostrum of the sphenoid, the nasal crest of the palatine, maxilla and premaxilla bones, nasal cartilage (quadrangular), the upper lateral cartilage, the membranous septum and the columella (Fig. 14.2) [17].

The septal cartilage (quadrangular) is the most important anatomical structure in nasal-septum operations. The caudal portion of the cartilage rests on the support of the anterior nasal spine and maxillary crest, linked to this structure by fibrous attachments. Posteriorly, the septal cartilage extends into the vomer and perpendicular plate of the ethmoid bone through osseocartilaginous junctions. These interosseous and osseocartilaginous junctions are very important for nasal, facial and cranio-facial growth. The superior and septal cartilages form a single anatomic structure. The junction between these two cartilages forms the nasal-valve angle, which is the narrowest portion of the nasal cavity, forming an angle of approximately 10–15° in Caucasians. Alteration of the nasal-valve angle may result in nasal obstruction [15].

The attachment of the nasal septum and upper lateral cartilages to the undersurface of the nasal bones forms an articulation of substantial strength, often termed the "keystone" area of nasal support [12]. The most anterior portion of the nasal septum, located between the columella and the caudal end of the septal cartilage, is called the membranous septum, and it is composed of subcutaneous tissue covered by vestibular skin on both sides. The columella is formed by the paired medial crura of the inferior lateral cartilages, transverse ligament, subcutaneous fat and skin.

The commonly observed deflection of the nasal septum from the medial sagittal plane is a consequence of both genetic and environmental factors. Septal deviations and unilateral or bilateral exostosis are very common and have been noted to be present before birth. The bilateral septal lining is composed of periosteum and perichondrium, mucosa and submucosa. It is very rich in vessels and nerves, especially in the anteroinferior region near the incisive foramens.

Blood Supply

Both the internal and external carotid arteries provide the blood supply to the interior of the nose and the paranasal sinuses. The external carotid artery contributes the maxillary (internal maxillary) artery and the facial (external maxillary) artery, which are the main vessels supplying the nasal interior. The internal carotid artery contributes the anterior and posterior ethmoid arteries.

The facial artery gives rise to the superior labial artery. This artery ascends into the nasal vestibule, splitting into numerous smaller branches that largely supply the nasal septum.

The maxillary artery is the larger of the two branches of the external carotid system. Two terminal branches of the maxillary artery, the septal and posterior lateral nasal arteries, originate in the pterygopalatine fossa very close to its medial wall and pass through the sphenopalatine foramen bordering its superior and inferior edges, respectively [24]. The arteries enter the nasal cavity through the sphenopalatine foramen located slightly posterior to the posterior edge of the middle turbinate. The septal artery lies more superiorly, close to the anterior wall of the sphenoid sinus, continuing toward the upper and posterior portion of the nasal septum. The posterior lateral nasal artery arises from the inferior edge of the sphenopalatine foramen and descends beneath the periosteum in a vertical plane. At the edge of the sphenopalatine foramen, the posterior lateral nasal artery splits, sending a large branch to the caudal end of the middle turbinate (the middle-turbinate artery; Fig. 14.3).

The mucoperichondrial and mucoperiosteal lining of the upper nasal septum is supplied by anastomotic branches of the anterior and posterior ethmoid arteries, while the inferior and posterior portions are supplied by the terminal branches of the maxillary artery. The columella and the nasal septum receive blood from septal branches of the superior labial artery.

The septal artery is encompassed by the septal periosteum and delivers branches throughout the septum; its ascending branches anastomose with terminal septal branches of the posterior and anterior

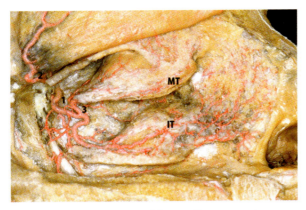

Fig. 14.3. Terminal branches of the maxillary artery, the posterior lateral nasal artery (*1*), septal artery (*2*) and middle-turbinate artery (*3*) entering the nasal cavity through the sphenopalatine foramen. *IT*, inferior turbinate; *MT*, middle turbinate (Courtesy of Prof. Navarro)

ethmoidal arteries. The convergence of these vessels in the region of the anterior septum and branches of the greater palatine artery is known as Little's area or Kisselbach's plexus.

Venous drainage is provided by the ophthalmic and facial veins and the pterygoid and pharyngeal plexus. Thus, it is partly intracranial and partly extracranial relative to the cavernous, coronary and transverse sinuses.

Innervation

The nasal cavity and paranasal sinuses have general-sensory and autonomic (secretory and vasomotor) nerve supplies in addition to special sensory fibers of the olfactory nerve. Branches of the ophthalmic and maxillary divisions of the trigeminal nerve provide the general-sensory nerve supply. The autonomic nerve supply to the nose is important in regulating vascular tone, turbinate congestion and the production of nasal secretions. Parasympathetic autonomic fibers first synapse in the sphenopalatine ganglion before entering the nasal cavity through the nasal mucosa. Sympathetic fibers pass through (but do not synapse) in the sphenopalatine ganglion before being distributed to the nasal vasculature. Branches of the sphenopalatine ganglion contain sympathetic and parasympathetic fibers and trigeminal fibers for pain, temperature and touch.

Sympathetic fibers leave the spinal cord by spinal nerves T1–T3, reaching the sympathetic trunk by white rami communicantes. The fibers travel up the cervical sympathetic trunk and synapse in the superior cervical ganglion. The post-ganglionic fibers accompany blood vessels to the mucosa of the nose and nasal sinuses. Some fibers pass (without synapsing) through the pterygopalatine ganglion.

Presynaptic parasympathetic fibers in the facial nerve exit at the geniculate ganglion in the greater superficial petrosal nerve. The greater superficial petrosal nerve unites with the deep petrosal nerve to form the nerve of the pterygoid canal (the vidian nerve). Through the vidian nerve, the parasympathetic fibers enter the sphenopalatine ganglion and synapse. Postsynaptic fibers are then distributed to all branches of the sphenopalatine ganglion, including the posterolateral nasal branches, the greater palatine (and its nasal branches) and nasal branches to the septum. Secretory and vasodilator fibers are then distributed to the glandular epithelium (Fig. 14.4).

Lymphatic System

The lymphatic drainage system consists of (1) an anterior system that collects lymph from the nasal pyramid and drains into the superficial, submandibular cervical lymph nodes and (2) a posterior system that drains the posterior part of the nasal cavity and the nasopharynx into the retropharyngeal lymph nodes and the jugular nodes.

Mucosa of the Nasal Cavity and Paranasal Sinuses

The internal lining of the nasal cavity consists of a thin layer of tissue, the respiratory epithelium. Histologically, it consists of a pseudostratified epithelium firmly attached to the basement membrane, which intimately adheres to either the periosteum or the perichondrium.

Glandular structures, particularly serous glands, are more numerous in the middle and superior turbinates than in the inferior turbinate. Their secretions provide a protective blanket for the mucosa. The thickness of the inferior turbinate is due to the presence of vascular channels or lakes and the degree to which they are empty or engorged.

Cavernous erectile tissue may also be found within the mucosa surrounding the ostia, and this can influence their patency. This variability of the opening of the ostia is also influenced by simultaneous variations in volume of the neighboring turbinates.

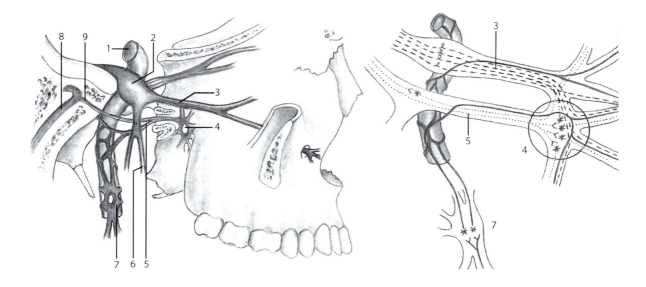

Fig. 14.4. Nerve supply of the nasal mucosa. *1*, Internal carotid artery with sympathetic plexus; *2*, gasserian ganglion; *3*, maxillary nerve; *4*, pterygopalatine ganglion; *5*, nerve of the pterygoid canal; *6*, mandibular nerve; *7*, superior cervical ganglion; *8*, facial nerve with nervus intermedius; *9*, greater petrosal nerve. The *inset* shows the locations of fibers in the pterygopalatine ganglion. *Continuous line*, sympathetic fibers; *dotted line*, parasympathetic fibers; *broken line*, trigeminal nerve [1]

Physiology

In addition to its aesthetic value, the nose serves many important functions: olfaction, as an airway to the lower respiratory tract, air conditioning (i.e., warming, humidification, filtration), protection against infection, self-cleansing (through coordinated mucociliary activity), speech resonation and as a chemosensor. The air we inspire daily contains billions of infectious, allergenic, irritative or toxic materials that must be filtered, neutralized and eliminated. Even extremely cold and dry air is instantly warmed to body temperature and completely humidified during nasal transit. Air-conditioning functions are largely modulated through changes in nasal blood flow, secretions and nasal resistance. The autonomic nervous system also plays a role in these processes.

Septal deformities can cause impairment of the nose's normal airflow patterns, creating turbulence, nasal obstruction, drying, crusting and bleeding on the side with increased flow. Anterior septal deviations usually cause more disturbance of the air flow than posterior deformities, because the anterior deviations often involve the nasal-valve region [15].

Treatment of Diseases of the Nasal Turbinates

The treatment of diseases of the nasal turbinates depends on a careful medical evaluation, taking into account all the factors that can contribute to their configuration.

Evaluation

- History
- Ear, nose and throat exam
  - Conventional anterior and posterior rhinoscopy
  - Endoscopy (flexible and rigid)
- Rhinomanometry and acoustic rhinometry [13]
- Imaging studies
  - Ultrasound (pregnancy)
  - Computed tomography (CT; Fig. 14.5)
  - Magnetic resonance (MR; Fig. 14.6)
- Laboratory tests
  - Nasal cytogram
  - Mucociliar tests

When either primary or secondary turbinate hypertrophy is determined to be the cause of nasal obstruction and rhinorrhea, the treatment may be medical or surgical, depending on whether the hypertrophy reflects mucosal (epithelial) or anatomical (structural) alteration, Overgrowth of the bony portions of the turbinates can also be a cause of nasal obstruction. The turbinate can grow toward the nasal septum, sometimes touching it, or toward the choana, eventually obstructing it completely (Fig. 14.7). Alterations of the middle

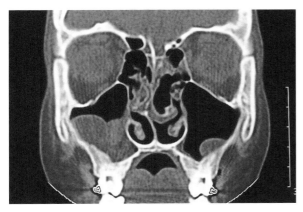

Fig. 14.5. Coronal computed tomography of a patient with septal deviation, concha bullosa and inflammatory process of the maxillary sinus

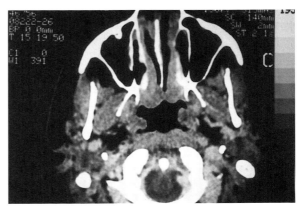

Fig. 14.7. Axial computed tomography of a patient with bilateral hypertrophy of the inferior turbinates, which completely obliterates the choana region

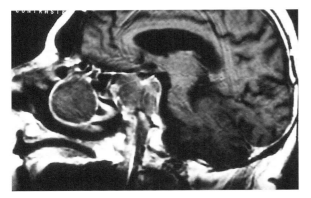

Fig. 14.6. Sagittal magnetic resonance of a patient with a mucocele of the middle turbinate

turbinate (such as paradoxical middle turbinate, concha bullosa, mucosal hypertrophy or polypoid turbinate) can also be responsible for nasal obstruction, frontonasal pain or narrowing of the middle-meatus region.

The rhinoscopic examination should be performed before topical vasoconstriction so that the level of congestion, extent of edema, presence and character of rhinorrhea and color of the mucosa can be evaluated. Once the above tests have been completed, a detailed examination is undertaken after topical application of 4% or 10% lidocaine (with 1:100,000 epinephrine for vasoconstriction). This can be done by spraying the solution or by applying cottonoid pledges. The condition of the mucosa is then compared to the pre-vasoconstriction state. In addition, the anatomy of the nasal septum and turbinates can be more accurately examined after a histamine-challenge test under experimentally induced nasal obstruction.

Medical Treatment

Pharmacological agents, such as decongestants, corticosteroids (oral or topical), antihistamines, cromolyn sodium, anticholinergics and immunotherapeutics, are all therapeutic agents that can be used in the medical treatment of diseases that affect the turbinates. The main drugs used in the medical treatment of diseases of the nasal turbinates are described next.

Decongestants

Oral decongestants commonly include pseudoephedrine, phenylephrine and phenylpropalamine. Since these drugs are sympathomimetic, the patient should be asked about any history of hypertension, cardiovascular diseases and glaucoma. Topical decongestants generally contain the same decongestants but in a drop, spray or inhaler form, and the same precautions apply. Additional reactions, unique to topical application to the nasal mucous membranes, include nasal irritation or burning and rhinitis "medicamentosa", which can be a cause of nasal obstruction.

Corticosteroids

Corticosteroids may be used systemically (orally or by injection) or topically. They are used alone or together with nasal decongestants or antihistamines in cases of allergic and inflammatory nasal obstruction. While the relief of nasal congestion/obstruction may be spectacular (especially in the obstruction of chronic, non-specific rhinitis), steroids should not be used routinely, but rather for selected patients. Their use may be contraindicated in patients with a history of diabetes, tuberculosis, pregnancy, peptic ulcer, renal disease, emotional disturbance or hypertension. In addition, their anti-inflammatory effects may mask the symptoms of infection and thus delay accurate diagnosis. In general, the lowest dosage of oral systemic steroids should be used for the shortest period of time, and the therapy should be individualized.

The same side effects and cautions apply for systemic steroids via the parenteral (usually intramuscular) injection route. Adrenocorticotropic hormone is the usual preparation, often in long-acting, repository form.

Intranasal topical steroids, usually beclomethasone (Beconase, Beclosol), fluticasone (Flonase, Flixonase), triamcinolone (Nasacort), budesonide (Rhinocort) and mometasone (Nasonex), exert a marked local anti-inflammatory effect on the nasal mucosa. They may be useful alone or in combination with nasal decongestants or antihistamines in the management of allergic and inflammatory nasal obstruction (nasal polyps). They are usually available in metered nasal sprays; the usual dose is two sprays per nostril after irrigation (since they are topically active, it is helpful to irrigate away excess mucus before application) one or two times per day.

Antihistamines

Antihistamines compete with histamine for H1-histamine receptor sites on the nasal mucosa and blood vessels and are most effective when taken prior to an allergen exposure. They relieve the wet symptoms of allergy (itching, sneezing and rhinorrhea) but have very little decongestant effect. Diphenhydramine, tripelennamine, meclizine, chlorpheniramine and promethazine are conventional or first-generation antihistamines. The primary side effects are sedation, excessive drying and potential aggravation of prostatism or narrow-angle glaucoma. Prolonged use of an antihistamine may produce a tolerance, necessitating a change to a different antihistamine class. New antihistamines include non-sedating systemic medications (loratadine, cetirizine, terfenadine, astemizole and ketotifen) and topical forms (levocabastine, azelastine, etc.).

Cromolyn sodium

Intranasal sodium cromoglycate has been shown to be effective in immunoglobulin E (IgE)-mediated allergic nasal obstruction and in various forms of eosinophilic rhinitis. It is usually used in combination with antihistamines, decongestants, and intranasal steroids. The clinical benefits are not apparent for 3–4 weeks.

Anticholinergics

Anticholinergics may be helpful for patients with nasal obstruction due to excessive nasal discharge. A side effect of the use of these drugs is excessive nasal dryness, and they are contraindicated in glaucoma. Iodide isopropamide and ipratropium bromide nasal spray (0.03%; Atrovent) have been proven safe and effective for short- or long-term treatment of non-allergic rhinitis.

Immunotherapy

Immunotherapy is indicated in patients who experience failure with the usual treatments and environmental control measures. It can be used in association with other pharmacological agents and works by doing the following:
1. Altering mast cells against IgE
2. Increasing the production of IgG, IgA and other lymphocytes
3. Decreasing the production of IgE

Other Drugs

Antibiotics, topical ointments, analgesics and diuretics can also be useful in appropriate situations.

General Care

Environmental modifications may be helpful at home, at work or at school. The patient should avoid dust, smoking, air conditioning and excessive humidity. The manner of sleeping can be important; in the supine position, there is a tendency for greater venous congestion. In these cases, we recommend the patient sleep with two pillows, elevating his head about 30° from the horizontal position. Irrigation with saline solutions is useful to decrease crusting and rhinorrhea of the nasal cavity and nasopharynx.

Surgical Treatment

Several surgical procedures of the inferior turbinates may be helpful in relieving nasal obstruction refractory to vigorous medical management. The surgical treatment of obstructing inferior turbinates was first reported in 1895 by Jones [14]. Widespread controversy over the safety of the inferior turbinectomy has continued despite its long history and the absence of documented evidence of adverse sequelae. Several procedures have been recommended depending on the etiology of the obstruction, including direct steroid injection of the turbinate, cryotherapy, laser vaporization (using CO_2, yttrium aluminum garnet (YAG) or argon lasers, etc.) mucous-membrane cautery, turbinate out-fracture, intra-turbinate stroma removal using a microdebrider [38], submucous resection of the inferior turbinate, turbinoplasty and partial or complete inferior turbinectomy.

Surgical treatment of the middle turbinate is indicated in cases of nasal obstruction or narrowing of the entrance of the middle meatus. The superior turbinate may be surgically approached when it is necessary to access the sphenoid sinus. When turbinate hypertrophy is determined to be the cause of nasal obstruction, the treatment should aim at the opening of the airway (without compromising nasal physiology) and preserving the local anatomy.

Although surgical procedures of the turbinates are thought to be one of the causes of atrophic rhinitis and "dry rhinitis", some authors, such as Ophir [25] and Martinez [21] (who defend the total removal of the inferior turbinate) and Mabry [20] and Saunders [30] [who are in favor of partial inferior turbinectomy and turbinoplasty (submucosal turbinectomy)] do not mention these complications.

Surgical Treatment of the Inferior Turbinate

The patient is placed in a supine position, with the head elevated approximately 30° and turned toward the surgeon. The surgery can be performed under local or general anesthesia. When local anesthesia is used, the nasal cavity is prepared with cottonoids soaked with lidocaine (4%) or neotutocaine (2%) with vasoconstrictor. After removal of the cottonoids, 1–2 ml of lidocaine (2%) with 1:100,000 epinephrine is slowly injected into the turbinate.

Usually, the surgery is accomplished using the 4-mm, 0° endoscope. The most utilized instruments are a Freer elevator, a suction elevator, microscissors, electric cautery (mono- and bipolar) and a shaver system.

Cryosurgery

Cryotreatment provokes submucosal destruction at the cost of surface destruction. This method is used in chronic rhinitis (mainly in allergic, "medicamentosa" and non-allergic rhinitis). This method fails to remove thickened bone and gives only partial, temporary relief. Complications, such as septal perforations, bleeding, crusting and infection, may occur [26]. Cryotreatment can be performed on an outpatient basis under local anesthesia with a cryosurgery unit that jets a stream of supercooled nitrogen gas or CO_2.

After local anesthesia is achieved, the cryoprobe is introduced into the nasal cavity in close contact with the turbinate, and the unit starts the cryotherapy. The probe should contact the turbinate for no more than 60 s and, to prevent removal of healthy tissue, it should not be removed suddenly. This procedure usually promotes partial, temporary relief (62% after 1 year and 35% later, according to Rakover et al. [28]).

Laser Vaporization

Laser vaporization has several rhinologic indications, including treatment of turbinate hypertrophy, telangectasia, papillomas, nasal polyps, intranasal adhesions, granulomas, choanal atresia and tumors. Due to the hemostatic properties of the laser, it has some advantages compared with other techniques, particularly in nasal diseases that bleed easily and in patients with coagulopathies. Lenz [19], using inferior-turbinate argon-laser treatment for vasomotor rhinitis, observed that 80% of his patients initially had good results, and 6% required repeat surgery.

Selkin [32] used the CO_2 laser for partial turbinectomies and noted initial improvement in 100% of his patients, but 6% had recurrent obstructions between 5 months and 1 year later. The KPT-532 or "New Laser" has also been used for polyp resection and photocoagulation of vascular lesions.

Figure 14.8 shows the preoperative (Fig. 14.8A) and 8 months postoperative (Fig. 14.8B) CT of a patient who underwent laser vaporization of the inferior turbinates for treatment of a nasal obstruction. After surgery, the patient developed both maxillo-ethmoidal sinusitis and necrosis of the medial wall of the maxillary sinus and the bony portion of the inferior turbinates. After local surgical debridement, a wide communication between the maxillary sinuses and the nasal cavity could be seen.

Electrical Submucosal Cauterization

Treatment of turbinate hypertrophy using electrical submucosal cauterization reduces the inferior-turbinate bulk and can be performed in patients with allergic or non-allergic rhinitis who have mild to severe nasal obstruction that may or may not be associated with adenoid or tonsil hypertrophy. The

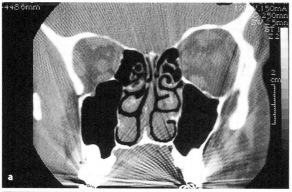
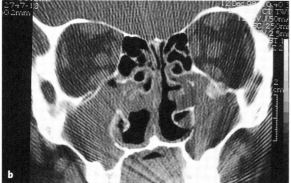

Fig. 14.8. Preoperative (**a**) and postoperative (**b**) coronal computed tomography of a patient with hypertrophy of the inferior turbinates who underwent laser vaporization for nasal obstruction. The patient developed osteonecrosis of the nasal wall of the maxillary sinus and inferior turbinates and, subsequently, maxillo-ethmoidal sinusitis

technique is performed either with a monopolar cautery using a special electrode or with radiofrequency. The needle is introduced into the submucosa of the inferior turbinate along its entire extent. As the needle is withdrawn, the electrical submucosal cautery creates an intensive reaction that reduces the turbinate bulk. Although this procedure can be repeated several times, it fails to remove thickened bone.

Care must be taken during this procedure in order to avoid direct contact of the electrode and the bony portion of the turbinate, thus preventing bone sequestration and osteitis. Postoperative bleeding is uncommon, and nasal packing is infrequently required; if needed, antibiotic gauze or a Merocel is left in place for 24 h. Sometimes a lateral out-fracture of the turbinate is performed with the electrical submucosal cautery in order to obtain a better result. Postoperative care includes daily nasal irrigation and weekly crust removal for 3–4 weeks.

Late complications, such as atrophic and infectious rhinitis are uncommon and may occur when the turbinate is too coagulated, leading to bone-necrosis sequestration. Intranasal adhesions are usually secondary to inadequate postoperative care and can be prevented by using nasal splints.

The patients usually have a good prognosis, with partial and temporary relief of the nasal obstruction. Some authors report limited success with submucosal diathermy, which has proven to be effective in reducing the airway resistance initially; however, no difference from the pre-surgical values was noted 1 year after surgery [27].

Turbinate Displacement (Out-Fracture)

Lateral out-fracture of the inferior turbinate can be useful for anatomical bony deviations but is not helpful for mucosal hypertrophy of dependent congestion. Turbinate displacement alone yields only short-term results but is very useful when associated with other procedures when micro-endoscopic surgery of the paranasal sinuses is required. After some time, the fractured turbinate returns to its original position [10, 18].

Intra-Turbinal Stroma Removal with a Shaver System

Intra-turbinal stroma removal with a shaver system initially consists of infiltration of the anterior portion of the inferior turbinate with lidocaine (2%) and epinephrine (1:100,000). At this region of the turbinate, an incision is made via monopolar cautery; through this incision, a Freer elevator creates a space between the bony part and the stroma along the turbinate extension. Next, the microdebrider is inserted into this space and works with an oscillatory rotation (800–1000 rpm), removing the excessive stroma (Fig. 14.9). The bony portion of the turbinate may be gently fractured laterally or partially resected. If the caudal portion of the turbinate presents severe hypertrophy, the microdebrider can be used directly on its surface [38]. At the end of the procedure, a Merocel packing is placed between the nasal septum and the inferior turbinate. Bielamowicz et al. [3], using a microdebrider with a 3.5-mm Jaguar blade, resected the inferior and lateral aspects of the inferior turbinate mucosa and removed the turbinate bone with long Stevens tenotomy scissors. Complications, such as bleeding, crusting and synechiae, can occur.

Inferior Turbinoplasty

Inferior turbinoplasty consists of a modified method of submucosal turbinectomy (the "mucosal-flap method"). The mucosa and the bone of the inferior

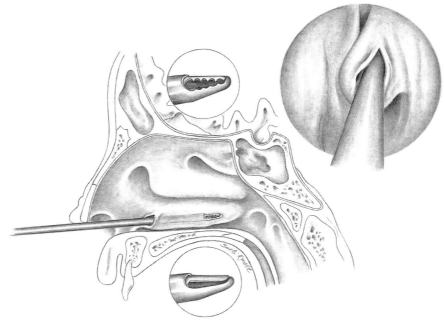

Fig. 14.9. Intra-turbinal stroma removal of the inferior turbinate with a 3.5-mm silver-bullet blade with or without teeth (shaver system; Xomed, Jacksonville)

turbinate are largely excised without causing any mucosal defects; thus, postoperative symptoms, such as crusting, dryness, scarring and bleeding, are minimal [7, 20].

Submucosal resection of the inferior-turbinate bone affords the removal of disordered tissue (thickened, spongy, space-occupying bone and hypervascular submucosa) while sparing the surrounding, physiologically normal tissue. Resection is indicated in patients with anatomically deviated obstructing inferior turbinates or with hypertrophic mucosa unresponsive to vigorous medical treatment.

After vasoconstriction and anesthesia are achieved, an incision is made in a posterior-to-anterior direction along the inferior edge of the inferior turbinate, continuing up the anterior aspect of the turbinate. Using a suction elevator, the mucoperiosteum is elevated off the medial and lateral aspects of the turbinate bone. The inferior-turbinate bone is fractured and removed subperiosteally with a Takahashi forceps or a Jansen–Middleton rongeur, carefully preserving the mucoperiosteal flaps. Excessive mucosa can be trimmed from the inferior portions of the mucoperiosteal flaps (principally the lateral one). The remaining mucoperiosteum is deflected laterally over the bare bone of the inferior turbinate remnant and is packed in place with antibiotic gauze or Merocel for 48–72 h (Fig. 14.10).

Postoperative crusting is rarely observed, because this method does not cause mucosal defects. Saline irrigation is prescribed, and patients usually do not complain of postoperative symptoms. Complications associated with submucous resection of the inferior turbinate include crusting, dryness, intranasal adhesions and bleeding but are very uncommon. Bleeding is observed less often than with other resection techniques and can be easily controlled. Fixation of lateral and medial flaps is difficult, because flap margins are loose and there is no firm base (such as bone) to which to fix them, and this can result in failure of the attachment of the mucosal flaps.

Mabry [20] reported good results in 25 patients after a 1 year follow-up with this technique. In 1999, Passali et al. [27] compared results from 382 patients presenting hypertrophy of the inferior turbinate and randomly treated with electrocauterization of the inferior turbinate (62 patients), cryotherapy (58), laser cautery (54), submucosal resection without lateral displacement (69), resection with displacement (94) and turbinectomy (45). Long-term results showed that submucosal resection with lateral displacement of the inferior turbinate resulted in the greatest increase of the air flow. According to the authors' experience, this technique provides quicker postoperative healing (2–4 weeks) and fewer office visits (two or three).

Partial Inferior Turbinectomy

Partial inferior turbinectomy using endoscopes provides complete visualization of the operative field and, thus, decreases risk of excessive or inadequate resection. This old technique has been criticized by many because of the potential risk of atrophic

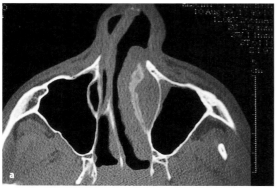 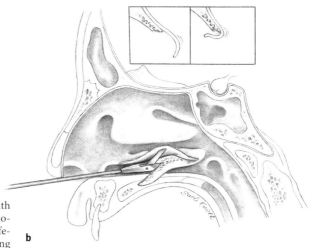

Fig. 14.10. a Axial computed tomography of a patient with hypertrophy of the left inferior turbinate. **b** Inferior turbinoplasty. Resection of the mucoperiosteum and bone of the inferior turbinate throughout its inferior extension, preserving the mucoperiosteum of the superior portion that covers the bare bone

changes and crusting. This is probably why it has not gained full acceptance, though some authors believe that partial inferior turbinectomy is effective and safe when properly performed [4]. It is usually indicated in well-established cases of chronic hypertrophic rhinitis that are refractory to medical treatment and may respond best to surgery of the nasal turbinates [33]. Chronic rhinitis frequently results from hyperactivity of the mucous glands of the inferior turbinates. Vasomotor disturbances and allergy are the most common causes of chronic rhinitis. Symptoms of nasal obstruction, rhinorrhea, postnasal drip, morning pharyngeal dryness, headaches, snoring and sleep apnea can be very distressing to the patient.

The inferior turbinate is gently out-fractured medially (Benfield maneuver), using a Freer elevator. The inferior third is resected along its entire anteroposterior length (under microscopic or endoscopic visualization) using a turbinate scissors after squeezing the portion to be resected and cauterizing the marked sulcus. Partial turbinectomy can be performed by using a diagonal excision and removing the anterior (anterior turbinectomy) or posterior (posterior turbinectomy) ends of the turbinate, depending on which part is more hypertrophic. Exposed turbinate bone and ragged mucosa are gently resected, and hemostasis is achieved with monopolar or bipolar cautery. The inferior turbinate is repositioned, and rayon or Merocel packing is placed for 48–72 h. The patient is usually discharged during the next day.

Postoperative care consists of a 7-day course of an oral antibiotic and nasal irrigation with saline solution. After removal of the nasal packing, a second visit is scheduled 7–10 days after the operation, when residual blood clots and crusts are removed. Follow-up is then scheduled at weekly intervals until complete healing is achieved.

Excessive postoperative crusting and bleeding are the most common complications. Bleeding may require prolonged packing and hospitalization, but the overall incidence of significant bleeding is very low if proper cauterization of denuded tissue is performed. Although uncommon, osteitis of exposed turbinate bone may occur and, because of that risk, meticulous debridement of any exposed bone should always be performed intraoperatively. According to Mabry [20], this problem is minimized by performing submucosal inferior turbinectomy with turbinoplasty, which gives similar results with respect to breathing and rhinorrhea but features faster postoperative healing and necessitates fewer office visits. This is probably due to mucosal covering of the resected area, promoting better and faster healing.

Total Inferior Turbinectomy

Total inferior turbinectomy has been condemned because it is believed to result in disturbance of nasal function due to turbulent airflow in an excessively enlarged nasal cavity, eventually leading to alteration of nasal breathing sensation, crusting, bleeding, chronic purulent infection, nasal dryness, atrophic changes and "foul odor" from the nose. Total turbinectomy may result in loss of functioning nasal membranes, which are needed to warm and moisturize air. The morbidity associated with complications of the operation is sufficient to warrant a cautious approach. In general, it should be reserved for tumors involving the inferior turbinate.

The inferior turbinate is fractured slightly medially and upward. Before resection of the inferior turbinate, its insertion into the lateral nasal wall is cauterized with a bipolar bayonet forceps. The next step is resection of the turbinate with a microscissors that is positioned as far posteriorly as possible, including as much turbinate mucosa and bone as surgical resection permits. After removal of the turbinate, if a hypertrophic, mulberry-like posterior turbinate tip or bone spicules remain, a debrider or through-cutting forceps can be used to complete the resection. The nasal cavity is then packed with rayon gauze or Merocel. The packs are removed after 48–72 h.

Postoperative care is basically the same as for partial inferior turbinectomy. Radical turbinectomies may produce more crusting, so Parson's nasal irrigation is recommended twice or three times per day.

For many years, total inferior turbinectomy has been condemned because of fears of subsequent complications, such as atrophic rhinitis (rhinitis sicca) and severe bleeding, but the literature has not established a link between turbinectomies and atrophic rhinitis. Many authors have reported studies of atrophic rhinitis, and none of them suggest that inferior-turbinate reduction and middle-turbinate resection are a cause of atrophic rhinitis [6, 23]. Other complications include postoperative hemorrhage, infection, pain, nasal dryness, crusting, adhesions and osteitis. Bleeding can be avoided with the use of monopolar cautery and nasal packing. If postoperative profuse bleeding occurs, eletrocoagulation of the nasal branches of the maxillary artery should be performed. Crusting, nasal dryness, infection and intranasal adhesions can be eliminated with good postoperative care. Intraoperative debridement of any exposed turbinate bone is essential to avoid osteitis. Epiphora is most likely due to damage to the opening of the nasolacrimal duct in the inferior meatus.

Although some authors [4, 21, 23] describe relief of airway obstruction in their patients submitted to total resection of the inferior turbinates, Moore et al. [23] describe extensive malodorous nasal crusting requiring daily crust removal in 88% of their series of 40 patients 3–5 years after surgery.

Surgical Treatment of the Middle Turbinate

The middle turbinate is an anatomically important part of the ostiomeatal complex and consists of an extension of the ethmoid sinus. It presents two surfaces: the meatal aspect (which is related to the middle meatus) and the septal aspect facing the nasal septum (Fig. 14.11). The middle turbinate may har-

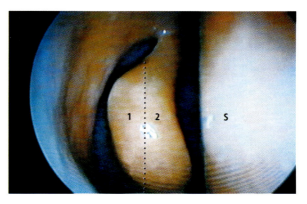

Fig. 14.11. Anterior endoscopic view of the right middle turbinate. The *broken line* separates the meatal (*1*) and septal (*2*) surfaces. S, nasal septum

bor and be the source of several conditions, including hyperplasia, excessive pneumatization (concha bullosa) and polyps. In addition, the middle turbinate can be associated with sinonasal diseases, blocking of the ostiomeatal complex or harboring of an inflammatory condition, such as mucocele or conchal sinusitis.

Surgical management of the middle turbinate is still controversial, especially during endoscopic sinus surgery, when the turbinate is thin and floppy and can obstruct the surgical field. Although, in this situation, partial middle-turbinate resection is indicated, it may lead to crusting, bleeding, hyposmia and frontal-duct stenosis [35].

Partial middle turbinate resection has been advocated by Wigand et al. [37], Biedlingmaier [2] and Stamm [34] as part of a complete ethmoidectomy. It is also recommended as part of a surgical approach to the sphenoid sinus [37].

Types and Indications for Middle-Turbinate Resections

- **Meatal surface.** When a concha bullosa is present, surgical resection of its meatal surface is performed by sectioning the middle turbinate with a straight microscissors, preserving the lateral lamella in order to maintain stability. Later, a small packing of Merocel or Merogel can be left in place (Fig. 14.12). This procedure is usually indicated for obstructive concha bullosa, conchal sinusitis, mucocele or as a part of endoscopic sinus surgery to facilitate exposure.
- **Anterior portion.** Resection of the anterior portion of the middle turbinate can be done with a curved microscissors; this can amplify the surgical access to the anterior ethmoidal cells and frontal recess. When the middle turbinate become

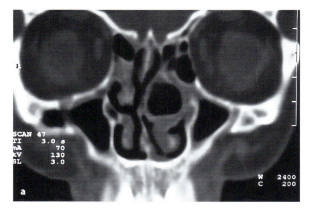

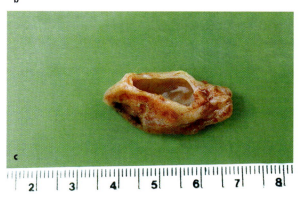

Fig. 14.12. a Coronal computed tomography of a patient with concha bullosa. b Surgical resection of its meatal surface with a microscissors. c Meatal surface of the middle turbinate after surgical resection

unstable during paranasal-sinus operations, anterior resection is recommended in order to avoid lateralization, leading to middle-meatus impairment (Fig. 14.13).

- **Posterior portion.** The posterior portion of the middle turbinate is usually resected with a curved microscissors after local electrocoagulation. This may be necessary in some cases of endonasal tumor resection or during an endonasal surgical approach to the sphenoid sinus (Fig. 14.14).
- **Total resection.** This surgical procedure is usually performed with a strong scissors, preserving only the insertion of the middle turbinate. It is indicated in special cases as part of a radical ethmoidectomy or in the resection of tumors when the middle turbinate is completely involved (stage-V recurrent polyposis).

When the middle turbinate is unstable because of previous endoscopic surgery, one should try to maintain the medial position, suturing the turbinate to the nasal septum or promoting adhesions between the septal face of the middle turbinate and the nasal septum. It should be noted that there are very few reports of significant complications from middle turbinectomy in the literature compared with several references reporting complications from inferior turbinectomy [2, 35].

Nasal-Septum Surgery

Surgery of the nasal septum utilizing the submucosal-resection technique was first performed by Freer [8] and Killian [16] at the beginning of the century. Recognizing the importance of the nasal-septum cartilage as a support structure, Killian [16] modified his technique in order to preserve the caudal and dorsal portions of the septal cartilage. The "swinging door" technique was proposed by Metzenbaum in 1929 [22] to address problems related to the caudal portion of the nasal septum.

In 1958, Cottle [5] introduced the concept of modern nasal-septum surgery with the "maxilla–premaxilla approach", which consisted of limited resections and reconstitution of the septal components. Conservative techniques partly or exclusively involving the nasal septum have been called "septoplasty" ever since.

When the septal deformity is limited, it can simply be resected; however, if a large region of the nasal septum is involved, reconstruction of the septal cartilage may be necessary. The most frequent causes of nasal-septum deformities are: intrinsic properties of septal cartilage, developmental deformity of the septum and traumatic (acquired) deformity.

Surgery to treat nasal-septum deformities are usually indicated when the septal deviation causes nasal obstruction or cosmetic deformities, as an adjunct for endonasal surgery for severe epistaxis or for non-obstructive snoring-apnea surgery. It is also indicated as part of the surgical approach to the pitu-

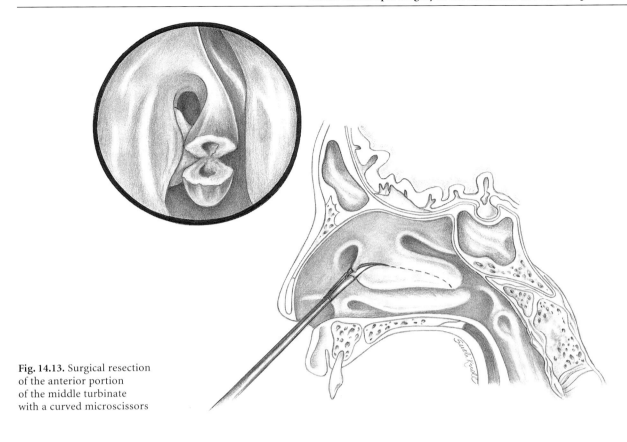

Fig. 14.13. Surgical resection of the anterior portion of the middle turbinate with a curved microscissors

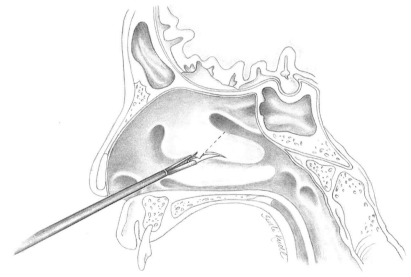

Fig. 14.14. Surgical resection of the posterior end of the middle turbinate

itary gland and micro-endoscopic surgery of the paranasal sinuses.

Depending on their nature, severity and location, nasal-septum deformities can be surgically treated by resection, structural realignment or camouflage. Surgery of septal spurs consists of ipsilateral elevation of the mucoperichondrium and/or mucoperiosteum followed by resection. The cartilaginous septal deviations are usually incised, shaved or morselized to align the septum on the midline. Large cartilaginous septal deformities can be resected and need to be repositioned if they are not part of the "L"-shaped supporting area. The cartilage of the nasal septum should always be preserved along the large "L" formed by its dorsal and caudal portions (Fig. 14.15). Surgical resections of the "L" components may result in a structural deficiency causing dorsal saddling, a crooked nose, loss of tip projection, retraction of the columella or nasal-valve collapse.

Surgical Technique – Basic Septoplasty

Although septoplasty can be performed under local or general anesthesia, we usually prefer general anesthesia associated with local infiltration of the anesthetic and vasoconstrictor. The nasal cavity is initially soaked with cottonoids embedded in vasoconstrictor solution. Infiltration of the region beneath the mucoperichondrium and mucoperiosteum is accomplished with lidocaine (2%) and epinephrine 1:100,000, producing a hydraulic dissection facilitating the surgical elevation. Before starting the surgical procedure, waiting 10 min is advisable to obtain satisfactory vasoconstriction and stabilization of the systemic hemodynamic condition.

The first step consists of a mucoperichondreal hemitransfixion incision at the caudal limit of the septal cartilage. When a significant anterior deviation is present, a full transfixation incision can be made. The choice of the side depends on the type of the deviation and the surgeon's experience. Although elevating the mucoperichondrium of the concave face is considered easier, most of the septal deformities need elevation of both sides in order to be corrected. Adequate identification of the sub-perichondreal plane is accomplished with a no. 15 surgical blade. Elevation of the mucoperichondrium and mucoperiosteum is made with the semi-cutting side of a suction elevator (Fig. 14.16). The surgeon should observe the peculiar bluish hue of the cartilage surface, advancing the suction elevator carefully in order to be sure the dissection continues beneath the mucoperichondrium or mucoperiosteum.

Disarticulation of the osseocartilaginous junction (septal cartilage, ethmoid plate and vomer) is performed with the same suction elevator, preserving the uppermost part of the osseocartilaginous junction to avoid dorsal saddling. When the deviation is located inferiorly and is associated with deformities of the maxillary crest, a strip of cartilage is resected along its extension, and the crest is mobilized. This maneuver helps identify the mucoperiosteum. The bony part of the nasal septum is dissected on both sides, followed by removal of the deviated region. If the deformity is located superiorly, it should be removed carefully, avoiding lateralization and preferably using instruments like Jansen–Middleton nasal forceps or heavy scissors in order to preserve

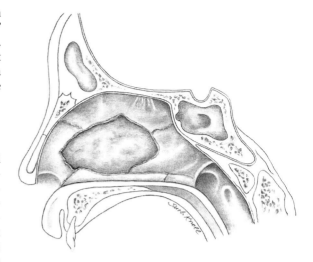

Fig. 14.15. Removal of a deviated nasal septum region, preserving the "L"-shaped support (*broken line*)

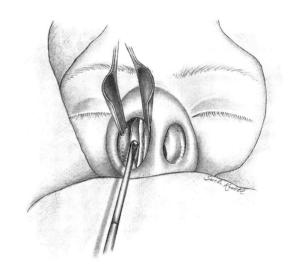

Fig. 14.16. Elevation of the mucoperichondrium of the anterior portion of the nasal septum with a suction elevator

the integrity of the cribriform plate. After adequate exposure of the nasal floor, any spurs along the maxillary crest are initially mobilized with a microosteotome; the mucoperichondrium is then elevated, and the fibrous tissue between the cartilage and the maxillary crest is incised.

Correction of deformities of the cartilage of the nasal septum usually involves the caudal region and angled parts. Caudal defects are usually treated by removal of a vertical band of cartilage posterior to the caudal edge. The position of the septal cartilage is maintained by suturing it to the nasal-spine periosteum. When the cartilaginous defect does not interfere with the "L" nasal support, it can be completely removed. An L-shaped strut of cartilage, at least 1.5 cm in width, is retained to support the nose [29]. Corrections of septal-cartilage deviations can be done with multiple incisions or by resecting, morselizing and replacing the cartilage. Stabilization and alignment of the dissected components of the nasal septum are obtained with a 4-gauge absorbable suture and a septal "splint". The septal "splint" is sutured to the septum with a non-absorbable 4-gauge suture and removed after 1 week. Nasal packing consists of a Merocel sponges inside a glove fingers, which are left in each nostril for approximately 48 h.

Complications of Septal Surgery

The most common complication consists of failure of the surgical correction. Other complications include infection, bleeding, adhesions, transient hypesthesia of the lip or palate and allergic reaction. Severe complications, such as damage to the cribriform plate and cerebrospinal-fluid (CSF) leak, olfactory-nerve injury, pneumocephalus and intracranial hemorrhage are rare and usually follow impulsive maneuvers at the level of the junction of the ethmoid plate and the cribriform plate. In order to avoid damaging this area, resection of superior septal deviations should be accomplished using Jansen–Middleton forceps or heavy scissors. Special care should also be taken to preserve the "L"-shaped support, to avoid damaging the mucoperichondrium and mucoperiosteum on both sides in the same region and to control intra-septal bleeding. Neglecting these important points may lead to dorsal saddling, septal perforation and intra-septal hematoma.

Limited Septoplasty Assisted by an Endoscope

The increasing utilization and success of endoscopic surgery of the paranasal sinuses has been followed by exploration, use and acceptance of endoscopic techniques for surgery of the nasal septum. Clinical evaluation of the nasal septum is usually accomplished with a 4-mm, 0° endoscope combined with imaging studies to identify isolated spurs, limited deviations of the nasal septum causing nasal obstruction, and impairment of surgical access to the paranasal sinuses. Endoscopic nasal-septum surgery can be an isolated procedure used to resect (1) septal spurs or ridges projecting into middle or inferior turbinates and (2) superior–posterior deviations or to perform revision operations. It can also be combined with surgery of the paranasal sinuses, CSF-leak repair, choanal atresia, tumor resection or surgery to access the sellar and parasellar regions [9, 11, 31, 37].

Technique

The most common surgical instruments utilized in endoscopic surgery of the nasal septum are the 4-mm, 0° endoscope, a no. 15 blade surgical knife, a suction elevator, Takahashi forceps, Jansen–Middleton septal forceps and a flat chisel. Before any septal surgical incision to correct septal deviations and spurs, local infiltration with lidocaine (2%) and epinephrine (1:100,000) is recommended.

Superior–posterior septal deviations are approached through a vertical incision just anterior to the deviation, followed by elevation of the mucoperichondrium and/or mucoperiosteum (using a suction elevator) on both sides of the deviated region (Fig. 14.17). The osseous deviation is then cut with a Jansen–Middleton forceps in order to avoid damaging the cribriform plate; the deviated part is removed with a Takahashi forceps, and the mucoperichondrial flap is repositioned. Septal spurs are basically resected in two manners. Localized spurs are usually accessed through a vertical incision anterior to the deviated part followed by elevation of the mucoperichondrium and/or mucoperiosteum and mobilization of the spur with a flat chisel. The spur is then removed with Takahashi forceps. If the septal spur is located inferiorly along the anterior–posterior plane of the septum, it is preferable to make an incision along and superior to the spur and, from this incision, to elevate the mucoperichondrium and/or mucoperiosteum inferiorly. Finally, the septal deviation is resected using a flat chisel (Fig. 14.18).

Usually, is not necessary to suture the free ends of the mucoperichondrial incision. When endoscopic septoplasty is combined with surgical intervention in the paranasal sinuses, a septal splint and Merocel packing may be used, helping to achieve perfect coaptation of the mucoperichondrium.

Final Remarks

Operations on the nasal septum and turbinates assisted by an operating microscope and endoscopes have decreased morbidity by making these procedures much more precise. Some additional advantages provided by these surgical techniques include: (1) complete identification of the involved region, which makes the surgical intervention more conservative, (2) limiting of the extent of dissection to the area of the septal deviation, (3) limiting of the

Fig. 14.17. a Limited endoscopic septoplasty using a 4 mm, 0° endoscope. **b** Removal of the deviated portion of the nasal septum with a Takahashi forceps

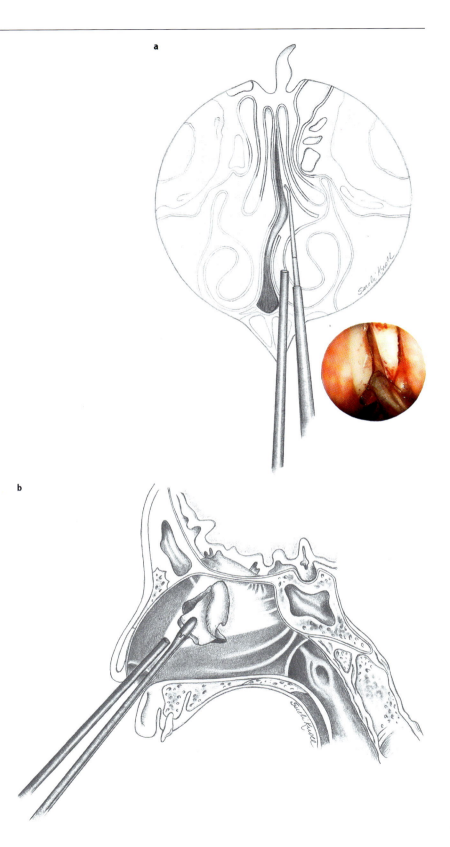

Fig. 14.18. a Coronal computed tomography of a patient with a nasal-septal spur compressing the inferior turbinate, with hypertrophy of the contralateral inferior turbinate. **b** Endoscopic resection of the nasal-septal spur by incising and elevating the mucoperichondrium along its superior extent

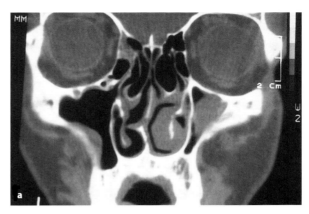

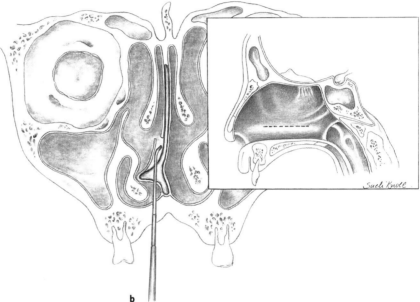

extent of resection of inferior-turbinate hypertrophy and deformities and diseases of the middle turbinate, (4) shortening of the surgical time, (5) lowering of the number of surgical instruments needed and (6) facilitation of teaching and documentation.

References

1. Becker W, Naumann HH, Pfaltz CR (1989) Ear, nose and throat diseases. Thieme, Stuttgart, pp 170–186
2. Biedlingmaier JF (1996) The middle turbinate window approach in endoscopic sinus surgery. Operative Tech Otolaryngol Head Neck Surg 7:275–277
3. Bielamowicz S, Hawrych A, Gupta A (1999) Endoscopic inferior turbinate reduction: a new technique. Laryngoscope 109:1007–1009
4. Courtiss EH, Goldwyn RM, O'Brien JJ (1978) Resection of obstruction inferior nasal turbinates. Plastic Reconstr Surg 62:249–255
5. Cottle MH (1958) The "maxilla–premaxilla" approach to extensive nasal septum surgery. Arch Otolaryngol Head Neck Surg 60:301
6. Dawes PJD (1987) The early complications of inferior turbinectomy. J Laryngol Otol 101:1136–1139
7. Elwany S, Hanison R (1990) Inferior turbinectomy: comparison of four techniques. J Laryngol Otol 104:206–209
8. Freer OT (1902) The correction of deflections of the nasal septum with a minimum of trauma. JAMA 38:636
9. Giles WC, Gross CW, Abram AC, et al. (1997) Endoscopic septoplasty. Laryngoscope 104:1507–1509
10. Goode RL (1978) Surgery of the turbinates. J Otolaryngol 7:262–268
11. Grossman J, Banov C, Boggs P (1995) Use of ipratropium bromide nasal spray in chronic treatment of nonallergic perennial rhinitis alone and in combination with other perennial rhinitis medications. J Allergy Clin Immunol 95:1123

12. Hinderer KH (1971) Fundamentals of anatomy and surgery of the nose. Aesculapius, Birmingham
13. Huygen PLM, Klaassen AB, de Leeuw TJ, Wentges RT (1992) Rhinomanometric detection rate of rhinoscopically assessed septal deviations. Rhinology 30:177–181
14. Jones M (1895) Turbinectomy. Lancet 2:879
15. Kasperbauer JL, Kern EB (1987) Nasal valve physiology: implications in nasal surgery. Otolaryngol Clin North Am 20:699
16. Killian G (1904) Die surnucose Fensterresektion der Nasenscheidewand. Arch Laryngol Rhinol 16:362
17. Lang J (1989) Clinical anatomy of the nose, nasal cavity and paranasal sinuses. Thieme, New York, p 144
18. Lanza DC, Kennedy DW (1993) Endoscopic sinus surgery. In: Bailey BJ (ed) Head and neck surgery. (Otolaryngology, vol 1) Lippincott, Philadelphia
19. Lenz H (1985) Acht Jahre Laserchirurgie an den unteren nasenmuscheln bei rhinopathia vasomotoria in form der laserstrich karbonisation. HNO 33:422–425
20. Mabry RL (1984) Surgery of the inferior turbinate: how much and when? Otolaryngol Head Neck Surg 92:571–576
21. Martinez SA, Nissen AJ, Stock CR, Tesmer T (1983) Nasal turbinate resection for relief of nasal obstruction. Laryngoscope 93:871–875
22. Metzembaum M (1929) Replacement of the lower end of the dislocated septal cartilage versus submucous resection of the dislocated end of the septal cartilage. Arch Otolaryngol Head Neck Surg 9:282
23. Moore GF, Freeman TJ, Orgren JP, Yonkers AJ (1985) Extended follow-up of total inferior turbinate resection for relief of chronic nasal obstruction. Laryngoscope 95:1095–1099
24. Navarro JAC (1997) Cavidade do nariz e seios paranasais. (Anatomia cirurgica, vol 1) All Dent, Baun, p 146
25. Ophir D, Schindel D, Halperin D, Marshak G (1992) Long-term follow-up of the effectiveness and safety of inferior turbinectomy. Plast Reconstr Surg 90:980–984
26. Ozenberger JM (1970) Cryosurgery in chronic rhinitis. Laryngoscope 80:723
27. Passali D, Anselmi M, Lauriello M, Bellussi L (1999) Treatment of hypertrophy of the inferior turbinate: long-term results in 382 patients randomly assigned to therapy. Ann Otol Rhinol Laryngol 108:569–575
28. Rakover Y, Rosen G (1996) A comparison of partial inferior turbinectomy and cryosurgery for hypertrophic inferior turbinates. J Laryngol Otol 110:732–735
29. Ridenour BD (1998) The nasal septum. In: Cummings CW, Fredrickson JM, Harker LA, et al. (eds) Otolaryngology, head and neck surgery. Mosby, St. Louis, pp 921–948
30. Saunders WH (1982) Surgery of the inferior nasal turbinate. Ann Otol Rhinol Laryngol 91:445–447
31. Scher RL, Gross CW (1990) Additional application for transnasal endoscopic surgery. Operative Tech Otolaryngol Head Neck Surg 1:84–88
32. Selkin SG (1985) Laser turbinectomy as an adjunct to rhinoplasty. Arch Otolaryngol 11:446–449
33. Spector M (1982) Partial resection of the inferior turbinates. Ear Nose Throat J 61:200
34. Stamm A (1995) Microcirurgia naso-sinusal. Revinter, Rio de Janeiro
35. Stammberger H (1991) Operative techniques. Saunders, Philadelphia, pp 290–296
36. Van Alyea OE (1939) Ethmoid labyrinth. Arch Otolaryngol Head Neck Surg 29:881–902
37. Wigand ME (1990) Endoscopic surgery of the paranasal sinuses and anterior skull base. Thieme, New York, pp 75–77
38. Yañez C (1998) A new use of the microdebrider in turbinate surgery: "intra-turbinate stroma removal with hummer". In: Stammberger H, Wolf G (eds) Congress of the European Rhinologic Society. Monduzzi, Bologna, pp 9–12

Surgical Approaches to Sinusitis: Overview of Techniques

Renato Roithmann, Ian Witterick, Michael Hawke, and Aldo Cassol Stamm

Introduction

In the early 1970s, Professor Walter Messerklinger began his investigations into the anatomy and physiology of the paranasal sinuses and the pathogenesis of sinusitis. He was the first to apply endoscopes routinely in the diagnosis of sinus disease and, subsequently, for the performance of surgical procedures on the paranasal sinuses.

Professor Messerklinger subsequently developed a specific surgical approach to chronic sinusitis; this approach has been called "the Messerklinger technique" or "functional endoscopic sinus surgery (FESS)". In the mid-1980s, Messerklinger's concept of the pathogenesis of chronic sinusitis and his surgical technique were rapidly introduced across the world, initially in the United States (by David Kennedy) and in Canada (by Michael Hawke).

In the intervening decade, surgeons have come to realize that the endoscope should be viewed as a key tool for visualization when performing surgical procedures on the paranasal sinuses. Nasal endoscopes have now been applied to a wide variety of different surgical approaches to the paranasal sinuses; as a result, we now refer to this type of surgery as "endoscopic sinus surgery". The following sections of this chapter will outline the principles of management of sinusitis and provide an overview of the techniques for the diagnosis and treatment of sinusitis in adult patients.

Principles of Surgical Management

Nasal/Sinus Anatomy

The anatomy of the paranasal sinuses and adjacent vital structures (like the orbit and the anterior cranial fossa) exhibits high variability from patient to patient. The surgeon must be able to recognize anatomical landmarks that provide surgical guidance for safe sinus surgery in each case. It is not recommended that a beginner start his/her learning curve by doing revision cases, where the anatomy may be very distorted and classical anatomical landmarks may not be available. Courses in endoscopic and/or microscopic sinus-surgery anatomy, especially those that include cadaver dissection, are essential to improve the surgeon's knowledge of nasal/sinus anatomy and the significant neighboring relationships. The anatomy of each individual patient should be clearly established by careful nasal endoscopy and radiological imaging prior to any proposed surgical procedure on the sinuses.

Diagnosis: Nasal Endoscopy/Computed Tomography

In our institutions, diagnostic nasal endoscopy complemented by computed tomography (CT) is a routine component of the preoperative management of every patient for whom sinus surgery is contemplated. Nasal endoscopy is performed in the outpatient setting with fiberoptic telescopes (rigid or flexible) and topical nasal decongestion, often in combination with a topical anesthetic agent. Rigid telescopes are available in a variety of diameters (4 mm and 2.7 mm) and angles of view (0°, 30°, 70° and 120°). The 0° and 30° 4-mm rigid telescopes provide an excellent field of view in most instances but may be too wide to access the narrow nasal cavities found in some patients. In these cases, a 2.7-mm rigid telescope or a flexible telescope may be useful. The goal of nasal endoscopy is to visualize and document the anatomy (normal and abnormal) and pathologic processes of the nasal cavity. Particular attention is paid to the lateral wall of the nose in the region of the middle turbinate and middle meatus, where the anterior ethmoid, frontal and maxillary sinuses drain (Fig. 15.1). Nasal endoscopy is able to delineate mucosal surfaces and identify middle-meatal soft-tissue abnormalities that are not diagnosed on imaging studies [63]. However, CT scans of the sinuses are indispensable for the further evaluation of the extent of disease and the clarification of anatomical relationships (Fig. 15.2). Plain sinus radiographs are of little value in assessment of the ostiomeatal complex (OMC), because they do not show this critical region in enough detail to differentiate subtle anatomical and mucosal changes [67]. CT scans clearly demonstrate and differentiate soft tissues, bone and air [47, 67] and are the imaging modality of choice for clinical evaluation and planning of FESS. Ideally, the CT examination should be performed when the patient is least symptomatic and after antibiotics have been administered to control acute exacerbation. The coronal plane (without contrast) is preferred, because the images are displayed in the anatomic plane encountered during endoscopic surgery, and this view affords optimal demonstration of the anterior ethmoid and ostiomeatal structures (Fig. 15.2a). The CT scans are scrutinized for anatomical variations, the relationship of the OMC to the skull base and lamina papyracea, evidence of inflammation, mucoperiosteal thickening and the patency of the ostia and ethmoidal prechambers. In previously operated patients or those for whom posterior ethmoid or sphenoid sinus surgery is contemplated, biplanar (coronal + axial) CT scanning is helpful in order to determine the relationships between important surrounding structures, including the optic nerve and the internal carotid artery (Fig. 15.2b).

Mucociliary Clearance (Messerklinger's Concepts)

In the past, conventional sinus operations for chronic infection were directed at removing diseased mucosa in the sinus. It was believed that chronic sinusitis was an infectious process arising primarily

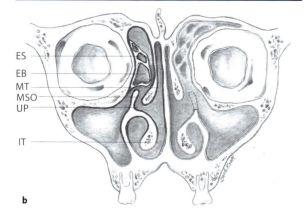

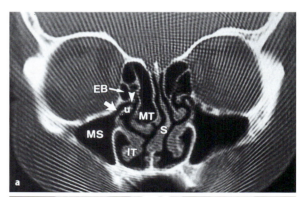

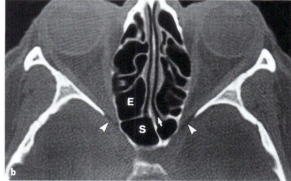

Fig. 15.1. a Endoscopic view of the middle turbinate and the area of the middle meatus. **b** Anatomy of the ostiomeatal complex. *EB*, ethmoid bulla; *ES*, ethmoid sinus; *IT*, inferior turbinate; *MSO*, maxillary sinus ostium; *MT*, middle turbinate; *UP*, uncinate process

Fig. 15.2. a Ostiomeatal complex: coronal computed tomography (CT) view. **b** CT (axial view). *E*, posterior ethmoid cell; *EB*, ethmoid bulla; *IT*, inferior turbinate; *MS*, maxillary sinus; *MT*, middle turbinate; *S*, nasal septum (coronal view) or sphenoid sinus (axial view); *U*, uncinate process [47]

from abnormalities within the affected sinus itself. Based on studies of mucociliary clearance, Messerklinger demonstrated that the etiology of most cases of chronic recurring sinusitis in the maxillary and frontal sinuses are related to anatomical variations or mucosal pathology obstructing the natural drainage pathways of these sinuses [39].

Surgically created drainage openings in a sinus at locations distant from the natural ostia are bypassed by the mucous blanket produced in the sinus, which is preferentially transported to the natural ostium by the cilia of the respiratory epithelia. The mucous produced in the anterior ethmoid, frontal and maxillary sinuses drains into the OMC, an anatomically complex region of the lateral nasal wall under the middle meatus (Fig. 15.1). The OMC is bounded by the middle turbinate medially, the roof of the ethmoid sinus superiorly and the medial wall of the orbit (lamina papyracea) laterally, and it opens into the middle meatus inferiorly.

Mucous from the anterior ethmoid, frontal and maxillary sinuses drains into the ethmoidal infundibulum and then into the middle meatus through the hiatus semilunaris. The hiatus semilunaris is located between the posterior edge of the sickle-shaped uncinate process anteriorly and the bulla ethmoidalis (the largest air cell of the anterior ethmoid sinus) posteriorly.

The OMC is an anatomically narrow region. Small variations in the size or shape of the structures or minor mucosal edema can cause opposing mucosal layers to come into contact with each other. This mucosal contact impairs the mucociliary clearance and ventilation of the sinuses draining into the OMC.

Messerklinger pioneered the concept that re-establishment of ventilation and drainage via the physiologic routes can bring about resolution of even extensive pathological changes in the dependent sinuses (maxillary and frontal) without touching mucosa in the sinus itself [25, 55]. Surgery to achieve this goal has been termed FESS by Kennedy [25]. Some sinus surgeons use the microscope and the endoscope (a technique termed micro-endoscopic endonasal surgery) to perform functional sinus surgery (see "Microscopic Endonasal Surgery") [1, 54, 65]. Restoration of mucociliary clearance and ventilation through the OMC is now considered essential in managing most patients with chronic recurring sinusitis [4].

Management of Systemic Disease

The pathogenesis and natural course of chronic hypertrophic inflammatory sinusitis is incompletely understood. In order to maintain control of their symptoms and to decrease the risk of recurrence of disease, patients with diffuse nasal polyposis, aspirin sensitivity and asthma require continued medical therapy following OMC surgery [24, 30, 58]. Surgery alone, performed by skilled surgeons, might improve ventilation through the natural pathways of drainage, but it will not solve the problem of recurring and chronic sinusitis in all circumstances, and it must be recognized that systemic host factors contribute to mucociliary dysfunction and nasal/sinus mucosal edema. These factors include allergies and immunological, environmental (pollution, smoking) and genetic influences [20, 24, 33].

Surgical Planning

Patient symptoms and endoscopic and imaging findings must all be taken into account in precise surgical planning. Factors considered in the choice of procedure include the type and extent of disease and the patient's physical condition and ability to withstand anesthesia. In addition, the expectations and wishes of the patient must be considered. In reality, for many patients with chronic sinusitis, the choice between a traditional, microscopic or endoscopic technique may largely depend on the training, experience and skill of the surgeon.

Whatever technique is used, the surgeon must remember the critical role of the OMC in the pathogenesis of chronic and/or recurring sinusitis. Current concepts of sinus surgery direct the surgeon to tailor the operation to the disease with maximal preservation of healthy tissue and to re-establish the sinus drainage system via the natural ostia (see "Mucociliary Clearance") [29, 57].

There are wide regional variations in the indications for surgery and the type of surgical procedure performed for chronic sinusitis; during the past decade, there has been increasing interest in and adoption of "minimal access surgery". In otolaryngology and head and neck surgery, the interest in FESS has grown exponentially since its introduction in North America in 1985 [25, 29, 35, 45, 50, 51, 56–58, 64, 66]. This does not mean that patients with chronic or recurrent sinusitis cannot be managed with equally good results by "traditional" or conventional sinus operations [12, 30].

Clear advantages of FESS include minimal trauma to normal, non-diseased tissue and the ability to remove often subtle pathology. Endoscopic techniques are well suited to opening the natural ostia of the maxillary and frontal sinuses. In addition, illumination and visualization with an endoscope are superior to illumination and visualization with a head light and nasal speculum, and endoscopes with deflected angles of view allow the evaluation and removal of disease from recesses that could not be seen by previously used methods. The microscopic technique also has advantages; for instance, it allows bimanual surgery, which may be helpful in cases with excessive bleeding (see "Microscopic Endonasal Surgery"). Informed consent is obtained after detailed discussion of the disease process, alternative management strategies and the proposed surgery, including the type of anesthesia (local or general), the risks of surgery (see "Complications"), realistic expectations and the importance of postoperative follow-up.

Fig. 15.3. Caldwell-Luc procedure

Conventional Techniques

The priorities of conventional sinus surgery include complete removal of all diseased tissue and opening of the natural drainage pathways. Operations were developed to create an opening into the sinuses (trephination) and to curette diseased mucosa out of every sinus either intranasally or via an external incision. If opening of the natural pathways is not feasible, wide communication into the nose at another site is created to permit gravitational drainage, or the sinus is obliterated to prevent ingrowth of mucosa and possible recurrence of disease. Cosmetic concerns with external procedures are also of concern, particularly to the patient.

Conventional Maxillary Sinus Operations

Antral irrigation or lavage is a procedure intended to clear infected purulent material from the maxillary sinus. The procedure is typically carried out in the physician's office with local anesthesia. A trocar is placed into the sinus either through the canine fossa or under the inferior turbinate. Sterile saline or some other solution is gently irrigated into the sinus. Antral lavage may help some patients with acute maxillary sinusitis that is not responding to medical management; such lavage is useful for obtaining culture material. Sinuses with thickened membranes are not appropriate for irrigation. In our experience, antral lavage is rarely helpful in managing patients with chronic sinusitis.

Intranasal antrostomy involves creating a large window into the maxillary sinus under the inferior turbinate. The window needs to be large to preclude subsequent closure during the healing phases. This procedure usually provides ventilation of the sinus but may not provide adequate drainage, as the cilia continue to transport mucous to the natural ostia, bypassing the antrostomy.

Prior to the development of FESS, the Caldwell-Luc was the standard procedure for patients with recurrent or chronic maxillary sinusitis (Fig. 15.3). Even with endoscopic techniques, it is still useful for removing fungal disease, massive polyposis and some mucoceles or cysts from the sinus. It can be performed under local or general anesthesia. An incision is made in the upper gingivobuccal sulcus over the canine fossa, and an opening large enough to inspect the entire sinus and remove diseased tissue is created in the anterior wall of the sinus. A nasoantral window is made under the inferior turbinate or the natural ostium is enlarged. The most common complication is paresthesia or anesthesia of the cheek, teeth and gingiva secondary to stretch injury of the infraorbital nerve; this may persist for 6 months. Other complications include persistent pain, bleeding, facial swelling, oroantral fistula and damage to the lacrimal system, the orbit and the cribriform plate.

Conventional Frontal-Sinus Operations

Trephination of the frontal sinus is performed for acute frontal sinusitis unresponsive to medical management. An incision is made in the superior and medial aspects of the orbit immediately below the brow. The sinus is entered through the inferior edge of the supraorbital rim, the purulent material is evacuated, and a drain is placed. Postoperatively, the sinus may be irrigated daily through the drain until the return is clear.

There are several external operations available for patients with chronic frontal sinusitis, including mucosa-preserving and mucosa-eliminating procedures. Both types of mucosa-preserving techniques begin with an external ethmoidectomy. The Lynch procedure removes a large part of the medial floor of the frontal sinus, and the Boyden procedure extends this to enlarge the nasofrontal duct and attempts to reconstruct it with a mucosal flap, usually pedicled from the septum. The disadvantages of these procedures are the required external incision and the potential for the orbital soft tissues to collapse medially and obstruct the nasofrontal duct once the stent is removed.

There are three basic types of mucosa-eliminating procedures, which remove variable portions of the frontal bone and all of the sinus mucosa. In the Reidel operation, the anterior wall and floor of the frontal sinus is removed, and the sinus is obliterated by re-draping the forehead skin over the posterior table, leaving a major cosmetic deformity.

The Killian procedure is similar but spares the supraorbital rim and approximately 1 cm of the adjacent inferior frontal bone, so the postoperative appearance is more acceptable. These two procedures are rarely performed but may be indicated for severe frontal sinusitis with osteomyelitis of the anterior table.

The third option is the osteoplastic flap, which is the external mucosa-eliminating procedure of choice in most circumstances. This is particularly suited to symptomatic mucoceles placed laterally in the frontal sinus distant from endoscopic access or in cases of stenosis of the frontal recess and/or ostia. The osteoplastic-flap procedure involves creating an anterior frontal-bone flap through a bicoronal or brow incision. The bicoronal incision is preferable, because it can usually be concealed by hair (which may recede later in life), whereas the brow incision leaves a noticeable scar and (usually) some postoperative numbness of the forehead (Fig. 15.4). A posteroanterior plain sinus X-ray is used as a template intra-operatively to define the frontal-bone flap, which is fractured inferiorly and attached by the periosteum. The mucosa of the entire sinus is removed, and several options exist for managing the frontal ostium/recess and the sinus cavity. Traditionally, the frontonasal ducts were plugged (for example, with fascia), and the sinus was "obliterated" by fat obtained from another site (abdomen, thigh, submental region). The fat is non-vascularized and may reabsorb over time. Another technique enlarges the frontal ostium/recess with a drill and stents the area for a variable period postoperatively. If one frontal recess is patent and functional, another option is to remove the inter-sinus septum so that mucous from both frontal sinuses can drain through the functional side.

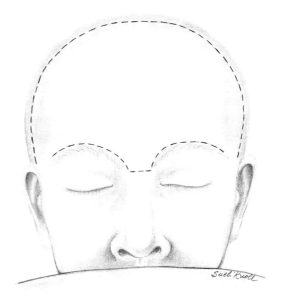

Fig. 15.4. Standard incisions to perform an osteoplastic flap include bicoronal or upper-brow and glabella ("seagull") incisions. A mid-forehead incision in a natural skin crease is also a possibility (not shown)

Conventional Ethmoid-Sinus Operations

The ethmoid sinus may be approached through the nose, through the maxillary sinus or by external incisions. A frequent indication for intranasal ethmoidectomy has been recurrent obstructive nasal polyposis in which a patent nasal airway could not be maintained by office polypectomy. The surgeon must recognize the ethmoid anatomy and the critical relationships with adjacent structures (such as the lamina papyracea, fovea ethmoidalis, cribriform plate and optic nerve) in order to avoid major complications during ethmoidectomy (Fig. 15.2).

The transantral approach to the ethmoid sinus gives good visualization of the floor of the orbit, which the surgeon can follow medially and superiorly as it becomes the lamina papyracea. If the only goal of surgery is to open the ethmoid sinuses, then this approach would not seem to be substantially preferable to intranasal ethmoidectomy, due to the requirements for a sublabial incision and a Caldwell-Luc-type approach. The transantral approach may be preferable when portions of the orbital floor are also removed for thyroid orbitopathy. External ethmoidectomy can be used for treatment of extensive polypoid sinus and nasal disease, chronic ethmoiditis and as an approach to neoplasms and cerebrospinal-fluid leaks of the frontoethmoid region. The advantage of external ethmoidectomy is excellent visualization, but at the price of an external incision and the removal of the lamina papyracea, which may be important as a barrier to protect the eye from future sinus disease.

Intranasal ethmoidectomy in the presence of extensive nasal polyposis can be a difficult and dangerous procedure because of problems associated with visualizing the anatomy with a headlight and mirror, bleeding associated with removal of the polyps and the distortion of landmarks due to the polyps themselves or due to previous polypectomies. Following removal of any nasal polyps, the anterior half of the middle turbinate is usually resected to gain access to the ethmoid sinuses, which are then removed. The associated bleeding can be profuse at times, so intermittent packing is required for control.

External ethmoidectomy is usually performed through a gently curved elliptical incision midway between the inner canthus and the dorsum of the nose (Fig. 15.5). Elevation of the periosteum extending from the lamina papyracea identifies the frontoethmoid suture line, which serves as a landmark for the level of the cribriform plate and the uppermost extent of the posterior ethmoidal cells. The anterior (and sometimes the posterior) ethmoid arteries are ligated or cauterized as they exit through the frontoethmoid suture line. The close proximity of the optic nerve to the posterior ethmoid artery must always be remembered. The lamina papyracea is removed, and the ethmoidal cells are exenterated.

Fig. 15.5. External ethmoidectomy incision

Conventional Sphenoid-Sinus Operations

The sphenoid sinus can be reached by trans-septal, transnasal and transethmoid approaches. The preferred procedure depends on the surgeon's experience, the local anatomy and the coexistence of ethmoid disease. If ethmoid disease is absent, a direct route (trans-septal or transnasal) is preferable but, if there is co-existent ethmoid disease, then one of the other approaches may be selected.

A sphenoidotomy is a relatively simple procedure once the surgeon recognizes the local anatomy of the posterior nasal cavity (the arch of the choana, the posterior attachment of the middle turbinate and the posterior portion of the nasal septum). The front face of the sphenoid in adults is usually at a 30° angle from the hard palate and is 7 cm posterior to the anterior nasal spine. The sphenoid sinus can be entered at this level, between the posterior attachment of the middle turbinate and the nasal septum, with a curette or a straight Blakesley forceps and the anterior wall removed with micro-Kerrison forceps. Bleeding from the sphenopalatine artery at the front face of the sphenoid can occur and is easily controlled with a suction coagulator.

Functional Endoscopic Sinus Surgery

The guiding concept of FESS is removal of tissue obstructing the OMC and facilitation of drainage while conserving normal non-obstructing anatomy and mucous membrane (see "Messerklinger's Concepts") [25, 56]. The rigid fiberoptic nasal telescope provides superb intraoperative visualization of the OMC, allowing the surgery to be focused and precise. Bleeding is minimized by vasoconstricting the mucosa, and atraumatic technique. The 0° (forward viewing) 4 mm telescope is the endoscope of first choice, and the 30° and 70° have special uses, such as examining the frontal recess or maxillary sinus through the natural ostium. The image may be projected to a television monitor through a small camera attached to the eyepiece of the endoscope. The surgeon has the choice of performing the operation while looking through the endoscope or at the monitor, or a combination of both. Proper equipment and training are required to identify the important anatomical landmarks and perform the directed surgical techniques. The development of powered, rotary soft-tissue shavers (microdebriders), has helped endoscopic sinus surgeons to accomplish the main goal of the functional sinus surgery, which is removal of pathologic tissue while preserving normal mucosa [17, 52]. Newer true-cutting instrumentation has also allowed precise resection of pathologic tissue, with less chance of mucosal stripping.

Indications and Contraindications

In our institutions, the main established indications for FESS include:
A. Continued chronic and/or recurrent sinusitis despite appropriate medical treatment or previous surgical treatment
B. Recurrent sinusitis secondary to massive obstructive nasal polyposis
C. Correction of anatomic variations predisposing the patient to chronic and/or recurrent sinusitis

The indications for surgical management of adult rhinosinusitis were recently reviewed in a paper published by Anand et al. [2]. In their classification, indication A is defined as relative for sinus surgery, which means surgery is required after medical treatment has failed. Indication B is classified as absolute, which means there is a clear reason for surgery. Although an association between anatomic variations of the lateral nasal wall and chronic/recurrent sinus infections has been observed in many patients, a cause–effect relationship has not been proved. These anatomic variations are found in many subjects without chronic or recurrent sinus infections. This indication is considered relative by the Task Force on Rhinosinusitis [3].

The ideal candidate for FESS is the patient who remains symptomatic despite appropriate medical treatment and whose OMC shows persistent inflammation and/or anatomic obstruction (associated or not with pathological CT findings) on nasal diagnostic endoscopy [23, 31]. Appropriate medical treatment should include prolonged courses of broad-spectrum antibiotics, topical nasal corticosteroids and, in some cases, systemic corticosteroids.

Approximately 30% of patients without sinus symptoms or endoscopic findings have been found to exhibit sinus "abnormalities" on CT scanning performed for reasons other than suspected sinus disease [5, 16]. Consequently, the significance of CT-scan abnormalities must be evaluated in conjunction with the patient's symptoms and endoscopic findings.

The vast majority of patients with acute sinusitis are managed medically, with surgery being reserved for those who fail to respond or who are threatened by complications (orbital, intracranial). Endoscopically placed cotton pledgets soaked with vasoconstrictor can be placed in the middle meatus for 20–30 min to reduce edema around the ostia and promote drainage of purulent secretions. This form of therapy is done in the office and, when repeated two to three times each day together with systemic antibiotics, may avert an impending complication and may even bring about the resolution of severe cases of acute sinusitis. Experienced surgeons have used endoscopic measures to drain periorbital abscesses caused by acute ethmoiditis [11, 57] and in the management of empyema of the frontal [43] or maxillary sinus [57].

Fungal sinusitis, mucoceles and nasal polyps have also been successfully managed through FESS [57]. Extended applications of the endoscopic technique include the management of cerebrospinal-fluid leak, nasal meningoceles and some benign neoplasms, orbital decompression for thyroid orbitopathy, refractory headache syndromes, choanal atresia, lacrimal obstruction, optic-nerve decompression and pituitary-tumor surgery [7, 21, 26, 38, 40].

The Messerklinger technique does not seem to be appropriate for the surgical approach to extensive, invasive disease involving the paranasal sinuses or skull base. Examples include osteomyelitis, malignant neoplasms and benign fibro-osseous lesions. It is also contraindicated in patients with an incipient central nervous system complication related to acute sinusitis. Endoscopic techniques may be unsatisfactory for accessing laterally positioned disease in the frontal or maxillary sinuses or for opening post-inflammatory stenoses of the ostium of the frontal sinus [29, 57].

Technique

FESS can be done under general or local anesthesia, depending on the patient's general condition, the extent of the disease and the surgeon's experience (see "Surgical Planning"). Patients undergoing surgery under local anesthesia tend to bleed less than those undergoing the same procedure under general anesthesia. Other advantages of local anesthesia include less risk of cardiorespiratory complications than with general anesthesia; also, intraoperative pain can provide the surgeon with an important early warning of impending injury to the roof of the ethmoidal sinus, orbit or optic nerve. In addition, if an orbital complication develops (orbital hematoma), it may be recognized earlier, and appropriate treatment and follow-up can be initiated [29]. In most adults, local anesthesia with intravenous sedation provides excellent hemostasis [14] and appropriate visibility.

Local anesthesia is generally associated with a shorter and easier post-operative recovery than general anesthesia. The anesthesiologist plays an important role in sedating and monitoring the patient during local anesthesia. Advantages of general anesthesia include immobilization of the patient and easier instrumentation because of the reduction of intraoperative pain and anxiety.

The medications used for topical anesthesia and vasoconstriction differ among centers. A single agent, such as topical cocaine (4–10%), or a combination of agents (2% tetracaine and 1:1000 epinephrine or 0.1% xylometazoline) may be used for these purposes. Although dosages based on patient weight must be adhered to, submaximal dosages of cocaine have been associated with cardiac arrhythmias, particularly when used in combination with epinephrine and some inhalational anesthetic agents. Cottonoids with the topical agents are placed in the OMC region for 5–10 min, followed by infiltration of the lateral nasal wall (uncinate process) with an anesthetic/vasoconstricting agent (1% or 2% xylocaine with 1:100,000 or 1:200,000 epinephrine) [46, 57]. The mucosa of the middle turbinate is injected when partial or total turbinectomy is contemplated. Inadvertent damage to the mucous membrane of the anterior nose by the tip of the needle must be avoided, and the number of injected sites should be minimized to limit oozing from puncture sites. If the patient reports pain during the course of the surgical procedure, topical anesthetic is reapplied. Injection of the sphenopalatine nerve occurs via either the intranasal or transpalatal route when total sphenoethmoidectomy is planned; this complements the local anesthesia protocol [24].

Bleeding is a major concern for the endoscopic surgeon. It is critical that surgeons see what they are doing and, although the rigid nasal telescope gives an excellent view of intranasal structures, even small quantities of blood can obscure landmarks, making the procedure more difficult and sometimes dangerous. Therefore, it is essential to maintain excellent hemostasis. Various techniques to control bleeding include topical mucosal vasoconstriction, infiltration of vasoconstricting drugs, hypotensive anesthesia and minimization of trauma to non-diseased tissue. Microdebriders have helped some sinus surgeons to perform endoscopic sinus surgery with less bleeding and less mucosal stripping in patients with large polyps [15, 17, 52]. In fact, it is important to prevent bleeding pre-operatively by instructing the patient to avoid aspirin and other non-steroidal anti-inflammatory drugs for at least 1 week before the procedure. Some patients with diffuse mucosal edema and nasal polyposis benefit from a short (7- to 14-day) preoperative course of systemic corticosteroids to reduce tissue edema (and possibly bleeding) at the time of surgery.

In order to minimize the risk of disorientation, experienced endoscopic surgeons recommend using the 0° (straightforward viewing) telescope as much as possible [24, 53, 57]. The 30° or 70° nasal endoscopes are more safely used after the important topographic landmarks have been identified and are usually required only in special situations (for viewing the frontal recess or the cavity of the maxillary sinus).

Although it is possible to perform total sphenoethmoidectomy by the Messerklinger technique, such radical procedures are avoided by most surgeons. According to Stammberger and Hawke [57], "under no circumstances should extensive bony surfaces be denuded of their mucosal covering, because this will increase the chance of postoperative osteitis". Regardless of the technique used, the CT scans are displayed in the operating room during FESS so the surgeon can easily review them during surgery.

Anterior-to-Posterior Dissection

The first step consists in opening the ethmoidal infundibulum by resecting the uncinate process in its entirety. The uncinectomy can be performed in an anterior-to-posterior direction with a sickle knife or Freer elevator or in a posterior-to-anterior direction with a backbiter. If a sickle knife is used for this purpose, it is important to keep the blade parallel to the lateral nasal wall to avoid injuring the lamina papyracea and orbital contents. Commonly, superior and inferior remnants of the uncinate process obscure access to the frontal recess and the maxillary sinus ostium and should be removed.

The next step depends on the preference and experience of the surgeon. Some surgeons prefer to preserve the ethmoidal bulla at this stage and to use it as a guide to the frontal recess and the maxillary-sinus ostium areas. Other surgeons routinely open the ethmoidal bulla at this stage, starting with its medial and inferior portions. The anterior ethmoid air cells are resected until the basal lamella is identified. The basal lamella is the lateral attachment of the middle turbinate and is also referred to as the "ground" or "grand" lamella. If the ethmoidal bulla and adjacent cells are completely removed, the operative field will be bordered medially by the middle turbinate, laterally by the lamina papyracea, superiorly by the roof of the ethmoid sinus and posteriorly by the vertical portion of the basal lamella of the middle turbinate. The posterior ethmoidal sinus is then approached (if diseased) through the basal lamella as far medially and inferiorly as possible. Removing the entire basal lamella should be avoided to prevent destabilizing the middle turbinate. Laterally, the optic nerve may lie in intimate contact with posterior ethmoid air cells (cells of Onodi). The angle of the endoscope must change (usually from 45° to 30°, relative to the palate) as the surgeon proceeds more deeply into the ethmoidal labyrinth. This maneuver allows the endoscopist to follow a path along the base of the skull. In the minority of cases in which it is necessary to open the sphenoid sinus, entry should be as far medially and inferiorly as possible to decrease the risk of injury to the optic nerve and internal carotid artery. In order to determine the anatomic relationships of these important structures, careful review of coronal and axial CT scans is essential before the sphenoid is approached surgically (Fig. 15.2b).

The surgical steps that follow or (in accordance to the surgeon's experience) precede the resection of the anterior ethmoid include inspection of the frontal recess and the maxillary-sinus ostium. The frontal recess should be cleared of disease but, in most cases, the sinus does not need to be entered, nor is the ostium of the frontal sinus routinely enlarged. The entrance to the frontal sinus is manipulated as little as possible in the Messerklinger technique and, although there are specific techniques described for the safe dissection of the frontal recess [32, 41], this is a technically challenging area to manage by endoscopic means. The frontal recess is anterior to the anterior ethmoidal artery, and many surgeons use this artery as a landmark for the posterior-to-anterior approach. One must be extremely careful, particularly medially, as the skull base is very thin and easily perforated at this point. The supraorbital ethmoid air cell that lies lateral to the frontal recess should be identified. It may be necessary to remove a superior remnant of the uncinate process to localize the frontal recess. If it is obscured by an enlarged agger nasi cell, we use Kuhn curettes (45° and 90°) and angled Blakesley-Weil forceps to explore this intricate region. Severe and long-standing sinus disease may cause frontal-recess contraction and bony obstruction (which can make endoscopic enlargement difficult or impossible [57]) and is often better managed through traditional external frontal-sinus procedures. If the frontal recess cannot be identified accurately from the nose, a small trephination in the anterior frontal sinus can be done to visualize the progress of the endoscopic procedure from above.

The ostium of the maxillary sinus is typically lateral to an inferior remnant of the uncinate process. Palpation with a bent spoon or curved suction along the bony insertion of the inferior turbinate will help to locate the ostium. If the natural ostium is stenosed, enlargement is performed posteriorly and inferiorly. The extent to which the surgeon enlarges the ostium depends on the severity of the condition, but excess anterior extension risks damage to the nasolacrimal duct and should be avoided. Although its benefit has not been conclusively established, surgical connection of accessory ostia with the natural ostium is routinely performed in an attempt to prevent the circular transportation of secretions shown by Messerklinger's studies [57]. Under the guidance of 30° or 70° scopes, the interior of the maxillary sinus is inspected through the natural ostium, but it is not routine to interfere surgically with mucosal disease.

Middle-turbinate surgery is not a routine part of functional endoscopic sinus surgery. Partial resection is done in cases where the turbinate prevents access to the middle meatus or when the middle turbinate is compromised by severe polyposis. Removal of the lateral half of a concha bullosa that contributes to ethmoidal disease or is itself diseased is, in many cases, the first surgical step of the endoscopic procedure. Total turbinectomy should be avoided, if possible, because the middle turbinate is an essential anatomical landmark [28] and has an important role in nasal physiology. Moreover, there is evidence indicating that resection of the middle turbinate might increase the risk of postoperative frontal sinusitis [62].

Turbinate surgery increases the risk of adhesion between the middle turbinate and the lateral nasal wall from denuded mucosal surfaces and weakens the support of the turbinate. A Merocel sponge moistened with beclomethasone, inserted into the middle meatus for 24–48 h, has been used to prevent adhesion, [57] but our current practice of placing silastic sheeting in the middle meatus and leaving it in position for at least 1 week postoperatively has shown promising results.

Following endoscopic surgery under local anesthesia, we do not routinely pack the middle meatus. In some cases, a strip of gauze soaked in vasoconstrictor is left in situ for 2–4 h and is removed before the patient is discharged. An overnight gauze pack impregnated with bismuth–iodoform–paraffin paste is a useful hemostatic measure when bleeding is severe during surgery or if, at the end of the procedure, there is persistent bleeding. In these cases, the patient's discharge from hospital is delayed. Petrolatum-based ointments are not used in the postoperative sites, because they might induce a submucosal subcutaneous inflammatory condition known as myospherulosis [27, 42].

Some surgeons successfully employ a "two-handed" video-endoscopic technique [31]. The principles of surgery are the same as in the procedure described above, but a well-trained assistant guides the endoscope, leaving the operating surgeon's hands free to manipulate the suction cannula while holding surgical instruments. Both surgeon and assistant monitor the procedure by video.

Posterior-to-Anterior Dissection (Wigand Technique)

The first step is to identify the sphenoid sinus by partial resection of the posterior end of the middle turbinate. The sphenoid ostium is then located, and the anterior wall of the sphenoid is penetrated; this is followed by dissection of the posterior ethmoid. The skull base is identified posteriorly, and dissection is completed in a posterior-to-anterior direction. The procedure offers some advantages in the endoscopic management of extensive sinus disease and is usually performed under general anesthesia [66].

Post-Operative Care

The early and long-term success of functional sinus surgery depends not only on the procedural expertise and competence of the endoscopic surgeon but on meticulous postoperative endoscopic follow-up and, in many cases, medical therapy. Long-term medical therapy for treatment of mucosal hyperreactivity is essential in many patients, such as those with cystic fibrosis or the ASA triad (asthma, polyposis and aspirin sensitivity).

The patient should be informed that nasal obstruction secondary to reactive swelling of the nasal mucosa is to be expected during the first few post-operative days; the patient should also be instructed to use acetaminophen for pain relief if necessary. The importance of the strict instruction not to blow the nose during the first 48 h after surgery is emphasized, and the risk of surgical emphysema of the facial tissues, the orbit or the cranial cavity is explained. These complications could ensue, especially if the lamina papyracea or bone of the skull base was breached during the procedure. A systemic antibiotic (penicillin or cephalosporin) is commonly prescribed for at least 1 week post-operatively. The thick crusts that form in the ethmoid cavity can be removed with greater ease if they are softened by saline spray and mineral oil, and it is our usual practice to have patients sniff a homemade solution consisting of 1 teaspoon each of sugar, salt and sodium bicarbonate dissolved in 240 ml of warm water. Between the first and fourth postoperative days, judicious removal of secretions, clots and crusts from the nasal cavity and middle meatus is undertaken and, in some cases, it may be possible to remove secretions from maxillary and frontal sinuses by means of a curved aspirator (Eustachian-tube catheter).

The healing phase, although not entirely predictable, requires at least 2–3 weeks. Careful, regular nasal endoscopy to clean the ethmoid cavity and to identify residual (remaining polyps, bone fragments) or developing problems (synechia, ostial stenosis) is essential on during the months following surgery. Topical nasal corticosteroids or short courses of systemic corticosteroids help to prevent or delay the recurrence of diffuse polypoid rhinosinusitis. Imaging is not routine during the first postoperative months unless there are complications requiring it.

Complications

The complications of endoscopic sinus surgery are similar to those associated with traditional techniques and depend largely on the surgeon's experience. Stankiewicz [59] has shown that, as the endoscopic surgeon's experience increases, the complication rate decreases. The beginner should start his/her learning curve by doing endoscopic surgery in patients with limited OMC disease (anterior ethmoid). As confidence with instrumentation and technique increases, surgery can be performed on more complicated cases (disease involving the posterior ethmoid or sphenoid sinuses). Revision cases, where conventional anatomical landmarks might not be easily recognized or are not present, and cases with scarring and fibrous tissue that can predispose the patient to excessive bleeding should be done only by surgeons having extensive experience with endoscopic sinus surgery.

Nevertheless, complications even occur in surgeries performed by competent physicians, and surgeons must be prepared to recognize and manage such problems at an early stage, especially problems involving the orbit and the brain [22, 37, 60, 61]. Major and minor complications are summarized in Table 15.1.

Computer-assisted endoscopic sinus surgery ("navigation sinus surgery") has been described, and it might be useful in reducing complications during difficult revision cases, where the anatomical landmarks are usually not available [6, 48]. This might be especially useful in providing accurate three-dimensional intraoperative localization in the anterior skull-base region and the frontal or sphenoid sinuses. The reader should refer to other sections of this book for a more detailed description of the complications of FESS and their management.

Table 15.1. Complications of endoscopic sinus surgery

Minor	Major
Synechiae	Intracranial injury and/or bleeding
Ostial stenosis	Cerebrospinal-fluid leak
Minor hemorrhage	Persistent diplopia
Periorbital ecchymosis	Blindness
Orbital emphysema	Carotid-artery injury
Transient diplopia	Orbital hematoma
Tooth numbness and pain	Severe hemorrhage
Nasolacrimal-duct injury	Meningitis and brain abscess
Disturbance of olfaction	

Results

Ventilation and drainage through the OMC are essential for the proper functioning of the sinuses. Ventilation is related to the local anatomy, and FESS is highly effective in correcting the anatomical abnormalities that obstruct the physiologic routes of drainage of the paranasal sinuses. Drainage, however, depends on both the local anatomy and on effective mucociliary function. Even when perfectly done, FESS might not improve the sinus drainage when the mucociliary disorder is related to a systemic disorder (cystic fibrosis) or a primary disorder of the cilia (Kartagener's syndrome) rather than a local anatomy problem.

It is our own and others' experience that the best results with the Messerklinger technique are achieved in patients with limited disease obstructing the OMC [24, 57]. Patients with diffuse nasal polyposis or hyperplastic sinusitis, allergic rhinitis, reactive airway disease (asthma), the ASA triad (aspirin, asthma, nasal polyps), immune disorders, cystic fibrosis and previous endonasal sinus surgery respond less well. They do not experience the same rate of symptom improvement as the group described above [51, 57] and require continued medical therapy.

It is difficult and probably inappropriate to compare results published from different centers, because there is no uniform rhinosinusitis staging system or reporting criteria for success and failure [13, 24, 35, 51]. The situation is further complicated, because subjective symptomatology correlates weakly with objective information (nasal endoscopy and/or CT) [5, 16, 36].

To date, the task force on rhinosinusitis [3] recommends the Lund-MacKay staging system to evaluate the extent of disease. The system is based on a simple numeric score derived from a CT scan obtained after an adequate trial of medical treatment [34]. Patient symptoms and the endoscopic appearance of the nose are also scored before surgery and at regular postoperative intervals [34].

In the absence of a standardized staging system or outcome measure for rhinosinusitis, the success rate is reported as greater than 80% in adult patients. Patients report substantial improvement in their symptoms and health status [8, 19, 44].

Microscopic Endonasal Sinus Surgery

Microscopic endonasal sinus surgery (MESS) and microscopic surgery of the nose was first used in the late 1950s by Heermann [18] in adults, and it was first reported for sinus surgery in children in 1976 by Dixon [9]. The philosophy of the MESS for the treatment of sinusitis is similar to that of FESS, as proposed by Messerklinger and others [25, 39, 55]. The target is ventilation and drainage through the OMC. Once these are accomplished, the diseased mucosa of the dependent sinuses should resolve without being removed or "touched".

In fact, most authors who perform MESS usually combine the advantages of the microscope and the endoscope and perform what is termed "microendoscopic" sinus surgery [1, 54, 65]. The main advantages of the microscopic technique are [10, 49]:
1. A three-dimensional binocular view with various magnifications, which allows a stable view and a wide surgical field
2. The use of a self-retaining speculum that allows bimanual surgery and convenient control of bleeding during the procedure
3. The ability to adapt video cameras without increasing the weight of the instrument of view

Some sinus surgeons report finding that the technique is especially advantageous when there is diffuse disease involving all paranasal sinuses, e.g., stage-IV nasal polyposis [1, 54]. The disadvantages of the microscope include:
1. The need for general anesthesia because of the pain caused by the constant pressure exerted by the self-retaining nasal speculum
2. A single, straight angle of view
3. Recording/documentation is not as good as with the endoscopic technique

The telescopes – especially the 30° and 70° telescopes – allow viewing of the so-called blind areas (such as the anterior and inferior walls of the maxillary sinus, the frontal recess and the frontal sinus) and nicely complement the microscopic technique.

Technique

The pre-operative considerations are the same as those described for the FESS procedure (see "Principles of Management"). The surgery is performed under hypotensive general anesthesia, with the patient in the supine position. The main instruments required for the microscopic stage of the procedure are a 250- or 300-mm objective lens microscope, a self-retaining nasal speculum and mono and/or bipolar cautery. The endoscopic stage of the surgery requires 4-mm telescopes with 30° and 70° angles of view. Both the microscope and the endoscope can be connected to video systems for recording/documentation; thus, other personnel in the operating room can watch the procedure. The surgical instruments used are more or less the same in both procedures (see "Functional Endoscopic Sinus Surgery"). Details of the procedure can be found in other chapters of this book, and a brief summary of the procedure is presented below.

Microscopic Stage

The nasal septum is addressed first in case a major deviation obscures the surgical field. Next, the middle turbinate is fractured toward the nasal septum, and a self-retaining nasal speculum is inserted, with one tine pressed against the meatal face of the middle turbinate and the opposite tine pressed against the inferior turbinate (Fig. 15.6). The region of the middle meatus is nicely exposed. To accomplish maxillary antrostomy, the self-retaining speculum is re-positioned in the direction of the lateral wall of the maxillary sinus. Using an angulated instrument, such as the curved curette, the surgeon palpates then penetrates the maxillary sinus just above the insertion of the inferior turbinate. The opening is enlarged anteriorly with Ostrom forceps. This maneuver must be delicate in order to avoid stripping the maxillary-sinus mucosa. The 30° and/or 70° telescopes are used to inspect the interior of the maxillary sinus. The ethmoidectomy begins with resection of the uncinate process, followed by resection of the ethmoidal bulla. Once the bulla is opened, the dissection continues in an anterior-to-posterior direction, mainly using true-cutting instruments to avoid mucosal stripping. The surgery is tailored to the diseased tissues and follows the functional concept (maximal preservation of normal sinus mucosa).

Endonasal microsurgery of the sphenoid sinus can be performed transethmoidally, trans-septally or transnasally (Fig. 15.7). Important landmarks for safe sphenoid entry (like the 90° angle between the anterior face of the sphenoid and the roof of the ethmoid, the tail of the middle and superior turbinates and the superior border of the choana) must be observed. The sphenoid-sinus normal mucosa must be preserved, as must the ethmoidal mucosa. Angled telescopes (30° and 70°) are used in sphenoid surgery in cases where extensive pneumatization is present and there is a need to inspect the lateral and posterior walls, where the optic nerve and the carotid artery are usually located.

Micro-endoscopic surgery of the frontal sinus is done less frequently, due to the lower incidence of frontal-sinus disease. The anatomic references for

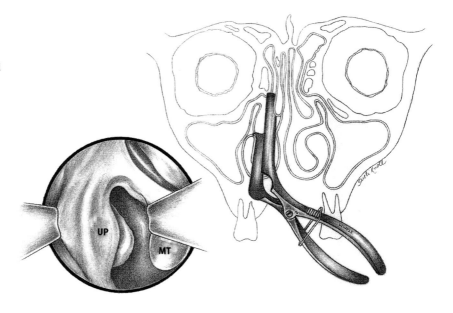

Fig. 15.6. Microscopic endonasal surgery of the paranasal sinuses (self-retaining speculum in place). The long blade is placed against the middle turbinate, and the short blade exposes the middle meatus. *MT*, middle turbinate; *UP*, uncinate process

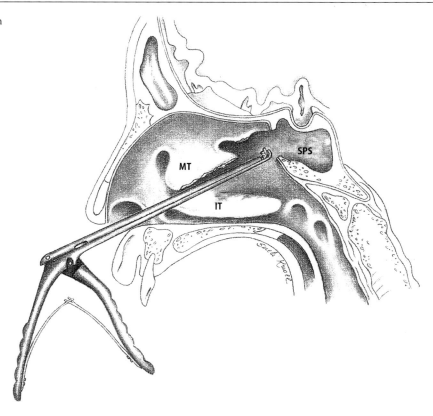

Fig. 15.7. Direct transnasal approach to the sphenoid sinus. *IT* inferior turbinate, *MT* middle turbinate, *SpS* sphenoid sinus

frontal-sinus surgery are: the uncinate process, the ethmoidal bulla, the insertion of the middle turbinate on the lateral nasal wall, the anterior ethmoidal "agger nasi" cells, the lacrimal sac and the anterior ethmoidal artery (when the dissection follows a posterior-to-anterior route). The procedure is started with microscope guidance. After identifying the insertion of the middle turbinate and the structures of the middle meatal complex, the uncinate process is resected, and the region of the agger nasi is dissected. Next, a 3- or 4-mm diamond drill is used to thin the lacrimal bone so that the lacrimal sac can be identified. The lacrimal sac is followed superiorly until the region of the nasofrontal recess is exposed. The head of the patient can be extended to help inspection of the frontal sinus. The mucosa of the posterior portion of the frontal recess must be preserved to avoid postoperative stenosis. To remove polyps and other lesions inside the frontal sinus, the telescopes and the Kuhn-Bolger giraffe forceps are used. In some cases, a diamond burr is used to remove the floor of the frontal sinus [66].

When safe access to the frontal sinus is not possible by the transnasal route, a 5- to 6-mm trephination can be performed on the anterior aspect of the frontal sinus, and a telescope can be introduced to inspect the inside of the sinus and its drainage to the nose. Next, a Nelaton catheter can be passed in the direction of the nose to endonasally clarify the location of the frontal recess.

Summary

Current sinus surgery techniques emphasize the importance of the OMC in the causation of disease in the maxillary, frontal and anterior ethmoid sinuses. It is the goal of FESS and microendoscopic surgical techniques to clear disease in this critical area. The choice of which type of surgery will be performed depends to a large extent on the training, experience and results of the treating surgeon. Traditional sinus-surgery techniques are still required in specialized circumstances, but many are being replaced by functional endoscopic and microendoscopic techniques. The essential step in carrying out any form of sinus surgery is accurate preoperative assessment with nasal endoscopy and CT scans. Surgery is considered after a thorough trial of medical therapy when the patient understands the risks, benefits and limitations of surgery.

Acknowledgement. The authors wish to acknowledge the support received from the Isabel Silverman Canada-International Scientific Exchange Program (CISEPO).

References

1. Amedee RG, Mann WJ, Gilsbach JM (1989) Microscopic endonasal surgery of the paranasal sinuses and the parasellar region. Arch Otolaryngol Head Neck Surg 115:1103–1106
2. Anand VK, Osguthorpe JD, Rice D (1997) Surgical management of adult rhinosinusitis. Otolaryngol Head Neck Surg 117:S50–S51
3. Anon JB (1997) Report of the rhinosinusitis task force committee meeting. Otolaryngol Head Neck Surg 117:S2–S68
4. Benninger MS, Anon J, Mabry RL (1997) The medical management of rhinosinusitis. Otolaryngol Head Neck Surg 117:S41–S49
5. Bolger WE, Butzin CA, Parsons DS (1991) Paranasal sinus bony anatomic variations and mucosal abnormalities: CT analysis for endoscopic sinus surgery. Laryngoscope 101:56–64
6. Carrau RL, Snyderman CH, Curtin HB, et al. (1994) Computer-assisted frontal sinusotomy. Otolaryngol Head Neck Surg 111:727–732
7. Clerico DM (1996) Pneumatized superior turbinate as a cause of referred migraine headache. Laryngoscope 106:874–879
8. Conte LJ, Holzberg N (1996) Functional endoscopic sinus surgery, symptomatic relief: a patient perspective. Am J Rhinol 10:135–140
9. Dixon H (1976) Microscopic antrostomies in children: a review of the literature in chronic sinusitis and a plan of medical and surgical treatment. Laryngoscope 86:1796–1814
10. Draf W, Weber R (1993) Endonasal micro-endoscopic pansinus operation in chronic sinusitis I. Indications and operation technique. Am J Otolaryngol 14:394–398
11. Elverland HH, Melheim I, Anke IM (1992) Acute orbit from ethmoiditis drained by endoscopic sinus surgery. Acta Otolaryngol (Stockh) 492:147–151
12. Friedman WH, Katsantonis GP (1990) Intranasal and transantral ethmoidectomy: a 20-year experience. Laryngoscope 100:343–348
13. Friedman WH, Katsantonis GP, Bumpous JM (1995) Staging of chronic hyperplasic rhinosinusitis: treatment strategies. Otolaryngol Head Neck Surg 112:210–214
14. Gittelman PD, Jacobs JB, Skorina J (1993) Comparison of functional endoscopic sinus surgery under local and general anesthesia. Ann Otol Rhinol Laryngol 102:289–293
15. Gross CW, Becker DG (1966) Instrumentation in endoscopic sinus surgery. Curr Opin Otolaryngol Head Neck Surg 4:28–33
16. Havas TE, Motbey JA, Gullane PJ (1988) Prevalence of incidental abnormalities on computed tomography scans of the paranasal sinuses. Arch Otolaryngol Head Neck Surg 114:856–859
17. Hawke WM, McCombe AW (1995) How I do it: nasal polypectomy with an arthroscopic bone shaver. The Stryker "Hummer". J Otolaryngol 24:57–59
18. Heermann H (1958) Endonasal surgery with the use of the binocular microscope. Arch Klin Exp Ohren Nasen Kehlkopfheilkd 17:295–297
19. Hoffman SR, Mahoney MC, Chmiel JF, et al. (1993) Symptom relief after endoscopic sinus surgery: an outcome-based study. Ear Nose Throat J 72:413–420
20. Huerter JR (1992) Functional endoscopic sinus surgery and allergy. Otolaryngol Clin North Am 25:231–237 1992
21. Jankowski R, Auque J, Simon C, et al. (1992) Endoscopic pituitary tumor surgery. Laryngoscope 102:198–202
22. Kainz J, Stammberger H (1990) The roof of the anterior ethmoid: a place of lease resistance in the skull base. Am J Rhinol 4:7–12
23. Kennedy DW (1990) First-line management of sinusitis: a national problem? Otolaryngol Head Neck Surg 103:847–888
24. Kennedy DW (1992) Prognostic factors, outcomes and staging in ethmoid sinus surgery. Laryngoscope 102:1–18
25. Kennedy DW, Zinreich SJ, Rosenbaum A, et al. (1985) Functional endoscopic sinus surgery: theory and diagnosis. Arch Otolaryngol Head Neck Surg 111:576–582
26. Kennedy DW, Goodstein ML, Miller NR, et al. (1990) Endoscopic transnasal orbital decompression. Arch Otolaryngol Head Neck Surg 116:275
27. Kyriakos M (1977) Myospherulosis of the paranasal sinuses, nose and middle ear. A possible iatrogenic disease. Am J Clin Pathol 67:118–130
28. Lamear WR, Davis WE, Templer JW, et al. (1992) Partial endoscopic middle turbinectomy augmenting functional endoscopic sinus surgery. Otolaryngol Head Neck Surg 107:382–389
29. Lanza DC, Kennedy DW (1992) Current concepts in the surgical management of chronic and recurrent acute sinusitis. J Allergy Clin Immunol 90:505–511
30. Lawson W (1991) The intranasal ethmoidectomy: an experience with 1,077 procedures. Laryngoscope 101:367–371
31. Levine H, May M (1993) Endoscopic sinus surgery. Thieme, New York
32. Loury MC (1993) Frontal recess dissection: endoscopic frontal recess and frontal sinus ostium dissection. Laryngoscope 103:455–458
33. Lund VJ (1990) Surgery of the ethmoids – past, present and future: a review. J R Soc Med 83:451–455
34. Lund VJ, Kennedy DW (1997) Staging for rhinosinusitis. Otolaryngol Head Neck Surg 117:S35–S40
35. Lund VJ, Mackay IS (1993) Staging in rhinosinus. Rhinology 31:183–184
36. Lund VJ, Holmstrom M, Scadding GK (1991) Functional endoscopic sinus surgery in the management of chronic rhinosinusitis. An objective assessment. J Laryngol Otol 105:832–835
37. Maniglia AJ (1991) Fatal and other major complications of endoscopic sinus surgery. Laryngoscope 101:349–354
38. Mattox DE, Kennedy DW (1990) Endoscopic management of cerebrospinal fluid leaks and cephaloceles. Laryngoscope 100:857
39. Messerklinger W (1978) Endoscopy of the nose. Urban and Schwarzenberg, Baltimore

40. Metson R (1991) Endoscopic surgery for lacrimal obstruction. Otolaryngol Head Neck Surg 104:473–479
41. Metson R (1992) Endoscopic treatment of frontal sinusitis. Laryngoscope 102:712–716
42. Moore DF, Grogan JB, Lindsey WH, et al. (1995) The myospherulotic potential of water-soluble ointments. Am J Rhinol 9:215–218
43. Perkins JA, Morris MR (1993) Treatment of acute frontal sinusitis: a survey of current therapeutic practices among members of the Northwest Academy of Otolaryngology. Am J Rhinol 7:67–70
44. Piccirillo JF, Edwards D, Haiduk A, Thawley SE (1995) Psychometric and clinimetric validity of the 31-item rhinosinusitis outcome measure (RSOM-31). Am J Rhinol 9:297–306
45. Rice DH (1993) Endoscopic sinus surgery. Otolaryngol Clin North Am 26:613–618
46. Riegle EV, Gunter JB, Lusk RP, et al. (1992) Comparison of vasoconstrictors for functional endoscopic sinus surgery in children. Laryngoscope 102:712–716
47. Roithmann R, Shankar L, Hawke M, et al. (1993) CT imaging in the diagnosis and treatment of sinus disease: a partnership between the radiologist and the otolaryngologist. J Otolaryngol 22:253–260
48. Roth M, Lanza DC, Zinreich J, et al. (1995) Advantages and disadvantages of three-dimensional computed tomography intraoperative localization for functional endoscopic sinus surgery. Laryngoscope 105:1279–1286
49. Rudert H (1988) Mikroskop- und endoskopgestützte chirugie der entzündlichen Nasennebenhölenerkrankungen. HNO 36:475–482
50. Schaefer SD, Manning S, Close LG (1989) Endoscopic paranasal sinus surgery: indications and considerations. Laryngoscope 99:1–5
51. Schaitkin B, May M, Shapiro A, Fucci M, Mester SJ (1993) Endoscopic sinus surgery: 4-year follow-up on the first 100 patients. Laryngoscope 103:1117–1120
52. Setliff RC, Parsons DS (1994) The "Hummer": new instrumentation for functional endoscopic sinus surgery. Am J Rhinol 8:275–278
53. Sillers MJ, Kuhn FA, Owen RG (1994) Surgery of the nose and paranasal sinuses. Curr Opin Otolaryngol Head Neck Surg 2:42–47
54. Stamm A (1995) Cirurgia micro-endoscópica naso-sinusal. In: Stamm A (ed) Microcirurgia naso-sinusal. Revinter, Rio de Janeiro, pp 183–214
55. Stammberger H (1986) Endoscopic endonasal surgery – concepts in treatment of recurring rhinosinusitis. Part I. Anatomic and pathophysiologic considerations. Otolaryngol Head Neck Surg 94:134–136
56. Stammberger H (1986) Endoscopic endonasal surgery: concepts in treatment of recurring rhinosinusitis. Part II. Surgical technique. Otolaryngol Head Neck Surg 94:147–156
57. Stammberger H, Hawke M (1993) Essentials of functional endoscopic sinus surgery. Mosby-Year Book, St. Louis
58. Stammberger H, Posawetz W (1990) Functional endoscopic sinus surgery: concepts, indications and results of the Messerklinger technique. Eur Arch Otorhinolaryngol 247:6376
59. Stankiewicz JA (1989) Complications in endoscopic intranasal ethmoidectomy: an update. Laryngoscope 99:686–690
60. Stankiewicz JA (1989) Blindness and intranasal endoscopic ethmoidectomy: prevention and management. Otolaryngol Head Neck Surg 101:320–329
61. Stankiewicz JA (1991) Cerebrospinal fluid fistula and endoscopic sinus surgery. Laryngoscope 101:250–256
62. Swanson PB, Lanza DC, Vining EM, et al. (1995) The effect of middle turbinate resection upon the frontal sinus. Am J Rhinol 9:191–195
63. Vining EM, Yanagisawa K, Yanagisawa E (1993) The importance of preoperative nasal endoscopy in patients with sinonasal disease. Laryngoscope 103:512–519
64. Vleming M, Middelweerd RJ, de Vries N (1992) Complications of endoscopic sinus surgery. Arch Otolaryngol Head Neck Surg 118:617–623
65. Weber R, Draf W (1992) Endonasale Mikro-endoskopische Pansinusoperaction bei chronischer sinusitis. Otorhinolaryngol Nova 2:63–69
66. Wigand ME (1990) Endoscopic surgery of the paranasal sinuses and anterior base skull bone. Thieme, New York
67. Zinreich SJ (1992) Imaging of chronic sinusitis in adults: X-ray, computed tomography, and magnetic resonance imaging. J Allergy Clin Immunol 90:445–451

High-Risk Areas in Endoscopic Sinus Surgery

Toshio Ohnishi

Introduction

From microscopic observations of the roof of the ethmoid sinus, Ohnishi [3] has identified the five main high-risk areas where surgical complications are likely to occur during sinus surgery. They are:
1. The medial wall of the ethmoid sinus or the lateral lamella of the cribriform plate
2. Along the course of the anterior ethmoidal artery/nerve
3. Around the anterior origin of the middle nasal turbinate or basal lamina
4. The antero-lateral aspect of the roof of the ethmoid sinus
5. Around the posterior ethmoidal artery/nerve

Figures. 16.1–16.5 show the operating microscopic findings for the roofs of five ethmoid sinuses with natural dehiscences, illustrating that such defects or dehiscences are not rare.

Due to violation of the barriers of the orbit and brain, relatively high rates of serious complications have been reported, even during the era of antibiotics [1, 2, 7, 8]. Awareness of each high-risk area during sinus surgery would be important for the prevention of such complications.

Figures. 16.1–16.5 show actual examples of vulnerable bony walls of the sinus cavities. One should remember that the bony walls of the upper sinus groups are not complete plates of bony walls, but instead are incomplete and have dehiscences and defects.

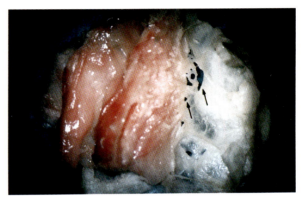

Fig. 16.1. Lateral lamella of the lamina cribrosa, viewed from the sinus side. Note the multiple dehiscences on the lateral lamella of the right cribriform plate (*arrows*). The top of the picture is anterior

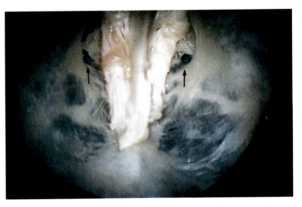

Fig. 16.2. Insertion of the anterior ethmoidal artery, viewed from the intracranium. Note the bone defect around the passage of the anterior ethmoidal artery on both sides (*arrows*). The top of the picture is anterior

High-Risk Areas in Endoscopic Sinus Surgery

Ohnishi [5] has described the following five high-risk areas that the surgeon encounters during endoscopic sinus surgery in the ethmoid sinus:
1. The lamina papyracea. The lateral wall of the ethmoid sinus consists of the lamina papyracea (the thin medial wall of the orbit), which is a convex bulge toward the cavity of the ethmoid sinus. Figure 16.6 shows an endoscopic view of the fully exposed lamina. This is the site of relatively frequent injuries during ethmoidal sinus operation,

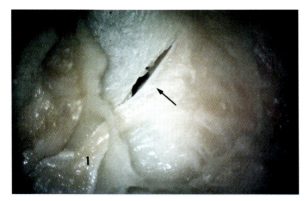

Fig. 16.3. Roof of the right ethmoid sinus, viewed from the sinusal side. Note the course of the anterior ethmoidal artery (*1*) from the right orbit, through the lamina papyracea, to the ethmoid sinus. Also note the defect of the bone along the entire course of the anterior ethmoidal artery within the ethmoid sinus. The top of the picture is anterior. The *arrow* shows dehiscence along the course of the anterior ethmoidal artery

Fig. 16.5. Roof of the ethmoid sinus after the dura mater and the mucoperiosteum have been removed. View from sinus cavity; note the thin bony roof of the right posterior ethmoid sinus and three small dehiscences (*arrows*)

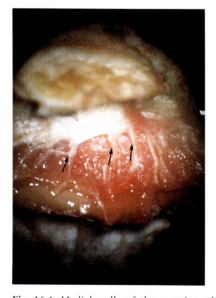

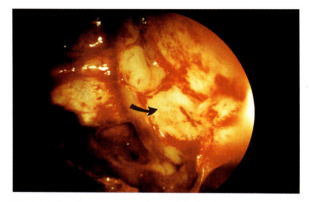

Fig. 16.4. Medial walls of the anterior ethmoid sinus; note abundant distribution of the olfactory fila (*arrows*) on the superior and middle nasal turbinates, as seen after resection of the roof of the ethmoid sinus. The right side of the picture is anterior

Fig. 16.6. Lamina papyracea (*arrow*), left ethmoid sinus

because the lamina is quite thin and fragile and protrudes into the operation area.

2. The roof of the ethmoid sinus around the anterior ethmoidal artery. The anterior ethmoidal artery usually travels along with the anterior ethmoidal nerve beneath the roof of the anterior ethmoid sinus. There are elevations of the bony walls near the medial and lateral ends of the bony canal containing the artery. These elevations are at risk of injury, and their injury can lead to orbital hematoma in the lateral wall and cerebrospinal-fluid leakage or intracranial infections in the medial wall.

Figure 16.7 shows a case where the left anterior ethmoidal artery and nerve have no bony canal and course unprotected below the roof of the ethmoid sinus. There are occasional defects of bone here, as seen in Fig. 16.3. The roof of the anterior ethmoid sinus is the area where surgical complications are likely to develop because of its unique anatomical features (especially the thin bony walls and the narrowness of the cavity, which tapers toward the ostium of the frontal sinus). Even minor injury to the thin bony walls may tear the dura mater and result in cerebrospinal-fluid leakage.

3. Lateral lamella of the cribriform plate. The medial wall of the anterior ethmoid sinus usually extends above the level of the lamina cribrosa,

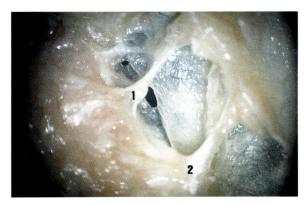

Fig. 16.7. Anterior (*1*) and posterior (*2*) ethmoidal arteries and nerves, right ethmoid sinus, dehiscence near the anterior ethmoid sinus

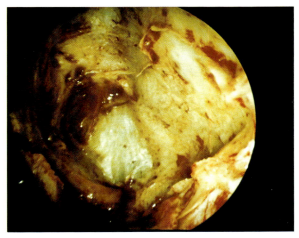

Fig. 16.9. Lateral lamina of the cribriform plate, left ethmoid sinus

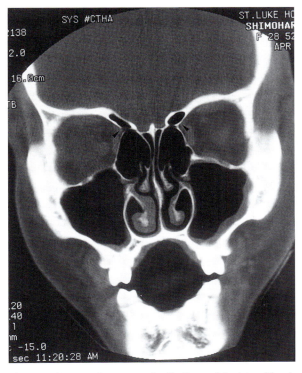

Fig. 16.8. Computed-tomography findings of the lateral lamina of the cribriform plate and the anterior ethmoidal arteries (*arrowheads*)

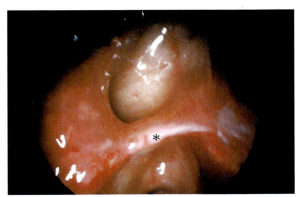

Fig. 16.10. Posterior ethmoidal artery (*), right ethmoid sinus

which is the lateral wall hiding the olfactory bulb. This is the lateral lamina of the cribriform plate. This wall often bulges into the sinus cavity, as shown in a CT scan in Fig. 16.8. This bulging is a likely site of surgical trauma during resection of disease in the frontal-recess area.

After penetrating the lamina cribrosa, the olfactory filaments distribute abundantly in the bony plates of the upper and middle nasal turbinates, which comprise the medial wall of the anterior ethmoid sinus. Because of these penetrations, the medial wall of the anterior ethmoid sinus occasionally has bony dehiscences, which may serve as a passageway for intracranial infections when damaged. Figure 16.9 shows an endoscopic view of the lateral lamina of the cribriform plate, and Fig. 16.1 illustrates a bony defect there.

4. The ethmoid roof near the posterior ethmoidal artery. The posterior ethmoidal artery travels in the roof of the posterior ethmoid sinus. The artery usually courses above the bony roof, which is thicker than the roof of the anterior ethmoid sinus and, in many cases, is hidden behind the bony plate. However, in some instances, the artery has only a thin bony canal around it and travels beneath the bony roof, where it can be damaged and can cause active bleeding. Figure 16.10 shows such a case, where the posterior ethmoidal artery travels underneath the sinus roof.

5. The area bordering the sphenoid and posterior ethmoid sinuses. In a number of cases, the bony bulge from the optic nerve can be seen in the walls of the sphenoid sinus during surgery (Fig. 16.11). However, injury to the nerve during the operation may occur at the area bordering the sphenoid and ethmoid sinuses, even when the optic canal is obscure. An attempt to remove the thick buttress between the two sinuses poses a risk of injury to the nerve if it lies superficially against the sinus wall. The surgeon should also be aware of Onodi cells where the optic canal protrudes into the posterior ethmoid sinus. The internal carotid artery courses along the lateral surface of the sphenoid bone within the cavernous sinus, causing a slight bone bulge into the sinus cavity. Injury to the artery may produce profuse and life-threatening bleeding into the sinuses and the intracranial space. Craniotomy and clamping of the artery may become necessary to prevent catastrophe.

Figure 16.12 shows a schematic drawing of the five high-risk areas in the left ethmoid sinus. Surgical injury of the lamina papyracea may cause a simple prolapse of fat, double vision from injury to the medial rectus muscle, intraorbital hemorrhage or infections, and in rare cases, blindness.

Complications are more severe when the lamina papyracea is traumatized in its posterior part, which is closer to the apex and the optic nerve. Intraorbital hematoma can cause blindness by extension of the optic nerve or compression of the central retinal artery. Injury to the lamina papyracea can be avoided by two important measures: (1) by preoperative study of coronal CT scans to determine the level of medial protrusion of the lamina, and (2) in surgery, by identifying the lamina at its most medial part and then exposing the entire lamina tangentially, along the wall.

The roof of the anterior ethmoid sinus is formed by the frontal bone and is quite thin, with occasional dehiscence. Inconsistency in the anatomical features in this area may confound the surgeon and lead to inadvertent complications. Identifying the anterior ethmoidal artery in the anterior ethmoid sinus is essential to prevent complications, because the careful technique required in approaching and preserving the artery should also be effective in preventing injury to the most vulnerable part of the roof of the ethmoid sinus.

The lateral lamella of the cribriform plate or the upper medial wall of the anterior ethmoid sinus is composed of a thin bony plate that bulges into the sinus cavity and which is prone to injury during surgery, resulting in cerebrospinal-fluid leakage. This area was referred to as the medial cranial wall by Ryo Takahashi [9] in 1944. Careful preoperative study of coronal CT scans should reveal the extent of protrusion in this area and should alert the surgeon prior to the operation.

The posterior ethmoidal artery leaves the ophthalmic artery proximal to the anterior ethmoidal artery and travels across and above the roof of the posterior ethmoid sinus. Occasionally, the artery appears beneath the roof of the ethmoid sinus and poses a risk. It should be preserved to avoid unnecessary bleeding, creation of a hematoma in the orbit and inadvertent injury to the base of the skull.

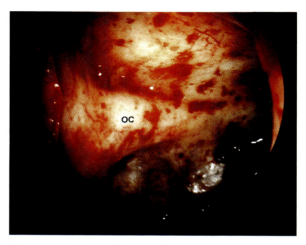

Fig. 16.11. Optic canal (*OC*) at the border area between the sphenoid and ethmoid sinuses, right sphenoid sinus

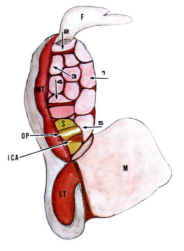

Fig. 16.12. Schematic drawing showing the high-risk areas in the left ethmoid sinuses. *1*, lamina papyracea; *2*, anterior ethmoidal artery; *3*, lateral lamella of lamina cribrosa; *4*, posterior ethmoidal artery; *5*, bordering area between sphenoid and posterior ethmoid sinuses. *F*, frontal sinus; *ICA*, internal carotid artery; *LT*, inferior nasal turbinate; *M*, maxillary sinus; *MT*, middle nasal turbinate; *OP*, optic nerve; *S*, sphenoid sinus

Direct injury to the optic nerve or internal carotid artery precipitates grave consequences. Surgeons must avoid such complications by every means possible. The optic canal can be seen bulging into the posterior-superior wall of the sphenoid sinus, while the internal carotid artery runs just below the canal.

The nerve is rarely injured when the optic canal is visible. The damage may occur when there is no visible optic canal, which is the border between the sphenoid and ethmoid sinuses. The buttress between the two sinuses often conceals the superficial optic nerve. Fractures of the optic canal may cause total blindness when the central retinal artery is ruptured, producing disruption of the blood supply to the retina. The use of endoscopes in sinus surgery has enabled visualization of minute details of the structures of the paranasal sinuses. At the same time, it allows more meticulous sinus surgery [4].

Our previous study indicated that serious surgical complications in endoscopic sinus surgery are likely to occur when the surgeon is hurrying to finish the operation because of other commitments. As Ohnishi et al. [6] indicated, a composed and calm attitude of the surgeon is an important requirement for sure and safe endoscopic sinus surgery. The most important point is that, in his or her mind, the surgeon should always visualize the important structures on the other side of the thin bony walls being addressed.

References

1. Freedman HM, Kern EB (1979) Complications in intranasal ethmoidectomy: a review of 1000 consecutive operations. Laryngoscope 89:421–434
2. Maniglia AJ (1989) Fatal and major complications secondary to nasal and sinus surgery. Laryngoscope 99:276–283
3. Ohnishi T (1981) Bony defects and dehiscences of the roof of the ethmoid cells. Rhinology 19:195–202
4. Ohnishi T (1991) Anatomy of the anterior ethmoid sinus. Operative Tech Otolaryngol Head Neck Surg 2:218–221
5. Ohnishi T, Esaki S, Iwasaki M, Tachibana T (1990) Endoscopic microsurgery of the ethmoid sinus. Am J Rhinol 4:119–127
6. Ohnishi T, Tachibana T, Kaneko Y, Esaki S (1993) High-risk areas in endoscopic sinus surgery and prevention of complications. Laryngoscope 103:1181–1185
7. Stankiewicz JA (1987) Complications in endoscopic intranasal ethmoidectomy. Laryngoscope 97:1270–1273
8. Stankiewicz JA (1989) Complications in endoscopic intranasal ethmoidectomy. An update. Laryngoscope 99:686–690
9. Takahashi R (1944) Clinicoanatomical studies of the canalis orbitocranialis and canalis orbit-ethmoidalis in relation to the ethmoid cells. J ORL Soc Jpn 50:224–240

Micro-endoscopic Surgery of the Paranasal Sinuses

Aldo Cassol Stamm

Introduction

The surgical treatment of inflammatory diseases of the paranasal sinuses has changed dramatically in the last few years with the development of new technology in equipment and operative techniques and better understanding of the pathophysiology of the ostiomeatal complex (OMC). These new techniques include transnasal micro-endoscopic surgery.

Transnasal microsurgery of the paranasal sinuses is based on the use of an operating microscope and self-retaining nasal speculums via the nasal cavity. The operating microscope can also be used as a complementary surgical instrument in external paranasal operations, such as external ethmoidectomy, osteoplastic surgery of the frontal sinus and the degloving approach.

Endoscopic surgery of the paranasal sinuses is an operative technique that employs endoscopes with visualization angles of 0°, 30°, 45° or 70° to visualize the nasal cavity and the paranasal sinuses during the surgical procedures. The surgical access is usually transnasal.

Micro-endoscopic surgery, therefore, is the association of the operating microscope and endoscopes. The development of the microscopic and endoscopic surgical techniques occurred simultaneously with advances in diagnostic techniques.

In 1958, Hans Heermann [20] introduced the transnasal microsurgery technique by using an operating microscope to perform an ethmoidectomy; he later added a controlled hypotensive anesthesia to minimize intraoperative bleeding. Since than, the microscope has been used in many different types of paranasal procedures and in surgery of the lacrimal sac and pterygopalatine fossa [1–3, 17, 22, 40, 51, 61].

After studying the physiology of the paranasal sinuses and making accurate clinical endoscopic observations, Messerklinger [44, 45] recognized the importance of the ethmoid sinus–middle meatus complex as the site of origin of sino-nasal diseases and envisioned the possibility of removing diseases of the middle meatus endoscopically, thus solving intrinsic problems of the paranasal sinuses. He standardized endoscopic-surgery techniques via the middle meatus, performing an antero-posterior ethmoidectomy beginning with resection of the uncinate process.

Special instruments were conceived and developed to facilitate the endonasal microsurgery procedures, leading some to utilize transnasal microsurgery techniques to treat paranasal-sinus diseases in children [6, 51]. The introduction of angled endoscopes in radical and extensive surgery involving sphenoethmoidectomy and the posterior-to-anterior technique was accomplished by Wigand et al. [74], who began the procedure at the ostium of the sphenoid sinus and proceeded anteriorly along the anterior skull base. Wigand also recommended general anesthesia and nasal-septum mobilization to facilitate surgical access [71, 72].

By this time, some authors considered surgeries of the ethmoid and sphenoid sinuses without the operating microscope to be blind procedures and associated them with increased risks of complication [7, 8]. The importance of the endoscope for clinical diagnosis and the value of maxillary sinusoscopy was also emphasized by Draf [9–11]. The combination of the operating microscope and angled endoscopes in extensive transnasal paranasal-sinus operations was soon recognized by many surgeons [1, 12, 52, 59, 62].

The concept of "functional endoscopic sinus surgery", which attempted to restore the aeration of the paranasal sinuses and to facilitate the drainage of secretions into the nasal cavity, was advocated by Kennedy et al. in 1985 [25, 30]. They further stressed the OMC concept, initially suggested by Messerklinger [43–45] and later emphasized by Stammberger [63–65], who indicated that anatomic variations were major factors causing obstruction of drainage of the paranasal sinus.

Straight endoscopes are currently recommended for the majority of the surgical dissections, and angled endoscopes are thought to work only in certain regions [29, 65]. May [41], for example, described ostioplasty of the frontal-sinus duct using angled endoscopes and a delicate diamond drill. Draf [12] described three basic techniques of frontal sinus drainage using a transnasal micro-endoscopic technique: simple drainage (type 1), enlargement of the natural ostium (type 2) and creation of a single, large median duct (type 3). The routine use of the endoscopes also allowed maxillary sinusitis to be classified according to characteristics of the mucosa and secretions found in the maxillary sinus [69].

Some authors recommended a combination of the operating microscope, angled endoscopes and a headlight [52, 77]. The importance of endoscopes in postoperative care and follow-up is widely recognized [29]. Setliff and Parsons [55] have recently recommended minimally invasive sinus surgery by using power instrumentation, such as microdebriders, in order to preserve the mucosa of the OMC as much as possible, avoiding wide openings between the sinuses and the nasal cavity.

The main advantages of the operating microscope are that the use of a self-retaining speculum allows magnification and bimanual work, which are advantageous when either bleeding or drilling and irrigation are required. However, endoscopes provide better photo and video documentation and greater surgical mobility. Both the excellent illumination and visualization provided by endoscopes and the opportunity to have angled fields of vision are especially helpful in areas that are impossible or difficult to see directly with the microscope, such as the frontal and maxillary sinuses.

Surgical Anatomy – Paranasal Sinuses and Lateral Nasal Wall

The paranasal sinuses are structures anatomically contained within the facial bones (maxillary, ethmoid, sphenoid and frontal). The mucosa lining the sinus cavities are histologically and physiologically similar under normal conditions of aeration and drainage.

Ethmoid Sinus

The ethmoid sinus is strategically located. It is the only sinus that is contiguous with all the other sinuses (frontal, sphenoid and maxillary) and with the nasal cavity, orbit and anterior skull base. Due to its multicellular composition, the ethmoid sinus has wide anatomic variations, and it is considered the key area of micro-endoscopic sinus surgery.

The cellular complex of the ethmoid sinus presents unique drainage areas that help to make it the most important of the sinuses from an inflammatory pathophysiological standpoint. Located in the middle meatus are the drainage ostia of the anterior ethmoid cells and ethmoid bulla. The concha bullosa has independent ostia. The drainage ostia of the posterior ethmoid cells are in the superior meatus, next to the sphenoethmoid recess, where the sphenoid sinus has its locus of aeration and drainage. The sites of the ostia of the drainage have singular clinical importance. If many cells drain to the same ostium, there is an increased chance of involvement by the same disease.

Ethmoid cells tend to remain small, their growth slowed by contact with compact bone, and they have greater growth in the directions of least resistance. The bony lamina that separates the ethmoid sinus from the orbit becomes thin, because the intraorbital pressure resists the ethmoid pressure and maintains the integrity of the orbital wall. This does not occur over the orbital globe, where the aeration progresses and supra-orbital ethmoidal cells develop. The ethmoid cells are usually divided into two groups, anterior and posterior, separated by the basal lamella of the middle turbinate [66].

The anterior ethmoid cells are very important pathophysiologically in inflammatory diseases, especially as they relate to the maxillary and frontal sinuses, because all these sinuses share the same areas of drainage. The anterior anatomical limit of the anterior ethmoid sinus is imprecise and controversial, and the OMC is considered part of it. Its lateral wall is the lamina papyracea, which is limited by the orbit; the posterior insertion of the middle turbinate is the posterior limit, and the roof or superior wall is the anterior portion of the fovea ethmoidalis, which forms the floor of the anterior skull base. The superior wall is positioned lateral to the cribriform plate and to the insertion of the middle turbinate. It is also formed by the frontal bone.

Keros [31] described three basic anatomic types of ethmoid-sinus roof according to the depth of the olfactory fossa, as measured by the extension of the lateral lamella of the cribriform plate. In this region, the anterior ethmoid artery and nerve can be observed coursing in a thin bony canal, and they are considered the superior limit of an ethmoidectomy; this area is at great surgical risk for CSF fistula and bleeding (Fig. 17.1). The anterior ethmoid artery is also a landmark used to access the frontal-sinus recess, which is always located anterior to the artery. The most anterior ethmoid cells are located anterior to the uncinate process around the frontal recess.

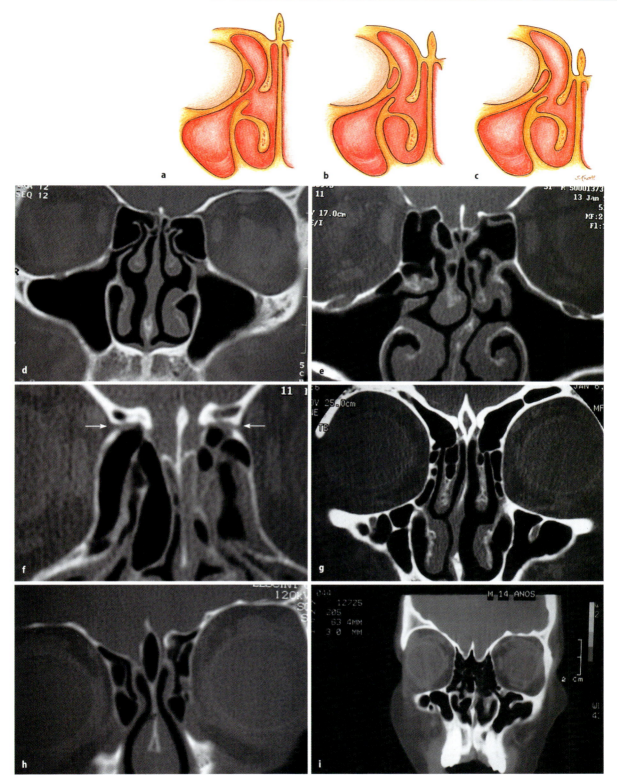

Fig. 17.1. a–c Configuration of the ethmoid roof according to the olfactory-fossa depth (Keros 1962) [31]. Type I: 1–3 mm; type II: 4–7 mm; type III: 8–16 mm. **d–f** Coronal computed tomography at the level of the crista galli process, showing examples of the three types of ethmoid roof. Anterior ethmoid-artery canals can also be seen (*arrows*). **g–i** Anatomical variant of the ethmoid roof and olfactory fossa

The frontal-recess area is better seen after resection of the antero-superior region of the uncinate process. Drainage of the anterior ethmoid cells preferentially occurs in the central region of the infundibulum, at the middle meatus.

The posterior ethmoid cells are very complex anatomically, varying greatly in size and number. The lateral wall of the posterior ethmoid cells is intimately related to the optic nerve, which may cross into the cells, especially when they are well pneumatized posteriorly and laterally or when they are superiorly related to the sphenoid sinus (sphenoethmoid cells/Onodi cells). The junction of the posterior ethmoid cells with the sphenoid sinus is a critical surgical area. The posterior ethmoid artery is located in the roof of the ethmoid sinus, posterior to the cribriform plate and anterior to the anterior wall of the sphenoid sinus.

The ostium of the sphenoid sinus is usually located between the superior turbinate and the nasal septum, in the sphenoethmoid recess. Identification of the anterior wall of the sphenoid sinus through transethmoid access is accomplished by observing an angle of approximately 90° that is formed between it and the roof of the posterior ethmoid sinus (Fig. 17.2). The sphenoethmoid recess is the region where secretions drain from the posterior ethmoid cells through the superior meatus and from the sphenoid sinus directly through its ostium in the superior meatus. Secretions proceed from this area directly to the choana and nasopharynx.

Fig. 17.2. Anatomic configuration of the 90° angle formed by the roof of the posterior ethmoid sinus and the anterior wall of the sphenoid sinuses (*arrows*). *IT,* inferior turbinate; *MT,* middle turbinate; *SS,* sphenoid sinus; *ST,* superior turbinate

face of the middle turbinate. Intracranially, it extends medially from the crista galli process and laterally to the lateral lamella of the olfactory fossa, accounting for any anatomic variations of this region (Fig. 17.1).

During an ethmoidectomy, the surgeon should work lateral to the middle turbinate but should remain in the middle-meatus region to minimize the risk of penetrating into the anterior cranial fossa through the cribriform plate, producing a CSF fistula, anosmia or damaging the brain tissue. The cribriform plate is inferior and thinner than the fovea ethmoidalis. The middle turbinate should be preserved if possible, because it is important as a surgical landmark in transnasal ethmoidectomy.

Basal Lamella of the Middle Turbinate

Portions of the middle turbinate are inserted in three different planes. The anterior portion is attached to the lateral region of the cribriform plate of the anterior skull base in a sagittal plane. The body portion is inserted in the lamina papyracea in a coronal plane, and the posterior part of the basal lamella joins the lamina papyracea or the medial wall of the maxillary sinus. In a horizontal plane, its stability depends mostly on the basal-lamella portion, especially its posterior and anterior insertions [66].

Sphenoid Sinus

The sphenoid sinus varies greatly in size, and it is surrounded by important anatomical structures. It is located medial and inferior to the posterior ethmoid cells; thus, the optic nerve and carotid artery can be damaged when the sphenoid sinus is entered via the posterior ethmoid cells.

The optic nerve lies dehiscent in the lateral wall of the sphenoid sinus in 6% of cases [34]. Siebert [56] presented a study of 50 cadavers showing that 57% of optic nerves bulge into the sphenoid sinus and 1% of optic nerves have no bony canal. The same study demonstrated that the horizontal portion of the intracavernous carotid artery extends prominently into the sphenoid sinus in 67% of cases and is dehiscent in 6%.

Cribriform Plate

The central part of the ethmoid bone forms the floor of the anterior skull base, the olfactory fossa, where the terminal branches of the olfactory nerve pass into the nasal cavity. It is located in the roof of the nasal cavity, between the nasal septum and the septal

A small sulcus area called the carotid–optic recess is located in the lateral and posterior walls of the sphenoid sinus, between the bony canal of the inter-

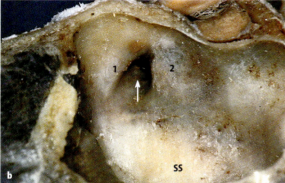

Fig. 17.3 a,b. Carotid–optic recess into the sphenoid sinus (*arrow*). S, nasal septum; SS, sphenoid sinus; *1*, Optic-nerve canal; *2*, internal carotid-artery canal

nal carotid artery and the prominence of the optic nerve, and another recess can be observed just above the optic nerve. Removal of diseased tissue in these areas requires special caution (Fig. 17.3).

Other anatomic structures neighboring the sphenoid sinus include the meninges, the pituitary gland and optic chiasm in the posterior and superior surfaces. The optic nerve and internal carotid artery lie in the lateral wall, and the cavernous sinus and III, IV, V and VI nerves lie farther laterally.

In addition to the structures already mentioned, the maxillary and pterygoid nerves can also be observed in the floor of the sphenoid sinus. The maxillary nerve is prominent in 48% of the sphenoid sinuses; it is dehiscent in 5% of the cases, whereas the pterygoid nerve is prominent 18% of the time [56].

Maxillary Sinus

The maxillary sinus is the largest of the paranasal sinuses and is anatomically related to the floor of the orbit (superiorly) and the infratemporal, zygomatic and pterygopalatine regions (posteriorly). Its inferior wall is part of the hard palate, while the medial wall is related to the nasal cavity. The most important area of drainage is the infundibulum, usually visible after surgical resection of the uncinate process. The fontanelle or posterior membranous wall is located posterior to the ostium, which connects the maxillary sinus and the nasal cavity. The ostia usually found in this membranous wall are called accessories and normally do not receive secretions from the sinus, because the mucociliary flow is directed toward the natural ostium. This is one of the reasons that maxillary-sinus antrostomy through the middle meatus is preferable, enlarging the principal drainage portal.

Frontal Sinus

The frontal sinus usually consists of two large cells separated by an intersinus septum, but small incomplete septa may extend into each large cell. The anatomic relationships of the sinus are mainly with the roof of the orbital cavity, agger nasi cells (ethmoid cells located anterior to the uncinate process) and (posteriorly) the anterior cranial fossa. Its drainage occurs through the frontal recess in the anterior part of the middle meatus, anterior to the ethmoid bulla.

Lang [34] found that the frontal sinus drained toward the semilunar hiatus in 67.7% of specimens and away from the semilunar hiatus in 33.3% of specimens. Most authors now believe that the most frequent site of frontal-sinus drainage is medial to the uncinate process in the middle meatus; less frequently, drainage is through the ethmoid infundibulum or through the superior portion of hiatus semilunaris [29].

Ostiomeatal Complex

The middle meatus is an anatomic space located in the lateral nasal wall, between the middle and inferior turbinates, where drainage and ventilation of the frontal, maxillary and anterior ethmoid sinuses occur. The OMC is the entire composite of anatomical structures (consisting of prominences and recesses) that can influence the ventilation and drainage system of the sinuses related to the middle meatus. The main structures of the OMC include the ethmoid bulla, the uncinate process, drainage ostia, the infundibulum, the hiatus semilunaris (inferior and superior) and anterior ethmoid cells (agger nasi). We also include the infraorbital ethmoid cell

(Haller's cell), the meatal face of the middle turbinate and the frontal recess as parts of the OMC (Table 17.1, Fig. 17.4).

Ethmoid Bulla

The ethmoid bulla is a differentiation of the ethmoid cell and is the largest and most regular of the ethmoid cells. It is located at the level of the middle meatus, superior to the inferior hiatus semilunaris, and forms the posterior wall of the frontal recess, serving as protection for the skull base in surgical procedures. It has its own drainage ostium and drains directly into the hiatus semilunaris or into the adjacent ethmoid cells. It is considered an important anatomical landmark in intranasal ethmoidectomies and in surgeries of the frontal recess and middle meatus. It varies in size, sometimes having aerated extensions called supra- and retrobulbar recesses. These recesses can be accessed through the superior hiatus semilunaris. A large ethmoid bulla usually adversely affects the drainage and ventilation of the related paranasal sinuses.

Uncinate Process

The uncinate process is a curved structure resembling a fingernail and is composed of mucosa, periosteum and a thin bony lamina. It is located in the middle meatus, at the medial wall of the maxillary sinus, joining the middle meatus with the inferior hiatus semilunaris and the infundibulum. The uncinate process varies in size and in its point of insertion. Usually, the attachment of its superior portion is lateral to the lamina papyracea; however, it may attach on the medial or central region of the skull base (Fig. 17.5). The free edge of the uncinate process is very important from a surgical standpoint; sometimes, it lies almost in the coronal plane covering the ethmoid bulla but, in other cases, it can be very small and may lie adjacent to the lamina papyracea.

Ethmoid Infundibulum

The ethmoid infundibulum is a three-dimensional space bordered medially by the uncinate process and laterally by the lamina papyracea. Posteriorly, the ethmoid infundibulum extends to the anterior face of the ethmoid bulla and opens into the middle meatus through the inferior hiatus semilunaris. The maxillary-sinus ostium can usually be found at the floor and lateral aspect of the infundibulum. Superiorly, the ethmoid infundibulum ends blindly exactly where the uncinate process is attached to the lamina papyracea. This recess is called the recessus terminalis. The ethmoid infundibulum is best visualized after the uncinate process is removed.

Table 17.1. Structures of the ostiomeatal complex

Structures of the middle meatus: ethmoid bulla and recess
Uncinate process
Ethmoid infundibulum
Hiatus semilunaris (inferior and superior)
Drainage ostia
Anterior ethmoid cells (agger nasi)
Infraorbital ethmoid cells (Haller's cell)
Meatal face of the middle turbinate
Frontal recess

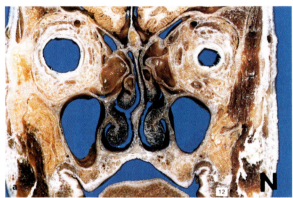
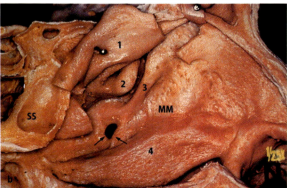

Fig. 17.4 a,b. Ostiomeatal complex. a Coronal view. b Lateral nasal wall (middle meatus after elevation of the middle turbinate). *1*, Middle turbinate; *2*, ethmoid bulla; *3*, uncinate process; *4*, inferior turbinate, showing an accessory ostium (*arrow*); (SS = sphenoid sinus, MM = middle meatus)

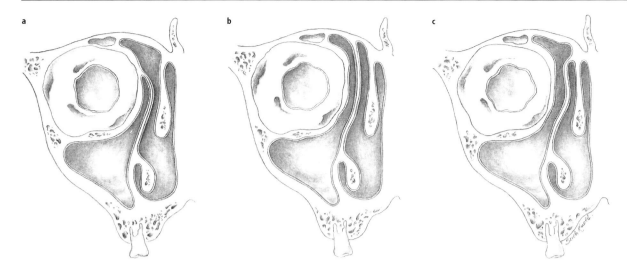

Fig. 17.5 a-c. Anatomic variations of the uncinate process, which most frequently is attached to the lamina papyracea (*a*) or to the skull base – centrally (*b*) or medially (*c*)

Anterior Ethmoid Cells (Agger Nasi)

The agger nasi cells are the most anterior ethmoid cells and are located above and anterior to the uncinate process. The agger nasi is the anterior limit of the frontal recess; thus, large agger nasi cells can impinge on the frontal recess. Endoscopically viewed, the agger nasi appears as a prominence superior to the insertion of the middle turbinate.

Hiatus Semilunaris Inferior

The hiatus semilunaris inferior is a space located between the uncinate process (antero-inferiorly) and the ethmoid bulla (postero-superiorly). This narrow space provides continuity between the middle meatus and the infundibula of the maxillary, ethmoid and frontal sinuses.

Hiatus Semilunaris Superior

This space is located between the ethmoid bulla and the insertion of the middle turbinate. It is the access to the supra- and retrobulbar recesses (sinus lateralis).

Infraorbital Ethmoid Cell: "Haller's Cell"

This cell of the ethmoid complex is located along the medial floor of the orbital cavity and is related to the roof of the maxillary sinus. When these cells are extensively developed, the ethmoid infundibulum region is often narrowed, leading to impaired drainage capacity of the maxillary sinus. This cell is also significant, because it can harbor residual disease after sinus operations.

Meatal Surface of the Middle Turbinate

The middle turbinate represents an anatomical differentiation of the ethmoid-cell complex. It may or may not be pneumatized and, when extensively aerated, is called the concha bullosa. It is considered one of the most important anatomical landmarks in transnasal micro-endoscopic surgery. Its insertion along the anterior skull base separates the olfactory region (cribriform plate) from the roof of the ethmoid sinus (fovea ethmoidalis), which is lateral to the middle-turbinate insertion. The posterior end of the middle turbinate is an important landmark for identification of the sphenopalatine foramen and the sphenoid sinus. The meatal surface of the middle turbinate is included in the OMC because, depending on its anatomical configuration, it can also compromise the middle meatus drainage and ventilation system (Fig. 17.6).

Frontal Recess

The frontal recess is the anterior–superior region of the anterior ethmoid complex. Its anatomical configuration usually resembles an inverted funnel connecting the frontal sinus with the anterior–superior aspect of the middle meatus. The frontal recess is bounded laterally by the orbital plane of the frontal

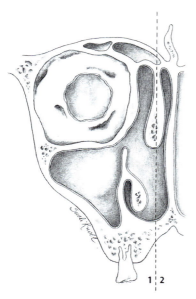

Fig. 17.6. A broken line divides the middle turbinate into two aspects: the meatal surface (*1*) and the septal surface (*2*)

Table 17.2. Preoperative assessment

Clinical history
Ear, nose and throat examination
Nasal endoscopy
Rigid
Flexible
Rhinomanometry
Axial, coronal and sagittal computed tomography
Magnetic resonance imaging
Patient counseling/medical decision

bone (which is medial to the orbital wall), medially by the vertical part of the middle turbinate and attachment to the skull base, posteriorly by the skull base, ethmoid roof, anterior ethmoid artery and ethmoid bulla, and anteriorly by the agger nasi cells and the supraorbital, frontal and suprabulbar cells. The frontal recess usually enters the middle meatus medial to the ethmoid infundibulum, either anterior (55% of cases) or superior (30% of cases) to the superior attachment of the uncinate process on the lateral nasal wall [70].

Preoperative Assessment

The success of paranasal-sinus surgery depends on careful preoperative evaluation, which includes a good clinical history, a good physician–patient relationship and appropriate information about possible complications and postoperative care (Table 17.2). The clinical history should establish the duration of the disease (chronic or recurrent), previous surgeries, complications, current medications (aspirin, anti-inflammatory agents, anti-hypertensives, etc.), changes of olfaction and vision, allergic manifestations, asthma and systemic diseases, such diabetes and hypertension.

Examination of the nasal cavity should be performed with the patient in a semi-sitting position. Initially, the general inspection is accomplished by using a headlight. The nasal cavity is prepared with a topical anesthetic solution containing a vasoconstrictor. Examination of the nasal cavity and paranasal sinuses is performed with rigid, 4.0-mm, 0°, 30° and 70° endoscopes. The flexible 3.2-mm endoscope is preferable with children and, occasionally, a straight 2.7-mm endoscope is utilized.

Inspection of the nasal cavity and recess begins at the vestibule and follows along the floor and inferior meatus to the choana. The sphenoethmoid-recess region is examined if it is accessible. Examination of middle-meatus structures is sometimes facilitated by gently displacing the middle turbinate medially, toward the nasal septum.

Angled endoscopes are used to evaluate the sphenoethmoid, frontal recess, olfactory and middle-meatal areas, where the ethmoid bulla and uncinate process can be identified. Posterior rhinoscopy with the 70° endoscope can be especially helpful in examining the choanal area and the nasopharynx. Endoscopes may also be useful in diagnosing maxillary-, frontal- and sphenoid-sinus diseases directly via sinusoscopy. Immunoallergic tests that help establish the diagnosis can be ordered.

In patients with histories of fracture or foreign bodies, we initially use plain radiographic studies (Caldwell, Waters, Hirtz and sagittal views) as a screening evaluation. In patients with incomplete resolution of acute sinusitis, complications, recurrent or chronic diseases or a history of previous paranasal-sinus surgery, high-resolution computed tomography (CT, with or without contrast) in axial, coronal and sagittal projections is preferred (Fig. 17.7).

High-resolution CT provides detailed information about the involved anatomical structures and helps the surgeon plan the micro-endoscopic surgical treatment. The most relevant aspects of CT images to be evaluated are [57]:

1. *Anatomical structures of the OMC*: ethmoid bulla, uncinate process, Haller's cell, anterior ethmoid cells, middle turbinate and frontal recess.
2. *Middle turbinate*: insertion, pneumatization, size and diseases.

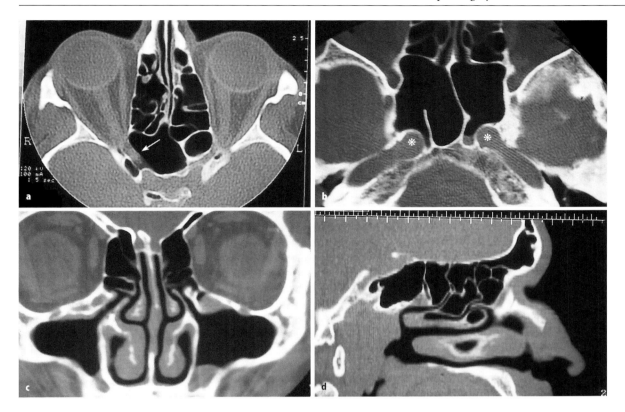

Fig. 17.7. a Axial computed tomography (CT), demonstrating the anatomic relationship between ethmoid cells and the orbit wall and between the optic nerve, posterior ethmoidal cells and the sphenoid sinus (*arrow*). b Axial CT showing the prominence of both internal carotid artery canals on the postero-lateral walls of the sphenoid sinus (*). c Ostiomeatal structures, lamina papyracea and the relationship between the fovea ethmoidalis and the cribriform plate, as seen in a coronal CT section. d Sagittal CT reconstruction showing the relationship between anterior ethmoid cells and the frontal recess and between posterior ethmoid cells and the sphenoid sinus

3. *Relationship between the roof of the ethmoid sinus and the cribriform plate*: this relationship and the course of the anterior ethmoid artery are best seen on the coronal CT.
4. *Orbital wall of the ethmoid sinus (lamina papyracea)*: coronal and axial projections provide good visualization. Special attention is needed in patients with ethmoid tumors, recurrent polyposis, fractures or a history of previous surgeries.
5. *Relationship between the superior edge of the choana and the sphenoid sinus*: the sphenoid sinus is located approximately 1.0 cm above the superior edge of the choana. Coronal CT gives a clear and indispensable picture of this relationship.
6. *Anatomical relationship among the posterior ethmoid cells, sphenoid sinus, optic nerve, internal carotid artery and cavernous sinus*: this is evident on both coronal and axial projections. The protuberances of the internal carotid arteries and optic nerves are well defined in axial projections. Identification of the cavernous sinus can be made in both basic projections, after the intravenous injection of contrast.
7. *Evaluation of extra-sinusal extension*: tumors or aggressive polypoid lesions may destroy the bony walls of the paranasal sinuses and compromise adjacent regions. Coronal and axial CT adequately define such alterations.
8. *Anatomical relationship among anterior ethmoid cells, the frontal recess, posterior ethmoid sinus and sphenoid sinus*: these anatomic relationships are extremely important in surgeries of the frontal recess and sphenoid sinus and are well identified in a sagittal CT section.

CT with three-dimensional reconstruction is particularly helpful in cases of tumors of the paranasal sinuses involving the anterior and middle skull base and fractures of the bones of the face, orbit and anterior skull base [29]. Using helical CT with reconstruction (employing the navigator system), it is possible to visualize the inside of the middle meatus and to enter an operated sinus the same way as with endoscopy (Fig. 17.8).

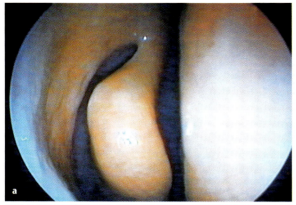

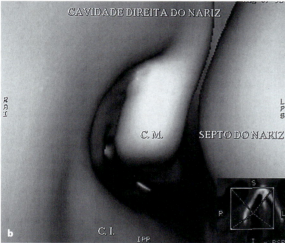

Fig. 17.8. a A view of the right middle turbinate by endoscopy. **b** The same view, using a three-dimensional reconstruction (virtual endoscopy)

Magnetic-resonance (MR) imaging effectively demonstrates the morphology of the soft tissues and the presence of liquid but is not helpful in visualizing bony architecture. It is helpful in the differentiation between neoplastic and inflammatory diseases and when a complication of a sinusitis is suspected, especially if paramagnetic contrast is used (gadolinium diethylene triaminopentaacetic acid). The absence of ionizing radiation is an advantage, and MR is the method of choice in pregnant women.

The preoperative assessment concludes with a detailed explanation from the physician to the patient (and family when appropriate) about the diagnosis, the natural course of the disease and the possible treatment options. When surgery is recommended, the type and rationale of the procedure are explained, as are the goals and potential complications. Any expected functional deficits or postoperative limitations are discussed. Surgery should not be performed without a solid physician–patient relationship.

Staging: Inflammatory Diseases of the Paranasal Sinuses

Preoperative staging of the disease helps determine the most appropriate operative technique, helps predict the results and allows comparison of results from different literature reports. Several different staging systems have been proposed for inflammatory and infectious diseases of the paranasal sinuses over the years [13, 14, 16, 26, 39, 42].

We believe staging of the inflammatory diseases of the paranasal sinuses – particularly nasal polyposis (Fig. 17.9) – should be based on the clinical endoscopic evaluation and radiographic findings from CT (and sometimes MR) examinations and should reflect the findings at surgery. Radiographically, it is sometimes impossible to differentiate between secretions and polyps in diseased sinuses, a distinction that has considerable prognostic importance; surgical correlation corrects for this deficiency.

The author's staging system consists of five stages (Fig. 17.10, Table 17.3):
I. Disease restricted to the OMC and middle turbinate (meatal and/or septal face)
II. Disease in the OMC, middle turbinate and ethmoid cells
III. Inflammatory disease involving the OMC, middle turbinate, ethmoid sinus and one more paranasal sinus (stage II plus one sinus)
IV. Involvement of the OMC, middle turbinate, ethmoid sinus and two more sinuses (stage II plus two sinuses)
V. Involvement of all paranasal sinuses

Using this clinical staging system (including the surgical findings), 65% of our 632 patients operated on due to sino-nasal polyposis had stage-III or stage-IV disease (Table 17.4). However, in the majority of

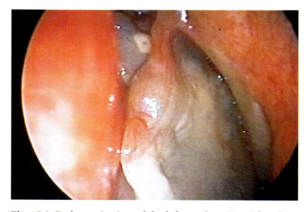

Fig. 17.9. Endoscopic view of the left nasal cavity, with polyps of the middle meatus

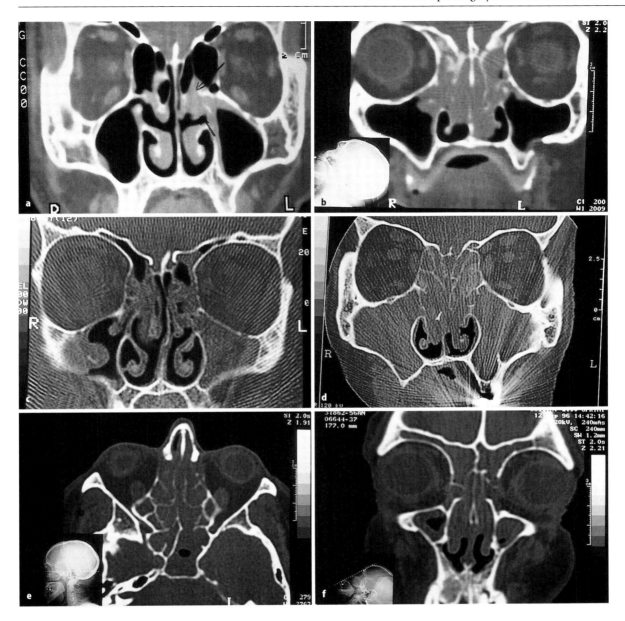

Fig. 17.10 a–f. Staging of nasal polyposis, according to the computed-tomography findings. a Stage I. b Stage II. c Stage III. d Stage IV. e, f Stage V

those cases, the disease was overestimated by CT and endoscopic findings alone. Approximately 30% of our patients had undergone previous nasal polyposis surgery, and 28% had associated asthma.

Surgical Prognostic Factors

Prognostic factors for micro-endoscopic surgery of the paranasal sinuses can be separated into pre-, intra- and postoperative factors (Table 17.5). The most important *preoperative* prognostic factors include an adequate endoscopic examination and careful analysis of the CT, noting any anatomic variations, the size of the paranasal sinuses, the surrounding structures (such as the optic nerve and internal carotid artery) and the character and extent of the disease (staging). A history of previous surgery may influence the final result, increasing the

Table 17.3. Sinonasal polyposis staging system by endoscopy, computed tomography and surgical findings [58, 60, 61]

Stage	Location
I	OMC/middle turbinate
II	OMC/middle turbinate/ethmoid sinus
III	Stage II plus one sinus
IV	Stage II plus two sinuses
V	All paranasal sinuses

OMC, ostiomeatal complex

Table 17.4. Sinonasal polyposis 1991–1999 (n=632)

Stage	Number of patients	%
I	62	10
II	107	17
III	218	34
IV	197	31
V	48	8

Table 17.5. Prognostic factors for micro-endoscopic surgery of the paranasal sinuses [60]

Preoperative
    Evaluation (endoscopy, computed tomography)
    Stage of the disease
    Previous surgery
    Associated systemic disease
    Asthma, allergy
    Use of antibiotics, steroids
    Smoking
Intraoperative
    Anatomic variations
    Amount of bleeding
    Anesthetic technique
    Choice of surgical technique
    Surgeon's experience
Postoperative
    Use and duration of antibiotics and steroids
    Endoscopic control during follow-up
    Allergic evaluation
    Adequacy of follow-up

risk of complications, especially because of the loss of the anatomic landmarks. Patients with systemic diseases (such as diabetes), hypertension or a history of chronic use of drugs that can cause problems with anesthesia or bleeding are at risk for complications. The association of sino-nasal polyposis with asthma is considered a negative prognostic factor by some authors, who cite a polyp recurrence rate of nearly 50% in such patients [35].

Patients with acute infection of the paranasal sinuses must receive preoperative antibiotics and systemic steroids in order to completely resolve the infection before surgery to minimize intraoperative bleeding and the chance of postoperative infection. Patients who smoke usually require a longer period of follow-up before complete mucosal healing takes place.

Intraoperative prognostic factors can be related to anatomic variations of the nose and paranasal sinuses. Small sinus cavities and narrow nostrils make surgical instrumentation more difficult. Other important factors relate to factors that affect bleeding in the operative field. Measures used to account for such factors include controlled hypotensive anesthesia, local infiltration of lidocaine (2% with 1:100,000 epinephrine) and topical vasoconstrictor, and elevating the dorsum and patient's head approximately 30°. The choice of the adequate surgical technique and the experience of the surgeon are decisive for the success of the surgery. Careful *postoperative* follow-up with periodic endoscopic examination and adequate control of infection with antibiotics and steroids are important positive prognostic factors.

Surgical Indications

Micro-endoscopic surgery of the paranasal sinuses attempts to re-establish aeration and drainage of the paranasal sinuses and to correct any alterations of the OMC anatomy. Surgery is indicated when mucosal disease persists even after adequate medical treatment of inflammatory and/or infectious processes that cause obstruction of the OMC (anatomic variations, polyps, fungae, etc.).

Micro-endoscopic surgery for the treatment of inflammatory and/or infectious diseases of the paranasal sinuses in children is more restricted. Treating predisposing factors (such as adenotonsillar hypertrophy, allergic rhinitis, choanal atresia or, occasionally, septal deviation) is often sufficient. The indications for micro-endoscopic surgery of the paranasal sinuses are tabulated for each of the individual sinuses below.

Ethmoidectomy

The role of the ethmoid sinus in the pathophysiology of sinusitis of all the other paranasal sinus is quite

established, and the most common surgical indications for ethmoidectomy are:
- Obstructive disease of the OMC (concha bullosa, polyps, hypertrophy of the ethmoid bulla and anatomic alterations of the uncinate process)
- Chronic maxillary sinusitis secondary to diseases of the OMC
- Sino-nasal polyposis
- Fungal sinusitis
- Acute, complicated, suppurative ethmoiditis
- Mucoceles or mucopyoceles of the ethmoid sinus
- Diseases of the middle turbinate
- Surgical access to the sphenoid and frontal sinuses
- Foreign bodies
- Trauma of the ethmoid sinus and anterior skull base
- Repair of a CSF leak
- Surgical access to the anterior and posterior ethmoid arteries
- Orbit and optic-nerve decompression
- Biopsy

Sphenoidotomy

Microendoscopic surgery of the sphenoid sinus may be performed through the nasal septum, directly through the nasal cavity or through the middle meatus, depending on the anatomy and the type of lesion. Surgical indications include:
- Chronic sphenoiditis
- Sphenoid-sinus empyema
- Polyposis
- Mucoceles or mucopyoceles of the sphenoid sinus
- Foreign bodies
- Surgical access to the petrous pyramid
- Fungal sinusitis
- Surgical approach to the pterygoid canal (vidian nerve)
- Repair of a CSF leak
- Surgical access to the pituitary gland
- Optic-nerve decompression
- Removal of benign tumors
- Biopsy

Maxillary Sinus

Surgical approaches to the maxillary sinus via the nasal cavity can be performed through the middle or the inferior meatus. Externally, it can be approached through the canine fossa. The majority of the inflammatory and infectious diseases of the maxillary sinus are secondary to diseases of the middle meatus. Surgical indications include:
- Chronic maxillary sinusitis
- Empyema of the maxillary sinus
- Recurrent bacterial sinusitis
- Sino-nasal polyposis
- Fungal sinusitis
- Surgical access to the pterygopalatine and zygomatic fossae
- Surgical access to the orbit
- Foreign bodies
- Biopsy

Frontal Recess and Frontal Sinus

Transnasal micro-endoscopic surgery of the frontal recess is indicated when the cause of the disease is in the nasofrontal recess (disturbing the sinus ventilation and drainage) or to treat intrinsic diseases of the frontal sinus itself. Surgical indications include:
- Disease confined to the frontal recess
- Recurrent frontal sinusitis with empyema
- Chronic hyperplastic frontal-sinus mucosal disease
- Sino-nasal polyposis
- Mucoceles or mucopyoceles
- Fungal sinusitis
- Benign lesions
- Foreign bodies
- Biopsy

Surgical Procedures

The most important transnasal micro-endoscopic surgical procedures of the paranasal sinuses can be seen in Table 17.6.

Table 17.6. Surgical techniques for micro-endoscopic surgery of the paranasal sinuses [60]

- Conservative surgery (uncinectomy, bullotomy, turbinoplasty and septoplasty)
- Antrostomy: middle, inferior, anterior and combined
  Ethmoidectomy: partial, total and radical
- Frontal sinus
  Frontal-sinus probing
  Frontal-recess surgery
  Intra-sinus procedures (Draf I, II and III procedures)
  Combined procedures
- Sphenoidotomy: transnasal, transethmoid and trans-septal
- Revision surgery

Conservative Surgery

These procedures all attempt to minimize mucosal contact in sinus-drainage areas, particularly in the OMC region. Conservation surgery includes surgical removal of the uncinate process (uncinectomy), partial or total resection of the ethmoid bulla (bullotomy), resection of the meatal surface of the middle turbinate and localized septoplasty [19, 60].

Maxillary Sinus

Surgical access to the maxillary sinus may be made by antrostomy via the middle meatus, inferior meatus or the anterior wall of sinus; access may also take place via a combination of these approaches. *Anterior antrostomy* is performed with a puncture of the canine fossa using a trocar and endoscope or through a mini-window (sufficiently large to introduce an endoscope and a microforceps) in the anterior wall of the maxillary sinus.

There is no consensus regarding how wide a *middle-meatus antrostomy* should be made. A large antrostomy permits drainage, aeration and easier access to the maxillary sinus, and it is usually indicated in cases of polyposis, fungal disease and chronic sinusitis with diseased mucoperiosteum. Ostioplasty or enlargement of the main drainage ostium is indicated in milder situations, such as acute, recurrent sinusitis. *Inferior-meatus antrostomy* is the opening of the inferior meatus, allows drainage of secretions by gravity and may be indicated to ensure successful removal of diseases of the maxillary sinus. *Combined antrostomy* (middle-to-inferior or middle-to-anterior) may be helpful in order to obtain a complete view of the sinus cavity.

Ethmoid Sinus

Partial ethmoidectomy is the resection of the anterior ethmoid cells, preserving the mucoperiosteum of the orbital wall and skull base. The middle turbinate is also preserved.

Total ethmoidectomy is an anterior and posterior ethmoidectomy, preserving the mucoperiosteum of the ethmoid-sinus walls (orbit, anterior skull base). It is indicated in cases of chronic sinusitis and/or polyposis with involvement of the entire ethmoid sinus. Middle turbinoplasty may be part of this surgical intervention.

Radical ethmoidectomy is the resection of the anterior and posterior ethmoid cells and the mucoperiosteum of the ethmoid orbital wall and skull base (except in the frontal-recess area). Maxillary-sinus mucoperiosteum is also usually preserved. The middle turbinate is often removed, preserving its insertion. This surgical procedure is mostly reserved for treating grade-IV and -V recurrent polyposis [58–60].

Frontal Sinus

Among the surgical procedures of the OMC, the most controversial involve surgery of the frontal recess; these are controversial principally because of the risk of postoperative stenosis from granulation-tissue formation, cicatrization and osteoneogenesis. For simple drainage or exploration of the frontal recess, the Ritter probe is indicated. The frontal recess can be widened by maintaining the integrity of the ethmoid bulla by working anterior to it. In cases of disease inside the frontal sinus, a larger opening that allows the surgeon to work inside the sinus is recommended, and the surgical techniques proposed by Draf (procedures I, II and III) are used. If the frontal recess is narrow and the surgeon has difficulty penetrating the frontal sinus, it may be advisable to combine an intranasal approach with external trephination, which permits simultaneous irrigation and endoscopic visualization.

Sphenoid Sinus

Endonasal surgery of the sphenoid sinus can be accomplished by three distinct approaches. The *transnasal direct approach* is used for diseases confined to the sphenoid sinus. In patients with concomitant disease in the posterior ethmoid sinus, the surgical access should be *transethmoid* or ethmoidectomy combined with the direct access. The *trans-septal* approach is rarely used for surgical treatment of sinusitis but is commonly used as a surgical approach to the pituitary gland. *Revision operations* of the paranasal sinuses are usually technically more complicated, especially because of the loss of anatomic landmarks.

After the optimum surgical technique has been selected, a careful assessment of the nasal cavity is essential to identify any nasal septal deviations and anomalies of the nasal turbinates, especially the middle turbinate. Even asymptomatic nasal septal deviations should be corrected, especially if they impair visualization of the middle meatus and the anterior insertion of the middle turbinate. They can also make postoperative care more difficult. The septoplasty can be restricted to a small area and can be performed endoscopically or in the conventional manner, with a headlight. Usually, septal splints are placed to avoid synechiae after septoplasty associated with paranasal-sinus surgery.

In patients with a large concha bullosa blocking the OMC, causing sinusitis and making the surgery more difficult, removal of the meatal face of the concha bullosa may be necessary. Surgery of the middle turbinate can be performed with micro-scissors, cutting forceps or microdebriders but, regardless of the technique, it is important to maintain the stability of the remaining turbinate and to avoid postoperative lateralization and stenosis in the middle-meatus region. In order to prevent this complication, a large portion of the middle turbinate can be resected or the remainder of the turbinate can be sutured to the nasal septum. Some authors even try to promote synechiae between the middle turbinate and the nasal septum. In such cases, a middle-meatal splint in conjunction with an expandable pack is recommended.

Perioperative Considerations

Positioning

In the operating room, the patient is positioned in a semi-sitting position. The patient's eyes are kept uncovered in order to permit periodic monitoring.

The surgical team is positioned as shown in Fig. 17.11, with the surgeon at the right side of the patient and the video equipment and anesthesiologist at the left. For lengthy procedures, a support for the arm that holds the endoscope may be helpful. When the procedure uses the operating microscope and a self-retaining speculum, the positioning of the surgical team is the same as for otologic surgery, with the patient supine and the surgeon on the patient's right side.

Instrumentation

The majority of paranasal-sinus surgical procedures are performed with the endoscopes attached to an endocamera and a video-monitor system

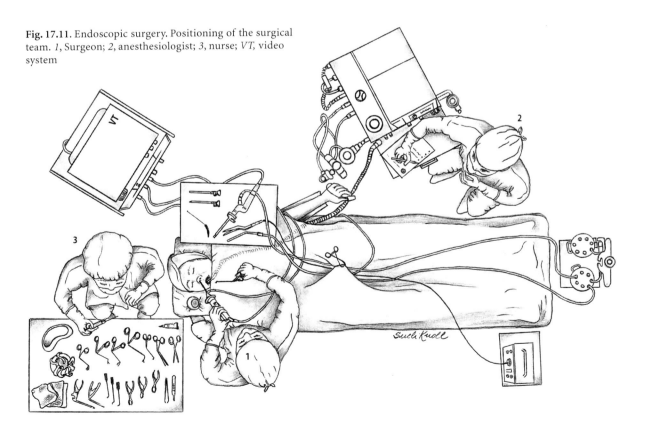

Fig. 17.11. Endoscopic surgery. Positioning of the surgical team. *1*, Surgeon; *2*, anesthesiologist; *3*, nurse; *VT*, video system

(Fig. 17.12). Although 0°, 30°, 45° and 70°, 4-mm endoscopes can be used, the 0° is used most often (Fig. 17.13).

Although conventional surgical instrumentation (such as nasal speculums, suction elevators and Freer elevators) can also be used, most of the micro-endoscopic surgical instrumentation has an articulation located at the edge, allowing adequate visualization of the operative field (Fig. 17.14). Suction tubes should be blunt edged in order to avoid unnecessary trauma and bleeding of the naso-sinusal mucosa. Their curved and straight shapes permit access to the maxillary and frontal sinuses. Conventional suction elevators usually used in septoplasty are also used in localized endoscopic septoplasty. They have multiple functions (including aspiration, elevation and cutting) and are especially helpful, because the surgeon can normally operate the elevator with one hand while holding the endoscope with the other.

Initial evaluation of the structures of the middle meatus and sphenoethmoidal and frontal recesses can be accomplished with a double-ended probe, also called a seeker/palpator (Fig. 17.42). A Freer elevator is also useful at the beginning of a surgical procedure. It has two ends: semi-sharp and blunt. The semi-sharp edge is used to incise and dissect the mucosa, while the blunt end can be used for palpation and displacement of structures of the nasal cavity, such as the middle turbinate. Removal of the uncinate process or any mucosa-cutting procedures can be done with a sickle knife (Fig. 17.18A).

Backbiting forceps (Karl Storz, Tuttlingen) are used to remove the uncinate process and anterior fontanelle. A shell-shaped modification in its cutting portion makes it more delicate, thus making removal of tissue easier and more precise. Blakesley's forceps, with straight and curved articulated ends, are frequently used to remove diseased tissue (such as polyps and bony fragments). A micro-Kerrison punch is normally used to open the anterior wall of the sphenoid sinus, to remove ethmoidal cells next to the medial wall of the orbit and to remove the optic-nerve bony canal (Fig. 17.44).

A 4-mm-diameter, angled forceps helps the surgeon to work in the frontal-recess region and in the frontal sinus (Fig. 17.28). The maxillary sinus is better accessed with an "antrum-grasping punch" (Heuwieser) because of the special angle of this punch, which was designed for use in this region (Figs. 17.20, 17.26A).

Micro-scissors, straight or curve-ended (right or left) are used for partial resection of the turbinates and tailoring of mucosa grafts. The "through-cutting" instrumentation (Gruenwald-Henke, Karl Storz, Tuttlingen; Fig. 17.38) represents an important development in paranasal-sinus surgery. It was designed to preserve the mucosa, avoiding unnecessary stripping and, consequently, leading to faster healing with less crusting. This diminishes the time needed for postoperative care. These forceps are especially useful in the frontal-recess region, medial orbital wall and skull base and in removing the posterior fontanelle while avoiding stripping of the mucosa and preventing postoperative stenosis.

Monopolar and bipolar eletrocautery permit one to control most of the bleeding that is a consequence of partial turbinectomy. Bleeding originating from branches of the anterior or posterior ethmoidal arteries is more effectively controlled by bipolar systems of coagulation.

Fig. 17.12. Video-endoscopic system (color monitor, cold-light fountain, endocamera, video recorder and videoprinter; Karl Storz, Tuttlingen)

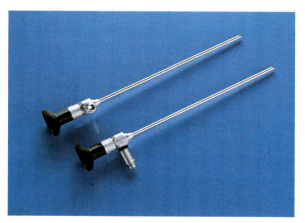

Fig. 17.13. Four-millimeter, surgical, rigid, angled endoscopes (Karl Storz, Tuttlingen)

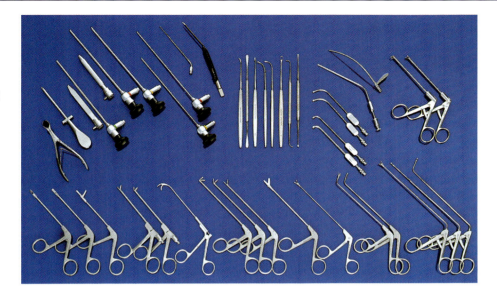

Fig. 17.14. Sinonasal endoscopic-instrument set (endoscopes, mono- and bipolar cautery, suction tubes, palpators, elevators and several types of articulated forceps; Karl Storz, Tuttlingen)

Powered instrumentation (i.e., microdebriders) initially developed for soft-tissue shaving represent a marked advance in endoscopic surgery of the paranasal sinus. They are instruments with multiple functions, including suction, cutting and irrigation. The first-generation equipment was deigned for the removal of soft tissue and was supplied with straight blades only. Currently, newer instruments including curved blades have been created; these are able to remove thin, bony laminas, such as ethmoidal cells. In addition, another advantage of the newer equipment is that it provides a precise cut of the diseased tissue, avoiding mucosal stripping and continuous irrigation; this considerably diminishes bleeding in the operative field (Fig. 17.15).

Fig. 17.15. Power instrumentation: microdebrider (Xomed, Jacksonville)

Anesthesia

Micro-endoscopic surgery of the paranasal sinuses can be performed under local or general anesthesia. Gittelman et al. [15] demonstrated that the complication rate for endoscopic surgery of the paranasal sinuses is similar in both local and general anesthesia. The choice of anesthesia depends on the experience and training of both the surgeon and the anesthesiologist, and their working relationship; general anesthesia with controlled hypotension is usually preferred. Topical vasoconstrictors and infiltration with lidocaine 2% and 1:100,000 epinephrine are also used. The local anesthesia is injected with a 25-gauge spinal needle. The key points to be infiltrated are the uncinate process, the anterior middle-meatus region, the sphenopalatine-foramen region, the superior insertion of the middle turbinate (anterior arch) and, if necessary, the sphenoethmoid recess. In order to obtain optimum mucosal vasoconstriction, at least 10 min should elapse after injection before making any incisions.

Operative Techniques

Maxillary Sinus

Transnasal access of the maxillary sinus was first performed via the middle meatus in 1761 by Jourdain [24], a Parisian dentist. In 1882, Zuckerkandl [78] already recommended direct access to the maxillary sinus by creating an opening just behind and below the infundibulum. In 1887, Mikulicz [46] performed an inferior meatotomy for the first time. Caldwell (in 1893) [5] and Luc (in 1897) [38] proposed a new method, currently known as the Caldwell-Luc method, consisting of creation of an opening in the anterior wall of the maxillary sinus and a secondary opening between the sinus and the nasal cavity (via inferior meatotomy). This method has been the most commonly used surgical procedure on the sinuses during the last 50 years, and it is still employed in many centers today. In the 1950s and 1960s, Heermann [20] began to utilize an operating microscope to approach the maxillary sinus surgically. Some years later, Messerklinger [43–45] recommended an infundibulectomy using an endoscope, and this method is still in use today. Maxillary sinus micro-endoscopic surgery is considered complex from the anatomic and physiologic standpoints. Mucociliary transport is known to ascend from the sinus floor to the ostium located in the middle meatus; thus, there is active drainage of the secretions from the sinus into the nasal cavity. Because of this, it is important to preserve as much maxillary-sinus mucosa as possible. Surgical access to the sinus follows three basic avenues: (1) the middle meatus, (2) the inferior meatus and (3) the canine fossa; sometimes, the approaches are combined (Fig. 17.16). Some aspects of the transansal maxillary-sinus approaches are still controversial (for example, whether the antrostomy should be performed through the middle or inferior meatus or both, and whether ostioplasty or antrostomy is preferable). According to Setliff [54] and Parsons et al. [49, 50], surgical procedures in the maxillary ostiomeatal area, regardless of the pathological condition, should attempt to keep the ostium undisturbed. Relief of this area is accomplished by partial or total resection of the uncinate process, particularly its inferior portion; this is called the "small-hole technique". Studies of maxillary-sinus mucociliary clearance concluded that mucosal damage in the region of the maxillary ostium and middle meatus decreases mucociliary flow from the sinus. Another area of disagreement concerns the anterior ethmoid cells, considered by some to be the key area and source of infection and re-infection of the maxillary sinus. For these reasons,

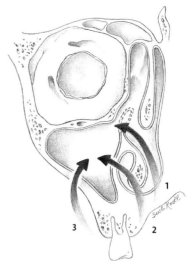

Fig. 17.16. Three basic avenues by which to surgically approach the maxillary sinus: via the middle meatus (*1*), via the inferior meatus (*2*) and via the canine fossa (*3*)

they recommend removing the anterior ethmoid cells and the infraorbital cells (Haller's cells) when maxillary-sinus surgery is performed. Any nasal-septum deviation is corrected first, usually through an endoscopic approach. The main reason to perform a septoplasty is to obtain complete visualization of the middle turbinate, especially its insertion into the lateral nasal wall (anterior arch), and to facilitate the use of surgical instruments in the middle-meatus region.

Middle-Meatal Antrostomy

The most important anatomical landmarks during middle-meatal antrostomy are the uncinate process, ethmoid bulla, superior insertion of the middle turbinate and superior margin of the inferior turbinate. If an operating microscope is used, the middle turbinate is gently displaced toward the nasal septum, and a number-2 self-retaining nasal speculum is inserted to retract (but preserve) the middle turbinate, exposing the middle-meatus area. The long blade of the speculum displaces the middle turbinate, while the short blade is set just anterior to the uncinate process (Fig. 17.17). If the procedure is performed with endoscopes, the middle turbinate is displaced with a Freer elevator and is maintained in position by the endoscope itself or with suction.

The next step is resection of the uncinate process (uncinectomy); however, before resecting the uncinate process, the CT should be rechecked to reassess

Fig. 17.17. Exposing of the middle meatus region by a self-retaining nasal speculum (a Rudert modified speculum)

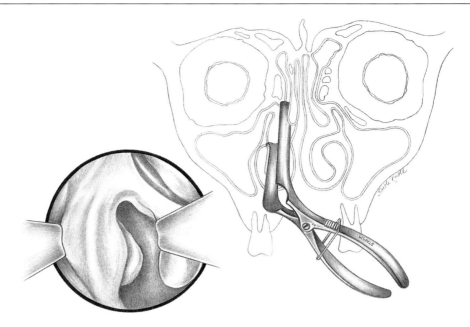

its position in relation to the orbit. The uncinectomy is similar in both the microscopic and endoscopic techniques. The uncinate process is gently sectioned with a sickle knife in a plane parallel to the orbit and lateral nasal wall and is displaced medially to avoid any chance of orbital penetration (Fig. 17.18A). Penetration of the orbit can result in eyelid ecchymosis or orbital complications. In order to avoid stripping the mucosa, the superior and inferior insertions of the uncinate process can be sectioned with a microscissors and removed with a micro-forceps. Another safe and effective technique is to remove the uncinate process with backbiting forceps, working in an inferior-to-superior direction (Fig. 17.18B). In revision cases, an effective method is use of a microdebrider or a Freer elevator for this step (Fig. 17.18C).

After uncinectomy, the ethmoid bulla is completely visible. The ethmoid bulla should initially be preserved as an anatomical landmark.

Next, the region of the natural ostium is identified with the help of a seeker or an atraumatic, curved aspirator and 30° or 45° endoscopes (Fig. 17.19). If the natural ostium cannot be identified, the maxillary sinus can be penetrated via the posterior fontanelle using a sharp instrument in order to avoid elevating the maxillary-sinus mucoperiosteum.

The natural ostium of the maxillary sinus should always be incorporated into the opening, because the mucociliary transport will continue in this direction and, if a second window is created, a recirculation pattern of mucociliary flow can develop, ineffectively moving the secretions around the maxillary sinus and leading to stagnation and infections. The size of the window varies from case to case. If needed, enlargement of the antrostomy initially incorporates the posterior fontanelle by using cutting forceps and avoiding the sphenopalatine-foramen region. If the antrostomy is enlarged anteriorly at the expense of the anterior fontanelle, the enlargement is done with backbiting forceps. Special care should be taken to avoid damage to the nasolacrimal duct and the orbit. The inferior limit of a middle-meatal antrostomy is the superior margin of the inferior turbinate.

Forty-five or 70° endoscopes permit adequate visualization of the interior of the sinus cavity (through the antrostomy) for drainage of an empyema or removal of polyps, fungus balls or cysts using an antrum punch (Fig. 17.20). If a Haller's cell blocks the infundibulum, it is removed first. The sinus mucoperiosteum is preserved, if possible, since its mucociliary function can recover after adequate ventilation and removal of the disease (Fig. 17.21). In many patients, and especially children, simple resection of the uncinate process (conservative surgery) is adequate treatment for chronic or recurrent sinusitis [54].

Inferior-Meatal Antrostomy

Inferior-meatal antrostomy is uncommon compared with middle-meatal antrostomy and is reserved for treatment of intrinsic lesions of the maxillary sinus unrelated to the OMC (such as cysts, isolated polyps, dental-apex disease or foreign bodies) or when

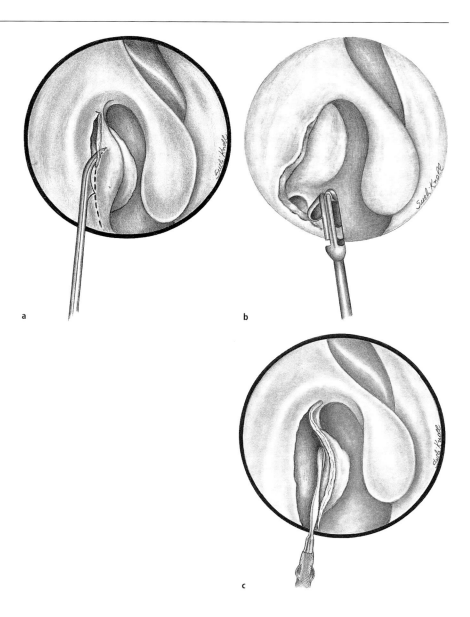

Fig. 17.18 a–c. Ways to perform an uncinectomy: by using a sickle knife (**a**), a backbiting forceps (**b**) or a Freer elevator (**c**)

repeated irrigations of the sinus are necessary [75, 76]. Inferior antrostomy may also be helpful in patients with ciliary dyskinesia, in order to facilitate drainage of the secretions (by gravity) [47, 69].

Because mucoperiosteum is removed from the antrostomy region, inferior antrostomy (or a combined antrostomy) can eventually result in disturbances of mucociliary transport from the sinus to the nasal cavity [28]. The surgical procedure is usually performed under local anesthesia, with installation of a topical anesthetic and vasoconstrictor solution followed by infiltration of the inferior-meatal mucoperiosteum with lidocaine (2%) and 1:100,000 epinephrine.

The inferior turbinate is displaced and gently fractured upward and is maintained in position by a number-2 self-retaining speculum or by the 0°, 4-mm endoscope itself. The mucoperiosteum is incised approximately 1 cm posterior to the head of the inferior turbinate and is displaced posteriorly and resected. The exposed bone is opened with a microosteotome or drill, and the window is enlarged with a micro-Kerrison punch. Care must be taken to avoid the insertion of the inferior turbinate and the ostium of the nasolacrimal duct (Hasner's valve). At this point, an excellent view of the entire sinus cavity is obtained with a 45° or 70° endoscope (Fig. 17.22). Pre- and postoperative CT scans of a patient operated on via an inferior-meatal antrostomy can be seen in Fig. 17.23.

In some situations, an antrostomy combining the middle- and inferior-meatal approaches is per-

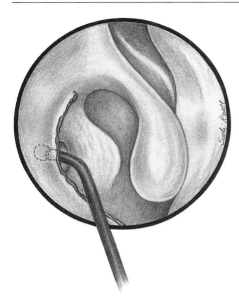

Fig. 17.19. Identification of the main maxillary-sinus ostium with an atraumatic curved aspirator

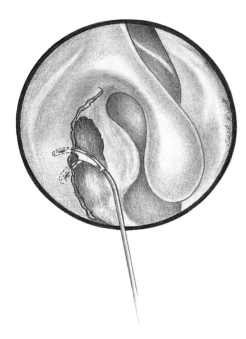

Fig. 17.20. Removal of diseases through a middle-meatal antrostomy can be accomplished using an antrum punch

formed. This combination is useful when a complete view of the inferior and alveolar-recess regions of the cavity is essential. In this situation, the inferior turbinate and its insertion are preserved by a bridge of bone between the two antrostomies.

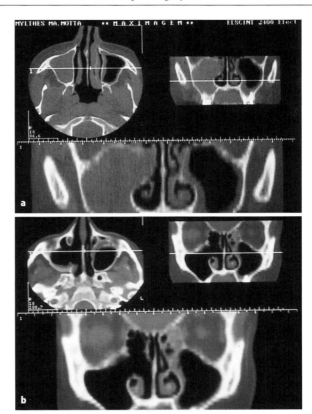

Fig. 17.21 a,b. Preoperative (a) and postoperative (b) computed tomography of a chronic-sinusitis patient operated on through a middle-meatal antrostomy

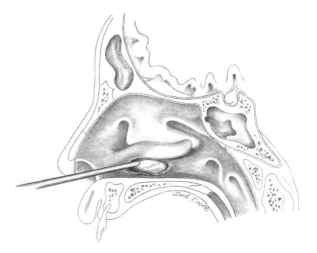

Fig. 17.22. Inferior meatal antrostomy assisted by 45° or 70° endoscopes

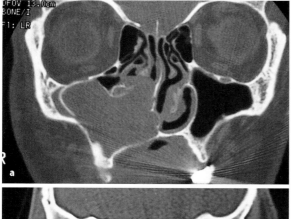
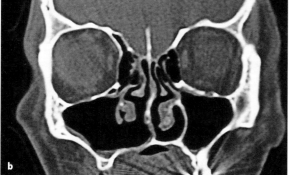

Fig. 17.23 a,b. Coronal computed tomography of a patient with odontogenic sinusitis operated on through an inferior-meatal antrostomy. **a** Before operation. **b** After operation

Anterior Antrostomy (Canine-Fossa Puncture or Mini-Anterior Antrostomy)

Endoscopic evaluation of the maxillary sinus through the canine fossa can be useful when a tumor is suspected, to remove polyps and cysts or to facilitate middle or inferior antrostomy from inside the sinus (Fig. 17.24). After sublabial infiltration of the canine-fossa region, a trocar is introduced into the maxillary sinus by rotation and gentle pressure. The trocar is then replaced by an endoscope. Biopsies and cyst removal can be performed by capturing the lesion with forceps attached to the endoscope. If the lesion cannot be removed through the cannula opening, combined access may be helpful (bimeatal endoscopy). Double access through the canine fossa or even slight enlargement of the anterior maxillary window (to approximately 1 cm in diameter) are feasible in order to allow simultaneous passage of an endoscope and a surgical instrument. These procedures can be useful in removing naso-antral polyps and foreign bodies and in correcting oro-antral fistulas. In such cases, the maxillary-sinus mucoperiosteum is always preserved (Fig. 17.25).

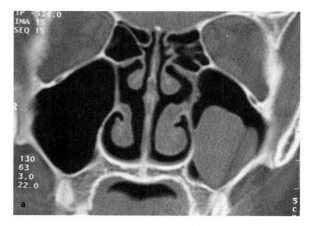
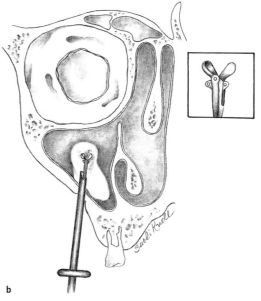

Fig. 17.24. a Coronal computed tomography shows a mucous retention cyst in the left maxillary sinus. **b** A forceps passed through the trocar can be used for biopsies and cyst removal

Combined Approach

Combining the middle and anterior mini-antrostomies or punctures is indicated in cases of extensive sino-nasal diseases. Hyperplastic sinusitis and polyposis may be surgically treated via the middle meatus by using angled forceps and microdebriders and may be guided by an endoscope introduced through the anterior maxillary wall (Fig. 17.26).

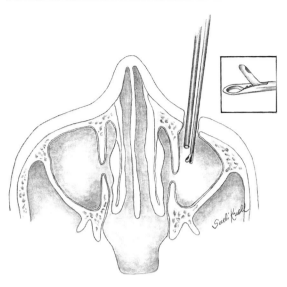

Fig. 17.25. Mini-anterior antrostomy, allowing simultaneous passage of a surgical instrument and endoscope

Causes of Failure

- Lateralization of the middle turbinate (obstructing the middle meatus)
- Incomplete resection of the uncinate process (leaving the natural ostium separate from the middle-meatal antrostomy, leading to recirculation phenomena)
- Middle meatus synechiae (blocking drainage and ventilation)
- Incomplete removal of a fungus ball (residual disease)
- Inadequate correction of nasal septal deviation (predisposing the patient to sinus infection, making postoperative care difficult)

Frontal Recess and Frontal Sinus

In the past, intranasal approaches to the frontal sinus and the frontal recess were used much less commonly than intranasal approaches to the other sinuses because of frequent anatomic variability and an inability to obtain sufficient visualization and exposure to perform safe and adequate surgery. Complications were frequent, especially postoperative stenosis. With the advent of angled telescopes, the frequency of the transnasal approach to the frontal sinus and frontal-recess region is now increasing, even though significant challenges remain.

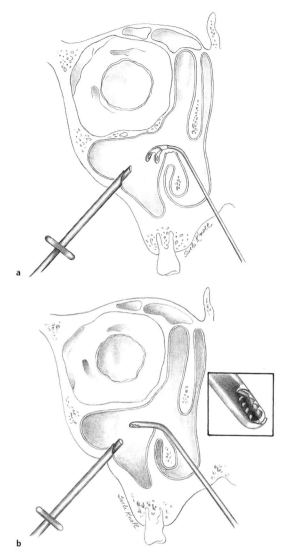

Fig. 17.26 a,b. A surgeon can work with an antrum-grasping punch (**a**) and an angled microdebrider (**b**) through the middle meatal antrostomy, guided by an endoscope introduced through the canine fossa

The goal of surgery of the frontal sinus and frontal recess is to obtain an open channel of communication with the anterior ethmoid and the middle meatus. The internal ostium of the frontal sinus (the "isthmus") is linked to the middle-meatus region via the frontal recess. Its width depends on the degree of pneumatization of the ethmoid bulla and "agger nasi" cells.

The uncinate process is often attached to the medial orbital wall. In these cases, the ethmoid infundibulum ends blindly, in the "recessus terminalis", while the frontal recess is medial to the uncinate process. The close positioning of these spaces may lead a surgeon to misinterpret them. Lee et al. [36], in a recent study, observed that the frontal sinus drains anteriorly to the uncinate process in 29.3% of cases and drains posteriorly in 68.3%; of those that drain posteriorly, 61% drain directly onto the ethmoid infundibulum.

The most important anatomical landmarks in frontal-recess and frontal-sinus surgery are:
1. The ethmoid bulla
2. The uncinate process
3. The insertion of the middle turbinate into the lateral nasal wall – the "anterior arch"
4. The anterior ethmoid artery and its canal
5. The lacrimal sac
6. Anterior ethmoid cells (agger nasi)
7. The lamina papyracea

There are several possible surgical techniques for approaching the frontal recess and frontal sinus, depending on the type of the lesion and the experience of the surgeon.

Currently, the most frequently used methods of surgical access are: an external approach with a coronal or supraciliary incision, a transnasal approach and a combined access. In this chapter, we emphasize the transnasal approach and its combination with an anterior trephination of the frontal-sinus wall.

External access to the frontal sinus can be obtained through a trephination of its anterior wall, external frontoethmoidectomy or osteoplastic surgery (with or without obliteration or cranialization of the sinus). Intranasal access was used extensively by several surgeons at the end of the last century and the beginning of this one. It was gradually abandoned because of the high incidence of orbital and intracranial complications.

It regained its popularity after Heermann [20] introduced the operating microscope in 1958. Draf [12] added the endoscope to the microscope to improve the approach to this area and described three different situations: type 1 (removal of bony partitions of the frontal recess and the area of the internal ostium), type 2 (widening of the frontal ostium) and type 3 (removal of the adjacent nasal septum and frontal intersinus septum to create a wide single opening). Gross et al. [18] called Draf procedure type III the "modified, transnasal, endoscopic Lothrop procedure". Wigand [72, 73], Stammberger [65], Schaefer [53], Kennedy [27], Kuhn [32], Lowry [37] and May [41] firmly established the use of endoscopes in transnasal access of the frontal sinus and frontal recess. The combined access is achieved by combining the transnasal approach with a trephination of the anterior frontal-sinus wall, through which endoscopes can be introduced and simple frontal-sinus irrigation can be carried out.

Surgical Technique

The surgical procedure begins with the resection of the uncinate process, especially its superior insertion. Initially, the ethmoid bulla is preserved for protection of the lamina papyracea, anterior ethmoid artery and the skull base; it is an excellent anatomical landmark, because the frontal recess is located anterior to the bulla [37, 52].

The frontal recess is first identified by gentle probing using number-3 or -4 Ritter probes (Fig. 17.27) or an atraumatic, curved aspirator. In some cases (for example, in an abscess of the frontal sinus), probing the frontal recess may be sufficient to resolve the disease [21, 52].

Next, the "agger nasi" cells that form the anterior wall of the frontal recess are removed. This can be done with an angled forceps or micro-curette in a

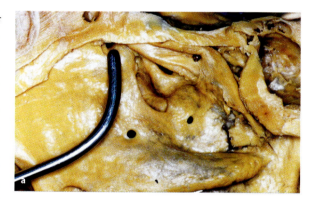

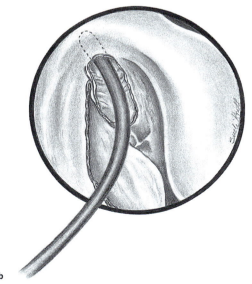

Fig. 17.27 a,b. Identification and probing of the frontal recess with a Ritter probe. a Lateral anatomical view. b Ritter probe entering the frontal sinus through the frontal recess

posterior-to-anterior and medial-to-lateral direction in order to avoid penetration of the anterior skull base. The surgeon should always try to preserve the mucoperiosteum of the frontal recess to minimize the risk of postoperative stenosis.

If the supraorbital ethmoid cells are well developed, it is possible to mistake them for the frontal sinus, which is normally located medial to them; in those cases, the bony wall between the ethmoid cells and the frontal sinus is also removed. Sometimes, it is necessary to partially resect the anterior part of the ethmoid bulla in order to clearly identify the anterior ethmoid artery, which is located in the posterior region of the frontal recess. If a pansinus operation is being performed, the ethmoid cells can then be removed in a posterior-to-anterior direction, working along the anterior skull base until the frontal recess and frontal sinus are encountered.

Identification of the internal ostium of the frontal-sinus "isthmus" is best accomplished with 30°, 45° or 70° endoscopes. Tumors and polyps are removed with an angled forceps specially curved for use in the frontal sinus (Fig. 17.28). Secretions are removed by periodic irrigation. If the visualization of the internal ostium is difficult because of a prominent nasofrontal "beak", it is removed with a curette (Fig. 17.29), diamond burrs [12, 32, 41, 52, 73] or high-speed bone-cutting drills that have an attached

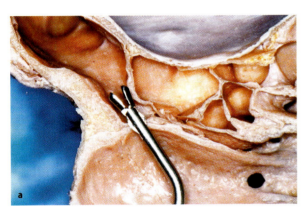

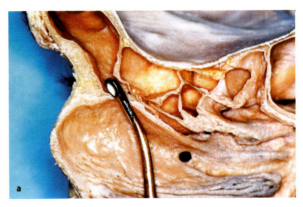

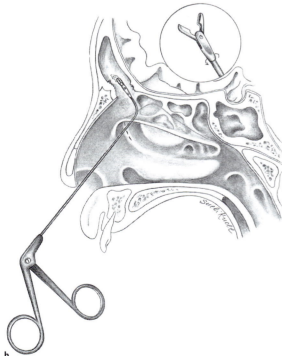

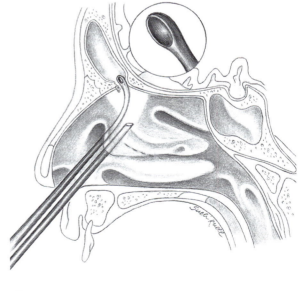

Fig. 17.28 a,b. Surgical manipulation of the frontal sinus, using an angled frontal forceps. a Lateral anatomical perspective. b An angled frontal forceps enters the frontal sinus, guided by angled endoscope

Fig. 17.29 a,b. Removal of the nasofrontal "beak" can be accomplished using a strong curette. a Lateral anatomical view. b A surgeon works with a surgical curette, guided by an angled endoscope

protective shield to avoid damaging the mucosa of the posterior wall of the frontal recess. Removal should be accompanied by simultaneous suction and irrigation (Fig. 17.30) [18].

If the anterior wall of the ethmoid bulla is absent (due to previous surgery or disease or because of an anatomic variation), the internal ostium can be identified at the insertion of the middle turbinate in the meatal face of the skull base, anterior to the anterior ethmoid artery and/or ethmoid dome. The anterior ethmoid artery crosses the skull base just posterior to the frontal recess. Care must be taken not to dislocate the middle turbinate from the anterior skull base, which promotes lateralization and blockage of the frontal recess. Pre- and postoperative CT aspects of a patient who underwent this procedure can be observed in Fig. 17.31. When these techniques fail to localize the "isthmus" of the frontal sinus, a combined access, adding external sinus trephination may be indicated [4, 59, 72].

The trephination is performed at the inferior medial eyebrow, below the supraorbital nerve. The trephination is approximately 5 mm in diameter to allow easy passage of a cannula/endoscope and a surgical dissector and is performed using a semi-cutting burr or using a mini-trephine set, which helps to locate the nasal frontal outflow tract and permits irrigation of the frontal sinus (Fig. 17.32). The transnasal approach, as described previously, should already be completed. The frontal sinus is irrigated through the trephination while observing the drainage point inside the nasal cavity. The drainage canal is then enlarged to reach the frontal sinus. The surgeon can also be guided by a red rubber catheter introduced from above, noting its emergence in the middle meatus. This catheter may be kept in place for a variable time after operation and may be sutured to the nasal

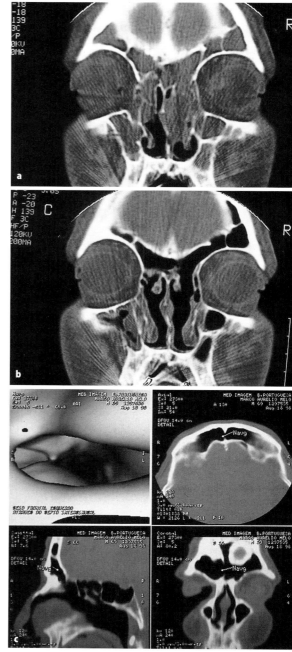

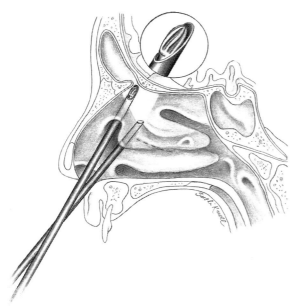

Fig. 17.30. The nasofrontal "beak" can also be removed with bone-cutting drills, specially protected in order to avoid frontal-recess-mucosa damage

Fig. 17.31 a-c. Preoperative (a) and postoperative (b) coronal computed tomography of a patient operated on for polyposis and chronic frontal sinusitis. c Postoperative three-dimensional reconstruction (virtual endoscopy)

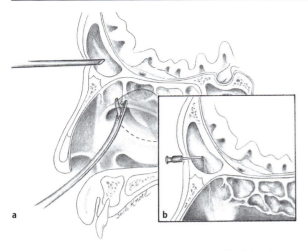

Fig. 17.32. a Trephination of the anterior wall of the frontal sinus, allowing passage of an endoscope. b Mini-trephination beneath the inferior medial eyebrow permits irrigation of the frontal sinus

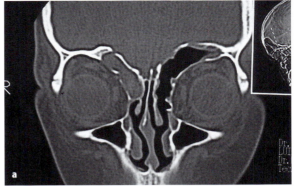

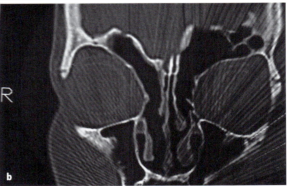

Fig. 17.34. a Postoperative cicatricial stenosis of the right frontal recess. b Coronal computed tomography after a re-operation that included reduction and medialization of the middle turbinate

Fig. 17.33. Ear, nose and throat image-guidance system (LandmarX; Xomed, Jacksonville)

septum. It maintains the patency of the communication and permits periodic irrigation. Another way to identify the frontal recess, especially in difficult cases with previous surgery, is through the use of image-guided technology (Fig. 17.33).

Causes of Failure

- Lateralization of the middle turbinate can be corrected by partial resection and the suturing of the remaining portion to the nasal septum. The Bolger method proposes scarification of the nasal septal mucosa and the septal surface of the middle turbinate in order to promote cicatricial attachment. In this case, middle-meatal spacers or septal middle-meatal splints are recommended.
- Agger nasi cells can lead to residual disease and must be removed with angled endoscopes and special forceps.
- Postoperative stenosis of the frontal recess is a third cause of failure. Cicatricial stenosis of the frontal recess is one of the most challenging situations. It may occur due to osteitis (osteoneogenesis) or because of excessive removal of mucoperiosteum from this area. It can be treated by introducing a "stent" and topical medications, with periodic removal of the debris until complete healing occurs. If nothing seems to work, an external revision-surgery approach should be considered.

Figure 17.34 shows pre- and postoperative CT scans of a patient who was re-operated due to lateralization of the middle turbinate and consequent stenosis of the right frontal recess.

Ethmoid Sinus

As described previously, an ethmoidectomy can be partial, total or radical. It may be accomplished in two basic ways: Messerklinger's anterior-to-posterior dissection and the Wigand posterior-to-anterior dissection.

The procedure proposed by Wigand [72, 74] is indicated in previously operated cases with loss of anatomical landmarks. The operation begins by opening the sphenoid sinus, followed by the dissection of the posterior ethmoid cells, progressing anteriorly through the ethmoid cells by following the skull base.

In contrast, the anterior-to-posterior dissection travels along the most inferior aspect of the ethmoid sinus cells until identification of the posterior ethmoid cells (including the Onodi cell, when it is present). Usually, exposing the roof of the ethmoid sinus (anterior skull base) in the posterior-to-anterior direction is safer, particularly for inexperienced surgeons.

Both techniques are based on the preservation of the mucoperiosteum of the orbital wall of the ethmoid, frontal recess and skull base. The middle turbinate is usually preserved but, if resection is needed, preservation of its insertion in the skull base is imperative as a surgical anatomical landmark.

Ethmoidectomy begins with resection of the uncinate process, as described previously (uncinectomy). It is followed by resection of the ethmoid bulla, which is an important anatomical landmark in procedures involving the maxillary and frontal sinuses.

Resection of the ethmoid bulla should begin at its inferior and medial aspects. The lumen can be entered with an atraumatic aspirator tip or a delicate forceps. Biting forceps may be used to remove the bulla and the smaller ethmoid cells as far as the orbital wall (laterally), the frontal recess (anteriorly) and the basal lamella (posteriorly; Fig. 17.35).

The most anterior ethmoid cells (agger nasi) should be removed if they are diseased. Due to their close relationship with the anterior skull base, a posterior-to-anterior dissection is recommended to avoid penetration into the orbit and cranium. The bony walls of the agger nasi cells and lacrimal fossae are usually extremely thin but occasionally are thick and sclerotic from chronic inflammatory diseases. In those cases, a micro-Kerrison punch or a drill is the best instrument for removal. Care should always be taken not to injure the frontal recess with sudden, abrupt movements. A *partial ethmoidectomy* is completed at this point (Fig. 17.36). Pre- and postoperative coronal CT scans of a patient who underwent partial ethmoidectomy are seen in Fig. 17.37.

If the posterior ethmoid cells need to be removed, it is done after inferior and medial penetration through the vertical portion of the basal lamella to preserve the horizontal portion and to avoid lateralization of the middle turbinate. During the resection of the posterior ethmoid cells, the surgeon must observe whether an Onodi cell is present; if it is, the surgeon should also know its relationship to the optic-nerve canal and the internal carotid artery. Usually, an Onodi cell can be seen on the preoperative CT scan. If there is uncertainty regarding whether an area is a large posterior ethmoid cell or the sphenoid sinus, one can try to identify the ostium of the sphenoid sinus first (via a direct transnasal view) before finishing the posterior ethmoidectomy. In some cases, use of an image-guidance system can be helpful.

At this point, complete removal of the ethmoid cells can be done in a posterior-to-anterior direction by removing the small, bony septations and remaining cells with a micro-curette or micro Kerrison punch and a biting forceps, thus completely exposing the skull base. The anterior and posterior ethmoid arteries can be identified during this stage. The posterior ethmoid artery is usually found anterior to the sphenoid sinus, while the anterior ethmoid artery crosses the roof of the ethmoid sinus posterior to the frontal recess.

If one of these arteries is injured, the bleeding can be normally be controlled with bipolar cauterization. In rare situations, the artery retracts into the orbital cavity, provoking an orbital hematoma, which requires an external access to coagulate the vessel.

Another significant ethmoidectomy complication is intracranial penetration producing a CSF fistula. The region most frequently injured is the junction of the anterior ethmoid artery with the lateral lamella of the cribriform plate, particularly when the fovea ethmoidalis is higher than the cribriform plate. In order to prevent this complication, the surgeon should always work lateral to the middle turbinate [68].

The lamina papyracea is the lateral limit of the ethmoid sinus and may occasionally be transgressed during surgery; it may also be found to be dehiscent or diseased. If this is suspected, gentle palpation of the globe will cause the orbital contents to bulge into the ethmoid sinus.

At this stage, if all the anterior and posterior cells of the ethmoid sinus have been removed and the mucoperiosteum of the orbit and skull base have been preserved together with the middle turbinate, a *total ethmoidectomy* has been performed (Fig. 17.38).

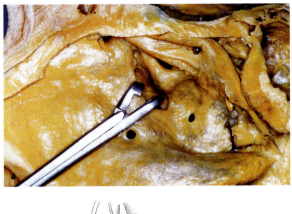
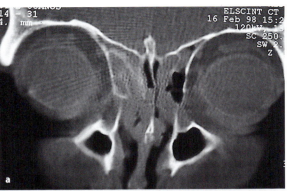
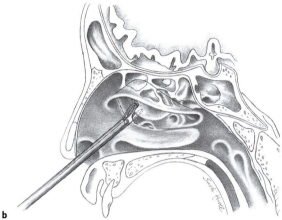
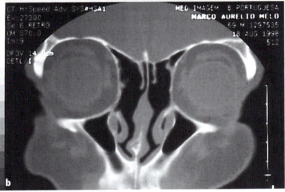

Fig. 17.35 a,b. Resection of the ethmoid bulla with biting forceps

Fig. 17.37 a,b. Preoperative (a) and postoperative (b) coronal computed tomography of a patient with stage-II nasal polyposis who underwent partial ethmoidectomy

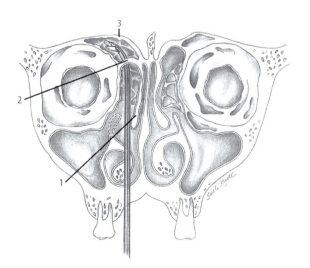
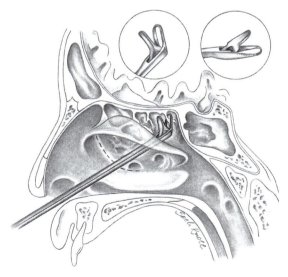

Fig. 17.36. A partial ethmoidectomy is completed after removal of the anterior ethmoidal cells. *1*, Middle turbinate; *2*, anterior "arcus"; *3*, ethmoid roof

Fig. 17.38. Total ethmoidectomy consists of complete removal of the anterior and posterior ethmoid cells, preserving the mucoperiosteum of the medial orbital wall and anterior skull base by using a "through-cutting" forceps. The middle turbinate is also preserved

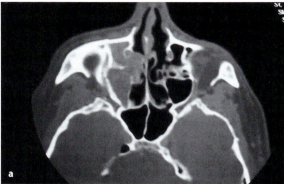
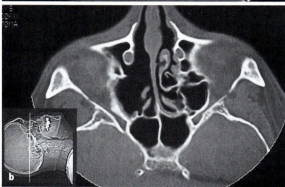

Fig. 17.39 a,b. Preoperative (a) and postoperative (b) computed tomography of a patient who underwent total ethmoidectomy for nasal polyposis

An example of this procedure can be observed on pre- and postoperative CT scans in Fig. 17.39.

Radical ethmoidectomy includes removal of the mucoperiosteum of the orbit and skull base (except in the frontal-recess region) and resection of the middle turbinate (preserving its superior insertion). It is only indicated in special revision cases (for example, after multiple previous operations when there is massive recurrence of polyps) and when anatomic landmarks are absent (Fig. 17.40) [23, 33].

Causes of Failure of Ethmoid-Sinus Surgery

- Lateralization of the middle turbinate and synechiae at the insertion of the middle turbinate in the lateral nasal-wall region ("anterior arch")
- Incomplete removal of the agger nasi cells
- Incomplete removal of polyps or fungal allergic disease, with blockage of the drainage (leading to a mucocele) [67]

When a radical ethmoidectomy is performed, the time necessary for healing is longer, and more intense postoperative care is required because of the removal of the mucoperiosteum.

Sphenoid Sinus

There are three principal approaches to the sphenoid sinus: the transethmoid approach, the trans-septal approach and directly through the nasal cavity (with or without partial resection of the superior turbinate). The choice of surgical access will depend on the extent of the disease, the degree of pneumatization of the sphenoid sinus, the relationship between the sinus, optic nerves and internal carotid artery and the experience of the surgeon.

The most important anatomical landmarks to help identify the anterior wall of the sphenoid sinus and the ostium are: the superior border of the choana, the superior and middle turbinates and the nasal septum. The ostium is usually circular or elliptical and is located in the superior portion of the anterior wall of the sphenoid sinus, approximately 8 mm above the superior choanal border.

The surgical opening of the anterior wall of the sphenoid sinus is usually made through a transnasal access, reserving the transethmoidal approach for special situations, usually when there is involvement of the ethmoid cells in the disease. If transethmoid access is performed, the surgeon should be wary of any protrusion of the optic nerve into the posterior ethmoid cells. If an Onodi cell overlaps the sphenoid sinus, one can easily penetrate the skull base (Fig. 17.41).

Transnasal Direct Approach

After identification of the middle and superior turbinates, the posterior region of the nasal septum and the choanal arch, the ostium of the sphenoid sinus can be probed with a seeker/palpator (Fig. 17.42). If diseased tissue, such as polyps, obstructs the access to the sphenoethmoid recess, a microdebrider can be helpful. After identifying the superior turbinate, it is removed to facilitate access (Fig. 17.43). When the surgical access is very narrow, the posterior portion of the middle turbinate can also be removed.

The initial opening of the sphenoid sinus is performed with a micro-Kerrison punch, beginning at the ostium, which is always incorporated into the sphenoidotomy opening (Fig. 17.44). Enlargement of the sphenoidotomy is initially directed inferiorly while watching for the septal artery (the terminal

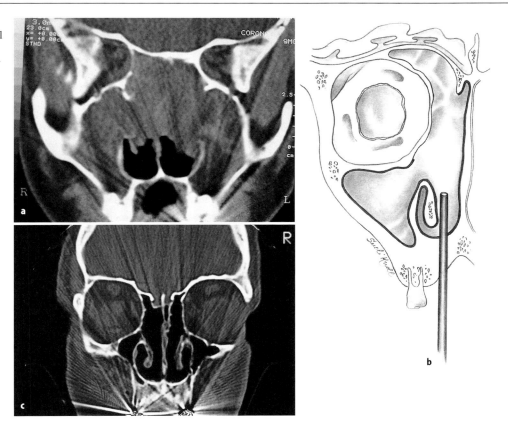

Fig. 17.40. a Massive recurrent nasosinusal polyposis, grade V. b Radical ethmoidectomy, which includes removal of the mucoperiosteum of the medial orbital wall of the anterior skull base and removal of the middle turbinate. c Postoperative computed tomography of a patient after radical ethmoidectomy

branch of the maxillary artery), which crosses the anterior wall of sphenoid sinus in that region. Any bleeding from this artery is controlled with eletrocoagulation. This step of the surgery can be done with an operating microscope or a 4-mm, 0° endoscope. Observing laterally, the bulge of the optic-nerve canal running along the posterolateral wall of the sphenoid sinus must be identified on each side. The nerve is dehiscent in 6% of cases [34]. The internal carotid artery is located in the inferolateral wall of the sphenoid sinus and may be dehiscent in approximately 25% of patients [65]. If the anterior wall of the sinus is quite thick, it can be opened with a diamond drill. The mucosa of the sinus should be preserved and all the bony fragments removed in order to avoid re-infection. The lateral extent of the sphenoid sinus is best visualized with 30° and/or 70° angled endoscopes. Figure 17.45 shows examples of patients operated on with this surgical approach.

Transethmoid Approach

First, an ethmoidectomy is performed. Next, the posterior ethmoid artery is identified in the anterior skull base, usually located anterior to the anterior wall of the sphenoid sinus, which makes an angle of approximately 90° with the roof of the ethmoid sinus.

The sphenoid sinus is first opened medially and inferiorly with a delicate curette or with an atraumatic aspirator. The initial opening is then enlarged with a micro-Kerrison punch, always incorporating the natural ostium (Fig. 17.46).

Another method of opening the sphenoid sinus during a transethmoid approach is to first identify and remove the superior turbinate and then identify, open and enlarge the natural ostium, proceeding as described previously.

At the termination of a direct transnasal access, an expander (such as a Merocel pack) is placed in the sphenoethmoid recess; if the surgical approach is transethmoidal, the expander is placed in the ethmoid sinus and middle meatus. Other common packing materials include rayon gauze or the finger of a glove filled with Merocel. Before insertion, the packs are usually impregnated with an ointment containing antibiotics and steroids.

Fig. 17.41. Exposure of the optic nerve at the posterior ethmoid cell/Onodi cell (*arrow*)

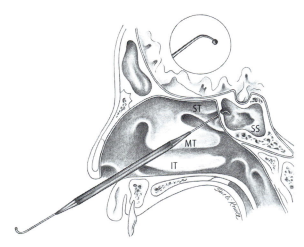

Fig. 17.42. Use of a seeker probe to identify and palpate the sphenoid-sinus ostium. *IT*, inferior turbinate; *MT*, middle turbinate; *SS*, sphenoid sinus; *ST*, superior turbinate

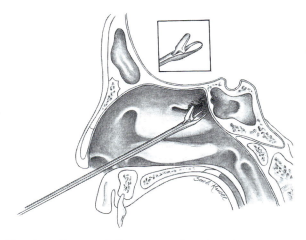

Fig. 17.43. Transnasal direct approach to the sphenoid sinus after removal of the superior turbinate

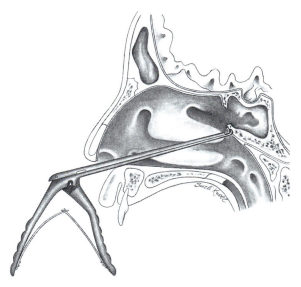

Fig. 17.44. Opening of the sphenoid sinus using a micro-Kerrison punch (beginning at the ostium) in a transnasal, direct approach

Causes of Failure of Sphenoid-Sinus Surgery

- Interpreting an Onodi cell to be the sphenoid sinus and never opening the sphenoid sinus itself
- Not incorporating the natural sphenoid ostium into the sphenoidotomy, thus creating a recirculation phenomenon analogous to that which occurs in the maxillary sinus
- Leaving residual disease in the cavity (such as a fungal ball)
- Postoperative stenosis of the sphenoidotomy

Usually, re-operation successfully resolves these failures.

Postoperative Care

A satisfactory postoperative result depends on both appropriate operative technique and meticulous postoperative care. The goal of the surgical treatment is to assure the best functional result and to avoid synechiae in the middle meatus, adhesions in the middle antrostomy and sphenoidotomy regions, and stenosis of the frontal recess.

Antibiotics and systemic steroids are usually given in the immediate postoperative period. Their duration is controversial and basically depends on the type of surgery and disease (sinusitis, tumor, polyps). Some factors related to the operative proce-

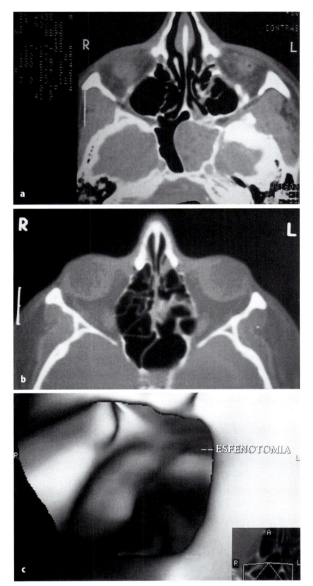

Fig. 17.45 a,b. Preoperative (a) and postoperative (b) computed tomography of a patient who underwent direct transnasal sphenoidotomy for sphenoid sinusitis. c Postoperative virtual endoscopy

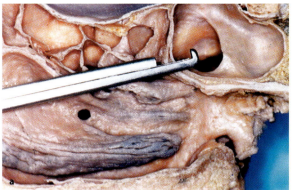

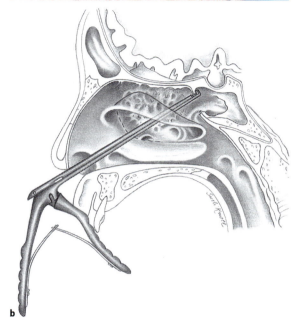

Fig. 17.46 a,b. Enlargement of the initial opening of the sphenoid sinus through a transethmoidal approach, using a micro-Kerrison punch. a Lateral anatomical view. b Schematic drawing

dure may interfere with postoperative care, such as the positioning and stability of the middle turbinate, denuded areas of bone and the presence of splints and packing.

Adequate postoperative care requires appropriate instrumentation, including 4-mm 0°, 30° and 70° endoscopes and 3.2- or 3.5-mm flexible endoscopes, straight and curved atraumatic aspirators, and straight and curved micro-forceps. Another important decision is the frequency of postoperative examinations.

Packing should be removed at the first postoperative visit, 2–4 days after surgery (earlier for uncomplicated cases and longer for more extensive procedures and with more aggressive diseases). The operated cavity is carefully suctioned, and any residual bony fragments are removed. The patient is asked to irrigate the nasal cavity with saline solution or sterile hypertonic seawater three or four times per day. This facilitates the expulsion of crusts, clots and secretions from the surgical cavity. Patients are also strongly advised to not to blow the nose vigorously and to avoid any drugs that may interfere with healing, such as aspirin.

A second visit is scheduled between the tenth and 12th day after surgery, when the operated cavity is again cleaned of crusts, granulation tissue, clots and secretions. At this time, the material inside the cavity is usually thick, gelatinous and quite dark. Any small, newly formed polyps or cysts that are detected are removed.

A third visit occurs between the third and fourth postoperative week. By then, the surgical cavity usually appears to be healing well. If purulent secretions are present, they are cultured. In patients operated for polyposis, topical steroids are prescribed for 4–6 months, and further periodic visits to the office are scheduled. Postoperative CT or MR is only indicated for documentation, if healing does not appear to progressing satisfactorily or if complications are suspected.

Final Considerations

Transnasal micro-endoscopic surgery is not the final and complete solution for all inflammatory/infectious diseases of the paranasal sinuses, although a recent review of the literature has shown that approximately 76–98% of the operated patients have improvement of the symptoms after surgery on follow-up after two or more years. However, re-operation is necessary in 18–25% of cases [48].

Despite advances in the medical field, including sophisticated equipment (such as image-guided technology), the rate of sinus-surgery complication remains between 0.36% and 1.1% [48]. Based on an experience of 20 years in micro-endoscopic surgery and taking in account all the pre-, trans- and postoperative surgical prognostic factors, a sinus-disease total cure is still a challenge. However, I strongly believe that it is possible to improve the quality of life of most patients.

References

1. Amedee RG, Mann WJ, Gilbach JM (1989) Microscopic endonasal surgery of the paranasal sinuses and parasellar region. Arch Otolaryngol Head Neck Surg 115:1103–1106
2. Bagatella F, Ferrrara AM (1980) Transansal microsurgical ethmoidectomy in nasal polyposis. Rhinology 18:25–29
3. Belal A Jr (1978) Surgical microcopy of the nose: new frontiers in nasal diagnosis and treatment. J Laryngol Otol 92:197–207
4. Bent JP III, Spears RA, Kuhn FA (1997) Combined endoscopic intranasal and external frontal sinusotomy. Am J Rhinol 11:349–354
5. Caldwell GW (1893) Diseases of the accessory sinuses of the nose and an improved method of treatment for suppuration of the maxillary antrum. N Y Med J 58:526
6. Dixon HS (1976) Microscopic antrostomies in children: a review of the literature in chronic sinusitis and a plan of medical and surgical treatment. Laryngoscope 86:1796–1814
7. Dixon HS (1983) Microscopic sinus surgery, transansal ethmoidectomy and sphenoidectomy. Laryngoscope 93:440–444
8. Dixon HS (1985) The use of the operating microscope in ethmoid surgery. Otolaryngol Clin North Am 18:75–86
9. Draf W (1978) Endoscokopie der Ansennebenhohlen – Technik Typische Befunde – Therapeutische Moghichkeiten. Springer, Berlin Heidelberg New York
10. Draf W (1982) Die chirurgische behandlung entzündlincher erkrankunger der ansennebenhohlen. Arch Otolaryngol 135:133–305
11. Draf W (1983) Endoscopy of the paranasal sinuses. Springer, Berlin Heidelberg New York
12. Draf W (1991) Endonasal microendoscopic frontal sinus surgery: the Fulda concept. Operative Tech Otolaryngol Head Neck Surg 2:234–240
13. Friedman WH, Katsantonis GP, Sivore M, et al. (1990) Computed tomography staging of the paranasal sinuses in chronic hyperplastic rhinosinusitis. Laryngoscope 100:1161–1165
14. Gaskins RE (1992) A surgical staging system for chronic sinusitis. Am J Rhinol 6:5–12
15. Gittelman PD, Jacobs JB, Skorina J (1993) Comparison of functional endoscopic sinus surgery under local and general anesthesia. Ann Otol Rhinol Laryngol 102:289
16. Gliklich R, Metson R (1994) A comparison of sinus computed tomography (CT) staging systems for outcome research. Am J Rhinol 8:291–297
17. Golding-Wood PH (1961) Observation on petrosal and vidian neurectomy in chronic vasomotor rhinitis. J Laryng Otol 75:232–247
18. Gross CN, Becker DG (1996) Power instrumentation in endoscopic sinus surgery. Operative Tech Otolaryngol Head Neck Surg 7:236–241
19. Gustafson RO, Bansberg SF (1993) Sinus surgery. In: Bailey BJ (ed) Head and neck surgery, otolaryngology. Lippincott, Philadelphia, pp 377–388
20. Heermann H (1958) Endonasal surgery with the use of the binocular Zeiss operating microscope. Arch Klin Exp Ohren Nasen Kehlkopfheilkd 171:295–297
21. Heermann J (1982) Endonasal microsurgery of the ethmoid sinus on a patient in a semi-sitting (Fowler position) with hypotensive anesthesia. HNO 30:180–185
22. Heermann J, Neues D (1986) Intranasal microsurgery of all paranasal sinuses, the septum and the lachrymal sac with hypotensive anesthesia. Ann Otol Rhinol Laryngol 95:631–638
23. Jankowski R, Pigret D, Decroocq F (1997) Comparison of functional results after ethmoidectomy and nasalization for diffuse and severe nasal polyposis. Acta Otolaryngol (Stockh) 117:355–361
24. Jourdain A (1778) Traite des maladies et des operations relellement chirurgicales de la bouche et des parties qui y correspondent, suivi des notes d'observations et de consultations interessantes tantes auciennes que modernes, vol 2. Valleyne l'aine, Paris
25. Kennedy DW (1985) Functional endoscopic sinus surgery. Technique. Arch Otolaryngol 111:643–649
26. Kennedy DW (1992) Prognostic factors, outcomes and staging in ethmoid sinus surgery. Laryngoscope 102:1–18

27. Kennedy DW, Senior BA (1997) Endoscopic sinus surgery. Otolaryngol Clin North Am 30:313–330
28. Kennedy DW, Shaalan H (1989) Reevaluation of maxillary sinus surgery: experimental study in rabbits. Ann Otol Rhinol Laryngol 98:901–906
29. Kennedy DW, Zinreich SJ (1989) Functional endoscopic surgery. Adv Otolaryngol Head Neck Surg 3:1–26
30. Kennedy DW, Zinreich SJ, Rosenbaum A (1985) Functional endoscopic sinus surgery: theory and diagnosis. Arch Otolaryngol 111:576–582
31. Keros P (1962) Uber die praktische bedeutung der niveauunter-schiede der lamina cribrosa des ethmoids. Laryngorhinootol 41:808–813
32. Khun FA (1996) Chronic frontal sinusitis: the endoscopic frontal recess approach. Operative Tech Otolaryngol Head Neck Surg 7:222–229
33. Klossek JM, Peloquin L, Friedman WH, et al. (1997) Diffuse nasal polyposis: postoperative long term results after endoscopic sinus surgery. Otolaryngol Head Neck Surg 117:355–361
34. Lang J (1989) Clinical anatomy of the nose, nasal cavity and paranasal sinuses. Thieme, New York
35. Lawson N (1991) The intranasal ethmoidectomy: an experience with 1077 procedures. Laryngoscope 101:367–371
36. Lee D, Broady R, Har-El G (1997) Frontal sinus out flow anatomy. Am J Rhinol 11:283–285
37. Lowry MC (1993) Endoscopic frontal recess and frontal sinus ostium dissection. Laryngoscope 103:455–458
38. Luc H (1897) Une nouvelle methode operatoire pour la cure radicale et rapide de lémpheme chronique du sinus maxillaire. Arch Laryngol 6:275
39. Lund VJ, Mackay IS (1993) Staging in rhinosinusitis. Rhinology 107:183–184
40. Masing H (1976) Surgery on the lateral nasal wall with the operation microscope. Rhinology 14:73–77
41. May M (1991) Frontal sinus surgery: endonasal endoscopic ostioplasty rather than external ostioplasty. Operative Tech Otolaryngol Head Neck Surg 2:247–256
42. May M, Levine HL, Schaitkin B, et al. (1993) Results of surgery. In: Levine HL, May M (eds) Rhinology and sinusology. Thieme, New York, pp 176–192
43. Messerklinger W (1970) Endoscopy of the nose. Moantsschr Ohrenheilkd Laryngorhinol 104:451–456
44. Messerklinger W (1972) Techniques and possibilities of nasal endoscopy. HNO 20:133–135
45. Messerklinger W (1978) Endoscope of the nose. Urban and Schwartzenberg, Baltimore
46. Mikulicz J (1887) Zur operativen Bechandlung des Empyems der Highmorehöhle. Arch Klin Chir 34:626
47. Muntz HR, Lusk RP (1990) Nasal antral windows in children: a retrospective study. Laryngoscope 100:643–646
48. Osguthorpe JD (1999) Surgical outcomes in rhinosinusitis: what we know. Otolaryngol Head Neck Surg 120:451–453
49. Parsons DS, Van Leeunen N (1996) Management of sinusitis in children. Adv Otolaryngol Head Neck Surg 11:23–36
50. Parsons DS, Setliff RC, Chambers D (1994) Special considerations in pediatric functional endoscopic sinus surgery: operative techniques. Otolaryngol Head Neck Surg 5:40–42
51. Prades J, Bosch J, Tolosa A (1977) Microcirurgia endonasal. Garsi, Madrid, p 7041
52. Rudert H (1988) Mikroskop und enedoskopgestützte chirurgie der entzündlichen ansennebenhohlenerkrankungen. HNO 36:475–482
53. Schaffer SD (1990) Endoscopic sinus surgery: posterior approach. Operative Tech Otolaryngol Head Neck Surg 1:104–107
54. Setliff RC (1996) Minimally invasive sinus surgery: the rationale and technique. Otolaryngol Clin North Am 29:1
55. Setliff RC, Parsons DS (1994) The "hummer": new instrumentation for functional endoscopic sinus surgery. Am J Rhinol 8:275–278
56. Siebert DR (1992) Estudo anatômico de seios esfenoidais em brasileiros adultos. Universidade de São Paulo, São Paulo
57. Simonetti G, Meloni F, Teatini G (1987) Computed tomography of the ethmoid labyrinth and adjacent structures. Ann Otol Rhinol Laryngol 96:239–2507
58. Stamm A (1992) A surgical staging system for sinonasal polyposis. 23rd Pan-American Congress of ENT, Head and Neck Surgery. Grune and Stratton, Orlando, p 115
59. Stamm A (1995) Microcirugia Nasossinusal. Revinter, Rio de Janeiro
60. Stamm A (1999) Círugia microendoscópica. Conceptos básicos. An ORL (Peru) 6:27–36
61. Stamm A, Burnier M Jr (1989) Tumores benignos nasosinusais. In: Brandão LG, Ferraz AR (eds) Cirurgia de cabeça e pescoço, vol 1. Roca, São Paulo, pp 465–491
62. Stamm A, Pignatari SS (1998) Transnasal micro-endoscopic surgery for CSF rhinorrhea. In: Stammberger H, Wolf G (eds) Congress of the European Rhinologic Society (Vienna, 1998). Monduzzi, Bologna, pp 329–335
63. Stammberger H (1986) Endoscopic endonasal surgery – concepts in treatment of recurring rhinosinusitis. Part I: anatomic and pathophysiologic considerations. Otolaryngol Head Neck Surg 94:143–147
64. Stammberger H (1986) Endoscopic endonasal surgery – concepts in treatment of recurring rhinosinusitis. Part II: surgical technique. Otolaryngol Head Neck Surg 94:147–156
65. Stammberger H (1991) Functional endoscopic sinus surgery. Mosby, St. Louis
66. Stammberger H, Kennedy DW, et al. (1995) Paranasal sinuses: anatomical terminology and nomenclature. Ann Otol Rhinol Laryngol 105:7–15
67. Stankiewicz JA, Donzelli JJ, Chow JM (1996) Failures of functional endoscopic sinus surgery and their surgical correction. Operative Tech Otolaryngol Head Neck Surg 7:297–304
68. Terrel JE (1998) Primary sinus surgery. In: Cummings CW, Fredricson J, Harker LA, et al. (eds) Otolaryngology, Head and Neck Surgery. Mosby, St. Louis, pp 1145–1172
69. Terrier G (1991) Rhinosinusal endoscopy. Diagnosis and surgery. Zambon, Osango
70. Van Alyea OE (1946) Frontal sinus drainage. Ann Otol Rhinol Laryngol 55:267–278
71. Wigand ME (1981) Transnasal ethmoidectomy under endoscopical control. Rhinology 19:7–15
72. Wigand ME (1990) Endoscopic surgery of the paranasal sinuses and anterior skull base. Thieme, New York
73. Wigand ME, Hosemann W (1991) Endoscopic surgery for frontal sinusitis and its complications. Am J Rhinol 5:85–89
74. Wigand ME, Steiner W, Jaumann MP (1978) Endonasal sinus surgery with endoscopic control: from radical operation to rehabilitation of mucosa. Endoscopy 10:255–260
75. Yanagisawa E (1995) Endoscopic excision of an antral lesion via inferior meatal antrostomy. Ear Nose Throat J 74:321–322

76. Yanagisawa E (1997) Inferior meatal antrostomy: is it still indicated? Ear Nose Throat J 76:368–370
77. Ziv M, Johnson JC, Prince L (1991) Combining the use of the microscope, fiberoptic, telescopes, and headlight in sinus surgery. Otolaryngol Head Neck Surg 105:269
78. Zuckerkandl E (1882) Normale und pathologiesche Anatomie der Nasennebenhohle und ihrer pneumatischen Anhange. Holder, Wien

Combined Microscopic and Endoscopic Surgery of the Ethmoid Sinuses

Heinrich H. Rudert and Christoph G. Mahnke

Introduction

The theoretical basis of modern paranasal sinus surgery was established by Messerklinger [20–24]. He systematically studied the clinical pathophysiology of paranasal sinus drainage using the Hopkins telescope (a new invention at that time) and then developed therapeutic principles for chronic sinusitis that are still currently valid. For his investigations, he used earlier anatomical and embryological studies of Zuckerkandl [39, 40] and Killian [17] who, at the turn of the century, had already inaugurated rather modern endonasal surgical techniques. Further proponents of endonasal surgery were Hajek [7, 8] and Halle [9–13] in Europe. It is not clear why this surgical approach was abandoned in the 1930s. One can only speculate that a high rate of severe endocranial complications was the reason. A quotation of Mosher from 1929 is well known: "Theoretically, the operation is easy. In practice, however, it has proved to be one of the easiest operation with which to kill a patient" [25].

In Germany, H. Heermann, who was educated by Halle, continued to follow the endonasal approach and improved its safety by introducing the operating microscope [14]. The systematic development of endonasal surgery, however, was started by the so-called endoscopists. These were, among others, Messerklinger [22, 24], Wigand [34–38], Stammberger [31–33] and Kennedy [15]. The most extensive use of microscopic endonasal surgery was made by Draf [2–6]. We apply both a microscope and an endoscope and also apply a self-holding nasal speculum. This allows us to operate with both hands when using the microscope, as is commonly done in microsurgery of the ear and the larynx [26–29].

Instruments

The self-holding nasal speculum (Fig. 18.1) has a short branch (with teeth attached), which is inserted

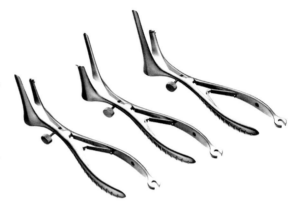

Fig. 18.1. Self-retaining nasal speculum. The short branch, with teeth attached, is inserted into the lateral wall of the pyriform aperture. Its long branch (which has three varying lengths) lies along the septum. The speculum is attached to the operating table by a flexible arm

into the pyriform aperture. Its long branch (which has three varying lengths) lies along the septum. The speculum is fixed to the operating table by a flexible arm.

In addition to the regular ethmoid-sinus forceps described by Blakesley, sharp cutting forceps described by Grünwald and Watson-Williams are used (Fig. 18.2). The latter cut the diseased mucosa at the desired location whereas, under some circumstances, the ethmoid forceps of Blakesley remove the complete mucosal covering of an ethmoid cell. Osseous cell septae can best be removed by a Kofler punch (Fig. 18.3). These minimize the risk of perforation at the skull base and the orbit, because they only cut backward.

For endonasal dacryocystorhinostomy and decompression of the optic nerve or the orbit, it is helpful to use modified tympanoplasty instruments (elevators, rasps, microforceps and microscissors) elongated by 2 cm. Drills are used to work on the osseous structures of the facial skeleton (for example, to reduce the os lacrimale in endonasal dacryocystorhinostomy or to remove the frontal process

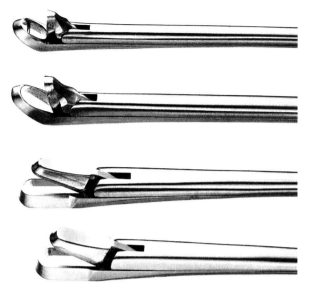

Fig. 18.2. Sharp cutting forceps, as described by Grünwald and Watson-Williams. These forceps cut the mucosa at the desired location, whereas the ethmoid forceps of Blakesley remove the complete mucosal covering of a cell

Fig. 18.3. Back-cutting punches, as described by Hajek and Kofler. Their use minimizes the risk of perforation of the skull base and the orbit

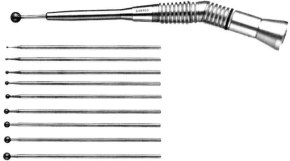

Fig. 18.4. Angulated burr hand piece with extra-long diamond burrs (12.5 cm) suitable for procedures using the microscope, allowing the left hand to be used for suction. An irrigation tube can be attached to the hand piece (not depicted)

and the spina nasalis frontalis in frontal-sinus surgery). The instrument best suited for this purpose is a drill with an angulated hand piece and extra-long diamond burrs, as used in transnasal surgery of the hypophysis (Fig. 18.4). Irrigation can be applied through a small tube attached to the hand piece of the drill. Since the application of the drill is controlled with the operating microscope, the second hand can hold the suction tube.

The use of so-called shavers should be left to experienced surgeons, because an unnoticed perforation of the skull base can lead to severe brain injuries, as we observed in one case at another institution. We have applied the new instrument beneficially in maxillary-sinus surgery. When using these shavers, a curved working unit is a requirement. It is introduced into the maxillary sinus through the enlarged maxillary ostium under the control of a 30° endoscope. The use of the shaver only helps the surgeon working purely endoscopically, because this instrument removes tissue and blood at the same time.

The operative procedure is aided by optical instruments, such as an operating microscope with a 300-mm lens and straight-forward and forward–oblique telescopes (0°, 30°, 45°, 70° and 120°). Endonasal-sinus surgery without optical instruments is not appropriate anymore, because the anatomical landmarks of the head of the middle turbinate, the uncinate process and the ethmoid bulla have to be located at the beginning of the procedure.

Anesthesia

Endonasal ethmoid-sinus surgery is almost always performed under general anesthesia and, if possible, with controlled hypotension. A neurosurgical gauze with a sympathomimetic (such as naphazolin or xylometazolin) is introduced into the nose for vasoconstriction and decongestion of the mucosa. It has a thread attached (as all things introduced into the nose should) to avoid accidental retention.

Positioning of the Patient

The upper body and the head of the supine patient should be slightly elevated. The skull base is thereby rotated away from the surgeon and is at a tangent rather than a perpendicular. Another advantage is a reduction of intraoperative hemorrhage. The position of the skull base can be envisioned by an imaginary plane going through the medial angle of the eye and the external auditory canal (Fig. 18.5a). Of further help is a second plane, parallel to the first, going through the floor of the nasal cavity (Fig. 18.5b). In

The Basics of Endonasal Ethmoid-Sinus Surgery

The anterior ethmoid sinuses are the key regions of the endonasal-sinus system, because blockage in this area causes a reduction in the ventilation and drainage of the maxillary sinus, the frontal sinus and the anterior ethmoid sinuses. It consists of the cells of the agger nasi, the uncinate process and the bulla ethmoidalis, which drain into the infundibulum and into the supra- and retrobullar recess. The border of the dorsal ethmoid sinus cells is marked by the basal lamella of the middle turbinate. Every rhinogenic sinusitis starts at the anterior ethmoid sinus and, from there, extends to other paranasal sinuses. Thus, the enlargement of the infundibulum and the opening of the anterior ethmoid sinuses is the minimal operative treatment in sinusitis and its complications.

Preoperative Imaging Studies

Computed tomography (CT) is essential in all endonasal procedures except in cases of emergency. The standard projection is a coronal view. When operating in the dorsal ethmoid sinus, the frontal sinus or the sphenoid sinus, additional axial planes are important. Additional images of these sections reformatted into sagittal planes are useful for operations of the frontal sinuses and should be obtained if modern CT technology with adequate software is available. Sections are obtained in 3-mm steps at the level of the frontal and sphenoid sinuses and in 2-mm steps at the ethmoid sinuses. The images are documented on X-ray films in the high-resolution bone window and in the soft-tissue window; the bone window is the more important one and must always be available [30]. The radiologist should comment on the following structures: connection of the uncinate process, high or low position of the ethmoid artery and the lamina cribrosa, bony dehiscences of the roof of the ethmoid sinuses or the orbital walls, and the presence of an Onodi cell or of other anomalies of the optic nerve or the internal carotid artery.

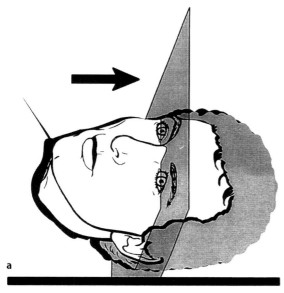

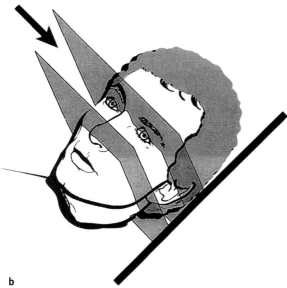

Fig. 18.5. a In the supine patient, the surgeon proceeds orthogonally toward the skull base. The location of the skull base can be visualized by an imaginary plane through the medial angle of the eye and the auditory canal. **b** An elevated upper body places the skull base in a direction more tangential to the surgeon. It also reduces intraoperative bleeding. Together with the first plane, a second plane at the floor of the nose indicates the space where one can operate without danger to the dura

the space between these two planes, one can operate without endangering the dura. Any necessary movement of the patient's head is managed by moving the whole operating table electrically, because the head is fixed by the self-retaining speculum.

Operative Technique

In endonasal surgery, we recommend a systematic progression from anterior to posterior. As mentioned above, the basic procedure is the enlargement of the infundibulum. This also often suffices to drain and ventilate the anterior paranasal sinuses. Only in isolated procedures of the sphenoid sinuses can one forego infundibulotomy.

Enlargement of the Infundibulum by Removal of the Uncinate Process

The ethmoid infundibulum is a bow-shaped sulcus that opens through the slit-shaped hiatus semilunaris toward the medial nasal duct. The infundibulum laterally borders the orbita, with its ascending anterior branch and with its dorsal horizontal branch the maxillary sinus (Fig. 18.6). Anteriorly, the infundibulum meets the agger nasi cells or, if these are missing, the saccus lacrimalis. The dorsal and cranial borders of the infundibulum are formed by the ethmoid bulla.

The opening or dilation of the infundibulum starts with the removal of the uncinate process, which forms the lower lip of the hiatus semilunaris (Fig. 18.6). The uncinate process is the descending crus of the first ethmoturbinal, which is only rudimentarily expressed in man. The agger nasi is the covered, ascending crus of the first ethmoturbinal, which is merged with the second ethmoturbinal (later the middle turbinate). It is very important to preserve this bow-shaped connection of the middle turbinate to the agger nasi because otherwise there is a great risk of a postoperative lateralization of the middle turbinate, with adhesions at the lateral nasal wall.

The removal of the uncinate process has to be applied with great caution, because the distance between the lacrimal sac and the infundibulum only measures 3 mm (range: 1–8 mm), according to Calhoun [1]. We perform this step with the operating microscope. First, an angulated probe is introduced into the infundibulum through the gap of the hiatus semilunaris to assess the anterior extension of the infundibulum and the connection between the uncinate process and the lateral nasal wall. Only then is the uncinate process removed with a sickle knife or a dorsally cutting Kofler punch. When removing a very slim uncinate process with a sickle knife, there is a danger of opening the lacrimal sac or the orbit. Possible sequels of this are an ecchymosis or, in some circumstances, a stenosis of the lacrimal duct.

With the removal of the uncinate process, the lower lip of the hiatus semilunaris is removed, and the sulcus of the infundibulum is exposed. The original position of the infundibulum is then only marked by the ethmoid bulla, which forms the upper margin of the infundibulum (Fig. 18.7). The ethmoid bulla should, therefore, be preserved as long as possible as an important landmark. In addition, its anterior wall

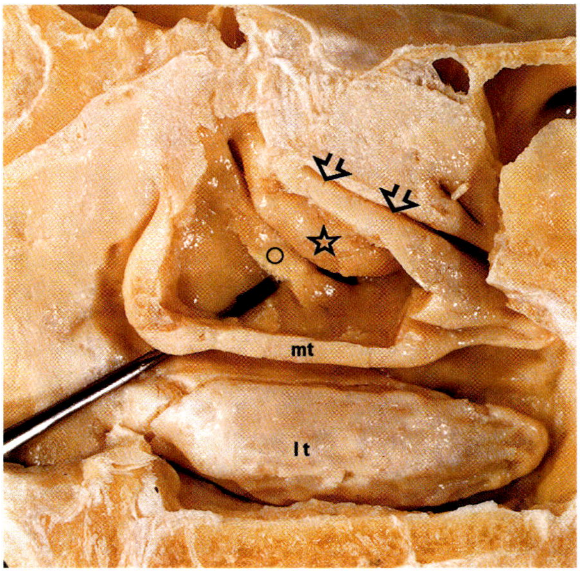

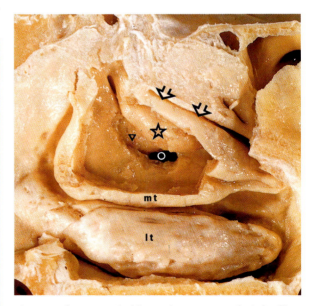

Fig. 18.6. The removal of the uncinate process (*open circle*) is the first step in infundibulotomy. *Open star*, ethmoid bulla; *open arrows*, basal lamella of the middle turbinate. *lt* Lower turbinate, *mt* middle turbinate. The middle turbinate is fenestrated to enable visualization of the middle nasal duct [26]

Fig. 18.7. After removal of the uncinate process, the infundibulum is opened wide. The ethmoid bulla (*open star*) should remain as long as possible, because it serves as an important landmark. *Open white circle*, ostium maxillare; *open triangle*, medial orbital wall (lamina papyracea); *open arrows*, basal lamella of the middle turbinate. *lt* Lower turbinate, *mt* middle turbinate [26]

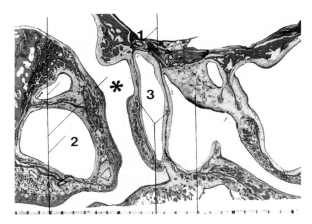

Fig. 18.8. Sagittal section through the ethmoid. *Star*, frontal recess; *1*, ethmoid artery and roof of the ethmoid; *2*, agger nasi cell; *3*, ethmoid bulla [18]

forms the dorsal wall of the frontal recess (Fig. 18.8). The entrance to the frontal recess can vary greatly, depending on the extent of the bulla ethmoidalis and the degree of pneumatization of the agger nasi.

Treatment of the Middle Turbinate

Next to the bulla ethmoidalis, the middle turbinate is the most important landmark in endonasal surgery. It should, therefore, be preserved; it should also be preserved because its large mucosal surface is important for the respiratory function of the nose. This is especially so for its bow-shaped origin at the agger nasi. If it is necessary to shift the middle turbinate medially for better exposure of the medial nasal duct intraoperatively, this should be done very carefully. The medial turbinate should never be fractured, as is sometimes recommended, as there is a risk of breaking its fixation at the lamina cribrosa, thus causing liquorrhea.

A prominent concha bullosa media is split in the longitudinal direction, and the lateral portion is resected. Again, one should carefully avoid injury of the lamina cribrosa. This can best be done by using sharp instruments and scissors. Only the lateral (never the medial) part may be removed. Particular attention should be paid to the origin of the middle turbinate at the base of the skull. According to Keros [16], the lamina cribrosa is in a low position in up to 70% of the patients. In these cases, the medial turbinate leads directly into the extremely thin lateral lamella of the lamina cribrosa at the base of the skull. One has to realize that the roof of the ethmoid in these cases is positioned higher than the lamina cribrosa and that the medial border of the ethmoid is

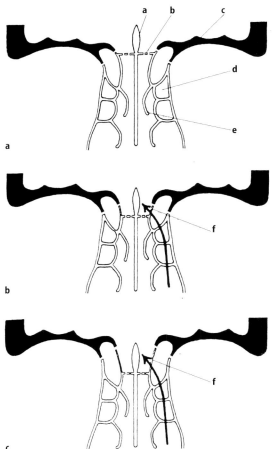

Fig. 18.9 a–c. Various types of the olfactory fossa (*a–c*). Note the thin lateral lamella of the cribriform plate in types 2 and 3 (f)[16]

formed only by the extremely thin lateral lamella of the lamina cribrosa (Fig. 18.9). Of course, the middle turbinate has to be resected partially (or sometimes even completely) if it shows polypoid changes as part of the ethmoid sinus' involvement with disease. In most cases, it is possible to preserve its attachment at the lamina cribrosa.

Sometimes, the anterior margin of the middle turbinate has to be removed if it is curved paradoxically. We have never seen a double middle turbinate. Any "doubling" was always imitated by a margin of the uncinate process protruding anteriorly.

Identification of the Maxillary Ostium

Following the removal of the uncinate process, the maxillary-sinus ostium is identified using a 30° telescope. Here again, the preserved bulla ethmoidalis offers an important landmark, because it indicates

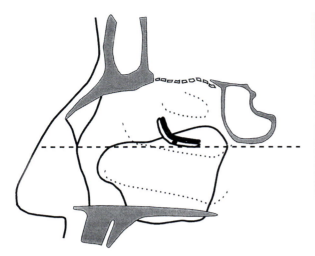

Fig. 18.10. The maxillary ostium is located in the upper region of the maxillary sinus, in close proximity to the orbit. It is found on an imaginary line that goes through the lower margin of the ethmoid bulla parallel to the floor of the nasal cavity. It is important to search for the maxillary ostium in the area of the dorsal horizontal branch of the infundibulum [29]

Fig. 18.11. The bulla (*open star*) is opened at its lowest point. *Open arrows*, basal lamella of the middle turbinate. *lt*, Lower turbinate; *mt*, middle turbinate [26]

the original position of the infundibulum (Fig. 18.10). In more than 90% of the cases [19], the mouth of the ostium maxillare is located at the dorsal two thirds of the infundibulum, i.e., at its horizontal branch. The safest way to search for the ostium is along an imaginary horizontal line (Fig. 18.10) parallel to the base of the nasal cavity and, therefore, to the skull base at the lowest point of the bulla. It is sometimes necessary to create a new foramen along this line by perforating the dorsal fontanelle. As we look for the ostium in a more anterior direction or as we leave this line cranially, the risk of damaging the orbit increases, because the distance between the ostium maxillare and the floor of the orbit only measures 1–3 mm in the cranial direction [18]. Incidental opening of the orbit most commonly happens if one tries to open the maxillary sinus at the anterior ascending portion of the infundibulum. This situation is depicted in Fig. 18.7. The already-dilated maxillary ostium is represented by a *white circle*. Anteriorly and above this is the medial wall of the orbit (*open triangle*). The orbit is injured when searching for the maxillary ostium, but this does not have major consequences as long as the fat of the orbit is not removed. The fat can easily be distinguished by its yellow color. The patient will wake up from anesthesia with a "blue eye". If one is not sure whether the orbit is open, the bulbus pressure maneuver (as described by Draf) should be applied. Mild pressure on the bulbus from outside causes movement of the fat or periorbita, which can easily be seen with the microscope or the endoscope. It is also useful to have the assisting nurse observe the eye by slightly lifting the eyelid during critical parts of the operation, since the eye moves when the orbit is opened accidentally.

The basic procedure of infundibulotomy is complete when the frontal recess and the ostium maxillare are identified. The operative procedure may end here if pathologic alterations are limited to the anterior ethmoid cells. Depending on the pathology, the next step is the opening of the bulla ethmoidalis and of the remaining anterior ethmoid sinus, including the frontal recess.

Opening of the Bulla Ethmoidalis

The bulla is perforated at its lowest point using ethmoid forceps (Fig. 18.11). The mouth of the ethmoid forceps is opened inside the bulla and is then withdrawn in an open position. This prevents damage to tissue in the space inside the bulla which, at this point, is not under direct optical control. This rule generally applies to all situations where closed compartments of the paranasal sinuses are opened. Another important rule is to always apply the instruments in movements parallel to the floor of the nasal cavity when working in the anterior–posterior direction. This assures that work proceeds parallel to the base of the skull (Fig. 18.12).

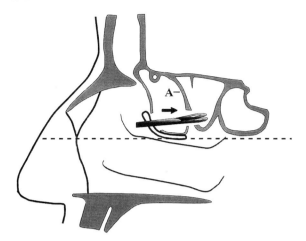

Fig. 18.12. It is important to open the cells of the anterior and posterior ethmoid at their lowest point parallel to the floor of the nasal cavity (and, therefore, parallel to the skull base) when working in an anterior–posterior direction. The basal lamella of the middle turbinate (*A*) and the skull base form a sharp angle that opens toward the posterior ethmoid

Dilation of the Frontal Recess

Dilation of the frontal recess (Fig. 18.13) is part of the surgical opening of the anterior ethmoid. It is achieved by removal of the anterior wall of the bulla ethmoidalis up to skull base. Thus, a wide space is created from the former recessus frontalis and the anterior ethmoid; this space is confined dorsally by the basal lamella of the middle turbinate and anteriorly by the agger nasi. Just anterior to the basal lamella of the middle turbinate lies the ethmoid artery, which runs from the lateral posterior direction to the medial anterior direction. Depending on the position of the lamina cribrosa, the artery either runs through the roof of the ethmoid or in an almost osseous mesentery to the olfactory fossa, which it enters. A high or low position of the ethmoid artery thus indicates the position of the olfactory fossa and, therefore, of the lamina cribrosa. The course of the anterior ethmoid artery through the roof of the ethmoid indicates a slight depression of the olfactory fossa. A low position of the artery in the ethmoid (indicating the presence of an osseous mesentery) marks a low position of the olfactory fossa (Fig. 18.9). An enlargement of the frontal recess anteriorly is necessary if a large agger nasi cell restricts the entrance (Fig. 18.8, and 18.13). Of all cases of frontal sinus inflammation or empyema, etc., 90% can successfully be treated by this simple dilation of the frontal recess. The frontal ostium can be detected with a 30° or 45° telescope in most cases. The described dilation of the frontal recess is frontal-sinus-drainage type I, according to Draf.

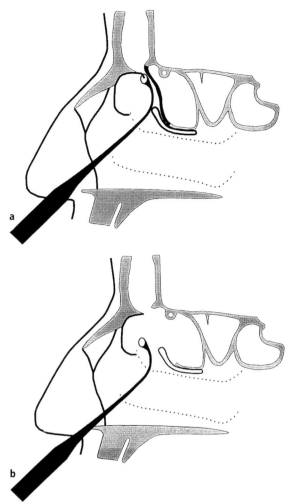

Fig. 18.13. a The frontal recess is enlarged by removing a possible agger nasi cell (*thin black line*) and the anterior wall of the bulla ethmoidalis (*thick black line*). The result is a wide space comprising the complete anterior ethmoid (**b**) [29]

A wider connection between the frontal sinus and the frontal recess (frontal-sinus-drainage type II by the Draf classification) has to be created only when it appears that the widening of the frontal recess does not suffice to relieve an empyema of the frontal sinus or to cure a chronic frontal sinusitis. Technically speaking, the wall of the agger nasi and the frontal process are removed in a cranial direction. The base of the middle turbinate is thereby repositioned more cranially and posteriexly. The frontal process is removed with a diamond burr above the lacrimal sac, the "blue line" of which can be exposed for orientation. The spina nasalis superior is also removed with the burr. The burr is always moved in an anterior direction, thus preventing injury to the skull base (Fig. 18.14). The dorsal margin of the frontal ostium

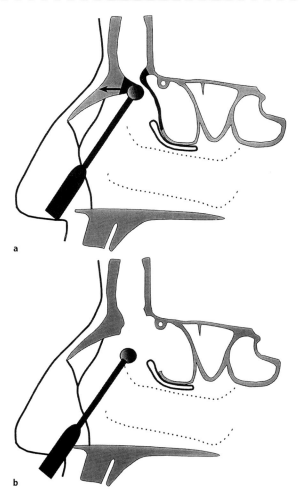

Fig. 18.14 a,b. The frontal process and the spina nasalis frontalis are removed with a burr only if the enlargement of the frontal recess (as shown in Fig. 8.13) is not sufficient to treat a frontal sinusitis. The skull base is protected as long as the burr is applied in an anterior direction (**a**). After removing any remaining anterior wall of the bulla, the frontal sinus is wide open (**b**) [29]

must be covered by mucosa to prevent shrinkage of the new foramen.

The Anteroposterior Opening of the Dorsal Ethmoid Cells

Proceeding in an anterior-to-posterior direction, the next structure reached is the basal lamella of the middle turbinate. It is an oblique lamella that separates the anterior and posterior ethmoid sinuses. When seen from the front, it forms a blunt angle with the skull base and forms a sharp angle dorsally

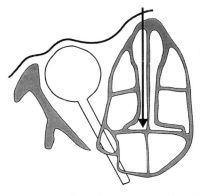

Fig. 18.15. The sphenoid sinus should always be opened transnasally and paraseptally, due to the high variability of the recessus sphenoethmoidalis. Otherwise, the optic nerve is endangered in cases where an Onodi cell is present. This approach also prevents damage to the internal carotid artery in cases where it reaches far into the sphenoid sinus [29]

(Fig. 18.12). It should, therefore, be perforated as caudally as possible and always parallel to the floor of the nasal cavity. Perforating it more cranially endangers the skull base, due to the oblique position of the basal lamella.

In most cases, the dorsal ethmoid consists of one large cell. Its dorsal border is the sphenoethmoid recess, the extension of which varies individually. The area of contact between the dorsal ethmoid and the sphenoid sinus can be quite small, depending on the lateral extension of the sphenoethmoid recess (Fig. 18.15).

The Opening of the Sphenoid Sinus

The sphenoid sinus should always be opened transnasally and paraseptally and not transethmoidally (Fig. 18.15) because of the great variability of the recessus sphenoethmoidalis among individuals. The sphenoid ostium can be found by careful exploration along the septum. The basal part of the ostium is dilated until a 30° telescope can be introduced; this allows visualization and localization of the contours of the optic nerve and the internal carotid artery.

Opening the sphenoid sinus by a transethmoidal approach is dangerous for two reasons. First, the optic nerve is endangered if an Onodi cell is present (Fig. 18.15). Second, in cases where the dorsal ethmoid is located above the sphenoid sinus (Fig. 18.16), the skull base may be damaged instead of entering the sphenoid sinus. In comparison, the transnasal approach is completely safe, because no important structures are located medially.

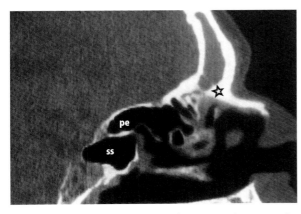

Fig. 18.16. Sagittal reconstruction of a computed tomography scan of the paranasal sinuses. The posterior ethmoid (*pe*) is located higher than the sphenoid sinus (*ss*). Any attempt to open the sphenoid sinus in a transethmoidal approach would endanger the skull base. The transnasal approach to the sphenoid sinus is, therefore, favored over the transethmoidal approach. *Star*, spina nasalis frontalis

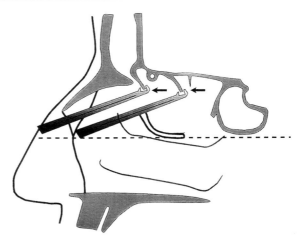

Fig. 18.17. The removal of the cranial ethmoid compartments is most safely done in a posterior–anterior direction after the ethmoid sinus has been opened caudally in an anterior–posterior direction. Cell septae inserting at the skull base are removed by Kofler punches or sharp forceps but not by blunt ethmoid forceps, such as Blakesley's forceps. In this way, breakage of the skull base and injury to the dura are prevented

The Retrograde Posterior–Anterior Preparation of the Ethmoid Near the Skull Base

To expose the dorsal frontobase after reaching the recessus sphenoethmoidalis the oblique dorsal wall of the ethmoid sinus is removed with the Kofler punch until the white, glimmering roof of the ethmoid is exposed. Subsequently, the remains of the basal lamella of the middle turbinate (up to the skull base) are removed in a posterior–anterior direction (Fig. 18.17). Next, the anterior ethmoid artery, which lies anteriorly in the recessus suprabullaris, is reached. Its osseous mesentery must not be mistaken for a cell septum. Approximately 1 cm before the anterior ethmoid artery, the roof of the ethmoid bends into the dorsal wall of the frontal sinus. This is where the frontal ostium is located. Cells of the suprabullar, infundibular or agger nasi region located in this area may easily be removed without endangering the skull base.

Exposure of the Optic Nerve Channel and the Tip of the Orbit

After opening and removing the anterior wall of the sphenoid sinus under direct view, the bony channel of the optic nerve is exposed. It forms an easily detectable bulge at the lateral wall of the sphenoid sinus. The optic channel and the tip of the orbit are decompressed with a diamond burr in a posterior–anterior direction.

Decompression of the Periorbita in Subperiostal Abscess

In cases of orbital complication with subperiostal abscess, the endonasal approach is favored over the extranasal approach. Exposure is through the bulla ethmoidalis, the lateral wall of which is formed by the lamina papyracea. It is anteriorly exposed by opening the bulla and is removed with a burr and elevators until the abscess is reached. It then drains into the ethmoid sinus.

Endonasal Transethmoidal Decompression of the Orbita in Graves Disease

The lamina papyracea is detected as described above (by starting at the bulla) and is thinned out with a diamond burr until it can be removed with an elevator. In the presence of a large supraorbital cell, the medial portion of the roof of the orbit is also exposed and removed in this way. Caudally, the lamina papyracea is followed to the roof of the maxillary sinus. The floor of the orbit can then be removed to the canalis infraorbitalis. The mucous membrane of the maxillary sinus remains intact. In this way, up to 180° of the orbital walls can be resected. The use of the operating microscope allows us to remove the bone with the burr until it can easily be taken away by elevators. Subsequently, the periorbita is slit open in an anterior-to-posterior direction, as though one were radially cutting an orange. The middle turbinate can thus be preserved. We have not applied the extranasal approach for several years, even though it is often recommended.

Important Key Points

1. Always guide your instruments with an awareness of the angle and distance to the floor of the nasal cavity. The skull base runs parallel to the floor of the nasal cavity.
2. Preserve the bulla ethmoidalis as long as possible, as it is an important landmark that can assist in locating the ostium maxillare and the recessus frontalis, which lead into the frontal sinus.
3. Remove the bulla and the dorsal ethmoid as basally as possible and always parallel to the floor of the nose in an anterior–posterior direction.
4. The roof of the ethmoid is most safely exposed retrogradely, in a posterior–anterior direction parallel to the floor of the nose, starting at the basally opened dorsal ethmoid cells.
5. The preserved medial portion of the middle turbinate protects the lamina cribrosa, the location of which is marked by the course of the ethmoid artery.
6. The sphenoid sinus should always be opened transnasally medially, not transethmoidally.

Acknowledgement. We appreciate the support of Prof. Tillmann, director of the Anatomical Institute of the University of Kiel, who donated the specimen in Figs. 18.6, 18.7 and 18.10.

References

1. Calhoun KH, Rotzler WH, Stiernberg CM (1990) Surgical anatomy of the lateral nasal wall. Otolaryngol Head Neck Surg 102:156–160
2. Draf W (1991) Endonasal micro-endoscopic frontal sinus surgery, the Fulda concept. Operative Tech Otolaryngol Head Neck Surg 2:234–240
3. Draf W (1992) Endonasale mikro-endoskopische Pansinusoperation bei chronischer Sinusitis. III. Endonasale mikro-endoskopische Stirnhöhlenchirurgie: eine Standortbestimmung. Otolaryngol Nova 2:118–125
4. Draf W, Weber R (1992) Endonasale Chirurgie der Nasennebenhöhlen – Das Fuldaer mikro-endoskopische Konzept. In: Ganz H, Schätzle W (eds) HNO Praxis Heute, vol 12. Springer, Berlin Heidelberg New York, pp 59–80
5. Draf W, Weber R (1993) Endonasal micro-endoscopic pansinus operation in chronic sinusitis. I. Indications and operation technique. Am J Otolaryngol 14:394–398
6. Draf W, Weber R, Keerl R, Constantinidis J (1995) Aspekte zur Stirnhöhlenchirurgie. Teil I: Die endonasale Stirnhöhlendrainage bei entzündlichen Erkrankungen der Nasennebenhöhlen. HNO 43:352–357
7. Hajek M (1915) Pathologie und Therapie der entzündlichen Erkrankungen der Nebenhöhlen der Nase. Deuticke, Leipzig
8. Hajek M (1923) Indikation der verschiedenen Behandlungs- und Operationsmethoden bei den entzündlichen Erkrankungen der Nebenhöhlen der Nase. Z HNO 4:511–522
9. Halle M (1906) Externe oder interne Operationen der Nebenhöhlen I. Berl Klin Wochenschr 42:1369–1372
10. Halle M (1906) Externe oder interne Operationen der Nebenhöhlen II. Berl Klin Wochenschr 43:1404–1407
11. Halle M (1911) Die intranasale Eröffnung und Behandlung der chronisch kranken Stirnhöhlen. Arch Laryngol Rhinol 24:249–265
12. Halle M (1915) Die intranasalen Operationen bei eitrigen Erkrankungen der Nebenhöhlen der Nase. Arch Laryngol Rhinol 29:73–112
13. Halle M (1923) Nebenhöhlenoperationen. Z HNO 4:489–522
14. Heermann H (1958) Über endonasale Chirurgie unter Verwendung des binokularen Mikroskops. Arch Ohren Nasen Kehlkopfheilkd 171:295–297
15. Kennedy DW (1985) Functional endoscopic sinus surgery: technique. Arch Otolaryngol Head Neck Surg 111:643–649
16. Keros P (1965) Über die praktische Bedeutung der Niveauunterschiede der Lamina cribrosa des Ethmoids. Laryng Rhinol Otol 41:808–813
17. Killian G (1895) II. Zur Anatomie der Nase menschlicher Embryonen. Die ursprüngliche Morphologie der Siebbeingegend. Arch Laryngol 3:17–47
18. Lang J (1988) Klinische Anatomie der Nase, Nasenhöhle und Nebenhöhlen. Thieme, Stuttgart
19. Lang J, Bressel S (1988) The hiatus semiluminaris, the infundibulum and the ostium of the sinus maxillaris, the anterior attached zone of the concha nasalis and its distance to the landmarks of the outer and inner nose. Gegenbaurs Morphol Jahrbb 134:637–646
20. Messerklinger W (1966) Über die Drainage der menschlichen Nasennebenhöhlen unter normalen und pathologischen Bedingungen. 1. Mitteilung. Monatsschr Ohrenheilkd 100:56–68
21. Messerklinger W (1967) Über die Drainage der menschlichen Nasenneben höhlen unter normalen und pathologischen Bedingungen. 2. Mitteilung: Die Stirnhöhle und ihr Ausführungssystem. Monatsschr Ohrenheilkd 101:313–326
22. Messerklinger W (1978) Endoscopy of the nose. Urban and Schwarzenberg, Baltimore
23. Messerklinger W (1982) Über den Recessus frontalis und seine Klinik. Laryngorhinootologie 61:217–223
24. Messerklinger W (1987) Die Rolle der lateralen Nasenwand in der Pathogenese, Diagnose und Therapie der rezidivierenden und chronischen Rhinosinusitis. Laryngorhinootologie 66:293–299
25. Mosher HP (1929) The surgical anatomy of the ethmoid labyrinth. Trans Am Acad Ophthalmol Otolaryngol 31:376–410
26. Rudert H (1988) Mikroskop- und endoskopgestützte Chirurgie der entzündlichen Nasennebebhöhlenerkrankungen. HNO 36:475–482

27. Rudert H (1995) Endonasal microscopic frontal sinus surgery. In: Tos M, Thomsen J, Balle V (eds) Rhinology: a state of the art. Kugler, Amsterdam, pp 59–62
28. Rudert H, Maune S (1997) Die endonasale Koagulation der Arteria sphenopalatina bei schwerer posteriorer Epistaxis. Laryngorhinootologie 76:77–82
29. Rudert H, Maune S, Mahnke CG (1997) Komplikationen der endonasalen Chirurgie der Nasennebenhöhlen. Inzidenz und Strategien zu ihrer Vermeidung. Laryngorhinootologie 76:200–215
30. Simmen D, Schuknecht B (1997) Computertomographie der Nasennebenhöhlen – eine präoperative Checkliste. Laryngorhinootologie 76:8–13
31. Stammberger H (1985) Unsere endoskopische Operationstechnik der lateralen Nasenwand – ein endoskopisch-chirurgisches Konzept zur Behandlung entzündlicher Nasennebenhöhlenerkrankungen. Laryngorhinootologie 64:559–566
32. Stammberger H (1986) Endoscopic endonasal surgery – concepts in treatment of recurring rhinosinusitis. Part I. Anatomic and pathophysiologic considerations. Otolaryngol Head Neck Surg 94:143–147
33. Stammberger H (1986) Endoscopic endonasal surgery – concepts in treatment of recurring rhinosinusitis. Part II. Surgical technique. Otolaryngol Head Neck Surg 94:147–156
34. Wigand ME (1981) Transnasale, endoskopische Chirurgie der Nasennebenhöhlen bei chronischer Sinusitis. I. Ein biomechanisches Konzept der Schleimhautchirurgie. HNO 29:215–221
35. Wigand ME (1981) Transnasale, endoskopische Chirurgie der Nasennebenhöhlen bei chronischer Sinusitis. II. Die endonasale Kieferhöhlenoperation. HNO 29:263–269
36. Wigand ME (1981) Transnasale, endoskopische Chirurgie der Nasennebenhöhlen bei chronischer Sinusitis. III. Die endonasale Siebbeinausräumung. HNO 29:287–293
37. Wigand ME (1982) Transnasal ethmoidectomy under endoscopic control. Rhinology 92:1038–1041
38. Wigand ME, Steiner W, Jaumann MP (1978) Endonasal sinus surgery with endoscopic control: from radical operation to rehabilitation of the mucosa. Endoscopy 10:255–260
39. Zuckerkandl E (1893) Normale und pathologische Anatomie der Nasenhöhle und ihrer pneumatischen Anhänge. Braumüller, Wien
40. Zuckerkandl E (1900) Atlas der topographischen Anatomie des Menschen. I. Heft: Kopf und Hals. Braumüller, Wien

Micro-endoscopic Surgery of the Maxillary Sinus

Rodolfo Arias, Hector Ariza, Ivan Correa, and Aldo Cassol Stamm

Introduction

The surgical treatment of chronic sinusitis has remained a controversial topic even though many years have passed since the first maxillary-sinus explorations were performed. The performance of the first endonasal antrostomy is attributed to Gooch [3] in the second half of the 18th century, but it quickly fell into disuse until it was reintroduced and described again by Mikulicz in 1866 [12]. Both the American Walter Caldwell (1893) and the Frenchman Henry Luc (1897) independently described the transcanine approach to the maxillary sinus, with opening of a nasoantral window through the inferior meatus [13]. From these two techniques, two separate surgical philosophies developed: the conservative one (which utilizes antrostomy) and the radical one (which was called transbuccal radical antrostomy, better known as the Caldwell-Luc operation). The first technique was not well received, primarily because techniques at that time gave poor lightning and visibility, hemorrhage was common, the procedure was quite traumatic and, in a number of cases, the new communication with the sinus closed, causing relapse of the disease.

In general terms, the Caldwell operation had a better acceptance, and it became so popular that it is still used today. However, a number of publications have argued against Caldwell-Luc for more than three decades [3, 16], citing frequent failures and postoperative sequelae, including trauma to the infraorbital and alveolar nerves, which causes dental sensory alterations, sometimes with dental-pulp death, permanent neuralgic pain, fibrous obliteration of the maxillary sinus, loculations and late abscess formation [6]. For these reasons, many authors still favor the surgically conservative treatment despite the risk of a relapse caused by obliteration of the antrostomy [5, 10].

With the introduction of the surgical microscope, borrowed from otologic and laryngeal microsurgery, a very important step was made in the treatment of chronic sinusitis, allowing access to the maxillary, ethmoid and sphenoid sinuses through an endonasal approach. In this respect, the work of Prades [18, 19], Dixon [3, 4], and Morgenstein [15] in this field is outstanding.

From the physiopathological studies of Van Alyea in 1945 [10] and the more recent work of Messerklinger [14], it has been established that the altered sinus mucosa can recover its function and restore adequate maxillary-sinus ventilation and drainage. Today, the concept of functional surgery is served by two techniques with a common purpose that developed in parallel: microsurgery and endoscopic surgery of paranasal sinuses.

Since Kennedy's publications in 1985 [8], the used of endoscopic surgery has been widespread, as detailed in numerous publications, including those of Stammberger [12], Wigand [25], Rice [20] and Levine [11]. However, the use of the microscope as an important part of bimanual surgery and binocular vision are significant advantages. Thus, many surgeons combine both procedures.

Surgical Anatomy of the Maxillary Sinus

The maxillary sinus is a hollow cavity within the maxilla, with a capacity of approximately 15 cm^3 and a quadrangular-pyramid shape. Its truncated vertex is in the malar region, and its base is in the inner wall, which also forms the external wall of the nasal cavity. From the surgical point of view, this wall is the most complex and important one, because it constitutes the intranasal approach to the paranasal sinuses (Fig. 19.1) [2, 17].

The lateral wall of the nasal cavity is formed by the union of several bone structures: the inferior turbinate, the ethmoid sinus, vertical palatal plate and the lachrymal bone. The inferior turbinate is considered an independent bone whose free edge is slightly curled outward and up, and its adhesive upper edge contributes to the sinus ostium through the maxillary or auricular process. By sectioning the inferior turbinate, the inferior meatus can be clearly observed; this turbinate is deeply linked to the maxillary sinus. The middle turbinate is part of the ethmoid sinus, extends laterally from its nasal origin and is often pneumatized (in which case it is called the concha bullosa); it is really an extension of the ethmoid sinus.

Important anatomic details that must be recognized are found in the middle meatus. The most important is a rounded formation called the bulla ethmoidalis. It constitutes an important surgical landmark for the ethmoid sinus. In front of and below it is the uncinate process, which is formed by the ethmoid uncinate-process articulation with the ethmoid process of the inferior turbinate. Between the two, a groove called the infundibulum or infundibular canal can be seen. Its hole is called the hiatus semilunaris. It communicates superiorly with the recessus frontalis, into which the nasofrontal duct flows. The ostium of the maxillary sinus is located inferiorly and is usually covered by the uncinate process, which obstructs direct visualization via the intranasal view. The ostium is formed by an osseous framework made up of the lateral masses of the ethmoid, the inferior turbinate, the unguis and the vertical portion of the palatine bone.

The framework is reduced to a diameter of approximately 3–4 mm it is surrounded by nasal mucosa. When approaching the maxillary sinus through the middle meatus, it is important to remember that the ostium is located on the medial and upper part of the maxillary sinus (near the orbit). There are important anatomic variations in the size of the maxillary ostium. From the physiological point of view, it is not exactly an orifice but is a canal called the nasoantral canal. Sometimes, another orifice, known as the accessory ostium or Giraldes orifice, can be observed behind or under the maxillary ostium. It is important to remember that the pharyngeal orifice of the Eustachian tube is located behind the inferior turbinate and that, on top of the tail of the turbinate, the palatal nerves and vessels are perpendicular and flow into holes also called pharyngeal orifices.

The upper wall or ceiling is the orbit floor, which is crossed by the infraorbital nerve (located within a very thin wall canal that is sometimes dehiscent). It is associated with the orbit and its contents (Fig. 19.2). The posterior wall is related to the pterigomaxillary fossa, making the maxillary sinus one way of approaching that fossa. The antero-lateral wall abuts the face and overlies the canine fossa; the lower portion of this wall is covered by the oral mucosa, through which the sinus can be surgically reached.

Fig. 19.1. Lateral nasal wall after upward displacement of the middle turbinate (*1*), exposing the middle-meatus region. *2*, Ethmoid bulla; *3*, uncinate process (being probed by a sickle knife); *4*, inferior turbinate. Accessory ostia (*arrows*)

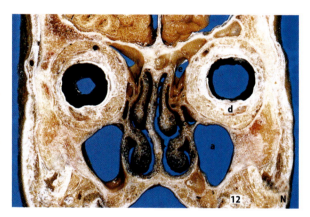

Fig. 19.2. Coronal anatomical section at level of the anterior ethmoid sinus, showing the relationship between the maxillary sinus (*A*) and the adjacent structures; *B*, inferior meatus; *C*, middle meatus; *D*, orbit; *E*, Anterior ethmoid (Courtesy of Prof. Navarro)

At birth, the maxillary-sinus floor is located over the nasal-cavity floor. As the sinus develops, it descends and, at the age of 8 years, it has nearly achieved adult height. With the subsequent development and emergence of permanent teeth, the sinus floor descends below the nasal-cavity floor. The most important relationship of the maxillary-sinus floor is with the roots of the premolar, molar and sometimes canine teeth, which may be covered only by a very thin osseous layer or, sometimes, only by mucosa. This explains how a suppurative dental process can cause infectious complications of this sinus.

Maxillary Sinusitis

Due to its relationships and location, the maxillary sinus is vulnerable to infectious and inflammatory processes, especially when predisposing factors, such as allergic rhinitis, septal deviation or dental alterations, are present. Maxillary sinusitis usually resolves satisfactorily with medical treatment; however, when predisposing factors are not properly handled, some recurrent episodes can occur, causing histopathological changes in the antral mucosa, with permanent obliteration of the maxillary-sinus ostium and irreversible lesions (such as cysts and polyps) that necessitate surgical treatment.

In many cases, chronic maxillary sinusitis extends its inflammatory pathology to the anterior ethmoid sinus, necessitating integrated management of both cavities. In fact, in many cases, access to the maxillary sinus through the middle meatus involves the anterior ethmoid sinus.

When studying a patient with chronic maxillary sinusitis, it is very important to perform a detailed endoscopic examination of the middle meatus after adequate vasoconstriction. A computed tomography (CT) scan should be obtained for any patient who is to be operated upon, because CT precisely delineates the extents of the pathological changes and gives detailed information on the anatomic characteristics of each patient. A complete review of CT in cases of sinusitis is presented in other chapters of this book.

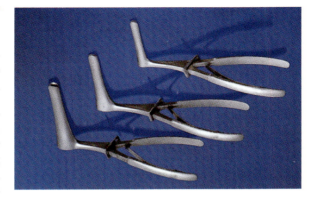

Fig. 19.3. Stamm's self-retaining speculum

Instruments

In microsurgery, the main instrument is a surgical microscope with inclined binoculars and pedal control; an objective lens with $f=250$ mm is preferred. As a second important element, self-retaining speculums are used. With them, it is possible to obtain good exposure and to free both hands for a proper bimanual work.

In our practice, we use Stamm speculums (Fig. 19.3), which can be adjusted to expose the middle meatus; 75- and 95-mm-long Prades speculums are also employed. With these speculums, the inferior meatus can be well visualized. Finally, suction cannulas and a set of varied instruments for dissection, cutting and extraction have to be available and are selected according to the surgeon's experience and preferences. For cutting bone, a universal handle with small Wagner edges in the dorsal and ventral directions work well.

In endoscopic surgery, Hopkins endoscopes with visualization angles of 0°, 30°, 70° and 120° and with optical-fiber cables and light sources are the fundamental instruments. The set of other instruments is very similar to that used in microsurgery; however, the forceps are shorter and thicker. For this reason, in combined microsurgery, we prefer to use microsurgical instruments.

The endoscopes are 4 mm in diameter and fit into the trocar's cannula, to which an additional piece or bridge is added to secure the endoscopes. For endoscopic work, round, flat handles have been designed to adapt to the endoscopes, allowing more comfortable and relaxed operation. For pre- and postoperative intranasal endoscopic examination, 2.7-mm-diameter endoscopes with 30° and 70° visualization angles are used at the office.

Maxillary-Sinus Microsurgery

Microsurgical techniques on the maxillary sinus are performed (under endoscopic control) by opening a 1- or 2-cm-diameter window at the middle or inferior meatus, depending on the pathology intended to be eradicated. In some cases, it is necessary to make the approach through both windows.

The choice of inferior or middle-meatus antrostomy is a very controversial topic, and depends on the existing pathology and the surgeon's skill and experience. Middle-meatus antrostomy is more common, because it is more physiological and is less likely to cause postoperative stenosis or obliteration.

The maxillary-sinus approach through this meatus is not used by all specialists, because of the oblique or tangential focus of the antrum and the poor visibility in some areas (especially in sinuses that are very well pneumatized in the superior and middle walls and the zygomatic recess). However, this limitation can be overcome by using endoscopes introduced through the microantrostomy.

Inferior-meatus antrostomy is less frequently used, especially as the surgeon gains experience in endoscopic surgery and can do a good job using middle-meatus antrostomy. However, we believe there is an important indication for inferior-meatus antrostomy. It should be chosen when there are isolated submucous cysts on the antrum floor without any pathology involving the maxillary ostium. A window and a mucoperiosteum flap can be made more easily with a microscope and a 75-mm Prades speculum than with endoscopes. We think that the ostiomeatal area should not be touched if it features no pathology.

Endoscopic Surgery of the Maxillary Sinus

Endoscopic surgery of the maxillary sinus was proposed by Messerklinger [14] and developed by Stammberger [21–23] and Wigand [25] in Europe and, recently, by Kennedy [7–9] in the United States, where it has become so popular that it is now the unavoidable theme of nearly every national and international convention. Every year, many publications and courses emphasize the advantages of this technique and describe its outstanding results. Other important writers on the topic include Rice [20], Levine [11] and Stankiewicz [24].

Two different approaches to endoscopic surgery of the maxillary sinuses have been recommended: the endonasal approach through the middle meatus and the transcanine approach through the oral vestibule. In the first, the middle turbinate is first endoscopically identified, then the ostiomeatal area in the middle meatus is found. Two important protuberances are immediately identified: the uncinate process and the ethmoid bulla. They form an angle at their bottom, where the maxillary-sinus ostium is found. The ostium is usually not visible, because it is covered by the uncinate process. To identify it, it is necessary to use a cannula with a ball-point explorer or with a curved suction cannula (Fig. 19.4). Depending on the pathology to be eradicated, it is common to first perform an uncinectomy and even a bullectomy before approaching the maxillary sinus. Once the ostium has been identified, it is enlarged by using Stammberger retrograde forceps or a Wagner curve terminal (complemented by straight and angled ethmoid forceps) until a window approximately 1 cm in diameter is obtained. Through this window, it is possible to explore the entire maxillary sinus with 30° and 70° endoscopes. If inflammatory tissue or polyps obliterate the ostium, tissue is removed and the window well is contoured, avoiding the remaining mucosa in order to reduce cicatrization. In this way, stenosis and postoperative obliteration are prevented. If the mucosa of the maxillary sinus is slightly hypertrophied, it can be preserved; if some polyps or cysts are found, they should be carefully removed to avoid disturbing the healthy mucosa. If the mucosa is necrotic, it should be completely resected (Fig. 19.5).

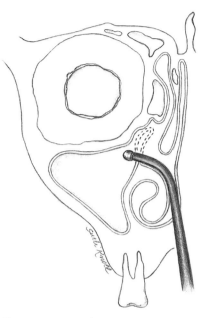

Fig. 19.4. A curved suction elevator entering the maxillary sinus via the middle meatus

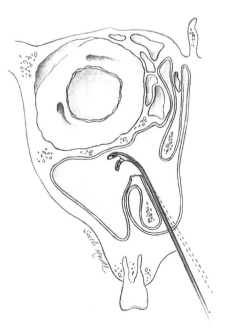

Fig. 19.5. Diseased tissue of the maxillary sinus can be removed by antrum-grasping forceps

The second option is basically an abbreviation of the Cadwell-Luc operation, in which trephination of the canine fossa is performed with a trocar, through which the maxillary sinus is explored with Hopkins endoscopes. There are two basic techniques: The first is enlargement of the contour of the entry hole until the introduction of the extraction forceps is allowed, but this has the disadvantage of being highly traumatic. In the second technique, the only light used is from the trocar, through which the 0° endoscope is usually introduced. Once the pathology is identified, the telescope is withdrawn, and the lesion is removed blindly. Some strongly advocate exploring the maxillary sinus as an office procedure in order to get a biopsy of any suspicious pathology or to remove (with local anesthesia) less extensive pathology, such as polyps or cysts. However, this is not a completely innocuous procedure, because hypoplasic sinuses frequently have a very thick anterior wall that makes the procedure difficult and can sometimes cause severe bleeding; local edema and pain can also occur.

Combined Technique for Microsurgical or Endoscopic Approach to the Maxillary Sinus

As in every surgical procedure, it is essential to have an accurate preoperative assessment of the pathology to be treated. It is very important to have a good case history and appropriate CT-scan studies. The endoscopic evaluation of the nasal cavities, especially at the middle-meatus level, is valuable; together with CT scan, endoscopic evaluation is considered the most effective method of detecting the incipient pathology at the maxillary-ostium level (Fig. 19.6).

The choice of general or local anesthesia depends on many factors, of which the general medical condition of the patient, the severity and degree of the maxillary-sinus inflammatory process and the concomitant pathology of the other paranasal sinuses are the most relevant. Due to the improvement in anesthetic techniques, almost all patients can undergo ambulatory surgery with general anesthesia without having to suffer the undesirable side effects of the prolonged sedation needed when using local anesthesia (including the poor patient tolerance of surgery that frequently occurs and disrupts the operation).

Surgical Technique

The patient lies on his back on the surgical table, with a slight dorsal flexion of the head. When using the microscope, the surgeon is positioned on the opposite side relative to the sinus that is to be operated upon. When endoscopes are used, the surgeon can position himself on the same side as the sinus or he can operate from the opposite side.

Once the patient is anesthetized, we use a local vasoconstrictor, usually cotton pledges soaked in 0.05% oxymetazoline. Using the headlight, these are placed on the floor and in the middle meatus of each nasal fossa. The pledges remain in place for 15 min. We make an endoscopic evaluation with the 0° endoscope, which offers a true portrait of the patient's conditions in relation to the pathology and condition of the middle turbinate, which is a very important reference point. Preferably under endoscopic control, the middle turbinate, agger nasi and uncinate process are infiltrated with 1% lidocaine and epinephrine. No more than 0.5 cm^3 is used at each location.

Middle-Meatus Antrostomy

If we are working with a microscope, we set the Stamm medium speculum after medial middle-turbinate displacement (Fig. 19.7). The large speculum is seldom used. The Prades 75-mm speculum can also be used; however, in our experience, the Stamm gives better exposure.

By opening the valves, it is possible to obtain a better visual field. The uncinate process, the ethmoid bulla, the infundibular groove and (through palpation) the maxillary-sinus ostium can be identified under microscopic control. As mentioned before, in most cases the ostium is generally covered by the uncinate process. The uncinectomy is usually performed first. This allows us to see the ostium, which

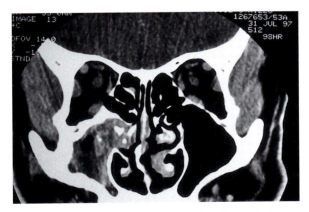

Fig. 19.6. Coronal computed tomography of a patient with fungal maxillary sinusitis

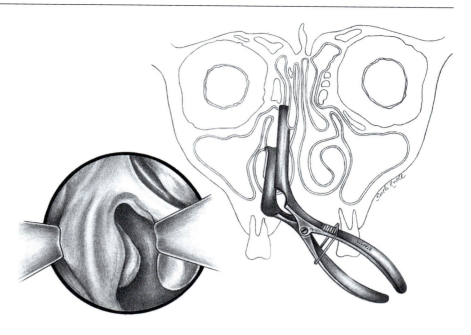

Fig. 19.7. Exposing of the middle meatus area by a self-retaining nasal speculum

is immediately enlarged backward and downward by using cutting forceps (such as Wagner terminals) or by using small straight or angular ethmoid forceps.

The sinus cavity is not easily observed under the microscope; thus, it is necessary to use 30° or 70° Hopkins endoscopes to complete the operation. If surgery is begun with endoscopes, the 0°, 4-mm-diameter one is used. The access route to the middle meatus is identified, avoiding fracturing the middle turbinate in order to prevent its destabilization. If the turbinate is very hypertrophied, half of the lateral part should be resected. This allows visualization of the ostiomeatal area, where it is easy to see the uncinate process, the ethmoid bulla and the ethmoid infundibulum, at the end of which it is possible to locate, through palpation, the maxillary sinus ostium by using a ball-point explorer. To enlarge the ostium, the Stammberger retrograde forceps is used, and the resulting window is enlarged using real cutting ethmoid forceps or Wagner cutting terminals. This window should have a diameter of no less than 1 cm, and its edges should be polished so that the mucosa will be free from fringes or festoons that, in the postoperative state, can cause stenosis, synechia or even obliteration. Due to the risk of causing trauma to the lachrymal–nasal canal, it is essential to avoid biting near the inferior turbinate at its anterior edge. In order to avoid orbital damage, it is also necessary to avoid enlarging the inferior turbinate superiorly.

When working with a microscope where there is very extensive maxillary-sinus pathology, surgery can be completed with an inferior microantrostomy operation. In addition, if the maxillary-sinus pathology is not clear, an antroscopy can be made through either the inferior meatus or the canine fossa in order to explore the maxillary sinus with 70° and 120° endoscopes without causing bleeding. This will help to more accurately determine the pathology and the intrasinusal state of the maxillary-sinus ostium. Pre and postoperative coronal CT. (Fig. 19.8 a,b).

Inferior-Meatus Antrostomy

After inferior-turbinate displacement, the Prades self-retaining 75-mm or sometimes 95-mm speculum is positioned. By opening it, one of the valves remains on the nasal floor and the other reflects the inferior turbinate upward and inward, giving a wide visualization of the inferior meatus. Posteriorly, it is possible to recognize the tail of the inferior turbinate, an important point of reference that the surgeon should avoid going beyond.

A posteriorly based mucoperiostial flap is made. When advisable, once the bone wall is denuded, an antroscopy can be made with the trocar. This will allow bloodless exploration of the maxillary sinus with 70° and 120° endoscopes. In this way, the entire cavity and recess can be observed. In addition, one can make a detailed examination of the maxillary ostium, the accessory ostium or Giraldes orifice (if it exists) and the pathology of the antrum (either diffuse or localized polyps and cysts), which frequently create valve mechanisms that alter the maxillary-sinus ostium function. We have found that many

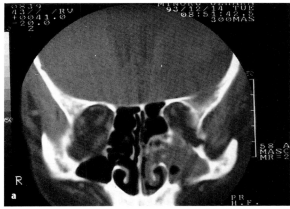
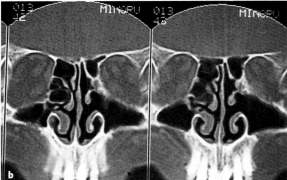

Fig. 19.8a,b. Coronal computed tomography of a patient with chronic maxillary sinusitis, operated on via a middle-meatal antrostomy, before (a) and after operation (b)

superiorly based cysts are purulent, and some small ones have not been detected by radiological studies. The orifice of the antroscopy is used as a guide to enlarge the window by using bone forceps (hollow punch), such as dorsal and ventral Wagner forceps. The forceps are mounted on a universal handle, which allows the user to direct them in the most convenient direction. Through this window, the pathology is removed during microscopic visualization. It is more difficult to make the window with endoscopes, but they are useful in detecting the remaining pathology in areas inaccessible to the microscope [1].

If the mucosa shows any significant necrotic or polypoid changes, it should be totally resected. However, if the mucosa is only edematous or hypertrophied, it is saved because, under these circumstances, we have observed complete recuperation.

The operation is completed by resecting the mucoperiostial flap into the maxillary sinus and by returning the turbinate to its anatomic position. Nasal packing is rarely necessary, because there is no bleeding. Sometimes, when the sinuses are very large and feature severe pathology, a middle-meatus antrostomy should also be made.

There has always been controversy in deciding which of the two meatal approaches offers better results. With the popularity of endoscopic surgery, the practice of inferior-meatus antrostomy has decreased considerably, because it is less physiological than middle-meatus antrostomy and is more difficult to execute with endoscopes. However, in the long term, the postoperative results are still good if the procedure is performed when the indications are correct.

The approach through the lower meatus has been criticized because of its adverse effect on function; the window remains at the floor of the nasal cavities, and the mucociliary flow goes toward the maxillary-sinus ostium. The mucoperiostial flap and the presence of the window alter the cilia movement. However, from a technical point of view, we think that the inferior-meatus approach is easier and intraoperative complications are less probable than in middle-meatus antrostomy, in which trauma to the sphenopalatine vessels and orbit penetrations can occur. In addition, in most of our patients, we have observed both clinical and radiological improvement in the operated maxillary sinus.

Finally, in both the short- and long-term postoperative periods, inferior-meatus-antrostomy permeability can be easily demonstrated through the introduction of a round-pointed probe; this avoids the use of endoscopes, which are necessary for postoperative middle-meatus-antrostomy evaluation. Through inferior-meatus antrostomies, exploration with 30° and 70°, 2.7-mm-diameter endoscopes can be practiced; in this way, we verified that, in several cases, a maxillary ostium (which, during the surgical operation, had been observed to have hypertrophied mucosa) recovered its functionality in approximately 3 months.

Conclusions

We believe that, because of the many advantages of endoscopic surgery, it is an excellent technique for surgical treatment of chronic sinusitis. However, microsurgery still plays an important role, because the endoscopic equipment is expensive, making microsurgery a cheaper procedure with the same results.

Summary

Traditionally, chronic maxillary sinusitis has a macroscopic surgery treatment with two ways to approach: the conservative, intranasal (commonly called antrostomy) and the other radical, through transcanine way (known as Caldwell-Luc). These operations have been practiced since the beginning of this century, causing development of two schools (one radical and the other conservative) that surgeons defend according to their experiences and results. However, both are controversial. Results in both have not been ideal, and there are frequent failures, with relapse of inflammatory pathology or sequelae that are difficult to manage.

In recent years, functional surgery via an intranasal approach has seemed to be the best alternative. It includes two procedures that arose in parallel and offer the possibility of obtaining better results than with conventional surgical treatment. The procedures are microsurgery (in which a surgical microscope is the main element) and endoscopic surgery (with Hopkins endoscopes as preferred instruments). Both techniques have positive and negative aspects. For this reason, many surgeons prefer to combine both procedures when approaching the maxillary sinus.

References

1. Arias R (1985) Antrostomía endonasal microquirúrgica en el tratamiento de la sinusitis maxilar crónica. Acta ORL Cir Cabeza Cuello 13:13–19
2. Ballenger JJ (1981) Enfermedades de la nariz, garganta y oído, 2nd edn. JIMS, Barcelona
3. Dixon H (1976) Microscopic antrostomies in children: a review of the literature in chronic sinusitis and plan of medical and surgical treatment. Laryngoscope 86:1796–1813
4. Dixon H (1985) The use of the operating microscope in ethmoid surgery. Otolaryngol Clin North Am 18:75–86
5. Eichell B (1973) Surgical management of chronic paranasal sinusitis. Laryngoscope 83:1195–1203
6. Harrison D (1971) Surgical anatomy of maxillary and ethmoidal sinuses. Laryngoscope 81:l658–664
7. Kennedy D (1985) Functional endoscopic sinus surgery. Theory and diagnostic evaluation. Arch Otolaryngol 111:576–582
8. Kennedy D (1987) Endoscopic middle meatal antrostomy: theory, technique and patency. Laryngoscope 97:1–9
9. Kennedy D (1987) Functional endoscopic sinus surgery. Technique. Arch Otolaryngol 111:643–649
10. Lavelle RJ, Harrison MS (1971) Infection of maxillary and ethmoidal sinuses. Laryngoscope 81:90–106
11. Levine HL (1993) Endoscopic sinus surgery. Thieme, New York
12. Litton W (1971) Acute and chronic sinusitis. Otolaryngol Clin North Am 4:25–37
13. MacBeth R (1971) Caldwell, Luc, and their operation. Laryngoscope 81:1652–1657
14. Messerklinger W (1978) Endoscopy of the nose. Urban and Schwarzenberg, Baltimore
15. Morgenstein K (1985) Intranasal sphenoethmoidectomy and antrostomy. Otolaryngol Clin North Am 18:69–74
16. Murray J (1983) Complications after treatment of chronic maxillary sinus disease with Caldwell-Luc procedure. Laryngoscope 93:282–284
17. Paparella MM, Shumrick DA (1982) Otorrinolaringología, 2nd edn. Médica Panamericana, Buenos Aires, pp 117–120
18. Prades J (1977) Microcirugía endonasal. Garsi, Madrid
19. Prades J (1980) Microcirugía endonasal de la fosa pterigomaxilar y del meato medio. Salvat, Barcelona
20. Rice DH, Schaefer SD (1988) Endoscopic paranasal sinus surgery. Raven, New York
21. Stammberger H (1986) Endoscopic endonasal surgery. Concepts in treatment of recurring rhinosinusitis. Part I. Anatomic and pathophysiologic considerations. Otolaryngol Head Neck Surg 94:143–146
22. Stammberger H (1986) Endoscopic endonasal surgery. Concepts in treatment of recurring rhinosinusitis. Part II. Surgical technique. Otolaryngol Head Neck Surg 92:147–156
23. Stammberger H (1991) Functional endoscopic sinus surgery. Decker, Philadelphia
24. Stankiewicz J (1987) Complications of endoscopic intranasal ethmoidectomy. Laryngoscope 97:1270–273
25. Wigand ME (1990) Endoscopic surgery of the paranasal sinuses and anterior skull base. Thieme, New York

Endonasal and External Micro-Endoscopic Surgery of the Frontal Sinus

Wolfgang Draf, Rainer Weber, Rainer Keerl, Jannis Constantinidis, Bernhard Schick, and Anjali Saha

Introduction

Transnasal surgery of the paranasal sinuses began (except for a couple of earlier reports) about a hundred years ago [14]. The development of frontal sinus surgery demonstrates clearly how difficult it is to achieve a practicable and reliable approach to the frontal sinus, to obtain a cosmetically satisfactory appearance and, above all, to prevent recurrent infection (which is, unfortunately, common) simultaneously in one operation.

In 1883, the frontal sinus was probed endonasally via its natural opening. Killian [48, 49] exposed the natural ostium and widened it via the nose. In order to reach this goal, he resected the uncinate process and the protruding ethmoid cells. With this technique, a small opening to the frontal sinus can be achieved. Ingals, at an annual conference of American laryngologists in 1905 [38] and Halle in 1906 [30] reported wide endonasal opening of the frontal sinus by removal of the nasal process of the frontal bone. The most important instruments in this operation were special burrs [31], which make it possible to fashion a wide communication between the frontal sinus and nose, with a relatively low risk (at that time, a microscope was not used).

In 1845, Dieffenbach [8,9] was the first to conduct transnasal removal of frontal sinus polyps and frontal-sinus opening after splitting of the nose. As described by Kuhnt [50], Voltolini [68] re-established patency of the frontanasal duct with galvanocautery.

By 1845, Dieffenbach had already noted the crown trephine, as described by Velpeau for trephination of the frontal sinus, but Dieffenbach preferred the path through the nasal cavity when treating inflammatory changes. It can be accepted that extranasal frontal sinus surgery for the treatment of frontal sinus empyemas was first conducted by Ogston in 1884 [2, 40, 41]. He used a trephine to make a 1-cm-diameter opening in the anterior wall of the frontal sinus. In a brilliant monograph, Kuhnt, an ophthalmologist in Königsberg, published a description of the obliteration of the frontal sinus by removal of the entire anterior wall (1895); therefore, he must be considered the pioneer of obliteration techniques. As an incision, Kuhnt chose the eyebrow incision. Marx (from the Strassburg Eye Clinic) had already successfully carried out free-fat transplantation (in 1910) in a second operation to improve the external appearance (as in the Kuhnt and Riedel operations).

Jansen [39] argued against removal of the anterior frontal sinus wall (as recommended by Kuhnt) and instead suggested total removal of the floor of the frontal sinus together with clearance of the ethmoid sinus via excision of the orbital and nasal walls. In 1906, in order to completely remove all mucosa without disfiguration in very high frontal sinuses, Ritter, mostly following Jansen, formed a small, oblique running canal to the frontal sinus by way of an additional skin incision. In 1898, Schenke [59] published the radical frontal sinus operation (as described by Riedel) which, in principle, removed the anterior wall and floor of the sinus in a manner similar to the procedure described by Kuhnt. Riedel also removed the frontal maxillary process.

Those considered as pioneers of osteoplastic frontal sinus surgery featuring temporary removal of a bony ridge are Brieger [4] and Czerny [6] (whose procedure included the formation of a skin-periostium bone flap), Schönborn [60] and Winckler [87]. In 1949, Tato and Bergaglio established a new procedure featuring closure of the frontonasal duct after osteoplastic opening of the frontal sinus [65]. Tato did not leave the frontal sinus to spontaneously obliterate but implanted fat, fibrin, gelatin or blood coagulum. Goodale and Montgomery [24–27] contributed to the popularity of this type of frontal-sinus operation in the US.

Naumann [57] preferred the coronal incision of Mygind [56] and Unterberger [67] for large frontal sinuses; for small sinuses, he used the incision of Bergara and Itoiz [1] and Tato et al. [66], which consists of an incision below the eyebrow, extending to a vertical paramedian incision if so required. Because of frequent serious complications, endonasal sinus surgery (especially for the frontal sinus) was not popular at that time.

The Renaissance of endonasal-sinus surgery began with the introduction of optical aids (such as the microscope [32] and endoscope [12, 13, 15, 52, 53]) and an improved understanding of the pathophysiology of inflammatory sinus disease [54, 83–85]. During the course of further developments, endonasal sinus surgery established itself as a frequently practiced surgical technique and as a routine operation with a low complication rate, when performed by experienced surgeons [19–21, 33, 34, 44–47, 62, 69, 71, 86].

Without doubt, the endonasal procedure using a microscope and/or endoscope for the operative treatment of inflammatory diseases of the frontal sinus is the most important procedure used today. Despite this, there are situations in which certain external operations on the frontal sinus are necessary (also using the endoscope and microscope). That is why, in this chapter, endonasal micro-endoscopic surgery of the frontal sinus and the role of external frontal sinus surgery are discussed in detail; in this way, both operative principles can be united into a single concept of frontal-sinus surgery. Modern imaging techniques permit precise analysis of individual patho-anatomical situations. Based on these analyses, an individually tailored operative treatment can be developed from the surgical approaches available, both endonasal and external; if necessary, a combination of both possibilities can be used.

Preoperative Investigations

The patient history is obvious in trauma cases. Inflammatory changes and tumors in the frontal sinus, head and facial pain, nasal obstruction, taste and smell disturbances, increased nasal discharge and frequent attacks of cold are characteristic.

After a complete ear, nose and throat (ENT) examination with the help of an examining microscope and flexible endoscope, radiological investigation is of great importance. The sinus skiagram in the occipito-dental view (or, for better demonstration of the frontal sinus and ethmoid sinuses, the occipito-frontal view), still has a place as a screening investigation when there is a suspicion of acute or subacute sinusitis, particularly when endoscopic findings are inconclusive. Due to the superimposition of other bony structures and anatomical-variation peculiarities, there is a relatively high rate of false positive and false negative findings [14]. To demonstrate the extent of the disease and the possible need for surgery and for exact anatomical orientation (especially for the internal carotid artery and optic nerve), a computerized tomographic (CT) scan of the paranasal sinuses (using 2-mm slices) should be carried out in each case of undefined paranasal sinus disease. We prefer primary transverse cuts, with reconstructions of the sagittal and coronal planes [29].

The advantages of this somewhat time consuming procedure are:
1. Less radiation exposure for the patient, with optimal three-dimensional orientation for the surgeon
2. Avoidance of artifacts in the coronal cuts when metallic teeth filings are present

We believe that, whenever there is a possibility that surgery will be necessary, a CT investigation should be carried out as a primary procedure. Conventional X-rays are then an unnecessary radiation exposure.

Magnetic resonance imaging (MRI) is not adequate as the only preoperative investigative method. To us, the information MRI provides on the individual anatomy of bony structures appears to be insufficient.

We next have a conversation with the patient concerning the indications for paranasal-sinus operation, based on the overall picture provided by the history, clinical aspects and CT findings. The following risks of the operation are mentioned during this conversation:

- Inflammation in the operative field
- Severe bleeding
- Injury to the meninges
- Loss of smell
- Blindness (described in scattered cases in world literature)
- Lid edema or periorbital hematoma formation
- Injury to the nasaolacrimal duct

Indications and Technique of Frontal-Sinus Surgery

There is general consensus that surgical treatment of acute or chronic inflammatory sinus disease is indicated only after exhausting all possible conservative treatments, including minor prophylactic surgical measures. By prophylactic surgery, we mean interventions that eliminate the need for major surgical measures on the sinuses or even simplify the surgical treatment, thus improving long-term results. These include adenoidectomy, operations on the nasal turbinates, endoscopic elimination of obstructions in the middle meatus, removal of single nasal polyps and septoplasty [7, 17, 18, 22, 78]

Operative Technique, Including Pansinus Surgery

The combined micro-endoscopic technique developed by us (Draf et al.) since 1979 is carried out under general intubation anesthesia, with neuroleptanalgesia after local anesthesia. This includes regional anesthesia of the infraorbital and nasopalatine nerves. We infiltrate the nasal mucosa with lidocaine and suprarenin 1:120,000 and insert strips soaked in 10% cocaine solution (maximum single dose 2 ml, equivalent to 20 mg). The blood pressure should be maintained between 80 mmHg and 100 mmHg. Placement of a hypopharyngeal pack is recommended.

For the first part of the operation, we use a Zeiss microscope with a 250-mm objective. Surgeons with very long arms should use the 300-mm lens. A Cholewa self-retaining speculum [5] is held in the left hand and, when work requires both hands, the speculum is fixed (Fig. 20.1). For video demonstration, a non-reflecting speculum should be used. We use the somewhat elongated diamond burr usually used in ear surgery, with an extra long hand piece. A large forceps and punches simplify the procedures. We avoid non-specific mucosal tearing with blunt instruments by using sharp cutting forceps of various sizes and angles (Fig. 20.2). To preserve as much mucosa as possible and to achieve the greatest surgical precision, we often make use of the Essential Shaver System (Richards HNO, Smith and Nephew GmbH, Hamburg, Germany), which sucks in and separates pathological tissue precisely (Figs. 20.3, 20.4) [28, 61].

The components of the system are: a power unit, a foot switch, hand piece, and a disposable tip. The power unit and the foot switch are not located in the operative field. The shaver system is controlled with a foot switch operated by the surgeon. Three settings are possible: rotating right, rotating left or oscillating. Experience shows that movement of the blade in only one direction does not allow effective grasping of the tissue for entry into the whirling blade. There-

Fig. 20.2. Cutting forceps (straight- and upward-cutting) in various sizes and angles, and a through-cutting punch (Stuemed, Würzburg, Germany)

Fig. 20.3. Essential Shaver System (Richards HNO, Smith and Nephew GmbH, Surgical Division, Schenefeld/Hamburg, Germany)

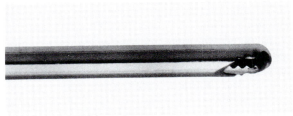

Fig. 20.4. Detail of the Essential Shaver System

Fig. 20.1. Cholewa's self-retaining speculum (Storz)

fore, the oscillating mode, which reverses the direction of the blade every 0.25 s, has been found to be most useful for functional endoscopic sinus surgery.

The hand piece is 2 cm×3 cm in diameter and is 15 cm long. It weighs only 142 g and fits comfortably in most surgeons' hands. The contours of the hand piece allow it to be placed alongside the telescope easily. A variable suction control is on the handle, but full suction is generally preferred. Rapid sterile turnaround of the hand piece is obtained with an autoclave.

The sterile, disposable tip adds 8 cm to the working length and is in two parts: an outer protective sheath with a window near the end and a rotating, hollow blade with a matched, serrated window. Tips with lengths of 2.5 mm and 3.5 mm are available in a variety of configurations. The larger-diameter tip with an aggressive toothed window is the preferred blade for most cases. The smaller-diameter tip is useful in pediatric cases and in narrow noses.

Blood is removed continuously, along with resected bone and soft tissue. By using irrigation to keep the telescope clean, there is less need to remove the instrument or telescope. When expertise with the "Hummer" is gained, the overall operating time for the procedure is reduced.

This instrument was originally designed for temporomandibular-joint surgery. The manufacturer has obtained regulatory approval from the Food and Drug Administration to market this instrument for sinus surgery.

The surgical head is a small, disposable, protected, spinning blade with continuous suction for precise removal of thin bone and soft tissue. This permits sharp trimming and excision of polyps, mucosa and thin bony septations without stripping mucous membranes (Fig. 20.4). The blade rotates within a protected sheath. The contour of the sheath tip is formed as a blunt end, greatly reducing the risk of violating the lamina papyracea or skull base. The risk of trauma to the remaining mucosa is consequently reduced. The blade's cutting edges are only exposed distally along one side. This allows the operator to reposition tissue with the protected tip, to selectively dissect and remove unwanted tissue, or to do both simultaneously. The center of the blade is hollow and, when connected to continuous high suction, enables uninterrupted evacuation of blood, excised tissue and irrigation fluid. This dramatically improves visualization and markedly reduces the frequency with which the instruments must be removed from the nose.

When the spinning blade touches soft tissue or thin bone, the whirling feature, combined with the suction, grasps the tissue and pulls it into the hollow center shaft. The tissue is immediately excised and propelled from the operative site into the suction canister.

The depth of tissue taken is controlled by the amount of pressure with which the surgeon holds the device against the tissue. Consequently, the surgeon can precisely trim small polyps or remove all the bony/mucosal septations of the ethmoid sinuses. As alternative systems. we use the Storz (Tuttlingen) Stammberger shaver system and the Xomed microresector.

We consider it very important to stand straight and relaxed at the operating table to avoid fatigue (which increases the risk of complications). This is why exact placement of the operating table is done before beginning the operation. During the operation, the table is adjusted to the surgeon and not vice versa.

Using the microscope, the middle turbinate is carefully mobilized medially with the septal scissors. After removing mucosa from the agger nasi, the lacrimal bone is drilled away with a diamond drill, and the lacrimal sac is identified. This is the anterior limit of the operative field. Following the lacrimal sac superiorly, a portion of the maxillary frontal process is drilled away by the diamond burr in order to gain access for removal of the anterior superior ethmoid cells and to identify the frontal-sinus infundibulum. This takes some time but improves visualization for the rest of the operation, and is particularly helpful during postoperative treatment, because the path to the frontal sinus is widened. Depending on the individual case, drainage of the frontal sinus is carried out by three quantitatively different techniques (Fig. 20.5; Table 20.1) [17, 18, 22, 78]. The three drainage types (I–III) have common goals: the frontal sinus outflow is bordered by bone on all sides and is covered with mucosa. This reduces the danger of secondary scarring and shrinkage of the drainage opening.

In simple drainage (type-I) cases, the natural opening of the frontal-sinus infundibulum is preserved after the anterior ethmoid cells (up to the skull base) are removed. The mucosa should be manipulated as little as possible in the frontal recess. An extended drainage (type II) is constructed when the floor of the frontal sinus (from the medial orbital border to the nasal septum anterior to the plane of the posterior wall of the frontal sinus) is removed. The resulting opening is positioned anterior to the anterior border of the olfactory fossa. The first olfactory fiber between the head of the middle turbinate and the nasal septum should be identified as an important landmark with a microscope. This is also necessary when creating a type-III drainage. While doing this, care is taken to disturb the mucosa at the transitional frontonasal zone as little as possible. An

Table 20.1. Indications for endonasal frontal-sinus drainage

Type I	Type II	Type III
Unopacified or opacified FS in pansinusitis without risk factors (aspirin hypersensitivity, asthma), no previous surgery	Opacified FS in chronic pansinusitis with large frontal sinus	Opacified FS in pansinusitis in case of special underlying disease (mucoviscidosis, Kartagener syndrome) and in cases with risk factors like aspirin intolerance and asthma
		Revision surgery in chronic pansinusitis
	Orbital complications of sinusitis	Orbital complications
	Endocranial complications (meningitis)	Endocranial complications (meningitis)
	Barotrauma of FS	
	Closure of CSF leak	Closure of CSF leak
	Removal of benign tumors (osteoma)	Removal of benign tumors (osteoma)

FS, frontal sinus

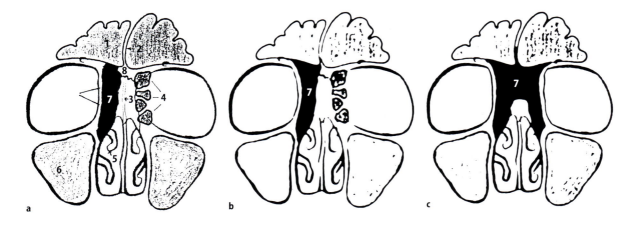

Fig. 20.5 a–c. Endonasal frontal sinus drainage. **a** Simple frontal sinus drainage (type-I). **b** Extended frontal sinus drainage (type-II). **c** Median drainage (type-III)

endonasal median drainage (type III) removes the upper nasal septum and the interfrontal septum that joins and includes the floor of the frontal sinus (on the same and the opposite sides), providing the greatest possible amount of endonasal drainage from the frontal sinuses into the nasal cavities.

Following adequate drainage of the frontal sinus, the anterior ethmoidal artery is identified. It generally lies 1–2 mm behind the angle formed by the posterior frontal sinus wall and the ethmoid roof. It should be noted that the anterior ethmoidal artery sometimes lies 1 mm below the anterior skull base and that the surrounding bone can be very thin. Occasionally, it lies above the visible bony skull base and cannot be identified by the surgeon. The anterior ethmoidal artery is an important landmark for exposure of the whitish-appearing anterior skull-base bone. In cases where there is doubt, demonstration of the anterior skull-base dura with the diamond burr may be necessary. With extensive polyposis, dissection proceeds along the anterior skull base, removing the superior posterior ethmoid cells until the anterior wall of the sphenoid sinus is reached. This is sometimes difficult to identify. The choana, which always lies underneath the anterior sphenoid-sinus wall, is a useful landmark. The opening into the sphenoid sinus should be made with a diamond burr. This reduces the risk of injury to important structures such as the dura mater, the optic nerve and the internal carotid artery. The opening in the anterior wall of the sphenoid sinus is widened with micropunch forceps, and its mucosa is inspected. A large branch of the sphenopalatine artery can run in the anterior wall of the sinus and cause severe bleeding. This bleeding can be stopped with the punch or the diamond drill without irrigation. Next, dissection in the posterior-to-anterior direction is completed by removal of the inferior posterior ethmoid cells. This completes the microscopic part of the operation. In general, the mucosa should be preserved as much as possible.

For endoscopy, we use equipment that allows the operation to be viewed on the monitor and permits video documentation and the printing of color prints (Fig. 20.6). With the 30° suction/irrigation endoscope of Wigand (Fig. 20.7) [83–85], the frontal sinus is inspected when a type-II or -III drainage has been performed, and the lateral ethmoid cell system and the lamina papyracea that could not be seen with the microscope are now identified and cleared. If there is difficulty identifying the lamina papyracea or the periorbit, external pressure over the eyeball and simultaneous endonasal visualization can help in identification (bulb-pressing test as described by Draf [16, 71]; Fig. 20.8). Stankiewicz recommended the same test independently in 1989.

Next, fenestration of the maxillary sinus is performed via the middle meatus and (in cases of very advanced disease) via the inferior meatus. In the middle meatus, an opening is made with 45°, upward-cutting nasal forceps or a diamond burr and is widened to about 2 cm×1 cm. While doing this, the lacrimal sac (located anteriorly), the sphenopalatine foramen (located posteriorly), the orbit (located cranially) and the attachment of the middle turbinate (located caudally) are carefully noted.

Fenestration of the inferior meatus is performed extremely rarely and is usually performed only as a second procedure if polypous maxillary sinusitis persists despite ethmoidectomy. A trocar is then introduced into the maxillary sinus after medially fracturing the inferior turbinate. This opening is widened to approximately 2 cm×1 cm with a punch and cutting forceps. The retrograde cutting forceps, as recommended by Stammberger and Messerklinger, are helpful. This antrostomy opening should not extend to the attachment of the inferior turbinate because, if it is extended too far superiorly, the nasolacrimal-duct opening and Hasner's valve can be damaged.

The opening is located approximately 1.5 cm behind the anterior end of the inferior turbinate, directly below the turbinate attachment on the lateral nasal wall. Polyps and cysts are usually removed using a 30° endoscope with a special suction/irrigation hand piece. In some cases, the 70° endoscope may also be used. The advantage of the 30° endoscope lies in its wider optical field. The removal of pathologically altered mucosa is carried out with forceps of different angles. Conchotomy of the inferior turbinate (corresponding to the opening formed in the inferior meatus [14]) guarantees long-term open maxillary sinus drainage into the inferior meatus.

At the end of the operation, the operative field is checked again using the endoscope and microscope.

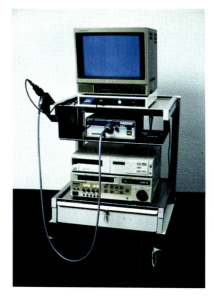

Fig. 20.6. Equipment for video endoscopy on a special trolley (Stuemer, Würzburg, Germany). From *top to bottom*: monitor; camera-control unit with programming board in a case and (on the left side) a charge-coupled-device camera with an endoscope in a telescope holder with a heating system; Xenon Video cold-light fountain; color video printer

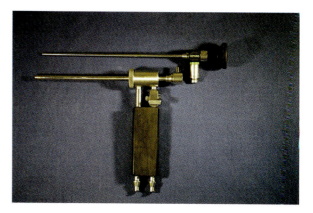

Fig. 20.7. Suction–irrigation Wigand endoscope (modification by the Storz Company)

Finally, a tamponade of three or four rubber finger packs filled with foam rubber (Rhinotamps; Vostra, Aachen, Germany; Fig. 20.9) is inserted into each nasal cavity. Silk threads and plaster are used to securely fix them to the nasal dorsum in order to prevent aspiration.

The finger packs are removed between the third and seventh postoperative days. This is followed by daily cleaning of the operated cavity. Using a special nasal douche, the patient rinses the mucosa with Emser brine (Rhinocare, Siemens and Company,

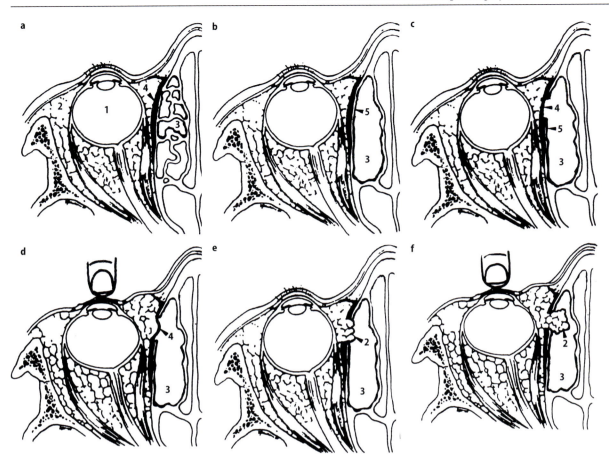

Fig. 20.8 a–f. Bulb-compression test [16, 64]. a Normal anatomy. b–f Various findings on the bulb-compression test. b Intact periorbit and lamina papyracea. c Intact periorbit with a defect in the lamina papyracea. d Bulging of the intact periorbit into the ethmoid cavity on pressure over the eyeball. e Injured periorbit. f Increased prolapse of fatty tissue into the ethmoid cavity on pressure over the eyeball following periorbital injury. *1*, Ocular bulb; *2*, periorbital fat; *3*, ethmoidal cell system, ethmoid cavity; *4*, periorbit; *5*, lamina papyracea

Germany). To prevent exuberant granulation formation and edematous protrusions of mucosa, we always apply the local corticosteroid budesonide (Astra Wedel, Germany). This is given for 3–6 months after the operation. Depending on the extent of the histologically proved mucosal eosinophilia, additional oral medical therapy with a non-sedating antihistamine is started or, in the event of high eosinophilia, a combination of corticosteroids and antihistamines is administered for approximately 6 weeks. We give a broad-spectrum antibiotic for 1–2 weeks if purulent sinusitis is seen during the operation. Depending on their local situations, the patients are discharged and placed in the care of their local ENT specialists on the sixth or seventh postoperative day. Local treatment for 4–6 weeks after the operation is as important as the operation itself [86].

The operation technique described here features various possible methods of endonasal drainage of the frontal sinus and the requisites of maximum endonasal operation for extensive polyposis: the so called endonasal micro-endoscopic pansinus operation. In circumscribed polypoid changes or an exclu-

Fig. 20.9. Rubber finger packs attached to the nasal bridge

sively purulent sinusitis, minor interventions on the ethmoid cell system are often sufficient, depending on the extent of the pathological–anatomical process. These are performed as described by the Graz school [54, 55, 62, 63], by Kennedy [46] in the USA and by many other authors.

Indications for Endonasal Frontal-Sinus Surgery

At present, indications for the three types (I–III) of endonasal frontal sinus opening are as follows [17, 18, 22, 72, 78, 79, 80, 81]. Simple drainage (type I) is coupled with circumscribed ethmoidectomy or endonasal pansinus operation when there is minimal frontal sinus disease pathologically and radiologically. We have changed our previous concept [17, 22] and now recommend type-I drainage for cases with severe opacification of the frontal sinus in patients without risk factors (such as aspirin intolerance, asthma or previous surgery). Our increasing experience indicates that most of these sinuses will be cured without a more extensive approach. However, it is important not to touch the mucosa of the frontal recess if possible, in order to avoid secondary obstruction by scarring.

Extended drainage (type II) is indicated when more drainage is necessary in patients without severe polyposis. The neo-ostium should have a minimum diameter of 5 mm in the anterior-to-posterior direction. Examples include: orbital complications of acute sinusitis and endocrinial complications (such as meningitis [23]). Recurrent barotrauma, which most frequently occurs after sudden extensive changes of height or depth without adequate pressure equilibration (flying, diving) is another ideal indication. The extended-drainage approach is also suitable for inspection of the frontal sinus after circumscribed trauma to the part of the ethmoid sinuses near to infundibulum. Modern high-resolution CT permits a precise preoperative analysis of the extent of injury and clear evidence of smaller, benign tumors close to the infundibulum (such as osteomas, osteofibromas or fibrous dysplasia), which can be removed via an endonasal ethmoidectomy with type-II frontal-sinus drainage.

Endonasal median drainage (type III) is recommended as the primary operation in the presence of extreme mucosal pathology of the frontal sinus or when the frontal sinuses are large. In our experience, the larger the frontal sinus is, the bigger the frontal sinus drainage should be. We also construct a median drainage primarily in patients with extensive polyposis of all the sinuses (including the frontal) who present with risk factors like aspirin intolerance and bronchial asthma. We do the same in patients with mucoviscidosis, the bone-destroying Woakes syndrome and Kartageners syndrome. It should not be performed in small frontal sinuses, where a neo-ostium of 5 mm is not possible. The median drainage is also indicated in orbital and intracranial complications (without bone destruction) originating in the frontal sinus and in frontal sinus mucoceles. If bone destruction of the sinus walls has occurred, an external approach seems safer to us. Smaller benign midline tumors are also removed via the median-drainage operation.

In frontal-sinus cases requiring revision after one or more previous operations, revision is always carried out in the form of median drainage so that the risk of a re-closure is reduced to a minimum. However, in cases following numerable previous unsuccessful frontal sinus operations, we prefer external frontal sinus osteoplastic obliteration using fat. We think the Jansen–Ritter and Lynch operations are no longer appropriate. This is discussed in detail below.

The decision of whether type-II or type-III drainage is constructed depends on the severity and extent of the prevailing pathology and individual anatomy. The smaller the existing frontal-sinus outflow, the more likely it is that primary type-III drainage is appropriate [35, 36, 78].

Technique and Indication of External Frontal-Sinus Surgery

Introductory Remarks

Today, most of the inflammatory frontal-sinus diseases requiring surgical treatment can be successfully treated endonasally [7, 14, 18, 74, 75, 82]. Figure 20.10 illustrates that the external approach to the frontal sinus is seldom indicated. However, there are a number of problem cases that cannot be successfully treated endonasally and require an external procedure.

There are also two very different operative techniques rarely carried out in our department: one is the procedure of Jansen [39] and Ritter [58], which uses an incision below the eyebrow (also associated with the Howarth [37] and Lynch [51] in Anglo-Saxon literature). Essentially, one eliminates the pathology via complete removal of mucosa from the frontal sinus, and the creation of wide drainage to the nose, which may become epithelialized. As a rule, ethmoidectomy is included. The approach includes the removal of the bony anterior and lateral ethmoid walls and the risk of scar contracture of the newly

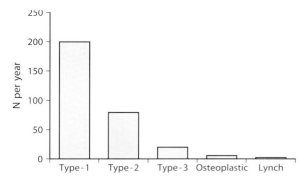

Fig. 20.10. Type and number of frontal-sinus operations [74]

laid drainage. The details of this "classical" operation are well known and are described in numerous operative textbooks. Therefore, it will not be dealt with in depth here.

For the "problem frontal sinus", the osteoplastic frontal-sinus approach has proved very successful during our experiences. This approach permits an optimal view encompassing the entire frontal sinus and permits complete micro-endoscopic removal of mucosa and obliteration of the sinus with abdominal fat. This operation is not commonly carried out, because of the great effort involved. Its success depends on numerous operative details. Thus, we will describe the technique comprehensively here.

In addition to high-resolution CT with primary axial cuts (with a maximum slice thickness of 2 mm) and reconstruction in the coronal plane (and, in special cases, in the sagittal plane), an occipito-frontal X-ray is necessary. From this image, the contours of the frontal sinus are cut out as a template and are preserved in disinfectant solution. Although blood transfusion is seldom required, two units of packed red cell concentrate should be kept ready for safety reasons (particularly if simultaneous endonasal ethmoidectomy is performed).

Several years ago, we stopped shaving the hair (as is done in cosmetic surgery) if a coronal scalp incision is used. Patients have welcomed this step. We have not yet seen an infection in the incision line because of retained hair. The hair must be washed with disinfectant solution the evening before the operation.

Operative Technique

Surgery requires general anesthesia, with orotracheal intubation and the insertion of a pharyngeal pack. In cases of acute inflammatory complications, antibiotic therapy with a second-generation cephalosporin preparation (cefuroxime) is used. After a few days, the parenteral administrations are changed to oral antibiotics for approximately 10–15 days total (Fig. 20.11).

The incision for a uni-or bilateral osteoplastic frontal-sinus operation is individually selected (Fig. 20.12). Selection should take into consideration aesthetic aspects, the prevention of complications and the prevailing anatomical and pathological conditions.

We use the bitemporal coronal incision [56, 67] for large frontal sinuses, in unilateral operations and in patients with normal hair growth. Young men must always be asked about the hair growth patterns of their father and brothers, because the coronal incision leaves an unaesthetic scar in bald persons.

The incision is made beginning at the attachment of the helix (or slightly ventral to it) on one side and is carried over the vertex (approximately 5 cm behind the hairline) to the corresponding point on the opposite side. A tightly stretched silk thread may be helpful as a marker. In the fronto-temporal region, care should be taken not to damage the superficial temporal vessels. We inject a local anesthetic with adrenaline (suprarenin) to lessen the bleeding from the scalp incision (which may be severe). The remaining bleeders are controlled by bipolar coagulation and the use of special scalp clamps on the edges. These clamps are left in place, covered by moist gauze until the end of the operation. Bleeding is usually minor after the clamps are removed. The residual bleeding can be stopped by bipolar coagulation.

In cases of large frontal sinuses and the presence of a risk of significant hair loss, we select a frontal skin crease for the incision [3]. A large frontal sinus can be adequately exposed by this approach, and sensory disturbances in the forehead region are avoided, because only peripheral branches of the supraorbital nerve are cut. Appreciably less bleeding occurs here than with bitemporal incision.

In small frontal sinuses (especially in unilateral operations) or when the supraorbital nerve was damaged by previous surgery, incision below the eyebrow (as described by Killian) is useful [14]. When the Killian incision is used during treatment for bilateral frontal-sinus disease, it is continued bilaterally as a spectacle or seagull incision, but forehead numbness usually results. Although the ethmoid cell system and the sphenoid sinus can also be exenterated by this approach, we recommend the endonasal ethmoid operation, with its respective indications, in order to preserve the bony surroundings in the region of the frontal-sinus floor and its drainage pathway.

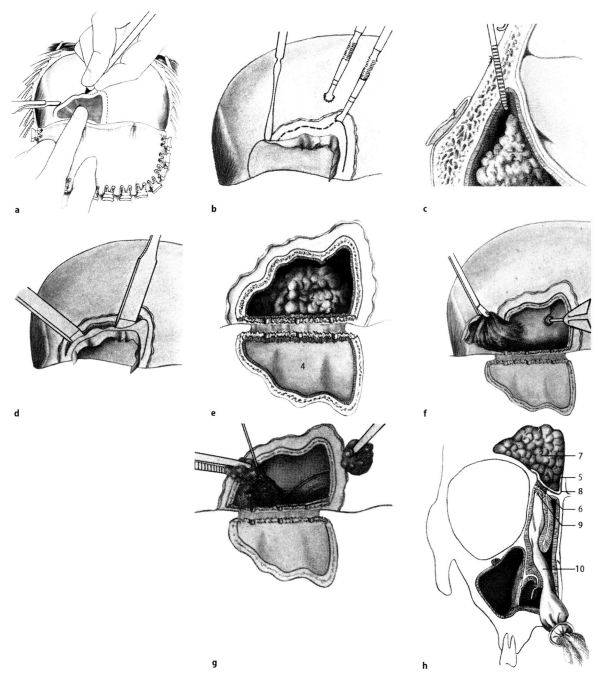

Fig. 20.11 a–h. Technique of the osteoplastic frontal sinus operation [18]. **a** The X-ray template is placed and the periosteal incision is marked; the supra-orbital nerves are preserved after scalp-flap plasty. **b** Formation of the bone flap with a saw corresponding to the limits of the frontal sinus. The periosteum is elevated from the area of bone incision. **c** An oblique incision into the bone enlarges the area for replacement of the bony flap. **d** After placing the bone incision and making drill holes in the base of the bony flap, it is lifted out with chisels. **e** The frontal sinus is opened. The bone flap hinges on the periosteum at the level of the supraorbital ridge. **f** The mucosa is completely removed and, using the operation microscope, the internal table drilled with the burr. **g** After blockage of the ostium and frontal sinus drainage by inverting the mucosa, insertion of cartilage and covering of the cartilage with preserved fascia (fixed with fibrin glue), the frontal sinus is filled with pieces of fat. These are made to cohere by fibrin glue. Finally, the bony flap is replaced, and closure of the wound begins. **h** Frontal section: condition at the operation's end [7]. *1*, Elevated galea/periosteum; *2*, pathologically altered mucosa (frontal sinus); *3*, drill holes; *4*, bony flap hinged on periosteum (anterior wall of the frontal sinus); *5*, preserved fascia; *6*, cartilage (shining through); *7*, transplanted fat with fibrin glue (*blue*); *8*, fibrin glue; *9*, resorbable sponge; *10*, rubber finger packs

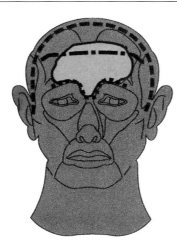

Fig. 20.12. Types of incision used in the osteoplastic frontal-sinus operation. *Dashed line*, coronal; *dashed/dotted line*, skin-crease incision; *dotted line*, spectacle or seagull incision

In cases secondary to trauma, soft-tissue injuries should be utilized as an approach and, if necessary, should be widened in the relaxed skin-tension lines or skin folds. The incision is extended to the periosteum without injuring it. Using partly blunt, partly sharp dissection extending to the supraorbital ridge and over the root of the nose, we pull the scalp flap caudally on both sides, leaving behind the periosteum and the bone. In this way, the entire frontal-bone area is exposed on both sides, thus preserving the supraorbital nerve.

Next, the template of the frontal sinus that was excised from the occipito-frontal X-ray is placed correctly (with respect to the side and orientation) on the *roof* of the nose so that the borders of the frontal sinus can be marked on the periosteum. The periosteum is then incised approximately 1.5 cm outside the template and is elevated slightly past the marking on the bone (Fig. 20.11a). Then the osteotomy is made in the bone (from which the periosteum has been elevated) a few millimeters inside the marked line. The oscillating saw is excellent for this and is made at an angle directed towards the frontal sinus (Fig. 20.11b, c). In this way, a surface that is as wide as possible is created for the later replacement of this bony lid. The bone incision reaches the supraorbital ridge on both sides. When opening the frontal sinus bilaterally, the intersinus septum must be separated from the anterior sinus wall with the help of a chisel angled over the surface. The fracture and elevation of the bony lid is done with a wide osteotome (Fig. 20.11d). During caudal fracture, the supraorbital ridge (slightly dorsal to the fracture, in the region of the sinus floor) is preserved. The bony lid is fashioned so that it hinges on the periosteal flap. The diseased tissue is then removed according to the pathological–anatomical findings. Fractures must be exposed to their full extent, repositioned and, if necessary, a dural lesion has to be treated with duraplasty. In cases secondary to trauma or osteoma, where there is healthy frontal-sinus mucosa, it must be decided whether the mucosa (especially that of the frontal nasal duct and infundibulum) is healthy enough to preserve the frontal sinus or whether an obliteration should be carried out.

A type-III median-drainage technique can be performed easily with the optimum exposure described above. If obliteration is indicated, the frontal-sinus mucosa can be removed macroscopically. In narrow niches, the inner layer must be drilled away with a burr under microscopic and endoscopic control, as only by doing so can the mucosa be completely removed (Fig. 20.11f). The mucosa in the region of the frontonasal duct is inverted nasally, and the drainage opening is obturated with bone or cartilage splints fixed with fibrin glue (Fig. 20.11h). If no autologous material is available, dehydrated human cartilage or preserved bovine cartilage can be used. The bone/cartilage is tightly sealed with solution-dried fascia. The fascia is similarly fixed with fibrin glue. The fascial piece and the bone or cartilage graft should securely seal the frontonasal duct opening caudally but must not cover a large area of the drilled bone of the frontal sinus because, if they do, no nourishing blood vessels can grow out of the bone into the fat used for obliteration. Through this three-layered closure (Fig. 20.11h), the frontal sinus lumen is securely isolated from the nasal cavity, and re-entry and growth of mucosa into the sinus with the associated risk of mucocele formation is prevented.

Next, abdominal fat is freshly harvested. If the patient has undergone appendectomy, the fat is taken from the right-sided scar. Otherwise, we place the incision in a left caudal skin crease in order to prevent confusion between the appendectomy scar and that of the incision for harvesting fat. Precise hemostasis and the insertion of a drain for about 3 days is necessary. The risk of hematoma formation is relatively high. The abdominal fat is temporarily preserved in isotonic saline solution and is then placed into the frontal sinus as larger pieces held together by fibrin glue, until the sinus cavity is completely filled (Fig. 20.11h).

Finally, the periosteum bone lid is replaced and wedged closed with a tap of the mallet, and the periosteum is sutured. Since the periosteal incision lies approximately 15 mm away from the bone incision, the sutures now lie away from the edges of the incision in the bone, so they are covered by periosteum. If the bone lid fractures during elevation, these bony fragments should be fixed together with absorbable thread or wire sutures during their replacement. To end the operation, the scalp flap is flipped back into place, a tube drain is inserted, and the coronal incision is closed with interrupted "O" sutures. The supraorbital incision or the incision in the frontal skin crease is closed in layers. Lastly, skin is apposed with five or six O sutures. A light pressure bandage suffices for better placement of the scalp flap. The tube drain is removed after 2–3 days.

If there is an extensive defect in the posterior frontal-sinus wall after clearance of the frontal sinus, it must be decided whether an obliteration or a cranialization [10] should be carried out [11]. In case of obliteration, abdominal fat is placed on the dura. In cranialization, the entire posterior sinus wall is excised, the mucosa is completely removed (as in obliteration), and the frontonasal duct is sealed in two layers. The frontal lobe of the brain expands into what was the frontal-sinus lumen. This decompression is especially useful in traumatic cases featuring brain edema and comminuted fractures of the frontal sinus.

Indications for External Frontal-Sinus Surgery

With small frontal sinuses, we carry out the "classical" external frontal sinus operation [37, 39, 51, 58] via an incision below the eyebrow so that the supraorbital neurovascular bundle and trochlea are not endangered by the exposure [72, 74–77]. In large frontal sinuses, we prefer the osteoplastic-flap frontal-sinus operation. The Jansen–Ritter operation is always done without obliteration. In the osteoplastic-flap operation, we routinely obliterate (except in unusual cases; for example, after trauma where there is good, healthy mucosa lining the frontal sinus). In the majority of our osteoplastic frontal sinus operations, particularly in inflammatory disease, obliteration is carried out with fat. External frontal-sinus surgery is indicated in laterally placed, localized mucoceles with orbital or intracranial complications of inflammatory origin, with destruction of the posterior or anterior frontal sinus wall, in large osteomas and (most commonly) in the treatment of frontal-sinus trauma.

When there have been one or more previous operations for frontal-sinus disease, especially in conjunction with an ethmoidectomy, the osteoplastic-flap operation with obliteration (i.e., the closure of the frontal sinus as an air-containing cavity via a secure seal from the nose) appears to be the method of choice. If the ethmoid sinus is not completely exenterated, we combine the endonasal ethmoidectomy and maxillary and sphenoid operations with the osteoplastic-flap sinus clearance as one surgical intervention.

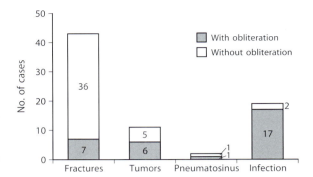

Fig. 20.13. Indications and number of osteoplastic frontal-sinus operations with and without fat obliteration [75]

In our series of sinus operations [73, 75] between 1979 and 1992, it can be seen that, from a total of 1500 patients whose sinuses were operated on, an osteoplastic-flap operation was carried out in 5% ($n=75$; Fig. 20.13). When the primary pathologic process was inflammatory, obliteration with fat was performed with the osteoplastic-flap procedure in nearly all cases; with other indications for osteoplastic flap, obliteration was seldom needed.

Fractures

The most common indication (51%) was frontal-sinus fractures. These could be clearly demonstrated by an osteoplastic approach and could be reduced and treated with a duraplasty if necessary. When the osteoplastic-flap operation is performed without obliteration, it is important to guarantee adequate frontal-sinus drainage in order to avoid late complications, such as mucoceles or pyoceles. Fractures that do not affect the region of the frontonasal duct or the frontal-sinus infundibulum do not require special measures. When scar-tissue formation in the drainage pathway is expected due to either the trauma or extent of surgery, the following operative techniques are indicated, depending on the individual situation:
- Construction of type-II or type-III drainage, with insertion of a silicon tube and a rubber finger pack for 2 weeks.
- Drilling out of the intersinus septum only, so that drainage can also take place from the opposite, healthy sinus (Fig. 20.14).
- Complete removal of mucosa. This gives the best chance of success and is mainly used in obliteration with freshly harvested abdominal fat, in cranialization in extensive, comminuted fractures of the infundibulum (Fig. 20.15) and in cases where chronic inflammatory frontal-sinus disease necessitates surgical trauma to the drainage pathway.

Fig. 20.14. Fracture of the anterior and posterior walls of the right frontal sinus with a CSF leak following an automobile accident. Treatment consists of osteoplastic frontal-sinus surgery with repositioning, duraplasty, use of a bony fragment to support the defect in the posterior frontal-sinus wall, and removal of the intersinus septum

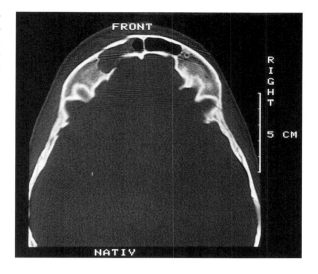

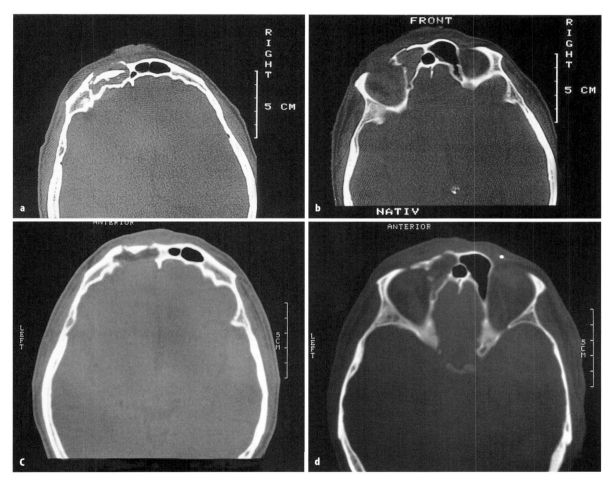

Fig. 20.15 a–d. Extensive right-sided fracture of the anterior and posterior wall of the frontal sinus with definite dislocation of the anterior wall and involvement of the infundibulum following an automobile accident. Treatment consists of osteoplastic frontal sinus surgery with fat obliteration. **a, b** Before operation. **c, d** Six weeks after operation

In cases of fractures involving the roof of the ethmoid sinuses, which extends to the basal region of the posterior frontal-sinus wall, endonasal treatment with repositioning, duraplasty (if necessary) and the construction of type-II or type-III drainage is often a practicable treatment option.

Tumors

Although we routinely remove small infundibular osteomas endonasally, the osteoplastic approach is used when osteomas arise laterally (from the anterior frontal-sinus wall) or are very large (Fig. 20.16) [76]. Similarly, the osteoplastic procedure is preferred in cases where the histology and character are not clearly defined. In addition, six osteomas of the frontal sinuses, one ossifying fibroma, one fibrous dysplasia, two inverted papillomas and one squamous cell carcinoma (the patient with squamous cell carcinoma cell also underwent resection of the external table, with reconstruction of the anterior sinus wall with methylmethacrylate) have been operated via the osteoplastic approach.

Pneumatosinus

Two patients with pneumatosinus dilatans frontalis were operated on, because the patients were concerned by the increasing bulge in their foreheads and desired cosmetic improvement. To remodel the anterior frontal-sinus wall, a horizontal strip of bone was removed from the area of maximum bulge or protrusion so that the bony lid hinging on the periosteum could be molded correspondingly.

Inflammatory Diseases

We have operated on 17 patients with mucopyoceles, one patient with subperiosteal abscess from acute frontal sinusitis and one patient with a high risk of infection following air entry into a frontal sinus that was previously obliterated via the osteoplastic approach. In five patients with mucopyoceles of the frontal sinus, an endonasal ethmoidectomy was also done because of the recurrence of a chronic polypoid ethmoid sinusitis. In 12 of 19 patients with mucopyoceles or related conditions, there had been previous sinus surgery. The mucopyoceles occurred after treatment of a frontal-sinus fracture occurring during a Jansen–Ritter operation for chronic sinusitis. In nine patients, the mucopyoceles occurred between 1 year and 22 years (average=10 years) after surgery. In two patients, mucopyoceles developed 2 years and

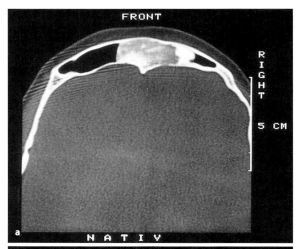
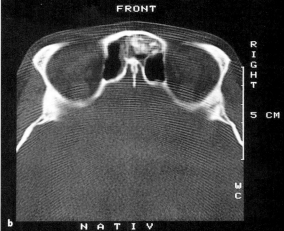
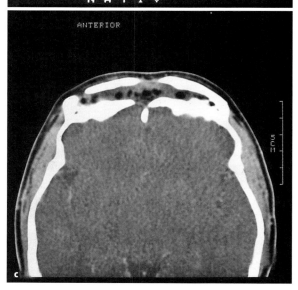

Fig. 20.16 a–c. Osteoma of both frontal sinuses with infundibular involvement. Treatment consists of osteoplastic frontal-sinus surgery with excision of osteoma and fat obliteration. **a, b** Pre-operative axial computed tomography (CT). **c** CT 1.5 years postoperatively

11 years after endonasal sinus surgery for chronic sinusitis. Mucopyoceles developed following transmaxillary surgery for chronic sinusitis in one patient. When mucopyoceles occurred after previous sinus surgery, these complications developed within the first two postoperative years in 63% of the cases, and occurred more than 5 years after operation in 26% of the cases.

Based on our experience with endonasal and external frontal-sinus surgery, the following are our current indications for an osteoplastic frontal-sinus operation in inflammatory disease. In almost all these indications, obliteration is indicated for permanent clearance.

- Recurrence of chronic frontal sinusitis after failed endonasal type-III (median) drainage, after numerous Jansen–Ritter operations (Fig. 20.17) or when concomitant ethmoid disease requires primary or revision endonasal ethmoidectomy
- Persistent or recurrent inflammatory processes in the lateral frontal sinus or supraorbital recess, e.g., mucopyoceles (Fig. 20.18)
- Complications of frontal sinusitis with bone destruction and abscess formation (forehead abscess, intracranial complications; Fig. 20.19)
- Complications of frontal sinusitis following a previous endonasal operation with construction of type-III (median) drainage (Fig. 20.20)
- Inflammatory complications following operative treatment of a frontal-sinus fracture with the introduction of foreign material (Fig. 20.21)

Results of Endonasal Micro-Endoscopic Drainage of the Frontal Sinus

We have carried out two studies. In the first retrospective study [22], we evaluated patients who underwent endonasal frontal-sinus drainage (471 type-I drainages, 125 type-II drainages and 52 type-III drainages) between 1979 and 1992. From these groups, random patients were examined; 42 patients with type-I drainage, 43 with type-II drainage and 47 with type-III drainage were entered into the study. In each patient, the indication was chronic polypoid sinusitis. In five cases with type-III drainage, an

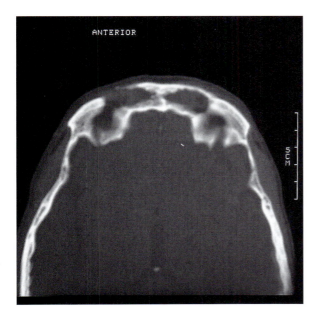

Fig. 20.17. Recurrence of a bilateral chronic polypoid frontal sinusitis with severe cephalgia. Three previous operations were performed (one endonasal, two Jansen Ritter operations), resulting in total absence of the frontal-sinus floor. Treatment consists of osteoplastic frontal-sinus surgery with fat obliteration

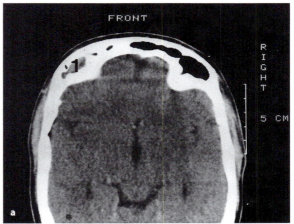
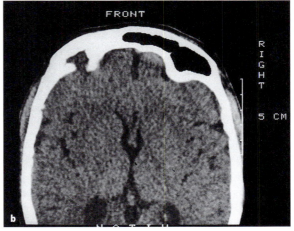

Fig. 20.18. Left laterally situated pyocele of the frontal sinus (1), condition after 18 surgeries (2) of the frontal sinus, with cranialization. a Pre-operative axial computed tomography (CT). b Axial CT after cranialization

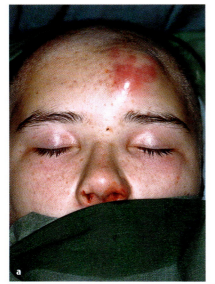

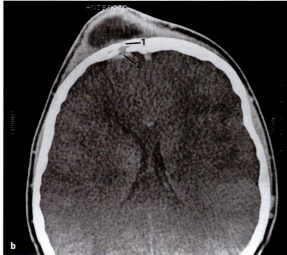

Fig. 20.19. Acute frontal sinusitis with destruction of the anterior (*1*) and posterior walls (*2*) of the frontal sinus with the formation of a forehead abscess (*3*). Treatment consists of osteoplastic frontal-sinus surgery with fat obliteration. **a** Macroscopically. **b** Axial computed tomography

orbital complication presented with the acute sinusitis. The follow-up time periods were between 1 year and 12 years, with a median of 5 years.

The subjective estimation of operative results by the patients is revealed in Fig. 20.22. Individual symptoms had improved to various extents (Fig. 20.23). It is notable that the tendency of increased secretion by the mucosa was not significantly improved after operation. Additional medical therapeutic trials should be started if required (topical corticosteroids). The fact that only a yes/no answer is possible must be taken into account. This

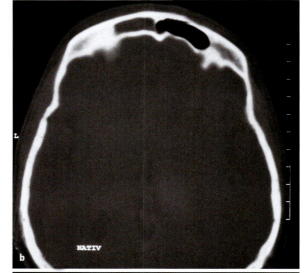

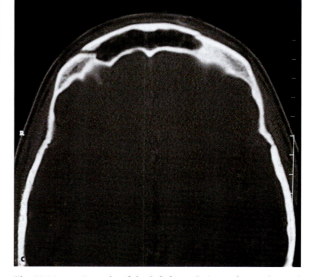

Fig. 20.20 a–c. Pyocele of the left frontal sinus after endonasal pansinus operation and type-III drainage of the frontal sinus. Treatment consists of osteoplastic frontal sinus surgery with fat obliteration. **a** Patient with preoperative swelling and congestion of the left upper lid. **b** Preoperative axial computed tomography with opacity in the left frontal sinus. **c** Condition after obliteration with fat

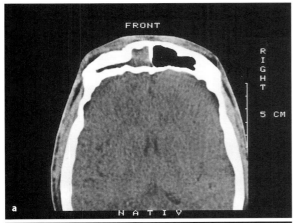

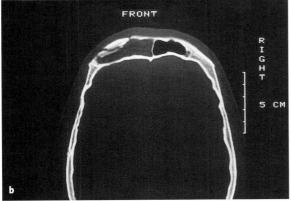

Fig. 20.21a,b. Pyocele of the left frontal sinus after neurosurgical treatment of an anterior and posterior frontal-sinus wall fracture with the insertion of a Pallacos plastic. Treatment consists of osteoplastic frontal-sinus surgery with fat obliteration and reconstruction of the anterior wall of the frontal sinus with a free parietal external table graft. **a** Preoperative axial computed tomography (CT). **b** Axial CT 3 months after operation

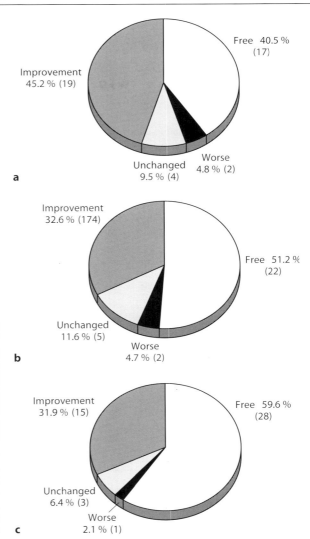

Fig. 20.22 a–c. Subjective judgement of results of frontal-sinus surgery 1–12 years after operation. **a** Type-I drainage. **b** Type-II drainage. **c** Type-III drainage

partly explains the relatively high degree of postoperative complaints.

If one applies subjective and objective criteria to evaluate the success of endonasal frontal-sinus drainage [grade 1 = endoscopically normal mucosa conditions, independent of the subjective picture of complaints; grade 2 = subjective freedom from symptoms, with endoscopically visible inflammatory mucosal changes still present; grade 3 = no subjective improvement, and pathological changes in the mucosa (failure)], we were able to achieve a success rate of 83.4% with type-I drainage, 83.7% with type-II drainage and 89.4% with type-III (median) drainage. This means that, despite the choice of prognostically unfavorable cases (Fig. 20.24), type-III drainage shows the best results. However, there is no statistically significant difference among the three groups.

In a second study, endoscopic and CT examinations were systematically carried out (Figs. 20.25–20.28) [78]. After 12- to 98-month follow-up of patients with type-II drainage, 58% of 83 frontal sinuses were ventilated and normal. A ventilated frontal sinus with hyperplastic mucosa was seen in 12%. Scar-tissue occlusion with total opacification on CT was evident in 15%. In 16%, total opacification was due to recurrent polyposis. Patients were free of symptoms or had only minor problems in 79%. Twelve to 89 months following type-III drainage, 59% of 81 frontal sinuses were ventilated and normal. A ventilated frontal sinus with hyperplastic mucosa was seen in 17%. Scar-tissue occlusion with total opacification on CT was obvious in 7% and, in

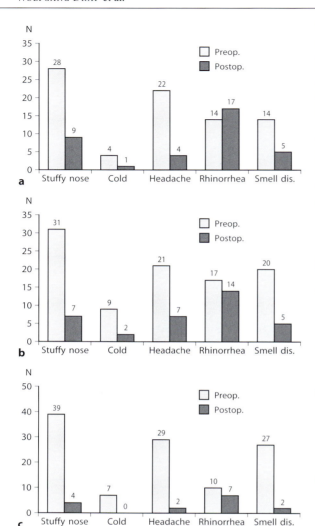

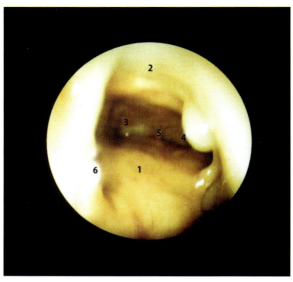

Fig. 20.24. View into the frontal sinus 4 months after type-III drainage. *1*, Posterior wall; *2*, anterior wall; *3*, right frontal sinus; *4*, left frontal sinus; *5*, remnant of the interfrontal sinus septum; *6*, nasal septum

Fig. 20.23 a–c. Pre- and postoperative complaints. **a** Type-I drainage. **b** Type-II drainage. **c** Type-III drainage

16%, there was total opacification due to recurrent polyposis. The patients were free of symptoms or had only minor problems in 95%.

Complications of Endonasal Micro-Endoscopic Frontal-Sinus Surgery

We have analyzed the complications of our endonasal micro-endoscopic pansinus operation in two studies [70, 71]. The significant complications were:
- Injury to the periorbit in 14%; this had no further consequences (except in one patient, who developed periorbital hematoma).
- Dural injury occurred in 2.3%. The subsequent course was uneventful and free from complications after immediate plastic closure of the defect with preserved dura and fibrin glue [78, 79]. Persistent CSF leakage or meningitis was not observed. No cases of blindness occurred.
- Eleven patients noted a postoperative disturbance in their sense of smell; this was confirmed by a smell test in only one case.

An evaluation with respect to endonasal frontal sinus surgery was not done. The operation itself can be classified as very safe when optical aids (such as the microscope and/or the endoscope) are used following the technique described by us [43–45]. According to this technique, the frontal process, the middle turbinate and the ethmoid roof (with the anterior ethmoidal artery) act as significant landmarks. The simplest way to open the frontal sinus is through the ethmoid roof, because the cells are removed with a punch or drill in the direction of the surgeon. In cases of doubt, the bone is gradually thinned by the diamond burr until it is possible to differentiate between mucosa and dura. When there is any uncertainty, operative measures taking place in an anterior inferior-to-posterior superior direction should be undertaken extremely cautiously.

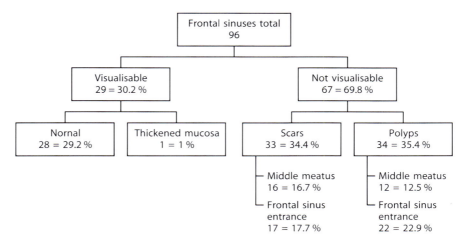

Fig. 20.25. Endoscopic findings 12–98 months following type-II drainage

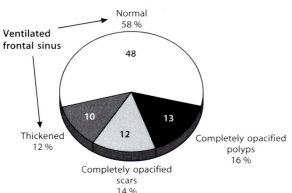

Fig. 20.26. Synopsis of CT and endoscopy 12–98 months following type-II drainage

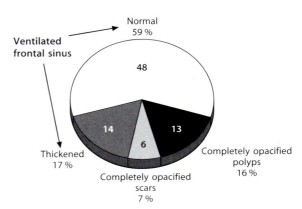

Fig. 20.28. Synopsis of CT and endoscopy 12–98 months following type-III drainage

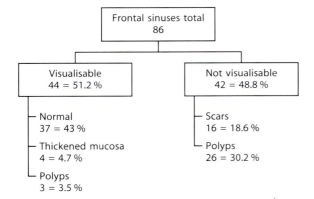

Fig. 20.27. Endoscopic findings 12–98 months following type-III drainage

Results and Complications of External Micro-Endoscopic Osteoplastic Frontal-Sinus Surgery

Overall, very good cosmetic and functional results were seen. No patient has complained about an unsightly scar. Permanent double vision or an unsightly alteration in the contours of the forehead were not observed. Severe complications, such as reduction in vision, dural injury or meningitis, did not occur. A permanent sensation of numbness in the forehead did not occur after coronal incisions or incisions in the frontal-skin creases. However, permanent numbness occurred in 86.7% of patients with incisions in the eyebrow area, and this was felt to be very disturbing by 20% of the patients.

Six patients required transfusion with erythrocyte concentrates. In five cases, this occurred in patients who also had other serious traumatic injuries, so the sum total blood loss made transfusion necessary. Only in one patient was the blood transfusion given on the basis of the frontal-sinus operation only.

In only one case (post-traumatic pyocele of the frontal sinus) was a revision required because of failure of the osteoplastic operative technique. The patient had forced air into the frontal sinus by nose blowing; infection threatened, and early revision surgery with repeat obliteration was necessary. We forbid nose blowing for a 4-week period after operation.

In two other cases, a hypersensitivity reaction occurred after a new type of glass–ionomer cement was used in remodeling the anterior frontal-sinus wall. After revision and removal of the offending material and use of the external table for reconstruction, the further course was uneventful. Because of this and other experiences with methylmethacrylate, great care is advised in using foreign materials for reconstruction of the anterior frontal-sinus wall. The foreign material should not come into contact with the mucosa and should never be placed in inflamed or infected tissue. No patient exhibited recurrence of the inflammatory process, tumor recurrence or delayed complications of fracture treatment.

Using MRI, it is possible to demonstrate the vitality of fat transplanted into the frontal sinus [43, 74, 75]. The fat implanted into the frontal sinuses of 11 patients who underwent an osteoplastic frontal-sinus operation with obliteration was examined by MRI after operation. Objectives were the time-dependent distribution of vital fat and connective tissue, the eventual development and recurrence of necroses or cysts, inflammatory complications and re-epithelialization of the frontal sinus 4–24 months after operation. Vital fat tissue was found in only 6 of 11 cases. Fatty necrosis occurred five times; in four of these cases, it was transformed into granulation tissue and, in one case, it was transformed into connective tissue. All 11 patients were complaint free. Long-term observations are needed to determine whether differences in the recurrence rate of frontal-sinus disease are dependent on whether the implanted fat remains viable or becomes necrosed and transformed.

Summary

A number of operative procedures for frontal sinus disease are at our disposal, including the various types of endonasal micro-endoscopic frontal-sinus drainage (types I–III), the so-called classical external frontal-sinus operation and osteoplastic frontal sinus surgery with or without fat obliteration (which also uses the aid of the microscope and, if necessary the endoscope). All allow eradication of frontal-sinus disease in nearly 90% of cases. Greater experience is necessary to indicate the operative technique promising the greatest chance of success in individual cases. In particularly problematic cases, the combination of endonasal sinus ablation – especially that of the ethmoid cell system – and osteoplastic frontal-sinus surgery with fat obliteration can be very successful. The well-versed sinus surgeon should master the external operative approaches and techniques in addition to the modern endonasal techniques in order to meet the often great demands of frontal-sinus surgery.

References

1. Bergara AR, Itoiz AO (1955) Present state of the surgical treatment of chronic frontal sinusitis. Arch Otolaryngol 61:616–628
2. Boenninghaus G (1923) Die Operationen an den Nebenhöhlen der Nase. In: Katz L, Blumenfeld F (eds) Handbuch der speziellen Chirurgie des Ohres und der oberen Luftwege vol 3, 3rd edn. Kabitzsch, Leipzig, pp 91–286
3. Bosley WR (1972) Osteoplastic obliteration of the frontal sinuses. A review of 100 patients. Laryngoscope 82: 1463–1476
4. Brieger (1894) Jahresber der Schles Ges für Vaterl Kultur. Arch Ohrenheilkd 39:213
5. Cholewa (1938) Selbsthaltendes Speculum Katalog der Fa. Aesculap, Hamburg
6. Czerny (1895) Zit nach Boenninghaus. Arch Klin Chir 50:217
7. Denecke HJ, Denecke MU, Draf W, Ey W (1992) Die Operationen an den Nasennebenhöhlen und der angrenznden Schädelbasis. Springer, Berlin Heidelberg New York
8. Dieffenbach JF (1845) Die operative Chirurgie, vol 1. Brockhaus, Leipzig
9. Dieffenbach JF (1848) Die operative Chirurgie, vol 2. Brockhaus, Leipzig
10. Donald PJ, Bernstein L (1978) Compound frontal sinus injuries with intracranial penetration. Laryngoscope 88:25–32
11. Donald PJ, Ettin M (1986) The safety of frontal sinus fat obliteration when sinus walls are missing. Laryngoscope 96:190–193
12. Draf W (1973) Wert der Sinuskopie für Klinik und Praxis. Laryngorhinootologie 53:890–896
13. Draf W (1978) Endoskopie der Nasennebenhöhlen. Springer, Berlin Heidelberg New York
14. Draf W (1982) Surgical treatment of the inflammatory diseases of the paranasal sinuses. indication, surgical technique, risks, mismanagement and complications, revision surgery. Arch Otorhinolaryngol 235:133–377
15. Draf W (1983) Endoscopy of the paranasal sinuses. Springer, Berlin Heidelberg New York
16. Draf W (1986) Kurs Endonasale mikro-endoskopische Chirurgie der Nasennebenhöhlen. Academic Teaching Hospital, Fulda

17. Draf W (1991) Endonasal micro-endoscopic frontal sinus surgery: The Fulda concept. Operative Tech Otolaryngol Head Neck Surg 2:234–240
18. Draf W (1992) Endonasale mikro-endoskopische Pansinusoperation bei chronischer Sinusitis. III. Endonasale mikro-endoskopische Stirnhöhlenchirurgie: Eine Standortbestimmung Otorhinolarygol Nova 2:118–125
19. Draf W, Weber R (1992) Die endonasale mikro-endoskopische Pansinusoperation bei chronischer Sinusitis. I. Indikation und Operationstechnik. Otorhinolarygol Nova 2:1–4
20. Draf W, Weber R (1992) Endonasale Chirurgie der Nasennebenhöhlen – Das Fuldaer mikro-endoskopische Konzept. HNO Praxis Heute 12:59–80
21. Draf W, Weber R (1993) Endonasal micro-endoscopic pansinus operation in chronic sinusitis I. Indications and operation technique. Am J Otolaryngol 14:394–398
22. Draf W, Weber R, Keerl R, Constantinidis J (1995) Aktuelle Aspekte zur Stirnhöhlenchirurgie I – Die endonasale Stirnhöhlendrainage bei entzündlichen Erkrankungen der Nasennebenhöhlen. HNO 43:352–357
23. Gammert C, Panis R (1977) Behandlung orbitaler Komplikationen bei Entzündungen der Nasennebenhöhlen. Dtsch Ärzteblatt 46:2737–2744
24. Goodale RL (1957) Trends in radical frontal sinus surgery. Ann Otol Rhinol Laryngol 66:369–379
25. Goodale RL (1965) The rationale of frontal sinus surgery. Laryngoscope 75:981–987
26. Goodale RL, Montgomery WW (1958) Experiences with the osteoplastic anterior wall approach to the frontal sinus. Arch Otolaryngol 68:271–283
27. Goodale RL, Montgomery WW (1961) Anterior osteoplastic frontal sinus operation. Five years experience. Ann Otol Rhinol Laryngol 70:860–880
28. Grevers G (1995) Ein neues Operationssystem für die endoskopische Nasennebenhöhlenchirurgie. Laryngorhinootologie 74:266–268
29. Haas JP, Kahle G (1986) Kurs Endonasale mikroendoskopische Chirurgie der Nasennebenhöhlen. Academic Teaching Hospital, Fulda
30. Halle M (1906) Externe oder interne Operation der Nebenhöhleneiterungen. Berl Klin Wochenschr 43:1369–1372, 1404–1407
31. Halle M (1915) Die intranasalen Operationen bei eitrigen Erkrankungen der Nebenhöhlen der Nase. Arch Laryngol 29:73–112
32. Heermann H (1958) Über endonasale Chirurgie unter Verwendung des binocularen Mikroskopes. Arch Klin Exp Ohren Nasen Kehlkopfheilkd 171:295–297
33. Hosemann W, Göde U (1994) Epidemiology, pathophysiology of nasal polyposis, and spectrum of endonasal sinus surgery. Am J Otolaryngol 15:85–98
34. Hosemann W, Wigand ME, Fehle R (1988) Ergebnisse endonasaler Siebbein-Operationen bei diffuser hyperplastischer Sinusitis paranasalis chronica. HNO 36:54–59
35. Hosemann W, Wigand ME, Wessel B (1992) Medico-legale Probleme in der endonasalen Nasennebenhöhlenchirurgie. Eur Arch Otorhinolaryngol Suppl 2:284–296
36. Hosemann W, Kühnel T, Held P, Wagner W, Felderhoff A (1997) Endonasal frontal sinusotomy in surgical management of chronic sinusitis – a critical evaluation. Am J Rhinol 11:1–10
37. Howarth WG (1921) Operations on the frontal sinus. J Laryngol Otol 36:417–421
38. Ingals EF (1905) New operation and instruments for draining the frontal sinus. Ann Otol Rhinol Laryngol 14:513–519
39. Jansen A (1894) Zur Eröffnung der Nebenhöhlen der Nase bei chron. Eiterung. Arch Laryngol Rhinol (Berl) 1:135–157
40. Jurasz A (1887) Zit. nach Boenninghaus. Berl Klin Wochenschr 24:34
41. Kasspariantz A (1900) Zit. nach Boenninghaus. Congress intern de med sect de Rhinol, Paris, p 90
42. Keerl R, Weber R, Draf W (1995) Operationsweiterbildung mittels Multimediatechnik am Beispiel der endonasalen mikro-endoskopischen Pansinusoperation. Laryngorhinootologie 74:361–365
43. Keerl R, Weber R, Kahle G, Draf W, Constantinidis J, Saha A (1995) Magnetic resonance imaging after frontal sinus surgery with fat obliteration. J Laryngol Otol 109:1115–1119
44. Keerl R, Weber R, Draf W (1996) Multimedia in der HNO-Chirurgie. Die endonasale Pansinusoperation. Ullstein Mosby, Wiesbaden
45. Keerl R, Weber R, Draf W (1998) Endonasal sinus surgery. Werner Giebel Verlag, Eiterfeld
46. Kennedy DW (1985) Functional endoscopic sinus surgery: technique. Arch Otolaryngol 111:643–649
47. Kennedy DW (1992) Prognostic factors, outcomes and staging in ethmoid sinus surgery. Laryngoscope 102[suppl]:1–18
48. Killian G (1894) Zit. nach Engelmann: Der Stirnhöhlenkatarrh. Arch Laryngol Rhinol 1:311
49. Killian G (1900) Zit. nach Boenninghaus. Heymanns Handbuch Laryngol 2:1062, 1150
50. Kuhnt H (1895) Über die entzündlichen Erkrankungen der Stirnhöhlen und ihre Folgezustände. Bergmann, Wiesbaden
51. Lynch RC (1921) The technique of a radical frontal sinus operation which has given me the best results. Laryngoscope 31:1–5
52. Messerklinger W (1970) Die Endoskopie der Nase. Monatsschr Ohrenheilkd 104:451–456
53. Messerklinger W (1972) Technik und Möglichkeiten der Nasenendoskopie. HNO 20:133–135
54. Messerklinger W (1978) Endoscopy of the nose. Urban and Schwarzenberg, München
55. Messerklinger W (1987) Die Rolle der lateralen Nasenwand in der Pathogenese, Diagnose und Therapie der rezidivierenden und chronischen Rhinosinusitis. Laryngorhinootologie 66:293–299
56. Mygind SH (1938) Herunterklappen des Skalps bei Ostitis frontis. Acta Otolaryngol (Stockh) 26:537
57. Naumann HH (1961) Gedanken zum gegenwärtigen Stand der Stirnhöhlen-Chirurgie. Laryngol Rhinol Otol (Stuttg) 40:733–749
58. Ritter G (1906) II. Eine neue Methode zur Erhaltung der vorderen Stirnhöhlenwand bei Radikaloperationen chronischer Stirnhöhleneiterungen. Dtsch Med Wochenschr 32:1294–1296
59. Schenke H (1898) Über die Stirnhöhlen und ihre Erkrankungen. Die Radikal-Operation nach Riedel. Friedrich-Schiller-Universität, Jena
60. Schönborn W (1894) Zit. nach Boenninghaus. Diss Würzburg 1:36
61. Setliff RC, Parsons DS (1994) The "Hummer": new instrumentation for functional endoscopic sinus surgery. Am J Rhinol 8:275–278

62. Stammberger H (1985) Unsere endoskopische Operationstechnik der lateralen Nasenwand- ein endoskopisch-chirurgisches Konzept zur Behandlung entzündlicher Nasennebenhöhlenerkrankungen. Laryngorhinootologie 64:559–566
63. Stammberger H (1991) Functional endoscopic sinus surgery. The Messerklinger technique. Decker, Philadelphia
64. Stankiewicz JA (1989) Complications of endoscopic sinus surgery. Otolaryngol Clin North Am 22:749–758
65. Tato JM, Bergaglio OE (1949) Surgery of frontal sinus. Fat grafts: new technique. Otolaryngologica 3:1
66. Tato JM, Sibbald DW, Bergaglio OE (1954) Surgical treatment of the frontal sinus by the external route. Laryngoscope 64:504–521
67. Unterberger S (1953) Kosmetische Schnittführung bei doppelseitiger Stirnhöhlen-Radikal-Operation. Monatsschr Ohrenheilkd 87:304
68. Voltolini A (1888) Die Krankheiten der Nase. S 344
69. Weber R (1987) Die endonasale chirurgische Therapie der chronischen Sinusitis. Inaugural dissertation. Philipps Universität, Marburg
70. Weber R, Draf W (1992) Komplikationen der endonasalen mikro-endoskopischen Siebbeinoperation. HNO 40:170–175
71. Weber R, Draf W (1992) Die endonasale mikro-endoskopische Pansinusoperation bei chronischer Sinusitis. II. Ergebnisse und Komplikationen. Otorhinolaryngologia Nova 2:63–69
72. Weber R, Draf W (1993) Endonasale Stirnhöhlenchirurgie bei chronischer Entzündung – derzeitiger Stand. HNO Aktuel 1:164–170
73. Weber R, Draf W, Constantinidis J (1994) Osteoplastic microscopic frontal sinus surgery. Am J Rhinol 8:247–51
74. Weber R, Draf W, Keerl R, Constantinidis J (1995) Aktuelle Aspekte zur Stirnhöhlenchirurgie II. Die externe Stirnhöhlenoperation – der osteoplastische Zugang. HNO 43:358–363
75. Weber R, Draf W, Keerl R, Constantinidis J (1995) Aktuelle Aspekte zur Stirnhöhlenchirurgie III. Indikation und Ergebnisse der osteoplastischen Stirnhöhlenoperation. HNO 43:414–420
76. Weber R, Draf W, Constantinidis J, Keerl R (1995) Aktuelle Aspekte zur Stirnhöhlenchirurgie IV. Zur Therapie des Stirnhöhlenosteoms. HNO 43:482–486
77. Weber R, Keerl R, Draf W, Wienke A, Kind M (1995) Zur Begutachtung: Periorbitales Paraffingranulom nach Nasennebenhöhlenoperation. Otorhinolaryngol Nova 5:87–90
78. Weber R, Draf W, Keerl R, Behm K, Schick B (1996) Langzeitergebnisse nach endonasaler Stirnhöhlenchirurgie. HNO 44:503–509
79. Weber R, Keerl R, Draf W, Schick B, Mosler P, Saha A (1996) Management of dural lesions during endonasal sinus surgery. Arch Otolaryngol 122:732–736
80. Weber R, Draf W, Keerl R, Schick B, Saha A (1997) Microendoscopic endonasal pansinusoperation in chronic sinusitis. Results and complications. Am J Otolayngol 18:247–253
81. Weber R, Hosemann W, Draf W, Keerl R, Schick B, Schinzel S (1997) Endonasal frontal sinus surgery with longterm stenting of the nasofrontal duct. Laryngorhinootologie 76:728–734
82. Weber R, Draf W, Kahle G, Kind M (1999) Obliteration of the frontal sinus: state of the art and reflections on new materials. Rhinology 37:1–15
83. Wigand ME (1981) Ein Saug-Spül-Endoskop für die transnasale Chirugie der Nasennebenhöhlen und der Schädelbasis. HNO 29:102–103
84. Wigand ME (1981) Transnasale, endoskopische Chirurgie der Nasennebenhöhlen bei chronischer Sinusitis. I. Ein biomechanisches Konzept der Schleimhautchirurgie. HNO 29:215–221
85. Wigand ME (1981) Transnasale, endoskopische Chirurgie der Nasennebenhöhlen bei chronischer Sinusitis. III. Die endonasale Siebbeinausräumung. HNO 29:287–293
86. Wigand ME (1989) Endoskopische Chirurgie der Nasennebenhöhlen und der vorderen Schädelbasis. Thieme, Stuttgart
87. Winckler (1898) Zit nach Boenninghaus. Arch Laryngol 7:347

Mini-anterior and Combined Frontal Sinusotomy and Drilling of the Nasofrontal Beak

Kanit Muntarbhorn and Sanguansak Thanaviratananich

Introduction

As an explanation of the title of this chapter, mini-anterior frontal sinusotomy (MAFS) is regarded as synonymous with "trephination". Frontal trephination is a procedure in which a trephine is used to make a small hole in the frontal bone. MAFS implies making a small hole (trephination) through an anterior wall (table) of the frontal sinus (AWFS) in order to enter a frontal sinus (FS) or both FSs. The trephine, the original instrument used for trephination, is not the only instrument used for trephination. Other instruments, e.g., a needle, a trocar, a chisel, a gouge or a drill, can also be used for trephination.

The approach for trephination/MAFS can be either an anterior approach or an antero-inferior approach. MAFS can provide access for frontonasal drainage surgery (FDS). Among the techniques of FDS, drilling of nasofrontal beak (DNB) is an important procedure for producing an enlarged frontonasal drainage passage (FDP). DNB can be performed via a MAFS or during an inferior transnasal frontal sinusotomy (ITFS) or during a combined frontal sinusotomy (CFS – MAFS and ITFS). In this chapter, an emphasis is placed on MAFS with DNB rather than ITFS alone.

Trephination

Ogston [26] was the first to publish a technique of median anterior trephination (Fig. 21.1); it involved the use of a trephine followed by the removal of a bony disc to reveal both FSs. Antero-inferior MAFS, the most common type of trephination, takes place at a different location (Fig. 21.1) and can be performed under local or general anesthesia [2, 18, 28]. Although acute frontal sinusitis is the common indication for frontal trephination [2, 18, 28], chronic frontal sinusitis is suggested as an important indication for frontal trephination/MAFS.

Fig. 21.1. Techniques of bony trephination: *1*, median anterior trephination (Ogston); *2*, antero-inferior trephination (Silcock); *3*, mini-anterior frontal sinusotomy type A (Muntarbhorn); *4*, an excessively large bony trephination

History of Anterior Frontal Sinusotomy

In 1884, Ogston [26] described the first trephination (Fig. 21.1, Table 21.1). In 1897, Silcock [31] described an antero-inferior trephination (Fig. 21.1, Table 21.1); Silcock's MAFS involved one frontal sinus per sinusotomy, whereas Ogston's MAFS involved two FSs per sinusotomy. Other early contributors (1884–1928) to MAFS included Schmidt [30], Lothrop [19, 20] and Kisch [16] (Table 21.1). Recent contributors to MAFS include Draf [3–5], Fry, Biggers and Fischer [8], Wigand [34], Hoffmann and May [14], Gerber, Myer and Prenger [9], Muntarbhorn [25] and Thawley and Deddens [3] (Table 21.2).

History of DNB

Nearly a century ago, Lothrop [19] mentioned the thick bony ring around the hiatus frontale, where the frontal bone articulates with the nasal bone and the frontal process of the maxilla, and suggested the removal of the nasal portion of the floor of the

Table 21.1. Trephination (mini-anterior frontal sinusotomy)

Name (year)	Incision	MAFS or trephination	Purpose and procedures
Ogston (1884) [26]	Vertical median incision from the root of the nose upwards (1.5 in.)	Trephination with a chisel	To gain access and to pass the trocar into the nasal cavity in the vicinity of the ostia of both frontal sinuses
Schmidt (before 1895) [30]	Incision along the eyebrow from the angle of the orbit, and reflection of flaps up and down	Trephination with a chisel	To drain pus
Silcock (1897) [31]	Incision below and under the cover of the eyebrow (~1 in.)	Trephination with a gouge (hole as big as a finger)	To probe, to irrigate, to apply a curette and to insert a catheter
Lothrop (1899) [19]	Curved incision on bone near frontonasal suture line in an upward direction for 15 mm, parallel to the fold of skin made by corrugator supercilli, curving outward towards the glabella	Trephination with a chisel	To explore, to curette the mucosa, to enlarge the opening and to remove the floor of the frontal sinus
Kisch (1928) [16]	Incision from just inside supraorbital notch to the inner canthus	Trephination with a chisel	To gain access, to remove bone, to expose the ostium of the frontal sinus and to remove the nasofrontal spine and polyps

MAFS, mini-anterior frontal sinusotomy

Table 21.2. Mini-anterior frontal sinusotomy

Name (year)	Incision	MAFS or trephination	Purpose and procedures
Draf (1974, 1983) [3–5]	Incision along the eyebrow	Trephination with a drill	To use an endoscope to examine the frontal sinus, to perform a biopsy and to instill topical therapy
Fry, Biggers and Fischer (1981) [8]	Incision in the eyebrow by puncturing with a needle	Percutaneous trephination	To irrigate the frontal sinus via a retained catheter
Wigand (1990) [34]	Incision in the eyebrow	Trephination by drilling	To examine the frontal sinus endoscopically and to perform an operation using two access ports
Hoffmann and May (1991) [14]	Curved incision (limited Lynch incision)	Trephination by drilling	To enter the floor of the frontal sinus, to drain the sinus and to inspect with the endoscope
Gerber, Myer and Prenger (1993) [9]	Stab incision below the eyebrow	Trephination by use of a trocar	To endoscopically examine the frontal sinus and to irrigate it
Muntarbhorn (1993) [25]	Peri- or intra-eyebrow incision	Trephination and optional drilling of the nasofrontal beak	See text
Thawley and Deddens (1995) [33]	Incision just superior to the medial portion of the eyebrow	Trephination by drilling (oval opening)	To identify and remove obstructions of the frontal outflow tract

MAFS, mini-anterior frontal sinusotomy

frontal sinus. In 1917, Lothrop [20] used a lateral roentgenogram to show the dense bone: the upper, thickened ends of nasal bones, nasal processes of maxillae and the region where the interfrontal septum and median plate of the ethmoid and frontal spine meet. Lothrop [19, 20] performed DNB via an external approach. DNB via a transnasal approach was described by Halle [11, 12]. Recent contributors to intranasal/transnasal/endonasal DNB include Draf [6], Close, Lee, Leach and Manning [1], May and Schaitkin [23] and Gross, Gross and Becker [10].

Indications

The indications for MAFS or CFS are (1) patients with persistent or chronic frontal sinusitis, (2) patients with frontal sinusitis after functional endoscopic sinus surgery (FESS) and (3) patients requiring frontal-sinus revision surgery. Among our 30 patients scheduled for revision surgery after failed FESS, 25 had frontal sinusitis (83.3%). The indications for ITFS are similar to those for MAFS. Indications for DNB include (1) bony stenosis of FDP, (2) bony narrowing of FDP, (3) obstructive FDP after FESS, (4) obstructive FDP after failed FDS and (5) obstructive FDP after failed frontal-sinus operation(s).

Planning the Bony Trephination

Although Ogston [26] used a trephine for trephination, drilling with a burr is a more common and modern technique [2, 4, 14, 18, 25, 28, 33, 34]. However, one should note that mislocated perifrontal sinus drilling can lead to a mini-craniotomy instead of a frontal trephination/MAFS.

For the *X-ray-template technique*, a 6-ft Caldwell radiograph is required; an X-ray plate is in contact with the forehead while the X-ray tube is 6 ft away [2]. This X-ray film of the FS can then be cut, sterilized and later used as a template during the marking of the outline of the FS (Figs. 21.2, 21.3). Requirements for *preoperative X-rays and computed tomography (CT) films* include: (1) radiographs of the FS (two films, Caldwell's view), (2) lateral X-ray film of the skull (for the anterior skull base and FSs), (3) coronal CT films of the paranasal sinuses, with particular attention to the FSs, ethmoid sinuses and anterior skull base and (4) axial CT films (though optional, these may yield important information about the antero-posterior distance to the chamber(s) of the FS).

Pre-operative antibiotic medication should be given parenterally, as MAFS with high trephination (MAFS type A) often involves entering the diploe. *Pre-operative information for the patient* should include the potential consequences of MAFS, e.g., a visible scar and sensory loss (due to damage to the supratrochlear and supraorbital nerves).

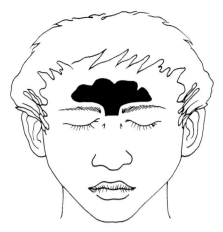

Fig. 21.2. An X-ray template is placed on the patient's forehead

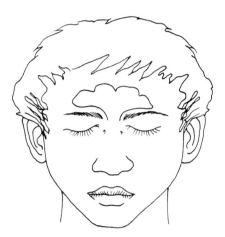

Fig. 21.3. An outline of the frontal sinuses is drawn

Surgical and Drilling Equipment

Surgical equipment includes a video system (camera and monitor), three endoscopes (4-mm rigid tele-endoscopes: 0°, 30° and 70°), a retractor (double or triple blunt-end hooks), a drilling motor (up to 30,000 rpm), an angled drill hand piece (main handle: 12.5 cm; shaft handle: 8.5 cm), six burrs (sizes 3–5 mm) 6–6.5 cm in length (mastoid burrs are usually too short), a set of FESS instruments, etc.

MAFS Type A

The positions of the patient, medical and nursing personnel and equipment in the operating room are shown in Fig. 21.4. The surgeon sits near the head of the table, and the table for the drilling equipment is on the surgeon's right. The monitor is placed near the foot of the table (approximately 1 ft away).

After planning the location of the incision (Fig. 21.5), 5 cm³ of 0.5% lignocaine/lidocaine with adrenaline (in a dilution of 1:200,000) is infiltrated into the soft tissue at the planned location while the patient is under general anesthesia. An intra-eyebrow or peri-eyebrow incision is then carried out. Electrocautery is used for hemostasis.

Soft tissues (muscles and pericranium) are incised and retracted to provide bone exposure (revealing the AWFS). A 4-mm cutting burr is used to perform a MAFS or a trephination of the AWFS (drill speed: approximately 15,000 rpm). If no entry is achieved after drilling 5 mm deep, one is advised to verify that the location is appropriate. Once the FS is entered, the hole is enlarged and shaped as required; the hole is usually round, oval, triangular with blunted edges or binocular-shaped (Fig. 21.6). An excessively large hole due to an excessively large bony trephination (Fig. 21.1) is likely to result in retraction and a dimpling deformity. In addition, it should be noted that, if the hole is placed too far medially, the DNB cannot be performed accurately, because the opening limits the direction and angle of the burr.

MAFS Type B

If the FS is very small, MAFS type B is chosen, because the location for drilling in a MAFS type-A procedure would be too high (causing the potential for a mini-craniotomy). The location for entry into the FS in the MAFS type-B procedure is medial and inferior (Fig. 21.7) to that for the MAFS type-A procedure. The template technique is recommended before drilling.

Drilling of the Nasofrontal Beak

The dense bone anterior to the apertura frontalis and the FDP is referred to as nasofrontal beak (NB), whose synonyms include the nasofrontal process and the nasofrontal spine (or bony spina nasalis interna). A line drawing of the NB is shown in Fig. 21.8.

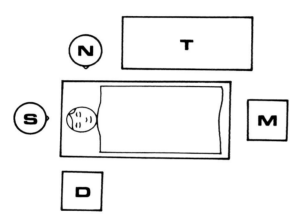

Fig. 21.4. Positioning of personnel and equipment in the operating room. *D*, table for the drilling equipment; *M*, monitor; *N*, nurse; *S*, surgeon; *T*, table for instrument

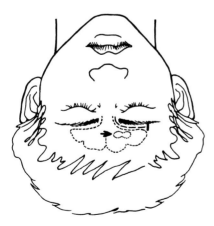

Fig. 21.5. A surgeon's view from the top of the table: the planned location of the incision (*I*), the planned location for mini-anterior frontal sinusotomy (*dotted binocular-shaped outline near the arrow*) and the outline (*dashed/dotted line*) of the frontal sinuses

Fig. 21.6. After skin incision, dissection and retraction of soft tissue, a bony trephination (binocular-shaped) is performed (near *arrow*). An endoscope (*E*) is connected to a video camera (*V*), and a burr secured to a drill hand piece (*D*) is inserted into the frontal sinus

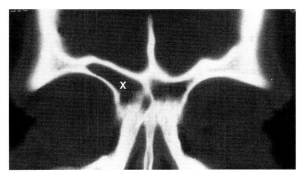

Fig. 21.7. Location (*X*) for entry of mini-anterior frontal sinusotomy type B

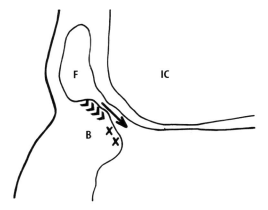

Fig. 21.8. The relationship between a frontal sinus (*F*), a frontal-sinus drainage passage (*arrow*), the nasofrontal beak (*B*) and an intracranial cavity (*IC*). The bony locations for drilling are shown (>>> and ××)

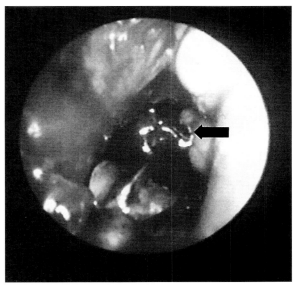

Fig. 21.9. An endoscopic photograph indicating a burr and a frontal sinustome (the *arrow* points to the edge of the sinustome)

Operative Complication

Although there was postoperative loss of sensation [25], it can be argued that this sensory loss at and near the wound should be regarded as an expected sequela rather than a true operative complication.

Comment

During the current decade, most reports have favored endoscopic transnasal/intranasal techniques for the surgical treatment of frontal sinusitis [1, 6, 10, 13, 15, 17, 21–24, 27, 29]. However, there are still opportunities for other procedures, especially in difficult cases (failed FESS cases). MAFS provides good visual and surgical access with optional DNB in an antero-medial and inferior direction (away from the ipsilateral brain and eye). MAFS, ITFS and CFS (all with optional DNB) can result in resolution of frontal sinusitis; examples of pre-operative and postoperative CT films are illustrated in Figs. 21.11 and 21.12.

In this chapter, MAFS with DNB highlights the difference between our approach and those of others. While we accept that many modern surgeons can successfully perform DNB via an intranasal/transnasal approach [1, 6, 10, 23], we have found DNB during ITFS alone to be more difficult than DNB via either MAFS or CFS. A possible explanation may be attributed to our experience on Thai patients, who

The NB can be drilled in one of three ways: (1) entirely via a MAFS, (2) entirely during an ITFS and (3) during CFS (i.e., partly via a MAFS and partly during an ITFS; ;Fig. 21.8). In Fig. 21.9, a burr and a frontal sinusotomy are shown as they appear during an ITFS.

Some operative steps in DNB via a MAFS are illustrated in Fig. 21.10. Suggestions for caution when drilling are listed in Table 21.3. It should be noted that there should be no anterior bony defect of the FDP, because such a defect can result in a space for potential soft-tissue herniation or collapse into the FDP.

Comparison of Various Endoscopic Methods of DNB

Some advantages and disadvantages of various endoscopic methods of DNB are listed in Table 21.4.

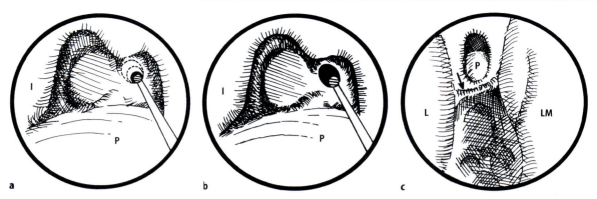

Fig. 21.10. a Drilling bone just anteromedial to the opening of the frontonasal drainage passage (FDP) of the frontal sinus: *I*, interfrontal septum; *P*, posterior wall (table) of the frontal sinus. The drilling is in an anteromedial and inferior direction (away from the ipsilateral brain and ipsilateral orbit). **b** Enlarged upper end of the FDP (*dark area*). **c** Enlarged lower end of the FDP, with a view of the posterior wall (table) of the frontal sinus (*P*), lateral nasal wall (*L*) and middle turbinate (*M*). The *arrow* indicates the position of the anterior ethmoidal artery

Table 21.3. Caution during drilling

Caution during DNB via MAFS
Do not enter the intracranial cavity
Do not drill in a posterior direction to the opening of the FDP (toward the meninges)
Do not drill in a lateral direction to the opening of the FDP (toward the lacrimal sac and orbit)
Avoid drilling out all the mucosa of the FDP
Avoid drilling too far anteriorly, thus drilling the soft tissue of the face
Caution during ITFS
Avoid traumatizing the frontal ostium unnecessarily
Avoid injuring the anterior ethmoidal artery
Avoid entering the orbit (except for drilling bone to identify the lacrimal sac for location of the anterior border of the operative field in accordance with the Fulda concept of endonasal micro-endoscopic frontal sinus surgery, as described by Draf [6])
Avoid aggressively removing agger nasi cells (removal of medial and posterior walls may be adequate)
Avoid operating in a posterior direction to bulla lamella because of the risk of (1) injury to the anterior ethmoidal artery and (2) penetration of the lateral lamella of the cribriform plate
Avoid removing mucosa from the posterior and lateral aspects of FDP

DNB, drilling of nasofrontal beak; *FDP*, frontonasal drainage passage; *ITFS*, inferior transnasal frontal sinusotomy; *MAFS*, mini-anterior frontal sinusotomy

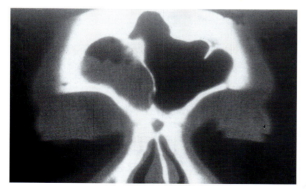

Fig. 21.11. Preoperative coronal computed-tomography film indicating unilateral frontal sinusitis

have Mongoloid skulls; measurements of Thai skulls are found to differ from those of Caucasian skulls [7, 32]. Lothrop's procedure has been performed both microscopically and endoscopically [6] and endoscopically only [1, 10, 23]; we prefer the external approach (via MAFS) to the transnasal/intranasal/endonasal approach.

For sinus surgeons without considerable FS-surgery experience, MAFS is a procedure that can provide good access to the FS, FDP and NB, and may be practiced on cadaver heads before performing it on patients. DNB can help to alleviate the problems of patients with chronic frontal sinusitis, an obstructed FDP and a history of failed FS surgery.

Table 21.4. Advantages and disadvantages of three approaches for endoscopic drilling of the nasofrontal beak

Advantages	Disadvantages
DNB via MAFS	
Easiest of the techniques	Incision
Good endoscopic view	Scar
Drilling away from meninges, CSF, brain and orbit	Sensory loss (at and near wound)
DNB during ITFS	
Good endoscopic view	Extra equipment, e.g., drill and sheath
No incision	Difficult access for burr and telescope
No scar	Difficult technique (for the authors)
	Drilling in a cranial direction (higher risk for the lesser experienced?)
DNB during CFS	
Double access ports	Incision
Good endoscopic view	Difficult access for drilling during ITFS
Double approaches for drilling	Drilling in a cranial direction during ITFS
	Scar
	Sensory loss (at and near wound)

CFS, combined frontal sinusotomy; *DNB*, drilling of nasofrontal beak; *ITFS*, inferior transnasal frontal sinusotomy; *MAFS*, mini-anterior frontal sinusotomy

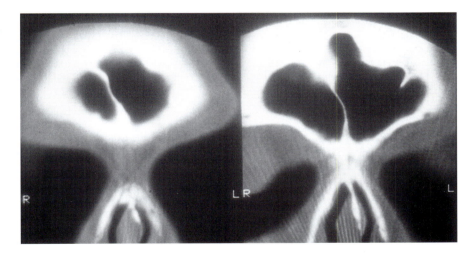

Fig. 21.12. Postoperative coronal computed-tomography film indicating resolution of sinusitis

References

1. Close LG, Lee NK, Leach JL, Manning SC (1994) Endoscopic resection of the intranasal frontal sinus floor. Ann Otol Rhinol Laryngol 103:952–958
2. Donald PJ. Surgical management of frontal sinus infections (1995) In: Donald PJ, Glickman JL, Rice DH (eds) The sinuses. Raven, New York, pp 201–232
3. Draf W (1974) Klinisch-experimentelle untersuchungen zur Pathogenese, Diagnostik und Therapie der chronisch-entzündlichen Kieferhöhlenerkrankungen unter Verwertung der direkten Beobachtung durch Sinuskopie. Postdoctoral dissertation. Johannes Gutenberg University, Mainz
4. Draf W (1983) Technique of paranasal endoscopy. In: Draf W (ed) Endoscopy of the paranasal sinuses. Springer, Berlin Heidelberg New York, pp 4–27
5. Draf W (1983) Therapeutic possibilities of endoscopy of the paranasal sinuses. In: Draf W (ed) Endoscopy of the paranasal sinuses. Springer, Berlin Heidelberg New York, pp 44–58
6. Draf W (1991) Endonasal micro-endoscopic frontal sinus surgery: the Fulda concept. Operative Tech Otolaryngol Head Neck Surg 2:234–240
7. Fooanant S, Kangsanarak J, Sorasuchart A (1995) Surgical anatomy of the nasal cavities and paranasal sinuses in Northern Thai cadavers. J Int Coll Surg Thai 38:26–30
8. Fry TL, Biggers WP, Fischer ND (1980) Frontal sinus trephination: a new technique for office procedure. Laryngoscope 90:838–841
9. Gerber ME, Myer III CM, Prenger EC (1993) Transcutaneous frontal sinus trephination with endoscopic visualization of the nasofrontal communication. Am J Otolaryngol 14:55–59

10. Gross CW, Gross WE, Becker DG (1995) Modified transnasal endoscopic Lothrop procedure: frontal drillout. Operative Tech Otolaryngol Head Neck Surg 6:193–200
11. Halle M (1906) Externe oder interne Operation der Nebenhöhleneiterungen. Berl Klin Wochenschr 42:1369–1372
12. Halle M (1906) Externe oder interne Operation der Nebenhöhleneiterungen. Berl Klin Wochenschr 43:1404–1407
13. Har-El G, Lucente FE (1995) Endoscopic intranasal frontal sinusotomy. Laryngoscope 105:440–443
14. Hoffmann DF, May M (1991) Endoscopic frontal sinus surgery: frontal trephine permits a "two-sided" approach. Operative Tech Otolaryngol Head Neck Surg 2:257–261
15. Hosemann W, Kühnel T, Held P, Wagner W, Felderhoff A (1997) Endonasal frontal sinusotomy in surgical management of chronic sinusitis: a critical evaluation. Am J Rhinol 11:1–9
16. Kisch H (1928) A new frontal sinus operation. J Laryngol Otol 43:407–409
17. Kuhn FA (1996) Chronic frontal sinusitis: the endoscopic frontal recess approach. Operative Tech Otolaryngol Head Neck Surg 7:222–229
18. Lawson W (1991) Frontal sinus. In: Blitzer A, Lawson W, Friedman WH (eds) Surgery of the paranasal sinuses. Saunders, Philadelphia, pp 183–218
19. Lothrop HA (1899) Anatomy and surgery of the frontal sinus and anterior ethmoidal cells. Ann Surg 29:175–217
20. Lothrop HA (1917) The treatment of frontal sinus suppuration. Laryngoscope 27:1–13
21. Loury MC (1993) Endoscopic frontal recess and frontal sinus ostium dissection. Laryngoscope 103:455–458
22. May M (1991) Frontal sinus surgery. Endonasal endoscopic ostioplasty rather than external osteoplasty. Operative Tech Otolaryngol Head Neck Surg 2:247–250
23. May M, Schaitkin B (1995) Frontal sinus surgery: endonasal drainage instead of an external osteoplastic approach. Operative Tech Otolaryngol Head Neck Surg 6:184–192
24. Metson R (1992) Endoscopic treatment of frontal sinusitis. Laryngoscope 102:712–716
25. Muntarbhorn K (1993) Anterior and inferior frontal sinusotomy by drilling with burrs. Am J Rhinol 7:191
26. Ogston A (1884) Trephining the frontal sinuses for catarrhal diseases. Med Chron 1:235–238
27. Rice DH (1993) Chronic frontal disease. Otolaryngol Clin North Am 26:619–622
28. Ritter FN, Fritsch MH (1992) Atlas of paranasal sinus surgery. Igaku-Shoin, New York, pp 174–189
29. Schaefer SD, Close LG (1990) Endoscopic management of frontal sinus disease. Laryngoscope 100:155–160
30. Lothrop HA (1899) Anatomy and surgery of the frontal sinus and anterior ethmoidal cells. Ann Surg 29:175–217
31. Silcock AG (1897) Distension by mucus and empyema of the frontal sinus, with illustrative cases. Practitioner 141:244–254
32. Thanaviratananich S, Sangsa-ard S, Tankongchumraskul C, Chaisiwamongkol K (1996) Surgical anatomy of lateral nasal wall in Northeast Thai cadavers. J Med Assoc Thai 79:177–184
33. Thawley SE, Deddens AE (1995) Transfrontal endoscopic management of frontal recess disease. Am J Rhinol 9:307–311
34. Wigand WE (1990) Operations. In: Wigand ME (ed) Endoscopic surgery of the paranasal sinuses and anterior skull base. Thieme, New York, pp 75–133

Recurrence of Polyposis: Risk Factors, Prevention, Treatment and Follow-Up

Pierre Rouvier and Roger Peynegre

Introduction

Nasal polyposis can be defined as a chronic inflammatory disease of the paranasal-sinus mucosa, leading to a protrusion of benign edematous polyps from the meatus into the nasal cavity [11]. It is generally acknowledged that polyps can recur, even after "radical" surgery and sometimes occur as late as 10–15 years after a previous procedure. There is now a consensus of most authors [2, 6, 7, 11, 15, 19, 22, 28, 35, 36] that treatment needs to involve both medical and surgical aspects: steroid therapy during inflammatory disease and surgery of polypoid disease.

The precise pathophysiology determining polyposis recurrence remains obscure and, despite long-term, sustained medical management, recurrence after surgery still occurs frequently. Because of this, three questions can be posed:
1. Preoperatively, is it possible to identify risk factors for polyp recurrences by screening?
2. Can the identification of certain clinical patterns of recurrence provide clues for successful treatment options of the recurrences?
3. Can a consensus be established regarding what constitutes appropriate follow-up of surgically treated nasal polyposis when recurrence is anticipated?

Risk Factors and Prevention

The precise risk of recurrence is difficult to determine from the literature. If the risk seems relatively modest (between 15% and 18%; Friedman calculated the risk as 15%, Kennedy calculated it as 16%, Eichel calculated it as 17.5% and Wigand calculated it as 18%), it is because polyposis manifests itself through symptomatology. The results are based on analysis of such symptoms and not on endoscopic findings.

The risk of recurrence is greater when it is assessed by authors using objective endoscopic evaluations. Their results are fairly congruent: a recurrence rate of 50–60% (although many recurrences are asymptomatic). Rouvier and Peynegre [28] calculated a 60% recurrence rate at 5 years in 100 ethmoidectomies (40 of these asymptomatic). Fombeur [8] calculated a 53% recurrence rate at 4 years on 132 cases (among which 43% were asymptomatic cases). Strunski [2] calculated a 63% recurrence rate at a mean follow-up of 2 years, on 116 ethmoidectomies (20% of these are "micro-recurrences"). Friedrich (as reported by Strunski) calculated a 53% rate of anatomic recurrences.

We know that nasal polyps represent the nasal manifestation of an unstable respiratory mucosa; the mucosa is the primary site of the disease, rather than a normal structure involved secondarily. This mucosa has no control over the inflammatory reaction, in which eosinophils seem to play a central role [10, 21, 30, 32, 33]. This pathologic process may involve the entire nasal and paranasal-sinus mucosa and may extend into the lower respiratory tract; polyps and asthma may be representations of the same disease. Nasal polyposis is a local manifestation of a general "inflammatory hyper-reactivity" and, if we understand this state of inflammatory hyper-reactivity, we may be able to understand why polyps recur despite medical or surgical treatment.

Causes of Recurrence

Edema is an integral part of the recurrence of polyps, but the pathogenic mechanisms responsible for its formation in respiratory mucosa are numerous and are not completely understood.

The Role of Eosinophils

The presence of eosinophils in the polyps is a dominant histopathologic feature. Their central role in polyposis seems to involve the classic responses of inflammation and edema and some effects (mediated by many factors) on the epithelium and extracellular matrix.

Initially, tissue eosinophilia was thought to be the result of immunoglobulin E (IgE)-mediated allergy, but now there is good evidence that IgE-mediated allergy plays only a minor role in the patho-physiology of nasal polyposis. IgE-mediated allergy certainly doesn't explain the accumulation of eosinophils. Indeed, several studies [10, 30] have shown that allergy occurs no more frequently in individuals with nasal polyps than in normal individuals, and the incidence of positive skin tests is approximately the same for both type of individual. In nine studies involving 287 patients, local specific IgE from polyp secretions or mucosa was detected in only 19% of the patients who did not exhibit systemic allergy.

Actually, IgE-mediated allergy seems to play only a minor role in eosinophil accumulation and, in 1979, Mullarkey and Jacobs introduced a new concept: non-allergic rhinitis with eosinophilia syndrome (NARES). This concept gives the eosinophil a central role in inflammatory processes associated with chronic rhinosinusitis. In it, the accumulation of eosinophils is postulated to be due to proteins (cytokines) and other molecules of adhesion of different origins that attract eosinophils via powerful chemotactic actions.

Among the various cytokines, GM-CSF (granulocyte/macrophage colony-stimulating factor) is the most important and has been detected in nasal polyp tissue (where activated eosinophils are especially numerous) but not in normal nasal mucosa. This cytokine has powerful biologic effects (including the regulation of the survival, proliferation and activation of granulocytes and the differentiation of hematopoietic cells) and could be responsible for eosinophil accumulation. Studies have shown that GM-CSF is mainly produced by eosinophils themselves (autocrine theory), leading to a hypothesis invoking some type of intrinsic eosinophilic inflammatory process.

The Role of Eosinophilia in the Formation of Polyps

The accumulation of activated eosinophils leads to the release of a wide array of cytokines responsible for inflammation and edema. However, eosinophils contribute to nasal-polyp formation and growth both through the classic mechanisms of inflammation and edema and by exerting effects (through the actions of growth factors) on the epithelium and extracellular matrix.

These peptides – insulin-like growth factor [26], TGF (transforming growth factor) [24] and PDGF (platelet-derived growth factor) – have already been detected in nasal polyps. They are produced by inflammatory cells and by epithelial cells themselves; they appear to be mitogenic factors for fibroblasts (by stimulating collagen synthesis) and epithelial cells [11]. Indeed, epithelial cell proliferation is higher in polyps than in "normal" mucosa. Morphological changes, such as secretory hyperplasia and squamous metaplasia, are also frequently observed in the epithelium lining nasal polyps [16, 25]. These findings suggest that modifications of normal epithelial differentiation and proliferation occur in nasal polyps. The epithelium itself could, therefore, play an active role in the pathogenesis of nasal polyposis, just as the inflammatory reactions of the lamina propria do.

There is a balance between upregulating and down-regulating factors, and the same cell type is able to produce both positive (PDGF) and negative (TGF) regulating factors [4, 12, 27]. It appears that epithelial cells and fibroblasts from inflamed tissues may be upregulated and may be causes of a perpetual inflammatory reaction that characterizes nasal polyposis.

The intrinsic factors that cause a predominance of upregulation are not well known but are probably present in diseases of mucociliary function, immune deficiencies, the vasculitides and allergy. Autonomic-nervous-system dysfunction and abnormalities of the degradation of membrane phospholipids (where steroids can block A2-phospholipase and stop the reaction at its initiation) may be even more important. Although external factors are still hypothetical, pollutants and infectious processes acting on the mucosa to damage the superficial epithelial layer could lead to abnormal stimulation of nerve endings in an already evolving neurogenic process.

In summary, many things can initiate the influx of eosinophils. Once present, the eosinophils induce epithelial damage [5, 23], which results in attempts at repair via increased cell proliferation [10–18]. This epithelial cell proliferation is stimulated by growth factors produced either by inflammatory cells or by epithelial cells; the proliferation also causes an increase in eosinophil accumulation. The growth factors secreted by the increased numbers of inflammatory cells (macrophages, eosinophils) and, subsequently, by the epithelial cells and the fibroblasts begin to reduce epithelial, connective and vascular proliferation in a self-regulating way. This leads to the formation and growth of polyps. The inflammatory reaction is upregulated, resulting in a "self-perpetuating" inflammatory state, so the abnormal nasal mucosa becomes a "tumor-like" structure.

Conclusions

We should do everything possible to minimize the accumulation of eosinophils in the sinonasal mucosa. This may justify more radical surgery (including removal of the ethmoid mucosa) in extensive polyposis to eradicate the "local eosinophil pool" and to reduce or slow eventual recurrences. This is also suggested by studies reporting recurrence rates in relation to the extent of polyposis [11, 14, 29]. The overall recurrence rate in a series of 120 patients was 12.3% [14], but the rate depended on the stage of the polyposis; it was 5.8% in stage-II polyposis patients, 16.2% in stage-III patients and 24.2% in stage-IV patients. However, the increasing rates of recurrence could be partially accounted for by increasingly difficult surgery and the increasingly greater likelihood of incomplete eradication of the disease.

However, cortisone decreases local eosinophilia, and it has been recognized that topical steroid instillation causes both polyp regression and a reduction of eosinophilic infiltration. Therefore, long-term steroid management is justified in fighting eosinophilic invasion.

Role of the Host Environment

The majority of authors recognize several environmental factors that affect the evolution of the disease, including heredity, sex, allergy, infection, asthma, aspirin intolerance and stress.

Heredity

An ongoing study by the "ORLI" group of French rhinologists (P. Rouvier, R. Peynegre, E. Serrano, J.M. Klossek, L. Crampette, D. Stoll and A. Coste) is evaluating survey results from 224 patients with polyposis. Of the 224 subjects, 80% had family histories of respiratory disease. Nasal polyposis was present in other family members in 48%: 39% with one member affected, 6% with two members affected and 3% with more than two members affected. A history of asthma was present in 45% of the families, with 34% having only one family member affected, 7% having two members affected and 3% having three members affected. A family history of aspirin intolerance was found in 13% of the subjects.

This strongly implicates heredity as a risk factor for the development of this inflammatory disease, but the study has not determined whether the familial factor modulates the severity of the disease. To date, the study identifies only an increased risk for the occurrence of polyps, not their severity or the risk of recurrence after treatment.

Age

In children, the recurrence of polyposis can be divided into two groups. Severe bilateral, continuously symptomatic polyposis occurs in children with disorders of the epithelium itself, such as mucoviscidosis and immotile-cilia syndrome, and occasionally occurs in those with a congenital immune-deficiency syndrome. The indications for surgery are rare in these cases (10–20%), with high rates of recurrence (especially in mucoviscidosis), leading some authors [1] to state that all treatments featuring surgery, though transiently beneficial, are never curative.

The other group of children have what is called primary polyposis, which is analogous to nasal polyposis in adults (except for a 24% incidence of associated asthma, which is higher than the incidence in adults [3]). In children, recurrences can be partly explained by the evolutive nature of polyp disease. In addition, endoscopic sinus surgery is technically more difficult, as are postoperative examinations and debridement, resulting in less control of healing [20, 31].

Sex

Although polyps occur more often in males than in females (in a ratio of 3:1), it is not clearly established whether the rate of recurrence is higher for one sex or the other. However, such a dramatic difference in occurrence by sex suggests a hormonal/environmental risk factor, as do the clinical observations that polyps seem to occur more frequently during menopause and during pregnancy in women and during periods of stress in men.

Stress

Stress could initiate dysfunction of the autonomic nervous system as an initial step in polyp formation. The isoprenaline test has demonstrated a hyperadrenergic state in the nasal mucosa in 50% of individuals. Such a state could induce local vasoconstriction, with tissue anoxia causing local mast cell degranulation. Ultimately, this could lead to release of histamine, substance P and eosinophil chemotactic factor A. Thus, dysautonomia would be followed by neurogenic inflammation, followed by a need for eosinophils capable of releasing new toxic substances.

Allergy

For many years, allergy has been considered to be the cause of primary polyposis, because allergic rhinitis, polyposis and asthma all feature hyper-eosinophilia and have some symptomatic analogies. Even though patients with nasal polyposis do not display a greater frequency of allergies than the general population (15–19%, depending on the authors), it is obvious that, in those patients who do have associated allergies, the inflammatory reaction (including the eosinophilic involvement) triggered by allergenic exposure on unstable respiratory mucosa can explain some relapses of active disease.

Asthma and Aspirin Idiosyncrasy

Asthma is the factor most frequently associated with nasal polyposis, with a 25–30% concordance cited by most studies, but the coincidence increases to 50% if only obstructive grade-IV polyposis (recurrences) is considered [28, 29]. This rate may even reach 75% in steroid-resistant cases. Asthma is also the most ominous associated factor, especially if associated with aspirin atopy. Larson and Toss [18] compared 96 patients without asthma, acute recurrent or chronic sinusitis, acetylsalicylic-acid intolerance or allergy with 84 patients who did have these characteristics. After a minimum follow-up of 4 years, the patients with the associated conditions required more polypectomies and topical-steroid treatments. We have noted [28] that, in two equivalent groups of patients operated on because of polyposis and followed for 5 years (51 patients had asthma and 49 did not), nasal symptoms and the use of systemic steroids were always higher in the asthmatic group. A greater number of steroid-dependent patients with the disease were also present in the asthmatic group.

Asthma, as a marker of the ultimate stage of inflammatory respiratory disease, thus appears to be an indicator of increased risk of recurrence. However, this risk does not constitute a contraindication for surgery [2]. Some results [11, 28–29] even seem to suggest some degree of improvement of asthma during the postoperative period, although it is difficult to distinguish how much improvement was related to the surgery and how much was due to postoperative steroid therapy and follow-up.

Infection

Long considered to be a principal risk factor, local infection is now largely regarded as the result of obstruction and sinusal retention. However, it plays an important part in cystic fibrosis and Kartagener syndrome, where numerous neutrophils (and, sometimes, intracellular bacteria within the neutrophils) can be found.

Work in progress from the "ORLI" group (mentioned above) demonstrates that, out of 224 cases of polyposis, 20% appeared to be triggered by an episode of upper-respiratory-tract infection. A direct study of 180 cases of newly-diagnosed polyposis, seen between 1984 and 1990 [12] and initially treated with surgical polypectomy and topical-steroid therapy, has demonstrated that 62 (34%) of cases required one or more further operations during the subsequent 5 years. These surgical procedures were almost always performed during winter months, when upper-airway infections are more common, and suggests that bacterial and viral infections have roles triggering factors in the recurrence of polyposis. Thus, to some degree, bacterial or viral infections are triggering factors in the recurrence of polyposis.

Role of Local Anatomical Factors

The majority of authors believe that nasal polyps originate in the ethmoid sinuses; in fact, all polyps are associated with some degree of ethmoid sinusitis. Surgical endoscopic observations consistently show that nasal polyposis almost always originates from the ethmoid mucosa; the maxillary, frontal and sphenoidal mucosa are less frequently involved (at least in the beginning).

Serial endoscopic and computed tomography (CT) scan exams of some patients have allowed reconstruction of their anato-clinical presentation. We have seen cases of "obstructive rhinitis without polyps" associated with a variable degree of opacification of the ethmoid sinuses (for instance, with NARES). Indeed, this constitutes the beginning of polypous disease. With the passage of time, the ethmoidal opacification progresses as the symptoms progress; this completes the presentation of a typical case of nasal polyposis.

We are of the opinion that the inflammatory edema initiated by a poorly controlled inflammatory reaction early on develops into a congestion of the intricate architecture of the ethmoidal labyrinth; the mucosa is thin and easily peeled off by edema, which quickly fills the small cells (particularly in the anterior ethmoid sinus). Although initially contained in the ethmoid itself, this edema eventually progresses via the natural ostia. Whether it is from mechanical strangulation at the level of the ostium, the weight of the diseased tissue, the Venturi effect (related to obstruction of individual ethmoid cells), infection or retention of secretions, a vicious cycle of "self-perpetuating polyposis" is established. The more elaborate the ethmoid cell system (Figs. 22.1, 22.2) [34], the more difficult it is for the body to contain the problem. Any developmental lesions (such as a deviated septum or a concha bullosa) or iatrogenic synechia and stenosis can aggravate the process of polyp growth.

This pathophysiologic concept emphasizes the importance of the initial surgical approach to the disease and argues in favor of complete ethmoid exenteration once the decision for surgery has been made after failure of medical management, especially in recurrent polyposis. However, without question, it is also true that the origin of the polyps is in the nasal mucosa of the anterior and middle meati in their lateral aspects, and polyps only secondarily invade the sinuses [17].

It is easy to understand that nasal inflammation, allergy or any inhaled particle that is transported to the osteomeatal complex can result in edema at the osteomeatal complex and can further reduce the space and perhaps completely obstruct the middle meatus. This will lead to stasis of secretions in the middle meatus and to infection of the sinuses, initiating a vicious cycle; the inflammation thus becomes chronic. Pathologic secretions, even in response to low-grade inflammation, can become inspissated in the osteomeatal complex and can result in further edema via a direct toxic influence on the mucosa. Epithelial necrosis and polyp formation can result. Such a pathophysiologic concept would argue in favor of surgery on an "as-needed" basis; surgery would be limited to obstructive lesions and would be performed as soon as irreversible lesions are detected (in order to avoid progression to extensive polyposis).

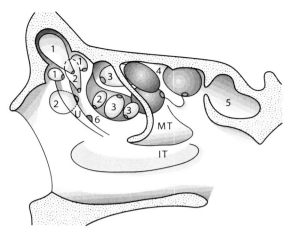

Fig. 22.1. Schematic arrangement of ethmoidal cells, according to Terrier [34]. *1*, Meatal group of cells; *2*, bullar group of cells; *3*, uncinate group of cells; *4*, posterior system; *5*, sphenoid sinus; *6*, maxillary ostium; *IT*, inferior turbinate; *MT*, middle turbinate; *U* uncinate process

Fig. 22.2. Sagittal section of the ethmoid sinus (cadaver dissection)

The Controversy

Supporters of the ethmoid origin of polyposis believe that surgery must be extensive and exhaustive. Supporters of the nasal origin of polyposis believe that surgery must be confined to irreversible lesions of the ethmoid and middle meatus and should be performed to preserve the mucosa and the middle turbinate.

The choice is difficult, but it seems that everyone recognizes the deleterious role of congestion (either at the level of the meatus or the sinus) due to inflammation at the nasal–ethmoid junction. The logical extension of that realization is that, if such congestive lesions cannot be reduced medically, only surgical debridement at this level will overcome the established vicious circle. It seems logical to believe that a limited surgical approach explains, at least in part, some cases of recurrence.

The Role of Surgery

Any surgeon confronted with a postoperative recurrence of polyp disease must eventually review his surgical intervention. Was it indicated? Was the exenteration too limited? Did surgery contribute to the recurrence? These questions tend to raise doubts about the efficacy of surgery and its possible role in the reactivation of the inflammatory process.

Surgery is always justified in extensive cases of polyposis that are refractory to vigorous medical treatment and are subjectively poorly tolerated. Proof of the appropriateness of surgery in such cases is provided by a study [28] that we conducted between 1982 and 1992 on 22 cases of debilitating polyposis referred for surgical intervention. Initially, and in a systematic manner, all the patients were treated medically with a 9-day pulse of steroids (prednisone 1 mg/kg/day), followed by topical beclomethazone (200 µg/day continuously). This was followed by systemic steroid treatment for 5 years, as needed, in cases of recurrence. At the end of 5 years, confronted with persistence of rhinologic symptomatology, all cases were operated on and followed for another 5 years (with the same continuous treatment with topical beclomethazone and systemic steroid treatment, as needed). We have discovered that (Fig. 22.3):
- After 5 years of medical management alone, rhinologic symptomatology was severe in 80% of cases. Persistence or recurrence of nasal polyposis was present in all cases and, in 90% of these, it reached grade IV. The consumption of steroids was an average of 160–300 mg/month in 50% of cases and was more 300 mg/month in 25% of cases.
- Five years after surgery, the nasal symptomatology was severe in 42% of cases, and the recurrence of nasal polyposis was 60%. The consumption of steroids was an average of 80–160 mg/month in 23% of cases and was 160–300 mg/month in 7% of cases.

We conclude that, even if surgical intervention does not resolve the problem of recurrence, it seems to delay it and clearly diminishes the need for systemic steroids.

We also believe that, from the onset, the surgery must be as complete as possible. In support of this idea is a retrospective study [11] of the functional results of two series of cases of diffuse nasal polyposis treated surgically by two different surgeons with the same background. A well-trained but less experienced surgeon treated 37 cases by classic ethmoidectomy, and 39 patients were operated on in an aggressive manner (nasalization) by an expert surgeon. The results (Fig. 22.4), assessed on a scale of ten points (10=no symptoms) demonstrated a mean score of 8.8 for the nasalization group and 5.9 for the ethmoidectomy group. The improvement in asthma noted in the two groups was similar, as was the consumption of steroids by the asthma patients. The need for steroids was significantly lower in the nasalization group than in the ethmoidectomy group. The conclusion that can be drawn, and which was clearly made by the author, R. Jankowski [11], is that "when dealing with nasal polyposis, the more radical the surgery, the better the functional results".

Some Inappropriate Techniques are the Source of Recurrences, Often in Atypical Fashions

If nasal polyposis affects the entire ethmoid sinus, any form of limited surgery will lead to disappointment. As mentioned above, classic ethmoidectomy is ultimately less effective than nasalization. For the same reason, a simple polypectomy or debridement of the anterior ethmoid will be even less effective. Furthermore, such a limited debridement usually preserves a non-stable middle turbinate, where synechia and stenoses frequently compound the situation restrict the middle meatus (Figs. 22.5, 22.6).

However, unnecessary surgery is to be avoided. Middle meatotomy is an indispensable part of the procedure, but inferior meatotomy offers no advantage, and the frequent protrusion of polyps at this level serves to prove this point (Fig. 22.7). Recurrence of nasal polyposis through the inferior meatus often causes early symptomatology that is more difficult to manage.

Inferior turbinectomy seems to represent an "unfortunate" gesture with serious and irreversible consequences. Not only does it have no impact on the recurrence of nasal polyposis, it induces iatrogenic sequela of permanent, crusty rhinitis, particularly when associated with wide ethmoid debridement. In rare cases, a "careful" amputation of the hypertrophic anterior aspect of the turbinates may prove useful in the improvement of the airway and for access of topical medication (Figs. 22.8, 22.9). In the vast majority of cases, however, one should avoid removing the inferior turbinate to any significant degree. If such a turbinectomy has previously been

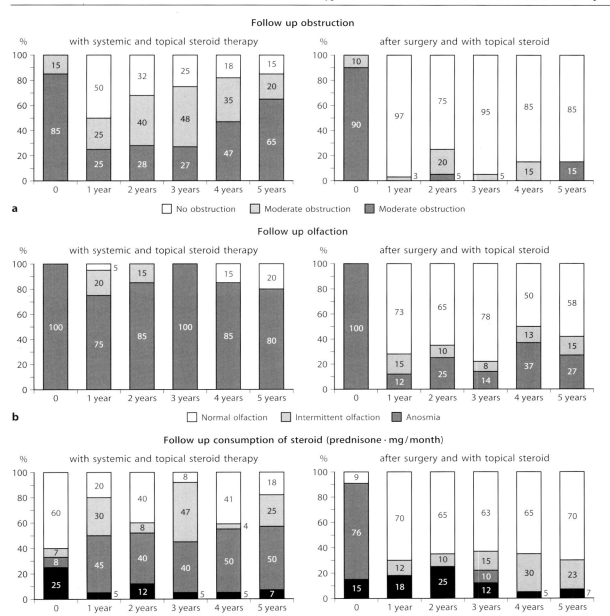

Fig. 22.3. Post-operative evolution of 22 cases of steroid-treated and postoperative nasal polyposis. **a** Evolution of nasal obstruction. **b** Evolution of olfactory function. **c** Evolution of steroid consumption

performed, there is a significant risk of development of a crusty type of rhinitis if the ethmoidectomy is too exhaustive. Such cases should only be treated through limited ethmoidectomy, with preservation of the middle turbinate, even if this increases the risk of recurrence.

The Role of Medical Management

All the statistics indicate that the only effective medical management is the use of steroids, either systemic or topical. A study (by Clement [36]) of 22 patients with massive nasal polyposis, treated over 4 days with 60 mg of oral prednisone and followed by progressively tapered doses (a reduction of 5 mg/day), demonstrated that 72% of the patients showed subjective improvement due to the involution of polyps in the nasal cavity. However, on CT

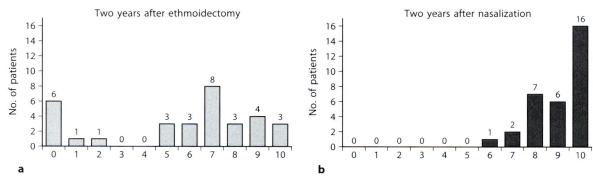

Fig. 22.4 a,b. Comparison of functional results [11]: **a** After classic ethmoidectomy surgery; **b** after a nasalization procedure

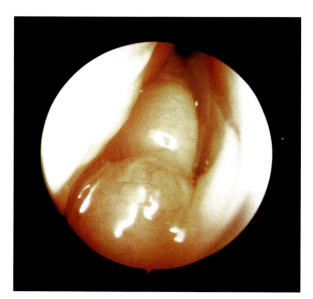

Fig. 22.5. Post-ethmoidectomy polypoid middle turbinate occluding the middle meatus

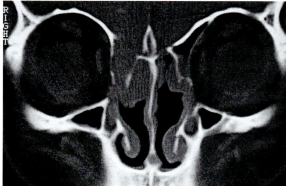

Fig. 22.6. Same patient as in Fig. 22.5. Coronal computed tomography

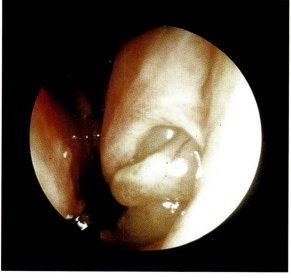

Fig. 22.7. Extrusion of polyps through a left inferior meatotomy

scans of the paranasal sinuses, only 52% showed definite improvement. The benefit is of short duration because, within 5 months after successful oral steroid therapy, there is a strong tendency of recurrence. That is when problems arise. Repeating the treatment at intervals of less than 4 months increases the risk of iatrogenic sequela due to the drug.

Lildholdt [19] reports on 126 cases of polyposis treated for a period of 1 month with 400 µg of budesonide powder or placebo. The increase in expiratory peak-flow index was about 60% in the actively treated group, as opposed to 16% in the placebo group. The overall assessment of treatment efficacy at 1 month showed a success rate of approximately 82%

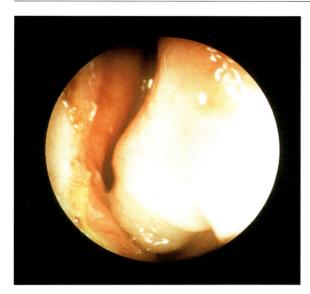

Fig. 22.8. Hypertrophy of the left inferior turbinate

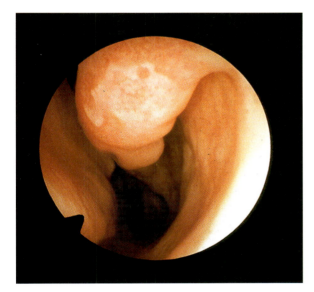

Fig. 22.9. Reduction via limited turbinoplasty

in the actively treated patients, as opposed to 43% in the placebo group. Kanai et al. [14] showed that the proportion of activated eosinophils was significantly lower in polyps than in patients treated with topical nasal steroids. The efficacy of topical-steroid therapy is undeniable, and this management option causes essentially no long-term iatrogenic consequences.

The combination of topical and systemic steroid therapies is even more effective. We followed (Fig. 22.3) [28] 22 patients with massive nasal polyposis who were treated for more than 5 years with 60 mg of oral prednisone over 7 days, followed by topical beclomethazone. Therapeutic efficacy, evaluated by means of symptom assessment, showed that, as early as 1 year after starting treatment, 50% of patients recovered normal nasal patency and (in 25% of patients) normal olfactory function. At 1 year, however, these rates drop to 50% and 10%, respectively.

A systemic "flash" of high doses of corticosteroids (which we call a "medical polypectomy") is used first and sometimes reduces polyposis enough to allow control of the disease through topical-steroid management. In cases of failure, preoperative systemic steroid therapy considerably facilitates the surgical procedure [36].

The combination of systemic and topical steroid management is certainly the most effective counterpart to surgery. Its use with surgery that opens widely the ethmoid sinuses, permitting access of topical agents, further enhances the results by delaying recurrences in a large number of cases. It should be considered an adjunct rather than an alternative to surgery.

Conclusions

Even if some progress has been made in understanding the mechanisms of the origin of nasal polyposis, we still are confronted by our ignorance of the "prime movers" of this state of inflammatory instability. We do not know why, at some specific date or age or under certain circumstances, the reaction develops and chronic inflammation begins. M. Wayoff postulated an evolution in two stages; the first ("sthenic") phase shows only edematous mucosa and is followed by a second ("asthenic") phase, where polyposis starts and, in some cases, recurs despite the treatment used.

We believe that, in many cases, it is possible to predict and possibly influence the evolution of recurrent nasal polyposis through a search (and if possible, an elimination) of all the possible risk factors and by a treatment that judiciously combines steroids and surgery. We admit, however, that there are some cases where it seems impossible to stop seemingly hopelessly recurring nasal polyposis.

Clinical Presentations of Recurrence and their Treatment

The recurrence of polyposis is characterized by the presence of polyps in the nasal cavity after previous surgical removal of polyps. The anatomic landmarks are often absent, particularly the important middle turbinate and uncinate process. However, even more significant is the severe damage that is sometimes imposed on the ethmoid cavity, particularly the medial wall of the orbit, the ethmoid roof, the inferior turbinate and the lateral wall of the inferior meatus. Loss of landmarks, breaks in the bony continuity and severe distortion are typical of recurrent polyposis and necessitate that, for any secondary surgical approach, one must have:

- Precise documentation of any data pertaining to previous surgical procedures (hematoma, hemorrhage, diplopia, headaches or clear rhinorrhea)
- A detailed assessment of previous operative records
- A high-resolution CT scan of the bony architecture of the ethmoid sinus, in two planes

The extent of recurrent disease, its clinical progression, and the information mentioned above, will allow distinction among three forms of recurrence, for which a specific treatment can be recommended:
- Limited recurrence
- Widespread recurrence
- Iatrogenic recurrence in the nasal fossa

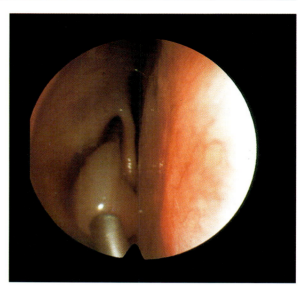

Fig. 22.10. Limited recurrence, with a single polyp in the right middle meatus

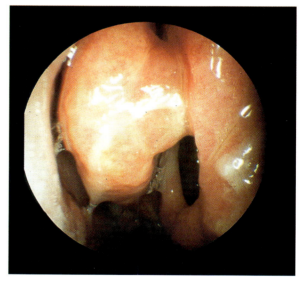

Fig. 22.11. Limited recurrence on the middle turbinate

Limited Recurrence

Clinical Presentations

Limited recurrence usually takes place early in the first two postoperative years, sometimes within a few months. They can be spontaneous and progressive (related to the natural history of the disease) or can be suddenly triggered by an aggressive inflammatory process (upper-respiratory-tract infection, exposure to an allergen, trauma, etc.).

When they are asymptomatic, these limited forms of recurrence are discovered during the systematic postoperative follow up. Sometimes, these polyps are uncomfortable and actually lead to the consultation. They present in two clinical patterns.

Localized polyposis (Figs. 22.10–22.12) can recur anywhere in the ethmoid sinus but most commonly presents anteriorly (sometimes on the head of the middle turbinate) and only rarely appears posteriorly or at the middle level of the ethmoid sinus. A specific but very uncommon form occurs when the polyps emerge from the middle meatotomy window or, occasionally, from an inferior meatotomy. Sometimes, there is a single polyp with an edematous, sessile base. In those cases, the maxillary sinus is the origin of dense polyposis with purulent secretions.

Polypoid edema (Figs. 22.13, 22.14) is a more common type of recurrence. It invades the ethmoid sinus from anterior side to the posterior side, extending toward the ostia of the anterior sinuses (maxillary and frontal). This edema is the first stage of widespread recurrence (which is discussed later).

Treatment

Even though symptoms may be minimal, these limited recurrences must be aggressively treated. However, there is a risk of being overly aggressive by surgi-

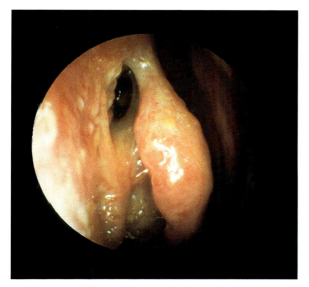

Fig. 22.12. Limited, multifocal recurrence in the right middle meatus

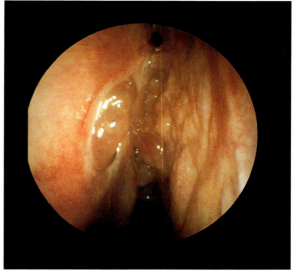

Fig. 22.14. Left anterior ethmoid polypoid edema threatening the frontal ostium

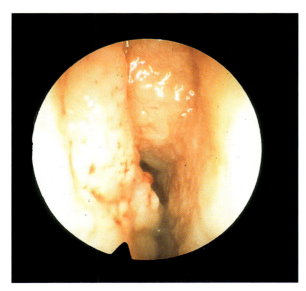

Fig. 22.13. Polypoid edema of the left ethmoid cavity

cally treating an asymptomatic lesion, thus aggravating the inflammatory process so that the diseased mucosa exerts control only poorly. The risk/benefit ratio, therefore, necessitates caution.

There are three therapeutic modalities:

- **Systemic steroid therapy ("medical polypectomy")**. The route of administration is unimportant. The treatment must be of short duration and must feature an "effective" dose. For this "steroid flash", we usually prefer prednisone because of its short-term action and oral administration. We use 1 mg/kg/day (in a single morning administration) for 8 days. Antibiotics are always given concomitantly because of the risk of associated infection.
- **Surgical polypectomy**. The polyps usually are removed under local anesthesia with a snare or forceps.
- **Laser polypectomy**. Those who use a laser in the surgical treatment of the naso-sinusal polyposis often find it most useful for the failures after classical surgical treatment.

When there is a small single polyp in the anterior ethmoid sinus, 200–400 J is enough to eradicate the recurrent polyp; local anesthesia is sufficient. When there are recurrences of large numbers polyps, as is frequent with Widal's disease and cystic fibrosis, the diseased mucosa is first removed with the forceps. Next, the bone is exposed, and all the sinuses are opened. Photocoagulation is then performed to destroy the remaining pieces of the mucosa. In the anterior ethmoid region, the energy released must be reduced to avoid damage to the adjacent dura mater.

Recommended Therapy

When there are no contraindications to systemic steroid therapy, the treatment should begin with oral steroid therapy for 1 week. If the lesion regresses well, the treatment is continued with topical beclomethazone steroid therapy; if a persistent large polyp will impede topical steroid therapy, we remove it with a laser.

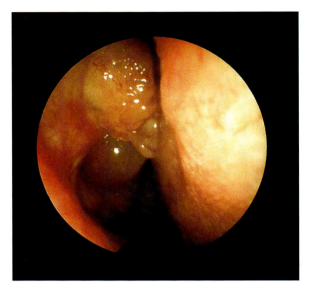

Fig. 22.15. Extensive recurrence in the right ethmoid cavity

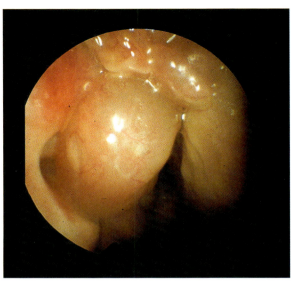

Fig. 22.16. Same case as in Fig. 22.15. Close-up of the posterior region

When there is a contraindication to systemic steroid management, we remove the polyps first (either with surgery or a laser) and proceed to topical steroid management.

Widespread Recurrence

In cases of widespread recurrence, the entire ethmoid cavity is filled by polyps, which overflow into the nasal cavity and obscure the anatomic landmarks (or what remains of them; Figs. 22.15–22.17). Edematous tissue, synechiae between the anatomic structures and the polyps, and purulent sinus secretions are frequent. The challenge of surgery in these cases relates to the difficulty in establishing the anatomic landmarks and the tendency of these recurrent lesions to bleed more than primary cases do.

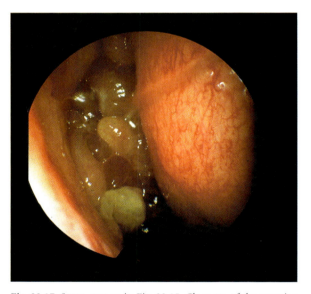

Fig. 22.17. Same case as in Fig. 22.15. Close-up of the anterior region

Treatment Approach

Such widespread recurrences of polyposis are best addressed by a thorough exenteration of the ethmoid cavity, resulting in its "nasalization", a concept defined and recommended by Jankowski [11] and by Katsantonis [15] under the name of "marsupialization". The aim of the nasalization procedure is to perform a total and radical ethmoidectomy, removing all the bony lamellae and mucosa within the labyrinth (Fig. 22.18). The procedure is performed in a systematic way and includes the following steps.

The Insertion of Lemoyne's Nail

These "frontal nails" are made of small, metallic tubes 20 mm in length and 1 mm in diameter, with a buttress that limits their penetration into the frontal sinus. They can be connected to a syringe (Figs. 22.19, 22.20).

An adequate radiologic assessment of the frontal sinuses (standard Caldwell view) is required (Fig. 22.21). It is preferable to obtain CT-scan images for this specific purpose, though they are difficult to interpret. If the frontal sinuses reach more than 7–8 mm above the supra-orbital ridge, the insertion of the nails should not cause problems, even in relatively small frontal sinuses. Access to the sinuses is achieved with a hand drill (which we prefer to an electric drill because of the better control it gives; Fig. 22.22) and a guarded bit of variable depth (to avoid injury to the posterior wall of the frontal sinus).

A line is drawn on the surface of the forehead between the medial ends of the eyebrows (Figs. 22.23, 22.24), which roughly correspond to the supra-orbital foraminae. A line in the midline represents the frontal-sinus septum. The point of penetration should be oriented on the horizontal line approximately 8–10 mm on each side of the midline.

A very typical sensation of "give" is perceived during the drilling when the anterior wall of the frontal sinus is penetrated and "gives" way; the soft tissues are stabilized and the drill bit removed before inserting the frontal nail. Each nail is then connected to a syringe filled with 20 cm^3 of sterile water, allowing intra-operative transcutaneous irrigation of the frontal sinus and easy identification of the frontal nasal duct during the endoscopic procedure.

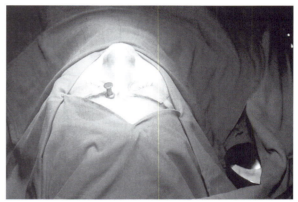

Fig. 22.20. Frontal nail in position in the left frontal sinus

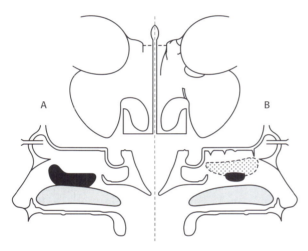

Fig. 22.18 a,b. Schematic representation of differences between nasalization (a) and ethmoidectomy (b) [10]

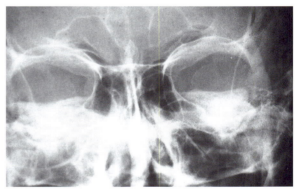

Fig. 22.21. Standard X-ray. Caldwell view

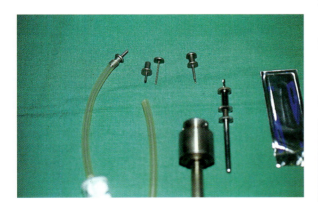

Fig. 22.19. Instrumentation for frontal-nail insertion. Trephination bit and drill, nails (with introducer) and a tubing-irrigation kit

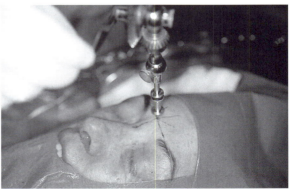

Fig. 22.22. Trephination with a hand-held drill

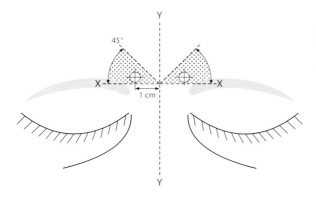

:::: Aire de projection des sinus frontaux
⊕ Point d'insertion du clou frontal
X---X Ligne inter-sourcilière
Y---Y Ligne médiane

Fig. 22.23. Diagram for site selection for insertion of the frontal nail. *Dotted area*, projection of the frontal sinus; ⊕, point of insertion of the frontal nail; *X–Y*, line between the eyebrows; *Y–Y*, median line

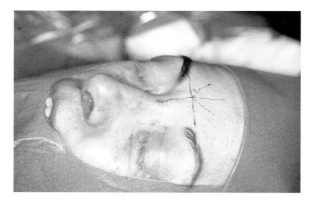

Fig. 22.24. Preoperative cutaneous landmarks

Cleaning of the Nasal Fossa and Middle Meatus

The surgery starts with a delicate polypectomy to find the middle turbinate or any remnant of it as the first key landmark. It is preserved until the end of the procedure to provide medial protection of the cribriform plate.

Middle Meatotomy

The second key landmark, the vertical contour of the lacrimal structures, is always intact and is fairly easy to distinguish anteriorly. Behind the lacrimal duct is the uncinate process, which is the third landmark. It can usually be found (especially its upright part).

Even in patients who were over-operated previously for severe polyposis and had removal of the uncinate process in an anterior-to-posterior fashion, a large middle meatotomy that enlarges the natural maxillary ostium is created. If no landmark is noted, the antrostomy is started in the fontanelle area. The middle antrostomy is enlarged as far as possible, reaching the shoulder of the inferior turbinate inferiorly, the palatine bone posteriorly, the bony lacrimal duct anteriorly and the roof of the maxillary sinus superiorly.

Anterior Ethmoidectomy

The roof of the maxillary sinus is the fourth key landmark and is used to find and dissect the medial wall of the orbit, which is the fifth key landmark. The anterior ethmoidal cells between the middle turbinate and the medial wall of the orbit can be exenterated safely until the ethmoid roof (the sixth key landmark) is identified. Because the anatomy of the anterior ethmoid is unpredictable [34], perioperative irrigation allows easy identification of the fronto-nasal duct (Fig. 22.25). Exenteration of the anterior ethmoid sinus is then completed, and the frontal ostium (the seventh key landmark) is opened widely into the cavity (fronto-meatotomy or frontotomy).

Posterior Ethmoidectomy

As many posterior ethmoidal cells as possible are removed by following the medial orbital wall, the ethmoid roof and the middle and superior turbinates. The risks posed by posterior ethmoid anatomy variations, however, are less if the dissection is restarted from the sphenoid-sinus lumen. To enter the sphenoid sinus safely, the endoscope is first placed into the rhinopharynx to identify the floor of the sphenoid bone (the eighth key landmark) and moved backward into the nose to identify the anterior face of the sphenoid (the ninth key landmark) and the septum (the tenth key landmark). The anterior face is then punctured close to the midline, and a large sphenoidostomy is achieved. The middle turbinate usually needs to be resected at this stage to allow better access to the sphenoid sinus and to complete posterior ethmoid dissection.

Finally, ethmoid exenteration is meticulously and patiently achieved by carefully removing all residual cells, bony lamellae and mucosa left on the orbital wall and ethmoid-sinus roof. Mucosa of the maxillary, sphenoid and frontal sinuses are left intact. Only irreversible lesions, such as intramucosal abscess, cholesterol cysts or large polyps, are removed (Figs. 22.26–22.30). At the end of these pro-

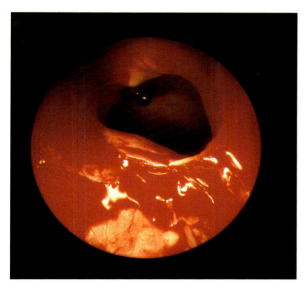

Fig. 22.25. Perioperative view of the tip of the nail through a frontotomy

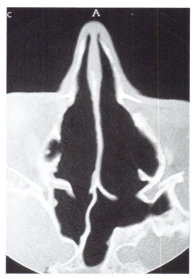

Fig. 22.27. Nasalization cavity (axial computed tomography view at the level of the ethmoid cavities)

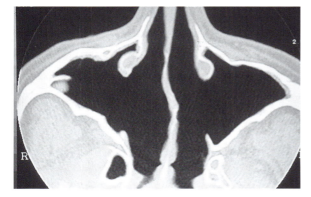

Fig. 22.26. Nasalization cavity (axial computed tomography view at the level of the maxillary sinuses)

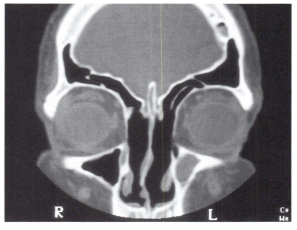

Fig. 22.28. Nasalization cavity (coronal computed tomography view at the level of the anterior ethmoid and frontal sinuses)

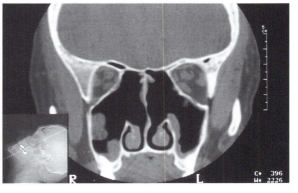

Fig. 22.29. Nasalization cavity (coronal computed tomography at the level of the posterior ethmoid sinuses)

cedures, all sinuses open into a large, reshaped nasal cavity; this is called "nasalization" of the paranasal sinuses.

For certain cases, the "external approach" remains an alternative to endonasal surgery. Since these are very well known, the description is not repeated in this chapter.

The use of microscopes and endoscopes in the approach to the frontal and maxillary sinuses makes it possible to reduce the size of the incision, and we currently perform "mini-Caldwell-Luc" or "mini-external frontotomy". Despite these improvements, the external approach to the ethmoid sinuses necessitates a large bone trephination without making the operating procedure substantially easier; this is why we believe the endonasal approach is still the best choice for these cases.

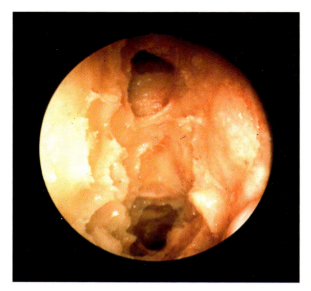

Fig. 22.30. Endoscopic view of left nasalization cavity. The frontal sinus is located superiorly and the sphenoid sinus is located inferiorly

Indications

Two situations are to be addressed. Most cases of widespread ethmoid-polyp recurrence do not feature significant disease in the maxillary or frontal sinuses. The goal of the traditional ethmoidectomy was never complete exenteration of the ethmoid cavity; therefore, it does not lead to large recurrences. Too often, it leaves a diseased middle turbinate, strands of diseased mucosa or non-exposed cells. This procedure is not as thorough in its exenteration of the ethmoid sinus and leads to the more pervasive and dangerous forms of recurrence. Only thorough debridement of the ethmoid sinus through "nasalization" can appropriately address diffuse recurrences, because this technique more completely removes the diseased mucosa and minimizes the risks associated with the debridement.

Occasionally, we encounter cases with "compounded" ethmoid disease, where there is also involvement of the maxillary and/or frontal sinuses. These situations are somewhat easier to approach via an external approach. There is no reason not to complete an endonasal nasalization procedure via a "mini-Caldwell" approach in order to clear a sinus filled by polyps, retention cysts or mycotic debris. The same is true for the frontal sinus, which can be accessed through a mini-external approach, especially for severe recurrent cases with stenosis or suppuration, which are often secondary to postoperative iatrogenic changes form the first surgery.

Recurrences Compounded by Iatrogenic Factors

Sometimes recurrences are associated with iatrogenic lesions, which compound the recurrence of polyp disease and subsequent surgical intervention. We distinguish three categories:
1. Synechia that form in the nasal fossae interfere with any surgical approach.
2. An inferior septo-turbinal synechia is always more extensive (in depth) than it appears, but it poses little risk (except for recurrence).
3. However, extensive synechia of the middle meatus (Figs. 22.31, 22.32) from postoperative lateralization of the middle turbinate is much more significant; it often extends to the frontal recess.

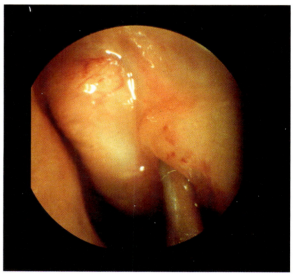

Fig. 22.31. Extensive synechia of the left middle meatus (from lateralization of the left middle turbinate)

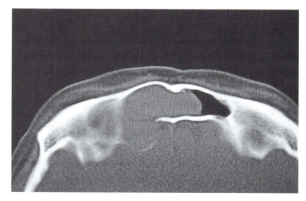

Fig. 22.32. Computed-tomography scan of the same case as that shown in Fig. 22.31. Left frontal mucocele

Stenosis

Stenosis of the frontal recess (Figs. 22.33–22.37) is the most common and the most serious. Its extent ranges from a simple low stenosis of the naso-frontal duct to complete obstruction with frontal mucocele formation. The most common clinical feature is dull, persistent frontal pain that is worse than the headaches that are usually encountered in non-complicated polyposis.

Such stenoses are not predictable but most often occur in cases that have had wide antero-superior ethmoid-sinus debridement in the area of the naso-frontal duct. Therefore, we always advise that the surgeon inspect the mucosal cuff at that site and avoid manipulation of the frontal sinus, even in procedures culminating in nasalization. In the anterior ethmoid sinus, the free flow of irrigating fluids from the frontal nail should identify the extent to which the naso-frontal duct has been reopened.

Although rarer, stenosis of the sphenoid ostium (Figs. 22.38, 22.39) must be suspected when there are severe retro-orbital headaches radiating to the vertex and in patients who have opacity of the sphenoid sinus. These generally result from incomplete dissection of the anterior aspect of the sphenoid sinus, which should always be widely opened. The possibility of progression to mucocele formation or more severe neurologic complications mandates early treatment.

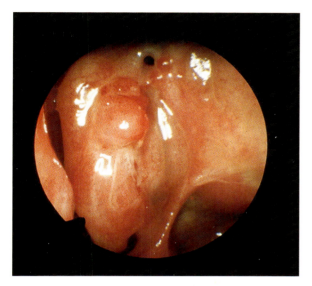

Fig. 22.33. Partial stenosis of the right frontal recess

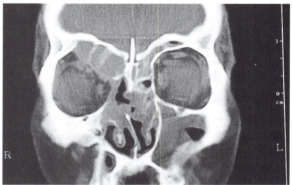

Fig. 22.35. Extensive recurrence. Computed-tomography scan demonstrating a right fronto-ethmoid mucocele

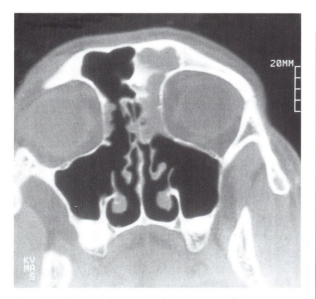

Fig. 22.34. Computed-tomography scan view of the same case as that shown in Fig. 22.33

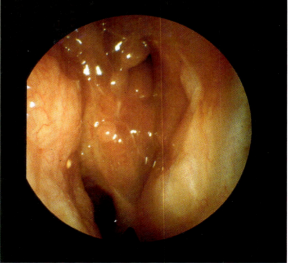

Fig. 22.36. Same case as in Fig. 22.35. Left nasal endoscopy

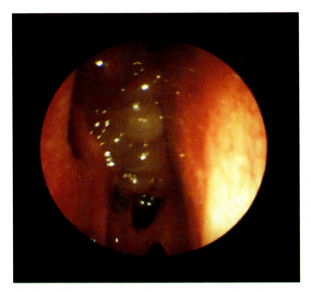

Fig. 22.37. Same case as in Fig. 22.35. Right nasal endoscopy

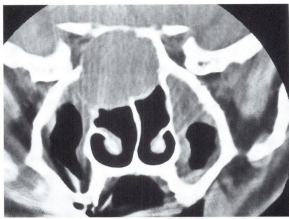

Fig. 22.39. Same case as in Fig. 22.38. Frontal view

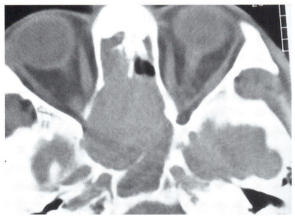

Fig. 22.38. Computed tomography, axial view. Sphenoid mucocele 3 years after ethmoidectomy

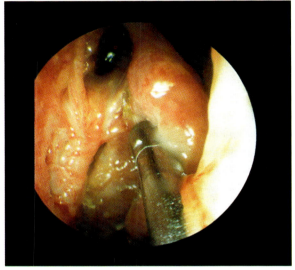

Fig. 22.40. Crusty rhinitis of iatrogenic origin following ethmoidectomy and inferior turbinectomy. Endoscopic view of the right nasal cavity

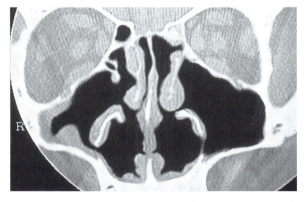

Fig. 22.41. Same case as in Fig. 22.40. Coronal computed-tomography view

Excessive endonasal bone removal is not uncommon in cases of recurrent polyposis, especially after multiple operations. The loss of bony borders laterally or superiorly in the ethmoid sinus are, unfortunately, intraoperative findings, and this should remind us of the need for careful study of the CT scan before any operation. Loss of bony borders should caution the surgeon against any radical surgery in that area. Amputation of the inferior turbinate (Figs. 22.40–22.42) is an extremely unfortunate surgical choice, because it leads to crusty rhinitis whenever ethmoidectomy is concomitantly performed or is subsequently needed.

Each case of iatrogenic problems is unique, and its treatment must be individualized. In cases of inferi-

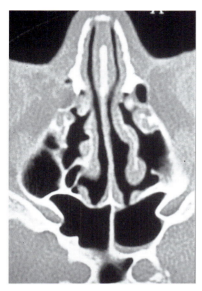

Fig. 22.42. Same case as in Fig. 22.40. Axial computed-tomography view

or septo-turbinal synechia, the treatment is usually easy with either electrocautery or laser photocoagulation. The most important technical point is to prevent a recurrence by the insertion of silastic splints on each side of the septum for 12–15 days and attentive postoperative care for 2–3 months.

Extensive Synechia of the Middle Meatus

Extensive synechia of the middle meatus is one of the most important indications for the "nasalization" technique. Careful debridement of the nasal fossa is followed by a wide meatotomy, carried out in an anterior direction from the thick, hard posterior edge of the palatine bone to the lacrymal bone. This must be done as low as possible, just above the inferior turbinate. Through this wide window, the inferior and medial orbital walls are easily outlined, allowing dissection of ethmoid cells along the lamina papyracea, which can then be safely opened. It is relatively easy to identify and remove the remnants of the middle turbinate superiorly.

Stenosis

At the level of the frontal recess, an external approach is often necessary to re-establish the patency of the naso-frontal duct or, at times, to obliterate the sinus, when there is severe recurrent suppuration. Stenoses of the sphenoid ostium are, however, quite easily accessible endoscopically and are amenable to a nasalization technique.

Excessive Bone Removal

The loss of bony portions of the ethmoid framework is not a specific contraindication for surgery, but it necessitates a careful technique of debridement that will leave behind mucosal lesions in the involved areas. Obviously, there is a greater risk of recurrence locally, but this is understood preoperatively. In the worst cases, when turbinates have been excised from the nasal fossa, it may be preferable to limit the intervention to "medical polypectomy" or to intervene through a very limited ethmoidectomy technique that preserves the middle turbinate, thus (unfortunately) creating the potential for future problems ("recurrences").

Post-Operative Follow-Up

All cases must be followed very closely, because any lapses in the monitoring or treatment of the condition can translate into problems. Early – even preventative – action is the only hope for optimal long-term results.

Postoperative preventive measures are especially important when risk factors for recurrence are present. Every attempt should be made to prevent any inflammatory or infectious insults, such as nasal-trauma surgery, epidemic or allergic rhinitis, and exposure to pollutants. Permanent topical-steroid therapy with beclomethazone is started, and systemic steroid therapy with short courses of oral prednisone is given as soon as recurrence is noted, but without exceeding three such courses per year (once every 4 months maximum) in order to avoid complications from the medication.

Methods of Follow-Up, or how are we to Identify Flare-Ups of the Disease?

The 50–60% rate of anatomic (objective) recurrence, as assessed by physicians, is clearly in excess of the 15–18% rate of symptomatic (subjective) recurrence reported by patients experiencing successful results. If early treatment recurrence is the only hope for good long-term results, we need to determine the "ideal" method of follow-up. Should we base this follow-up on nasal complaints, endoscopic findings or CT scans?
- The development of new *nasal symptomatology* is a practical criterion for screening for flare-up of polyp disease. Among all the subjective symptoms, the recurrence of anosmia (usually in a

transitory fashion at first) is the most important symptom. Nasal obstruction usually occurs too late to be helpful. Anosmia should be an alarm that prompts an appointment.

- *Endoscopic findings* of a "pre-polyp" or polypoid edema, though limited by the frequency of the visits, may document recurrence of polyp disease even earlier than is possible using nasal symptoms. This occurs more commonly in the anterior ethmoid region in the asymptomatic patient. It indicates the need for resumption of topical-steroid management (when it had been stopped) or a systemic steroid flash (if no improvement is observed 1 month later).
- *Postoperative CT scans* may not be as encouraging as improvements in symptomatology. Batteur and Strunski [2] noted an opaque anterior ethmoid sinus after half of their 116 ethmoidectomies. Radiographically, a distinction between polyposis and cicatricial and inflammatory reactions could not be made. Dense opacities partially filling the anterior ethmoid sinus were indicative of recurrence. In that situation, CT verified the finds obtained by endoscopy. Thus, for the experienced surgeon, clinical and endoscopic follow-up is appropriate and, for an inexperienced surgeon, CT-scan examinations, particularly 6 months after surgery, can be recommended [9].

Frequency of the Follow-Up

There is no standard for postoperative care and follow-up, but it is helpful to identify separate time intervals [37]. The first 6 months are characterized by the postoperative inflammatory reaction. Wound healing after nasalization takes place over a 3-month period. During the first 12 days, blood crusts cover the entire wound. Granulation tissue can be visible for 4 weeks. The increasingly edematous coverage of the wound reaches its maximum between the fourth and sixth weeks and gradually decreases between the seventh and 12th weeks. A macroscopically normal mucosa is generally observed after the 12–18th weeks. During this critical period, and up to the end of the first 6 months, local care must be consistently applied. Antibiotic and steroid treatments should be repeated as often as necessary to avoid an "inflammatory flare-up". As soon as the ethmoid operative site is suitable for topical steroid therapy, we replace systemic steroid management.

After the first 6 months, the frequency of the follow-up is modulated according to physical restrictions (distance, convenience, motivation, the social condition of the patient and the availability of the surgeon) and the findings that are present at the time. Our standard is one of full assessment every three months during the first year, followed by assessment every 6 months for the rest of the life of the patient. This is not an absolute rule, however, and the appearance of new symptomatology requires an earlier assessment.

Conclusion

Although we seem to partially understand the mechanisms of polyp formation, we do not know why the inflammatory respiratory disease becomes established and chronic. It is for this reason that the problem of recurrences is still unresolved and will continue to harass both the patient and the rhinologist.

We believe that it is possible to slow recurrences through better anticipation and management of the risk factors that trigger their development, through immediate extensive surgery (the nasalization procedure of Jankowski) when recurrences are resistant to medical management, and through sustained topical steroid therapy lasting the remainder of the patient's life.

Despite these precautions, some recurrences will occur (usually later, but occasionally relatively soon after operation). Such recurrences must be addressed even more vigorously than the initial presentations of polyp disease. Recurrence after non-invasive endoscopic surgery allows a more attractive surgical option to be offered to patients. It is up to surgeons to fulfil those expectations and to avoid causing iatrogenic sequelae (which are invariably symptomatic); this can be a difficult challenge.

Acknowledgement. The authors are grateful to Dr Guy Tropper (Cornwall, Ontario, Canada) for reviewing the manuscript and providing the English transl

References

1. Batsakis JG, El-Naggar AK (1996) Cystic fibrosis and the sinonasal tract. Ann Otol Rhinol Laryngol 105:329–330
2. Batteur B, Strunski V, Caprio D, Berthet V, Goin M (1994) Recurrence of nasosinusal polyposis after ethmoidectomy by endonasal approach. Functional, endoscopic, X-ray tomographic aspects and surgical implications. Ann Otolaryngol Chir Cervicofac 111:121–128
3. Bolt RJ, de Vries N, Middelweerd RJ (1995) Endoscopic sinus surgery for nasal polyps in children: results. Rhinology 33:148–151
4. Coste A, Rateau JG, Roudot-Thoraval F, Chapelin C, Gilain L, Poron F, Peynègre R, Bemaudin JF, Escudier E (1996)

Increased epithelial cell proliferation in nasal polyps. Arch Otolaryngol Head Neck Surg 122:432–436
5. Devalia JL, Sapsford RJ, Rusznak C, Davis RJ (1992) The effect of human eosinophils in cultured human nasal epithelial cell activity and the influence of nedocromil sodium in vitro. Am J Respir Cell Mol Biol 7:270–277
6. El-Naggar M, Kale S, Aldren C, Martin F (1995) Effect of beconase nasal spray on olfactory function in post-nasal polypectomy patients: a prospective controlled trial. J Laryngol Otol 109:941–944
7. Ferrara A, Stortini G, Bellussi L, Di Girolamo S, Zuccarini N, Passali D (1994) Furosemide long-term inhalation therapy in patients with nasal polyposis. Acta Otorhinolaryngol Ital 14:633–642
8. Fombeur JP, Ebbo D, Lecomte F, Simon D, Koubbi G, Barrault S (1993) Initial results of 132 ethmoidectomies by endonasal approach. Ann Otolaryngol Chir Cervicofac 110:29–33
9. Franzen G, Klausen OG (1994) Post-operative evaluation of functional endoscopic sinus surgery with computed tomography. Clin Otolaryngol 19:332–339
10. Jankowski R (1996) Eosinophils in the pathophysiology of nasal polyposis. Acta Otolaryngol (Stockh) 116:160–163
11. Jankowski R, Pigret D, Decroocq F (1997) Comparison of functional results after ethmoidectomy and nasalization for diffuse and severe nasal polyposis. Acta Otolaryngol (Stockh) 117:601–608
12. Jetten AM (1991) Growth and differentiation factors in tracheo-bronchial epithelium. Am J Physiol 260:1361–1373
13. Kanai N, Denburg J, Jordana M, Dolovich J (1994) Nasal polyp inflammation. Effect of topical nasal steroid. Am J Resp Crit Care Med 150:1094–1100
14. Karlson G, Runerantz GH (1982) A randomized trail of intranasal beclomethasone dipropionate after polypectomy. Rhinology 20:144–148
15. Katsantonis GP, Friedman WH, Bruns M (1994) Intranasal sphenoethmoidectomy: an evolution of technique. Otolaryngol Head Neck Surg 111:781–786
16. Krajina E, Zirdum A (1987) Histochemical analysis of nasal polyps. Acta Otolaryngol 103:435–440
17. Larsen L, Tos M (1991) Origin of nasal polyps. Laryngoscope 101:305–312
18. Larsen L, Tos M (1994) Clinical course of patients with primary nasal polyps. Acta Otolaryngol (Stockh) 114:556–559
19. Lildholdt T, Rundcrantz H, Lindqvist N (1995) Efficacy of topical corticosteroid powder for nasal polyps: a double-blind, placebo-controlled study of budesonide. Clin Otolaryngol 20:26–30
20. Lund VJ, MacKay IS (1994) Outcome assessment of endoscopic sinus surgery. J R Soc Med 87:70–72
21. Miszke A, Sanokowska E (1995) Cytology of nasal polyps. Otolaryngol Pol 49:225–230
22. Mostafa BE (1996) Fluticasone propionate is associated with severe infection after endoscopic polypectomy. Arch Otolaryngol Head Neck Surg 122:729–731
23. Nishioka K, Saito C, Nagana T, Okano M, Masuda Y, Kuriyama T (1993) Eosinophil cationic protein in the nasal secretions of patients with mite allergic rhinitis. Laryngoscope 103:189–192
24. Ohno I, Lea RG, Flanders KC, et al. (1992) Eosinophils in chronically inflamed human upper airway tissue express transforming growth factor β1 gene (TGF-β1). J Clin Invest l89:1662–668
25. Paludetti G, Maurazi M, Tassoni A, et al. (1983) Nasal polyps a comparative study of morphologic and etiopathogenic aspects. Rhinology 21:347–360
26. Petruson B, Hansson HA, Petruson K (1988) Insulin-like growth factor I is a possible pathogenic mechanism in nasal polyps. Acta Otolaryngol (Stockh) 106:156–160
27. Pierce GF (1990) Macrophages: important physiologic and pathologic sources of polypeptide growth factors. Am J Respir Cell Mol Biol 2:233–234
28. Rouvier P, Peynegre R (1997) Etude comparativre de l'évolution de 22 polyposes nasales invalidantes 5 ans après corticothérapie puis 5 ans après ethmoïdectomie. Communication faite à la réunion de la Sté de Laryngologie des Hopitaux de Paris. Ann Otolaryngol Chir Cervicofac 114:1
29. Rouvier P, Vandeventer G, El-Khoury J, de Lanversin H (1991) Les résultats à long terme (sur 5 ans) de l'ethmoïdectomie dans la polypose nasale invalidante. J Fr ORL 40:102–105
30. Rouvier P, Mondain M, El-Khoury J (1992) Eosinophilie et rhinite obstructive. Ann Otolaryngol Chir Cervicofac 109:264–271
31. Stankiewicz JA (1995) Pediatric endoscopic nasal and sinus surgery. Otolaryngol Head Neck Surg 113:204–210
32. Stoop AE, Van der Heijden HA, Biewenga J, Van der Baan S (1992) Clinical aspects and distribution of immunologically active cells in the nasal mucosa of patients with nasal polyps after endoscopic sinus surgery and treatment with topical corticosteroids. Eur Arch Otorhinolaryngol 249:313–317
33. Stoop AE, Van der Heijden HA, Biewenga J, Van der Baan S (1993) Eosinophils in nasal polyps and nasal mucosa: an immunohistochemical study. J Allergy Clin Immunol 91:616–622
34. Terrier G (1991) Rhinosinusal endoscopy. Diagnosis and surgery. Zambon, Milano
35. Tos M, Drake-Lee A, Lund V, Stammberger H (1989) Treatment in nasal polyps. Medication or surgery and which surgery. Rhinology 8:45–49
36. Van Camp C, Clement PA (1994) Results of oral steroid treatment in nasal polyposis. Rhinology 32:5–9
37. Weber R, Keerl R, Huppmann A, Schick B, Draf W (1996) Effects of postoperative care on wound healing after endonasal paranasal sinus surgery. Laryngorhinootologie 75:208–214

What is the Place of Endonasal Surgery in Fungal Sinusitis?

Valerie J. Lund

Introduction

Defining the role of endonasal surgery in the management of fungal sinusitis is an interesting problem, as the condition cannot be regarded as a single entity. Rather, it represents a spectrum of diseases, depending on the organism involved and the immunocompetence of the host. The immune status may improve or, deteriorate altering the patient's response, and we should never be lulled into a false sense of security by a disease entity that has a serious morbidity and mortality.

Classification

Fungal sinusitis can be broadly divided into non-invasive and invasive forms (Table 23.1). The term "invasive" relates to the presence of fungal hyphae within the sinus mucosa and not necessarily to the presence of bone erosion, which can be found in the non-invasive forms.

The range of organisms responsible for fungal infections is considerable, though *Aspergillus* and the dermatiaceous fungi probably account for the majority of cases (Table 23.2). There is also a clinical impression that the number of fungal sinusitis cases is increasing, though this may simply relate to greater awareness and consequent diagnosis. There certainly appears to be geographical clustering of cases, with reports of large numbers from centers with stable local referral patterns, though there seems to be little environmental linkage between, for example, the findings of Graz [13] and Poitiers [8] and findings in countries in the Middle East.

Table 23.1. Classification of fungal sinusitis

Non-invasive
Fungal ball (mycetoma)
Allergic fungal sinusitis
Invasive
Chronic, immunocompetent
Chronic, immunocompromised
Acute/fulminate, immunocompromised
Sclerosing
Granulomatous

Table 23.2. Range of fungal organisms responsible for sinonasal infection

Aspergillus fumigatus
Aspergillus flavus
Aspergillus niger
Dermatiaceous *Curvularia*
Dermatiaceous *Bipolaris*
Dermatiaceous *Alternaria*
Conidiobolus coronatus
Rhizopus oryzae (Mucor)
Rhinosporidium seeberi
Cryptococcus neoformans
Histoplasma capsulatum
Sporothrix schenckii
Candida

Non-Invasive

Mycetoma

The mycetoma or fungal ball is one of the most recognized forms of fungal infection occurring in normal, non-atopic individuals. It usually affects a single sinus (usually the maxillary sinus), though it is sometimes seen in a concha bullosa. It is symptomatically relatively inert, producing only mild symptoms or those associated with a superadded bacterial infection. Bone erosion and mucosal invasion do not occur, so the radiological appearances present an opaque or partially opaque sinus. The bony walls are normally intact, and the fungal mass may give a heterogeneous signal due to the chemical components retained by the fungus. The infecting organism is generally *Aspergillus fumigatus* [13], and the mass has a rather typical, clay-like appearance and is sometimes covered with obvious sporangia.

Removal of the mass and aeration of the sinus is almost always associated with complete cure, and an endoscopic approach is ideal for such cases. It is important, however, that the entire mass be completely removed, though it is noteworthy that the sinus mucosa is generally relatively normal despite the presence of fungal material (Fig. 23.1).

Allergic Fungal Sinusitis

Clinicians familiar with the condition of allergic bronchopulmonary aspergillosis recognized a similar manifestation in sinuses in the early 1980s. This association was first described by Millar in 1981 [10] and, subsequently, by Katzenstein et al. [7], who retrospectively identified seven patients from a group of 119 specimens taken from patients with chronic sinusitis. Although originally ascribed to *Aspergillus*, it is now recognized that the dermatiaceous fungi (*Curvularia*, *Bipolaris* and *Alternaria*) are more often responsible [1, 9].

Clinical Characteristics

The condition appears to affect young adults, though there does not appear to be any male or female preponderance. The majority of patients are atopic and asthmatic and, while some may give a positive skin-prick test for *Aspergillus*, many will not (presumably because their condition is due to one of the other fungi).

In allergic fungal sinusitis, the fungus provokes a polypoid reaction (sometimes unilateral or bilateral with one predominant side) but, notwithstanding this, the symptomatology is that of nasal polyposis, with nasal obstruction and loss of sense of smell as major symptoms. The amount of pain and headache probably relates to the presence of superadded bacterial infection (particularly with *Staphylococcus aureus*). Endoscopic examination of the nose will reveal rather unremarkable polyps, but mixed amongst these is the characteristic fungal mucin, which has a yellow–green–brown appearance (akin to that of axle grease) and has the consistency of chewing gum.

Diagnosis

The finding of fungal mucin, together with rather typical appearances on imaging, aids the clinician in diagnosis. Computed tomography (CT) scanning should be performed in the direct coronal and axial planes to assess the extent of the disease and frequently shows the typical heterogeneous pattern (Fig. 23.2) due to areas of high attenuation caused by

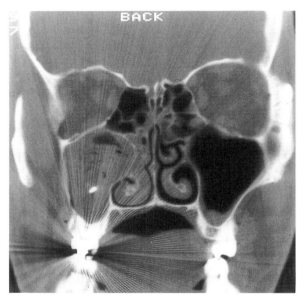

Fig. 23.1. Coronal computed tomography showing mycetoma of the right maxillary sinus

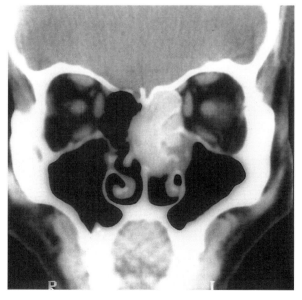

Fig. 23.2. Coronal computed tomography (CT) scan showing localized allergic fungal sinusitis of the left ethmoid complex

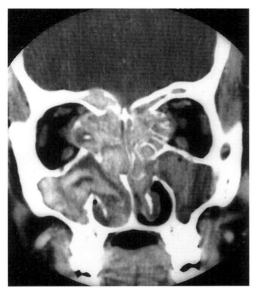

Fig. 23.3. Magnetic resonance image showing allergic fungal sinusitis

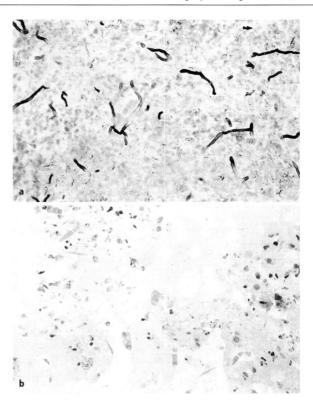

Fig. 23.4. a Smear of material taken from a patient with a case of allergic fungal sinusitis, showing fungal hyphae. b Smear from same case, showing eosinophils and Charcot-Leyden crystals

ferromagnetic elements (such as magnesium and calcium) within the fungal mass in 83%. Bone erosion of the adjacent sinus walls has been reported in 20–80% of cases. Thus, in particular with unilateral disease, the condition may need to be distinguished from chondrosarcoma. Erosion of the skull base and the lateral wall of the sphenoid may be present and demonstrates the clinical potential for what might be regarded pathologically as benign disease. The mechanism of bone resorption may be explained in part by the presence of bone-resorbing cytokines (such as interleukin 1 and tumor necrosis factor) at the mucosa–bone interface despite the lack of invasion. Under these circumstances a magnetic-resonance image may be of value. Again, the appearances are characteristic, as the ferromagnetic elements give a signal void on both T1 and T2 images, while the surrounding inflamed mucosa gives a high signal (Fig. 23.3).

The best method for confirming the diagnosis is histological examination of the fungal mucin, which demonstrates eosinophilia, Charcot-Leyden crystals and extramucosal hyphae. The Charcot-Leyden crystals simply represent the by-products of dead eosinophils and are characteristic of this condition. The adjacent mucosa should be examined to confirm that the fungal hyphae have not invaded the tissue. In some cases, the hyphae can be relatively few in number, and the pathologist should be asked to use special stains (such as Grocott) and to search carefully in cases where the clinician is highly suspicious (Fig. 23.4). Culture of the fungal material by conventional methods is only moderately successful, though this can be improved to almost 100% if care is taken with the selection and speed of submission of the specimen [2, 3]. The detection may be improved by immunofluorescence microscopy and enzymes linked immo-absorbant assays. Indeed recent studies suggest that fungal hypae may be found in the majority of patients with chronic rhinosinusitis with or without polyposis.

Treatment

Complete removal of the fungal material and sinus ventilation offer the best results for this condition and can be achieved by any one of a number of approaches to the sinuses [6], of which an endoscopic approach would be the first choice for many surgeons. Certainly, this offers a good way to clear the ethmoid, maxillary and sphenoid sinuses, though some difficulties may be experienced with the frontal sinus. Frontal-sinus secretions are particularly glutinous and can be resistant both to suction through a small frontal ostium and to trephination and irrigation. If fungal material is left behind, the condition will undoubtedly continue and, in these cases, the endoscopic approach may need to be combined with a small external approach to access the most lateral compartment of the frontal sinus.

Most authors agree that the use of intranasal steroid preparations (with or without alkaline or saline douching) is of considerable benefit postoperatively and, given the analogy with allergic bronchopulmonary aspergillosis, oral steroids have been employed in cases where resolution is not as quick as anticipated. It is our practice to employ oral prednisolone (30 mg for 5 days, followed by 20 mg for 5 days, then 10 mg for 5 days) in patients where there is no contraindication for systemic steroid use.

It has generally been stated that systemic antifungal agents (such as oral itraconazole, ketoconazole or, in particular, amphotericin) should not be used in allergic fungal sinusitis, as there is no evidence of mucosal invasion. This is true in the majority of cases. However, even in the absence of mucosal invasion, the presence of a systemic condition, such as diabetes, may compromise clinical response and may make the use of oral steroids undesirable. In cases such as these, I have employed long-term itraconazole with considerable success.

Prognosis

In 1986, Waxman et al. [16] retrospectively studied 15 patients, dividing them into immediate-recurrence (recurrence within months), delayed-recurrence (recurrence within years) and disease-free groups. In the absence of an endoscopic examination, three patients were reported as disease free for over 2 years following the surgery. In the literature, recurrence rates range from 32% [11] to 100% [12]. The methodology and length of follow-up may account for the disparity in some reports, and it may be that a more aggressive approach to the frontal sinus and/or the use of systemic antifungal agents could improve matters.

Invasive

Chronic

The clinical severity of this condition depends on the immunocompetence of the individual. Pathologically, it is characterized by mucosal invasion with fungal hyphae together with some degree of polypoid or granulomatous inflammation of the mucosa (often infiltrated with eosinophil plasma cells) and a variable degree of bone erosion. *Aspergillus* and some of the dermatiaceous fungi are most frequently implicated, and there are notable geographical "black spots" in the Middle East, Sudan [15] and India [5].

Surgical debridement and ventilation of the sinus is again the mainstay of the treatment, the extent of that surgery being indicated by the imaging. The frontal sinus, skull base and lateral extension from the sphenoid sinus into the cavernous sinus pose the greatest difficulties. Thus, an experienced endoscopic surgeon may feel confident in some cases, but it is our experience that more radical sinus surgery (up to and including craniofacial resection) has been required in extensive cases (Fig. 23.5). It should not be forgotten that these patients are potentially at risk of severe orbital and intracranial complications, and the disease should not be undertreated.

In the presence of mucosal invasion, an oral antifungal agent should be given for at least 8 weeks and, occasionally, if the disease cannot be completely extirpated (from the cavernous sinus) and/or if the fungus is resistant to oral antifungal agents, liposomal amphotericin may need to be used. Only a finite dose of this drug can be given, and close monitoring of liver and renal function must be instituted.

Acute Fulminate

This (fortunately rare) condition occurs exclusively in immunocompromised patients (such as poorly controlled diabetics), immunosuppressed patients (such as those from the oncology or transplant units) and patients with other infections, such as acquired immune deficiency syndrome (AIDS) or tuberculosis. *Aspergillus* can once again be an important pathogen, though this condition is often associated with one of the phycomycetes, *Rhizopus* or *Mucor*. In this disease, the fungus produces rapid necrosis of tissue due to ischemia of the vascular supply, with rapid destruction of the skull base and penetration of the orbit.

The condition requires emergency surgery, allowing only for the most urgent of CT scans, combined with systemic liposomal amphotericin and, where possible, correction of the immunosuppression. The necrotic tissue must be debrided back to normal tissue, which often necessitates a craniofacial resection together with clearance of the orbit. Nevertheless, the disease is fatal in 25–75% of cases (endoscopic endoscopy plays no role other than in the initial examination of the patient) [4].

Invasive Sclerosing

A rare form of fungal infection may be seen in patients from the Middle East. In contrast to the polypoid reaction associated with allergic fungal sinusitis, in this infection, the fungal hyphae provoke fibrosis, which results in a space-occupying mass (usually in the midline, affecting the ethmoid labyrinth, skull base and both orbits). Patients may

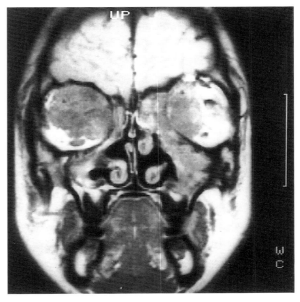

Fig. 23.6. Coronal magnetic resonance imaging scan showing a case of sclerosing fungal sinusitis invading both orbits after cranofacial resection (reproduced with kind permission of A.D. Cheesman)

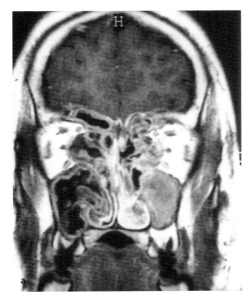

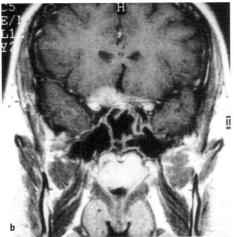

Fig. 23.5. a Coronal magnetic resonance imaging (MRI) scan (T1 sequence with gadolinium diethylene triaminopentaacetic acid) showing a case of invasive fungal sinusitis, with inflamed sinus mucosa, fungal material and mucus giving a heterogeneous signal. Ferromagnetic material on the right side appears as a signal void. **b** Coronal MRI scan from the same case, taken more posteriorly, with a signal void from fungal material in the sphenoid sinuses and extension of disease into middle cranial fossa

present with bilateral proptosis and visual loss. The exact cause of the condition is unknown and, as only an occasional fungal hyphae is found within the fibrous material, it has rarely been possible to identify the organism. To preserve any residual vision and excise the mass, a craniofacial approach is required, though it may be impossible to extirpate all the tissue, particularly from within the orbit (Fig. 23.6). Systemic amphotericin B has been used, but with limited success because of the problems of tissue penetration.

Fungal Sinusitis in Human-Immunodeficiency-Virus Infection

Although the patient with AIDS may present with acute fungal infection, it is generally felt that such patients are more at risk from chronic fungal sinusitis caused by *Cryptococcus*, *Histoplasma*, *Pseudallescheria boydii* [14] or, occasionally, *Candida*.

What is the Place of Endonasal Surgery in Fungal Sinusitis?

There can be no doubt that nasal endoscopy has made us suspect and diagnose a greater number of fungal infections than was hitherto possible. The role of sinonasal endoscopy in the follow-up of our patients is also invaluable in determining the therapeutic success (or failure) of our interventions and, given the propensity for recurrence, long term follow-up should be instituted wherever possible. As a therapeutic approach, endoscopic sinus surgery has much to commend it in the management of the mycetoma, allergic fungal sinusitis and selected cases of chronic invasive fungal sinusitis. However, we should always be prepared to recognize our own limitations and those of the technique when the location of the disease renders it suboptimal. We should never underestimate the little mushrooms!

References

1. Allphin AL, Strauss M, Abdul-Karim FW (1991) Allergic fungal sinusitis: problems in diagnosis and treatment. Laryngoscope 101:815–820
2. Bent JP, Kuhn FA (1994) The diagnosis of allergic fungal sinusitis. Otolaryngol Head Neck Surg 111:580–588
3. Bent JP, Kuhn FA (1997) Allergic fungal sinusitis and polyposis. In: Settipane G, Lund VJ, Tos M, Bernstein J (eds) Nasal polyps: epidemiology, pathogenesis and treatment. Oceanside, Providence, pp 25–30
4. Blitzer A, Lawson W (1993) Fungal infections of the nose and paranasal sinuses. Otolaryngol Clin 26:1007–1035
5. Chakrabarti A, Sharma SC, Chander J (1992) Epidemiology and pathogenesis of a paranasal sinus mycosis. Otolaryngol Head Neck Surg 107:745–750
6. De Shazo RD, Chapin K, Swain RE (1997) Fungal sinusitis N Eng J Med 337:254–9
7. Jonathen D, Lund VJ, Milroy C (1989) Allergic *Aspergillus* sinusitis – an overlooked diagnosis. J Laryngol Otol 103:1181–1186
8. Karpovitch-Tate N, Lund VJ, Smith E, Dewey FM, Gurr PA, Gurr SJ Detection of fungi in sinus fluid of patients suffering from allergic fungal rhinosinusitis. Acta Otolaryngologic (Stockholm) In press
9. Katzenstein AA, Sale SR, Greenberger PA (1983) Allergic *Aspergillus* sinusitis. A newly recognised form of sinusitis. J Allergy Clin Immunol 72:89–93
10. Klossek J-M, Peloquin L, Fourcroy P-J, Ferrie J-C, Fontanel J-P (1996) Aspergillomas of the sphenoid sinus. Rhinology 34:179–183
11. Lund VJ, Lloyd G, Savy L, Howard DJ (2000) Fungal rhinosinusitis: radiology in focus J Laryngol Otol 114:76–80
12. Manning SC, Schaefer SD, Close LG, Vuitch F (1991) Culture-positive allergic fungal sinusitis. Arch Otolaryngol Head Neck Surg 117:174–178
13. Millar JW, Johnston A, Lamb D (1981) Allergic aspergillosis of the maxillary sinuses. Thorax 36:710
14. Ponikau JU, Sherris DA, Kern DB, Homburg RHA, Frigass E, Gaffey TA, Roberts GD (1999) The diagnosis and incidence of allergic fungal sinusitis. Mayo Clin Proc 74:877–884
15. Schwietz LA, Gourley DS (1992) Allergic fungal sinusitis. Allergy Asthma Proc 13:3–6
16. Sher TH, Schwartz HJ (1988) Allergic *Aspergillus* sinusitis with concurrent allergic bronchopulmonary *Aspergillus*: a report of a case. J Allergy Clin Immunol 81:844–846
17. Stammberger H, Jakse R, Beaufort F (1985) Aspergillosis of the paranasal sinuses. Ann Otol Rhinol Laryngol 94[suppl 119]:1–11
18. Tami TA, Wawrose SF (1992) Diseases of the nose and paranasal sinuses in the human immunodeficiency virus-infected population. Otolaryngol Clin 25:1199–1210
19. Veress B, Malik OA, El Tayeb AA, El-Daoud S, El-Mahgoub S, El-Hassan AM (1973) Further observations on the primary paranasal *Aspergillus* granuloma in the Sudan. Am J Trop Med Hyg 22:765–772
20. Waxman JE, Spector JG, Sale SR, Katzenstein AA (1987) Allergic *Aspergillus* sinusitis: concepts in diagnosis and treatment of a new clinical entity. Laryngoscope 97:261–266

Combined Fiberoptic Headlight and Endoscopic Sinus Surgery with Adjunct Use of Middle-Meatal Stenting

Paul H. Toffel

Introduction

The most substantial recent advance in rhinologic management has been the advent of functional endoscopic sinus surgery [1, 2, 4]. Its efficacy in the treatment of chronic obstructive rhinosinusitis has been well established in European centers for 20 years. Two distinct approaches evolved. The first technique is that espoused by Wigand et al. [10], which is rhinologic in style and advocates broad fenestration of the sphenoid, ethmoid, frontal, and maxillary sinuses, partial resection of middle turbinates and obligatory septal correction. In contrast is the technique of Messerklinger, who advocates a much more limited approach suggesting that opening the narrow anterior ethmoid sinuses and osteomeatal complex will physiologically reverse sinus disease. Despite technical divergence, these two approaches share the basic principle that most sinus mucosal disease will resolve if aeration and drainage are established (Fig. 24.1).

Using an intermediate technique that incorporates elements of both European schools has allowed me to achieve a high success rate with few minor complications and no major complications [8]. I think that this technique offers other otolaryngologists a secure way to perform endoscopic sinus surgery as an adjunct to functional nasal surgery [6].

The first key point I discovered was the value of maintaining three-dimensional orientation by alternating between the macroscopic view of the headlight and the microscopic view of the endoscopes. As such, a fiberoptic headlight is worn throughout the procedure, a style similar to the Wigand et al. [10] approach. The alternating method is especially helpful to the beginning endoscopist, who is likely to become disoriented when using angled optical instruments and angled forceps. The macroscopic view allows checking of positioning and maintenance of perspective, whereas the endoscopic view allows precise visualization and removal of diseased tissue.

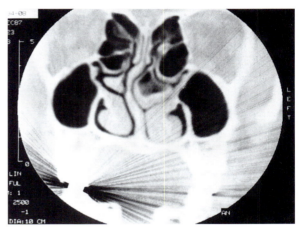

Fig. 24.1. Anatomic findings that narrow the osteomeatal complex: large ethmoid bulla, large concha bullosa, paradoxically shaped middle turbinate (convex laterally), large medially displaced uncinate process, wide or deviated septum, localized mucosal edema, granulation tissue or polyps

The most efficient use of time and energy is employment of the endoscope in four key areas of sinus surgery and use of the fiberoptic headlight between these areas. The endoscope is used for initial exploration through the uncinate process, exploring the frontal recess, finding and enlarging the natural ostium of the maxillary sinus and exploring the sphenoid sinus. Between these key areas, the combination of a narrow-beam fiberoptic headlight and tactile instrument feedback will facilitate the remainder of the sinus procedures and can be augmented by endoscopy as necessary.

I use suction–irrigation endoscopes, because they help maintain a clear operative field. The primary operating scope is a 25°, wide-angle, pistol-grip endoscope. The 70° endoscope is used for nasofrontal-duct examination and work on the natural ostia of the maxillary sinuses. In addition, the use of moderately sized straight and angled biting forceps affords optimal tactile appreciation, an important but rarely mentioned benefit.

Endoscopically guided ethmoidectomy is performed in an anterior-to-posterior direction, following the disease as found but always leaving a rim of mucosa. The ethmoid sinuses are entered at the level of the uncinate process opposite the leading edge of the middle turbinate (Fig. 24.2). The frontal recess is carefully cleansed and inspected. Ethmoid-sinus dissection is continued posteriorly throughout the ground lamella of the middle turbinate, along the medial corridor of the ethmoid sinuses, creating a neoinfundibulum. Lateral dissection is done with great care, especially in the posterior lateral corner, where the bony covering of the optic nerve may be dehiscent. The eyes are left uncovered so that dangerous movement of the lamina papyracea can be detected as orbital ballottement by the surgeon or assistant (Stankiewicz maneuver [5]). The middle turbinate, which is preserved at this point if it has not undergone polypoid degeneration, serves as key anatomic reference during ethmoidectomy.

After ethmoidectomy, the natural ostium of the maxillary sinus is found above the midportion of the inferior turbinate and is enlarged. If, on inspection of the antrum via the enlarged natural ostia, there exists significant mucosal disease, I use an additional access port via the inferior meatal route so that intramaxillary surgery can be performed securely by viewing through one opening and operating through the other. Sphenoid-sinus endoscopy is performed when indicated by findings of polypoid disease on computed tomography (CT); however, such surgery is not routine.

Similar to the Wigand et al. [10] style, partial middle turbinectomy with preservation of the olfactory rim is performed. However, Wigand et al. [10] performed middle turbinectomy as the initial step; I preserve the middle turbinate until sinusotomies are complete so that this important landmark is maintained throughout the endoscopic procedure.

Partial resection of the inferior, bulky portion of the middle turbinate is important for a number of reasons. First, the middle turbinate may be a source of pathological obstruction of the osteomeatal outflow tract. Second, roughened or injured middle-turbinate mucosa (especially of the anterior aspect) may be important in postoperative synechiae formation. As has been suggested elsewhere, the middle turbinate may have a tendency to drift laterally and abut the lateral nasal wall during the early postoperative period. To avoid this problem, I place a wedge-shaped spacer of thin, Merocel sponge soaked in antibiotic solution between the resected middle turbinates and the lateral wall for 5–7 days. The middle turbinate, when pneumatized or hypertrophied, may also impact the septum or lateral wall and, thereby, may be instrumental in the etiology of midfacial pain. For these reasons, limited resection of the bulky portion of the middle turbinate is performed in most cases.

Combined Functional Rhinologic and Endoscopic Surgical Technique

The combination of history, physical examination (including diagnostic endoscopy) and limited osteomeatal coronal CT will indicate those patients who might benefit from surgical correction of various anatomic functional deformities. In any given patient, the obstructive anatomy might vary from simple septum deviation to turbinate hypertrophy, polyposis, nasal pyramid deflection, osteomeatal unit obstruction, sinus polyposis [9] or any combination thereof. The specific abnormalities present determine the logical order of the correction.

The technical aspects of simultaneous endoscopic and rhinologic surgery are facilitated by the surgeon's use of a narrow-beam fiberoptic headlight throughout the endoscopic aspects of the procedure. This single feature enables easy alternation between endoscopic and normal rhinologic surgery at any time necessary during the overall procedure, allowing the surgeon to easily use all tools at his or her disposal.

Septoplasty, if indicated, is usually performed first, and conservative management of hypertrophied inferior turbinates may be accomplished by crushing, outfracturing or suction–shaver contouring and striation with a potassium titanyl phosphate,

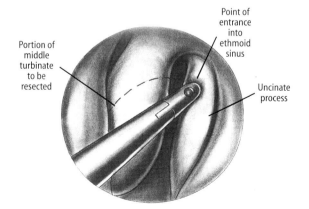

Fig. 24.2. Endoscopic view of the entrance to the anterior ethmoid sinus at the level of the uncinate process opposite the leading edge of the middle turbinate. The middle turbinate is preserved until sinusotomies are completed; the bulky portion is then partially resected along the line shown

532-nm laser as necessary. I am careful to be conservative with inferior-turbinate modification if partial resection of the bulky portion of the middle turbinate is preplanned during endoscopic sinus surgery.

The main point of the concept of secure endoscopic sinus surgery [6] as an adjunct to functional nasal surgery is that a competent rhinologic and endoscopic surgeon can do whatever needs to be done during one operation, and can do it in whatever order indicated by the specific anatomy and pathological condition. I have found the results to be facilitated by the appropriate use of middle-meatal packing or spacers.

Perioperative Management

Patients with chronic hypertrophic rhinosinusitis who are to be managed by endoscopic and rhinologic surgery are carefully prepared for anesthesia by their internist, allergist or pulmonologist. Special attention is given to the frequent concomitant asthmatic and sinobronchial-syndrome patients, who often have lower-airway sensitivity and rhinosinologic polypoid disease. These patients are brought to surgery under steroid and bronchodilator management. At the time of surgery, all patients receive a burst of corticosteroids [usually 8 mg intravenous decadron dexamethasone (Merck and Company, West Point)] and prophylactic antibiotic (typically 1 g cefazolin intravenously). This may be preceded by oral antibiotic therapy (such as cephalosporins) for several days to minimize infectious effects on polypoid mucosa.

Most operations are performed with the patient under relatively hypotensive general anesthesia. I found no complications in this 1463-case series, during which I operated with excellent, gentle anesthesiologists. Because my patients are operated on in the outpatient setting of the hospital but are allowed to stay overnight for recuperation, smooth and gentle emergence from general anesthesia is the rule, and the patients are typically comfortable when discharged the day after the outpatient procedure. Although patients receive general anesthesia, painstaking local anesthetic preparation is performed before surgery, with topical 4% cocaine solution on cottonoids slid into the middle meatus (if possible), infiltration of 0.5% lidocaine (Xylocaine; Astra Pharmaceutical, Westborough) and 1:200,000 epinephrine solution injected gently into all target areas with a tonsillectomy needle. Gentle periodic re-injection with dilute Xylocaine and adrenaline (Parke-Davis, Morris Plains) solution is continued throughout surgery before advancing to a new region of surgery.

Surgery is performed while either sitting or standing, whichever is more comfortable for the surgeon. I have found no difference between the two methods, in terms of my ability to gently handle tissues, provided the table height is adjusted carefully for close use of the endoscope. Also, the surgeon should rest his or her hands gently on the patient's nose or face when manipulating instruments, just as in microscopic otosurgery.

Packing System and Aftercare

At the conclusion of a combined endoscopic functional and rhinoplastic procedure, I use a light packing and respiration system. A special Toffel Merocel (polyvinyl acetal) Sinus-Nasal Pak expandable sponge system (Merocel, Mystic) is used to insert a small wedge into the osteomeatal cleft to keep the middle-turbinate remnant from adhering to the lateral nasal wall. The Merocel sponge is coated with water-soluble antibiotic ointment. A second Merocel sponge is inserted in the lower nose adjacent to the septum to keep the septum narrow and non-edematous during healing, and a respiration tube is inserted along the floor of the nose for patient comfort (Fig. 24.3). The system is maintained for 5–7 days with administration of cephalosporin antibiotic, yielding excellent results with narrow septa and ethmoid clefts without synechiae. My patients have

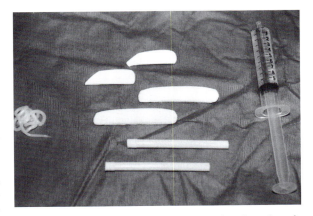

Fig. 24.3. Merocel (polyvinyl acetal) wedge-shaped, soft, expansile packing system for insertion into the mid-meatus, with complementary lower nasal pack and airway tube for use in a combination of endoscopic sinus surgery and functional nasal surgery. The packing system can be removed in a brief outpatient, anesthetic procedure 5–7 days after insertion, and cavities can be carefully inspected and debrided under endoscopic illumination

experienced no toxic shock episodes in more than 3000 nasal operations when the patients were treated with packs and cephalosporin antibiotic for several days, but scrupulous administration of antibiotics is necessary. The packing system is often removed under brief intravenous sedation in our hospital's outpatient department 1 week after surgery.

After removal of the packing system, cleaning is periodically accomplished in the office via phenylephrine (Neo-Synephrine, Sterling Health, New York) irrigation, and instrument hygiene is facilitated by endoscopy. With the open cavity provided by partial resection of the middle turbinate, inspection and care of the postoperative field are tremendously facilitated.

All helpful preoperative medical, allergy, immunologic and pulmonary therapy is continued indefinitely to minimize return of hypertrophic disease. The patients are postoperatively treated using a home saline irrigator and beclomethasone or triamcinolone aqueous sprays, and antibiotics are administered quickly any time a postoperative infection develops. The patients are followed particularly closely for 1 year, because I believe it takes at least one rotation through the four seasons for the nasal and sinus structures to be maximally healed.

Risks and Complications

The beautiful optics of endoscopes have markedly aided diagnostic ability and surgical control of the operative field in rhinosinus surgery. However, they afford a two-dimensional view, and it must be emphasized that disorientation can occur with tragic orbital or cerebral consequences. Lack of landmarks in extensive polypoid cases can also make pure endoscopic surgery difficult, even for an expert surgeon. I advocate the combined use of a narrow-beam fiberoptic headlight and endoscopes to provide three-dimensional referencing with binocular stereo vision any time the surgeon is in doubt. I believe this to be a minor precaution with high dividends of surgical security and efficiency. Frequent intraoperative referencing of the coronal CT scans also improves safety, because each patient has unique anatomic variants for which the surgeon can make adjustments if forewarned.

Troublesome bleeding during surgery is best prevented by meticulous atraumatic technique. If the anterior or posterior ethmoid arteries are violated, bleeding can be controlled with microfibrillar collagen slurry or bipolar cautery. If bleeding during surgery is so severe as to obstruct the surgeon's view, the procedure should be terminated and resumed at another time. Operating in a field where there is much blood creates a potential for orbital or cerebral penetration. Extreme care should be exercised when operating in the lateral postero-superior ethmoid sinuses and lateral sphenoid-sinus regions, because these are areas where the optic nerve, carotid artery and cerebral contents are at risk. In addition, cerebrospinal fluid (CSF) leaks are more likely medial to the frontal recess area, where the fovea ethmoidalis blends into very thin bone.

In case of orbital penetration, do not proceed further. If any packing is employed in these cases, it should be non-expansile. If expanding orbital hematoma does occur, high doses of intravenous corticosteroids and mannitol should be administered to control swelling. If the situation worsens, arrange an ophthalmologic consultation. Lateral canthotomy and orbital decompression may be necessary.

Small CSF leaks can be recognized and sealed with patches of nasal septal cartilage and turbinate mucosa, using fibrin glue. Larger penetration requires neurosurgical consultation for possible craniotomy procedures.

Results Using Mid-Meatal Stenting

Endoscopic sinus procedures have been performed in 1463 patients. Many of these patients had concurrent functional nasal surgery (i.e., septoplasty and turbinoplasty [3]). Some had concurrent rhinoplasty. Virtually all patients underwent packing of the middle meatus, as described earlier. All had limited coronal CT of the osteomeatal unit and sinuses. Patient selection was based on symptoms consistent with chronic or recurrent hypertrophic rhinosinusitis in conjunction with positive physical examination and radiologic findings (Figs. 24.4, 24.5).

The results presented here reflect a comparison between the preoperative and postoperative symptoms of the 1463 consecutive patients who underwent endoscopic sinus procedures by the secure method [6] between 1986 and 1995. Specific attention was directed at the evaluation of the primary symptoms of nasal obstruction [7], purulent postnasal drainage and midfacial cephalgia, similar to the analysis of Wigand [10]. The cases were also staged according to the severity of disease, (as determined by preoperative coronal CT, in accordance with the method of Friedman; Table 24.1). Of the 785 patients available for 1-year follow-up, a total of 90.75% reported complete resolution of the nasal obstructive symptoms, whereas purulent post-nasal drainage and midfacial cephalgia were eliminated in

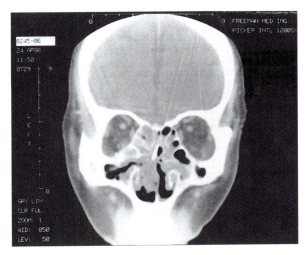

Fig. 24.4. Preoperative, limited, direct coronal osteomeatal-sinus computed-tomography scan of a patient with extensive ethmoid and maxillary sinusitis, obstruction of osteomeatal tracts, deviated septum and hypertrophic middle and inferior turbinates

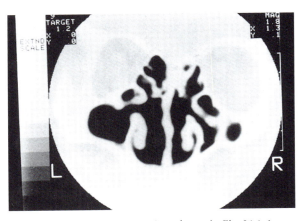

Fig. 24.5. View of the same patient shown in Fig. 24.4, 1 year after multiple endoscopic sinusotomies, enlargement of the natural ostia of the maxillary sinuses, septoplasty and modification of the middle and inferior turbinates

83.3% and 84.4% of patients, respectively. Patients who still had symptoms almost universally reported improvement (Table 24.2).

Two hundred and five patients were followed for 3 years. As indicated in Table 24.3, the data at 3 years indicate results consistent with those at 1 year.

Additional analysis was performed to use CT to subdivide the results according to Friedman's preoperative disease-severity staging categories. The results at 1 year and 3 years are shown in Tables 24.4 and 24.5. They also show that patients with stage-I, -II and -III disease have results similar to the overall case analysis but that, as would be expected, patients

Table 24.1. Friedman computed-tomography staging system for nasal/sinus hypertrophic disease

Stage I	Unifocal nasal and sinus hypertrophic disease
Stage II	Multifocal but discontinuous nasal and sinusal hypertrophic disease
Stage III	Multifocal, contiguous pan-polypoid nasal and sinusal disease
Stage IV	Multifocal, contiguous pan-polypoid disease with bony changes

Table 24.2. Symptoms in 785 patients 1 year after surgery, all stages

Symptom	Gone (%)	Decreased (%)	No change (%)
Midfacial cephalgia	84.4	14.8	0.8
Postnasal discharge	83.3	16.1	0.6
Nasal obstruction	90.7	9	0.3

Table 24.3. Symptoms in 205 patients 3 years after surgery, all stages

Symptom	Gone (%)	Decreased (%)	No change (%)
Midfacial cephalgia	80	19.5	0.5
Postnasal discharge	81.5	17.6	0.9
Nasal obstruction	88.3	11.2	0.5

Table 24.4. Symptoms in 785 patients 1 year after surgery, divided into Friedman preoperative stages I, II, III and IV

Symptom	Gone (%)	Decreased (%)	No change (%)
Stage I			
Midfacial cephalgia	87.4	11.8	0.8
Postnasal discharge	84.9	15.1	0
Nasal obstruction	94.1	5.9	0
Stage II			
Midfacial cephalgia	87.4	12.6	0
Postnasal discharge	87.1	12.9	0
Nasal obstruction	93.5	6.5	0
Stage III			
Midfacial cephalgia	84.0	14.7	1.3
Postnasal discharge	82.0	16.7	1.3
Nasal obstruction	88.0	11.4	0.6
Stage IV			
Midfacial cephalgia	71.4	28.6	0
Postnasal discharge	67.3	30.6	2.1
Nasal obstruction	85.7	12.2	2.1

Table 24.5. Symptoms in 205 patients, 3 years after surgery, divided into Friedman preoperative stages I, II, III and IV

Symptom	Gone (%)	Decreased (%)	No change (%)
Stage I			
Midfacial cephalgia	80.9	19.1	0
Postnasal discharge	89.1	14.9	0
Nasal obstruction	93.6	6.4	0
Stage II			
Midfacial cephalgia	86.5	13.5	0
Postnasal discharge	86.5	13.5	0
Nasal obstruction	94.2	5.8	0
Stage III			
Midfacial cephalgia	76.4	23.6	0
Postnasal discharge	77.5	21.3	1.2
Nasal obstruction	82.0	18.0	0
Stage IV			
Midfacial cephalgia	50.0	37.5	12.5
Postnasal discharge	50.0	37.5	12.5
Nasal obstruction	87.5	0	12.5

Table 24.6. Site of surgical intervention in 1463 consecutive patients

	Number of patients
Ethmoid sinuses	1458 right, 1451 left
Maxillary sinuses	1433 right, 1435 left
Frontal sinuses	957
Sphenoid sinuses	604
Partial middle turbinectomy	1430
Septoplasty	1268
Rhinoplasty	204

Table 24.7. Complications in 1463 consecutive cases

Complication	Number of patients	Percentage of patients (%)
Intraoperative hemorrhage	0	0
Postoperative hemorrhage with transfusion	3	0.20
Lamina papyracea defect without ocular injury	4	0.30
Ocular injury	0	0
Cerebrospinal-fluid leak	0	0
Synechiae	41	2.7

with stage-IV disease (i.e., multifocal, contiguous pan-polypoid disease with bony changes) have more dismal results at 1 year and 3 years. Therefore, stage-I, -II and -III patients can probably be counseled preoperatively with optimistic expectations, but stage-IV patients must be counseled that recurrent disease and repeated procedures may eventually be likely in nearly half the cases. Ninety-one patients of the series had recurrent disease that required repeat operations (~7%). Table 24.6 indicates site of surgical intervention for this series of 1363 patients. Note that most had bilateral ethmoid and maxillary management and partial middle turbinectomies. Also, 19% had concurrent rhinoplasties.

The charts of the same 1463 patients were reviewed for complications (Table 24.7). There were no cases of intraoperative hemorrhage requiring termination of the procedure or transfusion. Three cases of postoperative hemorrhage that required transfusion occurred in the early stages of the surgeon's experience; all three began bleeding more than 48 h postoperatively. Synechiae between the middle turbinate and lateral nasal wall in the ethmoid or natural maxillary-ostia areas occurred in 2.7% of patients. Fifteen of these 41 patients required revision procedures. Four patients experienced lamina-papyracea defects without ocular injury. In this series, there were no cases of CSF leak, brain injury or ocular injury. The overall low complication rates are attributed to the stereoscopic referencing with a fiberoptic headlight to augment the endoscopic views, and the regular use of Merocel nasal and sinusal compressed sponges in the middle meatus (4. declination) for 1 week to allow mucosalization of the surgical cavities before postoperative instrumentation.

References

1. Kennedy DW (1985) Functional endoscopic sinus surgery: technique. Arch Otolaryngol Head Neck Surg 111:643–649
2. Kennedy DW, Zinreich SJ, Rosenbaum AE, Johns ME, et al. (1985) Functional endoscopic sinus surgery, theory and diagnostic evaluation. Arch Otolaryngol Head Neck Surg 111:576–582
3. Mabry RL (1991) Inferior turbinoplasty. Otolaryngol Head Neck Surg 2:183–188
4. Stammberger H (1986) Endoscopic endonasal surgery: concepts in the treatment of recurring rhinosinusitis. II. Surgical technique. Otolaryngol Head Neck Surg 94:147–156
5. Stankiewicz JA (1987) Complications of endoscopic nasal surgery: occurrence and treatment. Am J Rhinol 1:45–49
6. Toffel PH (1989) Secure endoscopic sinus surgery as an adjunct to functional nasal surgery. Arch Otolaryngol Head Neck Surg 115:882–825
7. Toffel PH (1990) Chronic nasal obstruction. In: Gates GA (ed) Current therapy in otolaryngology head and neck surgery, 5th edn. Decker, Philadelphia, pp 283–287
8. Toffel PH (1992) Instruction course. Am Acad Otolaryngol Head Neck Surg 5:165–171

9. Toffel PH (1994) Nasal polyposis. In: Gates GA (ed) Current therapy in otolaryngology head and neck surgery, 5th edn. Decker, Philadelphia, pp 383–391

10. Wigand ME, Steiner W, Jaumann MP, et al. (1978) Endonasal sinus surgery with endoscopic control: from radical operation to rehabilitation of the mucosa. Endoscopy 10:255

Sinus Mucosal Wound Healing Following Endoscopic Sinus Surgery: Mucosal Preservation

Hiroshi Moriyama

Introduction

When treating chronic sinusitis, sufficient ventilation and drainage of the paranasal sinuses are essential for achieving physiological healing of the sinuses and rapid recovery of the ciliated mucosa lining in the sinuses. In order to achieve good postoperative aeration of the sinuses, endoscopic surgery must be performed to attain sufficient opening of the anterior and posterior ethmoid sinuses, maximum possible patency of the maxillary-sinus fontanelle and adequate patency of the frontal-recess region [2–4].

In order to obtain drainage of the sinuses, mucosal preservation is essential. Rapid postoperative recovery of mucociliary function in the sinuses depends on how quickly the cilia on the mucosal surface can be regenerated. A key factor in determining this recovery rate is the surgical treatment applied to the mucosa, i.e., the mucosal preservation of the sinuses [1].

During the surgical procedure, it is essential not to expose the bone by removing all diseased mucosa on the outer walls of the ethmoid sinus, i.e., the roof, medial wall of the orbit and lateral wall of the superior and middle turbinates. The most important requirement for successful surgery is achievement of early regeneration of normal mucosa composed of many ciliated cells.

Methods for Treatment of Sinus Mucosa and Postoperative Healing

A histological study [scanning electron microscopy (SEM) findings] evaluating different treatments of the mucosa was conducted at regular intervals after surgery to evaluate post-endoscopic sinus surgery (ESS) mucosa healing. Mucosal treatment appropriate for obtaining physiological wound healing will be discussed using ethmoid-sinus mucosa as an example.

Appropriate Treatment in Cases with Mild or Moderate Disease

If the degree of diseased sinus mucosa is mild or moderate, the lamella of the ethmoid sinus should be resected with through-cutting forceps to empty the entire cavity, and the mucosa should be preserved (Fig. 25.1). In these cases, mucosal disease is reversible with good ventilation. In intraoperative SEM findings (Fig. 25.2), scattered cilia were documented in moderate chronic sinusitis. One month after surgery, the number of ciliated cells gradually increased with good ventilation and drainage. Six months after surgery, abundant ciliated cells were observed, and the mucosa had become nearly normal.

Appropriate Treatment in Severe Cases

If the degree of diseased sinus mucosa is severe, with polypoid changes and irreversibly diseased mucosa, the inflamed mucosal surface and part of the subepithelial layer should be resected with through-cutting forceps (Fig. 25.3). However, we do not expose the bone surface. With this treatment, normalized epithelium will cover the diseased portion. Re-epithelization soon occurs from the surrounding tissue and covers the normalized connective tissue. Postoperative findings of resected, diseased mucosal surfaces were observed with SEM (Fig. 25.4). One month after surgery, flattened epithelium without cilia covered the whole lesion. However, 6 months after surgery, regenerated ciliated cells covered many parts of the lesion and, by 1 year after surgery, normalized ciliated cells were regenerated.

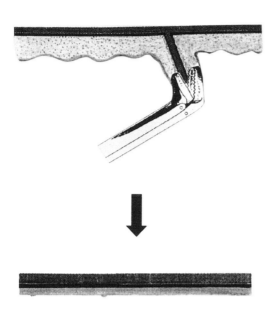

Fig. 25.1. Sinus mucosa is left intact (only the lamella is resected, using cutting forceps)

Inappropriate Treatment

If the diseased mucosa is completely removed and the bone surface is exposed widely with cup-type forceps (Fig. 25.5), regeneration of mucosa is delayed. By removing the entire layer of diseased mucosa, the exposed bone becomes covered with hypertrophic mucosa having flattened epithelium that is not ciliated. SEM (Fig. 25.6) inspection revealed that only a few cilia and abundant non-ciliated cells were present both 6 months and 2 years after surgery.

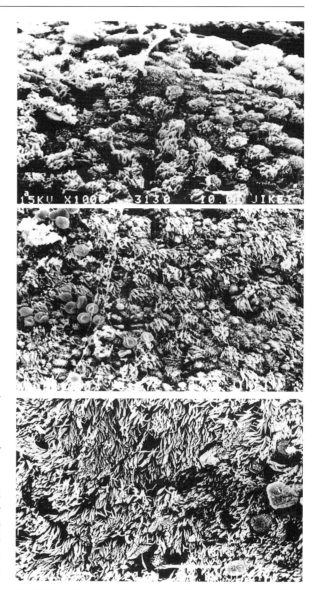

Fig. 25.2 a–c. Postoperative scanning electron microscope findings in the ethmoid sinus. **a** Intraoperative findings. **b** 1 Month after surgery. **c** 6 Months after endoscopic sinus surgery (all regions was covered with normalized ciliated epithelium)

Importance of the Selection of Forceps and Other Surgical Instruments

In order to preserve mucosa and prevent exposure of bone, appropriate selection of instruments (such as through-cutting forceps) is very important. The using of cup-type forceps should be avoided as much as possible.

Taking care to avoid exposing the bone surface results in rapid regeneration of mucosal epithelium, with resumption of the mucociliary functions necessary for healthy sinuses. Considering the importance of proper mucosal treatment to achieve the desired cure of sinusitis, it is clear that effective cutting instruments – such as cutting forceps and the microdebrider – are extremely useful in the surgical therapy of sinusitis.

Figure 25.7 shows the basic cutting forceps, which should be used to carry out the treatment of the pathological sinus mucosa described above. The microdebrider provides yet another option. In recent years, the range of applications of the microdebrider

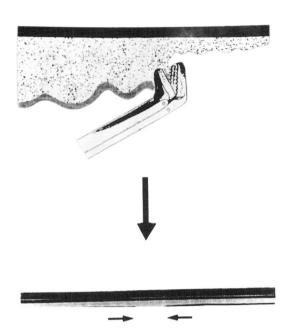

Fig. 25.3. The inflamed mucosal surface and part of the subepithelial layer is resected with cutting forceps

has rapidly expanded, and it is now being used for the treatment of the paranasal sinuses and of deep regions of the nasal cavity.

Importance of Postoperative Care

Appropriate postoperative care greatly improves the results of operations for chronic sinusitis. Recurrence of sinusitis is highly probable without appropriate postoperative treatment, such as periodic observation, cleaning of the sinus and drug therapy (low-dose, long-term macrolide therapy).

In order to prevent infection, an antibiotic (such as penicillin) is administered for 1–2 weeks. Next, low-dose, long-term macrolide (erythromycin, roxithromycin, clarithromycin) therapy is given for 3–6 months after ESS. The effects of this therapy are cell-activating actions, such as stimulation of migration of multinucleated leukocytes and activation of macrophages and lymphocytes (in addition to the macrolide's primary, anti-bacterial effect). The actions also include acceleration of mucociliary function, activation of natural-killer cells, suppression of mucus secretion and regulation of various immunological actions.

If the patient is allergic, an anti-allergic agent is simultaneously administered. Steroids are given in

Fig. 25.4 a–c. Postoperative scanning electron microscope findings. a 1 Month after surgery. b 6 Months after surgery. c 1 Year after surgery (the entire region was covered with normalized ciliated cells)

very small doses for 3–4 weeks in patients with severe allergy or in eosinophil-dominant cases. Asthma is involved in some patients showing poor postoperative healing; thus, systemic control of asthma is also an important aspect of postoperative care.

It is thought that patients should be followed-up for 2 years. Sinuses are to be inspected with an endoscope, and drug therapy for recovery of mucosa and corrective treatment for postoperative changes (such as adhesion and stenosis and obstruction of the sinus entrance) are performed. Postoperative

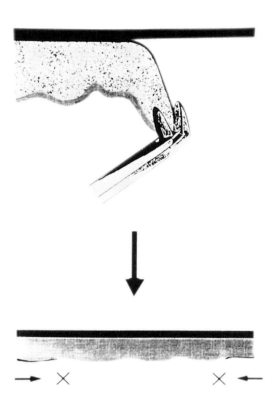

Fig. 25.5. When the mucosa is completely removed and bone surface is exposed, regeneration of mucosa is delayed

Fig. 25.6 a,b. Scanning electron microscope findings. Most of the mucosal surface of the bone-exposed region was covered with nonciliated epithelium (**a**, 6 months), and there were only scattered ciliated cells (**b**, 2 years)

observation of the sinus at regular intervals reveals that 3–5 months are required for stabilization of the mucosa in most patients, although the time depends on the preoperative severity of lesions. Therefore, postoperative care for at least a year is necessary.

Discussion

The basic principle of treatment of sinusitis is that sufficient aeration and drainage of the paranasal sinuses are essential. For achieving healing of the sinuses and improving the mucociliary functions, it is necessary to effectively treat the pathological mucosa in the sinuses.

The conventional approach to the treatment of chronic sinusitis has been complete removal of pathological mucosa in the ethmoid sinus. However, if the mucosa and periosteum are removed and the bone surface is exposed, the exposed bone surface is first covered by edematous tissue, which later undergoes cicatrization. The regeneration of

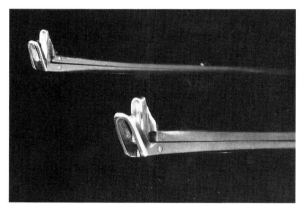

Fig. 25.7. Upward cutting forceps (regular size and slim type)

healthy mucosal epithelium on top of this scar tissue is slow; a long period of time is required before the surrounding mucosal epithelium grows out and covers the scar tissue. SEM observations show that, in some patients, normal ciliated epithelium did not regenerate until more than 1 year after the surgery. When the ethmoid-sinus mucosa is completely removed, the regenerated mucosal epithelium has only a small number of ciliated cells, and it can be surmised that the drainage function of this mucosal epithelium is poor.

Accordingly, the mucous membranes should be left intact as much as possible. These mucous membranes include the mucosa of the ethmoid-sinus roof, the lamina papyracea, the lateral wall of the middle turbinates, the maxillary sinus, the frontal recess, etc. When the mucosa of the ethmoid sinus is pathological and is observed to be edematous or polypous, the epidermis of the pathological mucosa and the subepidermal pathological tissues should be excised using a cutting forceps. A thin layer of connective tissue should be left on the bone surface, and the surface of the bone of the limiting walls should never be exposed. Performance of the excision of the pathological sinus mucosa in accordance with these guidelines will result in smooth epithelization of ciliated mucosa, and the mucociliary function will also be recovered. Based on these morphological observations, excising the mucosa with a through-cutting-type forceps without removing all of it is important in preserving the normal function of the mucosa after surgery.

Particularly in frontal recess region, mucosal preservation is essential for maintaining the postoperative opening. Even diseased mucosa around the frontonasal duct and frontal recess should be left intact in order to achieve good postoperative communication. Cells and lamella of frontal recess should definitely be excised to allow wide communication to the frontal sinus. If diseased mucosa is widely removed and the bone surface of frontal recess is exposed, granulation and edematous tissue occur soon after surgery. Communication will then be obstructed by fibrous tissue and scar tissue.

Conclusion

In the treatment of chronic sinusitis, the mucosa of the ethmoid sinus should be conserved; if the mucosa has been severely diseased, only the mucosal surface should be removed. In addition, in case of irreversibly edematous and thickened pathological mucosa, only the epidermis should be cut, and normal, ciliated mucosa should then be allowed to regenerate. Thus, it is important that the bone surface never be exposed due to excessive removal of the mucosa. Correct treatment of diseased mucosa in sinuses, appropriate selection of the forceps and proper postoperative care lead to good postoperative wound healing.

References

1. Forsgren K, Stierna P, Kumlien J, Carlsöö B (1993) Regeneration of maxillary sinus mucosa following surgical removal – experimental study in rabbits. Ann Otol Rhinol Laryngol 102:459–466
2. Moriyama H, Ozawa M, Honda Y (1991) Endoscopic endonasal sinus surgery. Approaches and post-operative evaluation. Rhinology 29:93–98
3. Moriyama H, Yanagi K, Ohtori N, Fukami M (1995) Evaluation of endoscopic sinus surgery for chronic sinusitis: post-operative erythromycin therapy. Rhinology 33: 166–170
4. Moriyama H, Yanagi K, Ohtori N, Asai K, Fukami M (1996) Healing process of sinus mucosa after endoscopic sinus surgery. Am J Rhinol 10:61–66

Wound Healing after Endonasal-Sinus Surgery in Time-Lapse Video: A New Way of Continuous in Vivo Observation and Documentation in Rhinology

Rainer Weber, Rainer Keerl, Andreas Huppmann, Wolfgang Draf, and Anjali Saha

Introduction

Following endonasal sinus surgery, the operated area becomes a more or less extensive wound. Surgical trauma activates wound-healing mechanisms, the understanding of which is extremely important for the analysis of endonasal procedures. From this knowledge of physiology and pathophysiology of wound healing, the best and most effective course of follow-up treatment arises. We know that the postoperative treatment significantly influences operative results [27, 38, 118].

Cutaneous Wound Healing

First, we will present a general overview of cutaneous wound healing [16–19, 21, 36, 39, 41, 52–54, 58, 59, 68, 72, 77, 88, 89, 92, 96, 105, 109, 110, 125]. Normal wound healing is characterized by a complex sequence of physiological processes after the organism is injured in some way. This is to protect against a repeated injury (whether caused by mechanical factors, heat or invasion of pathogenic microorganisms), to prevent the loss of important substances (for example, proteins or water) and to replace anatomical features and repair structural anatomical damage. Normal wound healing comprises three temporally overlapping stages: the inflammatory phase (the first 3–4 days), the proliferative phase (10–14 days) and the phase of contraction and remodeling (6–12 months).

The inflammatory phase begins with an initial hemostatic reaction to injury. Vessels contract for about 5–10 min after injury, and then dilate. Platelets adhere to collagen and aggregate with each other, forming a hemostatic plug. While the platelets are aggregating, the extrinsic and intrinsic coagulation cascades are activated, forming a fibrin network within the wound. This provides initial stability to the wound, increases hemostasis and helps form the scaffolding over which cellular migration occurs. A number of biologically active mediators are released, and the kinin pathway and complementary systems are activated. This results in vasodilation, increasing vascular permeability (with extravasation of fluids) and chemotactic, proliferative and activating signals for effector cells, such as polymorphonuclear leukocytes (PMNs), monocytes, macrophages and fibroblasts.

PMNs are the dominant cells during the first days of healing. They are attracted to the wound approximately 6 h after injury, reach their greatest number after 24–48 h and begin to disappear by 72 h after injury. PMNs act as phagocytes, removing debris and bacteria from the wound. After phagocytosis, these cells die within a few hours and release their intracellular contents, contributing to the local exudate; they may also be phagocytosed by macrophages.

Macrophages infiltrate the wound within 48–96 h after injury. They play the most important role in regulating further wound healing. This control is achieved through release of many biologically active mediators. The proliferative phase (lasting 10–14 days) consists of epithelialization, fibroblast proliferation (with deposition of a collagenous matrix) and angiogenesis. Resurfacing of cutaneous defects begins 12–24 h after injury. Epithelial cells near the wound edges undergo phenotypic changes; replication increases, and migration occurs across the wound. This migration involves leap-frogging of the cells and interaction with fibronectin and cellular receptors in a "ratchet" mechanism. The migration rate depends on the local humidity and oxygenation.

Proliferation of endothelial cells starts the second or third day after injury. Angiogenesis is first stimulated by growth factors from activated platelets. Later, macrophages secrete angiogenic factors in response to high lactate concentrations and low oxygen levels. The result of endothelial proliferation and migration is capillary-bud formation and the gradual appearance of anastomotic vascular channels. The increasing tissue PO_2 reduces the stimulus for additional angiogenic-factor release.

Fibroblasts appear 48–72 h after injury. A lag phase of approximately 48 h occurs before significant collagen synthesis. This results when blood flow returns and the oxygen level increases due to angiogenesis. Collagen synthesis reaches its maximum in the first 2 weeks. Wound collagen levels reach a maximum at 3 weeks after injury.

During the first few days, the integrity of the wound is determined by adhesion of the wound surfaces and the fibrin network and by the cohesion of the newly formed epithelium. After 2 weeks, it consists of only approximately 10% non-wounded skin. In the following weeks, it correlates with the rate of collagen synthesis and so-called remodeling, which begins after approximately 3 weeks. This consists of rearrangement of collagen (with removal of water and proteoglycans), replacement of type-III collagen with type-I collagen, formation of larger bundles of fibers and alteration of inter- and intramolecular networks ("cross-links") [7]. A scar rich in fibers and with few cells results; this scar shows poor vascularity. Maturation of the scar takes months or years [71, 75, 76, 108].

Wound Healing in the Region of the Respiratory Mucosa

Wound healing in the region of the respiratory mucosa has a course similar to that of wound healing in the skin [8, 61, 62, 70, 88, 89, 119]. Just a few hours after the wound settles, intact cells on the wound edges flatten out, lose their cilia and migrate into the wound area. In 8 h, up to 200 µm distance can be covered [119]. The elastic lamina serves as a guiding path. If this is disturbed, then all motion halts; this is followed by a more gradual migration. The rate of mitosis of the basal cells is highest 24–48 h after wound settlement. After 96 h [88, 89], organization and differentiation begin. Transitional cell epithelium becomes cuboid and regains its polarity. The regeneration is complete (*restitutio ad integrum*) when structures lying beneath the epithelium remain intact. Otherwise, poor differentiation and scar formation occur. Functional differentiation is partially controlled by these subepithelial mesenchymal structures [88, 89]. Boling observed a new ciliary attachment beginning on the 13th day [11]. Following denudation of the tracheal epithelium of a chicken, Battista and others found overwhelming regeneration after 29 days [9]. Keenan was able to observe almost normal epithelium after 120 h. Mesenchymal changes, however, took longer than epithelial ones.

Wound Healing in Nasal Mucosa

In both animal and clinical experiments, Hosemann systematically researched wound healing in nasal mucosa by both endoscopic and histological examination [47–51]. In the animal experiments, he observed histological changes after a standardized injury made on the medial maxillary-sinus wall of a rabbit. For this injury, a circular wound 4 mm in diameter was made, exposing the bone.

Twelve hours after this injury, exposed bone in this area was found to be covered with fibrin, detritus, erythrocytes and granulocytes. The epithelium on the wound edges began to flatten toward the wound and to be pushed onto its surface. After 48 h, the wound base was irregularly or completely covered by immature mesenchymal cells. Epithelial migration from the wound edges continued. The cells were loosened, enlarged and vacuolized. More frequent mitoses were observed at the wound edges and injured area. Almost all the epithelium at the wound edges was arranged in two layers. Most of the cells retained their cilia. A similar finding was seen after 72–96 h, with increasing closure of the wound. Bony ridges in the injured area are replaced by osteoclasts. After 120 h, a lamina propria was partially formed over the bone. The first addition of bone substance in the wound area was seen at that time. At 168 h after injury, distinct osteoid apposition of varying development was observed. Additionally, exuberant granulation developed on top of the surrounding intact mucosa. The migrating epithelium in this area did not possess cilia. After 192 h, the newly formed granulation tissue increased the thickness of the mucosa three- to sixfold. The frontline epithelium did not possess cilia, but those more lateral did. At 238 h following injury, the maximal thickness of granulation tissue was achieved, though marked inflammatory infiltration was absent.

The wound as a whole concentrically decreased in size by migration and proliferation. The rate of migration declined from 20 µm/h (in the first 24 h) to 4–5 µm/h (after 168 h). The regenerated epithelium usually retained parts of its ciliary attachments but often lost its multiple rows for the advantage of multilayerability. Mature metaplasia did not occur. Osteoclasts were noticed 48 h after operation. After 120 h, activated osteoblast seams were seen to be associated with increases in the osseous matrix. At the same time, tissues began to produce hyperplastic granulation. Glandular regeneration was not seen to occur during the observation period.

In summary, wound healing following endonasal ethmoidectomy (as observed by endoscopic observation) was completed in four phases [47, 48, 51]:

1. Blood crusting of the operated area (up to 10 days)
2. Obstructive lymphedema (up to 30 days)
3. Mesenchymal transformation (up to 3 months)
4. Scar formation (over 3 months)

A Dynamic Process Must be Analyzed by a Dynamic Method

Analysis of physiological and pathophysiological processes is usually carried out at a fixed point in time toward the end of the process under study, and it is accessible to visual observation [63, 112–114, 116]. However, any such measurement is only a snapshot of a continuously changing system. Such a measurement may be sufficient for functional analysis, provided it represents a constant quantity in a control circuit. A functional system that is continually changing over time, however, can be adequately recorded only by a measuring or visualization technique that takes proper account of the time function. However, in many cases, a longitudinal continuous measurement lasting several days or weeks is obviously not practicable. The presentation of sinus wound healing as a dynamic process in its totality has not been accomplished. Until recently, documentation was accomplished by instant endoscopic photos and histological examination [48, 101].

Methods

To document long-lasting mucosal changes in the dynamic state, we looked for a method that would be non-invasive and practical for use in a conscious patient [63, 112–114, 116]. Endoscopy has proven to be a suitable imaging method [23, 25, 66, 82, 100, 118].

Continuous videoendoscopy for the documentation and visualization of these mucosal changes is unsatisfactory because of non-compliance of the patient or the inconstancy of the image (due to wobbling). Replay involves the use of time-lapse techniques, and the picture disturbances due to the wobble effect become an unavoidable part of the picture content. Instead of presenting a continuous film, we processed "snapshots" and reconstructed the natural dynamics of a time-dependent process so that we could actually see and comprehend the process realistically. The method is based on computer-assisted "morphing" and the preparation of a time-lapse video. Morphing is a general term for the process of transforming shapes or forms into something different (for instance, turning a square into a circle) [87]. In computerized animation, it means the smooth, flowing metamorphosis of two objects. The switching of the system is shown in Fig. 26.1. Table 26.1 shows the necessary equipment.

Morphing enables us to make a smooth, continuous transition from one chronologically discrete measurement (snapshot) to another, thereby visualizing the natural dynamics in a manner not previously possible. The static measurements for the observation of changes in the nasal mucosa are obtained by videoendoscopy. The best shot is selected from a video sequence covering several seconds, is stored as a single take (called an "original", because it comes from the original film) and is loaded into the computer (Fig. 26.2).

If the process being studied is to be adequately recognized, it must not run too quickly or too slowly when adapted to the physiology of signal reception and the processing of optical stimuli. We have selected a time unit of 60 s as convenient and manageable. The events being analyzed have been compressed into these 60 s. Video film requires 25 frames/s. This means that 1500 frames are needed for 1 min of film. These are made up of originals (paired A and B points; Fig. 26.2) and the morphed images, which are placed between them. In this way, a time-lapse film of

Table 26.1. Equipment used for videoendoscopy and computer-assisted morphing

Videoendoscope	Rigid 4-mm, 0° and 30° endoscopes (Hopkins Optic) 0.5-in. Charge-coupled-device video camera 175-W xenon light source, color monitor S-VHS video recorder Screen divider [Panasonic Color Quad System WJ-450 (Stuemer, Würzburg, Germany)]
Hardware	PC (586, 90 MHz CPU, 16 MB RAM) 1-GB hard disc Graphics board [Elsa Winner 1000 Pro VESA VGA version 1.06 (Elsa, Aachen)] PC-compatible digital video studio [Video Machine Lite (Fast Electronics, München)] Single digital player/recorder (Fast Electronics, München) with a 1-GB internal hard disc
Software	Photomorph (North Coast Software, Barrington, USA)

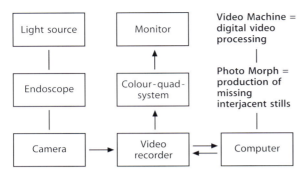

Fig. 26.1. Switching of the system

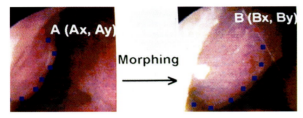

Fig. 26.2. Image processing by morphing single frames. Here, we see an endoscopic view of the right nasal cavity, with various states of swelling of the inferior nasal turbinate. The time difference between the first and the second image is 2 h. The inferior nasal turbinate has become swollen in the course of the nasal cycle, as shown in the right-hand picture. The two pictures, which are virtually identical except for the swelling, were downloaded from the computer's memory into the morphing program. Minor incongruities not related to the process can be removed by means of image processing programs in order to clarify what is actually going on. Mathematically speaking, the morphing used here is a two-dimensional interpolation. One generates any number of intermediate points T_i ($A_x + i \times (B_x - A_x)/N$; $A_y + i \times (B_y - A_y)/N$, where i=1 to N–1) between point A (A_x, A_y) of one image and the corresponding point B (B_x, B_y) of a second image. By defining as many paired points A–B (*black*) as possible and creating intermediate points by interpolating (i.e., morphing), the image changes. The greater the number of paired points defined, the more precise the smooth transition will be. If one then displays points A, T_i and B in sequence on a monitor, one will see a continuous movement of the points from A to B; in our example, this would be a swelling process in the inferior nasal turbinate

the process being analyzed is obtained. Using split-screen video techniques, the changes taking place on the two sides can be displayed simultaneously on one screen. Adding a time axis permits better orientation.

Patients and Postoperative Treatment

Wound healing after endonasal sinus operations on 23 sinuses in 12 patients was investigated over a time period of 6 months. Preoperatively, nasal endoscopy showed variably pronounced nasal polyps in all patients. Total or subtotal opacification of the ethmoid labyrinth was found in all cases on preoperative computed tomography (CT) of the paranasal sinuses. The maxillary, frontal and sphenoid sinuses showed varying degrees of associated involvement. In one patient, an osteoma of the anterior ethmoid sinus was found in addition to a unilateral subtotal opacification of the frontal and maxillary sinuses. With this degree of involvement, the endonasal pansinus operation was performed in 19 cases, with complete opening of the ethmoid sinus [27], wide fenestration of the maxillary sinus and widening of the frontal-sinus drain via type-II drainage [26, 27]. The frontal nasal-duct opening was not addressed in four cases (type-I drainage). The sphenoid sinus was opened in 12 cases. Three patients had previously undergone external sinus surgery (two had Caldwell Luc and transmaxillary ethmoidectomy, one had endonasal ethmoidectomy); one patient suffered from bronchial asthma and analgesic intolerance. For better comparison and standardization, all surgery was carried out by one surgeon.

Postoperative care consisted of (1) packing the operative cavity with rubber finger packs (Rhinotamp; Vostra, Aachen) after pack removal and (2) two or three daily rinses with a nasal douche (Rhinocare; Siemens and Company, Bischofsheim) and Ems brine (Siemens and Company, Bischofsheim) for 6 months. Manual removal of crusts was done only when severe crusting caused significant nasal-airway obstruction. In a few cases, excessive secretion had to be suctioned off for examination. We defined the wound healing in this type of postoperative treatment as mostly uninfluenced (or "normal") if the tamponade remained for only 3 days. In some cases, to examine the effect of a variably long-standing tamponade, it was left in situ for 1–3 weeks. To document any possible effects of topical-corticosteroid therapy, topical application of 50 µg of budesonide (Pulmicort Topinasal; Astra, Wedel) was carried out twice daily for 6 months [34]. In order to counteract the sometimes pronounced individual differences occurring during the course of wound healing, it seemed necessary and sensible to make some intra-individual comparisons. If the two sides are treated differently postoperatively in a patient with similar extents of disease and similar operative interventions, differences in the course of healing can probably be accounted for by the variable effects of the postoperative treatment procedure.

In total, ten operated sides with "normal" wound healing, ten operated sides with long-standing tam-

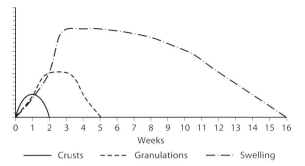

Fig. 26.3. Time-dependent distribution of mucosal changes during wound healing after endonasal sinus surgery

ponades (lasting 1 week in four cases, 2 weeks in four cases and 3 weeks in two cases) and eight operated sides receiving topical budesonide therapy were evaluated. Use of a long-term tamponade and corticosteroid application overlapped five times. Evaluation was done by analysis of time-lapse films and by taking into consideration an additional flexible endoscopy evaluation obtained at the end of the follow-up period.

Results

In uninfluenced wound healing (Figs. 26.3, 26.4), it was observed that blood crusts and eschars covered the operative field within the first 7–12 days. These freed themselves spontaneously in stages and could be removed with instruments or could be douched out or blown out by the patient. Granulations were seen about 1 week after operation and usually remained visible for approximately 2–4 weeks. The presence of blood clots and crusts prevented definite visualization of early proliferating granulation. At about the same time, an increasingly edematous swelling developed; it reached its maximum between the third and fifth weeks. Thus, the entire operated area may be completely swelled closed so that the entrance to the ethmoid cavity and other sinuses was sealed. Contact between the middle turbinate and lateral nasal wall (due to the pronounced swelling) can lead to the development of broad synechia at this stage. The swelling usually began to regress between the 7th and 12th weeks (in 78% of cases); in three cases, regression began during the fourth week (13% of cases) and, in a single case, onset of regression was the delayed until the 15th week. In one instance, no significant regression occurred during the observation period. The transformation of the more edematous, pale blue swelling into a more reddish mucosa (an expression of Hosemann's mesenchymal remodeling [48, 51]) was usually complete with the regression of the swelling. A macroscopically complete normal mucosa was achieved in 17 cases (74%). Residual swelling was documented in the paranasal-sinus systems of the rest of the patients. Normal mucosa could be observed between the fourth and 18th weeks, most frequently from the 12th to 18th week [in 12 of 17 cases (70%) of the clear paranasal-sinus systems]. In all ethmoid cavities, marked scars could be seen on the concave surfaces. These were first seen clearly during the stage of healing. Broad synechia developed between the middle turbinate and lateral nasal wall in three cases (Fig. 26.5), and smaller synechia developed between the middle turbinate and septum in two cases. The frontal-sinus interior could be visualized with the flexible endoscope in seven cases (30%) and was wide open. The frontal sinus could not be visualized endoscopically due to scarring in 11 cases (48%) and due to polypoid mucosal swelling in the region of the ethmoid- and frontal-sinus openings in three cases (13%). That does not necessarily mean that the frontal-sinus drainage was occluded. In two patients (9%), the entrance to the ethmoid sinus in the middle meatus had already closed due to major synechia.

The time courses of uninfluenced wound healing were so variable that a more exact detailed description is not possible. The range of the "natural" course varied between a very quick healing without marked swelling lasting 9–12 weeks, a medium course of 18 weeks and a prolonged course lasting 26 weeks, with pronounced swelling. Clearly variable courses ranging from relative symmetry to marked differences were found for each side with respect to the extent and duration of edema, even though the extents of disease, therapy and postoperative treatment were almost identical. These observations make it difficult to establish a precise "normal" course of wound healing.

With this background, the evaluation of the effects of a long-term tamponade and topical steroid therapy become problematic. As a group, patients treated with budesonide had a very variable healing course similar to the course that occurred without treatment with steroids. However, analysis of patients who had the same treatment (with respect to the tamponade and nasal care) on both sides and who also applied budesonide on one side showed a definite tendency for reduced amounts of granulation tissue, edema and swelling on the side receiving steroid application. Normal mucosa was re-established earlier on this side (Fig. 26.6).

The effect of a long-duration tamponade with an occlusive of rubber finger pack dressing was more

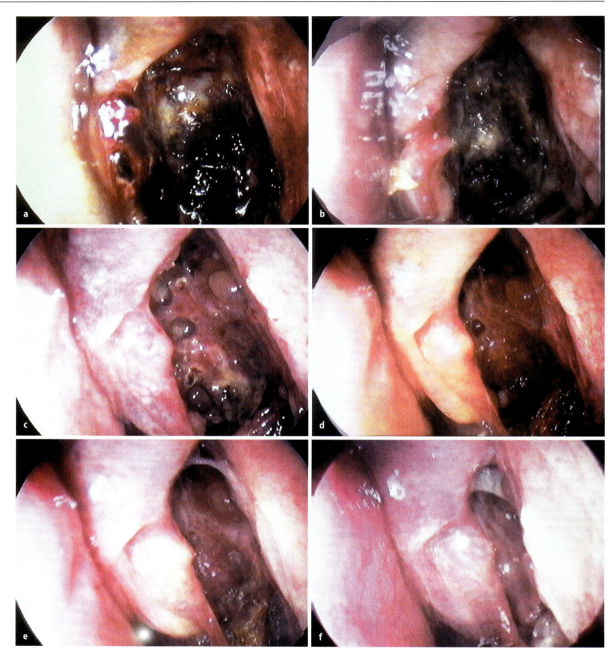

Fig. 26.4 a–j. Excerpted extracts taken from a time-lapse film of "normal" wound healing following an endonasal pansinus operation. An endoscopic view into the left nasal cavity with a 30°, wide-angle optic (Stuemer, Würzburg). **a** Four days after operation (1 day after pack removal). Blood crusts cover the operated cavity. **b** One week after operation. There is a definite decrease in blood crusts. In the wound base, separations of fibrin and secretions are found. Granulations have begun to sprout. **c** Two weeks after operation. There is a definite increase in granulation formation. An increasingly edematous swelling has begun to develop. **d** Three weeks after operation. There is more visible granulation and an increase in swelling. **e** Four weeks after operation. There is a further increase in swelling, with definite constriction of the entrance to the operative cavity. **f** Six weeks after operation. The swelling has reached its maximum. Only a slit for aeration of the cavity remains. **g** Eight weeks after operation. The extent of swelling remains similar; the mucosa is edematous and thickened. **h** Twelve weeks after operation. A gradual regression of the swelling has begun. The mucosa changes slowly and stepwise into a macroscopically normal condition. **i** Sixteen weeks after operation. Endoscopically, the mucosa appears normal. Despite this, there is thickening of the tissue as a whole. **j** Twenty-six weeks after operation. In the past 10 weeks, a further reduction of the mucosal thickening has occurred. The mucosa's color has become pale

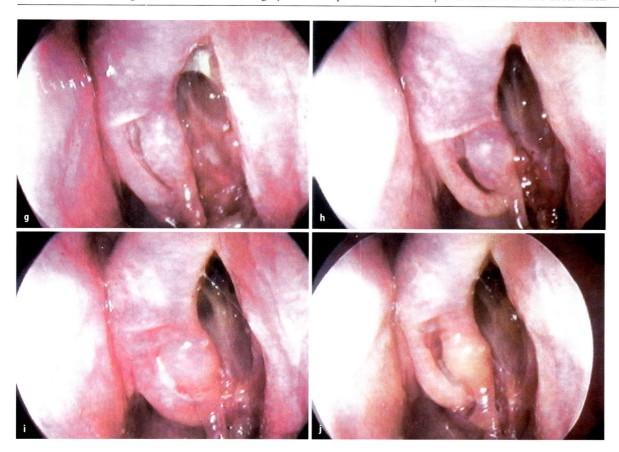

difficult to judge. However, video analysis showed that, in the anterior portion of the operated area, edema and granulations were generally less pronounced than in cases with short-duration tamponades. Sheet scarring in the ethmoid region seemed to be less marked in cases with long-duration tamponades than in other cases. Scarring and stenosis of the frontonasal duct opening seemed to be unaffected. This can be explained by the location of the packs; they lie away from the nasofrontal duct, in the nose and middle meatus or in the inferior ethmoid opening.

In our series, edema lasted much longer than described by Hosemann et al. [48, 51], and this could not be entirely explained by extravasation and increased vascular permeability resulting from the inflammatory response to injury. Another important factor was the accumulation of lymph (lymphedema) due to damaged lymphoid circulation. Furthermore, the diseased mucosa itself had a tendency for edematous swelling before surgery. Another possible causative factor is a more severe surgical trauma (complete pansinus operation with extensive removal of irreversibly diseased polypoid mucosa).

The dynamic-observation method allows us to study the time-related behavior of site-dependent edema for the first time. In the film, it can be seen that edema and swelling in the ethmoid entrance (the anterior end of the middle turbinate medially, the agger nasi laterally, the anterior ethmoid region and the frontal recess) are clearly more pronounced than in the posterior ethmoid. There are many possible explanations for this. This area of the operative field experiences the greatest surgical trauma, because it is the entrance point for the entire sinus surgery; therefore, it receives the maximum mechanical manipulation by instruments. Postoperative stress due to the drying and mechanical effects of air during respiration is also greatest in this region. Furthermore, the disease is often most extensive in this area, because the ostiomeatal complex has a key role in the development of chronic sinusitis [67, 84, 117]. Whether topographical variations in the mucosa and differences in its reaction to trauma also play a role remains unknown.

Four patients suffered from an upper-respiratory viral infection toward the end of the observation period. These always led to a clearly visible reaction in the nose and sinus mucosa, taking the form of a

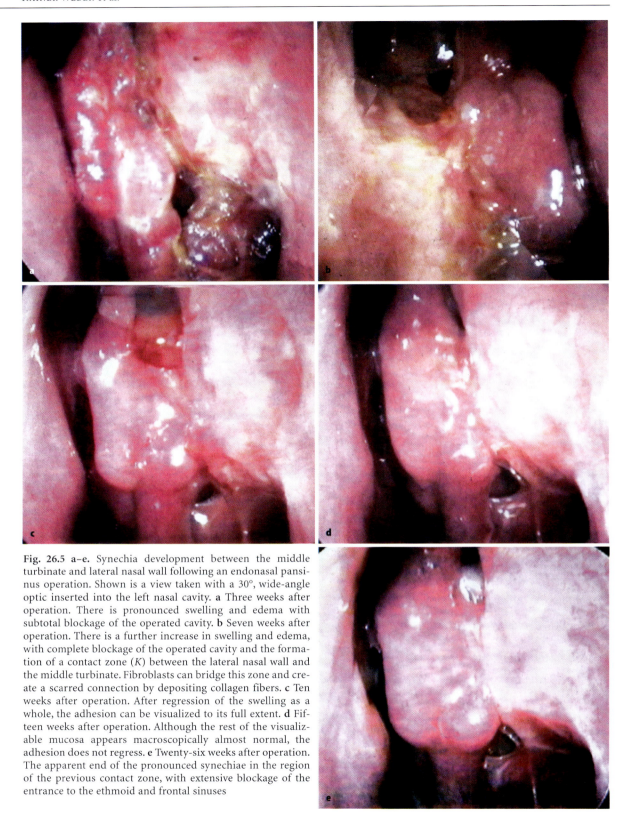

Fig. 26.5 a–e. Synechia development between the middle turbinate and lateral nasal wall following an endonasal pansinus operation. Shown is a view taken with a 30°, wide-angle optic inserted into the left nasal cavity. **a** Three weeks after operation. There is pronounced swelling and edema with subtotal blockage of the operated cavity. **b** Seven weeks after operation. There is a further increase in swelling and edema, with complete blockage of the operated cavity and the formation of a contact zone (*K*) between the lateral nasal wall and the middle turbinate. Fibroblasts can bridge this zone and create a scarred connection by depositing collagen fibers. **c** Ten weeks after operation. After regression of the swelling as a whole, the adhesion can be visualized to its full extent. **d** Fifteen weeks after operation. Although the rest of the visualizable mucosa appears macroscopically almost normal, the adhesion does not regress. **e** Twenty-six weeks after operation. The apparent end of the pronounced synechiae in the region of the previous contact zone, with extensive blockage of the entrance to the ethmoid and frontal sinuses

glassy, gray edema. This remained visible for 3–6 weeks and took a significantly longer time to regress than the phase of edema formation, whether after the primary operative trauma or after intercurrent infections.

What are the practical considerations we can learn from the physiology and pathophysiology of wound healing illustrated on the time-lapse films of different postoperative-treatment modalities?

1. In an extensive operation, a correspondingly longer period of time is to be expected before the desired normal mucosal lining is achieved. The patient must have it clearly explained to him that, in these cases, it usually takes many months. Even severe swelling or edema seen during this period should not be equated with a recurrence of polyposis and, therefore, must not cause discouragement. The examiner and treating ear, nose and throat specialist must evaluate the operative results with the same understanding. This final evaluation should take place after 12 months at the earliest.

2. The fundamental rule that the wound-healing reaction (comprising inflammation, edema and, ultimately, scarring) will be more pronounced the greater the trauma holds true here [53]. The demand for minimization of surgical trauma is banal but still relevant. The larger the fibronectin matrix based on the inflammatory reaction, the greater the probability that fibroblasts will wander along this pre-formed pathway, depositing collagen. Scarred synechia results. One should avoid:
   - Removal of tissue, thus creating extended areas with open wound surfaces and bare bone
   - Blunt avulsion of tissue, with bruising and stretching of remaining structures
   - Extended drilling, with mechanical and thermal damage to the surrounding tissue and production of tiny tissue particles that necrose
   - Aggressive placing of the speculum, damaging primary healthy mucosa in the anterior nasal part during the use of the microscope

The use of sharp cutting instruments (Fig. 26.7), whereby pathological tissue can be accurately removed and healthy tissue left behind, is recommended. The use of shaver systems (Fig. 26.8) also helps to minimize surgical trauma, by precise suction and cutting of polypoid tissue.

3. Postoperative tamponading of the surgical cavity is both helpful and necessary. Following an injury, the organism forms a so-called scab or crust in order to protect itself from heat, the invasion of pathogenic bacteria, a repeated injury or the loss of important substances (protein, water, etc.). A crust, however, results in drying out of the wound, thus negatively affecting wound healing (necrotic areas, difficulty in epithelial migration, hampering of neutrophil function, absence of growth factors).

Occluded, moist wounds provide a near-ideal environment for migration and fast resurfacing [36]. Almost all documented clinical experiences have shown that occlusively dressed wounds are less painful, less tender, less swollen and less red than open wounds [1–5, 12, 13, 28, 30, 33, 40, 46, 55, 60, 73, 78, 79, 81, 90, 91, 94, 120, 122, 123].

In addition, the eventual appearance of occlusively dressed wounds is superior. Beneath the film, epidermal cells begin to migrate sooner and complete their resurfacing more quickly. Connective-tissue regeneration begins about 3 days earlier [121], and the inflammatory response is reduced.

A similar occlusive dressing in the sinuses is, in reality, rather more complicated. Considering the type and location of such dressings, several factors must be considered for an ideal dressing. The dressing should be occlusive and ideally should remain in this position for a longer period of time than is necessary for many dressings, but must still be able to be removed without undue difficulty for the doctor or patient. Spontaneous dislodging of the dressing, creating a risk of aspiration, must be prevented. Furthermore, the breathing and smell functions of the nose should be retained as much as possible. It is also desirable that, during the immediate postoperative period, blood and secretions be absorbed or drained, and there should be free drainage from the opened sinus. An ideal dressing with these characteristics does not yet exist.

The use of rubber finger packs (Rhinotamp; Vostra, Aachen) is a compromise solution. Video analysis allows recognition of a general decrease in granulation, edema and scar areas (with a tamponade period of at least 1 week) and is usually well tolerated by patients. The finger packs permit occlusion but do not become adherent; i.e., no tight binding between the tamponade and the regenerating wound occurs. Removal of the tamponade does not lead to renewed tissue trauma (as was demonstrated to occur with adherent material by Kühnel et al. in 1995 [69]). However, only the nasal cavity itself, the middle meatus and the anterior inferior ethmoid sinuses can be packed with certainty. An exact and controllable placement in the paranasal sinuses themselves is not possible. Occlusion of the nasal cavity, with complete obstruction to the nasal airway over a longer period of time, is only poorly tolerated by

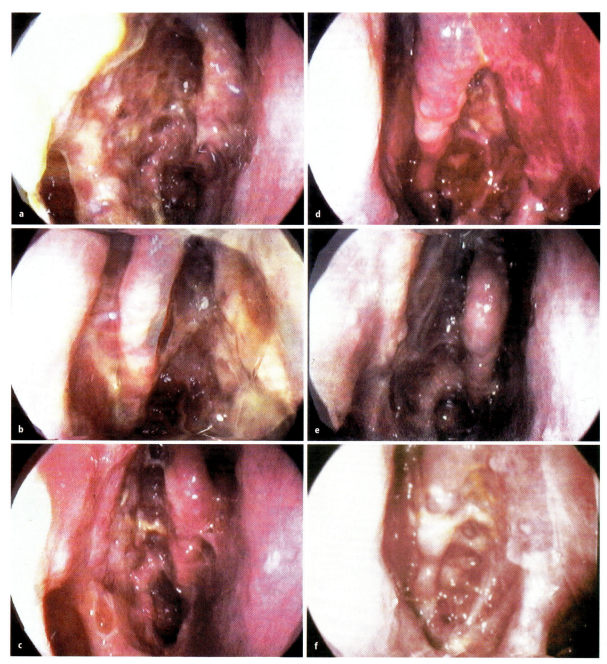

Fig. 26.6a–l. Effect of topical budesonide therapy (Pulmicort Topinasal; Astra, Wedel, Germany) following endonasal pansinus surgery. A view take with a 30°, wide-angle optic in the right and left nasal cavities of the same patient. Postoperative treatment of the right side consisted of pack removal after 6 days, douching with Ems brine and budesonide treatment (two doses of 50 µg each); postoperative treatment of the left side consisted of pack removal after 6 days and douching with Ems brine. **a** One week after operation. Right side. After pack removal, a moist wound base with granulations and low-grade swelling is generally found. **b** One week after operation. Left side. Findings similar to those seen on the right side. **c** Three weeks after operation. Right side. Only low-grade swelling and a well-visualized operated cavity are seen. **d** Three weeks after operation. Left side. Moderate swelling, with granulations and definite constriction of the operated cavity. **e** Seven weeks after operation. Right side. Macroscopically, the mucosa appears to be normal. **f** Seven weeks after operation. Left side. More definite swelling and constriction of the operated cavity are seen.

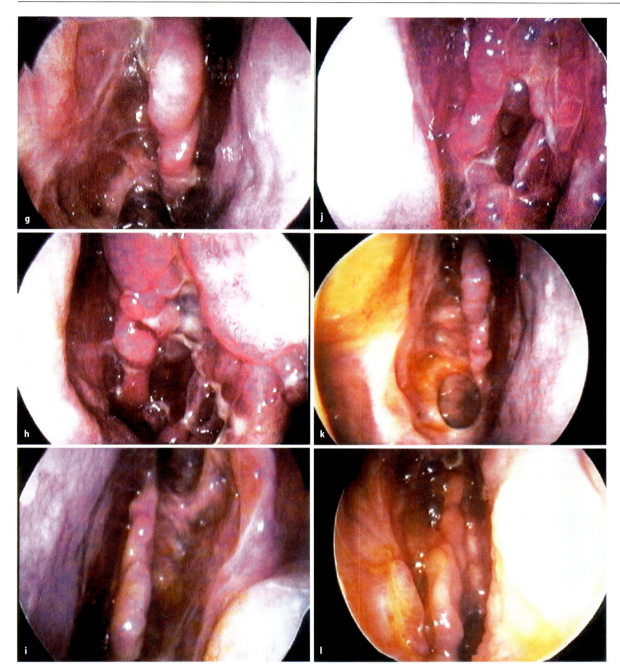

g Ten weeks after operation. Right side. Mucosa are still normal. Formation of small sheet scars is seen. **h** Ten weeks after operation. Left side. Definite swelling and constriction of the operated cavity are still seen. **i** Eighteen weeks after operation. Right side. Normal mucosa. A small area of edematous swelling after biopsy above the maxillary-sinus fenestra is seen. **j** Eighteen weeks after operation. Left side. An increase in edematous swelling, with complete blockage of the operated cavity, is seen. The patient experienced a subjective sensation of having "a cold". **k** Twenty-six weeks after operation. Right side. A completely normal, well-visualized operative field is seen. The yellow coloration is due to the application of fluorescein. **l** Twenty-six weeks after operation. Left side. Compared with the findings at week 18, there is definite regression of the swelling, but some edema remains. The yellow coloration is due to the application of fluorescein

the patient. In some cases, the tamponade has to be removed early because of the presence of an odor.

A 5- to 7-day-long tamponade (with insertion of a stent in the ethmoid sinuses and a tube in the nasal cavity to maintain nasal breathing) has been recommended by Toffel for 10 years (Toffel Sinus Pack) [106], and his experience agrees with our results. The use of a non-adherent packing material is important! Therefore, the often-used insertion of gauze strips is to be strictly avoided. During removal of such gauze packs, the surface tissue is peeled away, leading to renewed trauma, bleeding, initiation of inflammatory processes, fibrin exudation and consequent scar formation. This process is only desirable for debridement of necrotic surfaces and is not relevant for wound care after sinus surgery. The addition of ointment preparations to the packing can cause the development of a paraffin granuloma (Fig. 26.9) [64, 113–115]. We have seen paraffin-inclusion bodies in routine examinations of mucosa excised during revision operations, so it must be assumed that subclinical paraffin granulomas are not uncommon entities. The possibility that this foreign material, which cannot be broken down by the body, may be responsible for unexplained, chronic, postoperative, inflammatory reactions can not be ignored.

4. Further improvement of surgical results is to be expected due to dressing techniques using stents in the surgically enlarged approaches to the sinuses. It has been known for a long time that removal of mucosal strips over concave bony surfaces leads to sheet scarring of variable thickness [42, 43, 45]. Scarring can lead to drainage obstruction, especially when it lies in a direction opposite that of mucociliary transport. Circular resection of mucosa can result in a scar diaphragm that divides the frontal sinus into two halves, and bony transformation of the primary connective-tissue scar or diaphragm is possible [45]. Even with normal wound-healing processes, scarring and closure of the frontal-sinus ostium after resection of a circular piece of mucosa can occur and can lead to mucocele formation [42, 43, 98, 111], even within an intact bony framework [44]. Wound contracture with movement of existing tissue from the wound edges to the center begins after approximately 1 week. This is directed by myofibroblasts, which wander along the collagen fibers and guide the movement of the fibers toward each other [35, 37]. Remodeling occurs after 3 weeks and consists of a new arrangement of collagen, with the formation of larger fiber bundles and an alteration of inter- and intramolecular networks

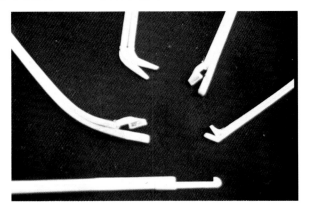

Fig. 26.7. Cutting forceps (straight and up-cutting) of various sizes and angles, and a through-cutting punch (Stuemer, Würzburg, Germany)

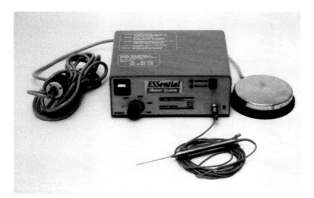

Fig. 26.8. Essential Shaver System (Smith and Nephew GmbH, Surgical Division, Schenefeld, Germany)

(cross-links). The scars take months or years to mature [108].

These processes apply mainly to the frontal sinus, which has a risk of mucocele formation when the drainage is blocked by sealing the exit path or the frontal recess. A permanent guarantee of frontal-sinus drainage has been a problem for many years. A number of mucous-membrane plasties were developed, and stenting with drainage tubes was also carried out [usually associated with the classical, external fronto-orbital approach (Jansen–Ritter)] to address this goal [24]. However, the results were unsatisfactory. Wigand [118] also often found the insertion of stents and silicon tubes to be disappointing. Even if they are left for 6–10 weeks in situ, remaining stenoses are frequently seen. Despite this, some authors use the insertion of stents [15, 20, 74, 85, 86, 95, 97, 99, 102, 106, 107].

Taking pathophysiology into account, an unsatisfactory result is very likely when the stents are

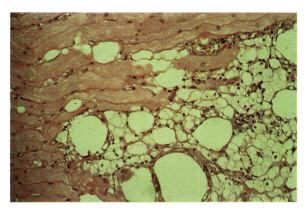

Fig. 26.9. Histology of a lower-lid paraffinoma.

Fig. 26.10. Prototype of silicone stents (Vostra, Aachen, Germany) for the maxillary sinus (*left*) and frontal sinus (*right*)

removed too early. A minimum of 3 months is necessary, and 6 months is preferred so that wound-healing processes are generally completed. In the stenting of choanal atresia or tracheal stenosis, we also leave the cylindrical stents for 6 months; this yields a high rate of success.

Primary attempts with specially designed silicon stents, which are left for 6 months in the maxillary and frontal sinuses, show encouraging results (Fig. 26.10). For continuing evaluation, the experiences obtained to date are insufficient. It appears to be of importance that a large portion of the lateral nasal wall be occlusively covered by the stent, therefore favorably influencing the healing process. This area is critical with respect to trauma and inflammatory reactions. Both stents similarly act as occlusive dressings in the ostial regions. Definitive criteria for the use of stents are still lacking.

5. The question of whether intensive mechanical cleansing should be done postoperatively or the wound cavity is self cleaning without external measures is very controversial. In an obstructed nose or sinus, when the patient has complaints that can be explained by occlusion of the ostial region by crusts, a mechanical cleaning must be done. However, since each cleaning leads to injury, freshly granulating tissue and partial removal of new epithelium (thereby inducing renewed inflammatory processes [69]), a rather controlled/conservative approach to instrument cleaning seems appropriate. After pack removal, excessive secretions should be suctioned out of the nose and sinuses (if they can be reached easily) without inducing bleeding. An atraumatic cleaning is achieved by rinsing the nose with a saline solution; this is done by the patient. The use of Ems brine [with special douching equipment (Rhinocare; Siemens and Company, Bischofsheim)] is generally thought to be superior to use of the normal salt solution [65]. Ems brine has no toxic effect on mucociliary clearance [6, 22, 124] and appears to favorably influence regenerative processes [83]. It is important that the solution be at body temperature. Cool or cold douches act as inflammatory agents, inducing trauma of the nasal mucosa [104].

If, after 2–3 weeks, pronounced tissue contacts between the middle turbinate and the lateral nasal wall constrict the ethmoid entry, these should be separated instrumentally in order to prevent later scar synechia. Limited surgical measures may also be necessary in persistent edema or granulations. Use of sharp cutting instruments is to be preferred.

6. Observation of wound-healing processes shows a favorable effect of topical budesonide application, with reduction of swelling and faster healing. Although corticosteroids generally have an inhibitory action on almost all stages of wound healing [31, 32, 80, 93], and even topical corticosteroids affect cutaneous wound healing [4, 10, 29], our results confirm common clinical experiences. This may be explained by undefined positive effects on the underlying pathologic process leading to chronic sinus inflammation and polyp formation. The very different wound healing courses in the group treated with corticosteroids can be explained by the assumption of different disease entities with correspondingly different corticosteroid sensitivities and the uniform appearance of polyposis [103]. Formation of sheet scars and synechia (with shifting of sinus openings) was not prevented by budesonide administration. This could simply be because this area is not reached by the medication and,

therefore, no effect can be achieved there [114]. Hosemann [49] also demonstrated the favorable effect of intramuscular prednisolone (2 mg/kg) on the healing process in the rabbit maxillary sinus. This led to quicker and more complete wound closure. A decrease in rate was seen only during the first 96 h. The tendency for hyperplastic granulation and bony apposition of the wound was lower.

The administration of newer preparations (budesonide, flunisolide, fluticasone, mometasone) is unequivocally preferred over administration of other products (such as beclomethasone) because of the higher potency and fewer systemic side effects of the newer drugs [14, 57]. Administration should be continued until macroscopic healing is achieved, i.e., for 6 weeks to 6 months.

Other applications of this new method of videoendoscopy, such as observation of gastrointestinal-mucosal changes [56] or the healing of bone fractures, pneumonia (imaging with X-ray), otitis media or tonsillitis (imaging with microscope or endoscope), are possible. Preconditions for application include the availability of a valid imaging method for documentation and measurement, generation of congruent images (if the process under analysis must be continuous without sudden accelerations) and adequate timing of single-measurement procedures.

Acknowlegements. We want to thank our pathologists [Dr. H. Arps, Dr. H. Kronsbein and Dr. M. Kind (Institute of Pathology, Hospital Fulda)] for their kind cooperation.

References

1. Alling P, North AF (1981) Polyurethane film for coverage of skin graft donor sites. J Oral Maxillofac Surg 39:970–971
2. Alper JC, Welch EA, Ginsberg M (1983) Moist wound healing under a vapor permeable membrane. J Am Acad Dermatol 8:347–353
3. Alvarez OM (1987) Pharmacological and environmental modulation of wound healing. In: Uitto J, Perejda AJ (eds) Connective tissue disease. Molecular pathology of the extracellular matrix. Dekker, New York, pp 367–384
4. Alvarez OM, Levendorf KD, Smerbeck RV, Mertz PM, Eaglstein WH (1984) Effect of topically applied steroidal and nonsteroidal anti-inflammatory agents on skin repair and regeneration. Fed Proc 43:2793–2798
5. Alvarez OM, Mertz PM, Eaglstein WH (1983) The effect of occlusive dressings on collagen synthesis and re-epithelialization in superficial wounds. J Surg Res 35:142–148
6. Badre R, Dirnagl K, Guillerm R, Hee J, Kummer A, Schnelle K (1970) Untersuchungen über die Wirkung von Bad Emser Quellprodukten auf das Flimmerepithel. Z Angew Bäder Klimaheilkd 17:40–58
7. Bailey AJ, Bazin S, Sims TJ, LeLeus M, Nicholetis C, Delaunay A (1975) Characterization of the collagen of human hypertrophic and normal scars. Biochem Biophys Acta 405:412–421
8. Bang FB, Bang BG (1977) Mucous membrane injury and repair. In: Brain P (ed) Respiratory defense mechanisms. Dekker, New York, pp 453–488
9. Battista SP, Denine EP, Kensler CJ (1972) Restoration of tracheal mucosa and ciliary particle transport activity after mechanical denudation in chicken. Toxicol Appl Pharmacol 22:59–69
10. Berliner DL, Williams RJ, Taylor GN, Nabors CJ (1967) Decreased scar formation with topical corticosteroid treatment. Surgery 61:619–625
11. Boling LR (1935) Regeneration of nasal mucosa. Arch Otolaryngol Head Neck Surg 22:689–724
12. Bolton LJ, Johnson CL, Rijswijk L van (1992) Occlusive dressings: therapeutic agents and effects on drug delivery. Clin Dermatol 9:573–583
13. Bothwell JW, Rovee DT (1971) The effect of dressings on the repair of cutaneous wounds in humans. In Harkiss KJ (ed) Surgical dressings and wound healing. Crosby Lockwood, London, pp 78–97
14. Brattsand R, Andersson PT, Edsbäcker S, Ryrfeldt A (1987) Development of glucocorticosteroids with lung selectivity. In: Godfrey S (ed) Glucocorticosteroids in childhood asthma. Excerpta Medica, Amsterdam, pp 9–25
15. Bumm P (1980) Eine Methode, das nasale Kieferhöhlenfenster offenzuhalten. Arch Otorhinolaryngol 227:643–645
16. Chapvil M, Koopman CF Jr (1984) Scar formation: physiology and pathologic states. Otolaryngol Clin North Am 17:265–272
17. Clark RA (1991) Cutaneous wound repair. In: Goldsmith LA (ed) Physiology, biochemistry and molecular biology of the skin, 2nd dn. Oxford, New York, pp 576–601
18. Clark RA, Henson PM (1988) The molecular and cellular biology of wound repair. Plenum, New York
19. Cohen IK, Diegelmann RF, Lindblad WJ (1992) Wound healing. Saunders, Philadelphia
20. Deitmer T, Rath B (1988) Befunde, Behandlung und Verlauf frontobasaler Frakturen. Laryngorhinootologie 67:13–16
21. Dineen P, Hildick-Smith G (1981) The surgical wound. Lea and Febinger, Philadelphia
22. Dirnagl K, Guillerm R, Hee J, Badre R, Schnelle K (1979) Untersuchungen über den Einfluss von Soleverdünnungen unterschiedlichen pH-Wertes auf die ziliäre Transportfunktion. Z Angew Bäder Klimaheilkd 26:5–14
23. Draf W (1978) Endoskopie der Nasennebenhöhlen. Springer, Berlin Heidelberg New York
24. Draf W (1982) Surgical treatment of the inflammatory diseases of the paranasal sinuses. Indication, surgical technique, risks, mismanagement and complications, revision surgery. Arch Otorhinolaryngol 235:133–305
25. Draf W (1983) Endoscopy of the paranasal sinuses. Springer, New York
26. Draf W (1991) Endonasal micro-endoscopic frontal sinus surgery: the Fulda concept. Operative Tech Otolaryngol Head Neck Surg 2:234–240

27. Draf W, Weber R (1993) Endonasal pansinusoperation in chronic sinusitis. I. Indication and operation technique. Am J Otolaryngol 14:394–398
28. Eaglstein WH (1985) Experiences with biosynthetic dressings. J Am Acad Dermatol 12:434–440
29. Eaglstein WH, Mertz PM (1981) Effect of topical medicaments on the rate of repair of superficial wounds. In: Dineen P, Hildick-Smith G (eds) The surgical wound. Lea and Febinger, Philadelphia, pp 150–170
30. Falanga V (1988) Occlusive wound dressings. Arch Dermatol 124:872–877
31. Fauci AS (1979) Immunosuppresive and anti-inflammatory effects of glucocorticoids. Monogr Endocrinol 12:449–65
32. Fauci A, Dale D, Balow J (1976) Glucocorticosteroid therapy: mechanisms of action and clinical considerations. Ann Intern Med 84:304–315
33. Fisher LB, Maibach HI (1972) The effect of occlusive and semipermeable dressings on the cell kinetic of normal and wounded human epidermis. In: Maibach HI, Rovee DT (eds) Epidermal wound healing. Year Book Medical, Chicago, pp 113–122
34. Fuller GC, Cutroneo KR (1992) Pharmacological interventions. In: Cohen IK, Diegelmann RF, Lindblad WJ (eds) Wound healing. Saunders, Philadelphia, pp 305–315
35. Gabbiani G, Ryan GB, Majne G (1971) Presence of modified fibroblasts in granulation tissue and their possible role in wound contraction. Experientia 27:549–550
36. Gibson FB, Perkins SW (1993) Dynamics of wound healing. In: Bailey BJ, Johnson JT, Kohut RI, Pillsbury HC III, Tardy ME Jr (eds) Head and neck surgery. (Otolaryngology, vol 1) Lippincott, Philadelphia, pp 187–198
37. Grinnell F (1994) Fibroblasts, myofibroblasts, and wound contraction. J Cell Biol 124:401–404
38. Gross CW, Gross WE (1994) Postoperative care of functional endoscopic sinus surgery. Ear Nose Throat J 73:476–479
39. Harris DR (1979) Healing of the surgical wound. I. Basic considerations. J Am Acad Dermatol 1:197–207
40. Helfman T, Ovington L, Falanga V (1994) Occlusive dressings and wound healing. Clin Dermatol 12:121–127
41. Hernandez-Richter HJ, Struck H (1970) Die Wundheilung. Thieme, Stuttgart
42. Hilding AC (1933) Experimental surgery of the nose and sinuses. III. Results following partial and complete removal of the lining mucous membrane from the frontal sinus of the dog. Arch Otolaryngol Head Neck Surg 17:760–768
43. Hilding AC (1933) Experimental surgery of the nose and sinuses. II. Gross results following the removal of the intersinus septum and of strips of mucous membrane from the frontal sinus of the dog. Arch Otolaryngol Head Neck Surg 17:321–327
44. Hilding AC (1950) Physiologic basis of nasal operations. Calif Med 72:103–107
45. Hilding AC, Banovetz J (1962) Occluding scars in the sinuses: relation to bone growth. Laryngoscope 73:1201–1218
46. Hinman CD, Maibach HI (1963) Effect of air exposure and occlusion on experimental human skin wounds. Nature 200:377–378
47. Hosemann W (1990) Klinische und experimentelle Untersuchungen zur Wundheilung in den Nasennebenhöhlen. Habilitationsschrift, Erlangen
48. Hosemann W, Wigand ME, Goede U (1991) Normal wound healing of the paranasal sinuses: clinical and experimental investigations. Eur Arch Otorhinolaryngol 248:390–394
49. Hosemann W, Dunker I, Göde U, Wigand ME (1991) Experimentelle Untersuchungen zur Wundheilung in den Nasennebenhöhlen. III. Endoskopie und Histologie des Operationsgebietes nach einer endonasalen Siebbeinausräumung. HNO 39:111–115
50. Hosemann W, Göde U, Länger F, Röckelein G, Wigand ME (1991) Experimentelle Untersuchungen zur Wundheilung in den Nasennebenhöhlen. I. Ein Modell respiratorischer Wunden in der Kaninchenkieferhöhle HNO 39:8–12
51. Hosemann W, Göde U, Länger F, Wigand ME (1991) Experimentelle Untersuchungen zur Wundheilung in den Nasennebenhöhlen. II. Spontaner Wundschluss und medikamentöse Effekte im standardisierten Wundmodell. HNO 39:48–54
52. Hunt TK (1982) Wounds and wound healing. Dis Colon Rectum 251:1–5
53. Hunt TK (1983) Physiology of wound healing. In: Burke JF (ed) Surgical physiology. Saunders, Philadelphia, pp 1–13
54. Hunt TK, Dunphy JE (1979) Fundamentals of wound management. Appleton-Century-Crofts, New York
55. James H (1994) Wound dressings in accident and emergency departments. Accid Emerg Nurs 2:87–93
56. Jaspersen D, Keerl R, Weber R, Huppmann A, Hammar C-H, Draf W (1996) Video time lapse endoscopy. Gastrointest Endosc 45:516–518
57. Johansson S-A, Andersson K-E, Brattsand R, Gruvstad E, Hedner P (1982) Topical and systemic glucocorticoid potencies of budesonide, beclomethasone dipropionate and prednisolone in man. Eur J Respir Dis Suppl 122:74–82
58. Jurkiewicz MJ, Morales L Jr (1983) Wound healing, operative incisions, and skin grafts. In: Hardy JD (ed) Hardy's textbook of surgery. Lippincott, Philadelphia, pp 108–122
59. Kanzler MH, Gorsulowsky DC, Swanson NA (1986) Basic mechanisms in the healing cutaneous wound. J Dermatol Surg Oncol 12:1156–1164
60. Katz S, McGinley K, Leyden JJ (1986) Semipermeable occlusive dressings: effects on growth of pathogenic bacteria and re-epithelization of superficial wounds. Arch Dermatol 122:58–62
61. Keenan KP, Combs JW, McDowell EM (1982). Regeneration of hamster tracheal epithelium after mechanical injury. II. Multifocal lesions: stathokinetic and autoradiographic studies of cell proliferation. Virchows Arch 41:215–229
62. Keenan KP, Combs JW, McDowell EM (1982) Regeneration of hamster tracheal epithelium after mechanical injury. I. Focal lesions: quantitative morphologic study of cell proliferation. Virchows Arch 41:193–214
63. Keerl R, Weber R, Huppmann A (1995) Darstellung zeitabhängiger Veränderungen der Nasenschleimhaut unter Einsatz modernster Morphsoftware. Laryngorhinootologie 74:413–418
64. Keerl R, Weber R, Draf W, Kind M, Saha A (1996) Periorbital paraffin granuloma following paranasal sinus surgery. Am J Otolaryngol 76:137–141
65. Keerl R, Weber R; Müller C, Schick B (1997) Effectiveness and tolerance of nasal irrigation following paranasal sinus surgery. Laryngorhinootologie 76:137–141

66. Kennedy DW, Zinreich SJ, Rosenbaum AE, Johns ME (1985) Functional endoscopic sinus surgery. Theory and diagnostic evaluation. Arch Otolaryngol Head Neck Surg 111:576–582
67. Kennedy DW, Gwaltney JM Jr, Jones JG (1995) Medical management of sinusitis: educational goals and management guidelines. Ann Otol Rhinol Laryngol 104[suppl 167]:22–30
68. Knapp U (1981) Die Wunde. Thieme, Stuttgart
69. Kühnel T, Wagner W, Göde U, Hosemann W (1995) Wie traumatisierend ist die mechanische Nasenpflege nach Nebenhöhleneingriffen? Eine histologisch-immunhistochemische Untersuchung. Deutschen Gesellschaft für HNO-Heilkunde, Kopf- und Halschirurgie, Karlsruhe
70. Lane BP, Gordon R (1974) Regeneration of rat tracheal epithelium after mechanical injury. I. The relationship between mitotic activity and cellular differentiation. Proc Soc Exp Biol Med 145:1139–1144
71. Levenson S, Geever EG, Crowley LV (1965) The healing of rat skin wounds. Ann Surg 161:293–308
72. Lindner J, Huber P (1973) Biochemical and morphological basis of wound healing and influences upon it. Med Welt 24:897–911
73. Linsky CB, Rovee DT, Dow T (1981) Effect of dressings on wound inflammation and scar tissue. In: Dineen P, Hildick-Smith G (eds) The Surgical Wound. Lea and Febiger, Philadelphia, pp 191–205
74. Lusk RP, Muntz HR (1990) Endoscopic sinus surgery in children with chronic sinusitis: a pilot study. Laryngoscope 100:654–658
75. Madden JW, Peacock EE Jr (1968) Studies on the biology of collagen during wound healing. I. Rate of collagen synthesis and deposition in cutaneous wounds of the rat. Surgery 64:288–294
76. Madden JW, Peacock EE Jr (1971) Studies on the biology of collagen during wound healing. III. Dynamic metabolism of scar collagen and remodeling of dermal wounds. Ann Surg 174:511–520
77. Maibach HI, Rovee DT (1972) Epidermal wound healing. Year Book Medical, Chicago
78. Mandy SH (1983) A new primary wound dressing made of polyethylene oxide gel. J Dermatol Surg Oncol 9:153–155
79. May SR (1984) Physiology, immunology and clinical efficacy of an adherent polyurethane wound dressing OPsite. In: Wise DL (ed) Burn wound coverings, vol 2. CRC, Boca Raton, pp 53–78
80. McCoy BJ, Diegelmann RF, Cohen IK (1980) In vitro inhibition of cell growth, collagen synthesis, and prolyl hydroxylase activity by triamcinolone acetonide. Proc Soc Exp Biol Med 163:216–222
81. Mertz PM, Marshall DA, Eaglstein WH (1985) Occlusive wound dressings to prevent bacterial invasion and wound infection. J Am Acad Dermatol 12:662–668
82. Messerklinger W (1978) Endoscopy of the nose. Urban and Schwarzenberg, München
83. Michel O, Charon J (1991) Postoperative Inhalationsbehandlung nach Nasennebenhöhleneingriffen. HNO 39:433–438
84. Naumann HH (1965) Pathologische Anatomie der chronischen Rhinitis und Sinusitis. (International Congress series, vol 113) Excerpta Medica, Amsterdam
85. Neel HB III, Whicker JH, Lake CF (1976) Thin rubber sheeting in frontal sinus surgery: animal and clinical studies. Laryngoscope 86:524–536
86. Neel HB III, McDonald TJ, Facer GW (1987) Modified Lynch procedure for chronic frontal sinus disease: rationale, technique and long-term results. Laryngoscope 97:1274–1279
87. Oliver D, Anderson S, Zigon B, McCord J, Gumes S (1993) Tricks of the graphics gurus. Sams, Indianapolis
88. Peacock EE Jr (1984) Wound healing and wound care. In: Schwartz SI (ed) Principles of surgery, 4th edn. McGraw-Hill, New York, pp 289–312
89. Peacock EE Jr (1984) Wound repair, 3rd edn. Saunders, Philadelphia
90. Pollack SV (1979) Wound healing: a review. II. Environmental factors affecting wound healing. J Dermatol Surg Oncol 5:477–481
91. Pollack SV (1982) Wound healing: a review. IV. Systemic medications affecting wound healing. J Dermatol Surg Oncol 8:667–672
92. Porras-Reyes BH, Mustoe TA (1994) Wound healing. In: Cohen M, Goldwyn RM (eds) Mastery of plastic and reconstructive surgery, vol 1. Little, Brown, Boston, pp 3–13
93. Reed BR, Clark RA (1985) Cutaneous tissue repair: practical implications of current knowledge. II. J Am Acad Dermatol 13:919–941
94. Rovee DT, Kurowsky CA, Labun J (1972) Local wound environment and epidermal healing. Arch Dermatol 106:330–334
95. Rubin JS, Lund VJ, Salmon N (1986) Frontoethmoidectomy in the treatment of mucoceles. Arch Otolaryngol Head Neck Surg 112:434–436
96. Sahl WJ Jr, Clever H (1994) Cutaneous scars: part I. Int J Dermatol 33:681–691
97. Schaefer SD, Close LG (1990) Endoscopic management of frontal sinus disease. Laryngoscope 100:155–160
98. Schenck NL (1974) Frontal sinus disease. II. Development of the frontal sinus model: occlusion of the nasofrontal duct. Laryngoscope 84:1233–1247
99. Shikani AH (1994) A new middle meatal antrostomy stent for functional endoscopic surgery. Laryngoscope 104:638–640
100. Stammberger H (1986) Endoscopic endonasal surgery – concepts in treatment of recurring rhinosinusitis. Part II. Surgical technique. Otolaryngol Head Neck Surg 94:147–156
101. Stammberger H (1991) Functional endoscopic sinus surgery. Decker, Philadelphia
102. Stammberger H (1993) Komplikationen entzündlicher Nasennebenhöhlenerkrankungen einschliesslich iatrogen bedingter Komplikationen. Eur Arch Otorhinolaryngol Suppl 1:61–104
103. Stammberger H (1995) Nasal polyposis: attempting classification. American Rhinologic Society, Palm Desert, pp 13–15
104. Stark WB (1928) Irrigations with aqueous solution. Their effect on the membranes of the upper respiratory tract of the rabbit. Arch Otolaryngol Head Neck Surg 8:47–55
105. Timberlake GA (1986) Wound healing: the physiology of scar formation. In: McSwain NE (ed) Current concepts in wound care. Macmillian, Chicago, pp 4–14
106. Toffel PH (1995) Secure endoscopic sinus surgery with middle meatal stenting. Operative Tech Otolaryngol Head Neck Surg 6:157–162
107. Toffel PH, Aroesty DJ, Weinmann RH IV (1989) Secure endoscopic sinus surgery as an adjunct to functional

nasal surgery. Arch Otolaryngol Head Neck Surg 115:822–825
108. Verzar F, Willenegger H (1961) Das Altern des Kollagens in der Haut und in Narben. Schweiz Med Wochenschr 41:1234–1236
109. Vogt PM (1993) Kutane Wundheilung. Habilitationschrift, Bochum
110. Wahl LM, Wahl SM (1992) Inflammation. In: Cohen IK, Diegelmann RF, Lindblad WJ (eds) Wound healing: biochemical and clinical aspects. Saunders, Philadelphia, pp 40–62
111. Walsh TE (1943) Experimental surgery of the frontal sinus. The role of the ostium and nasofrontal duct in postoperative healing. Laryngoscope 53:75–92
112. Weber R, Keerl R (1996) Einsatz moderner Bild-Datenverarbeitung in der klinisch rhinologischen Forschung. Eur Arch Otorhinolaryngol Suppl 1:271–296
113. Weber R, Keerl R, Huppmann A, Draf W, Saha A (1995) Wound healing after paranasal sinus surgery by video time lapse sequences. Operative Tech Otolaryngol Head Neck Surg 6:237–240
114. Weber R, Keerl R, Huppmann A, Schick B (1995) Nasenpolypen und topische Kortikoidtherapie. HNO Aktuell 3:224–228
115. Weber R, Keerl R, Draf W, Wienke A, Kind M (1995) Zur Begutachtung: Periorbitales Paraffingranulom nach Nasennebenhöhlenoperation. Otorhinolaryngol Nova 5:87–90
116. Weber R, Keerl R, Jaspersen D, Huppmann A, Schick B, Draf W (1996) Computer-assisted documentation and analysis of wound healing of the nasal and esophageal mucosa. J Laryngol Otol 110:1017-1021
117. Wigand ME (1981) Transnasale endoskopische Chirurgie der Nasennebenhöhlen bei chronischer Sinusitis. I. Ein biomechanisches Konzept der Schleimhautchirurgie. HNO 29:215–221
118. Wigand ME (1989) Endoskopische Chirurgie der Nasennebenhöhlen und der vorderen Schädelbasis. Thieme, Stuttgart
119. Wilhelm DL (1953) Regeneration of tracheal epithelium. J Pathol 65:543–550
120. Winter GD (1962) Formation of the scab and the rate of epithelialization of superficial wounds in the skin of the young domestic pig. Nature 193:293–294
121. Winter GD (1971) Healing of skin wounds and the influence of dressings on the repair process. In: Harkiss KJ (eds) Surgical dressings and wound healing. Crosby Lockwood, London, pp 46–60
122. Winter GD, Scales JT (1963) Effect of air drying and dressings on the surface of a wound. Nature 197:91–92
123. Wisemann DM, Pharm MR, Rovee DT, Alvarez OM (1992) Wound dressings: design and use. In: Cohen IK, Diegelmann RF, Lindblad WJ (eds) Wound healing. Biochemical and clinical aspects. Saunders, Philadelphia, pp 562–580
124. Wolf G, Koidl B, Pelzmann B (1991) Zur Regeneration des Zilienschlages humaner Flimmerzellen. Laryngorhinootologie 70:552–555
125. Zitelli JA (1987) Wound healing for the clinician. Adv Dermatol 2:243–268

Pediatric Endoscopic Endonasal Sinus Surgery

Gerald Wolf

Introduction

Endonasal sinus surgery for the treatment of acute and chronic sinusitis was already well established by the first third of this century [6]. It was based on the brilliant anatomical studies of Zuckerkandl, Onodi and Grünwald, which represented the high standard of anatomical knowledge of that time [5, 26, 42, 43].

Before the introduction of penicillin, the pathology that was seen was different. Diseases of the sinuses were one of the major causes of intracranial inflammatory and infectious complications [9]. Thus, endonasal methods were much more radical than they are today. The techniques and outcomes were limited by the lack of appropriate imaging modalities [such as computed tomography (CT)], adequate light sources, optical instruments with the possibility to deflect the angle of view and endotracheal intubation for general anesthesia.

In 1929, Mosher advocated caution in ethmoid surgery, stating that the transnasal approach to the ethmoid sinus "should be simple, but it has proven one of the easiest ways to kill a patient" [25]. Because of the potential risks, endonasal techniques were abandoned, and replaced by radical external techniques.

The development of new optical instruments and the new understanding of the physiology and pathophysiology of the nose and paranasal sinuses led to the rebirth of endonasal surgical methods and modern rhinology. In 1960, Karl Storz presented extracorporal cold light and light transmission by glass fibers. In 1966, Harald Hopkins developed the rod/lens system, with its excellent resolution, high contrast, large field of view and perfect fidelity of vision. This was the start of effective nasal and paranasal-sinus endoscopy.

With the help of the new optical devices, Messerklinger (of Graz, Austria) investigated physiology and pathophysiology of the sinus mucosa. He developed and established a systemic endoscopic diagnostic approach to the lateral wall of the nose in combination with conventional tomography [18–24]. Messerklinger's endoscopic findings led to the development of an endoscopic endonasal surgical technique primarily directed toward key areas in the anterior ethmoid sinus. His operation was directed specifically at the primary site of disease in the narrow gaps of the anterior ethmoid sinus, trying to preserve as much mucosal lining as possible. Limited, circumscribed endoscopic surgical procedures result in the recovery of mucosal pathology in the adjacent sinuses without actually operating on these areas [18, 22–24]. In Germany, Draf (of Fulda) and Wigand (of Erlangen) performed endoscopic endonasal sinus surgery but with a technique and instrumentation different from Messerklinger's [2, 3, 37, 38].

In 1984, Kennedy became interested in the concepts of Messerklinger and Stammberger [32–35]. The technique was first disseminated throughout the US, then became known worldwide. Kennedy coined the term "FESS", functional endoscopic sinus surgery [8, 10]. James Zinreich, the Johns Hopkins neuroradiologist, developed the parameters for coronal CT scanning used worldwide today [41].

Based on anatomical studies of Onodi and Ritter, Messerklinger began with pediatric endoscopic endonasal sinus surgery by the early 1970s at Graz University Hospital [27, 30]. Stammberger and Wolf followed, adapting the technique to pediatric patients and their needs [39, 40].

At that time, adenotonsillectomy, antral lavages and inferior-meatal antrostomies were the common surgical methods used to treat uncomplicated chronic inflammatory diseases in children resistant to medical therapy. These methods did not pay any special attention to clearing the key-areas of the anterior ethmoid. External ethmoidectomy and transantral approaches to maxillary-sinus diseases were the standard of care for treatment of complicated sinus disease; these approaches have the risk of traumatizing developing teeth and disturbing the development of the facial skeleton.

At that time, intranasal ethmoidectomy in children was not commonly performed. It was not recommended because of the small size of the sinuses and the increased potential for complications.

Rodney Lusk (of St. Louis) modified the technique for the pediatric population and developed pediatric instruments. He defined indications, published results and started his first pediatric sinus courses in the United States in 1990 [14, 15, 17].

Specific Features of Pediatric Patients

The Developing Sinus System

The paranasal sinuses in infants and children are different (in size, shape and proportion) from those found in adults. Each age has a specific character and anatomic feature [39].

In newborn children, the ethmoid-sinus system is well developed compared with the other sinuses (Fig. 27.1). The Eustachian-tube ostium is found behind the posterior end of the inferior turbinate and ascends to the level of the posterior end of the middle turbinate by the age of 4 years. The ethmoidal infundibulum is already well developed at birth and is a constant landmark. At birth, ethmoidal cells are already present in a definite number but not with a definite size. They are surrounded by connective tissue, into which they expand. Between the age of 8 years and 12 years, ethmoid cells are already completely developed, and hardly any connective tissue is left between the cells.

The maxillary, frontal and sphenoid sinuses have a slower development. The maxillary sinus is present at birth but is only a mucosal indentation into the lateral nasal wall at the floor of the ethmoid infundibulum. Until the age of 4 years, it expands laterally to the infraorbital canal and inferiorly to or below the attachment of the inferior turbinate. At the age of 8 years, pneumatisation has extended laterally to the infraorbital canal and has descended to the middle of the inferior meatus.

At birth, the sphenoid sinus is a blindly ending mucosal sac, that has not yet reached the sphenoid bone or cartilage. Pneumatisation of the sphenoid bone starts at the age of 1 year.

In infants, the frontal sinus is only an anterior ethmoidal cell ("frontal cell") draining into the middle meatus. Pneumatisation of the frontal bone can be found at the age of 4 years. Acceleration of pneumatisation is seen between the age of 8 years and 12 years.

Paranasal-sinus development is directly linked to the dentition and development of the facial part of the skull. Adult proportions are usually developed before the age of 12 years. After this, there are only minor changes in the paranasal-sinus system. These mainly concern pneumatisation of the sphenoid sinus, pneumatisation of the maxillary sinus into the alveolar recess, and extension of height of the nasal cavity. Knowledge of the unique anatomy and pneumatisation of pediatric sinuses is important to understand the pathophysiology and complications of sinusitis. Furthermore, it is important to evaluate radiographs and to avoid complications during surgery.

Signs and Symptoms of Chronic Recurrent or Persistent Sinusitis

Sinus diseases may vary in their signs and symptoms. Particularly in children, the patient's history is of

Fig. 27.1. Sagittal section (right side) of a newborn cadaver. *O*, orbit (exenterated); *T*, tooth buds; * maxillary sinus

great importance and may give relevant information about the underlying disease. However, only older children will be able to describe their problems. In order to make a proper diagnosis, it may be necessary to see and examine a child repeatedly, both when the patient is symptomatic and when he is asymptomatic. The predominant symptoms of chronic sinusitis are nasal obstruction, rhinorrhea, headache, cough, low-grade fever and persistent or recurring upper airway infections. Sometimes, changes in behavior and halitosis are noted [4, 14, 15, 17, 29, 40].

Pathophysiology

The pathogenesis of chronic sinusitis in children is multifactorial and, because of the developing immunologic system and anatomy, is different from that in adults. In younger children, the ability to produce antibodies against Haemophilus influenzae and Streptococcus species is lower than for older children. Attention should be paid to the possible presence of predisposing factors and/or sinus involvement in systemic diseases. The most common predisposing factors are allergies, frequent viral infections of the upper respiratory tract, environmental factors (tobacco smoke) and hypertrophy of the adenoid tissue [14, 36]. Sinus involvement is frequently found in cystic fibrosis and in systemic diseases (like immunodeficiency syndromes) when ultrastructural changes of cilia, such as cilia dyskinesia (as in Kartagener's syndrome; Fig. 27.2), are present [14]. With the developing anatomy, anatomic variations that can predispose the patient to sinus infection may occur.

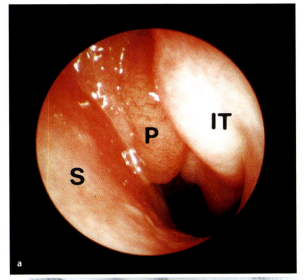

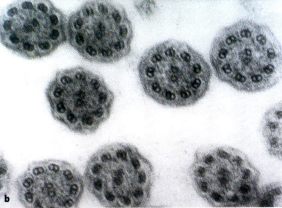

Fig. 27.2. a Kartagener's syndrome in a 10-year-old girl. Left nasal cavity. *IT*, inferior turbinate; *P*, polypoid mucosa; *S*, septum. **b** Electron microscopy of nasal mucosa from the same patient as in **a**. Typical ultrastructural features of Kartagener's syndrome with primary ciliary dyskinesia (missing dynein arms of the microtubules) are seen

Preoperative Assessment

Findings from inspection and palpation of the skin, nasal pyramid, orbit and periorbital region, sensory abnormalities of the infra- and supraorbital nerve and tenderness to percussion the frontal and maxillary sinuses may indicate the presence of sinus disease. Examination of the nasal cavity and the nasopharynx in children can be difficult due to narrow anatomical conditions and the lack of compliance. Anterior rhinoscopy with a speculum and headlight sometimes only allows viewing of the head of the middle turbinate and the anterior part of the septum. In smaller children, inspection of the nasopharynx with the headlight and mirror to evaluate the size of the adenoid pads may not be possible at all. Thus, the method of choice for precise examination of the nasal cavity and the nasopharynx is nasal endoscopy. We usually use rigid Hopkins endoscopes (2.7-mm diameter, 0° and 30°; Fig. 27.3). Flexible endoscopes can also be of help to get an overall view of the nasal cavity and adenoid tissue if a child does not allow introduction of a rigid telescope. In chronic sinusitis, anatomic abnormalities, mucosal edema or polyps, nasal discharge, crusts and the size of the adenoids (Fig. 27.4) must all be noted. In addition to findings from radiological information, nasal endoscopy and laboratory examinations (such as allergy tests, microbiologic and immunologic investigations and sweat electrolyte tests) may be needed to ascertain whether surgical treatment is indicated [14].

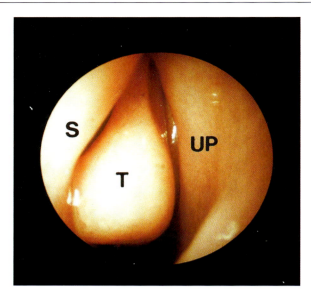

Fig. 27.3. Endoscopic view of a left nasal cavity of an 8-year-old child (as seen with a 2.7-mm, 0° endoscope). *S*, septum; *T*, middle turbinate; *UP*, uncinate process

Plain Radiographs, CT Scans, Magnetic-Resonance Imaging

Plain radiographs in the standard projections are readily available, inexpensive and (usually) adequate for the diagnosis of acute sinusitis. However, in cases with complications of acute sinusitis, tumors or chronic sinusitis, additional anatomical information about the ostiomeatal complex, bony erosions or variations and mucosal changes is needed (Fig. 27.5). For these situations, or for chronic recurrent or persistent sinusitis, CT is the modality of choice. In sinus diseases, CT has become the radiologic diagnostic standard. In a prospective study, Alister and Lusk demonstrated that plain sinus radiographs and coronal CT scans fail to correlate in 74% of pediatric patients [1, 16].

CT scans perfectly display the anatomic details, bony structures and normal and diseased mucosa. CT displays the maxillary and ethmoid air cells and, if they are already present, the frontal and the sphenoid sinuses and shows any orbital and intracranial complications. The coronal plane is preferred, because it gives excellent visualization of the ostiomeatal complex. Furthermore, it is the plane in which the surgeon approaches the sinuses. Care has to be taken to choose the correct window settings and center point and to choose a slice thickness of 3–4 mm [41]. Younger children usually have to be sedated to get adequate-quality pictures. Since the introduction of spiral CT, the time required for taking the scans and the necessity of sedation has been significantly reduced. To avoid overestimation of persistent chronic sinusitis, CT scans should not be taken during an acute episode.

Magnetic-resonance imaging in T1- and T2-weighted images does not demonstrate bone but gives superior soft-tissue contrast. It can distinguish the extent of soft-tissue expansion, fluid levels, inflammatory mucosa thickening and fungal infection. It can help to augment information from CT in selected cases, such as neoplasia and encephaloceles.

Most of the children we see for chronic sinus diseases have small adenoids or underwent adenoidectomy before. Except in these cases, adenoidectomy is our first surgical recommendation. Surgical treatment of the sinus system for chronic recurrent or persistent sinusitis should only be the last choice of treatment, after repeated attempts at medical treatment and adenoidectomy fail. When there is doubt, the child should be examined repeatedly. The decision for surgery should be based on the history, nasal endoscopy and CT scan and should not be made during an acute episode of rhinosinusitis. Surgical patients have to be selected carefully. The procedure itself has to be done by experienced surgeons who have gained sufficient experience in adult patients in order to perform the surgery safely under the difficult and narrow conditions present in children, including the presence of tooth buds.

Surgical Technique

The surgery is done under general anesthesia with endotracheal intubation. Preoperatively, the mucosa is decongested. To obtain better conditions, vasoconstricting agents should be either sprayed or applied by cotton-wool pledgets prior to endoscopy. Oxymetazolines (0.025% or 0.05%), such as suprarenin (1:1000 in older children), have proven to be well tolerated. Usually, 0.5 cm^3 of 1% xylocaine plus epinephrine (1:100,000) is injected into the lateral nasal wall anterior to the uncinate process to reduce diffuse mucosal bleeding. Care has to be taken not to apply excess xylocaine in order to avoid unexpected side effects, such as increased blood pressure.

CT scans are displayed in the operating room for frequent reference. The endoscopes used at our University hospital for surgery are primarily the 4-mm endoscopes (0° and 30°). In narrow anatomic situations, 2.7-mm endoscopes are sometimes used, providing a smaller view and less light. The instruments used are the Karl Storz adult and pediatric instru-

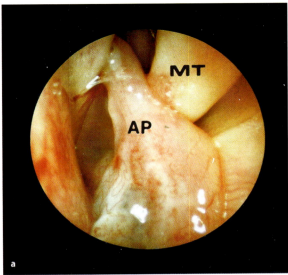
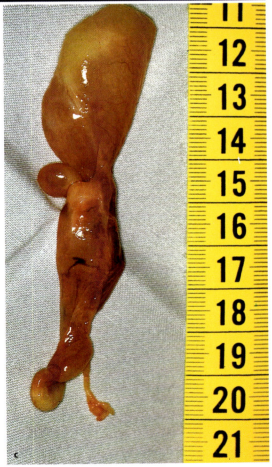

Fig. 27.4. a Antrochoanal polyp in the right middle meatus of a 12-year-old girl. *AP,* antrochoanal polyp; *MT,* middle turbinate. **b** Antrochoanal polyp. Same patient as in **a**. The polyp derives from a right maxillary sinus via an accessory ostium. *A,* accessory ostium in the lateral nasal wall; *P,* stalk of the polyp. **c** Antrochoanal polyp after resection (length in centimeters)

ment sets. Microdebriders, introduced by Settliff, have proven to be quite useful [31].

The concept of the surgical technique refers to the technique widely described by Messerklinger, Stammberger, Kennedy and Lusk [8, 10, 14, 22, 35]. It has to be emphasized that, in children, we try to be even more conservative than in adults, trying to preserve as much mucosa as possible.

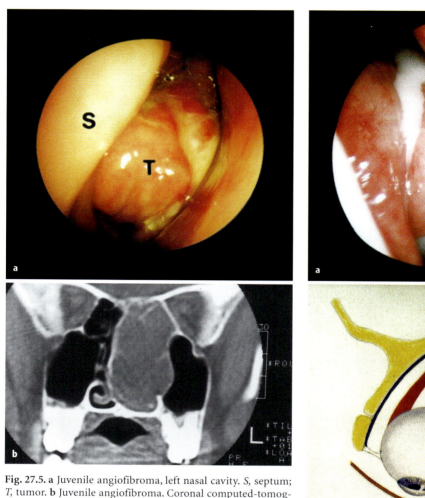

Fig. 27.5. a Juvenile angiofibroma, left nasal cavity. *S*, septum; *T*, tumor. **b** Juvenile angiofibroma. Coronal computed-tomography scan of same patient as in **a**

Usually, the children are hospitalized for two nights after the surgery. Perioperatively, antibiotics and topical steroids are administered. We do not use stents, nor do we perform follow-up endoscopy under general anesthesia to clean the ethmoid cavity. A Merocel sponge soaked in steroid drops is placed into the middle meatus overnight, without any further packing.

Complications of Acute Sinusitis

Complications resulting from acute sinusitis are diseases of different character than chronic sinus disease. They concern the orbit (edema, subperiosteal abscess, abscess of the orbital tissue, orbital cellulitis, thrombosis of the cavernous sinus) or cerebrum (meningitis, epi- or subdural abscesses, intracerebral abscesses). They are life threatening and have signif-

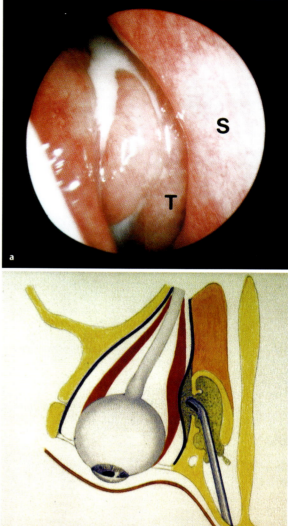

Fig. 27.6. a Acute sinusitis with orbital complication (subperiosteal abscess). Endoscopic view of the right nasal cavity, with pus protruding from the middle meatus. *S*, septum; *T*, middle turbinate. **b** Schematic depiction of endoscopic endonasal drainage of a subperiosteal abscess (courtesy of Dr. Heinz Stammberger)

icant morbidity. The character of the disease and its course are dramatic. The surgery is technically demanding regardless of whether endonasal or external approach is chosen. Abscesses must be drained immediately. The approach depends on the extent and location of the structures involved. Experienced surgeons can use the endonasal approach in cases of ethmoiditis with orbital edema or subperiosteal abscesses located lateral to the lamina papyracea (Fig. 27.6). We treat abscesses at all other locations (especially intracranial ones) and frontal sinusitis with osteomyelitis (Pott's puffy tumors) with a combined approach (endoscopic endonasal drainage of the sinuses combined with an external approach).

Results

From a review of the literature, it is difficult to objectively assess the success of pediatric endoscopic endonasal sinus surgery, because there are no standardized methods of diagnosis or staging, follow-up is too brief, and different percentages of underlying systemic diseases are included in the patient groups. However, the outcomes are very positive in the current literature, and no major complications were reported by the (very experienced) surgeons.

Küttner, in 57 children aged 4–15 years, found a significant improvement in the headache, asthma, bronchitis, hypersecretion, nasal obstruction and susceptibility to upper-respiratory-tract infections associated with sinusitis [11]. He also documented his results by anterior rhinomanometry.

Lazar reports improvement of obstruction in 85% of cases, of cough in 86%, of discharge in 83% and of headache in 84% [12, 13]. Parsons, in 52 children between 7 months and 17 years of age reported improvement or relief of obstruction in 92% of cases, of discharge in 88%, of postnasal drip in 61%, of cough in 84%, of halitosis in 75%, of headache in 96% and of asthma in 96% [29].

Jones and Parsons found a significant improvement in long-term quality of life in pediatric patients with cystic fibrosis. FESS led to a decline in the frequency of nasal discharge, obstruction and postnasal drip and resulted in a high level of patient satisfaction [7, 28].

Gross investigated 57 children between 3 years and 15 years of age. The follow-up period was 3–13 months. In a subjective assessment, 92% of the patients found the procedure to be helpful; in 64% the problems were improved. In 28%, the problems were solved, and 88% would agree to a surgical procedure again, should it become necessary [4].

Lusk compared other methods (adenotonsillectomy, antral lavage, and inframeatal antrostomy) with pediatric FESS and found FESS to be superior. One year after surgery, 71% of the patients were considered normal by their parents (80% if systemic diseases are excluded [15]). A recent personal communication described a significant improvement of rhinorrhea, nasal obstruction, headache, irritability and cough in 406 ethmoidectomies over a period of 4.5 years.

A review of 124 pediatric patients undergoing endoscopic endonasal sinus surgery at our university hospital over an 11-year period showed a successful outcome and a high level of patient satisfaction [40]. The patient's ages ranged from 3 years to 16 years. Seventy-one patients were operated for chronic sinus disease without polyposis; 53 patients presented polyposis (including antrochoanal polyps). All patients had undergone extensive medical therapy preoperatively and had not responded well to medical management. Of 124 patients, 40 underwent preoperative adenoidectomy without success. Predisposing factors leading to chronic recurrent sinusitis were inhalational allergies (25%), bronchial asthma (4%), immunodeficiency syndromes (3%), cystic fibrosis (3%) and Kartagener's syndrome (2%). In 33 patients (27%), the procedure was considered difficult (mainly due to narrow and difficult anatomical situations and diffuse mucosal bleeding). Of the patients with preoperative persistent rhinorrhea, we achieved complete resolution in 53% and improvement in 40%. Nasal obstruction was resolved in 31% of cases and was improved in 57%. Recurrent infections were eliminated or decreased in frequency in 91% of patients. Forty-one percent of patients were relieved of headache; 44% experienced improvement. Pulmonary symptoms were relieved or improved in 58% of patients; cough was improved in 72%. In a subjective patient or parent assessment, 41% respondents said they were very satisfied, and 46% were satisfied; 83% would agree to have the surgery again. Re-operation was necessary in 20 children (16%); 15 children were re-operated because of recurring polyposis, three because of antrochoanal polyps and one because of synechia and cystic fibrosis.

Current Standard

Endoscopic endonasal surgery is very useful for the many cases of chronic inflammatory sinus diseases in children that are uncomplicated and resistant to therapy. It is a minimally invasive method that re-establishes ventilation and drainage of the paranasal sinuses. This technique helps to change chronic sinusitis into a medical problem. Long-term follow-up studies show a successful outcome and a high level of patient satisfaction.

No scarring of the face or damage of tooth buds occurs. There is no evidence of deleterious effects on facial growth and development, though further follow-up is needed. We regard the Caldwell–Luc operations and external approaches for chronic sinusitis in children to be obsolete.

Extended applications of endoscopic endonasal-sinus surgery – such as identification and endonasal repair of CSF leaks (the fluorescein technique), optic-nerve decompression (for treatment of traumatic loss of vision, endonasal treatment of tumors, mucoceles and choanal atresia) and endonasal drainage of subperiosteal abscesses – are expanding the usefulness of this technique (Figs. 27.7–27.9). Pediatric endoscopic endonasal-sinus surgery should only be performed by well trained surgeons after sufficient experience in adult surgery. It is a safe and effective approach for skilled surgeons. In the future indications, for endoscopic endonasal-sinus surgery in children must be clearly defined, and the results must be critically followed.

References

1. Alister WH, Lusk RP, Muntz HR (1989) Comparison of plain radiographs and coronal CT scan in infants and children with recurrent sinusitis. AJR Am J Roentgenol 153:1259–1264
2. Draf W (1978) Endoskopie der Nasennebenhöhlen. Diagnostische und therapeutische Möglichkeiten. Springer, Berlin Heidelberg New York
3. Draf W (1982) Die chirurgische Behandlung entzündlicher Erkrankungen der Nasennebenhöhlen. Arch Otorhinolaryngol 235:133–305
4. Gross C, Gurucharri M, Lazar R, Long T (1989) Functional Endoscopic Sinus Surgery in the Pediatric Age Group. Laryngoscope 99:272–275
5. Grünwald L (1925) Deskriptive und topographische Anatomie der Nase und ihrer Nebenhöhlen. In: Denker A, Kahler O (eds) Handbuch der Hals-Nasen-Ohrenheilkunde, vol 1. Springer, Berlin Heidelberg New York, p 1
6. Hajek M (1926) Pathologie und Therapie der entzündlichen Erkrankungen der Nebenhöhlen der Nase, 5th edn. Franz Deuticke, Leipzig
7. Jones J, Parsons D, Cuyler J (1993) The results of endoscopic sinus surgery on the symptoms of patients with cystic fibrosis. Int J Pediatr Otorhinolaryngol 28:25–32
8. Kennedy DW (1985) Functional endoscopic sinus surgery: technique. Arch Otolaryngol Head Neck Surg 111:643–649
9. Kennedy DW (1997) Sinus surgery: a century of controversy. Laryngoscope 107:1–5
10. Kennedy DW, Zinreich SJ, Rosenbaum AE, Johns ME (1985) Functional endoscopic sinus surgery. Theory and diagnostic evaluation. Arch Otolaryngol Head Neck Surg 11: 576–582
11. Küttner K, Siering U, Looke G, Eichhorn M (1992) Funktionelle endoskopische Siebbeinrevision bei entzündlichen Nasennebenhöhlenerkrankungen im Kindesalter, HNO 40:158–164
12. Lazar RH, Younis RT (1990) Functional endonasal sinus surgery in the pediatric age group: In: Lazar RH, Younis RT (eds) Advances in otolaryngology year book, vol 4. Mosby, St. Louis, p 1
13. Lazar RH, Younis RT, Long TE (1993) Functional endonasal sinus surgery in adults and children. Laryngoscope 103:1–5
14. Lusk RP (1992) Pediatric sinusitis. Raven, New York
15. Lusk RP, Muntz H (1990) Endoscopic sinus surgery in children with chronic sinusitis: a pilot study. Laryngoscope 100:654–658
16. Lusk RP, Muntz HR, McAllister WH (1989) Comparison of paranasal sinus radiographs and coronal CT scans in children. 8th International Symposium on Infection and Allergy of the Nose. ISIAN, Baltimore
17. Lusk RP, Lazar RH, Muntz HR (1989) The diagnosis and treatment of recurrent and chronic sinusitis in children. Pediatr Clin North Am 36:1411–1421

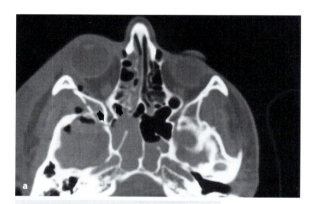

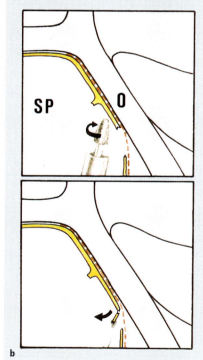

Fig. 27.7. **a** Post-traumatic compression of the optic nerve in its canal (*arrows*), resulting in optic-nerve ischemia and blindness. **b** Schematic depiction of decompression of the optic nerve in its canal, with the help of a drill. *O*, optic nerve; *SP*, sphenoid sinus (courtesy of Dr. Heinz Stammberger)

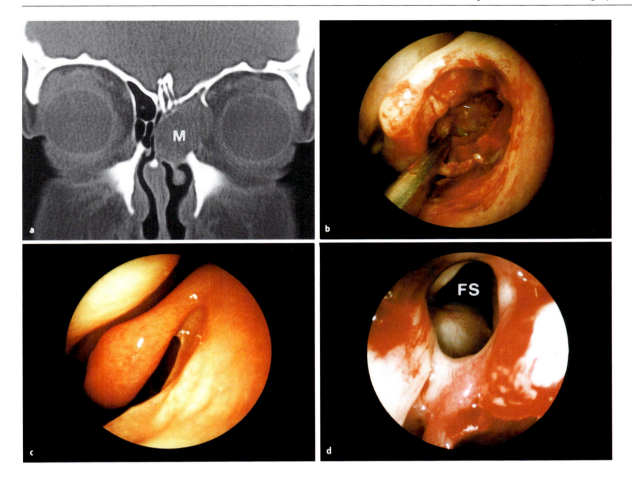

Fig. 27.8. a Computed-tomography scan of a mucocele of a right ethmoid and frontal sinus. *M*, mucocele. b Intraoperative feature of the same patient as in a; endoscopic endonasal approach. c Postoperative control of the left middle meatus after re-epithelialization. Same patient as in a. d Postoparative endoscopic control of the frontal sinus. Same patient as in a. *FS*, frontal sinus (courtesy of Dr. Heinz Stammberger)

18. Messerklinger W (1966) On the drainage of the human paranasal sinuses under normal and pathological conditions. 1. Monatsschr Ohrenheilkd Laryngorhinol 100:56–68
19. Messerklinger W (1966) On the drainage of the human paranasal sinuses under normal and pathological conditions. 2. Monatsschr Ohrenheilkd Laryngorhinol 100:313
20. Messerklinger W (1972) Nasenendoskopie: Der mittlere Nasengang und seine unspezifischen Entzündungen. HNO 20:212–215
21. Messerklinger W (1972) Technik und Möglichkeiten der Nasenendoskopie. HNO 20:133–135
22. Messerklinger W (1978) Endoscopy of the nose. Urban and Schwarzenberg, München
23. Messerklinger W (1979) Das Infundibulum ethmoidale und seine entzündlichen Erkrankungen. Arch Otolaryngol Head Neck Surg 222:11–22
24. Messerklinger W (1982) Über den Recessus frontalis und seine Klinik. Laryngorhinootologie 61:217–223
25. Mosher HP (1929) The symposium of the ethmoid. Trans Am Acad Ophthalmol Otolaryngol 34:376
26. Onodi A (1910) Die topographische Anatomie Eröffnung der Nasenhöhle und ihrer Nebenhöhlen. Curt Kabitzsch, Würzburg
27. Onodi A (1911) Die Nebenhöhlen der Nase beim Kinde. Curt Kabitzsch, Würzburg
28. Parsons DS (1992) Sinusitis and cystic fibrosis. In: Lusk RP (ed) Pediatric sinusitis. Raven, New York, p 65
29. Parsons DS, Phillips SE (1993) Functional endoscopic surgery in children: a retrospective analysis of results. Laryngoscope 103:899–903
30. Ritter FN (1973) The paranasal sinuses, anatomy and surgical technique. Mosby, St. Louis
31. Setliff RC III (1996) The hummer: a remedy for apprehension in functional endoscopic sinus surgery. Otolaryngol Clin North Am 29:95–104
32. Stammberger H (1986) Endoscopic endonasal surgery – concepts in treatment of recurring rhinosinusitis. Part I. Anatomical and pathophysiological considerations. Otolaryngol Head Neck Surg 94:143–147
33. Stammberger H (1986) Endoscopic endonasal surgery – concepts in treatment of recurring rhinosinusitis. Part II. Surgical technique. Otolaryngol Head Neck Surg 94:147–156
34. Stammberger H (1986) Nasal and paranasal sinus endoscopy – a diagnostic and surgical approach to recurrent sinusitis. Endoscopy 6:213–218

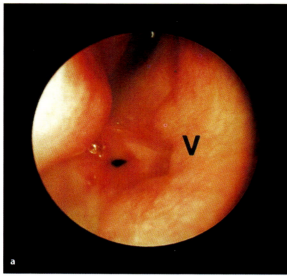

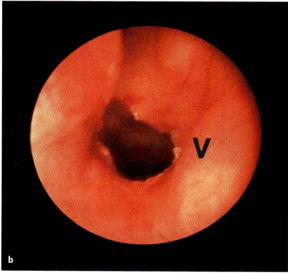

Fig. 27.9. a Membranaceous choanal stenosis of a newborn child, right side. *V*, vomer. **b** Postoperative feature. Same patient as in **a** (courtesy of Dr. Heinz Stammberger)

35. Stammberger H (1991) Functional endoscopic sinus surgery. Decker, Philadelphia
36. Terrahe K, Potrafke T (1992) Die Wirkung von inhalativen Umweltschadstoffen auf die Schleimhaut der oberen Luftwege. HNO 40:153–157
37. Wigand ME (1981) Transnasal ethmoidectomy under endoscopic control. Rhinology 19:7–15
38. Wigand ME, Steiner W, Jaumann MP (1978) Endonasal sinus surgery with endoscopic control: from radical operation to rehabilitation of the mucosa. Endoscopy 10:255–260
39. Wolf G, Anderhuber W, Kuhn F (1993) The development of the paranasal sinuses in children. Ann Otol Rhinol Laryngol 9:705–711
40. Wolf G, Greistorfer K, Jebeles JA (1995) The endoscopic endonasal surgical technique in the treatment of chronic recurring sinusitis in children. Rhinology 33:97–103
41. Zinreich SJ, Kennedy DW, Rosenbaum AE, Gayler BW, Kumar AJ, Stammberger H (1987) CT of nasal cavity and paranasal sinuses: imaging requirements for functional endoscopic sinus surgery. J Radiol 163:769–775
42. Zuckerkandl E (1882) Normale und pathologische Anatomie der Nasenhöhle und ihrer pneumatischen Anhänge. Wilhelm Braumüller, Wien
43. Zuckerkandl E (1892) Normale und pathologische Anatomie der Nasenhöhle und ihrer pneumatischen Anhänge. II. Wilhelm Braumüller, Wien

# Micro-endoscopic Sinus Surgery in Children

Shirley Shizue Nagata Pignatari and Aldo Cassol Stamm

Introduction

Advances in micro-endoscopic surgery of the paranasal sinuses in conjunction with development in image studies have changed the traditional concepts of the diagnosis and treatment of paranasal sinuses diseases in children. Although paranasal-sinus surgery is still considered a controversial subject, especially in children, the micro-endoscopic technique has assumed a dominant position during the past few years. It has been shown that, when correctly utilized, this type of surgery has low morbidity and encouraging results. In this chapter, the authors emphasize practical clinical and surgical aspects of the use of micro-endoscopic surgery in children. Table 28.1 lists our micro-endoscopic surgical experience with 504 children (15 years of age and younger, between 1983 and 1999) for a variety of lesions of the nose and paranasal sinuses.

Table 28.1. Frequency of micro-endoscopic surgery in children, by type of lesion (1983–1999); $n=504$

Lesion	Number
Congenital malformation	
Choanal atresia	32
Meningoencephalocele	2
Nasal glioma	4
Inflammatory/infectious process	
Sinusitis	75
Nasal polyposis	46
Mucocele	12
Mucous retention cyst	19
Antrochoanal polyp (Killian)	15
Benign tumors	
Angiofibroma	34
Meningioma	3
Fibrous dysplasia	2
Traumatic lesions	
CSF fistula	8
Microseptoplasty	94
Microsurgery of the turbinates	148
Severe epistaxis	10

Anatomic Development of the Paranasal Sinuses

Knowledge of the anatomic development of the paranasal sinuses is essential for any rhinologic surgeon, especially for those involved in treating children. Only the ethmoid and maxillary sinuses are present at birth and, during childhood (during the development of the skull and middle facial region), they grow and surround the orbital cavity. The newborn maxillary sinus is approximately 6–8 cm³ in size (Fig. 28.1) [16]. It grows rapidly until the age of 3 years and grows more slowly until age 7 years, after which there is another growth spurt, followed by slower growth until the sinus reaches maximum pneumatization after dental eruption is completed during adolescence [10]. Because the orbital cavity is almost completely developed at birth, all paranasal sinuses have a very close relationship with it. At birth, the ethmoid sinus is already large enough to present clinical significance and will become fully developed by age 12–14 years [10]. The frontal sinus arises from the anterior ethmoid in the frontal-recess region. This ethmoidal cell migrates to the frontal bone and starts growing at approximately 6 months of life, becoming radiographically perceptible as a frontal sinus at approximately 4–6 years of age. At birth, the sphenoid sinus is an evagination of the sphenoethmoid recess into the sphenoid bone; it begins to grow rapidly between the ages of 5 years and 7 years; its pneumatization also accelerates. Its aeration continues until late in the second decade of life, or even until adulthood (Fig. 28.2) [10, 23].

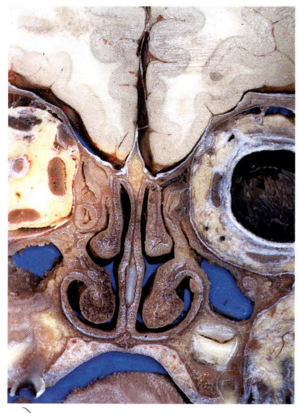

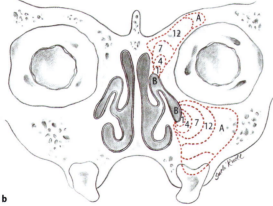

Fig. 28.1. a Coronal section of a 2-year-old patient at the level of the maxillary and ethmoid sinuses. b Schematic drawing of the development of the maxillary and ethmoid sinuses at *1, 4, 7* and *12* years of age (*broken line*). *A* Adult, *B* newborn

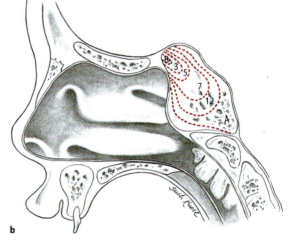

Fig. 28.2. a Anatomic sagittal view of the sphenoid bone in a 2-year-old patient. b Schematic drawing of the pneumatization of the sphenoid bone at *1, 4, 7* and *12* years of age (*broken line*). *A* Adult, *B* newborn

Congenital Malformations

Choanal Atresia

Choanal atresia may be unilateral or bilateral (Figs. 28.3, 28.4). Its incidence is estimated to be approximately one per 5000–8000 newborns [21]. In 50% of the cases, other associated congenital anomalies, such as microtia, developmental retardation, heart defects or facial paralysis, may be present [3, 22]. Recent review has shown that 70% of the atresic plates are mixed (bony-membranous) and about 30% are purely bony [6].

The diagnosis of choanal atresia is normally suspected in the nursery due to severe respiratory distress even after the upper airway is aspirated. When the choanal atresia is bilateral, acute respiratory insufficiency may occur quickly and is sometimes lethal. Diagnostic evaluation is based on the clinical history and is confirmed by endoscopic and computed tomography (CT) exams.

Although the respiratory reflex in newborns is physiologically nasal, and the treatment of the choanal atresia is always surgical, the surgery is not always considered an emergency [27, 29]. Unilateral cases are rarely an emergency, and the repair is usually delayed for at least 6 months.

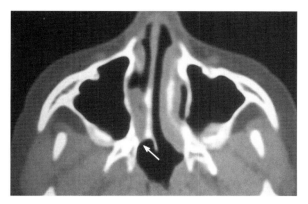

Fig. 28.3. Axial computed tomography of a newborn with unilateral choanal atresia. *Arrow* indicates the bony atresic plate

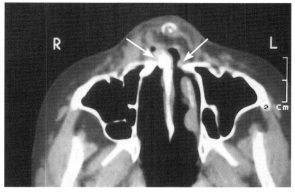

Fig. 28.5. Axial computed tomography of a child with stenosis of the pyriform aperture (*arrows*)

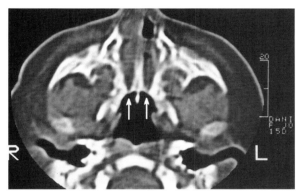

Fig. 28.4. Axial computed tomography of a newborn with bilateral choanal atresia. *Arrows* indicate the atresic bony plates

If the atresia is bilateral, the eventual treatment may be postponed for some months and, during this period of time, an orthodontic nipple with an opening in both sides may be used. Gastric or enteric feeding and a Guedel cannula may be necessary for immediate care of a newborn [27]. In some of these cases, the acute respiratory distress may lead to a severe asphyxia or even immediate death after birth. The authors' series of 32 patients with choanal atresia, submitted to surgical treatment by using the transnasal micro-endoscopic technique, is presented in Chap. 33.

Pyriform Aperture Stenosis

The pyriform aperture is the portion of the nasal cavity bounded below and laterally by the maxilla, above by the nasal bones and is separated medially by the nasal septum. Stenosis of this aperture has been described as an anomaly that occurs secondary to the excessive growth of the nasal process of the maxilla, and occurs during the first months of neonatal life [15, 17].

The signs and symptoms are similar to those found in bilateral choanal-atresia patients. The nasal cavity may be so narrow that it does not allow passage of a flexible endoscope; in such cases, diagnosis can be confirmed by CT (Fig. 28.5).

Conservative measures, such as a McGovern nipple, gavage feeding and apnea monitor, can be sufficient in mild cases. When the airway obstruction is more significant, surgical treatment is preferable. Transnasal and sublabial approaches have been described and, in both approaches, the excessive amount of bone can be removed with drills under endoscopic or microscopic visualization [1, 5].

Meningoencephalocele

The meningoencephalocele, encephalomeningocele and encephalocele, are herniations of the cerebral contents through a defect in the skull base during embryonic development. They may consist of meninges only (meningocele) or may involve brain and meninges (meningoencephalocele). They are classified as occipital, sincipital (nasofrontal) or basal. Although the occipital type is by far the most common, nasal or sinciptal encephaloceles (arising between the frontal and nasal bones or through the foramen cecum and protruding into the nasoethmoidal region) and basal encephaloceles (confined exclusively to the nasal cavity) represent approximately 10% of the cases. They can be associated with other malformations in 40% of the cases [12, 22]. A nasal meningoencephalocele, which is visible externally, presents as a bulge on the glabella region and is

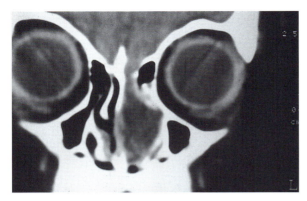

Fig. 28.6. Coronal computed tomography of a 4-year-old patient with a meningoencephalolocele, showing cerebral tissue protruding into the nasal cavity

soft, pulsatile and distendable with jugular-vein compression (Furstenberg's sign). CSF may be present [4, 24]. The intranasal meningoencephalocele is seen medially on the roof of the nasal cavity (adjacent to the nasal septum) as a soft, bluish, compressible lesion. CT and magnetic resonance (MR; Fig. 28.6) are essential to define the bony structures, the roof of the nasal cavity and the soft tissue of this region, both preoperatively and postoperatively. Differential diagnosis includes dermoid cysts, lachrymal-duct cysts, hemangiomas, rhabdomyosarcoma and lymphangioma.

Meningitis is a potential complication if the lesion is inadvertently manipulated. The current recommended surgical approach for removal of meningoencephaloceles is transnasal micro-endoscopy, though multiple surgical approaches (such as coronal flap, lateral rhinotomy and frontal craniotomy) can be considered.

Nasal Glioma

The nasal glioma is not a real neoplastic lesion and can be considered a type of encephalocele. Like encephaloceles, nasal glioma consists of ectopic neural tissue (glial cells) and fibrous and vascular connective tissue, but with no intracranial connection. Twenty percent of the gliomas have a persistent fibrotic band that extends intracranially. Typically, the lesion appears reddish, firm and non-compressible [22].

Although the literature reports that approximately 60% of gliomas are extranasal, 30% are intranasal and 10% are mixed [15], in our series of four patients, three were exclusively intranasal and one was mixed. Clinically, they may be confused with

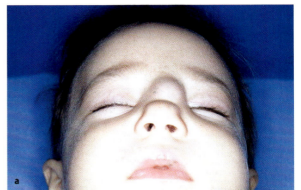
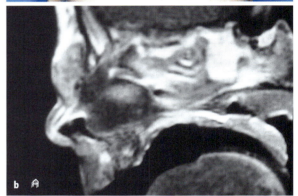
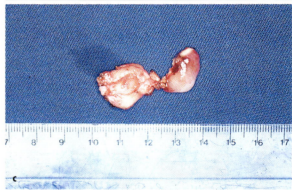

Fig. 28.7. a A 9-month-old child with a nasal glioma that appears as a bulge in the glabella region. **b** Sagittal magnetic resonance showing the internal and external portions of the lesion. **c** Nasal glioma after surgical removal

nasal polyps, although gliomas are (in general) harder, larger and less translucent. They normally cause nasal obstruction and nasal deformity, and the first signs and symptoms normally appear before the first year of life. There is no evidence of familial predisposing factors or gender predominance. The diagnosis is based on clinical and CT- and MR-image findings). Treatment is basically surgical [2]. All four patients of our series were operated on via the transnasal micro-endoscopic approach (Fig. 28.7).

Nasolacrimal-Duct Cysts and Dacryocystitis

Nasolacrimal-duct cysts are congenital malformations that occasionally lead to nasal obstruction. Approximately 30% of full-term infants are born with nasolacrimal-duct obstruction. When the proximal or distal portions of the system are occluded, liquid accumulates and a cyst is formed [34]. The clinical picture in children can include epiphora, nasal obstruction and respiratory distress.

A nasolacrimal-duct cyst is diagnosed (via anterior rhinoscopy) as a mass in the inferior meatus and is confirmed by CT. In 85% of cases, these cysts resolve spontaneously before approximately 9 months of age [8, 13]. Surgical treatment is indicated if resolution does not occur or when the child presents with recurrent infections, respiratory obstruction and feeding difficulties.

Surgery consists of endonasal endoscopic marsupialization. It is advisable to consult an ophthalmologist, because these patients may require intraoperative nasolacrimal probing and placement of nasolacrimal stents.

Inflammatory and Infectious Processes

The maxillary and ethmoid sinuses are the sinuses most frequently involved in inflammatory diseases in childhood. In children, the frontal and sphenoid sinuses are less frequently involved in isolation and are usually part of a pansinusitis; they may have greater clinical significance after puberty [23].

It is very difficult to determine which inflammatory disease of the paranasal sinuses needs surgical intervention, especially in children. There is a lack of specificity of signs and symptoms and a poor correlation between the clinical picture and image findings. Depending on the extent of the disease, the signs and symptoms may include nasal obstruction, aqueous, mucous or purulent rhinorrhea, headache, hyposmia and cough [18, 19].

Rhinosinusitis

It is believed that, in general, the development of chronic or recurrent rhinosinusitis is related to ostiomeatal-complex alterations (Fig. 28.8) and that, in most cases, the involvement of those structures can be more important than the extent of the process in developing the symptomatology. In children, however, despite a lack of consistent studies or good understanding of the pathophysiology of this clinical entity, other factors, such as systemic diseases (cystic fibrosis, immotile-cilia syndrome, immunodeficiencies; Fig. 28.9) and the immaturity of the immune system, are considered more important in development. Severe immunodeficiencies are well recognized as predisposing factors in the rhinosinusitis process [18, 19].

Selective immunoglobulin-G deficiency, although diagnosed more frequently today, is not always easy to identify. In a child with recurrent sinus infection, predisposing factors should always be considered, particularly if systemic manifestations of infection are also present [23, 32].

Nasal allergy produces disorders of the sino-nasal mucosa and ostiomeatal complex, leading to rhinosinusitis. The adenoids, in addition to acting as an obstructing factor, may serve as a reservoir of pathogenic microorganisms. Air pollution, air condition-

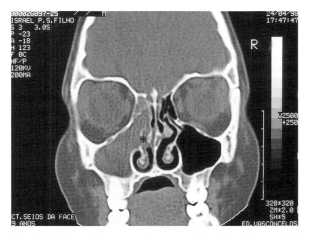

Fig. 28.8. Coronal computed tomography of an 11-year-old boy with chronic rhinosinusitis, showing obstruction of the maxillary- and ethmoid-sinus drainage ostia (*)

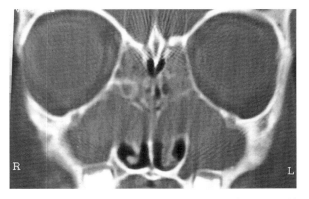

Fig. 28.9. Coronal computed tomography (CT) of a 7-year-old patient with cystic fibrosis and nasal polyposis. CT shows opacification of all paranasal sinuses

ing and swimming may also produce edema of the mucosa and disturbance of ciliary function [20].

Surgical intervention is very controversial and, currently, most authors agree that it should be reserved for special cases, when medical treatment has been exhaustively tried. Radical procedures, such as total ethmoidectomy and removal of the mucoperiosteum, are rarely indicated. Indirect procedures, such as adenoidectomy, septoplasty, puncture and partial turbinectomy, should be considered first. According to national and international consensus, surgery in the treatment of rhinosinusitis in children is strongly recommended in the following situations: mucocele, mucopyocele, polyposis, antrochoanal polyp, fungal sinusitis and complications. However, surgery of the ostiomeatal complex may be part of the treatment in cases of persistent infection after failure of indirect procedures and extensive medical treatment, particularly in patients with coexistent diseases, such as immunodeficiencies or severe asthma [7, 20].

Sino-Nasal Polyposis

Sino-nasal polyposis is a disease of the mucosa of the nasal cavity and paranasal sinuses and is relatively infrequent in children (Fig. 28.10). Due to its association with cystic fibrosis, especially during childhood, clinical investigation of mucoviscidosis must be considered, because 7–32% of children with cystic fibrosis have nasal polyps [31]. Nasal endoscopic examination usually shows the nasal cavity to be filled completely by polyps; sometimes, there is an isolated polyp in the middle-meatus region. Extension and staging of the disease are best evaluated through high-resolution CT [24]. Although clinical treatment may be an option (particularly in the pediatric population), surgical treatment is preferable, and the choice of the technique and approach will vary according to the stage of the disease.

Surgical Technique

Micro-endoscopic surgery in children is always performed under general anesthesia. The nasal cavity is first prepared with topical vasoconstrictors. CT images (coronal and axial projections) are mandatory in the operating room. Surgical instruments should always be very small. Surgical technique varies according to the type and extent of the disease.

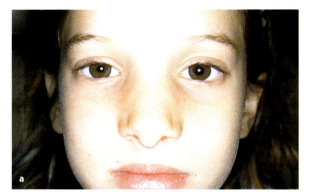
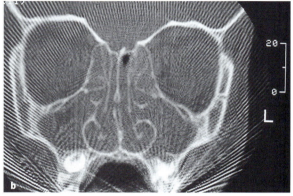

Fig. 28.10. **a** A 10-year-old girl presenting with deformity of the nasal dorsum due to extensive nasal polyposis. **b** Coronal computed tomography of the same patient, showing involvement of all paranasal sinuses

Maxillary Sinus

Surgical access to the maxillary sinus is usually achieved via a middle-meatus antrostomy after partial or total uncinectomy. The main ostium is then probed with a delicate seeker and is best visualized with an angled endoscope. If the disease is chronic rhinosinusitis, relieving the middle-meatus region by an uncinectomy is normally sufficient (Fig. 28.11).

If an antrostomy is needed, it is performed with a modified backbiting forceps. The size of the window depends on the type and extent of the disease. The posterior anatomic limit is the sphenopalatine foramen, and the anterior limit is the nasolacrimal duct.

Ethmoidectomy

In general, the procedure is the same as for adults. Special care must be taken regarding the operative field, which is considerably smaller. Previous knowledge of the level of the development of the paranasal sinuses is mandatory and should be always evaluated via CT.

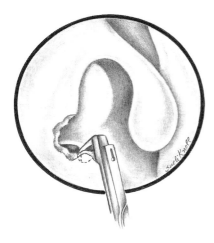

Fig. 28.11. Partial removal of the uncinate process with a backbiting forceps, exposing the ostium of the maxillary sinus

Although the ethmoid sinus is frequently involved in inflammatory processes during childhood, surgery is rarely indicated. In chronic and recurrent processes, when medical treatment is no longer effective, conservative procedures (such as transnasal endoscopic "bullectomy") may be considered. The anatomic relationship between the ethmoid sinus and the anterior skull base and cribriform plate is very delicate, and the structures are very close together; the risk of damaging these structures during a surgical procedure is greater than in adults.

To access the ethmoidal cells, it is usually necessary to perform a preliminary uncinectomy, which can be achieved with a microknife or a special (mini) backbiting forceps. If an operating microscope is used, the middle turbinate is gently displaced toward the nasal septum at the beginning of the procedure; the middle turbinate is then kept in position by a number-one self-retaining nasal speculum, exposing the middle-meatus region and the structures of the ostiomeatal complex.

The ethmoidectomy itself, whether using an operating microscope or the 0°, 4-mm endoscope, is performed by removing the anterior wall of the ethmoid bulla, which can be done with a straight, delicate forceps. The anterior ethmoidal cells are removed by using 30° and 45° endoscopes and curved forceps. If necessary, manipulation in the frontal-recess region must be done carefully in order to avoid stripping the adjacent mucosa, which usually leads to stenosis.

Sphenoid Sinus

Although the sphenoid sinus can be approached in three basic ways (trans-septal, transnasal and transethmoidal), in inflammatory and infectious processes, the transnasal and transethmoidal approaches are used most. The choice of route depends on the type of the lesion, the involvement of the surrounding structures and the degree of pneumatization of the sinus. Due to the difficulties of the postoperative care in children, splints are always recommended to prevent synechia formation and should be kept in the nasal cavity and sinus for 7 days.

Postoperative Care

Usually, nasal and paranasal-sinus packing is removed 24–48 h after surgery. Antibiotics active against Staphylococcus aureus are usually given for 7 days, and the parents are instructed to instill topical saline solution into the nasal cavity. In special cases, such as polyposis, topical steroids are also prescribed after the first postoperative week.

Cleaning of crusts and secretions should be scheduled once or twice per week until complete healing of the surgical cavity occurs; preferably, cleaning should be done while using endoscopes. When adequate postoperative care cannot be accomplished due to inability of the patient to cooperate, general anesthesia may be employed.

Mucocele

Mucoceles are benign cystic lesions limited by the mucosa of the paranasal sinus itself. In general, the frontal sinus is the most frequently involved paranasal sinus. In our series of 12 children, mucoceles were mostly found in the ethmoid and sphenoid sinuses.

The signs and symptoms are related to the mucocele's location; however, because of its slow development, it may remain asymptomatic for a long period of time. In some of our patients, the diagnosis was late because there were no specific complaints. Clinical manifestations, such headache and proptosis, were the most common complaints with ethmoid mucoceles.

CT and MR are important to provide details about the extent and bone involvement of the mucocele and for surgical planning (Fig. 28.12). The treatment of the mucocele is surgical, and the approach is specific for each paranasal sinus (usually using a micro-endoscopic technique, except in cases with involvement of the lateral region of the frontal sinus, in which case a combined external and transnasal micro-endoscopic approach is indicated). In gener-

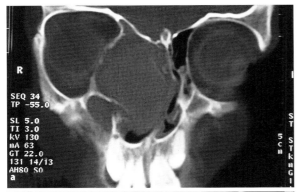

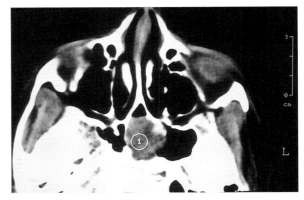

Fig. 28.13. Axial computed tomography of a 6-year-old boy shows a retention cyst in the sphenoid sinus (*1*)

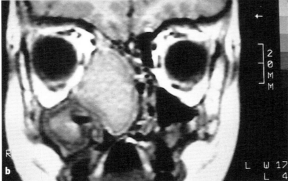

Fig. 28.12 a,b. Ethmoid-sinus mucocele in a 7-year-old boy, as seen with coronal computed tomography (a) and magnetic resonance (b)

al, the technique is the same as for adults, as previously described.

Mucous-Retention Cysts

Cysts of the paranasal sinuses can be considered a type of mucocele. They are most frequently localized in the floor of the maxillary sinus and assume a domed shape. Usually, the symptomatology is minimal or absent and, frequently, they are an incidental radiological finding (Fig. 28.13).

Treatment of mucous-retention cysts is controversial, especially in children. Surgical removal depends on how symptomatic the cyst is. Very large cysts obstructing the drainage ostium may cause headaches or lead to recurrent sinusitis. If resection of the cyst is indicated or if analysis of its content is desirable, the surgical approach should be as conservative as possible. A simple puncture of the canine fossa or inferior meatus is sufficient in most cases.

Antrochoanal Polyp (Killian Polyp)

The benign antrochoanal or Killian polyp should always be considered in a child with unilateral nasal obstruction accompanied by mucous or purulent secretion [30]. They are most frequent in patients less than 10 years of age and in adolescents. In our series of 22 patients, 12 were under age 13 years.

Anterior rhinoscopy usually shows a single polyp arising from the middle-meatus region and filling the posterior part of the nasal cavity. Endoscopic examination identifies the origin of the polyp in the nasal cavity. It usually protrudes from the accessory ostium of the maxillary sinus, projecting itself to the choanal region.

CT-image studies typically show complete opacification of the maxillary sinus and homolateral nasal cavity (Fig. 28.14). Differential diagnosis in children includes mucous-retention cysts, mucocele, tumors, meningoencephaloceles and maxillary sinusitis.

Treatment consists of endoscopic surgical resection of the polyp's implantation in the maxillary sinus (via a mini-fenestration at the anterior wall of the maxillary sinus or via a middle-meatus antrostomy) under general anesthesia (Fig. 28.15). If the child is too young, one may prefer to partially resect the polyp (intranasally, from the accessory ostium) in order to relieve the nasal airway.

Benign Neoplasms

Benign tumors of the nasal cavity and paranasal sinuses are relatively uncommon in children. They usually grow slowly, producing nonspecific symptoms, such as rhinorrhea and nasal obstruction. They include hemangiomas, squamous papilloma,

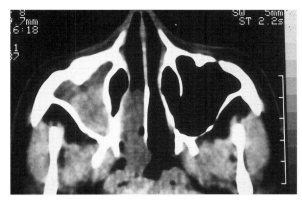

Fig. 28.14. Axial computed tomography of an 11-year-old boy with an antrochoanal polyp, showing typical opacification of the ipsilateral maxillary sinus

serosanguineous or purulent rhinorrhea, facial deformities, rhinolalia, hypoacusis secondary to otitis (with effusion) and exophthalmos. The diagnosis is made by a combination of the clinical picture, otolaryngological exam and CT. The treatment is surgical resection of the tumor. Some authors recommend preoperative embolization, attempting to diminish the tumor and minimize operative bleeding [33]. The surgical approach depends on the extent of the tumor and staging. In our series of 25 children, ages 8–15 years, the most utilized surgical approach was a mid-facial degloving approach combined with use of operating microscope and bipolar cautery (Fig. 28.16) [24, 29].

Meningioma

Meningiomas of the nasal cavity and paranasal sinuses are extremely rare, and about 44% occur in patients under 20 years of age [14]. Intracranial meningiomas with extension to the nasal cavity and paranasal sinuses are also very uncommon. These tumors originate from the embryonic arachnoid cells. They are usually single and grow slowly. Malignant transformation is rare, but the tumor can be locally invasive. Signs and symptoms include nasal obstruction, exophthalmos, facial deformity, epistaxis and loss of visual acuity. A clinical exam shows an uncharacteristic tumor filling the middle upper part of the nasal cavity. CT and MR are essential for evaluating the origin and extent of the disease and planning the surgical approach (Fig. 28.17) [24].

Treatment is always surgical. If the tumor is confined to the nasal cavity, the surgical approach can be the transnasal by using the micro-endoscopic technique. If the tumor involves the paranasal sinuses, the best approach may be a combination of a degloving approach and micro-endoscopy [29].

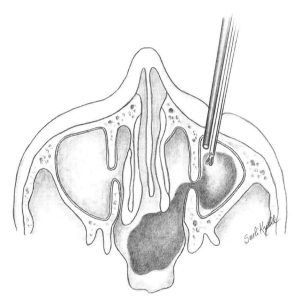

Fig. 28.15. The antrochoanal polyp is separated from the maxillary-sinus mucosa through an anterior mini-aperture

craniopharyngioma, angiofibroma and meningioma. The two most important are angiofibroma and meningioma.

Angiofibroma

Angiofibroma is a histologically benign tumor and is most frequent in young, prepubescent males. It usually originates from the sphenopalatine foramen, extending to the nasal cavity, paranasal sinuses and anterior skull base [11]. Symptoms include nasal obstruction, severe and intermittent epistaxis,

Traumatic Lesions: CSF Fistula

CSF leakage into the nasal cavity and paranasal sinuses may be congenital, iatrogenic or post traumatic. The usual presentation is continuous or intermittent watery rhinorrhea. Because children often accidentally hit their heads and commonly have rhinorrhea, the diagnosis may not be evident, and frequently it is only suspected after a bout of meningitis. High-resolution pneumocisternography (coronal and axial projections, with contrast) and MR are essential to define the diagnosis, localization and extent of the fistula (Fig. 28.18).

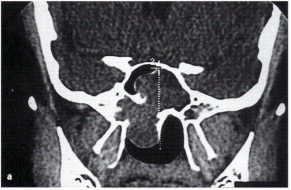

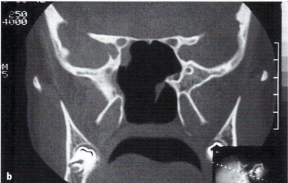

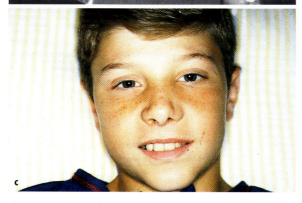

Fig. 28.16. a Coronal computed tomography (CT) of a 13-year-old boy with a nasal angiofibroma involving the right nasal cavity, pterygopalatine fossa and sphenoid sinus. b Postoperative CT after removal of the tumor through a degloving surgical approach. c Postoperative appearance of the patient 1 day after surgery

The surgical approach to seal the fistula depends on its location and extent. The transnasal micro-endoscopic technique is normally utilized to treat fistulas in the ethmoid sinus and cribriform-plate regions [25, 28]. The trans-septosphenoidal route can be employed to treat fistulas in the sphenoid sinus. Craniotomy should be reserved for large fistulas located in the frontal sinus or in patients with sinonasal infection.

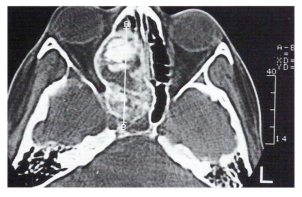

Fig. 28.17. An axial computed tomography of a 14-year-old girl with a meningioma involving the ethmoid and sphenoid sinuses, displacing the right orbit

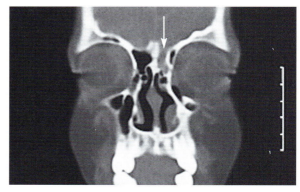

Fig. 28.18. Coronal computed tomography of a 12-year-old boy, with a traumatic cerebrospinal-fluid leakage in the left cribriform plate (*arrow*)

Microseptoplasty and Surgery of the Turbinates

Although septoplasty and turbinoplasty are controversial procedures in children, they are indicated in specific situations. Nasal-septum deviation and hypertrophy of the turbinates are responsible for nasal obstructions in a large number of children (Fig. 28.19). These structural anomalies may lead to functional alterations of the ostiomeatal complex, predisposing the patient to development of chronic sinusitis. In addition, nasal obstruction and continuous mouth breathing can also influence maxillofacial development.

The micro-endoscopic surgery of the turbinates can be particularly helpful in cases of concha bullosa (Fig. 28.20) and hypertrophy of the inferior turbinate in allergic patients. A partial turbinectomy or turbinoplasty may be performed under micro- or endoscopic control (Fig. 28.21) [29].

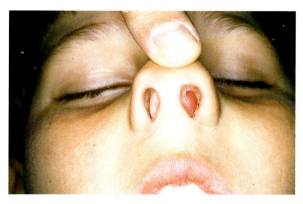

Fig. 28.19. An 8-year-old boy with an anterior nasal-septum deviation

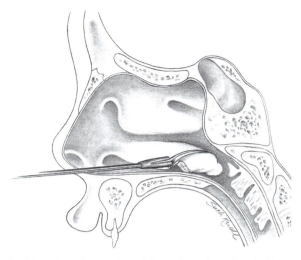

Fig. 28.21. Partial resection of the end portion of the inferior turbinate, using microscissors

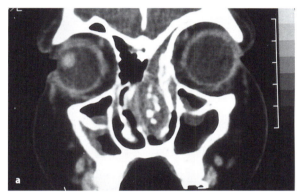

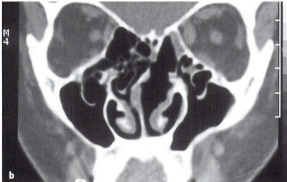

Fig. 28.20 a,b. Preoperative (a) and postoperative (b) coronal computed tomography of a 10-year-old child with a mucocele of the middle turbinate

Microseptoplasty and surgery of the turbinates in children are usually performed as indirect procedures to treat or prevent diseases (such as chronic or recurrent sinusitis) or disturbances of maxillofacial growth. They may also be performed in association with other surgical procedures, as part of the treatment or to facilitate surgical access to specific regions (for example, in a sphenoethmoidectomy or transnasal middle antrostomy).

If septoplasty is indicated in a child, the surgeon must be as conservative as possible, respecting the growth areas, particularly the anterior portion of the cartilage of the nasal septum. Nasal splints should always be used after surgery, in order to prevent synechia formation.

Epistaxis

Epistaxis is infrequent in the general population. Some authors believe that it occurs preferentially during cold seasons [9].

Most of the nasal-bleeding episodes in children are self limited, originating in the anterior nasal-septum area (Kisselbach or Little), and the usual predisposing factors are inflammatory conditions of the nasal mucosa, such as allergy and viral infections [5]. However, systemic diseases (including coagulopathies) and hematopoietic disorders must be considered as causes of epistaxis. When the bleeding persists (despite conventional care) and surgical procedures are warranted, they may be accomplished using the same guidelines as for adults [26, 29].

Foreign Body

Introduction of a foreign body into the nasal cavity and nasopharynx is a very common situation in children less than 5 years of age. Typically, the symptoms and signs include nasal obstruction and uni-

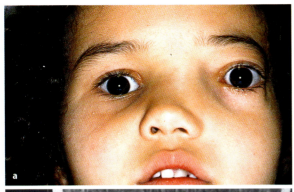
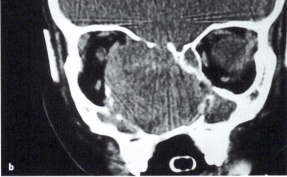

Fig. 28.22. a A 5-year old girl presenting with deformity of the face, caused by a chondrosarcoma of the nasal cavity and paranasal sinuses. **b** Coronal computed tomography of the same patient shows a tumor occupying the nasal cavity, maxillary and ethmoid sinuses and invading the orbital cavity

lateral, malodorous rhinorrhea, sometimes accompanied by epistaxis.

The diagnosis is easily made through an adequate inspection of the nasal cavity. Usually, the foreign body is localized on the floor of the nasal cavity or is impacted between the inferior turbinate and the nasal septum. In these cases, the foreign body can be removed after adequate immobilization of the child.

When the diagnosis is delayed, the foreign body can become attached to the adjacent tissue via a local inflammatory reaction. It may become infected, spreading the infection to the paranasal sinuses, or it may become a large mass (with edema, secretion and granulous tissue) and may be misinterpreted as a real tumor necessitating removal via a surgical procedure [29].

Malignant Sino-Nasal Tumors

There are a large variety of malignant lesions of the nasal cavity and paranasal sinuses. In childhood, the most common are lymphoma, squamous cell carcinoma, rhabdomyosarcoma and esthesioneuroblastoma. This region can also be a site of metastasis, most commonly from neuroblastoma. In general, the symptoms consist of nasal obstruction, rhinorrhea and epistaxis. Cervical adenopathy may or may not be present [22].

Regardless of the characteristic appearance of bone destruction and invasion seen on CT, the diagnosis is essentially made with a biopsy. Figure 28.22 shows a coronal-section CT scan of a 5-year-old girl; the scan reveals a large tumor occupying the nasal cavity and maxillary and ethmoid sinuses, destroying the ethmoid-sinus orbital wall and invading the orbital cavity.

References

1. Austin MB, Mills ES (1986) Neoplasms and neoplasm-like lesions involving the skull base. Ear Nose Throat J 65:25–50
2. Blumenfeld R, Skolnik EM (1965) Intranasal encephaloceles. Arch Otolaryngol Head Neck Surg 82:527
3. Brown K, Brown OE (1998) Congenital malformations of the nose. In: Cummings CW, Frederickson JM, Harker LA (eds) Otolaryngology, head and neck surgery. Mosby, St. Louis, pp 92–103
4. Brown OE (1995) Current management of pediatric nasal airway obstruction. Curr Opin Otol Head Neck Surg 3:402
5. Brown OE, Myer CM, Manning SC (1989) Congenital nasal pyriform aperture stenosis. Laryngoscope 99:86
6. Brown OE, Pownell P, Manning SM (1996) Choanal atresia: a new anatomic classification and clinical management implications. Laryngoscope 106:97
7. Clement PAR, Bluestone CD, Gordts F, Luzk RP, Otten FWA, Goossens H, et al. (1998) Management of rhinosinusitis in children. Consensus meeting, Brussels, September 13, 1996. Arch Otolaryngol Head Neck Surg 124:31–34
8. Edmond JC (1991) Congenital nasolacrimal sac mucocele associated with respiratory distress. J Pediatr Ophthalmol Strabismus 28:287
9. Friedman WH, Rosenblum BN (1987) Epistaxis. In: Goldman J (ed) The principles and practice of rhinology. Wiley Medical, New York, pp 375–383
10. Graney DO, Rice DH (1998) Anatomy. In: Cummings CW, Frederickson JM, Harker LA (eds) Otolaryngology, head and neck surgery. Mosby, St. Louis, pp 1059–1064
11. Harrison DFN (1987) The natural history, pathogenesis and treatment of juvenile angiofibroma. Arch Otolaryngol Head Neck Surg 113:936–942
12. Hengerer AS (1996) Congenital malformations of the nose and paranasal sinuses. In: Bluestone CD, Stool SE (eds) Pediatric otolaryngology. Saunders, Philadelphia, pp 718–728
13. Hepler KM (1995) Respiratory distress in the neonate. Arch Otolaryngol Head Neck Surg 121:1423
14. Ho KL (1980) Primary meningiomas of nasal cavity and paranasal sinuses. Cancer 46:1442–1447
15. Hui Y, Friedberg J, Crysdale WS (1995) Congenital nasal pyriform aperture stenosis as a presenting feature of holoprosencephaly. Int J Pediatr Otorhinolaryngol 31:263

16. Kasper KA (1936) Nasofrontal connections: a study based on one hundred consecutive dissections. Arch Otolaryngol Head Neck Surg 23:322
17. Lang J (1989) External nose skeleton. In: Lang J (ed). Clinical anatomy of the nose, nasal cavity and paranasal sinuses. Thieme Medical, New York, p 8
18. Lusk RP, Stankiewicz JA (1997) Pediatric rhinosinusitis. Otolaryngol Head Neck Surg 117:53–59
19. Parsons DS, Wald ER (1996) Otitis media and sinusitis: similar diseases. Otolaryngol Clin North Am 29:11–25
20. Pignatari SSN, Weckx LLM, Solé D (1998) Rhinosinusitis in children. J Pediatr 74:31–36
21. Pirsig W (1986) Surgery of choanal atresia in infants and children: historical notes and updated reviews. Int J Pediatr Otolaryngol 11:153
22. Potsic WP, Wetmore RF (1987) Pediatric rhinology. In: Goldman J (ed) The principles and practice of rhinology. Wiley Medical, New York, pp 801–845
23. Reilly JS (1990) The sinusitis cycle. Otolaryngol Head Neck Surg 103[suppl]:856–862
24. Stamm A, Burnier M Jr (1989) Tumores benignos naso-sinusais. In: Brandão LG, Ferraz A (eds) Cirurgia de cabeça e pescoço. Roca, São Paulo, pp 465–491
25. Stamm A, Pignatari S (1998) Transnasal miro-endoscopic surgery for CSF rhinorrhea. In: Stammberger H, Wolf G (eds) European Rhinologic Society and International Symposium on Infection and Allergy of the Nose meeting, 1998 (Vienna). Monduzzi, Bologna, pp 329–335
26. Stamm A, Pinto JA, Neto AF, Menon AD (1985) Microsurgery in severe posterior epistaxis. Rhinology 23:321–325
27. Stamm A, Mekhitarian L, Pato C (1988) Microcirurgia transnasal no tratamento da atresia coanal. Rev Bras Otorhinolaryngol 54:43–50
28. Stamm A, Mekhitarian LN, Braga FM, Souza HL (1992) Microcirurgia transnasal – tratamento da rinoliquorréia. Rev Bras Otorhinolaryngol 58:176–184
29. Stamm AC, Pignatari SSN, Pozzobon M (1995) Cirurgia microendoscópica naso-sinusal na infância. In: Stamm AC (ed) Microcirurgia naso-sinusal. Revinter, Rio de Janeiro, pp 417–427
30. Stammberger H (1997) Rhinoscopic surgery. In: Settipane GA, Lund VJ, Bernstein JM, Tos M (eds) Nasal polyps: epidemiology, pathogenesis and treatment. Oceanside, Providence, pp 165–176
31. Stern RC, Boat TF, Wood RE, Matthews LRW, Doershuk CF (1982) Treatment and prognosis of nasal polyps in cystic fibrosis. Am J Dis Child 136:1067–1070
32. Teufert KB, Aita FS, Pignatari SSN (1998) Sinusitis, rhinitis and systemic diseases. Rev Bras Otorhinolaryngol 64:6–15
33. Waldman SR, Levini HL, Astor F (1981) Surgical experience with nasopharyngeal angiofibroma. Arch Otolaryngol Head Neck Surg 107:677
34. Yee SW (1994) Congenital nasolacrimal duct mucocele: a cause of respiratory distress. Int J Pediatr Otorhinolaryngol 29:151

Revision Endoscopic Sinus Surgery

Jean-Michel Klossek

Introduction

Since the development of endonasal endoscopic sinus surgery, many studies have attested to its therapeutic effectiveness [5, 6, 8, 15, 17, 22]. Nevertheless, for some patients, endoscopic surgery fails to relieve persistent or recurrent sinus disease. This subset of patients may require revision surgery. The management of patients previously operated on with an endonasal or external approach remains a challenge for the surgeon, and some guidelines (Table 29.1) are helpful before deciding to perform revision surgery [1, 7, 12]:

- Past medical history
- Type of previous surgery
- Preoperative management
- Technical approach

Table 29.1. Non-exhaustive features to control in case of a revision procedure

Before the procedure
Type of pathology involving the sinus cavities
Type of previous surgery: endonasal, external, combined
Preoperative radiological exploration: CT scans, MRI
Anatomic variations: thickness of the skull base and the medial orbital wall, bone defects, position of the internal carotid artery and optic nerve
Preparation of the mucosa
During the procedure
Presence of the middle turbinate
Presence of adhesions
Access to the choana and the sphenoethmoidal recess
Access to the maxillary sinus
Presence of the uncinate process, the bulla ethmoidalis and the basal lamella of the middle turbinate
Frontal irrigation
At the end of the procedure
Prevention of adhesions and stenosis

CT, computed tomography; *MRI*, magnetic resonance imaging

First Step: Study of the Past Medical Story

Although the results of previous studies are difficult to analyze (because of a lack of standardization of the pathologies included [14]), inconsistent medical treatment and misdiagnosis are suggested as explanations for some failures. Before deciding on a new procedure, meticulous nasal endoscopy and analysis of the patient's complaints are required to avoid a second failure. Analysis of the literature leads to the conclusion that results also depend on the pathology involving the sinuses. For example, failure of surgery in cases of mucoceles or a fungus ball is extremely rare compared with the results in cases of polyposis [9–11]. Two mechanisms may be suggested to explain a failure: technical considerations and recurrence or persistence of the pathology. Postoperative compliance of the patient must also be evaluated, because inadequate follow-up has been given as a reason for surgical failure [6].

Second Step: What was the Previous Procedure?

Nasal endoscopy is the key to analysis of the results of the previous procedure(s). Any scarring of the middle meatus, medialization of the middle turbinate or persistence of diseased mucosa is noted. The middle meatus is inspected to try to identify any remaining anatomic landmarks, such as the uncinate process, bulla or the middle turbinate. The presence and patency of a middle antrostomy or a sphenoidotomy are also noted. Edema, polyps and secretions are located. At the completion of nasal endoscopy, anatomic variations and potentially difficult areas are reviewed and explained to the patient [2, 8, 18].

Third Step: Preoperative Management

If adequate medical treatment and evaluation suggest that revision surgery is indicated, computerized tomographic (CT) scanning is systematically performed. Although some authors consider coronal sections an adequate evaluation for primary sinus procedure, in revision cases, axial and sometimes sagittal sections are also recommended for analysis of anatomic variations resulting from prior surgery [4]. The main landmarks (the middle turbinate, uncinate process, orbital wall and the roof of the ethmoid sinus) are inspected (Figs. 29.1, 29.2). All abnormalities, such as defects or thinned areas of bone, are meticulously recorded to help avoid complications. If necessary preoperative treatment of the mucosa is initiated with oral antibiotics and steroids to reduce inflammation.

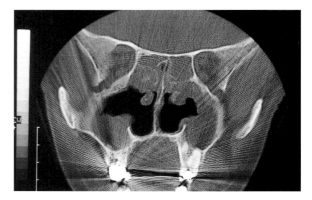

Fig. 29.1. Coronal computed tomography section/revision ethmoidectomy. Previous procedure: inferior turbinectomy and partial ethmoidectomy. There is persisting disease in the posterior ethmoid cells. Note the presence of the superior turbinate

Fourth Step: Technical Approach

Regardless of the extent of the surgical procedure, some rules must be considered to avoid new postoperative failures. After evaluation of the extent of the pathology (as determined from the history, the nasal endoscopy and the CT findings), preparation of the nasal fossa is suggested in order to reduce infections and the inflammation of the mucosa. In case of asthma or diffuse inflammation (polyposis), many authors give preoperative systemic steroids to reduce bleeding and facilitate identification of the main structures. In all cases, revision includes the following steps, depending on the procedure scheduled.

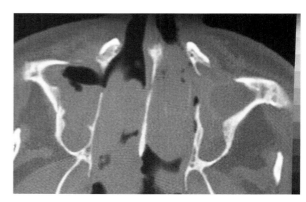

Fig. 29.2. Axial computed tomography view. Revision ethmoidectomy. Previous procedure: trans-maxillary ethmoidectomy. There is a partial defect of the anterior wall of the maxillary sinus

Access to the Operative Field

If a nasal septal deviation is thought to compromise airway patency or does not allow adequate visualization of the middle or superior meatus and the anterior wall of the sphenoid, submucosal resection (via an endoscopic or external approach) is added to the procedure. This may also improve the ease of postoperative care and may reduce the risk of adhesions (Fig. 29.3).

Revision Middle-Meatal Antrostomy

Dysfunction of mucus transport is the most common problem observed after a middle antrostomy and is due to the non-incorporation of the ostium into the antrostomy. In such cases, the inferior part of the uncinate process (which comprises part of the inferior border of the maxillary ostium) has been left in place and must be removed.

Closure of the antrostomy can also occur, especially if the pathology has not been completely eradicated (mycosis) or if the ostium wasn't incorporated into the antrostomy. In such cases, the complete removal of the uncinate process must be performed to re-open the antrostomy. When there is persistent suppuration, dental evaluation is indicated (if it has not been performed already).

Although middle-meatus scarring may sometimes be totally asymptomatic, it often obstructs the ostiomeatal complex and can be a cause of failure. Controversy remains concerning the removal of the anterior free part of the middle turbinate. Nevethe-

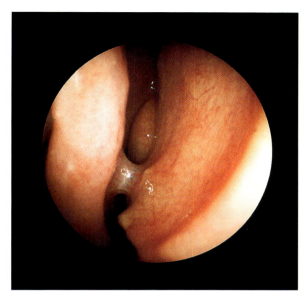 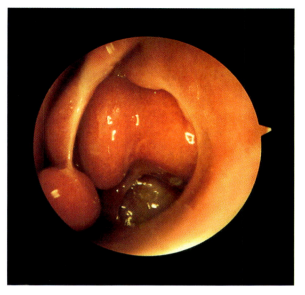

Fig. 29.3. Postoperative left nasal endoscopy (0°). There are adhesions between the inferior turbinate and the nasal septum

Fig. 29.4. Right revision middle-meatal antrostomy. Removal of the anterior part of the middle turbinate to increase access to the antrostomy and reduce the risk of synechia

less, resection of a pneumatized or poorly medialized middle turbinate is necessary to prevent new adhesions and to facilitate the postoperative care of the antrostomy. This technique prevents postoperative crusting (Fig. 29.4).

Ethmoidal Revision Surgery

In these situations, analysis of preoperative CT scans is essential. The principal landmarks (the middle turbinate, persisting bone separation, uncinate process, bulla, basal lamella, etc.) are studied. These landmarks are also located during the preoperative endoscopy. In a majority of cases, the procedure is performed in an anterior-to-posterior direction. At the beginning of the procedure, potential difficulties are evaluated. If no landmarks are easily recognized, the identification of the maxillary sinus is recommended before opening the ethmoid. If the uncinate process is still present, the removal of its inferior portion can facilitate the identification of the maxillary ostium and the posterior enlargement of the middle-meatal antrostomy. If the uncinate process has been previously resected, the use of a curved, non-traumatic aspirator to puncture just above the back of the inferior turbinate allows safe entry into the maxillary sinus through an accessory ostium. The enlargement of the antrostomy is continued anteriorly to incorporate the natural ostium in the fenestration. After this step, the roof of maxillary sinus becomes the main landmark used to open the anterior ethmoid sinus safely. With an aspirator or a blunt instrument, the roof is palpated to identify its medial limit, which corresponds to the level of the inferior aspect of the bulla ethmoidalis. Thus, it is easier and safer to penetrate the anterior ethmoid between the middle turbinate medially and the beginning of the roof of the maxillary sinus laterally. The penetration is performed exclusively in a plane parallel to the orbital wall, strictly avoiding any lateral penetration.

After this step, the orbital wall is easily identified as the first vertical plate of bone lateral and parallel to the plane of the middle turbinate. If necessary, the mucosa can be carefully removed in order to identify the character of the bone, which can be very thin. This dissection is performed parallel to the plane of the orbital wall until reaching the basal lamella of the middle turbinate. It is not necessary to identify the roof of the ethmoid sinus at this time, because this step is more easily performed in the posterior ethmoid, where the cells are larger and there usually is no superposition of cells. To avoid complications, penetration of the basal lamella of the middle turbinate is carried out medially and inferiorly. The walls of the cells are located with a blunt aspirator, recognizing that the superior wall corresponds to the roof of the ethmoid sinus and the lateral wall corresponds to the orbital wall. In cases of diffuse inflammation, careful removal of the mucosa makes identification of the bones of these walls easier. Once this landmark is identified, the procedure may be extended posteriorly and (eventually) anteriorly to remove all bony septa. If sphenoidotomy is not part of the procedure, management of the middle turbinate concludes the procedure. If the inferior part of the basal lamella of the middle turbinate has been removed, the partial resection of its free edge, preserving the superior attachment, is recommended to avoid scarring and lateralization (Fig. 29.5). In a partial ethmoidectomy, the middle turbinate may be preserved or partially removed (i.e., the head of the turbinate may be removed) unless doing so reduces access to the postoperative cavity. Regardless of the technique, crusting and the healing process are identical [19].

Sphenoid Revision Surgery

Failure after sphenoid surgery is uncommon, and the technique for primary surgery is now well standardized. Nevertheless, closure of the sphenoidotomy occasionally does occur, and revision is then needed. In such cases, CT scans and magnetic-resonance imaging (depending on the pathology) are performed to estimate the size of the cavity and the positions of the internal carotid artery and the optic nerve. A direct approach is recommended when possible. The first step of the revision procedure is localization of the ostium at the anterior wall of the sphenoid. It is located between the nasal septum and the superior turbinate. Locating it with a smooth aspirator or palpator is recommended. If all turbinates have been previously resected, it is necessary to identify the choana and the posterior vomer. From this area, the anterior wall of the sphenoid sinus is inspected, and a perforation is made medially and inferiorly, close to the nasal septum. Since strategic structures are lateral or superior, the risk of complications is lower. After the puncture of the anterior wall, an angled palpator is introduced to estimate the volume of the cavity and to control the enlargement of the opening with minimal risk. Enlargement is done either with a Blakesley forceps (which is placed in the ostium, opened and then rotated) or with a punch forceps (to remove the anterior bony wall). The edges of the enlarged opening are then smoothed, trying to minimize any injury of the mucosa of the sphenoethmoidal recess, which could result in postoperative adhesions. If the sphenoethmoidal recess is narrow, partial resection of the inferior portion of the middle turbinate is necessary. Once the resection is performed, the posterior ethmoid sinus may be reached and opened after opening the inferior part of the basal lamella of the middle turbinate. At this stage, the inspection of the anterior wall of the sphenoid sinus is facilitated via its trepanation. The sphenoidotomy is enlarged as much as possible to prevent closure of the opening. No stent or packing is required if the dissection has been meticulously performed; otherwise, a non-obstructive, small piece of Merocel® is placed in the sphenoethmoidal recess.

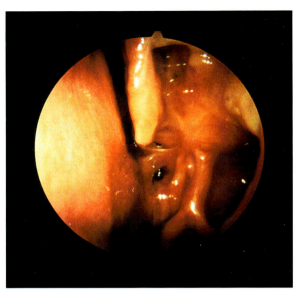

Fig. 29.5. Right revision sphenoethmoidectomy with subtotal resection of the middle turbinate

Revision Frontal-Sinus Surgery

The frontal sinus is probably the most difficult area to manage initially, because it has the highest incidence of postoperative stenosis [3, 13, 16, 20, 21]. The prevention of such problems requires respect for the mucosa surrounding the frontal ostia and accurate dissection of the frontal recess. Even with these precautions, failures occasionally occur, and revision surgery is then necessary. Postoperative stenosis of the frontal sinus was the most frequent complication encountered in our experience. Its management requires careful preoperative evaluation to determine the appropriate surgical technique. Accurate analysis of the CT scan and nasal-endoscopy findings are necessary to estimate the size of the frontal recess and the state of the mucosa at the origin of the stenosis (edema, fibrosis, polyps). The patency of the naso-frontal duct may be estimated by using the minimal-trephination technique, which allows analysis of any retained secretions; when lavage is also utilized, mucosal inflammation in the cavity can be treated. Sometimes the stenosis is secondary to persistent inflammatory disease and resolves after lavage. Unfortunately, sometimes either the sinus becomes blocked again or the stenosis recurs days or weeks after the local treatment. In these cases, revision surgery is advisable, either through an external or an endonasal approach (or both). If an endonasal approach is selected, several different techniques are available, and a single optimal technique has not been clearly defined. The first step of the procedure is analysis of the existent anatomy. Identification of the anterior portion of the middle turbinate is essential, because the frontal ostium is frequently located just lateral to it. Removal of the surrounding mucosa must be carefully avoided, because damaging it can lead to development of postoperative adhesions. Utilization of frontal irrigation is helpful, especially if no landmarks are recognized. With irrigation, the frontal ostium can be accurately localized; postoperative lavage is also possible (Fig. 29.6).

Once the frontal recess is reached, meticulously careful removal of the residual bony septae (upper part of the uncinate process) is necessary to completely open the frontal sinus, taking care not to injure the mucosa of the ostia. Management of the cicatrization is difficult because of the risk of recurrent stenosis. Several techniques have been proposed, but considerable controversy remains (concerning, for example, how large a frontal sinus neoostium must be to remain permanently open, and whether stenting improves the results). When to use a stent and what material it should be made of are still debatable. Some authors advise leaving a stent in place for days or months after the procedure, but others argue that such a foreign body leads to a secondary inflammatory response. Whatever technique is chosen, close follow-up is necessary to prevent recurrent stenosis. We advise frontal irrigation for some days after operation to remove secretions and reduce the postoperative edema induced by the surgery. If, because of the anatomy, access to the frontal sinus requires drilling of the bone below the ostium, postoperative reactions in the bone can occur, increasing the risk of postoperative stenosis. This can occur even if the opening of the frontal recess is made as large as possible through complete removal of the bottoms of both frontal sinuses and resection of the superior part of the nasal septum. In such cases, long-term stenting of the frontal opening is recommended in order to reduce the risk of recurrent stenosis of the frontal neo-ostium. In all cases, the opening must be sufficiently large to avoid a recurrence of the stenosis (both in the short-term and the long-term), because stenosis can occur many months or even years after the procedure. If the

Fig. 29.6. Frontal irrigation with a single-use Teflon needle (Micro-France)

frontal sinus can not be adequately accessed endonasally, an external or combined approach must be performed.

References

1. Corey JP, Bumsted RM (1989) Revision endoscopic ethmoidectomy for chronic rhinosinusitis. Otolaryngol Clin North Am 22:801–808
2. Dessi P, Castro F, Triglia JM, Zanaret M, Cannoni M (1994) Major complications of sinus surgery: a review of 1192 procedures. J Laryngol Otol 108:212–215
3. Draf W (1991) Endonasal micro-endoscopic frontal sinus surgery. The Fulda concept. Operative Tech Otolaryngol Head Neck Surg 2:234–240
4. Ferrie JC, Vandermarcq P, Azais O, Klossek JM, Drouineau J (1993) High resolution CT: pre operative assessment of chronic and recurrent rhinosinusitis. Eur Radiol 3:150–155
5. Hoseman W, Gode U, Wigand ME (1993) Indications, technique and results of endonasal endoscopic ethmoidectomy. Acta Otorhinolaryngol Belg 47:73–83
6. Kennedy DW (1992) Prognostic factors, outcomes and staging in ethmoid sinus surgery. Laryngoscope 102:1–18
7. King JM, Caldarelli D, Pigato JB (1994) A review of revision functional endoscopic sinus surgery. Laryngoscope 104:404–408
8. Klossek JM, Fontanel JP, Dessi P, Serrano E (1995) Chirurgie endonasale sous guidage endoscopique, 2nd edn. Masson, Paris
9. Klossek JM, Peloquin L, Fourcroy PJ, Ferrie JC, Fontanel JP (1996) Aspergillomas of the sphenoid sinus: a series of 10 cases treated by endoscopic sinus surgery. Rhinology 34:179–183
10. Klossek JM, Peloquin L, Friedman WH, Ferrie JC, Fontanel JP (1997) Diffuse nasal polyposis: postoperative long-term results after endoscopic sinus surgery and frontal irrigation. Otolaryngol Head Neck Surg 117:355–361
11. Klossek JM, Serrano E, Peloquin L, Percodani J, Fontanel JP, Pesey JJ (1997) Functional endoscopic sinus surgery and 109 mycetomas of paranasal sinuses. Laryngoscope 107:112–117

12. Lazar RH, Younis RT, Long TE, et al. (1992) Revision functional endonasal sinus surgery. Ear Nose Throat J 71:131–133
13. Loury MC (1993) Frontal recess dissection: endoscopic frontal recess and frontal sinus ostium dissection. Laryngoscope 103:455–458
14. Lund VJ, Mackay IS (1993) Staging in rhinosinusitis. Rhinology 31:183–184
15. Lund VJ, Mackay IS (1994) Outcome assessment of endoscopic sinus surgery. J R Soc Med 87:70–72
16. Metson R (1992) Endoscopic treatment of frontal sinusitis. Laryngoscope 102:712–716
17. Stammberger H (1991) Functional endoscopic sinus surgery. Decker, Philadelphia
18. Stammberger H, Kennedy DW (1995) Paranasal sinuses: anatomic terminology and nomenclature. The anatomic terminology group. Ann Otol Rhinol Laryngol Suppl 167:7–16
19. Weber R, Keerl R, Huppmann A, Schick B, Draf W (1996) Effects of postoperative care on wound healing after endonasal paranasal sinus surgery. Laryngorhinootologie 75:208–214
20. Weber R, Hosemann W, Draf W, Keerl R, SchickB, Schinzel S (1997) Endonasal frontal sinus surgery with permanent implantation of a place holder. Laryngorhinootologie 76:728–734
21. Weber R, Schauss F, Keerl R, Draf W (1997) La chirurgie ostéoplastique du sinus frontal: indications, procédures et résultats à propos de 75 cas. Rev Laryng Otol Rhinol (Bord) 118:91–94
22. Weber R, Draf W, Keerl R, Schick B, Saha A (1997) Endonasal microendoscopic pansinusoperation in chronic sinusitis. II. Results and complications. Am J Otolaryngol 18:247–253

Powered Instrumentation in Rhinologic and Endoscopic Sinus Surgery

Daniel G. Becker and David W. Kennedy

Introduction

Experience with the endoscopic approach to sinus and nasal disorders has led to the widespread recognition of the critical importance of mucosal preservation. The aim of functional endoscopic sinus surgery (FESS) should be to open diseased sinuses at the natural ostium and, wherever possible, to completely remove the underlying bony partitions in areas adjacent to chronic disease. At the same time, an intact mucoperiosteum-lined cavity should be left in the areas of the medial orbital wall, skull base and the frontal recess. While not always technically feasible at this time, this approach leads to more rapid healing with less crusting, reduced incidence of persistent disease and a more favorable postoperative course. Experience has shown that, when bone is exposed, healing is slowed and more intensive postoperative care is required. Scarring can be formidable in this setting. Indeed, it would appear that regenerated mucosa never re-develops normal ciliary density [22]. New bone growth may occur in areas of stripped mucosa and may provide a nidus for chronic infection. This becomes even more critical in the all-important area of the frontal recess, where mucosal scarring and new bone growth can lead to complete stenosis and persistent infection.

The initial instrumentation that was available for endoscopic sinus surgery consisted primarily of manual instruments without suction, such as Blakesley forceps and cupped forceps, which tended to strip mucosa from the underlying bone. It soon became evident that poor intraoperative visualization (as a result of bleeding) was frequently a limiting factor in the surgical procedure. Reduced visualization interfered with the necessary precision in this confined anatomic area and was a significant underlying factor in intraoperative complications. The introduction of suction forceps was an advance in this regard but was not a complete solution, as early versions tended to be bulky and suction was not available for all instruments. Devices like the Endo-scrub (Xomed, Jacksonville) and other similar devices have been helpful in keeping the tip of the endoscope clean and thereby improving visualization (even when some mucosal oozing is present).

A significant development for soft-tissue removal was the development of "through-cutting" instrumentation (Fig. 30.1). Low-profile heads and ergonomic designs have added to their usefulness. Furthermore, convenient shapes for instruments used in the frontal recess (Fig. 30.2), where the importance of this "through-cutting" instrumentation is critically important, have been a significant advance. However, it has been difficult to incorporate suction into these instruments in a completely satisfactory manner. Also, these instruments tend to become blunt over time.

Powered instrumentation using soft-tissue shavers offer another significant advance to the endoscopic sinus surgeon. There were several devices used in the late 1960s and early 1970s that formed the basis for today's soft-tissue-shaving systems. These devices may seem recent to most otolaryngologists, but a similar device was used by

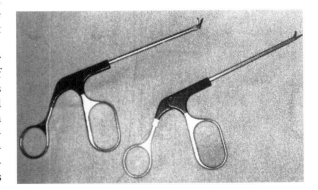

Fig. 30.1. "Through-cutting" instruments with narrow cutting heads and ergonomic design have added significantly to the ability to preserve mucosa during surgery, leading to subsequent quicker healing, less crusting and less intensive postoperative care (Black Star; Xomed, Jacksonville)

William House in the early 1970s for morselizing tissue associated with acoustic neuroma and is still described in some otologic texts [4]. The original patent was held by Dr. J.C. Urban (it was filed March 6, 1969 under the name "vacuum rotary dissector"); in the late 1970s, orthopedic surgeons developed a system based on similar principles for arthroscopy. These instruments achieved widespread use in orthopedic surgery before their potential application in sinus surgery was recognized. In 1994, Setliff and Parsons [30] were the first to report the use of soft-tissue shavers in endoscopic sinus surgery. Soft-tissue shavers have subsequently become widely accepted for the removal of polyps and soft-tissue masses. All currently available instruments work according to a similar principle, with an electrically powered, disposable cutting blade open only through a small window on the side and/or tip. Oscillation of the blade is accompanied by continuous suction through a hollow shaft; this removes soft tissue, blood and debris from the operative field [10].

Soft-tissue shavers and new bone-cutting drills offer potential advantages in the endoscopic treatment of a number of pathologies. As ongoing modifications have altered these instruments to more accurately reflect the needs of specific procedures, these instruments have become more useful and more widely used and have found utility in an increasing number of procedures. In this report, we will review the use of powered instrumentation in rhinologic and endoscopic sinus surgery.

Instrumentation

Soft-Tissue Shavers for Functional Endoscopic Surgery

A number of companies manufacture suction drills and soft-tissue shavers. The drill systems available are similar in several features. The hand pieces are generally small, light and ergonomic. The soft-tissue cannulas have a blunt cannula tip and a lateral or end port. The proximal and distal edges of the aperture are smooth; the lateral aspects of the aperture are smooth or serrated. An oscillating or rotating inner cannula with a similar port cuts and extracts soft tissue as it is suctioned through the port of the cannula (Fig. 30.3). The inner blade cuts and extracts soft tissue as it is suctioned into the side port.

The inner blade can oscillate, or it can rotate in forward or reverse directions. The oscillation mode appears to decrease the incidence of pulling or tearing of tissue. In the oscillation mode, some shavers

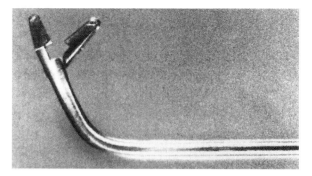

Fig. 30.2. Close-up of a cutting instrument with a fine head angled for work in the frontal recess, where stripping of mucosa may result in disastrous consequences of scarring and chronic obstruction

rotate one arc of rotation before reversing, while others rotate as many as five times in one direction before reversing. Whether this causes a significant difference in the amount of mucosal tearing has not been documented. While the maximum speeds of the shavers differ, all of the soft-tissue shavers can oscillate at least 1000 rpm [10].

The soft-tissue shavers are available in various shapes and sizes. Generally, surgeons select a size between 3.5 mm and 5.5 mm in diameter. The 3.5-mm blade removes smaller pieces of tissue and is preferred by many. Analogous to the use of the largest possible mastoid burr as the safest and most efficacious approach in mastoid surgery, however, some surgeons prefer the 4.2- to 5.5-mm shavers in a majority of cases.

The 3.5-mm blades made by all the manufacturers are generally 8 cm long, while the larger blades are up to 13.3 cm long. Some surgeons prefer the shorter length due to the theoretically decreased risk to vital structures beyond the front wall of the sphenoid. Some surgeons who operate while looking directly through the endoscopic eyepiece find that the additional length of the larger shavers allows the surgeon's hand to be further from the patient and allows an increased facility of movement. In all cases, critical emphasis is placed on a detailed knowledge of the anatomy (with known landmarks, such as the sphenoethmoid recess), including a careful review and constant referral to the patient's computed tomography (CT) scan [10].

Clogging is a problem cited by many sinus surgeons using this instrumentation. From a technical standpoint, there are several factors that affect the performance of the shavers [1]. The design of the shavers has been modified to maximize their cutting efficiency and minimize clogging. The configurations of the inner- and outer-blade cutting windows are designed to excise tissue in pieces that will be small enough to flow through the inside diameter of the inner tube/suction line to the suction canister. The size of the tissue piece can be affected by the size of the mouth of the tube and the speed of rotation of the inner blade relative to the outer, stationary blade. Built-in irrigation systems have also been included in some shaver systems in an effort to decrease the incidence of clogging.

Some of the shaver systems have a variable suction-control mechanism. Some of the shavers have pre-bent cannulas available, while some have the ability to be bent at the surgical site as desired by the surgeon. Although some of the more powerful units seem too expensive for the average office, some of the small units are more affordable for this setting and are quite adequate for office polypectomies. Hawke and McCombe describe 50 office polypectomies using a soft-tissue shaver without the need for nasal packing [14]. Undoubtedly, later generations of these drill systems will share many of these favorable characteristics.

Bone-Cutting Drills

The bone-cutting drills adapted from orthopedic arthroscopic surgery, with suction at the site of resection, allow excellent visibility at the operative site. They have proven beneficial in a variety of clinical settings. The bone-cutting drills consist of a spherical, tapered or oval-shaped burr affixed to a sheathed or guarded shaft. The "back" side of the burr is typically guarded by a beveled sheath. The sheath guards the shaft and the "back" side of the burr, and suction is applied through the sheath so that material can be suctioned from the site of the burr (Fig. 30.4).

The speed of rotation affects both the rate of resection and the controllability of the cut. Both tend to improve with increased rotational speed. High torque and rotation speeds (0–6000 rpm) allow effective and precise drilling. The one concern is that,

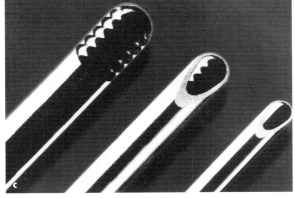

Fig. 30.3 a–c. Examples of power instrumentation for endoscopic sinus surgery. A wide variety of shaver blades and bone-cutting burrs are available. **a** This handle is long and slender enough to be held adjacent to an endoscope and contains both suction and irrigation (Xomed, Jacksonville). **b, c** These hand pieces are also light and slender and maneuver easily. These hand pieces have a variable suction-control mechanism (Linvatec, Largo)

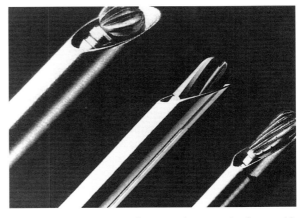

Fig. 30.4. A resection site. Shown are bone-cutting burrs with suction. A beveled outer sheath guards the posterior aspect of the drill (Linvatec, Largo)

with high-speed burrs, irrigation is critical to avoid thermal necrosis, and the burr may resonate or chatter at certain speeds within their range. Bone-cutting suction drills are available with a non-beveled sheath when end cutting is desired.

Specific modifications have been made to the bone-cutting burrs to tailor them for specific uses [1]. The fewer the number of flutes, the more aggressive a drill will be if all other factors are held constant. Burrs with two to six flutes tend to be very aggressive but are sometimes difficult to control. Burrs with eight or more flutes are less aggressive but are usually easier to control. The burrs for the treatment of choanal atresia must be small, but they must be aggressive to drill through a thick, bony atresia plate. These burrs are on the order of 2.9 mm in diameter and have as few as two flutes.

The burrs for frontal drilling must also be relatively aggressive to cut through the hard nasofrontal beak; consequently, they have a relatively low number of flutes (typically five flutes) for increased aggressiveness. In contrast, bone-cutting burrs designed for modifying the nasal dorsum require *less* aggressiveness and increased precision. In their current design, these drills have ten flutes.

Bendability

Soft-Tissue Shavers

Malleable and pre-bent soft-tissue shavers have application in select surgical circumstances. Pre-bent blades have a fixed angle of bend at a fixed distance from the distal end of the blade. Malleable blades offer the surgeon the opportunity to modify the bend of the blade at the surgical site. Advantages include an ability to bend the cannula at a wide range of points along its length. Bending of up to 30° is achievable.

Blades with a bend in them may have an increased tendency toward clogging. The ability to remove and then re-insert the inner cannula facilitates the unclogging of a bent soft-tissue shaver.

Blade heating may occur in some malleable or pre-bent blades, due to frictional heat between the inner and outer cannula in the bend area. With irrigated blades, the heat is removed from this area by the irrigating fluids; without irrigation, blade heating may cause thermal damage to susceptible mucosa. Decreased clogging may also be an advantage of irrigated blades in this setting.

The engineering principles behind the pre-bent and malleable blades are essentially the same [1]. The outer cannula undergoes a "plastic" deformation, meaning the blade does not return to a straight configuration after the bending stress is released. The inner cannula must conform with this bend; therefore, it is constructed to bend with an elastic deformation, which is a bend that results in return of the cannula or tube to its original shape after the bending stress is released.

The inner cannula or tube must, therefore, be able to bend elastically to conform to the outer cannula while also being able to withstand the motor's torque during operation. Polymeric tubing with high strength and high flexibility provides the physical properties necessary to create such an inner cannula and appears to be the most effective approach.

Bone-Cutting Burrs

A curved bone-cutting burr may be advantageous in certain surgical situations. For example, curved bone-cutting burrs that are tailored specifically for the frontal-sinus region should facilitate drilling in this confined anatomic area. One engineering approach has been the use of interlocking, ball-and-socket "links" that yield a multiply jointed inner shaft (Linvatec, Largo). Another solution has been the use of a high strength, flexible composite tube linked to a distal burr (Xomed, Jacksonville). The bend that can be created approaches 55°.

Powered Instrumentation: Clinical Applications

General Principles

The concept of functional sinus surgery emphasizes the potential of diseased mucous membrane to recover from removal of only irreversibly diseased tissue. While the term "microdebrider" is commonly applied to power instrumentation, this is a misnomer. The power instruments do not distinguish between healthy and unhealthy tissue, as the word "debride" may imply. It is the experienced sinus surgeon, not the power instrument, who determines what tissue must be removed and what will return to normal. The power instruments do offer precision, but this precision relies on the surgeon. The soft-tissue shavers provide a "true" cut, but equivalent precision can generally be achieved with conventional instrumentation in a capable surgeon's hands. The power instruments do provide specific advantages for specific circumstances for most surgeons; it appears they are being used by a great number of surgeons in a wide range of surgical situations [10].

We have observed that the increased popularity of power instrumentation has been accompanied by a re-emphasis of the important underlying principles of functional sinus surgery. Certainly, the precise nature of excision does allow and perhaps encourage the surgeon to proceed in a meticulous way. However, a surgeon can readily perform a radical removal of normal tissue with powered instrumentation. *It is not the instrument but the surgeon who must bring a functional philosophy to the operating room.*

In general, better visibility allows safer functional sinus surgery. In situations in which power instruments facilitate the removal of blood and debris from the surgical field, they clearly provide increased safety. Such situations include patients with polypoid sinusitis, sinusitis with subperiosteal abscess and some cases of revision sinus surgery. A better surgical result may also be attainable due to improved visualization. With regard to safety, it must always be pointed out that there is no substitute for a detailed knowledge of anatomy and careful, precise surgical technique.

Great care must be taken when using powered instruments. Knowledge of anatomy is essential, as these instruments can be aggressive. May has pointed out that there is a learning curve in the use of these instruments [11]. In cadaver dissections, May has demonstrated that these instruments can suction periorbital fat if care is not taken. Setliff has received a report of exposure of orbital fat without postoperative sequelae in one revision case [11], and we are aware of cases of both orbital entry and intracranial entry with a soft-tissue shaver. One disadvantage of using a shaver is that the ability to palpate that is available with conventional instruments is lost to some degree whereas, with conventional up-biting instruments, it is possible to palpate behind a bony partition in order to ensure that removal is safe. This is not possible with a shaver. Shavers may, therefore, be inappropriate for the less-experienced endoscopic sinus surgeon.

Due to the unique properties of the soft-tissue shaver, tissue removed is generally in significantly smaller pieces than with conventional instrumentation. The use of a suction trap to collect the shaved-tissue specimens is strongly recommended so that unexpected pathology will not be inadvertently missed. We also recommend the use of conventional instrumentation in obtaining biopsy specimens, e.g., for suspected malignancies, benign tumors (inverting papilloma) or other suspicious tissue. Also, biopsies for ciliary dyskinesia should be performed with conventional instrumentation.

Specific Applications

Polyps

Polypoid sinusitis and other situations in which increased surgical bleeding can be expected pose a challenge to the endoscopic surgeon. Lens-cleaning and -protecting equipment, such as Endoscrub and Endosheath, are designed to keep the lens clean longer to avoid the frequent need to remove the scope from the nose. Conventional non-powered instrumentation that has built in suction exists. These have tended to be bulky and have had problems with clogging [10, 11], although recently designed suction instruments may be an improvement. The sharp oscillating blade of the soft-tissue shaver allows for precise cutting of polyps and mucosa while leaving adjacent tissue intact (Fig. 30.5). However, perhaps the most significant advantage of the soft-tissue shaver is its continuous suction and ability to maintain a bloodless field. Soft-tissue shavers evacuate tissue from the surgical site without the need to remove the instrument, providing potentially continuous suction of blood from the field with the opportunity for improved visualization and precision and for less frequent interruptions during the procedure. This improves visualization (and, potentially, safety) during the procedure, particularly in the setting of massive nasal polyposis, where the use of these instruments is a dramatic advance in our ability to reduce bleeding and identify the anatomy. Additionally, since the blades are disposable (and, therefore, are always sharp), they tend to cut more cleanly than hand-held "through-cutting" instruments, which become significantly blunted with time.

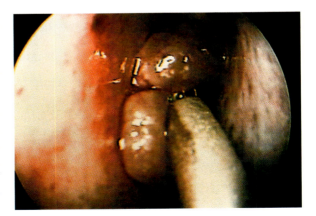

Fig. 30.5. Patient with massive intranasal polyposis, illustrating the optimal application of powered instrumentation. The bloodless field allows excellent visualization and mucosal preservation

Soft-tissue shavers offer the opportunity to debulk polypoid disease with relative ease in the clinic. Hawke and McCombe describe 50 office polypectomies using a soft-tissue shaver without the need for nasal packing [14].

"Small-Hole" Surgery

Setliff has described the use of powered instruments in the approach to sinus surgery called the "small-hole" technique [27–29]. He postulates that the uncinate process, the postero-medial wall of the agger nasi, the medial wall of the ethmoid bulla and possibly the basal lamella are the critical anatomic components in recalcitrant sinus disease. He asserts (1) that surgical treatment of these "transition spaces" may be all that is required in cases of sinusitis requiring surgical treatment and (2) that powered instrumentation provides a degree of precision that enables the surgeon to more readily perform this approach. However, others question whether patients with minor disease really require surgical intervention or whether, in many cases, they could also be handled appropriately with medical therapy.

Krouse and Christmas reported a non-randomized, non-blinded study of 250 patients who had undergone surgery with the soft-tissue shaver and compared them with 225 earlier patients who had undergone "traditional" endoscopic surgery [17]. Bleeding was not a significant problem in either group, and differences in operating time were not noted. No synechiae were identified in the powered-instrument group, while four patients in the traditional group had synechiae. One patient experienced ostial re-occlusion in the powered-instrument group, compared with seven in the traditional group. The authors felt that the patients in the powered-instrument group healed faster. Significantly, an essentially identical percentage of patients (86% in powered-instrument group, 87% in the traditional group) were symptom free at least 6 months after surgery. Most authors agree that powered instruments offer a precise cutting action; however, whether the soft-tissue shavers are *more* precise in the excision of tissue (with the avoidance of stripping of mucosa) in comparison to conventional "true-cutting" instruments is not known.

Ethmoid Bone

Several authors have demonstrated the ability to perform "mini-FESS" procedures with excellent mucosal preservation entirely with powered instruments [23, 27–29]. However, at least in the adult population, we believe that the underlying bone may play a significant role in the pathogenesis of sinusitis. Tetracycline-labeling studies provide evidence that the rate of bone turnover is similar to that in osteomyelitis. The histomorphology of ethmoid-bone specimens from patients with chronic sinusitis is compatible with a diagnosis of osteomyelitis [25]. Although we have not been able to demonstrate the presence of bacteria or bacterial DNA in the underlying bone, clinical evidence suggests that such DNA may be an active participant in the disease process. We believe that the utility of currently available soft-tissue shavers is limited by their preferential cutting of soft-tissue rather than bone. Preferential removal of the bony intercellular partitions should, therefore, be a primary aim and cannot be readily achieved with currently available powered instruments.

Partial Inferior Turbinectomy

Davis and Nishioka have described the use of soft-tissue shavers as an alternative instrument for partial inferior turbinectomy with removal of erectile soft tissue from the lateral and inferior borders of the turbinate [7]. In their report, turbinate-bone removal is undertaken with conventional cup forceps.

Pediatrics

Many consider the soft-tissue shavers to be particularly useful in the smaller anatomic confines of the pediatric patient. Setliff and Parsons [30] reported 345 patients with ages ranging from 3 years to 85 years who underwent surgery with soft-tissue shavers. Ninety of the patients were children. Mendelsohn and Gross [20] have reported that a clearer surgical field due to suction at the resection site combined with the precision "through-cutting" provided by the soft-tissue shavers decreased the risk of denuding mucosa. Parsons [23] has also described the use of soft-tissue shavers for other rhinologic uses in children, including adenoidectomy, choanal-atresia repair and correction of an obstructing septal spur. Smith, Boyd and Parsons [31] have described the usefulness of soft-tissue shavers for pediatric sphenoidotomy.

Adenoidectomy

Especially in view of the prominent role of adenoid hypertrophy in pediatric rhinologic disorders, the application of powered instrumentation for adenoidectomy warrants discussion here. Yanagisawa and Weaver [36] have described a transnasal endoscopic adenoidectomy using a soft-tissue shaver. After applying nasal and nasopharyngeal topical and infiltrative anesthesia, they pass a soft-tissue shaver transnasally for removal of adenoid tissue. They employ suction electocautery transnasally under endoscopic visualization to achieve hemostasis. They advise great caution around the Eustachian-tube orifice and cite disadvantages, including difficult maneuverability, especially in the presence of a deviated septum. Difficulty in achieving complete removal via a transnasal approach (due to the typical lateral location of the cutting port and problems with clogging) were also discussed. They felt that this approach should be reserved for select cases only.

Parsons [23] describes a transnasal endoscopic adenoidectomy performed using soft-tissue shavers. Indications for this approach include abnormal growth of adenoids into the nasal cavity anterior to the posterior choanae, unsuspected adenoid tissue identified during FESS, children with choanal atresia and patients with velopharyngeal flaps and obstructive apnea and/or otitis media with effusion.

Koltai et al. [16] reported on a trans-oral adenoidectomy using a soft-tissue shaver with an approximate 45° bend. They reported that the angle of the blade allows easy trans-oral placement and maneuverability. The blade orientation is favorable for end-on removal of the adenoid tissue.

Choanal Atresia

Powered instrumentation has demonstrated utility in choanal-atresia repair [23, 33]. The presence of suction at the drill site seems to clear debris away more quickly and has allowed improved visualization at this small operative site. Randall, Kang and Mohs [24] report favorable results from the application of a 3.5-mm mini-arthroscopic drill system via a transnasal endoscopic approach in three cases of choanal atresia. Even the small nostrils of neonates were able to accommodate the mini-arthroscopic drill and a 0°, 2.7-mm endoscope. The suction at the resection site and a protective sheath were described as two particularly favorable attributes of the drill system.

Septum

The advent of endoscopic septoplasty, first described by Lanza et al. [18] and later by Giles et al. [9], has significantly changed our handling of limited septal deformities. By utilizing a septal incision made anterior to an area of deviation and under endoscopic visualization, the extent of mucoperiosteal elevation and postoperative edema can be markedly reduced. After transecting the cartilage or bone and elevating the contralateral mucosal flap, the deviated portion of the cartilage or bone is resected with scissors or through-cutting forceps. The Endoscrub (Xomed, Jacksonville) and a suction elevator are very useful in these endoscopic intra-septal procedures. Bone-cutting drills have allowed a further modification in which the mucoperichondrial or mucoperiosteal flap may be raised (on one side) over the offending spur, which may then be drilled away while the raised flap is protected by the beveled sheath. Thus, there is no need to raise a flap on the opposite side.

Nasal Burr

Becker et al. have described the use of a burr with a protective sheath and suction at the resection site for precise modification of the bony nasal dorsum during rhinoplasty [1, 3]. The protective sheath covers all but the active part of the drill, protecting the skin/soft-tissue envelope (Fig. 30.6). Designed to offer a precise alternative to the rasp, the drill may be less

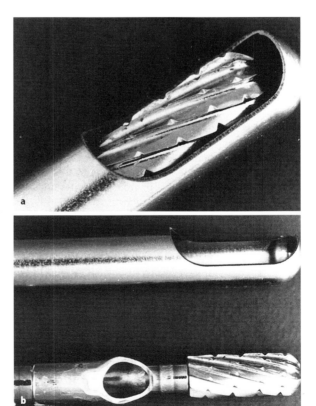

Fig. 30.6 a,b. A resection site; drilling of nasal dorsum. Shown is a burr with a protective sheath and suction

traumatic to the skin/soft-tissue envelope, because it does not rely on the potentially bruising back-and-forth motion typical of rasping. The authors have also described a powered reciprocating rasp with a back-and-forth excursion of 5 mm and speeds up to 6000 rpm. The nasal-dorsum drill and the powered reciprocating rasp may decrease the incidence of bony dorsal irregularities after rhinoplasty.

Frontal Sinus

With the advent of CT scans and endoscopic techniques, several centers around the world developed a renewed interest in re-establishing drainage pathways from the frontal sinus. Draf, May, Wigand, Close et al. and Gross et al. are among those who have used a drill transnasally to enlarge frontal-sinus drainage [2, 6, 8, 12, 13, 19, 35]. In the endoscopic era, Draf [8] first reported the transnasal use of a drill to address the frontal sinus. In a series of 387 patients requiring frontal-sinus surgery, 12 cases required "median drainage", which appears to be very similar to a transnasal Lothrop procedure. After a 5- to 8-year follow-up, Draf reported a closure rate of approximately 20% at the April 1995 Rhinologic Society spring meeting in Palm Desert, California.

Close et al. [6] reported a frontal drilling procedure. They used a conventional cutting burr adapted with an "extender" to drill away the floor of the frontal sinus in 11 patients. Five patients required a small infrabrow incision and frontal sinusotomy. They reported one CSF leak in this series.

Gross et al. report that the availability of advanced drill technology with a protective sheath and suction at the surgical site has been instrumental in allowing a reassessment of the intranasal techniques [2, 12, 13]. The beveled sheath protects the posterior table and its mucosa, helping to avoid both circumferential mucosal injury and violation of the posterior table and has enabled surgeons to achieve the widest possible anatomic opening. Advanced drill technology and endoscopic techniques have allowed this procedure to be performed safely and expeditiously [2, 12, 13].

In the modified transnasal-endoscopic Lothrop procedure (or frontal drillout), one common opening that includes both natural frontal openings is created. Creating one common opening that includes both natural frontal openings avoids the potential for circular re-flow. Theoretical advantages of this procedure over frontal-sinus obliteration include decreased morbidity, improved cosmesis and the ability to endoscopically evaluate postoperative patients for recurrent disease. The modified transnasal Lothrop procedure is generally reserved for surgical management of patients with diseases similar to those previously treated with an osteoplastic-flap approach [2, 12, 13].

Endoscopic Dacrocystorhinostomy

We prefer to perform endoscopic dacrocystorhinostomy endoscopically, using drills (as has been reported by other authors [21]). The advantages of an endoscopic approach using drills rather than a laser include the ability to create a very wide exposure of both sac and duct without heating the surrounding bone at the edge of the opening. The presence of built-in suction at the resection site in bone-cutting drills may increase visibility and maneuverability in these procedures. Irrigation also helps minimize bone heating. After the sac is opened, anteriorly and posteriorly based flaps of the sac mucosa can then be rotated over the bony edges and secured in place with a small, bipolar "weld".

Sphenoid

Setliff [27], Smith, Boyd and Parsons [31], Christmas and Krause [5] and others have described the use of soft-tissue shavers for sphenoidotomy. The availability of suction at the surgical site is, at times, of benefit when working in this area; soft-tissue shavers provide advantages similar to those of conventional suction/through-cutting instruments in this setting.

Optic-Nerve Decompression

Optic-nerve decompression is probably best performed endoscopically, using drills [32]. In cases where traumatic visual loss secondary to blunt trauma does not improve with high dose steroids, the bone may be removed from the optic canal with a diamond burr under excellent visualization of 180° or more. However, an adequate suction or suction/irrigation drill has not been described. The potential addition of built-in suction at the resection site in bone-cutting drills may prove to increase visibility and maneuverability in these procedures, but currently available suction/irrigation drills probably do not have sufficient irrigation to allow safe drilling in this region without supplemental irrigation at the site.

Pituitary

Endoscopic approaches to pituitary tumors were first reported by Jankowski in 1992 [15]. The use of the endoscope in this situation allows more careful evaluation for residual tumor, particularly when supra-diaphragmatic or lateral tumor extension is present. As neurosurgeons continue to gain familiarity with the use of the endoscope, it is now possible to treat selected lesions transnasally, avoiding the necessity for a trans-septal or trans-ethmoid approach. Sethi has recently reported excellent results utilizing endoscopic techniques in 80 patients [26].

While the neurosurgical literature has an increasing number of reports on the endoscopic trans-septal approach to the sphenoid sinus for pituitary surgery, there are no reports of the use of powered instruments for pituitary surgery. However, it seems possible that a very small soft-tissue shaver may prove helpful in tumor removal and drills are clearly helpful in removing a thick sella or the posterior bony septum.

Skull-Base Lesions

The ability to close CSF leaks and skull-base defects with high reliability and minimal morbidity has led us to revise traditional thinking with regard to conservative management of many cases of CSF rhinorrhea and is currently leading us to a new frontier of the potential applications of endoscopic techniques. Specifically, since 1989, one of the authors (Kennedy) has developed experience in managing benign lesions of the anterior skull base and orbit, including mucoceles, inverted papilloma, fibro-osseus lesions and juvenile nasopharyngeal angiofibroma.

Endoscopic approaches to benign tumors of the nose, sinuses, anterior cranial fossa and orbit are now becoming widely utilized. The role of endoscopic techniques for definitive resection is more controversial. One of the more common lesions encountered in this area is inverted papilloma. Traditional approaches to inverted papilloma usually involve a 10–15% recurrence rate and, in view of the malignant potential of the lesion, some skepticism regarding the potential role of a minimally invasive procedure initially appears to have merit. However, few could argue against resection of small, localized lesions where a clear margin can be obtained endoscopically, at least by skilled surgeons. Similarly, it has been our choice to approach larger lesions with skull-base involvement via an endoscopic approach, drilling the underlying bone with a diamond bur. Once involvement of the skull base has occurred, only a craniofacial approach will improve the margin more than is possible endoscopically, and most surgeons do not recommend a primary craniofacial approach for this tumor. The role of endoscopic techniques in inverting papillomas with other sites and degrees of tumor involvement is controversial, although it is increasingly being accepted as the treatment of choice among skilled endoscopists. Although the important principle is that the approach must not compromise the initial resection, there is increasing evidence that the recurrence rate of inverting papilloma following skilled endoscopic resection is similar to that seen in more radical procedures. Waitz and Wigand performed a retrospective evaluation of 51 patients; six of 35 (17%) who were treated endoscopically experienced recurrence, while three of 19 (19%) who were treated extranasally experienced recurrence [34]. In this series, however, the endoscopically managed tumors tended to be smaller.

In general, endoscopic surgery for inverted papilloma should include resection or drilling down (via a burr) of the underlying bone at the site of attachment (medial orbital wall) and preservation of the underlying periosteum or dura. Long-term endoscopic follow-up is essential, and a smooth, widely patent cavity should be created so that all areas can easily be examined during follow-up for early evidence of recurrent or persistent disease.

Conclusion/Summary

Powered instrumentation has gained widespread use in endonasal endoscopic surgery. The exacting nature of soft-tissue and bone removal in a confined anatomic area with adjacent important anatomic structures lends itself to consideration of anatomic contouring with these instruments. The soft-tissue-shaving cannulas and new bone-cutting drills embody a potentially less traumatic and more precise method of soft-tissue extraction and bone removal. The increased precision, the ability to keep the endoscopic field clear of blood and debris and the ability to precisely and safely drill bone endonasally have allowed widespread application of minimally invasive approaches to a variety of clinical problems.

References

1. Becker DG (1997) Technical considerations in powered instrumentation. Otolaryngol Clin North Am 30:421–434
2. Becker DG, Moore D, Lindsey WH, Gross WE, Gross CW (1995) Modified transnasal endoscopic Lothrop procedure: further considerations. Laryngoscope 105:1161–1166
3. Becker DG, Gross CW, Tardy ME, Toriumi DM (1997) Powered Instrumentation for Dorsal Reduction. Facial Plast Surg 13:291–297

4. Benecke JA, Stahl BA (1994) Otologic instrumentation. In: Brackmann DA (ed) Otologic surgery. Saunders, Philadelphia, pp 19–94
5. Christmas DA, Krouse JH (1996) Powered instrumentation in functional endoscopic sinus surgery. I. Surgical technique. Ear Nose Throat J 75:33–38
6. Close LG, Lee NK, Leach JL, Manning SC (1994) Endoscopic resection of the intranasal frontal sinus floor. Ann Otol Rhinol Laryngol 103:952–958
7. Davis WE, Nishioka GJ (1996) Endoscopic partial inferior turbinectomy using a power microcutting instrument. Ear Nose Throat J 75:49–50
8. Draf W (1991) Endonasal micro-endoscopic frontal sinus surgery: the Fulda concept. Operative Tech Otolaryngol Head Neck Surg 2:234–240
9. Giles WC, Gross CW, Abram AC, Greene WM, Avner TG (1994) Endoscopic septoplasty. Laryngoscope 104:1507–1509
10. Gross CW, Becker DG (1996) Power instrumentation in endoscopic sinus surgery. Operative Tech Otolaryngology Head Neck Surg 7:236–241
11. Gross CW, Becker DG (1996) Instrumentation in endoscopic sinus surgery. Curr Opin Otolaryngol Head Neck Surg 4:20
12. Gross CW, Gross WE, Becker DG (1997) Endoscopic frontal sinusotomy (Lothrop procedure). In: Schaefer SD (ed) Common problems in rhinology. Mosby, St. Louis, pp 19–97
13. Gross WE, Gross CW, Becker DG, Moore D, Phillips CD (1995) Modified transnasal endoscopic Lothrop procedure as an alternative to frontal sinus obliteration. Otolaryngol Head Neck Surg 113:427–434
14. Hawke WM, McCombe AW (1995) How I do it: nasal polypectomy with an arthroscopic bone shaver: the Stryker "Hummer". J Otolaryngol 24:57–59
15. Jankowski R, Auque J, Simon C, Marchal JC, Hepner H, Wayoff M (1992) Endoscopic pituitary tumor surgery. Laryngoscope 102:198–202
16. Koltai PJ, Kalathia AS, Stanislaw P, Heras HA (19) Power-assisted adenoidectomy. Arch Otolaryngol Head Neck Surg 123:685–688
17. Krouse JH, Christmas DA (1996) Powered instrumentation in functional endoscopic sinus surgery. II. A comparative study. Ear Nose Throat J 75:42–44
18. Lanza D, Rosin D and Kennedy DW (1993) Endoscopic septal spur resection. Am J Rhinol 7:213–216
19. May M (1991) Frontal sinus surgery: endonasal endoscopic ostioplasty rather than external osteoplasty. Operative Tech Otolaryngol Head Neck Surg 2:19–91
20. Mendelsohn MG, Gross CW (1997) Soft tissue shavers in pediatric sinus surgery. Otolaryngol Clin North Am 30:443–450
21. Metson R (1995) Endoscopic dacrocystorhinostomy: an update. Operative Tech Otolaryngol Head Neck Surg 6:217–220
22. Moriyama H, Yanagi K, Ohtori N, Asai K, Fukami M (1996) Healing process of sinus mucosa after endoscopic sinus surgery. Am J Rhinol 10:61–66
23. Parsons D (1996) Rhinologic uses of powered instrumentation in children beyond sinus surgery. Otolaryngol Clin North Am 29:105–114
24. Randall DA, Kang R, Mohs DC (1997) Use of the miniarthroscopic drill in choanal atresia repair: how we do it. Otolaryngol Head Neck Surg 116:696–697
25. Senior BA, Gannon FH, Montone KT, Hwang PH, Lanza DL, Kennedy DW (1998) Histology and histomorphometry of ethmoid bone in chronic rhinosinusitis. Laryngoscope 108:502–507
26. Sethi D, Pillay P (1996) Endoscopic pituitary surgery: a minimally invasive technique. Am J Rhinol 10:141–148
27. Setliff RC (1996) Minimally invasive sinus surgery: the rationale and the technique. Otolaryngol Clin North Am 29:115–129
28. Setliff RC (1996) The Hummer: a remedy for apprehension in functional endoscopic sinus surgery. Otolaryngol Clin North Am 29:93–103
29. Setliff RC (1997) The small-hole technique in endoscopic sinus surgery. Otolaryngol Clin North Am 30:341–354
30. Setliff RC, Parsons DS (1994) The "Hummer": new instrumentation for functional endoscopic sinus surgery. Am J Rhinol 8:275–278
31. Smith WC, Boyd EM, Parsons DS (1996) Pediatric sphenoidotomy. Otolaryngol Clin North Am 29:159–167
32. Stankiewicz JA, Chow JM (1997) Powered instrumentation in orbital and optic nerve decompression. Otolaryngol Clin North Am 30:467–478
33. Vickery CL, Gross CW (1997) Advanced drill technology in treatment of choanal atresia. Otolaryngol Clin North Am 30:457–466
34. Waitz G, Wigand M (1992) Results of endoscopic sinus surgery for the treatment of inverted papillomas. Laryngoscope 102:917–922
35. Wigand M, Hoseman W (1991) Endoscopic surgery for frontal sinusitis and its complications. Am J Rhinol 5:85–89
36. Yanagisawa E, Weaver EM (1997) Endoscopic adenoidectomy with the microdebrider. Ear Nose Throat J 72–73

Image-Guided Surgical Navigation in Functional Endoscopic Sinus Surgery

Winston C. Vaughan and Frederick A. Kuhn

Introduction

Functional endoscopic sinus surgery (FESS) has become the standard and most frequently performed surgical procedure for sinus disease. After the introduction of this technique, as described by Messerklinger [10–12], several publications have demonstrated it to be highly effective and safe [5, 6, 9, 13, 14, 16–18].

There are, however, major complications reported with endoscopic sinus surgery, as reflected by the fact that FESS has become the most common reason for lawsuits against otorhinolaryngologists in the USA [2, 6]. This is due to the learning curve surgeons experience when they start to perform a new procedure. This was also recently experienced by abdominal surgeons with the introduction of the laparoscopic cholecystectomy [8]. Surgeons need to continually assess the risks, benefits, applications and costs of newly introduced technology [7].

In 1929, the close anatomic relationship between sinuses and neighboring vital structures was said by Mosher to "often lead to tragedy… [it is] one of the easiest ways to kill a patient" [13]. This close relationship, lack of familiarity with the anatomy and disorientation due to bleeding or disease, may lead to complications.

The development of better surgical instrumentation and surgical expertise has led to more extensive endoscopic procedures, such as excision of tumors, orbital decompression and endoscopic treatment of recurrent sphenoid and frontal-sinus disease. With this extension of FESS into more difficult surgical scenarios, better technology was needed to keep these surgeries safe and effective. The development of newer modalities with direct application to FESS, including steerable, through-cutting and powered instruments, newer medications, surgical procedures for recurrent disease and image-guided surgical navigation will hopefully satisfy surgeons' desire to continue to extend FESS. The avoidance of complications should remain of the utmost importance if endoscopic sinus procedures are to remain acceptable to patients and third-party providers. One should also continue to improve the efficacy, thoroughness and efficiency of these operations.

The Instrumentation and Technology

Image-guided surgical navigation is definitely the most important technological advance with respect to both safety and improved surgical efficacy. This is also referred to as frameless, stereotactic surgical navigation, computer-assisted FESS, three-dimensional computed tomography (CT) intraoperative localization, intraoperative CT guidance and image-guided surgery. There are several systems currently available, including the Instatrak system (VTI, Boston), the Viewing Wand (ISG, Ontario) and the Stealth system (SNT, Broomfield).

Several studies using these systems have been published [1, 3, 4, 15]. The authors of these studies routinely state that these systems improve technique, safety and the surgeon's confidence in difficult cases. Most state that they believe that the surgical time is reduced when using these systems, as it leads to directed, safe tissue identification and removal. Some authors advocate its use in altered anatomy or difficult cases, while others feel that it should be available for all cases [1, 3, 15].

The Georgia Rhinology and Sinus Center has used the Instatrak surgical navigation system in over 200 patients during a 2-year period. This system, as others have reported, has proven to be very valuable in the surgical treatment of sinus disease at our center. The system is used in all revisions and in some primary cases of chronic sinusitis.

The system has been previously described [3]. The device utilizes a flexible headset (Fig. 31.1) rather than a rigid frame and consists of an armless pointer attached to an electromagnetic receiver. The pointer at present is a series of suction elevators (Fig. 31.2); eventually, surgical instruments will also

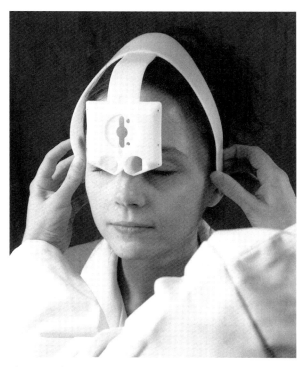

Fig. 31.1. The Instatrak surgical navigation system headset

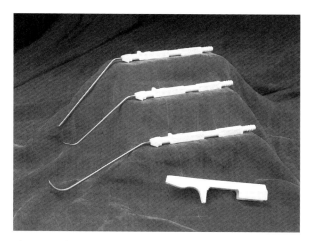

Fig. 31.2. The Instatrak surgical navigation system pointer

be used as pointers. These suction elevators are used in combination with the plastic headset, which has an electromagnetic transmitter attached. The pointer and sensor work in conjunction with a microcomputer, which provides instantaneous intraoperative CT positioning on a large computer screen (which also displays the video image; Fig. 31.3). The position of the pointer in the surgical field is represented by crosshairs on three images: axial, sagittal and coronal reconstructions from the preoperative CT scan. The patient wears a headset that contains fiducial markers during CT-scan acquisition. The axial CT scan, obtained at 1-mm intervals, is used to reconstruct sagittal and coronal images. This CT data set is stored and transferred to the Instatrak system for later use. It can often also be used for future revision surgeries. It will not of course, show the intrasinus surgical changes from the previous procedure, but it can be used to review constant, bony landmarks.

In the operating room (OR), the data set is retrieved by the computer and displayed on the monitor. This is compared with the actual hard films to ensure that the patient and date are correct before the patient is brought into the OR. The images are reviewed to plan the surgical approach and to note unusual findings and obstructive cells. The ability to scroll through images is an invaluable asset during surgical planning, as a continuous relationship between different images can be obtained. This is very similar to the scroll function of the rollerball on a CT scanner, which technicians and radiologists employ. This is also the point at which a senior surgeon should carefully review the findings with the resident or assisting colleague. Anatomical teaching and clinical correlation are facilitated by these functions prior to the operation.

For patient comfort and to avoid excessive pressure, the operation is started without the headset in place. Once navigation is needed, the circulating nurse places the headset on the patient, and a sterile drape is used to re-drape the field. The pointer and headset transmitter are calibrated. This is followed by the selection of a surgical reference point. This is performed by placing the suction pointer on a firm intranasal surface (such as the superior insertion of the middle turbinate) and verifying that this is seen on the imager. One should routinely return to this or other reference points during the course of surgery to reconfirm accuracy.

During FESS, there are several transition points at which the navigation system is routinely used. These include the identification of the basal lamella of the middle turbinate, the posterior ethmoid cells, the lamina papyracea, the sphenoid face and the junction of the sphenoid face and the skull base. The navigation system is also used during the posterior-to-anterior dissection along the skull base toward the frontal recess, and during the identification and removal of obstructing cells from the frontal recess.

Once the need for navigation has ceased, the boundaries of the surgical field are reconfirmed, and the headset is removed. A few patients have noted postoperative tingling or numbness around the earlobes. Therefore, we have decreased the length of time that the headset is on the patient to reduce the possibility of this problem. The headset is placed

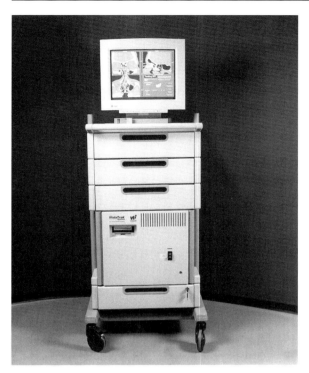

Fig. 31.3. The Instatrak surgical navigation system video output. The position of the pointer is shown relative to a preoperative computed-tomography image on a computer screen

when navigation will be needed for an extended time and is removed as soon as the majority of the operation is completed. The headset can be removed and replaced multiple times during the procedure; however, this necessitates re-calibration and re-draping.

Advantages of the System

Stereotactic surgical navigation may be invaluable in cases in which the anatomy has been distorted by previous surgery or recurrent disease. This is especially important in patients with nasal polyposis or those who have had middle-turbinate resections. The identification of potentially complex obstructing cells near the frontal recess and in the posterior ethmoid sinuses has been greatly improved with this device and allows more complete removal.

Most importantly, this instrument allows continuous CT-data correlation with the surgical field. There is no substitute for knowing the anatomy, but this instrument provides a level of reassurance needed during long and difficult cases, especially if bleeding persists [3].

The ability to re-verify the position of the suction is another main advantage. This allows for a continuous re-checking of accuracy in an efficient manner. Others have noted that this instrument may also improve surgical confidence in difficult cases, resulting in a more complete surgery [3]. We have found this to be the case. Lastly, the teaching effectiveness of this instrument must be emphasized.

In a teaching setting where students are attempting to overcome a learning curve, it is desirable to avoid complications. This instrument allows teaching before, during and after surgery in a logical and relational manner. The residents and fellows who have been trained with this device have found it to be invaluable in their education. Surgical navigation with this instrument has allowed residents and fellows to perform most of the operation (under supervision) while maintaining patient safety and clinical efficacy. We consider it a required instrument when teaching revision cases or cases with difficult anatomy.

Disadvantages of the System

The main disadvantage of these systems is the initial capital investment. This may be higher than $250,000 for some systems. As with every new technology, however, this should decline with increased availability and less expensive hardware. The potential cost savings experienced by a medical center or third-party provider with this technology has not been reported but could be significant. The potential complications that are avoided by using the instrument and the medical and personal impact of such complications seem to make the instrument well worth the investment. The advantages listed would appear to outweigh the costs, but no prospective, controlled study has been published to justify this statement.

There are also concerns about increased operating-room time due to device setup and use. We have found setup and use to be simple and have found the device to lead to more efficient surgery.

Criteria Recommended When Choosing a System

There are several systems available; therefore, one should have some criteria to assist in choosing the most appropriate system. Roth et al. have listed such criteria [15]. These include accuracy within 2–3 mm, lack of the need for a second CT scan, ability for the system to compensate for head movement, sensors on suction elevators and dissecting instruments and the ability of the device to be operated by the surgeon without need for an assisting technician.

The capital investment in a system, the instruments and the conversion from radiology are usually the primary determining factors when this new technology is being evaluated. However, the following criteria should also be carefully considered:
1. Ease of technical conversion with current the CT system
2. User-friendliness of the system for radiology and nursing staff
3. Reliability of the system in terms of mechanical defects
4. Ability of system to identify structures in the OR
5. Availability of company representatives to solve problems
6. Upgradability of the system, and the cost of upgrades and maintenance
7. Current regional distribution of systems, and the company's reputation
8. Patient comfort during the CT scan and operation and after surgery
9. Whether a repeat scan will be needed for revision surgeries
10. Ability to use the system in cadaver-dissection labs and during resident teaching
11. Potential use by other specialties, e.g., ophthalmology, otology, orthopedics and neurosurgery

Future Developments Needed

The current systems will hopefully continue to improve as newer technology and more experience are gained with their usage. Since the cost of these systems will determine their regional availability, we hope to see several models with different price ranges to meet varying needs. Hopefully, these systems will go through a pricing phase similar to that experienced by camcorders and lasers, with resultant increased accessibility. It will be interesting to see if radiology centers, in conjunction with manufacturers, will offer leased services for providers in the future.

These systems need to maintain and improve their accuracy. Early-warning systems and improved three-dimensional navigation are in the early phases of development. Advanced instrumentation, including the attachment of sensors on the new and powerful endoscopic debriders, drills, seekers and through-cutting forceps that are becoming widely used, would be beneficial. Addition of memory and surgical documentation by the navigation system in concert with the endoscopic video recorder would also be useful in some settings. Lastly, a "morphing" of the endoscopic image and navigation data set on a single video display may allow improved relational surgery.

Whenever new technology has been introduced, there is a learning curve for the surgeon, the patient and the third-party insurance company/payor. The patient or consumer needs to be educated realistically about this technology and must realize that it is still only an assisting tool, albeit a very critical one in some cases.

All parties need to realize that these devices are no longer considered experimental and represent some of the best equipment available in the modern care of our patients. This needs to be made abundantly clear to payors, who may be in a cost-containment mode and may not appreciate the major advantages previously mentioned. Continued cost effectiveness and risk-management documentation should be published. Most importantly, the acceptance of these devices as "a standard of care" in certain settings should be supported by the subspecialties that utilize them. This may necessitate the formulation of new Current Procedural Terminology and International Classification of Diseases codes, and guidelines for reimbursement and utilization.

Case Presentation

A 42-year-old female underwent trans-sphenoidal pituitary removal in 1990. She subsequently developed deep, recurrent retrorbital pain at least three times per day during mid-1991. She was treated with multiple courses of antibiotics for sinusitis without improvement. She eventually underwent three endoscopic sinus procedures between July 1991 and October 1993. Her symptoms returned shortly after each operation. In June 1995, she was treated with 6 weeks of intravenous ceftazidime and vancomycin after cultures grew methicillin-resistant *Staphylococcus aureus* (MRSA) and *Pseudomonas* species. Her retrorbital pain persisted and, in October 1996, she had recurrent *Pseudomonas* and MRSA infections. A consultation was sent to the senior author (Kuhn), and a fine-cut axial CT scan with coronal and sagittal reconstructions was obtained. These showed sphenoid sinusitis with possible bony sequestrum.

She then underwent stereotactic surgical navigation with sphenoidotomy (Fig. 31.4). Intraoperative findings included bony and methylmethacrylate obstruction (which required drilling) and entrapped mucus. She did well; her last visit was in January 1998, at which time she exhibited normal sphenoid mucosa, no purulence and resolved retroorbital pain (Fig. 31.5).

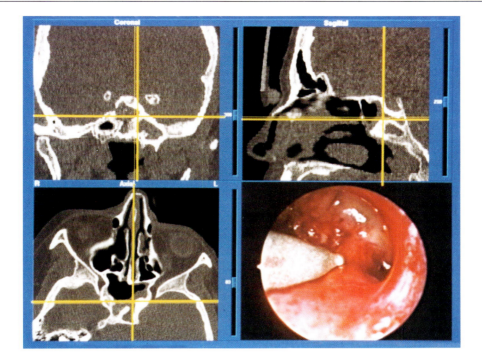

Fig. 31.4. Surgical navigation with sphenoidotomy of a 42-year-old female with sphenoid sinusitis

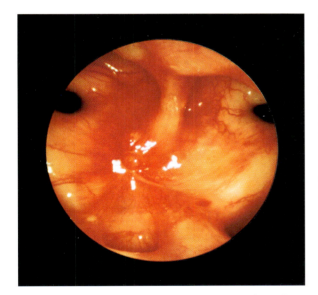

Fig. 31.5. Same patient as in Fig. 31.4, following sphenoidotomy

Summary

Image-guided surgical navigation systems have proven to be very valuable in the surgical management of sinusitis. They allow direct correlation of objective evidence (the CT scan in three views) with the view the surgeon has of the surgical field. It assists the surgeon in continuously navigating through complex anatomy in small spaces and next to vital structures in an organized manner. The benefits of these systems appear to outweigh their costs, especially if reduced complication and recurrence rates are obtained. These systems are becoming a standard of care in selected FESS cases, just as nerve monitoring has become standard in other areas of otolaryngology.

References

1. Anon J, Lipman S, Oppenheim D, Halt R (1993) Computer-assisted endoscopic sinus surgery. Laryngoscope 103: 1174–1176
2. Bolger W, Kennedy D (1993) Complications of surgery of the paranasal sinuses. In: Eisele D (ed) Complications of head and neck surgery. Mosby Year Book, St. Louis, pp 458–470
3. Fried M, Kleefield J, Gopal H, Reardon E, Ho B, Kuhn F (1997) Image-guided endoscopic surgery: results of accuracy and performance in a multi-center clinical study using an electromagnetic tracking system. Laryngoscope 107:594–601
4. Fried M, Kleefield J, Taylor R (1998) New armless image-guided system for endoscopic sinus surgery. Otolaryngol Head Neck Surg 119:528–532
5. Kennedy D, Zinreich S, Rosenbaum A, Johns M (1985) Functional endoscopic sinus surgery. Theory and diagnostic evaluation. Arch Otolaryngol Head Neck Surg 111:567–582

6. Kennedy D, Shaman P, Wei H, Selman H, Deems D, Lanza D (1994) Complications of ethmoidectomy: a survey of fellows of the American Academy of Otolaryngology, Head and Neck Surgery. Otolaryngol Head Neck Surg 111:589–599
7. Lanza D (1996) New technology. Otolaryngol Head Neck Surg 115:200–205
8. Legoreta A, Silber J, Costantino G, Kobylinski R, Zatz S (1993) Increased cholecystectomy rate after introduction of laparoscopic cholecystectomy. JAMA 270:1429–1432
9. Levine H (1990) Functional endoscopic sinus surgery: evaluation, surgery and follow-up of 250 patients. Laryngoscope 100:79–84
10. Messerklinger W (1967) Uber die Drainage der menschlichen Nasennebenhöhlen unter normalen und pathologischen Bedingungen II: Die Stirnhöhle und ihr Ausführungssystem. Monatsschr Ohrenheilkd 101:313
11. Messerklinger W (1978) Endoscopy of the nose. Urban and Schwarzenburg, Baltimore
12. Messerklinger W (1985) Endoskopische Diagnose und Chirurgie der rezidivierenden Sinusitis. In: Krajina Z (ed) Advances in nose and sinus surgery. Zagreb University, Zagreb, pp 31–34
13. Mosher HP (1929) The symposium of the ethmoid. Trans Am Acad Ophthalmol Otolaryngol 34:376
14. Rice D (1989) Endoscopic sinus surgery: results at two-years follow-up. Otolaryngol Head Neck Surg 101:476–479
15. Roth M, Lanza D, Zinreich J, Yousem D, Scanlan K, Kennedy D (1995) Advantages and disadvantages of three-dimensional computed tomography intraoperative localization for functional endoscopic sinus surgery. Laryngoscope 105:1279–1286
16. Stammberger H (1986) Endoscopic endonasal surgery – concepts in the treatment of recurrent rhinosinusitis. Part I. Anatomic and pathophysiologic considerations. Otolaryngol Head Neck Surg 94:143–147
17. Stammberger H (1986) Endoscopic endonasal surgery – concepts in the treatment of recurrent rhinosinusitis. Part II. Surgical technique. Otolaryngol Head Neck Surg 94:147–156
18. Wigand ME (1990) Endoscopic surgery of the paranasal sinuses and anterior skull base. Thieme Medical, New York

Part III

Transnasal Micro-Endoscopic Advanced Surgery

Chapter 32
Severe Epistaxis: Micro-endoscopic Surgical Techniques 393
Aldo Cassol Stamm, Glaura Ferreira,
João A. Caldas Navarro and Luiz A. Silva Freire

Chapter 33
Choanal Atresia: Transnasal Micro-endoscopic Surgery 405
Aldo Cassol Stamm, Levon Mekhitarian,
and Shirley Shizue Nagata Pignatari

Chapter 34
Endoscopic Transnasal Dacryocystorhinostomy 415
Gustavo A. Riveros-Castillo and Alfredo Campos

Chapter 35
Endonasal Surgery of the Lacrimal System 425
Joachim Heermann and Ralf Heermann

Chapter 36
Endoscopic Transnasal Orbital Decompression 433
Peter H. Hwang and David W. Kennedy

Chapter 37
Endoscopic Optic-Nerve Decompression in Traumatic Optic Neuropathy 441
Luis Alfonso Parra Duque

Chapter 38
Cerebrospinal Fluid Rhinorrhea – Transnasal Micro-endoscopic Surgery 451
Aldo Cassol Stamm and Luiz A. Silva Freire

Chapter 39
Endoscopic Repair of Cerebrospinal-Fluid Rhinorrhea 465
Alfredo Herrera and Emiro Caicedo

Chapter 40
Endonasal Micro-endoscopic Surgery of Nasal and Paranasal-Sinuses Tumors 481
Wolfgang Draf, Bernhard Schick, Rainer Weber,
Rainer Keerl, and Anjali Saha

Chapter 41
Micro-endoscopic Surgery of Benign Sino-Nasal Tumors 489
Aldo Cassol Stamm, Cleonice Hirata Watashi,
Paulo Fernando Malheiros, and Lee Alan Harker

Part III

Chapter 42
Juvenile Nasopharyngeal Angiofibroma – Transantral Microsurgical Approach 515
ALEJANDRO E. TERZIAN

Chapter 43
Endoscopic Approach to Lesions of the Anterior Skull Base and Orbit 529
DHARAMBIR S. SETHI

Chapter 44
**Endonasal, Microscopic, Trans-Septal, Sphenoidal Approach
to Sellar and Parasellar Lesions** 543
RICARDO SERGIO COHEN, ALDO CASSOL STAMM, and ANDRÉ BORDASCH

Chapter 45
Transnasal Endoscopic Surgery of the Sella and Parasellar Regions 555
ALDO CASSOL STAMM, ANDRÉ BORDASCH, EDUARDO VELLUTINI,
and FELIX PAHL

Chapter 46
Midfacial Degloving – Microsurgical Approach 569
ALDO CASSOL STAMM, MOACIR POZZOBON,
and OSWALDO L. MENDONÇA CRUZ

Chapter 47
Complications of Micro-endoscopic Sinus Surgery 581
ALDO CASSOL STAMM

Severe Epistaxis: Micro-endoscopic Surgical Techniques

Aldo Cassol Stamm, Glaura Ferreira,
João A. Caldas Navarro, and Luiz A.S. Freire

Introduction

Epistaxis is a frequent emergency in medical practice. Approximately 10% of the normal population has suffered at least one significant episode of nasal bleeding [5]. In most cases, the bleeding is anterior (90%), from the anterior portion of the nasal septum (Little's area, or Kiesselbach's plexus), and the diagnosis is made by anterior rhinoscopy. It is stopped by local chemical cautery, electrocautery or anterior packing with gauze or other agents.

Different methods of treatment are usually necessary for severe epistaxis. The classical treatment is anteroposterior nasal packing using different materials, such as gauze, a Foley catheter, inflatable nasal packs, a sponge, Avitene or Merocel, depending on the physician's experience. Alternative treatments include pterygopalatine-fossa injection, septoplasty, posterior endoscopic cautery, vessel ligation and angiography with selective embolization.

In our department, surgery is used to manage posterior epistaxis unresponsive to conservative treatment, including properly placed packs. To be effective, surgical management of epistaxis requires precise identification of the bleeding sites and adequate knowledge of the vascular anatomy of the nose.

Anatomic Features of the Nasal Blood Supply

The sources of the nasal blood supply are the branches of the internal and external carotid arteries (ECAs). The anterior and posterior ethmoidal arteries, both of which are branches of the ophthalmic artery of the internal carotid system, provide blood to the nasal bones, septal cartilage, roof of the nose, nasal areas related to the pituitary region and the lateral and medial superior nasal mucosa. The nasal blood supply from the external carotid-artery system is provided by two terminal branches of the maxillary artery: the posterior lateral nasal artery and the septal artery [21, 23].

The anterior ethmoidal artery leaves the orbit by passing through the anterior ethmoidal canal to reach the area adjacent to the cribriform plate [17] near the crista galli, where it turns inferiorly to supply the middle and anterior ethmoid cells, the infundibulum of the frontal sinus, the anterior nasal cavity (septum and lateral walls) and the skin over the cartilaginous part of the nose. The posterior ethmoidal artery passes through the posterior ethmoidal canal to the region of the cribriform plate and has branches that descend to the posterior upper part of the nasal cavity. There is anastomosis between the ethmoidal arteries and the septal and lateral posterior nasal arteries and between both ethmoidal arteries. The posterior ethmoidal artery is 4–7 mm anterior to the optic nerve (Fig. 32.1) [18].

The maxillary artery is one of the two terminal branches of the ECA. It arises behind the neck of the mandible and is divided into three parts: the mandibular portion (between the mandibular neck and the sphenomandibular ligament), the pterygoid portion (which passes in an upward and forward direction lateral and medial to the lateral pterygoid muscle) and the pterygopalatine portion (which passes into the homonymous fossa) [23]. Branches from the first division supply the tympanic membrane, the auricle, the temporomandibular joint, the mandible, the teeth and the dura. Branches from the second division supply the masticatory muscles and, close to the tuberosity of the maxilla, the artery gives rise to the superior and inferior alveolar branches, the infraorbital branches and the descending palatine branches. After a short medial course, the maxillary artery enters the pterygopalatine fossa, where its third division is located (Fig. 32.2). The third division gives rise to orbital branches, branches to the foramen rotundum, and the pterygopalatine canal (Boeck's canal). Passing medially through the fossa, the third division splits into the posterior lateral nasal artery and the posterior septal artery, both entering the nasal cavity through the sphenopalatine

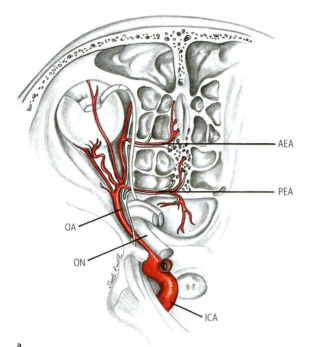

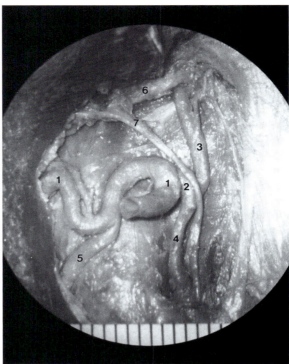

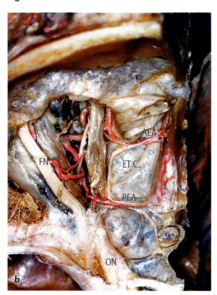

Fig. 32.1 a,b. Intracranial view of the anterior and posterior ethmoidal arteries running through the ethmoidal roof and cribriform plate. **a** Schematic drawing. **b** Anatomic dissection. *AEA*, anterior ethmoidal artery; *ETC*, ethmoid cells; *FN*, frontal nerve; *ICA*, internal carotid artery; *OA*, ophthalmic artery; *ON*, optic nerve; *PEA*, posterior ethmoidal artery

Fig. 32.2. The maxillary artery at the tuberosity of the maxilla. *1*, Maxillary artery; *2*, infraorbital/posterosuperior alveolar branch; *3*, infraorbital artery; *4*, posterosuperior alveolar artery; *5*, descending palatine artery; *6*, maxillary nerve; *7*, posterosuperior alveolar nerve

nasal cavity. The septal artery passes through the sphenopalatine foramen and courses along the roof of the nasal cavity toward the sphenoid rostrum and divides into a number of vessels that extend to the septum and the superior nasal walls. The posterior lateral nasal artery splits in the lateral wall of the nasal cavity, thus providing arterial supply to the turbinates and meatal spaces (primarily the middle and the inferior meati; Fig. 32.3).

The descending palatine artery, as a terminal buccal branch of the maxillary artery, courses between the middle, posterior and anterior walls of the fossa, divides into major and minor branches (which enter their respective canals) and reaches the buccal cavity through the homonymous foramen. The greater palatine artery courses through the palate to the incisive foramen, where it joins the nasopalatine artery, which comes from the septum. Anastomosis between the contralateral branches of these arteries can occur.

The anterior and inferior septa are supplied by the superior labial branch of the facial artery. An area of multiple arterial anastomosis of the three sources of vascular supply to the nose is found superficially in

foramen, just above the caudal, bony end of the middle turbinate. In 6.6% of the cases, Navarro et al. [21] describe an accessory foramen (located a little superior the sphenopalatine foramen) through which a branch of the posterior lateral nasal artery enters the

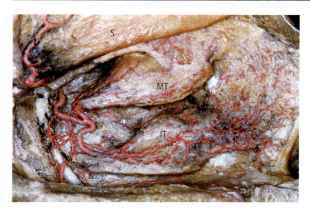

Fig. 32.3. Lateral nasal wall, demonstrating the terminal branches of the maxillary artery. The posterior lateral nasal artery and the septal artery are identified close to the sphenopalatine foramen. *IT,* inferior turbinate; *MT,* middle turbinate; *S,* nasal septum

the anterior nasal septum (Kiesselbach's plexus or Little's area). This area is responsible for most anterior nosebleeds.

Classification of Epistaxis

Epistaxis can be classified according to its location or its etiology. According to Montgomery [18], epistaxis is best classified according to its location: anterior, superior or posterior. Anterior bleeding occurs in Kisselbach's plexus (or Little's area) from a branch of the anterior ethmoidal artery, the septal branch of the superior labial artery, the septal branch of the maxillary artery and/or the nasal branch of the greater palatine artery. Superior bleeding occurs from either the anterior and/or posterior ethmoidal arteries or the septal artery, a branch of the maxillary artery. Posterior bleeding occurs from either the posterior lateral nasal artery or the septal arteries, both branches of the maxillary artery.

There are many different causes of epistaxis, and they can be divided into four groups, according to Younkers et al. [39]: (1) local causes, (2) causes associated with neoplastic disorders, (3) causes associated with hematological disorders and (4) other causes. Local causes include nasal allergy, trauma, infection, ulcers, nasal surgery, intranasal medications, nose-picking habits and foreign-body reactions. Epistaxis-associated with neoplasms include mainly angiofibroma and malignant tumors. Epistaxis associated with primary hematological disorders include leukemia, anemia, purpura, polycytemia, hemophilia, lymphoma and Hodgkin's disease. Other causes include hypertension, smoking, familial telangiectasia, liver disorders, chronic nephritis, sudden atmospheric-pressure changes, psychopathies and medications, such as aspirin or chemotherapeutic agents.

Treatment of Severe Epistaxis

Several modalities of treatment have been proposed to control epistaxis. Their use is based mainly on the location and etiology of the bleeding and on the physician's experience (Table 32.1).

During the initial evaluation, a brief history must be collected while initial measures to control the bleeding are taken. The examination of the nasal cavity requires a good source of illumination (headlight or telescopes), an appropriate nasal speculum and strong suction. It is best done using topical anesthesia containing a vasoconstrictor. This examination is of prime importance in order to locate the bleeding site and to choose the most appropriate treatment. In case of anterior nasal bleeding, anterior nasal packing and/or chemical cautery or direct electrocautery of the bleeding site are the usual procedures. Severe nasal bleeding of posterior, posterosuperior or superior origin requires posterior nasal packing (gauze, inflatable nasal packs, sponge or other packing; Fig. 32.4). The use of antibiotics is advisable, because the impaired sinus drainage caused by the pack can lead to sinus infection. The packing remains in place for 48–72 h. Many patients with an anterior–posterior packing require hospitalization, consultation with an internist and hematological evaluation. Because of the risk of sedation with posterior packing, only cautious use of sedatives and analgesics is advisable.

Nasal packing can produce local, regional and general complications. The most common general complication is the modification of arterial pO_2 and pCO_2, leading to serious hypoxemia and hemodynamic changes in patients with decreased cardiopulmonary reserves [4, 8, 22, 35]. Sleep apnea can occur as a complication of nasal packing [36].

One of the principal local/regional complications is sinusitis, occurring because of the impairment of drainage of the paranasal sinuses [16]. Swallowing problems, aspiration, septal perforation, tubal obstruction, alar necrosis, sepsis and secretory otitis media have all been described [8, 16, 35, 37].

Inflatable nasal packs are easy to place, both for the patient and the physician, compared with conventional posterior packs. The inflatable pack consists of two balloons of different sizes. The smaller of them, which has a volume of 10 ml, is placed at the level of the choana. The second is bigger and oblong shaped and has a volume of 30 ml; it is placed inside the nasal cavity. Once positioned

Table 32.1. Severe epistaxis: topographic diagnosis and surgical treatment

Origin	Arterial system	Surgical treatment
Anterior	Branches of anterior ethmoidal artery, septal branch of the superior labial artery, septal branch of the maxillary artery and nasal branch of the greater palatine artery	Cauterization (chemical or electrical) Septoplasty Dermoplasty
Superior	Anterior and posterior ethmoidal arteries	Eletrocoagulation of the ethmoidal arteries, with or without partial ethmoidectomy Bipolar coagulation by an external approach
Posterior	Posterior lateral nasal and septal arteries (branches of the maxillary artery)	Electrocoagulation of the nasal branches of the maxillary artery through a transnasal approach

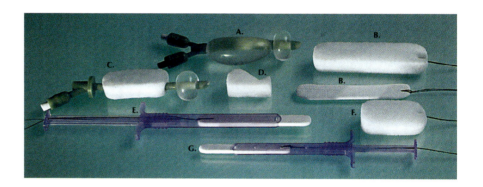

Fig. 32.4. Examples of nasal packing (Xomed, Jacksonville). *A*, Epistat nasal catheter; *B*, Pope Flex-Pak nasal packing; *C*, Epistat II nasal catheter; *D*, Staxi-Stat pack; *E*, Fast Pack with applicator; *F*, Weimert epistaxis packing; *G*, Fast Pack with small applicator

inside the nasal cavity, the catheters are filled with liquid (saline solution, distilled water). Filling the balloons with air should be avoided, because the packing can then lose volume in 24–48 h, thus becoming ineffective [2]. These catheters are usually better tolerated by the patient than the traditional tampon. The disadvantages of inflatable nasal packs are: (1) constant pressure on the same place, resulting in local necrosis; (2) possible rupture of the balloon because of sustained pressure and aspiration of the fluid; (3) loss of pressure of the balloon; and (4) persistence of bleeding [2].

Merocel is a soft, porous material resembling a pressed sponge; it expands in the presence of liquids and is used for nasal packing. Its advantages are easy introduction and removal. However, it does not always control severe bleeding and can be difficult to place in the presence of a septal deviation.

Classical packing sometimes fails to control intense and persistent bleeding from the nose. Montgomery [19] reported a 25% failure rate with posterior packing; Stamm et al. [32, 33] reported a 20% failure rate, and Procino [25] reported a 14% failure rate. Marks [16] treated 103 patients with posterior packing; 55% of them required further selective arterial ligation.

Digital subtraction angiography is indicated for diagnosis in patients treated repeatedly without success or when an embolization procedure is anticipated. Embolization can be used in patients in poor clinical condition or in patients who do not tolerate posterior packing or surgical treatment [6, 8, 15]. One complication of embolization is aching of areas of the scalp, which can occur from occlusion of the superficial temporal artery [15, 16]. Younkers et al. [39] describe the possibility of retrograde migration of the embolus to the internal carotid arterial (ICA) system, sometimes producing ischemia in areas of the brain. According to Golding-Wood [8], facial paralysis, trismus and facial pain can occur, as can diverse central nervous system lesions. Another possible outcome is recurrence of bleeding.

Myssiorek and Lodespoto [20] perform selective embolization of the maxillary artery as a second treatment option after failure of artery ligation (14 cases) or posterior packing (two cases). Two minor and transient complications (diplopia and disturbed facial sensitivity) were described in this series [20]. These authors believe that this treatment modality allowed earlier removal of nasal packing and shorter hospitalization [20]. In cases of ethmoid-artery bleeding, angiography should be used only for diagnostic purposes because of the high risk of ophthalmic complications from embolization [1, 20].

Pterygopalatine-fossa injection of vasoconstrictor substances through the greater palatine foramen has also been advocated as a method for the treatment of epistaxis [15]. This injection can result in infraorbital anesthesia, visual losses of variable extent and recurrence of bleeding after dissipation of the vasoconstrictor. The use of cryosurgery [15] at the site of bleeding is a secondary option because, although it can be effective when performed by experienced surgeons, it can also lead to crust formation and atrophic rhinitis. Septoplasty can be used to control recurrent bleeding from the nasal septum. Septal dermoplasty is considered for the treatment of familial telangiectasia [15, 27].

Arterial ligation is indicated when non-surgical methods of treatment fail [32–34, 37], especially when nasal bleeding persists or recurs after properly placed nasal packing. Montgomery [18] advocates early surgical intervention for epistaxis [26]. Maris and Werth [15] propose some general principles for the surgical treatment of epistaxis.

1. Patients in poor clinical condition who have pulmonary or cardiovascular problems or disorders of the central nervous system should not be treated with nasal tamponade. Early vessel ligation is indicated in these cases.
2. Young patients should be treated with conservative measures, with vessel ligation indicated only in cases of failure.
3. Packing is preferred in children because of the size of the nasal cavities and maxillary sinus.

According to Small and Maran [29], surgical treatment should not be delayed for a long time. The morbidity of vessel ligation is not much greater than that of posterior nasal packing and causes less discomfort.

External carotid-artery ligation was used by Hide [12] for the treatment of epistaxis, but now the use of the microscope, endoscope and other technological improvements has displaced it, especially because of the frequency of anastomosis between the ECA and ICA systems and the distance between the site of bleeding and the location of the ligature, which severely reduces its effectiveness [15, 17].

Transmaxillary ligation of the maxillary artery in the zygomatic fossa is frequently effective in controlling bleeding arising from nasal branches [19, 31]. Using the surgical microscope, the artery is dissected in the zygomatic fossa close to the tuberosity of the maxilla. Difficulty identifying the artery and all its branches is usually due to poor technique [21, 37], but there is also considerable variation in the number and size of the branches arising from the maxillary artery [25, 37]. This can result in an ineffective ligation and persistence of bleeding [8, 29]. Complications have included oroantral fistula and hemifacial anesthesia [19], and the method itself sometimes fails to control bleeding. This technique was initially proposed by Seiffert in 1928 [26] and was improved by Chandler and Serrins in 1965 [3]; Montgomery [18, 19] deserves credit for its popularization in 1980. Macery, in 1984 [14], proposed the transoral ligation of the maxillary artery in the zygomatic region without opening the maxillary sinus. Postoperative trismus, dental or facial dysesthesia and the distance between the ligature and the bleeding point are disadvantages of this method.

As implied by the etymological meaning of the word epistaxis (bleeding that comes from above) bleeding can originate in the ethmoid arterial system and, although less frequent, can be remarkably severe and impressive [8, 10, 11]. According to Golding-Wood [8], ethmoidal bleeding occurs most often in young people following trauma to the face. The external ligation of the anterior ethmoidal artery and septal artery, as recommended by Silverblatt [28] in 1955, appeared to be effective in controlling epistaxis but did not gain popular favor because of technical difficulties. External ligation can result in permanent or transient loss of vision because of the close relationship of the posterior ethmoidal canal and the optic canal (3–8 mm) [7]; it can also lead to diplopia [4, 16]. These complications are uncommon, and an external approach using a surgical microscope is an excellent approach to the ethmoidal artery, with a high rate of success [34]. Snyderman and Carrau [30] performed simultaneous ligation of the anterior ethmoid and sphenopalatine arteries in 67% of their patients due to difficulty localizing the site of bleeding.

Heermann (1986) [10, 11] reported treating all his patients with severe epistaxis by monopolar electrocoagulation of the anterior and posterior ethmoidal arteries in their osseous canals in the roof of the ethmoid sinus, using a microscopic intranasal approach. The most significant problem of the intranasal arterial electrocoagulation is the possibility of a CSF fistula when cautery is applied to the interior of the ethmoidal canals because the surgeon does not properly identify the vessels [10, 11, 34].

Intranasal electrocoagulation of the nasal branches of the maxillary artery, combined with bipolar electrocoagulation of the anterior and posterior ethmoidal arteries (with or without partial ethmoidectomy using an operating microscope or endoscopes) appears to be an effective method for the treatment of severe epistaxis [9–11, 13, 22, 30, 32–34, 38]. Advantages of this method are the proximity of the arterial ligation and the bleeding site, and direct visualization of the involved vessels [5, 10, 11, 13, 32–34, 38]. Associated disorders, such as polyps, sinusitis, septal deformity and tumors can be treated with the same approach.

Nasal endoscopy can also be used in the treatment of posterior epistaxis. Authors using nasal endoscopy as an adjunct report a success rate between 82% and 90%, depending on the surgeon's experience [5, 24, 30, 38]. Sedation and the use of topical anesthesia are advised. The patient should be in a semi-recumbent position with the head elevated, allowing visualization of the bleeding point without the use of a vasoconstrictor. After identifying the bleeding site, local vasoconstriction and local and regional anesthesia must be used. Wurman [38] recommends this procedure, using 4-mm-diameter endoscopes of 0° and 30°. Prechamandra, using the endoscope [24], places small nasal packs at the bleeding area, in addition to using local electrocautery or chemical cautery. Snyderman and Carrau [30], using a transnasal endoscopic technique, approached the sphenopalatine foramen. Resecting the anterior inferior portion of the middle turbinate can also be performed to maximize surgical access to the middle meatus. A large middle-meatus antrostomy is created, and the sphenopalatine foramen (with its artery) are identified. The terminal branches of the maxillary artery are dissected and may be coagulated or ligated.

Transnasal Micro-Endoscopic Electrocoagulation of the Nasal Branches of the Maxillary Artery

The surgery can be performed with either the operating microscope or endoscopes using general anesthesia. The patient is placed in a supine position, with the dorsum elevated approximately 30°.

The nasal cavity is prepared with cottonoids soaked with a vasoconstrictor solution to reduce the bleeding and the size of the turbinates. The middle turbinate is fractured toward the nasal septum very carefully in order to avoid damaging its arterial supply. When the operating microscope is used, a number-two self-retaining speculum is placed to maintain good exposure of the surgical field. Usually the surgery is done with a 4-mm, 0° endoscope attached to a video camera system. The caudal end of the middle turbinate is an important anatomic landmark because of its close relationship with the sphenopalatine foramen, through which the nasal branches of the maxillary artery arise. Another landmark is the posterior wall of the maxillary sinus. Using microcautery or a microknife, a vertical incision is made approximately 1 cm anterior to the caudal end of the middle turbinate through the mucosa and periostium (Fig. 32.5). A mucoperiosteal flap is elevated until the edges of the sphenopalatine foramen are exposed, identifying the posterior lateral nasal artery and the septal artery, which arise from the maxillary artery and enter the nasal cavity from the pterygopalatine fossa (Fig. 32.6). Once identified, both arterial branches are isolated by blunt dissection and are coagulated with a monopolar or bipolar

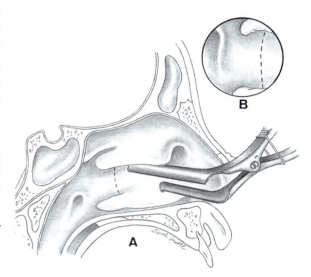

Fig. 32.5. Location of the incision at the middle meatus (mucosa and periosteum), approximately 1 cm anterior to the caudal end of the middle turbinate. A Self-retaining speculum number two in place, exposing the middle meatus and prechoanal area. B Endoscopic view

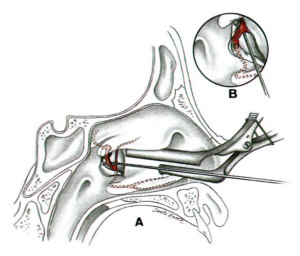

Fig. 32.6. The sphenopalatine foramen and its arteries after elevation of the mucoperiosteal flap. A Lateral view of the surgical procedure, as provided by a self-retaining speculum and operating microscope. B Endoscopic view of the surgical field

system (Fig. 32.7) as far from the edges of the foramen as possible (to avoid any possible retraction of the proximal segment into the pterygopalatine fossa; Fig. 32.8). At the end of the procedure, the mucoperiostial flap is replaced and packed in place with pieces of Gelfoam or Surgicel. The packing also minimizes oozing from the surgical area. The middle and inferior turbinates are replaced, and loose gauze packing inside a glove finger or a Merocel pack is placed into the nose for 24 h to prevent any minor bleeding caused by surgical manipulation. When it is necessary to ligate the terminal branches of the maxillary artery by the transnasal, transmaxillary approach, it is desirable to make a wide middle-meatus antrostomy to help identify the sphenopalatine foramen (Fig. 32.9A). A Kerrison punch is then inserted into the sphenopalatine foramen to remove the bone of the posterior wall of the maxillary sinus (Fig. 32.9B). Finally, the maxillary artery and its branches are identified, individually isolated and bipolarly electrocoagulated or clipped (Fig. 32.9C).

Transnasal Microendoscopic Electrocoagulation of the Anterior and Posterior Ethmoidal Arteries

The nasal cavity is prepared with a vasoconstrictor solution, as described above. The patient is placed in a supine position, and the head is hyperextended in order to have the ethmoidal complex in a plane parallel to the light from the microscope. Using a self-retaining, properly placed speculum, the surgical microscope and continuous aspiration, the bleeding point is identified at the confluence of the roof of the nose and the superior portion of the septum. This technique can also be performed using endoscopes (mostly 0°, 4-mm-diameter endoscopes).

Anterosuperior bleeding comes from the lateral and medial branches of the anterior ethmoidal artery or from a branch of the posterior ethmoidal

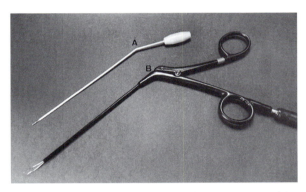

Fig. 32.7. Electrically shielded monopolar cauteries. *A*, Plain. *B*, Alligator forceps handle

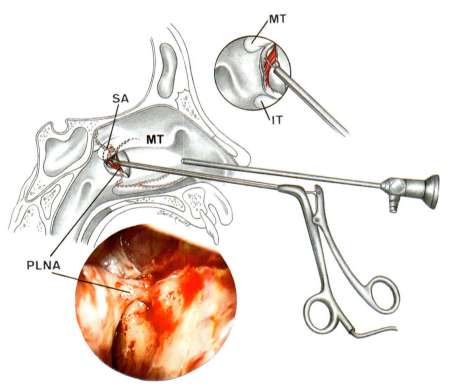

Fig. 32.8. Terminal branches of the maxillary artery arising from the sphenopalatine foramen and being coagulated separately with monopolar forceps under endoscopic control. *IT*, inferior turbinate; *MT*, middle turbinate; *PLNA*, posterior lateral nasal artery; *SA*, septal artery

artery. Posterosuperior bleeding comes from branches of the posterior ethmoidal artery or the septal artery. Posterosuperior bleeding from both arterial systems (internal and external carotid) is controlled by direct coagulation of the bleeding point and the surrounding nasal mucosa (Fig. 32.10).

If there is any persistence of bleeding after this procedure, coagulation of the ethmoidal arteries should be performed at the level of their osseous canals in the roof of the ethmoid sinus, by approaching them through a transnasal partial ethmoidectomy. To perform the transethmoidal approach, the uncinate process and the ethmoidal bulla are identified and removed to gain complete exposure of the ethmoidal complex. The ethmoidectomy is performed until the level of the floor of the anterior cranial fossa and the ethmoidal canals is reached. The next step is coagulation of the arteries when they emerge from their canals (Fig. 32.11). Bipolar coagu-

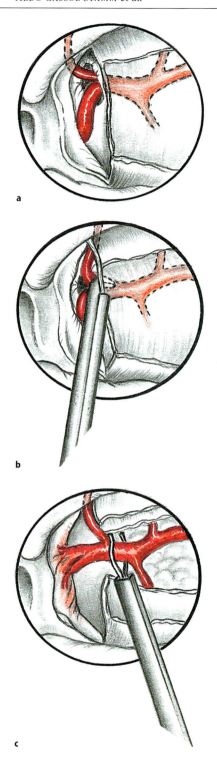

Fig. 32.9 a–c. Endoscopic, transnasal, transmaxillary eletrocoagulation of the maxillary artery after a wide middle-meatus antrostomy. **a** Identification of the sphenopalatine foramen and its arteries. **b** Enlargement of the sphenopalatine foramen toward the posterior maxillary-sinus wall. **c** Coagulation of the maxillary artery

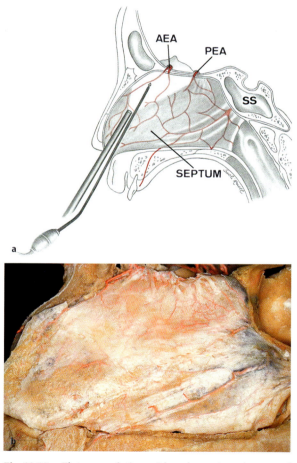

Fig. 32.10. **a** Eletrocoagulation, with endoscopic assistance, of the branches of the anterior ethmoidal artery (*AEA*) located at the junction of the upper septum and the roof of the nose. *PEA*, posterior ethmoidal artery; *SS*, sphenoid sinus. **b** Arterial supply of the nasal septum

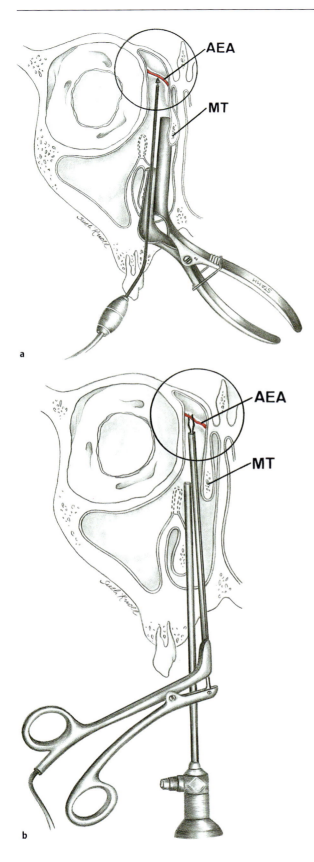

lation is used, if possible. Pieces of Gelfoam or Surgicel are placed close to the area. At the time of coagulation of the ethmoidal arteries, the introduction of electrocautery deep in the canals must be avoided to prevent the development of a CSF fistula [10, 11, 33].

Microsurgical External Electrocoagulation of the Anterior and Posterior Ethmoidal Arteries

When bleeding from the ethmoidal arteries system cannot be controlled transnasally, the use of the external approach under a surgical microscope is necessary. An incision is made close to the inner canthus, similar to that for an external ethmoidectomy. A subperiostial dissection of the orbital contents from the internal osseous wall is performed until reaching the level of the fronto-ethmoidal junction, where the anterior ethmoidal artery can be found and coagulated with a bipolar system after adequate dissection and isolation of the vessel (Fig. 32.12). To keep the orbital contents away from the soft tissues of the inner canthus and the nose, a Stamm self-retaining retractor, which has a blade on one side and a hook on the other side, is used (Fig. 32.13).

To coagulate the posterior ethmoidal artery, the elevation of the soft tissues of the orbit is continued posteriorly for approximately 1 cm. The close relationship of the posterior ethmoidal artery to the optic nerve (4–7 mm) should be recalled during the bipolar electrocoagulation of this vessel. As a practical landmark, the "two–one–one half rule" is useful; the anterior ethmoidal artery is found approximately 2 cm posterior to the skin incision, the posterior ethmoidal artery is found 1 cm behind the anterior, and the optic nerve is 0.5 cm from the posterior ethmoidal artery.

Results and Complications

Between 1985–1999, 173 patients were treated surgically to control severe epistaxis. The techniques consisted of selective microscopic or endoscopic electrocoagulation of the nasal branches of the maxillary

Fig. 32.11 a,b. Electrocoagulation of the anterior ethmoidal artery (*AEA*) after an ethmoidectomy. **a** Performed with an operating microscope, self-retaining speculum and plain monopolar cautery. **b** Performed with an endoscope and monopolar-forceps-type cautery. *MT*, middle turbinate

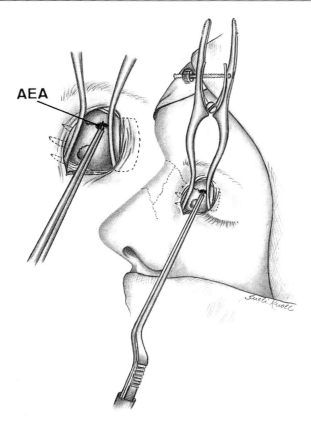

Fig. 32.12. External approach for bipolar coagulation of the anterior ethmoidal artery (*AEA*)

Fig. 32.13. Stamm's self-retaining retractor with two arms, a blade in one side and a hook in the other side

artery transnasally or monopolar or bipolar electrocoagulation of the ethmoidal arteries through an intranasal or external approach (Table 32.2).

Because of an error in the topodiagnosis of the bleeding sites, failure to control the bleeding occurred in six patients (3.4%) during the immediate postoperative period. Failure also occurred during the late postoperative period in four patients (2.3%) because of inadequate surgery (one of the terminal branches of the maxillary artery was not coagulated).

Complications have been infrequent and were usually related to either failure in the localization of the site of bleeding or inadequate surgical technique [33, 34]. According to Legent et al. [13], satisfactory results were obtained with intranasal ligation of branches of the maxillary artery alone in 30 patients, and only eight patients required ligation treatment of the ethmoidal arteries. Snyderman and Carrau [30] had recurrences of epistaxis after transnasal endoscopic ligation of the terminal branches of the maxillary artery in 13% of their cases.

Adhesion between the middle turbinate and the lateral nasal wall or nasal septum, transient or permanent unilateral hyposmia or anosmia, crust formation and local infection are complications related to surgical trauma or extensive coagulation of the nasal mucosa. Orbital ecchymosis and edema were the most common complications of the external approach to anterior and posterior ethmoidal-artery ligation.

Conclusions

The results of surgical treatment for severe epistaxis by intranasal micro-endoscopic ligation of the nasal branches of the maxillary artery and/or intranasal or external ligation of the ethmoid arteries lead to the following considerations:

1. Micro-endoscopic surgical technique results in minimal surgical trauma, decreased surgical time and morbidity and improved visualization.
2. Electrocoagulation of the main nasal branches of the maxillary arteries must be selective, thus avoiding the development of collateral circulation.
3. Electrocoagulation of the ethmoidal arteries must be performed at the sites of bleeding into the nasal cavity.
4. Bipolar coagulation of the ethmoidal arteries through an external approach using a surgical microscope is effective in controlling severe epistaxis.
5. The surgical treatment of severe epistaxis can result in shorter hospitalization time, is well tolerated by patients and can also be performed in children.

References

1. Breda SD, Choi IS, Persky MS, Weiss M (1989) Embolization in the treatment of epistaxis after failure of internal maxillary artery ligation. Laryngoscope 99:809–813
2. Cannon RC (1993) Effective treatment protocol for posterior epistaxis. A 10-year experience. Otolaryngol Head Neck Surg 109:722–725
3. Chandler JR, Serrins AJ (1965) Transantral ligation of the internal maxillary artery. Laryngoscope 75:151–159

Table 32.2. Distribution of patients with severe epistaxis by the involved arteries and surgical approach (n=173)

Arterial supply	Surgical approach	Number	Percentage
Maxillary-artery branches/ethmoid arteries	Intranasal/external	92	53
Maxillary-artery branches/ethmoid arteries	Intranasal/intranasal	38	22
Maxillary-artery branches only	Intranasal	27	16
Ethmoid arteries	External	10	6
Ethmoid arteries	Intranasal	6	3

4. Cooke ETM (1985) An evaluation and clinical study of severe epistaxis treated by artery ligation. J Laryngol Otol 99:745–749
5. El-Silimy O (1993) Endonasal endoscopy and posterior epistaxis. Rhinology 31:119–120
6. Felker CE III (1981) Angiography and embolization of epistaxis. Ear Nose Throat J 60:57
7. Friedman WH, Rosenblum BN (1987) Epistaxis. In: Goldman JL (ed) The principles and practice of rhinology. Wiley, New York, pp 375–383
8. Golding-Wood PH (1983) The role of arterial ligation in intractable epistaxis. J Laryngol Otol Suppl 8:120–122
9. Heermann H (1958) Endonasal surgery with the use of the binocular Zeiss operating microscope. Arch Klin Exp Ohren Nasen Kehlkopfheilkd 171:295–297
10. Heermann, J (1986) Intranasales mikrochirurgikes vorgehen bei epistaxis der riechspalte und weitere eingreffe mit hypotension. HNO 34:208–215
11. Heermann J, Neues D (1986) Intranasal microsurgery of all paranasal sinuses, the septum, and the lacrimal sac with hypotensive anesthesia. Ann Otol Rhinol Laryngol 35:631–638
12. Hide FT (1925) Ligation of the external carotid artery for control of idiopathic nasal bleeding. Laryngoscope 35:899
13. Legent F, Boutet JJ, Wesolouch, Viale M, Galiba J, Beauvillain C (1986) Treatment chirurgical des épistaxis interét de la micro-chirurgie endo-nasale. Rev Laryngol Otol Rhinol (Bord) 107:31–33
14. Maceri DR, Makielski KH, Arbor A (1984) Intranasal ligation of the maxillary artery for posterior epistaxis. Laryngoscope 94:734–741
15. Maris, C; Werth, JL (1981) Surgical management of epistaxis. Ear, Nose Throat J 60:463–466
16. Marks HW (1980) Complications of posterior epistaxis. Ear Nose Throat J 59:39–42
17. Mercurio GA Jr (1981) Anatomic considerations of nasal blood supply. Ear Nose Throat J 60:443–446
18. Montgomery WW (1979) Surgery of the upper respiratory system, 2nd edn. Lea and Febiger, Bostin, p 321
19. Montgomery WW, Reardon EJ (1980) Early vessel ligation of control of severe epistaxis. In: Snow JB (ed) Controversy in otolaryngology. Saunders, Philadelphia, pp 315–319
20. Myssiorek D, Lodespoto M (1993) Embolization of posterior epistaxis. Am J Rhinol 7:223–226
21. Navarro JAC, Toledo Filho JL, Zorzetto NL (1982) Anatomy of the maxillary artery into the pterygomaxillopalatine fossa. Anat Anz 152:413–433
22. Nicolaides A, Gray R, Pfleiderer A (1991) A new approach to the management of acute epistaxis. Clin Otolaryngol 16:59–61
23. O'Rahilly R (1978) Anatomia de cabeça e pescoço. In: Gardner E (ed) Anatomia. Guanabara-Koogan, Rio de Janeiro, pp 656–658
24. Prechamandra DJ (1991) Management of posterior epistaxis with the use of fiberoptic nasolaryngoscope. J Laryngol Otol 105:17–19
25. Procino ND (1978) Treatment of the posterior epistaxis. Ear Nose Throat J 57:305–308
26. Seiffert A (1928) Under bindung der arteria maxillaris interna. Z HNO 22:323–325
27. Sherrerd PS, Teet TJ (1981) Diagnosis and management of less common causes of epistaxis. Ear Nose Throat J 60:59–66
28. Silverblatt BL (1955) Epistaxis. Evaluation of surgical care. Laryngoscope 65:431
29. Small M, Maran AG (1984) Epistaxis and arterial ligation. J Laryngol Otol 98:281–284
30. Snyderman CH, Carrau RL (1997) Endoscopic ligation of the sphenopalatine artery for epistaxis. Operative Tech Otolaryngol Head Neck Surg 8:85–89
31. Spafford P, Durham JS (1992) Epistaxis: efficacy of arterial ligation and long-term outcome. J Otolaryngol 21:252–256
32. Stamm AC, Pinto JA, Neto AF, Menon AD (1985) Microsurgery in severe posterior epistaxis. Rhinology 23:321–532
33. Stamm AC, Ferreira GMP, Navarro JAC (1988) Microcirurgia transnasal no tratamento da epistaxe severa. Fed Med (Bras) 96:315–322
34. Stamm AC, Ferreira GMP, Navarro JAC (1995) Epistaxe severa – microcirurgia transnasal. In: Stamm AC (ed) Microcirurgia naso-sinusal. Revinter, Rio de Janeiro, pp 289–297
35. Stemm RA (1981) Complications of nasal packing. Ear Nose Throat J 60:45–46
36. Vaartjes M, Striges RLM, Devries N (1992) Posterior nasal packing and sleep apnea. Am J Rhinol 6:71–74
37. Ward PH (1980) Routine ligation of the internal maxillary artery is unwarranted. In: Snow JB (ed) Controversy in otolaryngology. Saunders, Philadelphia, pp 320–326
38. Wurman LH, Garry-Sack J, Flannery JV, Paulson O (1988) Selective endoscopic eletrocautery for posterior epistaxis. Laryngoscope 98:1348–1349
39. Younkers AJ, Glessman TM, Mercurio GA Jr, Werth JL, Blattner RE (1981) Etiology and management of epistaxis. Ear Nose Throat J 60:453–456

Choanal Atresia: Transnasal Micro-endoscopic Surgery

ALDO CASSOL STAMM, LEVON MEKHITARIAN, and SHIRLEY SHIZUE NAGATA PIGNATARI

Introduction

Choanal atresia is a congenital malformation of the posterior portion of the nasal cavity and can be unilateral or bilateral. The unilateral form is more frequent, but a bilateral presentation occurs in 30–40% of cases, and the incidence of the disorder is estimated to be one in 5000–8000 live births [26]. There is a female-to-male predominance of five to one in Caucasians [6].

The atresia plate contains bone in 80–90% of cases and is composed only of mucosa and fibrous connective tissue in approximately 10% [5]. However, recent reviews have failed to show purely membranous atresias, and approximately 30% are pure bony, whereas 70% are mixed bony/membranous [4]. The thickness of the plate is usually 4–6 mm [5, 14].

The first description of the abnormality was made by Roederer in 1755 [23]. Emmert, in 1853 [10], was the first to surgically approach choanal atresia when he passed a trocar through the obstruction. In 1880, Ronaldson [28] established the importance of the condition when he reported autopsy findings of newborns who had died from asphyxia.

Other congenital anomalies can be associated with choanal atresia; the most common are deviation of the nasal septum, malformation of the maxillary sinus and deformity of the ipsilateral Eustachian-tube orifice. In the series of 17 patients with choanal atresia described by Hall in 1979 [13], several anomalies were encountered, including growth retardation, small ears, cardiac defects, microcephaly, hypogenitalism, hearing loss, micrognathia, and facial paralysis. Pagon et al. [25] first described the "CHARGE" association in 1981. Choanal atresia is one of the six possible categories of associated malformations; the others are choloboma, heart defects, retarded growth and development of central nervous system, genito-urinary tract anomalies in male patients, and ear anomalies.

The etiology of choanal atresia is still inadequately understood. Several embryological theories have been proposed over the years, but there are now four that are the most prominent [7, 15, 24]:
1. Persistence of the superior bucopharyngeal membrane
2. Persistence of the buconasal membrane of Hochstetter
3. Proliferation of abnormal mesodermal tissue forming adhesions in the choanal region
4. Obliteration of the choana by abnormal growth of bone secondary to local factors

In addition to the more common congenital form, choanal atresia or stenosis can also occur as an acquired condition secondary to local trauma or surgery [27].

Diagnosis

The diagnosis of congenital bilateral choanal atresia is suspected soon after birth when a newborn infant presents a clinical picture of asphyxia and cyanosis [2, 14, 18]. Infants with bilateral atresia can be separated into two distinct groups. In the first group are infants who have recurrent bouts of significant asphyxia that is usually alleviated during crying when the child opens its mouth. Infants in this group are at significant risk of developing sudden asphyxia and death [8, 34]. In the second group, the breathing difficulty is less severe but become worse (and potentially life threatening) during eating. Because the child is not able to eat and breathe through the mouth at the same time, an acute obstructive picture with cyanosis and the necessity for artificial respiration develops. The most critical period for newborns with bilateral choanal atresia is during the first 3 weeks of life, after which respiratory emergencies are less common [18].

In unilateral choanal atresia, the picture is similar to that for unilateral nasal obstruction and secretions and can be confused with other lesions, eluding definitive diagnosis for several years. The respiratory difficulty in newborns with unilateral atresia is less dramatic. When unilateral or bilateral atresia is suspected, the diagnosis can be confirmed at the bedside by the inability to pass the nelaton sound or probe through the nose into the nasopharynx [30, 31]. Chronic unilateral choanal (or nasal) obstruction results in the accumulation of secretions on the involved side, which increases the chance of infection. In the clinical evaluation, it is important to remember to search for associated congenital abnormalities [3].

Otorhinolaryngologic Examination

An appropriate age-related otorhinolaryngology examination is performed in the usual manner; attention is then directed to the nasal cavity. After aspiration of any secretions and preparation of the cavity of the nose, examination in very small children is performed with the 3.2-mm-diameter flexible endoscope. When possible, we prefer the 0° and 30° rigid endoscopes with a diameter of 4.0 mm or 2.7 mm (Fig. 33.1). We also use the operating microscope to establish the diagnosis of choanal atresia, though this is more difficult in infants and small children than in adult patients [30].

Ancillary Examinations

Radiological evaluation can be accomplished using conventional X-rays taken in the position of Hirtz, with the introduction of radio-opaque contrast material into both nasal fossae; there will be retention of the contrast material in the involved side or sides (Fig. 33.2). Computed tomography (CT) in the axial and coronal projections is the method of choice for the evaluation of choanal atresia and the adjacent structures. The axial views supply important information, including the location of the obstruction, whether the lesion is unilateral or bilateral and the thickness of the atresia plate. This is helpful, because transnasal correction of atresia is easier if the bony portion of the atresia is thin. The demonstration of abnormal anatomy is of vital importance in surgical planning and in determining the choice of the operative technique to be used; it also permits identification of other congenital malformations in the middle

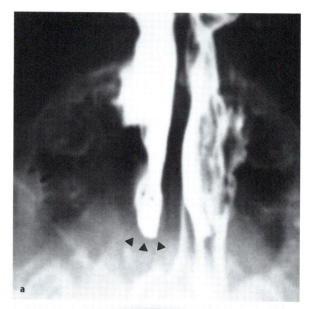

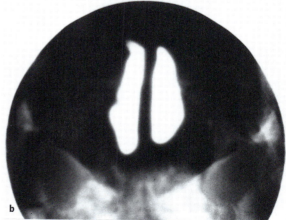

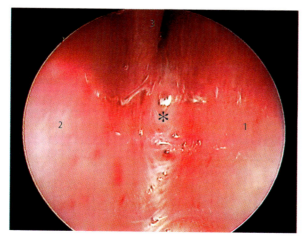

Fig. 33.1. Endonasal endoscopic view of a right choanal atresia. *1*, Nasal septum; *2*, lateral nasal wall; *3*, posterior portion of the inferior turbinate. An *asterisk* indicates the atresia plate

Fig. 33.2. a Plain X-ray (taken in the Hirtz position) of a right choanal atresia after radio-opaque contrast has been instilled into the nasal cavities. **b** Bilateral choanal atresia demonstrated by the same method

third of the face, such as septal deviation and sinus anatomical variations. (Fig. 33.3). The presence of associated congenital abnormalities was quite frequent in our series of patients, as shown in Table 33.1. Postoperative CT and endoscopic examination are utilized to detect any partial or complete re-stenosis of the choana.

Treatment

Once the diagnosis is established, it is important to realize that even bilateral choanal atresia is not always a surgical emergency (despite the presence of an innate reflex for nasal breathing), because it is possible to establish and maintain a patent airway with the use of a McGovern nipple [20] or a cannula of Guedel (Fig. 33.4). In exceptional cases, orotracheal intubation can be performed, thus stabilizing the condition until surgical correction can be done. The exact timing of surgery varies depending on whether the condition is bilateral or unilateral (unilateral cases can nearly always be delayed); the timing also depends on the age of the patient and other general factors. The younger the patient, the higher the risk of intraoperative and postoperative complications and re-stenosis. In unilateral cases, it is ideal to delay surgical treatment until the infant is at least

Table 33.1. Associated congenital anomalies seen in 32 cases of choanal atresia

Anomalies	Number
Ear malformation	4
Facial asymmetry	4
Retardation of growth	2
Cardiac defect	2
Coloboma	2
Treacher-Collins syndrome	1
Down syndrome	1
Total	16

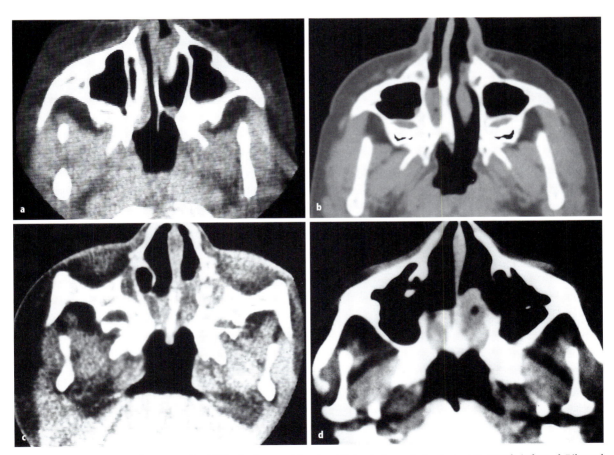

Fig. 33.3. a Axial computed tomography (CT) of a 5-year-old child with left membranous choanal atresia. **b** Right mixed choanal atresia in a 6-year-old child. **c** Axial CT demonstrating bilateral choanal atresia in a 7-month infant. **d** Bilateral choanal atresia in a 22-year-old patient

6 months old, when the operative field is larger and the opportunity for control during surgery is better [30]. If the choanal atresia is bilateral, the condition is much more urgent and, when not promptly diagnosed, can lead to severe asphyxia or death immediately after birth. Since Emmert's initial trocar perforation in 1853 [10], several approaches to correct choanal atresia have been described [1, 22, 35], including (1) transnasal, (2) transpalatal, (3) transseptal, (4) sublabial trans-septal, (5) transantral and (6) external rhinoplasty.

The transpalatine approach offers excellent exposure and provides one with the opportunity to suture the mucosal flaps into position, but the operation usually takes longer than the other techniques. It also has a higher risk of injury to the greater palatine artery. We have sometimes used this approach in children when the dimensions of the nasal cavity were too small to allow adequate intranasal exposure or when a transnasal technique had achieved an unsatisfactory result. Currently, this approach is used more in adults because, in children, it can impair palatal growth [29].

The trans-septal approach is effective but is used very little because of the resultant impairment of facial growth. However, this approach is effective in adolescents and adults. An important technical point is the need to remove the posterior portion of the vomer.

The sublabial, trans-septal access has advantages and disadvantages similar to those of the trans-septal approach, and it further reduces the likelihood of postoperative granulation-tissue formation and stenosis by avoiding surgical trauma to nasal mucosa [17]. Transantral access is the least-used approach, because it requires a well-developed maxillary sinus that is not present in infants and many older children and adults with choanal atresia [11].

Correction of choanal atresia using an external rhinoplasty approach is usually restricted to unilateral cases when the child is 1 year of age or older. The medial crura are separated, and the cartilaginous edge of the nasal septum is exposed. The mucoperichondrium and the mucoperiosteum are elevated to the region of the atresia. The atresia plate is than removed, and the posterior portion of the vomer is resected [16].

The transnasal access is recommended by many authors, including Whinther [34] Maniglia [19], Prades [27], Stamm [30, 31] and Koltai [17], in whose opinion the use of the operating microscope and/or endoscopes are essential. We use a self-retaining nasal speculum that reaches the atresia region without causing trauma to the nasal vestibule but still provides adequate visualization of the operative field. We used an operating microscope with a self-

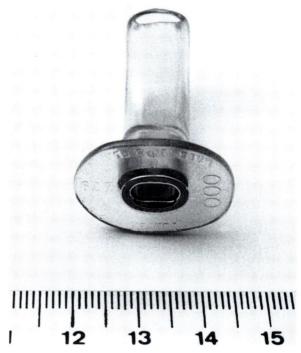

Fig. 33.4. Guedel cannula, which can be used in newborns with bilateral choanal atresia

retaining speculum during choanal-atresia repair before we began using endoscopes, and we find the excellent illumination and visibility of the endoscopes to be a very valuable addition. The technical aspects of endoscopic and microscopic endonasal repair can be facilitated by a microdebrider with different types of tips. [33]

Stankiewicz [32] repairs choanal atresia with otologic instruments, including a drill and diamond burrs; he also uses endoscopes and, sometimes, a laser. The rigid endoscope produces better illumination and resolution; an angled view of the atresia plate and the posterior nasal septum allows much more precise surgery.

Microendoscopic Technique

General anesthesia is employed, using techniques that include controlled hypotension. An operating microscope with a 250-mm objective lens is utilized, as is a 4-mm-diameter endoscope with 0° and 30° angles. The surgical instrumentation includes a self-retaining nasal speculum, dissectors, backbiting forceps, a microdebrider, a micro-Kerrison punch and conventional nasal-surgery instruments. The nasal cavity is prepared with cotton soaked in a topical vasoconstrictive solution to decrease blood loss and shrink the turbinates. Using a 4-mm, 0° endoscope, the site of the atresia plate is subperiostially infiltrated with a 1:100,000 solution of epinephrine. Using the operative microscope, the maneuver of Benfield (elevation of the inferior turbinate) is gently performed in order to avoid mucosal bleeding. Elevation of the inferior turbinate is maintained by one of the blades of a self-retaining speculum modified by the senior author; the other blade is secured in the floor of the nasal cavity (Fig. 33.5). Using a microsurgical scalpel, an incision is made through the mucosa and periosteum of the floor of the nasal cavity and the septum until reaching the lateral wall of the nose.

This flap is elevated so that it is permanently pedicled (with its vascular supply) on the lateral nasal wall (Fig. 33.6).

The bony plate is then visualized, and removal is begun, commencing with a micro-osteotome (Fig. 33.7a), microdebrider or diamond drill (Fig. 33.7b) and continuing with the modified micro-Kerrison punch forceps to complete the removal (Fig. 33.7c). Following this, mucosa of the rhinopharynx (which, up to this point, had been preserved) is incised, and the bony portion of the posterior septum is removed (utilizing a backbiting forceps) in order to create a larger-than-normal opening. This removal is technically easier when it is

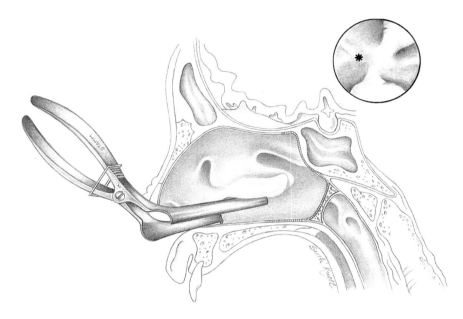

Fig. 33.5. Exposing of the atresia plate after displacement of the inferior turbinate (Benfield maneuver)

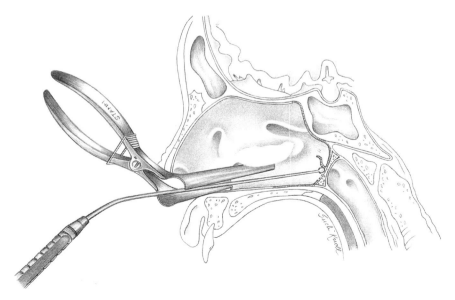

Fig. 33.6. The mucoperiosteum is incised in order to create flap pedicles on the lateral nasal wall

Fig. 33.7. a Opening the atresia plate using a micro-osteotome. **b** Diamond burr. **c** Enlargement of the neo-choana with a micro-Kerrison punch

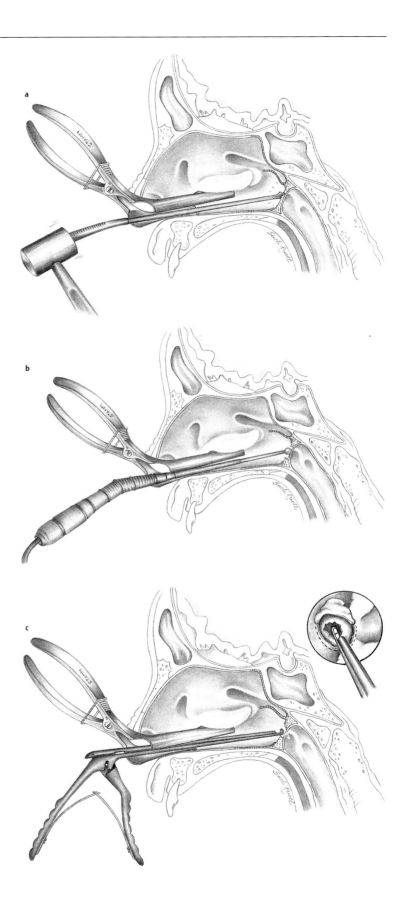

performed via the contralateral nasal cavity, which provides a better angle of vision for the microscope and endoscopes (Fig. 33.8). The removal of the atresia plate and the posterior third of the osseous nasal septum can also be visualized using 30° and 70° endoscopes in the nasopharynx. The new flap is then fixed to the lateral nasopharyngeal wall using fibrin glue. It is important that this flap does not cover the area of the orifice of the Eustachian tube (Fig. 33.9). If necessary, removal of adenoid tissue in children (or nasal septoplasty in adults) is performed at the same time.

At the end of the surgical procedure, a stent can be used to maintain the flap in position and to maintain patency of the neo-choana. This stent can be unilateral or bilateral and usually remains in the nasal cavity for 1–2 weeks (Fig. 33.10). After analysis of cases in which stents were used, our most common postoperative problems were edema, accumulation of secretions, infection and the formation of crusts. During the postoperative period, we try to avoid the accumulation of secretions in order to prevent secondary infections. For this purpose, nasal irrigation with saline solution containing steroids and antibiotics is recommended. If a stent is used, its removal should be performed under direct endoscopic vision. In children, general anesthesia is preferable so that any granulation tissue in the operative area can also be removed.

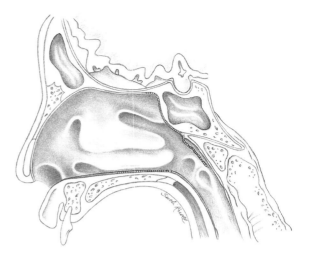

Fig. 33.9. Mucoperiosteal flap rotated to the nasopharynx, covering the lateral wall of the neo-choana

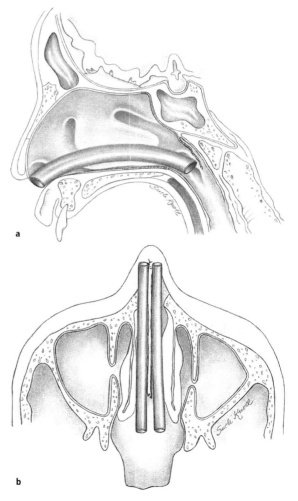

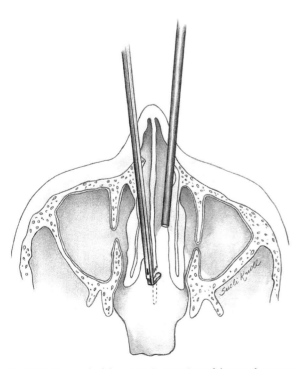

Fig. 33.8. Removal of the posterior portion of the nasal septum with a backbiting forceps via endoscopic visualization

Fig. 33.10. a Stent made with a Foley catheter (non-inflatable portion) sutured to the nasal septum. **b** Bilateral stent

Morgan and Bailey [21] employ an endotracheal tube as a stent, leaving it in position for 6 weeks. Duncan and Sulek [9] prefer a silastic tube as a stent for 4–6 weeks and recommend using a carbon dioxide or potassium-titanyl-phosphate laser to remove any granulation tissue. Koltai [17] prefers to leave the stent in the nasal cavity for only 2 weeks. According to Freng [12], the new choanal opening decreased in size by half in 84% of cases, due to inadequate postoperative care and excessive formation of scar tissue. An important aspect in the prevention of postoperative re-stenosis is the systematic removal of the posterior third of the nasal septum.

Postoperative care is given weekly during the first month and every 3 months for the first year. It is mandatory that endoscopes be used to check the passage of the neo-choana and to remove any crusts and granulation tissue during follow-up of these patients. CT is indicated at approximately 6 months after surgery to access the anatomical condition at the surgical site and to document choanal patency.

Results and Complications

Analysis of the results is based on clinical improvement, microscopic and endoscopic examinations and CT findings. The follow-up period should be at least 5 years (Fig. 33.11). The results of our 32 patients, operated on with micro-endoscopic techniques and followed for at least 5 years after operation, are shown in Tables 33.2 and 33.3.

The principal factors resulting in failure in micro-endoscopic surgery for the treatment of choanal atresia were: (1) surgery performed too early, (2) inadequate surgery exposure (usually when associated with other malformations), (3) inappropriate surgical technique with insufficient removal of the atresial plate and the posterior nasal septum, and (4) inadequate postoperative follow-up.

Table 33.4 shows results achieved by micro-endoscopic surgery for choanal atresia. The complications observed in the 32 operated patients include one case of stenosis of the nasal vestibule subsequent to trauma to the nasal speculum that necessitated correc-

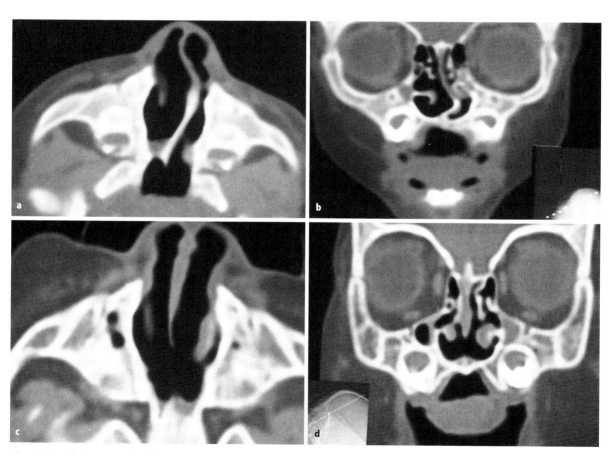

Fig. 33.11 a–d. A 10-month-old child with unilateral choanal atresia and nasal-septum deviation. **a** Axial computed tomography (CT). **b** Coronal CT. **c** Postoperative axial CT. **d** Postoperative coronal CT

tion via rotation of a flap of adjacent skin (Fig. 33.12). Infection, facial edema, crusts and re-stenosis were also observed (Fig. 33.13). Local infection secondary to retention of secretions and inadequate postoperative follow-up occurred in 30% of the operated patients. The local infections were treated with systemic and local antibiotics. In children, edema of the vestibule of the nose occurred mostly as a result of surgical trauma and normally resolved during the first few postoperative days. The appearance of the crusts is directly related to the presence of raw, bloody areas and the excessive use of eletrocautery during the operation. The crusts must be removed in order to avoid secondary infection and increased fibrous scar-tissue formation. Fibrous adhesions in the area of the new choana must be removed at the first postoperative visit to prevent the formation of synechiae. This is only a consideration

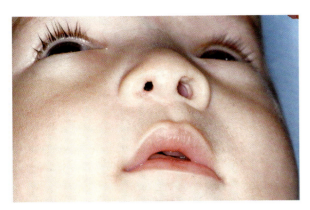

Fig. 33.12. Right nasal vestibule stenosis secondary to trauma from a self-retaining speculum

Table 33.2. Frequency of choanal atresia by gender and involved side (n=32)

Gender	Side			Number
	Right	Left	Bilateral	
Female	7	5	8	20
Male	6	3	3	12
Total	13	8	11	32

Table 33.3. Types and frequency of atresia plates (n=32)

Type	Number	Percentage
Mixed bony and membranous	25	78
Membranous	7	22
Total	32	100

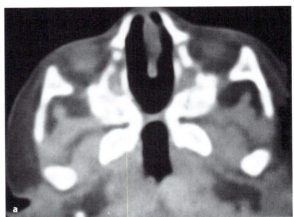
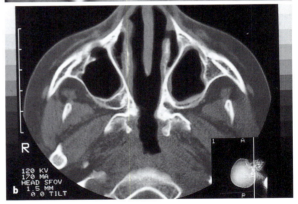

Fig. 33.13. a Postoperative stenosis 6 months after surgery. The removal of the posterior portion of the nasal septum can also be observed. b Axial computed tomography after surgical correction of the stenosis

Table 33.4. Results of the surgical repair after 1 year of follow-up (n=32)

Nasopharyngeal opening Side	Type	Normal	Slight narrowing	Re-stenosis
Bilateral	Mixed	6	–	2
	Membranous	2	–	1
Unilateral	Mixed	11	3	3
	Membranous	3	1	–
Total		22	4	6

in patients who did not have an intranasal stent at surgery.

Conclusions

After analysis of 32 patients with choanal atresia, the following clinical and therapeutic conclusions can be made:
1. Endoscopy and CT are the methods of choice for both diagnosis and follow-up
2. Transnasal micro-endoscopic surgeries utilizing a nasal self-retaining speculum or 0° and 30° endoscopes are the techniques of choice in the treatment of the choanal atresia
3. Micro-endoscopic transnasal surgery utilizing a mucoperiosteum flap pedicled on the lateral nasal wall (with removal of the posterior osseous nasal septum) diminishes the incidence of re-stenosis
4. Utilization of intranasal stents is not obligatory in all cases

References

1. April MM, Ward RF (1996) Choanal atresia repair: the use of powered instrumentation. Operative Tech Otolaryngol Head Neck Surg 7:348–351
2. Benjamin BF (1985) Evaluation of choanal atresia. Ann Otol Rhinol Laryngol 94:429–432
3. Brown K, Brown OE (1998) Congenital malformations of the nose. In: Cummings CW, Frederickson JM, Harker LA, et al. (eds) Otolaryngology, Head and Neck Surgery. Mosby, St. Louis, pp 92–103
4. Brown OE, Pownell P, Manning SM (1996) Choanal atresia: a new anatomic classification and clinical medical implications. Laryngoscope 106:97
5. Carpenter RJ, Neel HB (1977) Correction of congenital choanal atresia in children and adults. Laryngoscope 87:1304–1311
6. Cherry J, Bordley JE (1966) Surgical correction of choanal atresia. Ann Otol Rhinol Laryngol 75:911–920
7. Craig DH, Simpson NM (1959) Posterior choanal atresia with a report of ten cases. J Laryngol Otol 73:603–611
8. Crockett MD, Healy BG, McGill JT, Friedman ME (1987) Computed tomography in the evaluation of choanal atresia1 in infants and children. Laryngoscope 97:174–183
9. Duncan NO, Sulek M (1994) Management of congenital nasal malformations. Operative Tech Otolaryngol Head Neck Surg 5:12–17
10. Emmert C (1853) Bilateral bony atresia. Lehrbuch Chirurgie, Stuttgart
11. Ferguson JL, Neel HB (1989) Choanal atresia: treatment trends in 47 patients over 33 years. Ann Otol Rhinol Laryngol 98:110–112
12. Freng A (1978) Surgical treatment of congenital choanal atresia. Ann Otol Rhinol Laryngol 87:346–350
13. Hall DB (1979) Choanal atresia and associated multiple anomalies. J Pediatr 95:395–398
14. Healy GB, McGill T, Jako GJ, Strong MS, Vaughan CW, et al. (1978) Management of choanal atresia with the carbon dioxide laser. Ann Otol Rhinol Laryngol 87:658–662
15. Hengerer AS, Strome M (1982) Choanal atresia: a new embryologic theory and its influence on surgical management. Laryngoscope 92:913–921
16. Koltai PJ (1991) The external rhinoplasty for the correction of unilateral choanal atresia in young children. Ear Nose Throat J 70:450–453
17. Koltai PJ (1994) The surgical management of choanal atresia. Operative Tech Otolaryngol Head Neck Surg 5:5–11
18. Lantz HJ, Birck HG (1981) Surgical correction of choanal atresia in the neonate. Laryngoscope 91:1629–1634
19. Maniglia AJ, Goodwin WJ Jr (1981) Congenital choanal atresia. Otolaryngol Clin North Am 14:167–173
20. McGovern F, Fitz-Hugh GS (1961) Surgical management of congenital choanal atresia. Arch Otolaryngol 73:627–643
21. Morgan DW, Bailey CM (1990) Current management of choanal atresia. Int J Pediatr Otolaryngol 19:1–13
22. Muntz HR (1987) Pitfalls to laser corrections of choanal atresia. Ann Otol Rhinol Laryngol 96:43–46
23. Otto AW (1830) Lehbuch der pathologie anatomy des menschen and der thiere. Rucker, Berlin, p 181
24. Owens H (1951) Observations from treatment of seven cases of choanal atresia by transpalatal approach. Laryngoscope 61:304–314
25. Pagon RA, Graham JM, Zonana J, et al. (1981) Coloboma, congenital heart disease and choanal atresia with multiple anomalies: CHARGE association. J Pediatr 99:223–227
26. Pirsig W (1986) Surgery of choanal atresia in infants and children: historical notes and updated reviews. Int J Pediatr Otolaryngol 11:153
27. Prades J, Bosch J, Tolosa A (1977) Microcirurgia endonasal. Garsi, Madrid, pp 257–266
28. Ronaldson TR (1880) Bilateral membranous atresia. Edinburgh Med J 26:1035
29. Schwartz ML, Savestky L (1986) Choanal atresia: clinical features. Surgical approach and long-term follow-up. Laryngoscope 96:335–339
30. Stamm A, Mekhitarian L (1995) Atresia coanal: cirurgia microendoscópica transnasal. In: Stamm A (ed) Microcirurgia naso-sinusal. Revinter, Rio de Janeiro, pp 279–287
31. Stamm A, Mekhitarian L, Pato C (1988) Microcirurgia transnasal no tratamento da atresia coanal. Rev Bras ORL 54:43–50
32. Stankiewicz JA (1989) The endoscopic repair of choanal atresia. Otolaryngol Head Neck Surg 101:183
33. Vickery CL, Gross CW (1997) Advanced drill technology in treatment of congenital choanal atresia. Otolaryngol Clin North Am 30:457–465
34. Winther LK (1978) Congenital choanal atresia. Arch Otolaryngol 104:72–78
35. Wright W, Shambsugh G (1947) Congenital choanal atresia. A new surgical approach. Ann Otol Rhinol Laryngol 56:120–126

Endoscopic Transnasal Dacryocystorhinostomy

Gustavo A. Riveros-Castillo, Alfredo Campos

Introduction

The lacrimal drainage system has provoked great deal of investigation throughout history. There are reports of lacrimal-drainage-system studies in the literature since the time of Celsus, at the beginning of Christendom [18, 23]. Three main stages of development of surgical treatment of the pathological lacrimal system have been described. During the first stage, surgery was restricted to the lacrimal gland; epiphora was treated exclusively with probing and, in suppurative dacryocystitis, aseptic and caustic lavage and thermocautery were used. In the second stage, during the last century, extirpation of the lacrimal sac was initiated as a unique solution for suppurative dacryocystitis. The third stage began in Italy in 1904, when Toti started draining the lacrimal sac directly into the nasal fossa through the unguis [external dacryocystorhinostomy (DCR)]. Simultaneously, West-Poliak and Mosher performed the same drainage procedure, but did so from the nasal cavities (endonasal DCR). This surgical procedure required extensive septal dislodgment or perforation to reach the lacrimal recess in the nasal lateral wall [4, 8, 10, 18–20]. With the advent of the microscope, illumination and magnification improved, but the lateral exposure was not sufficient to allow adequate identification of the lacrimal sac and duct within the lateral nasal wall. It was only after the introduction of Hopkin's small-diameter (4-mm) rigid endoscopes (which could be maneuvered in the nose to give lateral visualization) and the increased magnification provided by video-documentation equipment that surgery under total visual control was possible.

Once it was established that an obstructed lacrimal system could be surgically treated by re-establishing drainage into the nasal fossa, it became necessary to introduce measures to maintain the patency of the drainage pathway. Early attempts involved intubating the lacrimal duct with stents or molds allowing drainage of tears by capillary action through the external walls [2, 8, 10, 18–20, 25].

Henderson pioneered the insertion of synthetic material into the lacrimal system in 1950. He inserted a rigid tube made of polyethylene, securing it to the skin with transfixion sutures and removing it after 45 days. Although initially successful, the technique failed relatively quickly, because the polyethylene was rigid and difficult to manipulate and, more importantly, it induced an intense, inflammatory tissue reaction [12]. To overcome these disadvantages, Quicker and Dryden introduced the more flexible silicone tube, which continues to be used today [1, 12].

Although several methods and materials of intubating the lacrimal duct have been described, the technique of Crawford probably has the greatest popularity. It has three advantages:
1. Manufacturers provide silicone tubes joined by adhesive to the metallic guides; these do not become dislodged easily during intubation.
2. The distal ends of the guides are in an olive form, which permits easy removal with a Crawford hook.
3. Once both ends of the silicone tube have been passed into the cavity, they are tied together in various knots to avoid displacement [11, 12, 18, 25].

Anatomy

Tear production occurs in the lacrimal gland, where the racemic elements secrete the tears and disperse them onto the conjunctival surface. The gland is located on the supero-external portion of the orbit, in a bony recess; it is firmly bound to the periosteum by several ligaments [5, 24]. It is divided into two parts by an expansion of the levator muscle of upper eyelid: a voluminous orbital portion and a smaller palpebral portion [15, 16]. The lobules of the gland disperse the tears to the eyeball through three to five ducts running downward and forward at the supero-external region of the conjunctival "cul-de-sac". Lacrimal dispersion throughout the conjunctiva is uniform and carries the tears toward the internal angle of the eye; this process is facilitated by blinking. The internal canthal tendon acts as a pump mechanism, pressing the lacrimal sac to mobilize the tears toward the nose.

In normal conditions, the tear pellicle consists of a trilaminar liquid stratum: a layer of mucus coating the corneal epithelium, a second aqueous central layer and a more external lipid layer. These components are produced by the caliciform conjunctival cells, the lacrimal glands and the Meibomian glands, respectively.

The lacrimal canaliculi and pores act by capillarity. They are two small outlets (an upper one and a lower one) that are 0.1–0.4 mm in diameter. Each has a 2.0- to 2.5-mm vertical portion that joins with a 7- to 10-mm horizontal portion at a right angle. The horizontal limbs of the two ducts then enter the lacrimal sac, either independently or after joining together to form a common duct. The horizontal portions have fascicles of Horner's muscle on their posterior aspects; these fascicles run parallel to the duct and form loops that encircle the lower portion of the ampoule formed by the vertical duct, thus creating a sphincter that narrows and even occludes the distal portion. This sphincter functions together with movements of the orbicular tendon and the above-mentioned pumping mechanism [3, 5, 15]. The canaliculi are 0.6 mm long, and the common duct is 1–3 mm, but they can easily be distended to 1.5 mm by Bowman probes, after which they return to their normal diameters.

The lacrimal sac, located beneath the orbicular tendon, is 12 mm long and 3 mm wide, and its walls are 0.6 mm thick; it merges with the nasolacrimal duct formed by the ascending apophysis of the maxilla. It also merges with the unguis on its posterior aspect [9, 15]. The unguis is up to 20 mm long, with a diameter between 3 mm and 7 mm; it tapers inferiorly. It inclines 12° obliquely, posteriorly and medially [4]; its terminal portion is variable and sometimes has a submucosal tract several millimeters long in the inferior meatus, 8–15 mm from the head of the inferior turbinate. Its shape is oval, and it is 2–6 mm in size [3–5, 8].

The lining of the nasolacrimal duct and sac have columnar pseudostratified epithelium similar to that found in the upper respiratory tract. The walls feature many elastic fibers, collagen and lymphoid tissue (Fig. 34.1) [4, 5].

Throughout the lacrimal system, there is a series of valves (mucosal folds that occupy a portion of the lumen) that help govern the lacrimal flow. They are:
1. Boschdalek's valve, at the pore level
2. Rossenmüller's valve, at the common duct mouth in the sac
3. Krausse's valve, between the sac and the duct
4. Taillefer's valve, at the medial portion of the nasolacrimal duct
5. Hasner's valve, at the lower nasal outlet (Fig. 34.2)

Endoscopic Surgical Anatomy

Reference points for the lacrimal system's location at the nasal wall are commonly found in the literature [3, 4, 8, 9, 13, 17–21], are based on the location of the middle turbinate head and are very inconstant because of the anatomical variability of that structure. These variations include bullous turbinates, paradoxical curvatures, frequent middle-turbinate pathology (such as polypoidal degeneration, severe hypertrophies and allergy-induced mucosal changes, which are discovered because of the increasingly frequent use of endoscopes during routine examination of the nose).

The difficulties encountered in endonasal DCRs performed under microscopic control since 1986 (and under endoscopic control since 1990) and the findings from dissections and exposures of the whole lacrimal system (performed under endoscopic control without disturbing the adjacent areas of the ethmoid or frontal sinuses and without causing cosmetic deformities) motivated the analysis of six fixed sagittal sections and the study of 40 lacrimal systems in fresh cadavers. These investigations yielded the following findings:

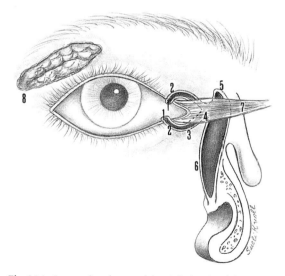

Fig. 34.1. Anatomic scheme of the right lacrimal duct, with the orbicularis tendon and lacrimal gland. *1*, Lacrimal canaliculi; *2*, vertical duct (2 mm); *3*, horizontal duct (8 mm); *4*, common duct (2 mm); *5*, lacrimal sac (15 mm); *6*, nasolacrimal duct (17 mm); *7*, orbicularis tendon; *8*, lacrimal gland

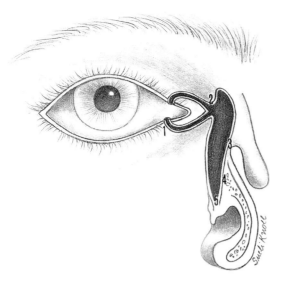

Fig. 34.2. Valves of the lacrimal system. *1*, Boschdalek; *2*, Rosenmüller; *3*, Krause; *4*, Taillefer; *5*, Hasner

1. The lacrimal-duct/middle-turbinate length ratio is an inconstant finding because of the frequent variances and the above mentioned pathology.
2. The ratio of the length of the lacrimal duct to that of the vertical portion of the unciform apophysis is more constant. The lacrimal duct splits at the beginning of the horizontal portion of the apophysis; the right branch ends at Hasner's valve in the inferior meatus.
3. In the maxillary crest and the maxillary line, the duct is always immediately anterior to the uncinate process.
4. The bony wall covering the system is usually very strong, and it protects the duct during resection of the uncinate process and makes its exposure difficult during surgery; this structure is soft and thin in 8% of individuals.
5. Wide exposure of the lacrimal sac and duct using a 0° endoscope provides a clear anterior pathway, which always emerges beneath the middle turbinate at the "keel". The course of the system was determined via the analysis of six sagittal sections in which the true posterior orientation relative to the Frankfurt plane could be seen. Optical distortion was due to several factors:
   a. The duct frequently occurs at a distance of 2–3 mm behind the head of the middle turbinate.
   b. It parallels the middle turbinate head in 30% of cases.
   c. The form of the middle-turbinate head and its clearly posterior orientation can cause distortion.
   d. The patient is in a supine position, with a hyper-extended head.
   e. The optics are always located inferior to the operative field.
6. The average distance to the maxillary ostium is only 7 mm, which makes the lacrimal system vulnerable to injury during a middle-meatus antrostomy, when the meatus is widened anteriorly, starting from the ostium.
7. The distance between the nasal vestibule and Hasner's valve is 21.2 mm; the distance between the nasal vestibule and the inferior-turbinate head is 12.7 mm, and that between the inferior-turbinate head and Hasner's valve is 8.5 mm (Fig. 34.3).
8. There is a close relationship between the duct and the orbit. Caution must be observed when the uncinate process has been removed, as it is separated only by the lamina papyracea; the possibility of fat herniation is high and is avoided by staying close to the maxillary line.
9. There is also a close relationship between the frontal sinus and the upper portion of the duct, because it is part of the recessed anterolateral wall and the lamina papyracea.

Pathology

The pathology of the lacrimal system is dominated by the cardinal symptom of dacryostenosis: epiphora. Dacryostenosis is the partial or total blockade of

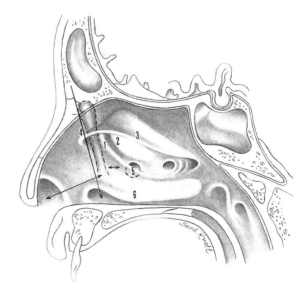

Fig. 34.3. Scheme of the lateral nasal wall. *1*, Uncinate process; *2*, bulla ethmoidalis; *3*, middle-turbinate head; *4*, lacrimal sac and duct (maxillary line); *5*, maxillary ostium; *6*, inferior turbinate

the lacrimal duct, pre- or post-saccal. The most common cause of dacryostenosis described in the literature is a lack of canalization in the terminal portion of the duct at the inferior-meatus level. The second most common cause is a lack of canalization at the sac/nasolacrimal-duct junction. Septal deviation and middle-turbinate impaction are also described as common causes [18].

Among nasolacrimal-system pathologies are the following:

- Infections
  - Suppurative dacryocystitis secondary to various infections presents with a dilated sac and a history of frequent facial inflammation and constant epiphora.
  - Canaliculitis by *Streptomyces* or *Actinomyces israelii*.
  - Syphilis and fungi.
- Congenital defects
  - Absence of valves, especially Hasner's valve, can facilitate formation of pneumatoceles.
  - Saccal abnormalities, which can generate fistulae and diverticuli.
  - Abnormalities of the lacrimal pore, including congenital atresia, double or supernumerary outlets or altered locations.
  - Canaliculus abnormalities or atresias.
  - Agenesia of the proximal lacrimal system (pore, canaliculus).
- Trauma
  - Trauma associated with craniofacial injury
  - Trauma secondary to osteotomies in rhinoplasty
  - Trauma secondary to resection of lesions involving the lateral nasal wall, such as inverted papillomas, etc.
  - Trauma secondary to functional surgery on paranasal sinuses for inflammatory pathology, etc.
  - Trauma secondary to multiple probing, resulting in stenotic areas or false non-functional routes
  - Tears of the proximal lacrimal system
  - Radiotherapy
- Tumors
  - Associated skin cancer, (base cell carcinoma in 90% of cases and squamous cell carcinoma in 5%)
  - Miscellaneous (such as adenocarcinomas or hemangiomas)
  - Lymphomas
- Granulomas
  - Rhinoscleroma
  - Sarcoidosis
  - Non-specific granulomas

Diagnosis

Diagnosis of lacrimal-system pathology begins with a good clinical history, including a complete examination of the lacrimal pore, observing its position relative to the upper and lower eyelids.

Fluorescein Test

- **Jones procedure 1.** Two or three drops of 2% fluorescein are applied in the conjunctival sac after the application of a nasal anesthetic and a decongestive agent. A dry applicator is located ipsilaterally at the inferior meatus and is removed 5 min later; the response to the colorant is then determined.
- **Jones procedure 2.** After dilating the lacrimal pore, a cannula is inserted for fluorescein infusion. It is sometimes necessary to use pressure to insert the cannula and to allow the sample to be taken at the nasal level [19, 20].

This test (contraindicated in acute cases) does not determine location of the stenotic site. It only corroborates the obstruction.

Dacryocystography

Dacrycystography had been more or less abandoned because of the risk of making false passages and extravasating the contrast medium, which occur because the radiologist has to canalize the lacrimal system. This useful procedure is presently being revived, and we use it intraoperatively in disruptions of the duct in orbital trauma if dacryocystography with ^{99}Tc does not reveal a passage to the proximal or distal ducts, causing doubt regarding the existence of the sac. It is essential to use an image intensifier to support the procedure.

Dacryocystogammagraphy requires injection of ^{99}Tc into the conjunctival sac. The patient is then taken to the γ-camera in order to see the marker traverse the lacrimal system. Its rate of travel is observed, and the obstruction level is determined (Fig. 34.4) [22]. *Subtracted dacryocystotomography* has recently been introduced as a diagnostic method for the study of various pathologies of the lacrimal system. The excellent images this method offers are an important aid for precise diagnostic evaluation (Fig. 34.5) [7].

Endoscopic Transnasal Dacryocystorhinostomy

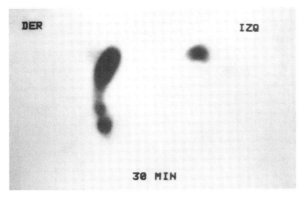

Fig. 34.4. Dacryocystogammagraphy showing blockage at the level of the left sac

The procedure is performed under general or local anesthesia. The area of the middle meatus is infiltrated 20 min after vasoconstrictor application via rolled cotton.

First Stage – Nasal Phase

Under endoscopic visualization, the uncinate process and its posterior edge at the semilunar hiatus are identified; by palpating its soft complexion, we continue anteriorly to the firmer maxillary line, where a mucosal window is created at the level of the middle third of the uncinate process. A rectangular incision of approximately 6 mm is made, and the mucosal membrane is removed (Fig. 34.6), creating a window through which we can observe the bony wall to be removed. This wall is removed using either the curved chisels, the Stammberger drill or the microdebrider. It is important to realize that the use of these high-velocity instruments can expose the lacrimal sac in a few seconds. Once the bony

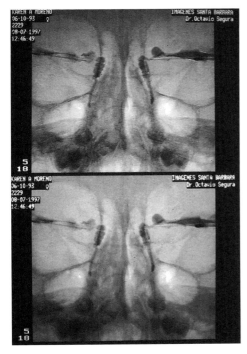

Fig. 34.5. Subtracted dacryocystotomography in a 4-year-old girl presenting with bilateral agenesia of Hasner's valve

Surgical Technique

Apparatus

1. A Bowman dilator set for the lacrimal system
2. Optics (0–25°, 4 mm long)
3. Video documentation equipment
4. Basic set of instruments for nose surgery
5. Crawford's set of instruments
6. Curved chisels (4 mm and 6 mm long)
7. Stammberger's microdebrider or drill (optional)

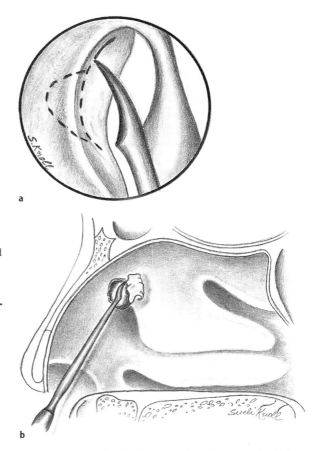

Fig. 34.6. a Surgical technique: creation of a mucosal window. b Mucosal window: sagittal view.

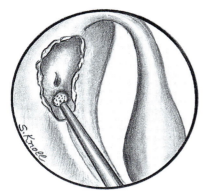

Fig. 34.7. Creation of a bony window. Note the lacrimal sac

window is formed (Fig. 34.7), the lacrimal sac, gray in color, is clearly identified with the aid of the magnification provided by the video documentation equipment. The sac is then opened via its most inferior portion.

Second Stage – Canalization of the Lacrimal System

The pore dilator is passed vertically through the inferior lacrimal pore, then it is moved horizontally 3–4 mm; the insertion of dilators is then initiated (beginning with the no. 00 dilator) until dilators can be gently passed into the entire lacrimal system. The dilators are first vertically inserted to a depth of 2 mm, then they are directed horizontally until they gently hit the nasal bone. During this step, external traction of the lower eyelid is required; when the dilator hits the nasal bone, the level of the lacrimal sac has been reached. The dilator is then manipulated until the sac can be entered with light pressure; the dilator is then removed. Further dilation is then performed, using no. 00, 1 and 2 dilators (and usually a no. 3 dilator); the largest is left in the sac. The endoscope is then used to make an incision in the most inferior part of the sac, preserving most of the sac. Then the dilator is removed, and the Crawford catheter is passed through the lower lacrimal pore, placing the metal end at the nasal level and ensuring its insertion along the route created. The other end of the catheter is inserted through the upper pore, and the same operation is repeated. Once the catheter is in place, the metallic ends are gently pulled and knotted (Fig. 34.8), avoiding applying too much pressure at the lacrimal pores. To further fix the catheter in place, its ends are knotted with a non-absorbable suture, preferably a nylon suture (Fig. 34.9).

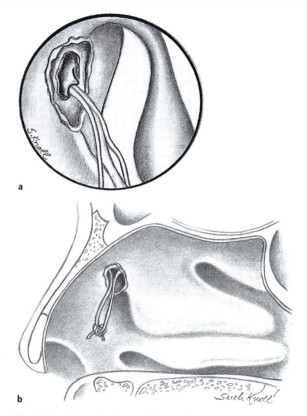

Fig. 34.8 a, b. Canalization of the lacrimal system

Surgical Management of Agenesia of the Proximal Lacrimal System

The surgical technique we use for the management of congenital defects of the lacrimal system is based on long hours of cadaver dissection and the knowledge and surgical expertise gained after 12 years of experience practicing DCR procedures. In our experience, we have found that two different types of clinical situations require certain variations in the surgical technique. These are:
- **Type I.** A lacrimal sac with normal size and good function provided by a normal medial canthal tendon
- **Type II.** An oversized lacrimal sac with poor function due to the disproportionate sizes of the sac and the medial canthal tendon, leading to diminished pump action (Fig. 34.10)

In any case, the first step is to localize the lacrimal sac transnasally, using a 0° or 25°, 4-mm endoscope, as described previously. A wide exposure is desirable in these cases. Normally, we create a bony window with a size of at least 7 mm×7 mm to access the lacrimal sac. We then proceed to open the sac. In cas-

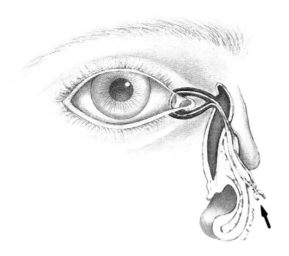

Fig. 34.9. Final position of a Crawford catheter, avoiding too much pressure at the lacrimal pores

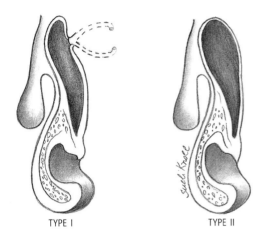

Fig. 34.10. Agenesia of the proximal lacrimal system. The lacrimal sac has a normal size (type I). Also shown is an oversized lacrimal sac (type II)

es with a type-II clinical presentation, we try to ensure wide marsupialization of the sac by suturing the borders of the open sac mucosa to the nasal mucosa using 10–0 nylon sutures.

Our next step is the creation of a proximal lacrimal system, including superior and inferior pores and corresponding superior and inferior lacrimal ducts. This is achieved using a no. 11 surgical blade to create lacrimal pores in the areas where they are normally found. We must keep in mind that the newly created superior lacrimal pore should not contact the inferior pore during blinking of the eye; in this way, we can recreate normal anatomy, stimulate normal physiology and avoid possible synechiae formation.

We will then proceed to create the superior and inferior canaliculi using a lacrimal-pore dilator followed by a no. 2 Bowman dilator. The Bowman dilator is first placed vertically into the newly created pores, then is inserted medially (in a horizontal fashion) in search of the lacrimal sac, in the same manner used for conventional endoscopic DCR. Once the lacrimal sac is reached through the newly created superior and inferior lacrimal canaliculi and the location of the sac is confirmed endoscopically, we insert a Crawford silicone stem (in the same manner described previously), which is kept in place for at least 12 months.

Surgical Technique for the Management of Traumatic Transection of the Lacrimal Canaliculi

We have seen two different types of clinical presentation in cases of traumatic transections of the lacrimal canaliculi. They are:
1. Recent presentation, which refers to cases in which the patient consults and can be managed during the first 48 h following the traumatic event
2. Delayed presentation, which refers to cases in which patients are treated more than 48 h following the traumatic event

We make a clear distinction between these two presentations. We have seen that, during the first 48 h following trauma, we can usually find the proximal and distal ends of the canaliculi that have been sectioned and can re-canalize them by inserting a Crawford silicone stem into the lacrimal sac and the nose, as described previously (Fig. 34.11).

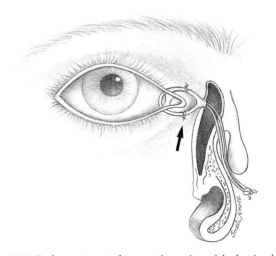

Fig. 34.11. Early treatment of traumatic section of the lacrimal canaliculi

After 48 h, it is very difficult to find and re-canalize the distal canaliculi; in these cases, we make a 2-mm incision in the superior part of the sac and pass a Crawford silicone catheter through the proximal canaliculi, directly into the sac. From there, we pass the catheter into the nose in a conventional manner. The Crawford stem is kept in place for as long as possible (at least 1 year) to ensure fistulization (Fig. 37.12).

Indications

Indications for dacryostenosis are summarized in Table 34.1.

Results

Between 1986 and 1998, 121 endonasal DCRs have been performed by the same surgeon (Table 34.2). Thirteen (10.7%) of them were performed under microscopic control before the introduction of the Crawford tubes; at that time, fixation of the catheter was still a problem. At that time, peridural catheters were used and were only inserted through the lower pore; sometimes they were fixed to the nasal mucosal membrane with silk sutures, a knot at the conjunctival level or anti-allergic adhesive tape at the level of the facial skin. At that time, the maximum duration of catheterization was 15 days, which was believed to be an appropriate length of time. Satisfactory results were achieved in 65% of cases, proving that DCR is a useful technique for preservation of the sac in cases of suppurative dacryocystitis.

The other 108 (89.2%) patients were operated on using Crawford catheters under endoscopic control. The endoscopes facilitated the identification of the lacrimal sac's gray color. Initially, the tubes were left

Table 34.1. Indications for dacryostenosis

Indications
• **Primary dacryostenosis** Congenital blockade not responding to dilation with Crigler or Bowman dilators Congenital blockade before and after the lacrimal sac Agenesia of the proximal lacrimal system • **Secondary indications** Trauma Craniofacial trauma Facial injury Trauma secondary to resection of tumors Trauma after dilation with a Bowman dilator Osteotomies due to rhinoplasty • **Secondary dacryostenosis** Chronic dacryocystitis Radiotherapy External dacryocystotomy Secondary mucoceles After Crawford catheterization External dacryocystostomy the Jones fluorescein test

Table 34.2. Distribution of patients according to the etiology of the lacrimal-system obstruction (*n*=121)

Etiology	Number	Percentage
Dacryostenosis	50	41.3
Dacryocystitis	25	20.6
Previous surgery (Crawford catheterization)	20	16.5
Maxillary/facial trauma	13	10.7
Previous surgery (external DCR)	2	1.6
DCR plus Jones tube	2	1.6
DCR plus mucoceles	1	0.8
Via lacrimal section	1	0.8
Crawford catheterization plus canthal pexis	1	0.8
Septorhinoplasty	1	0.8
Agenesia (pore, canaliculus)	4	3.3
Radiotherapy	1	0.8
Total	121	100

DCR, dacryocystorhinostomy

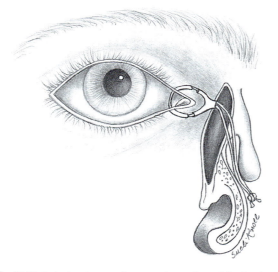

Fig. 34.12. Late treatment of traumatic section of the lacrimal canaliculi. Distal canaliculi not found

in place for 3 months, then for 6 months; there is now a debate among members of the American Academy of Ophthalmology regarding whether to leave the tubes in place for 6 months or permanently. It is important to realize that the tears flow along the external wall of the catheter.

Among the most frequent pathologies encountered were:
- Twenty-five (20.6%) patients with suppurative dacryocystitis
- Fifty (41.3%) patients with dacryostenosis and a history of failed attempts to dilate the lacrimal ducts
- Two (1.6%) patients with a history of external DCRs; one of these patients developed an ethmoidal mucocele (anteriorly located)
  Twenty (16.5%) patients with previous catheterization (with Crawford tubes) for dacryostenosis. These patients were admitted with the catheter in place but with their symptoms unresolved. One of them had a laceration of the lacrimal pore and the vertical and horizontal tracts of the lacrimal system
- One (0.8%) patient with history of epiphora secondary to septorhinoplasty
- Thirteen (10.7%) patients with a previous history of maxillary/facial trauma

The use of these catheters led to the initial belief (among some ophthalmologists) that, even if the dilators had to be forced through the lacrimal sac/nasolacrimal-duct junction, they were the definitive solution for this important pathology; the catheters were expected to provide lacrimal drainage. In one case, a catheter was found in the inferior turbinate, obviously with no control of the epiphora. This critical pathology must be handled by a team, and the otorhinolaryngologist must become more familiar with this technique, which is in accordance with tendency of the modern surgery to intervene in the most atraumatic way, preserving both structure and function. The use of a laser in creating the nasal window and opening the lacrimal sac, guided by the course of a luminous probe until the sac is reached, facilitates the procedure; however, use of the laser is not as significant as it seems to be [6, 14, 16]. By contrast, the introduction of Crawford catheters was certainly significant.

Conclusions

1. Our parameters for locating the lacrimal system and its anatomy have been constant and were corroborated by formolized samples of sagittal sections and by all the endoscopic procedures performed on fresh cadavers. Furthermore, the parameters have been applied in the last 56 endoscopic surgical interventions practiced on patients at the Hospital Clinic of San Rafael.
2. Anatomic variations observed are related to the middle-turbinate head more often than to the uncinate process and maxillary line because, even if there are exaggerated "pneumatizations" of the uncinate process, its relationship to the maxillary line is constant.
3. Thus, use of the reference universally accepted for creating the mucosal and bony windows (5 mm in front of the middle-turbinate head) is not safe, nor is the suggestion of creating a wider window.
4. In our opinion, methods and aids used for location (such as luminous probes or the movements of a dilator previously located along the lacrimal system) do not avoid the risk of creating a false route adjacent to the sac or the duct. These aids are not as reliable as adequate exposition of the system, which ensures that the procedure is performed under continuous visual control.
5. Good training in endoscopic surgery of paranasal sinuses is required, given the close relationships between the lacrimal system and the maxillary, ethmoid and frontal sinuses.
6. Otorhinolaryngologists must be increasingly familiar with the lacrimal system, as most of it is located within the lateral nasal wall.
7. The results are comparable to those of external dacryocystostomy: the procedure is efficacious in 90% of cases.
8. Preservation of the internal canthal tendon (the main propellant of tears) and the lacrimal sac and duct is considered the most important advantage of the technique.
9. The silicone tubes provide an outlet approximately 2 mm long, long enough for adequate lacrimal drainage.
10. Use of a laser is neither necessary nor justifiable in creating mucosal or bony windows, because secondary fibrosis can result.

References

1. Angrist RC, Dortzbach RK (1985) Silicone intubation for partial and total nasolacrimal duct obstruction in adults. Ophthal Plast Reconstr Surg 1:51–54
2. Arruga H (1946) Cirugia de las vias lagrimales. In: Arruga H (ed) Cirugia ocular. Salvat, Barcelona, pp 203–289
3. Bartley GB (1991) Acquired lacrimal drainage obstruction: an etiologic classification system, case reports and a review of literature. Part 2. Ophthal Plast Reconstr Surg 8:243–248

4. Botek AA, Goldberg SH (1993) Margins of safety in dacryocystorhinostomy. Ophthalmic Surg Lasers 24:320–322
5. Castillo IG (1994) Avances en el tratamiento de la obstruccion de la via lacrimal. Hospital Universitario Clinica San Rafael, Bogota
6. Echeverry GJ, Ortega DM (1994) Dacriocistorrinostomia endonasal con laser de CO_2. Ver Soc Colombiana Oftalmol 10:64–68
7. Holt JE, Holt GR (1988) Nasolacrimal evaluation and surgery. Otolaryngol Clin North Am 21:119–134
8. Katowitz J, Cahill K (1988) Anatomy of the orbit and ocular. Lacrimal gland and excretory system. Fundamentals and principles of ophthalmology. American Academy of Ophthalmology, pp 48–49
9. Kraft SP, Crawford JS (1992) Silicone tube intubation in disorders of the lacrimal system in children. Am J Ophthalmol 94:290–299
10. Latarjet M, Ruizliard A (1983) Anatomia humana. Panamericana, Buenos Aires, pp 468–477
11. Lee KJ (1995) Interrelacion con oftalmologia. In: Lee KJ (ed) Lo esencial en la otorrinolaringologia. Appleton and Lange, Norwalk, pp 347–350
12. Lusk J (1990) External disease and cornea, specific inflammatory diseases, lacrimal system. American Academy of Ophthalmology, San Fransisco, pp 133–135
13. Martinez R, Novoa J (1992) Dacriocistorrinostomia externa. trabajo de promocion Hospital Militar Central, Servicio de ORL. Santafe, Bogota
14. Massaro BM, Gonnering RS, Harris GJ (1990) Endonasal laser dacryocystorhinostomy. Arch Ophthalmol 108: 1172–1176
15. Metson R (1995) Endoscopic dacryocystorhinostomy: primary and revision. In: Stankiewicz JA (ed) Advanced endoscopic sinus surgery. Mosby, St. Louis, pp 127–135
16. Metson R, Woog JJ, Puliafito CA (1994) Endoscopic laser dacryocystorhinostomy. Laryngoscope 104:269–274
17. Neubauer H (1982) Operations of the eyelids, the lacrimal apparatus and the orbit. In: Naumann H (ed) Head and neck surgery. Saunders, Philidelphia, pp 149–162
18. Raymond E (1991) The conjuntive and lacrimal system. In: Tasman W, Jaeger EA (eds) Foundations of clinical opthalmology. Lippincott, Philadelphia, pp 14–18
19. Riveros GA (1988) Dacriocistorrinostomia endonasal. Hospital Clinica, San Rafael, pp 1–30
20. Riveros GA (1995) Dacriocistorrinostomia endonasal endoscopica. Acta Otorhinolaryngol Cirug Cabeza Cuello 23:117–121
21. Rubenfield M, Wirtschaffer J (1994) Oftalmologia en relacion con la practica de la otorrinolaringologia. In: Paparella M (ed) Otorrinolaringologia. Panamericana, Buenos Aires, pp 3536–3540
22. Sekhar GC, Dortzbach RK (1991) Problems associated with conjuntivodacryocystorhinostomy. Am J Ophthalmol 112:502–506
23. Taembaum M, McCord CD (1991) The lacrimal drainage system. In: Tasman W, Jaeger EA (eds) Foundations of clinical ophthalmology. Lippincott, Philadelphia, pp 1–13
24. Takeo I, Jakoblec FA (1991) Lacrimal glands. In: Tasman W, Jaeger EA (eds) Foundations of clinical ophthalmology. Lippincott, Philadelphia, pp 19–94
25. Tos M, Balle V, Andersen R (1986) Dacryocystorhinostomy. Ann Otol Rhinol Laryngol 94:352–355

Endonasal Surgery of the Lacrimal System

Joachim Heermann and Ralf Heermann

Introduction

In this century, the first external lacrimal-sac surgery was reported by the Italian rhinologist Toti [25] in 1904 (Fig. 35.1). In 1901, the German rhinologist Passow [22] recommended endonasal surgery for the lacrimal sac and, in 1909 [28], the American ophthalmologist Randolph at the John Hopkins Hospital induced his rhinologist J.M. West "to make the window big and carry it high up" into the lacrimal sac (Fig. 35.2) by the easier endonasal approach [28, 29]. Since 1912, four Heermann generations [5–20] at the Krupp Hospital of Essen, Germany, have performed endonasal surgery on the lacrimal system (Fig. 35.3) in up to 98 patients annually. Since 1958, all endonasal surgery has been done with the aid of a binocular Zeiss operating microscope [6]. Most of these cases (97%) have been referred by ophthalmologists.

Indication for Surgery

The indication [11] for surgery is epiphora of 4 months duration [15, 17], treated unsuccessfully by ophthalmologists with two irrigations of the inferior lacrimal duct and via a Bungerter probe in young children. Frequent irrigations can be followed by a presaccal stenosis and poor prognosis [20]. A short improvement after irrigation for only a few days is often observed [9] in cases with dacryoliths. Although dacryocystography (Fig. 35.8a) is of scientific interest, this does not change the surgical procedure [11] and can cause lacrimal-system damage [19].

Half-Sitting Semi-Fowler's Position

The half-sitting semi-Fowler's position of the patient (Fig. 35.4), with the head placed in a movable specially designed headrest [15], can be adjusted (by a nurse) into the optimal position for the surgeon. There are

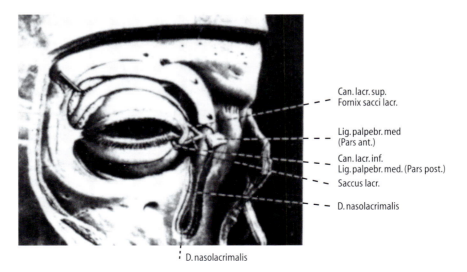

Fig. 35.1. In external Toti's surgery [25], the medial palpebral ligament and the muscle apparatus is incised [23]

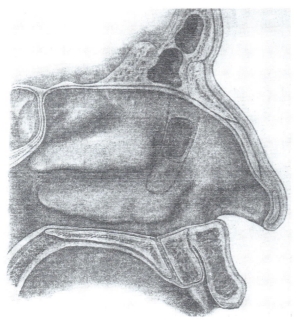

Fig. 35.2. In the original report of West [28], it is evident that the ophthalmic surgeon Randolph (of the John Hopkins Hospital, Baltimore) developed this technique, and West (an otorhinolaryngologist) was the first to perform it

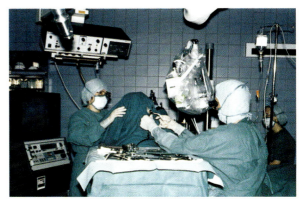

Fig. 35.4. Semi-Fowler's position of the patient. The surgeon can relax his elbows and focus the balanced movable operating microscope by head motions. The anesthetist adjusts hypotensive anesthesia with the aid of a video monitor

several advantages of using this position: (1) the surgeon can relax his elbows and can sit in a convenient position; (2) the view into the anatomy is the same in the office as in the operating room; (3) the operating microscope, in a balanced movable position, can be focused by head motions of the surgeon (depth of focus is immensely improved); (4) less bleeding occurs; and (5) fluids drain from the skull base, allowing better observation of dangerous spots [31].

Hypotensive Anesthesia of Systolic Pressure

Medication (midazolam or flunitrazepam) is given orally 1 h before surgery. In 80% of cases, patients do not remember being transported into the operating room.

To maintain a systolic blood pressure of 55–90 mmHg, intravenous propofol is given prior to intubation. Enflurane gas insufflation, fentanyl or alfentanyl hydrochloride, and nitroglycerine were also administered in approximately 25% of elderly patients, and α (clonidine) and β (esmolol hydrochloride) blockers were administered in approximately 20% of younger patients [31]. Remifentanil perfusors are also in use.

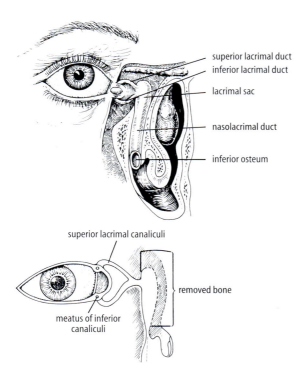

Fig. 35.3. We remove the medial bony covering of the lacrimal sac and the frontal process over the fornix [9, 12]

Operating Microscope or Endoscope

In 1958, Hans Heermann [6] introduced the operating microscope for removal of the uncinate process and ethmoid cells and for surgery of the lacrimal sac. If exposure is not sufficient for a binocular view, we use monocular endoscopes [2, 3, 24].

Endonasal Lacrimal-Sac Surgery

In infants, we never perform surgery of the posterior upper septum [18, 19] or remove the posterior ends of the lower turbinates. In elderly patients, however, this is indicated in patients with severe anatomic deviations [4, 10]. The anterior ethmoid cells are removed. This is necessary to see the distance from the fornix of the lacrimal sac to the skull base (Fig. 35.5) and to prevent postoperative adhesions. A beginner may then use a bayonet forceps (Fig. 35.6), placing one section externally (in the medial angle of the eye) and the other section in the nasal cavity to demonstrate [15–17] the lacrimal fossa and the bony covering of the lacrimal sac. Endonasal surgery is also indicated in acute abscesses (Fig. 35.7), but Toti's surgery should be performed later, after long antibiotic treatment. We use a hollow chisel with a 90° handle (Fig. 35.6). Starting in the lower part of the sac [13, 14], we chisel into the lacrimal fossa approximately three times, until reaching the frontal process over the fornix. By twisting the chisel's handle, we remove the bony covering (Fig. 35.8B, C) of the sac in one piece without hurting the sac itself. If the bone is still connected, it is relatively safe to use a larger punch forceps (Fig. 35.6). All the bone over the lacrimal sac and the fornix is removed (Fig. 35.8C) to expose the complete medial half of the sac and to prevent plugging of the common duct by later bone growth. This bone removal is also possible with a burr. We use Bowman probes (No. 0 and No. 1), passing them gently through the inferior lacrimal duct into the sac (Fig. 35.8D). In some cases, the use of a dilator is necessary. The probe pushes the mobile medial half of the sac into the nasal cavity. The lateral half of the sac firmly adheres to the bone. We incise the medial half of the sac with a neurosurgical blade (Figs. 35.6, 35.8D) from below until reaching the fornix, preserving the ostium of the common duct. The sac is then opened like a book (Fig. 35.8E) and, using a fine forceps (Fig. 35.6), we cut the posterior half to achieve complete removal of the medial half of the sac. Stitching the sac to the mucosa is not necessary [16], because the lateral half of the sac adheres to the untouched lateral bone and is not mobilized (as in Toti's technique).

If possible, we gently dilate the duct using No. 3 or No. 4 Bowman probes (Fig. 35.6). However, even in cases where we can dilate only with No. 0 or No. 1 probes, we never insert silastic tubes postoperatively unless there is congenital aplasia or a traumatic lesion [27, 30, 31] of the lacrimal ducts. Removal of the nasal packing is performed during the next day.

Stones

In the USA, Jones [21] observed dacryoliths in 30% of his cases; we have removed them in only 20% of our cases [20]. We also removed them in the common duct in seven cases (Fig. 35.9).

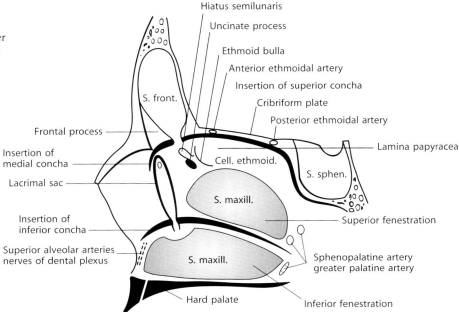

Fig. 35.5. Position of the lacrimal sac relative to upper and lower fenestrations and the ethmoid artery [10]

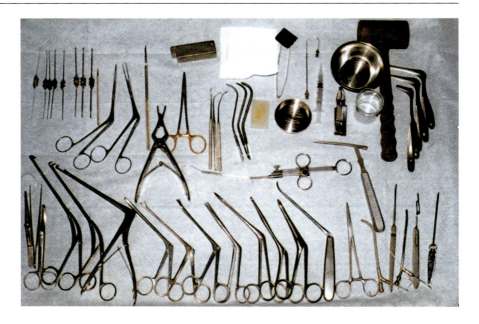

Fig. 35.6. Instruments: top row – Bowman probes No. 00-8, drill, Bruening forceps, neurosurgical knife, Luer power forceps, bent ear currettes, Ritter probes, Bungerter- and bent iffigation needles, hollow chisel with handle, hammer and speculas bottom row – angulated Heermann punchs, Hajek punch, fine double spoon forceps, suction, septal elevator and knife

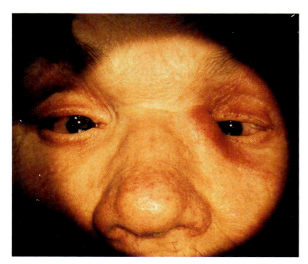

Fig. 35.7. Abscess of a lacrimal duct is an ideal indication for endonasal surgery, but Toti's surgery should be performed later, after long antibiotic treatment [18]

Presaccal Stenosis

For cases with a presaccal stenosis [17–20] in the common duct, we developed a new endonasal technique. A Bowman No. 1 probe is inserted into the inferior lower lacrimal duct, pushing the stenosis into the nose. Using a neurosurgical blade, we cut the complete stenosis in a circle around the common duct. The 360° retraction scars will keep the common duct open during subsequent healing and will pull the duct into the nose (Fig. 35.10).

Prosthesis for Lacrimal Ducts

The insertion of a plastic tube is not necessary in primary surgery unless there is congenital aplasia of the horizontal lacrimal ducts. Even in cases of epiphora lasting more than 20 years, where dilation is only possible with a No. 0 Bowman probe, we never insert a tube. Most cases heal well if the tear pump is working.

In this century, the first glass prosthesis for the tear duct was developed by Josef Heermann in 1925 [8]. Hans Heermann [5] reported a stainless-steel prosthesis in 1930. Jones [21] published a report of another glass prosthesis in 1962. In 1966, Joachim Heermann used a funnel-shaped plastic [9] prosthesis (Fig. 35.11) and, for cases with congenital aplasia, he proposed a funnel on both ends of the duct [18–20] to prevent loosening of the prosthesis while the patient blows his nose.

To reduce movement of the prosthesis by the eyelids, we insert the tube, reinforced by a cut needle (Fig. 35.11), into the lacrimal lake by incision of the horizontal inferior duct with a neurosurgical blade. Transplantation of veins or cartilage with mucosa [26] give unsatisfactory results. Bowman probes up to No. 8 are available. We have designed larger probes (up to No. 14) for easier insertion of the tubes (Fig. 35.6).

Disadvantages of the External Toti Technique

The external Toti technique is no longer used in otorhinolaryngologists but is still occasionally performed by ophthalmologists. Some disadvantages of the technique are:

1. The medial palpebral ligament commonly has to be temporarily dislodged [1].
2. The bony frame for the fixation of the lacrimal ducts and the muscle apparatus is partially removed. Scarring can subsequently disturb the tear pump: probing is possible, but the function is unsatisfactory.
3. Mobilization of the lacrimal sac necessitates sutures [1].
4. A non-cosmetic, visible scar remains.

Advantages of Endonasal Lacrimal-Sac Surgery

1. The horizontal lacrimal ducts (including the bony frame) are not touched, and the muscle apparatus is fixed.
2. In infants, the nose is short and so is the distance to the sac. This makes endonasal surgery easier. The septum is not touched. After removal of the packing (only in one side), further treatment by ophthalmologists is necessary.
3. In adults, a deviation of the septum or trouble with the nasal sinuses may contribute to the tearing problem and can be treated during the same operation.
4. Removal of the anterior ethmoid cells and the frontal process reduces postoperative problems.
5. Postoperative granulations or adhesions in the mouth of the common duct can be easily removed endonasally.
6. In most cases, postoperative control of the ostium of the common duct is possible using a binocular operating microscope without anesthesia.

Evaluation of 659 Cases Followed for Over 5 Years

During a 10-year surgical period, we followed 659 cases for 5 years. The prognosis was good in cases of functioning horizontal lacrimal ducts. All cases in young children healed well (there were few irrigations). Excellent results were achieved in cases with acute abscesses (Fig. 35.7) in which the canaliculus and the tear sac were enlarged by pus. These were ideal cases for endonasal surgery but not for the external Toti procedure.

In six cases (two tuberculosis cases and one case each of malignant lymphoma, basal cell papilloma, basalioma and lipid-vacuole-rich tumor), the inferior ducts were patent, as determined with Bowman's probes, but the tear pump did not work sufficiently and satisfactorily. Postoperative adhesions to the middle turbinate and bony growths in the fornix region are removed endonasally.

We inserted 17 prostheses (Fig. 35.11) in a cohort of 659 operated cases and followed these cases for 5 years (the prostheses were for four congenital aplasias, burn aplasia, stenosis after tumor and 11 cases of severe stenosis). Three cases refused a prosthesis. Only in two patients we could remove the prosthesis successfully after a 6-month period. In all other cases, we had to replace the prosthesis. By chance, we observed one case in which the patient had a stainless-steel prosthesis for 45 years without any problems. The canaliculus seemed to be very large but, after the removal of the prosthesis, the patient developed another stenosis within a few weeks. We had to replace the prosthesis. We have achieved satisfactory results – including re-operations after Toti's surgery – in 94% of cases.

Complications

For residents and novices in endonasal surgery, a thorough training on 50 cadavers is essential, as is frequent observation (through the observer tubes of surgical operating microscopes or on a video monitor). All surgery in the ear or nose (including septum surgery) should start with binocular visualization. When the above suggestions have been followed, we have never seen postoperative persistent disturbance of vision or ocular mobility because of the surgery.

In 1963, outpatient surgery on a purulent lacrimal sac was followed by meningitis, which was resolved with the administration of antibiotics. By using antibiotics routinely, we have not encountered meningitis for more than 30 years.

We observed a case of temporary blindness [12] after injecting local anesthesia into the ethmoid sinus. This condition was reversed with a stellate ganglion block [12].

During the past 37 years, more than 14,000 ethmoidectomies (947 in 1993) have been performed; only five cases of iatrogenic CSF leakage (0.03%) have occurred. All leakages were observed during operation and were treated with temporal fascia (acquired via the endaural incision described by Hans Heermann, 1930) without the need for re-operation.

Bleeding could be controlled intranasally in all cases. We use a fine monopolar needle for cauterization [15–17] of the ethmoid arteries through the

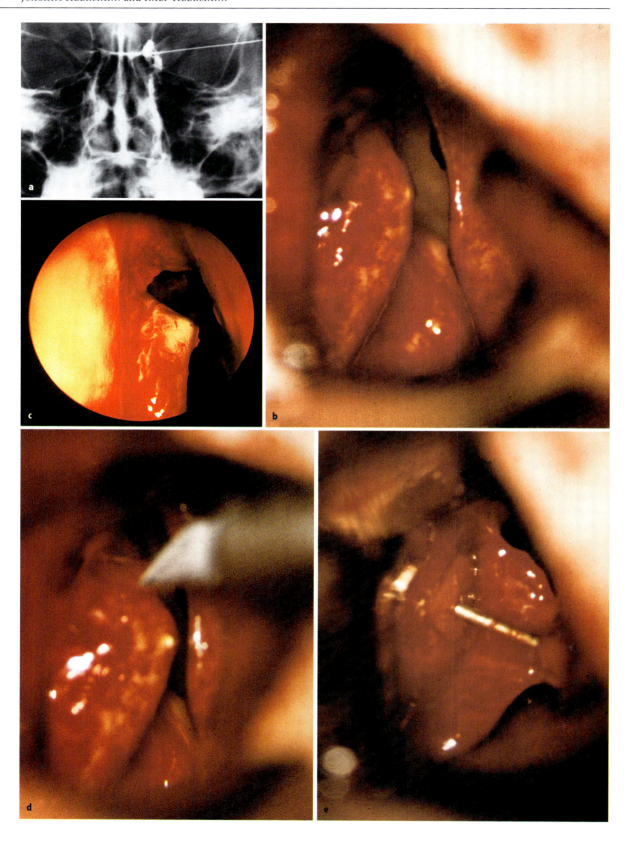

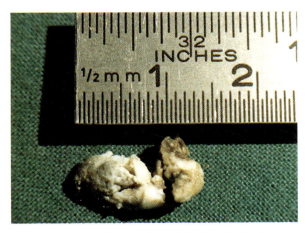

Fig. 35.9. A 1.5-cm-diameter dacryolith of the lacrimal sac

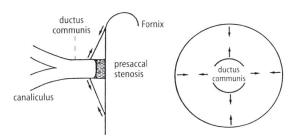

Fig. 35.10. A presaccal stenosis is resected circularly. Scar retractions of 360° pull the common duct open and into the nasal cavity [18]

Fig. 35.11. *Top*: a plastic, funnel-shaped prosthesis (reinforced with a cut needle) placed in the lacrimal lake during the treatment of severe stenosis of the common duct [9]. *Bottom*: funnels on both ends of the lacrimal duct during the treatment of congenital aplasia or complete stenosis [18]

◀──

Fig. 35.8. **a** Left dacryocystography with stone. **b** The medial bony covering of the lacrimal sac is removed. **c** The fornix, after removal of the frontal process. **d** A Bowman probe in the lacrimal sac, and a neurosurgical blade. **e** The sac is opened like a book; the Bowman probe is visible in the common duct bony canal in cases of upper bleeding. Blood transfusions were not necessary, due to hypotensive anesthesia.

Summary

The external Toti operation is performed by penetrating the suspension apparatus of the lacrimal ducts. Functional damage to the tear pump is possible. However, using the endonasal technique developed by Passow in 1901 and Randolph and West in 1909, only the medial half of the lacrimal sac is removed. Since 1912, this technique (performed on up to 98 patients annually) has been performed by four generations of the Heermann family at the Krupp Hospital in Essen. Most of the patients are referred by ophthalmologists (97%). In 1958, Hans Heermann introduced the binocular Zeiss operating microscope for removal of the uncinate process and the ethmoid cells and for lacrimal-sac surgery. In 1925, Josef Heermann introduced a glass prosthesis. In 1930, Hans Heermann used a stainless-steel prosthesis, and Joachim Heermann used a funnel-shaped plastic prosthesis in 1966, also designed with two funnels for the treatment of congenital aplasia of the lacrimal horizontal ducts. We have developed a new technique for resection of a presaccal stenosis. Endonasally, the stenosis is circularly resected so that scar contractions pull the common duct into the nose, achieving an open canaliculus. In a 10-year period, we were able to evaluate 659 cases 5 years after surgery. Including re-operations after Toti's surgery, satisfactory results were obtained in 94% of cases. Only 17 cases received a prosthesis. Three patients refused a prosthesis. In 38 cases, intranasal resection of a presaccal stenosis was successful. In more than 14,000 ethmoidectomies, we only observed five cases of iatrogenic CSF leakage (0.03%), which were treated using temporal fascia from an endaural incision at the time of the ethmoidectomy. There was no persistent visual or ocular disturbance due to surgery. In 1963, one patient developed meningitis in an outpatient procedure; the meningitis healed well on antibiotic administration. After local anesthesia in the ethmoid sinus, temporal blindness occurred and was reversed with a stellate ganglion block.

References

1. Busse H, Hollwich F (1978) Erkrankungen der ableitenden Tränenwege und ihre Behandlung. Enke, Stuttgart, pp 1–178

2. Draf W (1978) Die Endoskopie der Nasennebenhöhlen. Springer, Berlin Heidelberg New York
3. Draf W (1992) Operationen an den Tränenwegen. In: Kirschner M (ed) Allgem und Spez Operationslehre, vol 5. Springer, Berlin Heidelberg New York, pp 269–402
4. Halle M (1912) Modifikation der Westschen Operation. Verh Laryngol Gesamte Berl 26:1
5. Heermann H (1930) Seven years experience with the glass prosthesis in the inferior lacrimal duct. Congress II. Z HNO 27:570–576
6. Heermann H (1958) Endonasal surgery with the use of the binocular operating microscope. Arch Klin Exp Ohren Nasen Kehlkopfheilkd 171:295–297
7. Heermann J (1922) Endonasal surgery of the lacrimal ducts. Z Laryngol Rhinol 11:67–69
8. Heermann J (1925) Permanent connection (glass prosthesis) of the lacrimal ducts and the nose after failure of Toti and West operation or tuberculosis. Klin Monatsbl Augenheilkd 74:192–195
9. Heermann J (1966) Funnel plastic prosthesis after lacrimal sac surgery and severe stenosis of the horizontal common lacrimal duct. Laryngorhinootologie 45:842–947
10. Heermann J (1974) Endonasal microsurgery of the maxillary sinus. Laryngo-Rhinol-Otol. 53:938–942
11. Heermann J (1980) Endonasal microsurgical dacryo-cysto-rhinostomy. Deutsche plastische und Wiederherstellungschirurgie Kongress Murnau, 1977. Springer, Berlin Heidelberg New York
12. Heermann J (1980) Temporary amaurosis in endonasal microsurgery of the ethmoid and lacrimal sac in local anesthesia. Laryngo-Rhinol-Otol. 59:433–437
13. Heermann J (1982) Endonasal microsurgery of the ethmoid in a half-sitting position of the patient under hypotensive anesthesia. HNO 30:180–185
14. Heermann J (1985) Intranasal microsurgery of all paranasal sinuses and the lacrimal sac on a patient in a half-sitting position under hypotensive anesthesia – 25 years experience. In: Mayer EH, Zrunek M (eds) Aktuelles in der Otorhinolaryngologie (1984 Congress). Thieme, Stuttgart, pp 65–69
15. Heermann J (1986) Intranasal microsurgery of all paranasal sinuses, the septum, and the lacrimal sac with hypotensive anesthesia. 25 years' experience. Ann Otol Rhinol Laryngol 95:631–638
16. Heermann J (1986) Intranasal microsurgery in cases with epitaxis of the cribriform plate and further surgery under hypotensive anesthesia. HNO 34:205–215
17. Heermann J (1989) Intranasal surgery of the septum, paranasal sinuses (epistaxis, polyps, mucocel) and lacrimal sac under hypotensive anesthesia. World Congress, Madrid
18. Heermann J (1991) Rhinosurgical aspects of stenosis in the lacrimal passages. Otolaryngol Nova 1:227–323
19. Heermann J (1991) 33 years of intranasal microsurgery of the septum, paranasal sinuses and the lacrimal sac with resection of a presaccal stenosis. West German HNO Congress, Minden, 1991. Zentralbl HNO 140:682
20. Heermann J (1993) Intranasal microsurgery of the lacrimal ducts in 659 cases and resection of presaccal stenosis. Nova Acta Leopoldina 284:155–168
21. Jones LT (1962) The cure of epiphora due to canalicular disorders, trauma and surgical failures on the lacrimal passages. Trans Am Acad Ophthalmol Otolaryngol 66:506–524
22. Passow A (1901) Zur chirurgischen Behandlung des Tränenkanals. Munch Med Wochenschr 48:1403–1404
23. Pernkopf E (1963) Atlas der topographischen und angewandten Anatomie des Menschen, vol 1. Urban and Schwarzenberg, Munich
24. Ptok A, Draf W (1987) Operative Behandlung der Tränenwege. HNO 35:188–194
25. Toti A (1904) La restpocta dei fatti agli appunti mossi dal sott. Strassa al mio metodo conservatore di cura radicale delle dacriocistiti croniche (dacriocuistorhinostomia). Clin Moderno Ital 10:33–34
26. Walter C (1980) Reconstruction of the tear duct system. Arch Otolaryngol Head Neck Surg 106:118–119
27. Weber R, Draf W (1993) Zur Rekonstruktion der Tränenwege. Laryngo-Rhinol-Otol. 72:445–449
28. West JM (1909) A window resection of the nasal duct in cases of stenosis. Trans Am Ophthalmol Soc 12:654–658
29. West JM (1926) The intranasal lacrimal sac operation. Its advantages and its results. Arch Ophthalmol 55:351
30. Wielgosz R (1993) Intranasal microsurgery of the lacrimal sac. Proceedings of the 15th Ear, Nose and Throat World Congress, Istanbul. World Congress, Istanbul, pp 540–543
31. Wielgosz R (1995) The Heermann modification of intranasal microsurgery in lacrimal duct stenoses. Laryngo-Rhinol-Otol. 74:112–117

Endoscopic Transnasal Orbital Decompression

Peter H. Hwang and David W. Kennedy

Introduction

Dysthyroid orbitopathy, also known as Graves' ophthalmopathy, is an autoimmune disorder characterized by the deposition of anti-thyroglobulin immune complexes in the extraocular muscles [1, 18, 19, 28]. The immune complexes induce an inflammatory response that results in fibrosis, edema and hypertrophy of the extraocular muscles and orbital fat. Furthermore, stimulation of myogenous fibroblast activity results in orbital volume, manifested primarily as proptosis. Given the relatively small volume of the orbit (26 cm^3), an increase of only 4 cm^3 of orbital volume (16%) will result in 6 mm of proptosis [10].

Dysthyroid orbitopathy presents with hyperthyroidism in approximately 90% of patients, though euthyroidism or even hypothyroidism may be observed. In most cases, the orbitopathy is self limited and is not associated with visual impairment [11, 31, 34]. However, progressive cases of proptosis can potentially jeopardize vision in three ways [3, 4, 31, 34]. First, exposure keratopathy can result from an inability to completely close the eyelids. This may be due to a proptotic globe or to fibrosis of the levator palpebrae superioris muscle. Second, diplopia may result from impaired extraocular motility owing to hypertrophy and fibrosis of extraocular muscles. Third, an ischemic optic neuropathy may develop from chronically elevated intraorbital pressure. The hypertrophied extraocular muscles may contribute to optic neuropathy either by direct compression or by general elevation of intraorbital pressure. Optic neuropathy may manifest as diminished visual acuity, impaired color vision or scotomata.

Non-Surgical Treatment

Non-surgical treatments of dysthyroid orbitopathy are varied; none has proven to have longstanding therapeutic efficacy. Systemic steroids may provide significant relief of orbital signs and symptoms and may improve visual acuity. However, high doses of steroids, often between 40 mg and 100 mg of prednisone per day for many months, may be required. Even after aggressive doses, there is a strong possibility of recurrence or progression of symptoms once the steroid dose is tapered [7, 31, 34]. In addition, chronic steroid use as a primary treatment modality is impractical due to the multiple undesirable side effects of long-term steroid use, including accelerated osteoporosis, cataract formation and adrenal suppression.

External-beam radiotherapy is efficacious in the treatment of optic neuropathy symptoms, but proptosis and diplopia show much less favorable responses [2, 14, 21, 30]. In addition, although relatively low doses of radiation are administered (20 Gy to both orbits, divided over ten treatments), the patient is subjected to finite risks of radiation-induced malignancy and radiation injury to the retina or lens. Immunosuppressive therapy, alone or in combination with plasmapheresis, remains experimental and may be associated with potentially serious side effects [5, 9, 32, 35].

Surgical Treatment

The mainstay of therapy for dysthyroid orbitopathy is surgical decompression of the orbit [20, 23]. In 1911, Dollinger described a lateral-orbitotomy approach for treatment of exophthalmos; this treatment was modeled after that of Kronlein [8]. However, the degree of decompression attainable through this approach is minimal.

In 1931, Naffziger advocated a superior decompression of the orbit into the anterior cranial fossa [27]. However, this approach required a craniotomy and subjected the patient to risks of meningitis and cerebrospinal-fluid leak. Furthermore, a common postoperative complaint was the sensation of cerebral pulsations via the orbit.

Sewall continued the development of orbital decompressive techniques by introducing medial orbital-wall decompression via external ethmoidectomy [29]. Hirsch advocated inferior decompression by removal of the orbital floor through a Caldwell-Luc approach [12]. In 1957, Walsh and Ogura modified Hirsch's technique to include medial-wall decompression in conjunction with orbital-floor decompression [33]. Since then, the Walsh-Ogura technique has been the standard external approach to orbital decompression.

Endoscopic techniques have been successfully applied to orbital decompression surgery, achieving excellent surgical outcomes while avoiding the morbidity of a Caldwell-Luc antrotomy or external ethmoidectomy. Kennedy et al. first described an endoscopic transnasal technique for orbital decompression in 1990 [17] and, since then, the endoscopic technique has gained broad acceptance as a viable alternative to external surgical decompression procedures.

Preoperative Evaluation

Patients with dysthyroid orbitopathy are best managed by an interdisciplinary team including an endocrinologist, an ophthalmologist and an otolaryngologist. A complete endocrinologic assessment is required, particularly with respect to identification and treatment of thyroid dysfunction, including thyrotoxicosis. Normalization of thyroid function is important in two respects. First, control of thyroid function is critical in minimizing perioperative morbidity from the physical stress of surgery and general anesthesia. In addition, the hyperthyroid patient may have lid lag unrelated to the Graves' associated ophthalmopathy. If present, hyperthyroid lid lag can enhance scleral show and make the assessment of proptosis more difficult.

Next, a complete ophthalmologic exam is performed by the ophthalmologist. This examination should include a complete assessment of visual acuity, color vision and visual field. Slit-lamp examination, tonometry and motility exams should also be performed. Proptosis is measured using a Hertel exophthalmometer. The involvement of the ophthalmologist should continue throughout the postoperative course to measure clinical responses to surgical therapy and to assess the need for postoperative adjunctive therapies, such as eye-muscle correction.

The otolaryngologist's evaluation should include a complete head and neck exam. Nasal endoscopy is performed with particular attention to the anatomy of the middle meatus. During the preoperative counseling session, the patient is informed of all potential surgical risks, including sinusitis, visual loss and cerebrospinal-fluid leak. Even when diplopia is not present preoperatively, the orbital muscles are always involved in the disease process, and postoperative double vision is very common. Therefore, all patients should be counseled in this regard and advised that there is a good chance that they will need a second operation to correct the double vision 6–8 weeks following the decompression.

Preoperative radiologic evaluation consists of computed-tomography scans of the paranasal sinuses; these scans should be obtained in 3.0-mm sections in both coronal and axial planes. Images in the coronal plane allow for assessment of the ethmoidal anatomy in relation to the skull base and medial orbital wall. In addition, the thickness of the orbital floor can be evaluated. An unusually thick orbital floor or the presence of acute sinusitis are contraindications to an endoscopic approach. Axial images are useful in delineating the anatomy of the sphenoid and posterior ethmoid cells, particularly with respect to the internal carotid artery and optic nerve.

Surgical Technique

To elevate the head, the patient lies supine on the operating-room table in a slightly reversed Trendelenburg position. While we generally favor local anesthesia with sedation for most functional endoscopic sinus surgeries, orbital decompression is best performed under general anesthesia, due to the sensitivity of the periorbita and orbital contents. As with other endoscopic procedures, surgery is performed seated, with the surgeon sitting to the right of the patient and the surgeon's legs positioned beneath the operating-room table.

Preoperatively, the patient's nose is sprayed with oxymetazoline to initiate mucosal decongestion and to reduce subsequent cocaine absorption. Further intraoperative vasoconstriction is achieved by applying cocaine granules topically to the nasal mucosal surfaces with a cotton-tipped nasal applicator. In addition, 1% lidocaine with 1:100,000 epinephrine is infiltrated submucosally and through a greater-palatine-foramen block.

The surgical approach to orbital decompression begins with a complete intranasal endoscopic sphenoethmoidectomy in the method of Kennedy [15]. Particular care must be taken to thoroughly skeletonize the medial orbital wall without violating it, since premature exposure of orbital fat can compromise the surgical dissection. The skull base is likewise thoroughly dissected and skeletonized. A large sphenoidotomy is performed to provide visualization of the orbital-apex region. The position of the optic canal along the supero-lateral wall of the sphenoid provides a useful surgical landmark. Furthermore, a wide sphenoidotomy facilitates removal of the optic tubercle during decompression of the posterior orbit.

Frontal-recess dissection is not critical to orbital decompression. However, pre-existing frontal-sinus disease should probably be addressed surgically if present; postoperatively, there is a heightened risk of obstruction of the frontal recess from prolapsed orbital fat.

A maximally sized middle-meatal antrostomy is required in order to gain adequate surgical exposure to remove and decompress the orbital floor [16]. Anteriorly, the antrostomy is extended to the margin of the nasolacrimal duct. Superiorly, the antrostomy is extended to the level of the orbital floor. The posterior limit of the antrostomy is the posterior wall of the maxillary sinus. Inferiorly, the antrostomy is extended below the root of the inferior turbinate, requiring partial excision of the root of this structure.

Next, the medial orbital wall is removed (Fig. 36.1). Gentle fracturing of the lamina papyracea can be performed with a J-shaped curette followed by piecemeal removal of the bony fragments. The periorbita should not be violated during the bony dissection, because premature herniation of orbital fat can markedly impair surgical exposure and dissection of the posterior orbit. In addition, it is helpful to preserve the most superior aspect of the bony medial orbital wall adjacent to the frontal recess in order to prevent herniated fat from obstructing the frontal recess, which might predispose the patient to postoperative frontal sinusitis.

After complete removal of the medial orbital wall, attention is directed to decompression of the orbital apex. We advocate surgical treatment of the orbital-apex region in order to decompress the optic nerve where the hypertrophied extraocular muscles are most closely apposed to the nerve. In cases where vision is severely compromised due to the disease process, the decompression may be extended into the first part of the optic canal. Decompression of this area requires drilling of the thicker bone in the region just proximal to the annulus of Zinn, using a diamond drill and constant irrigation. The annulus of Zinn represents a fibrous condensation of periorbita in the orbital-apex region and is easily recognized by its dense circular fibers. Infero-medially, there is a firm buttress of bone located just superior to the pterygoid plates. Ophthalmologists differ in their opinions regarding whether this bone should

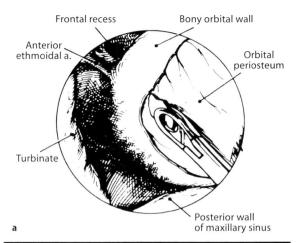

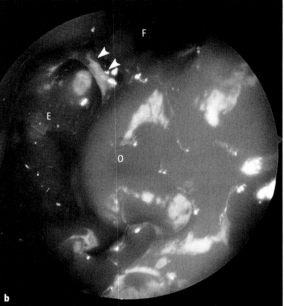

Fig. 36.1 a,b. Removal of the left medial orbital wall. **a** Artist's rendition of the use of Blakesley forceps to remove medial orbital-wall bone. **b** Endoscopic photograph (taken using a 30° endoscope directed laterally), depicting the orbital periosteum (*O*) following removal of the medial orbital wall. Note the anterior ethmoid artery (*arrows*) along the skull base, between the ethmoid cavity (*E*) and the frontal sinus (*F*) [17]

be removed. It adds to the level of decompression that can be achieved but may significantly increase the risk of postoperative diplopia. In general, we remove this bone if we want to achieve maximal decompression.

After complete decompression of the medial orbital wall and orbital apex, attention is directed to dissection and decompression of the orbital floor (Fig. 36.2). The lateral limit of dissection is the infraorbital nerve, which is visible along the roof of the

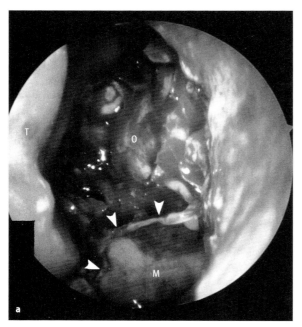

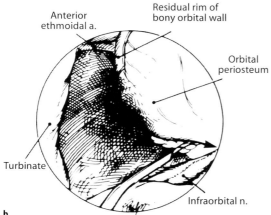

Fig. 36.2 a,b. Removal of the left orbital floor and incision of the orbital periosteum. **a** Endoscopic photograph depicting the creation of a widely enlarged antrostomy (*arrows*) of the maxillary sinus (*M*), providing surgical exposure of the orbital floor. *O* orbital periosteum, *T* middle turbinate. **b** Artist's rendition of incision of the inferior orbital periosteum using an angled arachnoid knife [17]

maxillary sinus with an angled endoscope. The bone of the orbital floor is fractured in a medial-to-lateral direction with a curved frontal-recess curette. Bony fragments may then be removed with fine 70° or 110° giraffe forceps, using the 30° or 70° endoscope for visualization. More anteriorly, the bone may be removed with a backbiting forceps, taking care not to traumatize the nasolacrimal duct.

After meticulous resection of the bony medial-orbital wall, orbital floor and orbital apex, the periorbita is incised. The periorbita of the orbital floor is incised first, using a disposable arachnoid knife that has been bent into a nearly 90° angle. Three or four parallel incisions are made, each running in a posterior-to-anterior direction. The first incision is best made at the most lateral aspect of the dissection, adjacent to the infraorbital nerve, progressing more medially with each successive incision. This prevents herniated fat from obscuring the field of dissection.

Next, the medial orbit is decompressed via periorbital incisions made with a straight Beaver arachnoid knife or a no.-12 knife blade. Again, to avoid obscured visualization due to prolapsed fat, the first incision should be made superiorly; next, two or three additional, parallel incisions should be made inferiorly. Repeat passes with the knife may be necessary for each periorbital incision in order to completely lyse the fibrous bands associated with the periorbita. This allows for extensive herniation of orbital fat into the sinus cavities. However, care must be taken to avoid injury to the medial rectus muscle during the dissection.

The extent of decompression can be assessed immediately by gently palpating the orbit and observing the herniated orbital contents endoscopically. Additional herniation of the orbital contents may be achieved by dividing any residual fibrous bands and by using slightly firmer palpation of the orbit to coax additional fat to prolapse. However, orbital fat should not be pulled directly through the periorbital incisions, as this may cause orbital bleeding or undue trauma to the extraocular musculature.

At the conclusion of the procedure, Gelfilm coated with mupirocin ointment is placed between the prolapsed orbital and the middle turbinate in order to reduce the tendency for adhesions to form. Some surgeons recommend the routine resection of the middle turbinate in orbital decompression, but this is not our recommendation. The patient is placed on a postoperative course of a broad-spectrum oral antibiotic, usually a cephalosporin that protects against β-lactamase-producing organisms. The maxillary sinus is suctioned under endoscopic visualization in the early postoperative period, and the Gelfilm is removed 1–2 weeks following surgery.

Results

Endoscopic approaches to orbital decompression can achieve excellent outcomes that compare favorably with those of traditional methods of orbital decompression (Figs. 36.3, 36.4). Using the endoscopic transnasal approach, Kennedy et al. reported

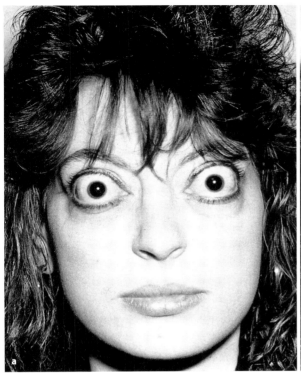

Fig. 36.3 a,b. Preoperative (a) and postoperative (b) photographs of a patient with dysthyroid orbitopathy treated with endoscopic orbital decompression

an average decompression of 4.7 mm, as measured using Hertel exophthalmometry [17]. This value compared favorably to an average of 4.5 mm of decompression for a comparison group of patients undergoing decompression via a Walsh-Ogura approach. When performed in conjunction with lateral orbitotomy, the procedure achieved an average of 5.7 mm of proptosis regression. Metson et al. reported an average decompression of 3.2 mm by an endoscopic approach [24].

There are several keys to performing satisfactory endoscopic decompression. First, the endoscopic sphenoethmoidectomy must be performed with minimal trauma and bleeding; otherwise, visualization will be compromised during the decompression phase of the procedure. Second, it is essential to create an extremely large antrostomy that extends into the root of the inferior turbinate. Although it might be possible to remove the orbital floor through a more limited access, the orbital fat continues to protrude into this area postoperatively; unless a mega-antrostomy is performed, the patient will develop postoperative maxillary-sinus obstruction. Finally, significant care is needed in the area of the frontal recess if postoperative frontal sinusitis is to be avoided. The bone of the frontal recess should be left intact during the dissection, and care should be taken to avoid incision of the orbital periosteum, even in the area adjacent to the frontal recess. Again, the fact that the fat continues to herniate postoperatively makes it important to perform less interoperative surgery in this region than might appear to be required.

The endoscopic transnasal approach to decompression provides distinct advantages. First, one can achieve regression of proptosis comparable to that possible with traditional approaches; nevertheless, the transnasal approach avoids the morbidity of an external ethmoidectomy or Caldwell-Luc approach [6, 7, 26, 34]. In addition, the endoscopic technique allows superior visualization of the posterior ethmoid sinus and the orbital-apex region. Improved surgical exposure of these areas allows more complete (and, therefore, more efficacious) decompression.

The primary limitation of the endoscopic technique is the decompression of the infero-lateral orbital floor. The relative inaccessibility of this area via current endoscopic instrumentation generally limits the lateral dissection of the orbital floor to the level of the infraorbital nerve. More thorough decompression of the lateral orbit may be achieved by supplementing the endoscopic transnasal approach with additional decompression via a later-

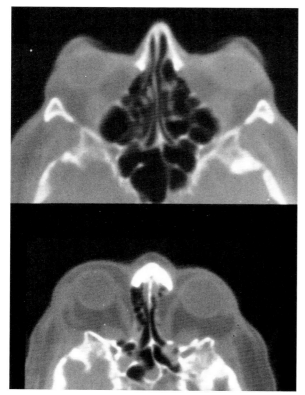

Axial computed-tomography scans depicting preoperative (*top*) and postoperative (*bottom*) views of a patient treated with endoscopic orbital decompression

al orbitotomy [22, 25]. The extent of orbital-apex decompression that can be achieved through this approach is better than the decompression possible via any of the alternative methods. Accordingly, our ability to improve both visual acuity and color vision in patients with visual compromise is excellent. No visual loss has occurred in any patient undergoing surgery.

The postoperative sequelae of endoscopic, transnasal orbital decompression are similar to those of standard approaches, but our studies indicate that, although the degree of decompression is slightly greater than with other approaches, the incidence of postoperative diplopia may be greater. During the immediate postoperative period, diplopia is best managed with prismatic lenses. If diplopia persists after several months of observation, eye-muscle corrective surgery may be warranted. A less common sequela of orbital decompression, regardless of approach, is the development of sinusitis or mucoceles secondary to obstruction by herniated orbital fat. Mucoceles arising after decompression surgery may often be managed endoscopically, as described by Hoffer and Kennedy [13].

In summary, the endoscopic approach to orbital decompression offers an effective alternative to standard external approaches. When executed carefully, endoscopic techniques can yield excellent decompression results with a low rate of morbidity.

References

1. Barbosa J, Wong E, Doe RP (1972) Ophthalmopathy of Graves' disease. Arch Intern Med 130:111–113
2. Brennan MW, Leone CR Jr, Janaki L (1983) Radiation therapy for Graves' disease. Am J Ophthalmol 96:195–199
3. Calcaterra TC (1993) Management of exophthalmos. In: Cummings CW, Fredrickson JM, Harker LA, Krause CJ, Schuller DE (eds) Otolaryngology, Head and Neck Surgery. Mosby, St. Louis, pp 2472–2479
4. Calcaterra TC, Thompson JW (1980) Antral-ethmoidal decompression of the orbit in Graves' disease: ten-year experience. Laryngoscope 90:1941–1949
5. Dandona P, Marshall NJ, Bidey SP, Nathan A, Havard CWH (1979) Successful treatment of exophthalmos and pretibial myxoedema with plasmapheresis. BMJ 1:374–376
6. DeFreitas J, Lucente FE (1988) The Caldwell-Luc procedure: institutional review of 670 cases: 1975–1985. Laryngoscope 98:1297–1300
7. Desanto LW (1980) The total rehabilitation of Graves' ophthalmopathy. Laryngoscope 90:1652–1678
8. Dollinger J (1911) Die drickentlastung der Augenhokle durch entfurnung der aussern Orbitalwand bei hochgradigen Exophthalmos und Koneskutwer Hornhauterkronkung. Dtsch Med Wochenschr 37:1888–1890
9. Glinoer D, Etienne-Decerf J. Schrooyen M, et al. (1986) Beneficial effects of intensive plasma exchange followed by immunosuppressive therapy in severe Graves' ophthalmopathy. Acta Endocrinol 111:30–38
10. Gorman CA (1978) The presentation and management of endocrine ophthalmopathy. Clin Endocrinol Metab 7:67–96
11. Hedges TR Jr, Rose E (1953) Hyperophthalmopathic Graves' disease. Arch Ophthalmol 50:479–490
12. Hirsch O (1950) Surgical decompression of exophthalmos. Arch Otolaryngol Head Neck Surg 51:325–331
13. Hoffer ME, Kennedy DW (1994) The endoscopic management of sinus mucoceles following orbital decompression. Am J Rhinol 8:61–65
14. Hurbli T, Char DH, Harris J, Weaver K, Greenspan F, Sheline G (1985) Radiation therapy for thyroid eye diseases. Am J Ophthalmol 99:633–637
15. Kennedy DW (1985) Functional endoscopic sinus surgery technique. Arch Otolaryngol Head Neck Surg 111: 643–649
16. Kennedy DW, Zinreich SJ, Kuhn F, Shaalan H, Naclerio R, Loch E (1987) Endoscopic middle meatal antrostomy: theory, technique, and patency. Laryngoscope 97[suppl 43]:1–9
17. Kennedy DW, Goodstein ML, Miller NR, Zinreich SJ (1990) Endoscopic transnasal orbital decompression. Arch Otolaryngol Head Neck Surg 116:275–282
18. Kodama K, Sikorska H, Bandy-Dafoe P, Bayly R, Wall JR (1982) Demonstration of a circulating autoantibody against soluble eye-muscle antigen in Graves' ophthalmopathy. Lancet 2:1353–1356

19. Konishi J, Herman MM, Kriss JP (1974) Binding of thyroglobulin and thyroglobulin–antithyroglobulin immune complex to extraocular muscle membrane. Endocrinology 95:434–446
20. Lee AG, McKenzie BA, Miller NR, Loury MG, Kennedy DW (1995) Long-term results of orbital decompression in thyroid eye disease. Orbit 14:59–70
21. Leone CR Jr (1984) The management of ophthalmic Graves' disease. Ophthalmology 91:770–779
22. Leone CR Jr, Piest KL, Newman RJ (1989) Medial and lateral wall decompression for thyroid ophthalmopathy. Am J Ophthalmol 108:160–166
23. McCord CD Jr (1985) Current trends in orbital decompression. Ophthalmology 92:21–33
24. Metson R, Shore JW, Glicklich RE, Dallow RL (1995) Endoscopic orbital decompression under local anesthesia. Otolaryngol Head Neck Surg 113:661–667
25. Miller NR, Iliff WJ (1984) Surgery of the orbit. In: Rice TA, Michels RG, Stark WJ (eds) Ophthalmic surgery, 4th edn. Mosby, St. Louis, pp 383–398
26. Murray JP (1983) Complications after treatment of chronic maxillary sinus disease with Caldwell-Luc procedure. Laryngoscope 93:282–284
27. Naffziger HC (1931) Progressive exophthalmos following thyroidectomy: its pathology and treatment. Ann Surg 94:582–586
28. Riley FC (1972) Orbital pathology in Graves' disease. Mayo Clin Proc 47:975–979
29. Sewall EC (1936) Operative control of progressive exophthalmos. Arch Otolaryngol Head Neck Surg 24:621–624
30. Threlkeld A, Miller NR, Wharam M (1999) The efficacy of supervoltage radiation therapy in the treatment of dysthyroid optic neuropathy. Orbit 18:1
31. Trobe JD, Glaser JS, Laflamme P (1978) Dysthyroid optic neuropathy. Arch Ophthalmol 96:1199–1209
32. Wall JR, Strakosch CR, Fang SL, Ingbar SH, Braverman LE (1979) Thyroid binding antibodies and other immunological abnormalities in patients with Graves' ophthalmopathy: effects of treatment with cyclophosphamide. Clin Endocrinol 10:79–91
33. Walsh TE, Ogura JH (1957) Transantral orbital decompression for malignant exophthalmos. Laryngoscope 67:544–549
34. Warren JD, Spector JG, Burde R (1989) Long-term follow up and recent observations on 305 cases of orbital decompression for dysthyroid orbitopathy. Laryngoscope 99:35–40
35. Weetman AP, Ludgate M, Mills PV, et al. (1983) Cyclosporin improves Graves' ophthalmopathy. Lancet 2:486–489

Endoscopic Optic-Nerve Decompression in Traumatic Optic Neuropathy

Luis Alfonso Parra Duque

Introduction

Nuhn reported the first optic-nerve injury in 1845. By 1892, Callan had published 80 cases of optic-canal fracture. Surgical decompression in the optic nerve via the lateral orbitotomy approach was described by Kronlein in 1911. Optic-nerve decompression via the craniotomy approach was described by Pringle in 1916 [17]. Decompression of the nerve medially through the lateral wall of the ethmoid and sphenoid sinuses was originally described by Sewall in 1926 and was developed by Niho in the 1960s [6]. Transantral ethmoid access to the medial optic canal has been reported by Kennerdell [9]. Takahashi reported success using microscopic endonasal optic-nerve decompression [31].

Endoscopic optic-nerve decompression is probably the approach indicated most often when surgical intervention is needed. Performing the procedure endoscopically reduces morbidity by eliminating the necessity for external incisions. At the same time, this procedure allows optimal visualization of the operative site [8].

Although infrequent, decreased visual acuity is one of the most disastrous complications of craniofacial trauma. Approximately 2% of all closed head trauma affects some portion of the optic nerve, usually at the level of the optic canal (Fig. 37.1) [22].

Optic-nerve decompression is controversial but may be indicated in cases of traumatic and/or inflammatory compression of the optic nerve at the apex of the orbit, its canalicular portion or the portion located in the walls of the posterior ethmoid and sphenoid sinuses. Surgical decompression of the intracanalicular segment of the optic nerve can be accomplished by either intracranial or extracranial approaches.

The endoscopic technique offers several advantages over other surgical approaches. It provides excellent visualization of the surgical site, avoids the morbidity associated with intracranial and external extracranial approaches and offers the best option for recovery of useful vision.

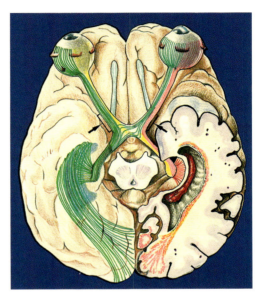

Fig. 37.1. The intracanalicular segment (*arrows*) of the optic nerve

Endoscopic endonasal surgery was originally introduced as a minimally invasive surgical therapy for inflammatory sinus disease, but it is now considered an alternative therapy for a wide variety of surgical problems. The endoscopic approach has been successfully applied in the closure of cerebrospinal-fluid rhinorrhea [14, 32], dacryocystorhinostomy, tumors of the anterior skull base, pituitary surgery, orbital decompression [8] and optic-nerve decompression.

Pathophysiology

Intracanalicular optic-nerve damage after craniofacial trauma results from many causes and is multifactorial. The common mechanisms of optic-nerve injury may be reversible or irreversible. Irreversible mechanisms include direct avulsion of the optic nerve and disruption of its blood supply, which leads to immediate or sudden blindness. Reversible mechanisms include compression from reactive edema or hemorrhage, microvascular spasm and thrombosis. While significant bone displacement within the optic canal produces disruption of the optic nerve and immediate loss of vision, most canal fractures with minimal displacement affect vision through hemorrhage, vascular spasm or edema and produce incomplete, progressive or delayed loss of vision (Fig. 37.2). The pathophysiology of traumatic optic neuropathy has remained unclear since the time of Hippocrates [5]. Trauma to the intracanalicular portion of the optic nerve can produce progressive axonal degeneration, possibly leading to permanent blindness.

There is no consensus about the treatment of traumatic optic neuropathy, but the patients are usually seen by an otolaryngologist, ophthalmologist and/or neurosurgeon, and very large doses of corticosteroids are given. If vision does not respond to medical therapy, surgery is indicated.

Optic-nerve decompression is analogous to facial-nerve decompression after head injury, where hemorrhage, edema or a fracture compresses the nerve in a tight osseous canal. Optic-nerve decompression is controversial but, when performed within a critical period of time, there is evidence to suggest that it may be beneficial and that it can lead to a reversal of edema, neuropraxia and compression of microcirculation, resulting in abatement of visual loss [3].

Surgical Anatomy

The physician who deals with traumatic optic neuropathy must be familiar with the anatomy and physiology of the optic canal and its surrounding structures. The optic nerve and the ophthalmic artery enter the apex of the orbit through the optic foramen. The optic foramen is somewhat eccentrically placed (medial and superior to the apex). The optic canal is 5.5–11.5 mm in length and is bounded by the lesser wing of the sphenoid bone. The proximal end of the optic canal measures 5.0–9.5 mm in diameter, and the distal end is narrower, measuring 4–6 mm in diameter [4]. Within the cranial cavity, the optic nerve is surrounded only by the pia mater. At the entrance to the optic canal, the optic nerve is surrounded by the pia mater, the arachnoid and dura. At the exit of the optic canal, the dura splits into two layers: the outer layer connects with the periorbita, and the inner layer forms the dural covering of the optic nerve. Thus, within the orbit and optic canal, the nerve is surrounded by all three layers. The subarachnoid space within the optic canal is narrow, making the vessels susceptible to shearing or thrombosis. The bone of the optic canal is thickest (0.6 mm) at the optic foramen and is thinnest (0.2 mm) proximal to the lateral sphenoid wall. In 4% of all cases, this bone is dehiscent [2]. The optic-nerve tubercle is a thick bulge of the medial aspect of the bone that surrounds the optic foramen.

Depending on the degree of pneumatization of the sphenoethmoid cells (Onodi cells), the optic tubercle can be seen in the transition area between the posterior ethmoid cells and the sphenoid sinus or in the sphenoid sinus itself [27]. The optic nerve

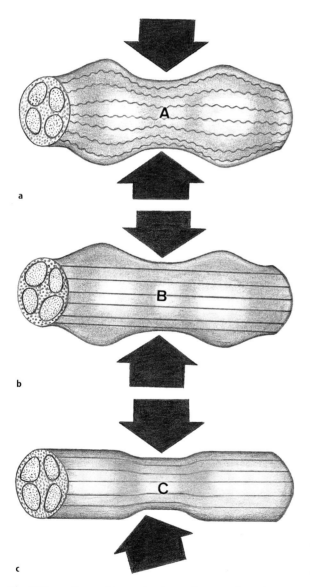

Fig. 37.2 a–c. Reversible mechanisms of traumatic optic neuropathy. **a** Compression from edema. **b** Compression from hemorrhage. **c** Compression from minimal bone displacement

and the carotid artery may be exposed in a sphenoethmoid cell when this cell pneumatizes laterally and superiorly to the sphenoid sinus (Fig. 37.3). It is very important to understand that the sphenoid sinus is medial and inferior to the furthest posterior ethmoid cell, because serious damage to the optic nerve or the carotid artery may result during an attempt to locate the sphenoid sinus just behind this last ethmoid cell.

Three structures indent the lateral wall of the sphenoid sinus: the optic nerve, the carotid artery and the maxillary division of the trigeminal nerve (Fig. 37.4). The optic nerve is usually found behind the superior portion of the lateral wall of the sphenoid sinus; it has the appearance of a mucosal prominence. Nevertheless, in approximately 25% of all cases, this characteristic landmark cannot be seen endoscopically, and the surgeon must rely on the position of the internal carotid artery (ICA) to localize the optic nerve [2]. The carotid artery lies inferolateral to the optic nerve. Extreme care must be exercised to prevent injury to the ICA. The bone adjacent to the vessel is less than 0.5 mm thick [2] and is dehiscent in the sphenoid sinus in 23% of patients [7]. The maxillary nerve indents the inferolateral sinus wall approximately 5–10 mm below the optic canal.

The main blood supply of the orbit is the ophthalmic artery, the first branch of the ICA. The ophthalmic artery is adjacent to the optic-nerve sheath and exits the dura at the anterior end of the canal. The ophthalmic artery enters the optic canal inferior and lateral to the optic nerve and occasionally may be positioned medially along the optic nerve (Fig. 37.5). In order to prevent further trauma to the nerve or the ophthalmic artery, the surgeon must be familiar with these anatomic variances when approaching the nerve. Most branches of the ophthalmic artery originate in the posterior third of the orbit. The central retinal artery splits from the ophthalmic artery and enters the optic nerve in the posterior third. The optic nerve is also nourished by a coaxial array of arterioles from the pia mater.

The anterior and posterior ethmoid arteries split from the ophthalmic artery, traverse the orbit medially and exit through the anterior and posterior ethmoidal foramina of the medial orbital wall. These foramina are in the frontoethmoid suture line at the level of the cribriform plate. The optic nerve traverses at the same level as the roof of the ethmoid sinuses and the anterior and posterior arteries. The average distance from the posterior ethmoidal foramen to the optic foramen is 6 mm; the distance from the posterior to the anterior foramen is 12 mm, and the distance from the anterior foramen to the anterior lacrimal crest is 24 mm (Fig. 37.6) [13, 20]. However,

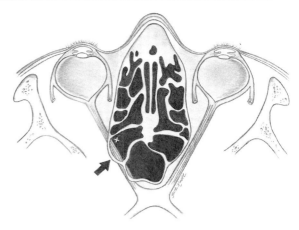

Fig. 37.3. This schematic drawing shows the optic nerve (*arrow*) exposed in a sphenoethmoidal cell (*asterisk*)

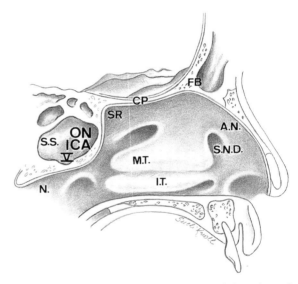

Fig. 37.4. Schematic drawing showing the left lateral nasal wall. *AN*, agger nasi; *CP*, cribriform plate; *FB*, frontal bone; *IT*, inferior tubinate; *MT*, middle turbinate; *N*, nasopharynx; *SND*, site of the nasolacrimal duct; *SR*, sphenoethmoid recess; *SS*, sphenoid sinus. Three structures indent the lateral wall of the sphenoid sinus: the optic nerve (*ON*), the internal carotid artery (*ICA*) and the maxillary division of the trigeminal nerve (*V*)

the distances are highly variable. The anterior ethmoid foramen is absent 16% of the time and, in over 30% of all cases, the orbit has multiple foramina for the posterior ethmoid artery.

The degree of pneumatization of the sphenoid sinus is highly variable; a conchal type of sinus with minimal extension is relatively rare, the presellar type of sinus (which extends posterior to the anterior sellar wall) is found in 40% of adults, and the postsellar type (which extends below the sella and may

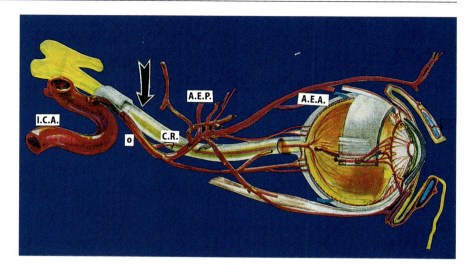

Fig. 37.5. Schematic drawing showing the relationship between the optic nerve (*arrow*) and the ophtalmic artery (*O*). The ophthalmic artery enters the optic canal inferior and lateral to the optic nerve. The central retinal artery (*CR*), posterior ethmoid artery (*AEP*) and anterior ethmoid artery (*AEA*) branch from the ophthalmic artery

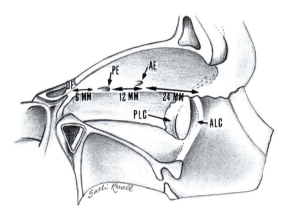

Fig. 37.6. The average distance from the posterior ethmoidal foramen (*PE*) to the optic foramen (*OF*) is 6 mm. The average distance from the posterior (*PE*) to the anterior foramen (*AE*) is 12 mm. The average distance from the anterior ethmoidal foramen (*AE*) to the anterior lacrimal crest (*ALC*) is 24 mm

extend further) occurs in 60% of all adults [24]. Within the cranial cavity, the optic nerve measures approximately 10 mm; within the intracanal portion, the nerve measures 9–10 mm. At the optic foramen, the optic nerve is fixed. When the head is accelerated rapidly backward by a frontal blow, injury to the optic nerve or its blood supply can occur.

The inertia of the globe may tear the optic nerve where it is fixed to the optic foramen. The orbital portion of the optic nerve measures approximately 30 mm until it enters the sclera. The annulus of Zinn is the thick, fibrous part of the periosteum from which the four rectus muscles originate. These muscles and the orbital fat provide substantial protection to the orbital portion of the optic nerve and act as shock absorbers to diminish nerve stretching and avulsion.

Diagnosis

The incidence of ocular and orbital injuries in the trauma population is significant. Ocular and orbital injuries may be the result of either blunt or penetrating trauma or both. The characteristic clinical presentation of traumatic optic neuropathy includes visual loss, a history of blunt trauma to the forehead and absence of reaction to direct light in the ipsilateral pupil. A complete ophthalmologic examination should be performed and properly recorded if possible. This may be very difficult, because the patient will be unconscious in some cases, and visual loss may be rapidly progressive after the blunt trauma. An orderly routine evaluation of the adnexa, ocular motility, pupillary function, ocular tension, anterior segment and fundus should be performed. It is important to exclude other causes of visual loss, such as hyphema, corneal or scleral laceration, retinal detachment, vitreous hemorrhage, orbital hematoma or corneal subluxation. With a cooperative patient, a Snellen chart can document visual acuity. With an uncooperative or unconscious patient, a Marcus-Gunn pupil is the most reliable indicator of total disruption of the affected optic nerve. The pupil of the affected side is equal in size to the contralateral pupil and reacts consensually, but when light is moved from the intact to the affected eye, a "paradoxical" dilation is seen (Fig. 37.7) [10]. The optical disk is usually normal in the early post-traumatic period, but pallor or papilledema may be present. After 1 month, optical-disk atrophy may appear.

Visually evoked-response testing is unnecessary if decompression is planned but is very useful in the unconscious patient lacking consensual pupillary

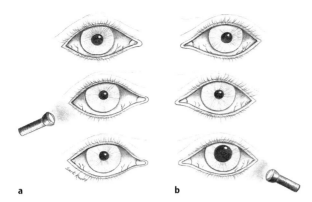

Fig. 37.7 a,b. Schematic drawing showing the Marcus-Gunn pupil phenomenon: the pupil of the affected side is equal in size to the contralateral pupil and reacts consensually but, when light is moved from the intact eye (**a**) to the affected eye (**b**), a "paradoxical" dilation is seen

constriction on light stimulation [22]. All patients with deterioration of vision following head trauma should receive thorough neurologic, ophthalmologic and maxillofacial evaluation. If a subdural hematoma or a significant fracture of the anterior skull base with intracranial bleeding, spinal-fluid leakage or brain and meningeal herniation are present, optic-nerve decompression should be performed via the craniotomy approach.

A coronal and axial computed tomographic (CT) scan (including soft tissue and bony windows) with a 1- to 2-mm slice thickness is performed in order to evaluate the ethmoid and sphenoid sinuses, the orbital apex and its adjacent structures. Precise imaging increases diagnostic accuracy. A CT scan is essential for localization of the injury and may identify a displaced fracture impinging on the optic nerve, although a non-displaced fracture may not be detectable with radiographic study [19]. Absence of an optical-canal fracture does not exclude optic neuropathy, because edema and hemorrhage alone can produce visual loss.

The surgeon must be cognizant of the many anatomic variants of this region. The CT scan is also obtained in order to evaluate the integrity of the skull base, the type and extent of sphenoid pneumatization (including whether or not a sphenoethmoid cell is present), the location of the intersinus septum and its relationship to the optic nerve and carotid artery, the dehiscence of the carotid artery (if any) and the location of and relationship between the optic nerve and carotid artery [27]. When a fracture of the optic canal is present, the proximity of the carotid artery to the optic canal and the possibility that a carotid cavernous-sinus fistula may develop over a period of weeks to months must be considered. A magnetic-resonance imaging or angiographic scan may be necessary in these cases.

Medical Therapy

Patients who present with immediate, documented loss of vision and radiographic evidence of injury to the nerve should be considered for immediate endoscopic nerve decompression [16]. If loss of vision is delayed following craniofacial trauma, aggressive medical and/or surgical decompression of the optic nerve will improve visual acuity. Once diagnosis of traumatic optic neuropathy is established, treatment is indicated. High-dose corticosteroid therapy and surgical optic-nerve decompression are more effective than observation alone. Begin treatment with very large doses of intravenous steroids (for example, methyl prednisolone in an initial dose of 30 mg/kg followed by 15 mg/kg every 6 h for 48 h, or an initial bolus of 0.75 mg/kg of dexamethasone, followed by 0.3 mg/kg every 6 h for 48 h). Matsuzaki and associates [12] advocate medical management alone for all optic-nerve injuries. Other studies have also reported significant restoration of vision with the use of steroids in traumatic optic neuropathy [21, 26]. When vision improves due to megadose steroids, therapy for five additional days (with gradual tapering) is recommended.

The decision to proceed with surgery is often difficult. Variable rates for recovery of vision have been reported after surgical optic-nerve decompression performed within a critical period. If vision does not respond to a 48-h trial of intravenous megadose corticosteroids [18] or if vision fails to continue improving or deteriorates after an initial response, prompt surgical optic-nerve decompression is indicated [25, 31].

Endoscopic Optic-Nerve Decompression Technique

Surgery is performed under general anesthesia. Meticulous technique is paramount. The nasal mucosa is decongested with oxymetazoline and is then injected with 1% lidocaine plus 1:100,000 epinephrine in a fashion similar to that used during endoscopic total sphenoethmoidectomy. The patient is positioned as if for conventional functional-sinus surgery with endoscopes. The cornea is coated with a bland ointment and is protected by a clear scleral cap. No tarsorraphy suture is needed. To permit intraoperative examination of the pupils and palpation of the globes, the eyes should not be taped. The operation can be performed with a video camera attached to the endoscope so that the assistant surgeon can observe the entire procedure on a video monitor; however, the surgeon's eye should always be on the endoscope, and he or she should always move with it. Most of the operation is performed using a 4-mm 0° telescope, although a 30° scope is occasionally required. A septoplasty may be required to ensure adequate access to the sphenoid sinus. Normally, a total endoscopic sphenoethmoidectomy is performed, with preservation of the middle turbinate (Fig. 37.8). The skull base should be thoroughly identified via meticulous hemostasis. This is essential for surgical success. The medial orbital wall and the ethmoid roof should be skeletonized throughout surgery. The maxillary-sinus natural ostium is identified, and (only if necessary) a large middle-meatal antrostomy is performed to identify the medial-orbital wall. The natural sphenoid ostium is identified. The ostium should be between the inferior border of the superior turbinate and the septum. If the sphenoid ostium cannot be identified, resection of the inferior edge of the superior turbinate may be required. The sphenoid sinus is entered, and the anterior wall is removed using a Stammberger circularly cutting sphenoid punch (Fig. 37.9). The sphenoid sinus is entered through the inferior medial aspect of the furthest posterior ethmoid cell, where the anterior wall can usually be seen as a bulge.

The posterior septal artery of the sphenopalatine artery may be encountered in the inferior aspect of the anterior sphenoid wall, just above the choana. If bleeding occurs, bipolar cautery can be used in this region. The optic nerve is identified in the superolateral aspect of the sphenoid sinus, and the ICA is located inferior and posterior to the nerve. The mucosa of the lateral wall of the sphenoid sinus is elevated and carefully removed. Caution should also be exercised in order to avoid stripping the sphenoid mucosa, because such stripping may result in considerable bleeding. Dissection must be meticulous and slow in order to avoid optic-nerve or carotid-artery injury. The optical tubercle is then visualized. Removal of the thick bone at the optic ring is facilitated with a cutting bur. Simultaneous irrigation is needed to prevent thermal injury to the nerve and devitalization of the surrounding bone. Approximately 1 cm before the optical tubercle, the lamina papyracea is carefully elevated away from the periorbita. The dissection proceeds posteriorly until reaching the optical tubercle. The annulus of Zinn can usually be identified there. The dissection must be carefully performed to avoid violation of the periorbita and injury to the medial rectus muscle. With a

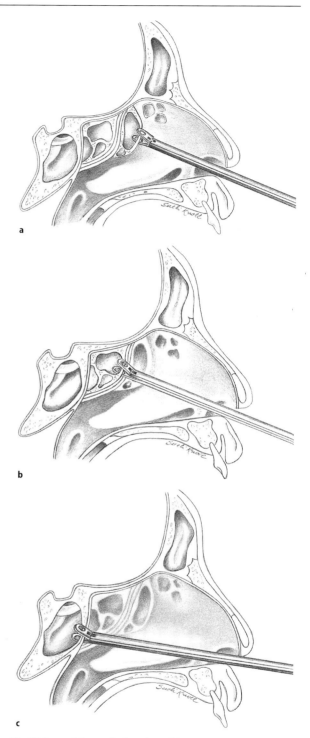

Fig. 37.8 a–c. Schematic drawing of the steps involved in endoscopic total sphenoethmoidectomy. **a** Anterior ethmoidectomy. **b** Posterior ethmoidectomy. **c** Sphenoidotomy

diamond bur, the bone above the optical tubercle and further along the optic canal is now carefully

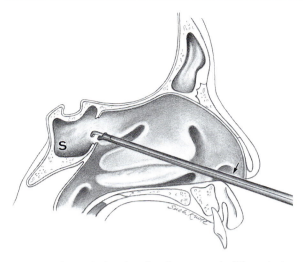

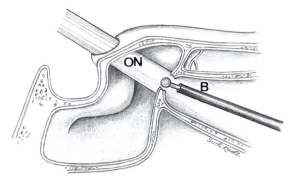

Fig. 37.9. Schematic drawing showing removal of the anterior wall of the sphenoid sinus (S) using a punch forceps (arrow). The ostium of the sphenoid sinus is enlarged, and the inferior and medial aspects of the anterior wall can be safely removed using the Stammberger circularly cutting sphenoid punch

Fig. 37.10. Schematic drawing showing optic-nerve decompression using a diamond bur (B). The bone of the optic canal is carefully thinned, and the optic nerve (ON) is exposed medially nearly 180° (from the orbit apex to the optic chiasm)

thinned from the anterior end to the posterior end, and the eggshell-thin remnants are removed completely using small curettes. The optic nerve and its sheath are exposed medially almost 180° (from the orbit apex to the optic chiasm; Fig. 37.10). During this part of the procedure, dissecting too far superiorly can lead to penetration of the roof of the ethmoid sinus and entrance into the cranial cavity. Using a disposable arachnoid knife, the incision of the optical-nerve sheath is carried out in a back-to-front direction until reaching the annulus of Zinn. The incision is made on the upper edge of the circumference in order to avoid risk of damaging the ophthalmic artery. This maneuver has been the subject of controversy in the literature, because it can lead to a cerebrospinal-fluid leak as a result of perineural extension of the subarachnoid space. If this occurs, it should be repaired with a free mucosal graft, fibrin glue or both [23].

Postoperative Care

After the operation, the patient is observed in the surgical intensive-care unit for 24 h. The patient is maintained on very high doses of steroids during the early postoperative period; steroid doses are then gradually tapered. Administration of a broad-spectrum antibiotic with good cerebrospinal-fluid penetration is started prior to surgery and continues postoperatively for 7 days. Initial postoperative care may also include complete bed rest for 3–5 days, 30° of head elevation for 1 week, analgesics as needed and stool softeners for 3–5 days.

The main concern during this period is the determination of the visual and ocular statuses. Periorbital swelling, proptosis, double vision and changes in visual acuity, should be documented. The patient is also requested not to blow his or her nose for at least 2 weeks in order to prevent orbital emphysema.

When the optic-nerve sheath is split, the incision must be covered immediately with fibrin glue in order to avoid a cerebrospinal-fluid fistula. In this case, no lumbar drain is required but, when a large skull-base defect featuring cerebrospinal rhinorrhea caused by the trauma is repaired, a 2- to 3-day lumbar drain is indicated.

The nose is packed to control bleeding. Vigorous sports and all activities that may lead to sudden intracranial hypertension should be avoided for 4–6 weeks. Nasal packing is removed after 3–5 days.

Complications

Endoscopic optic-nerve decompression is a relatively safe procedure when performed by an experienced surgeon. Although complications are infrequent, they can occur. Serious and catastrophic complications can be prevented by careful preparation and meticulous surgical technique [28, 30]. Entering the orbit or the anterior cranial fossa during endoscopic optic-nerve decompression creates risks. Penetration of the orbit and inadvertent dissection of orbital fat can injure the extraocular muscle or the optic nerve and can result in diplopia or blindness (Fig. 37.11) [1]. The most frequent problem is indirect injury to the orbital contents, caused by an

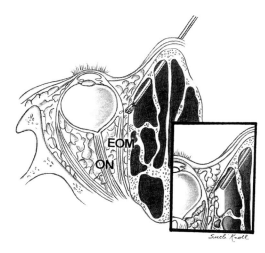

Fig. 37.11. Schematic drawing showing penetration of the orbit. Dissection of the orbital fat can inadvertently cause injury to the extraocular muscles (*EOM*) or the optic nerve (*ON*) and can result in diplopia or blindness (Adapted from [33])

expanding orbital hematoma. Bleeding within the orbit can abruptly increase the intraorbital pressure and, if this is not recognized immediately and treated appropriately, loss of vision can occur [29]. Intraoperative signs of a significant orbital hematoma include ecchymosis, proptosis and marked tension in the eye. Administration of mannitol and orbital massage should begin immediately. If pressure cannot be reduced, then a lateral canthotomy (with or without cantholysis) is performed. Failure to relieve pressure necessitates orbital decompression [15]. Further loss of visual acuity can result from direct surgical injury to the optic nerve. Cerebrospinal-fluid rhinorrhea, brain injury and carotid or cavernous-sinus bleeding are other possible complications. If a cerebrospinal-fluid leak occurs, it must immediately be repaired with fibrin glue or a free-mucosal graft. Massive hemorrhage from direct injury to major vessels can occur. Iatrogenic injury of the cavernous portion of the ICA is a rare and often fatal complication of sphenoid-sinus surgery [11]. When intraoperative injury of the ICA occurs, there is immediate (and usually massive) hemorrhage, delayed hemorrhage, a carotid-cavernous fistula or an intracavernous aneurysm. Massive hemorrhage must be rapidly controlled by ligature or by sphenoid-sinus and nasal packing and carotid-artery compression in the neck. If bleeding is controlled without carotid ligature, a carotid angiography is performed, and an elective balloon occlusion must be considered.

References

1. Edmund J, Gottfriedsen E (1963) Unilateral optic atrophy following head injury. Acta Ophthalmol Scand 41:693–697
2. Fujii K, Chambers SM, Rhoton AL (1979) Neurovascular relationships of the sphenoid sinus: A microsurgical study. J Neurosurg 50:31–39
3. Girard BC, Bouzas EA, Lamas G, Soudant J (1992) Visual improvement after transethmoid-sphenoid decompression in optic nerve injuries. J Clin Neuroophthalmol 12:142–148
4. Habal M, Maniscalco J, Rhoton A (1977) Microsurgical anatomy of the optic canal: correlates to optic nerve exposure. J Surg Res 22:527–533
5. Hughes B (1962) Indirect injury to the optic nerve and chiasma. Bull Johns Hopkins Hosp 111:98–126
6. Joseph M, Lessell S, Rizzo J, Momose K (1990) Extracranial optic nerve decompression for TON. Arch Ophthalmol 108:1091–1093
7. Kennedy D, Zinreich SJ, Hassab M (1990) The internal carotid artery as it relates to endonasal sphenoethmoidectomy. Am J Rhinol 4:7–12
8. Kennedy DW, Goodstein ML, Miller NR, Zinreich SJ (1990) Endoscopic transnasal orbital decompression. Laryngoscope 116:275–282
9. Kennerdell JS, Amsbaugh GA, Myers EN (1976) Transantral–ethmoidal decompression of optic canal fracture. Arch Ophthalmol 94:1040–1043
10. Kline LB, Moravetz RB, Swaid SN (1984) Indirect injury of the optic nerve. Neurosurgery 14:756–764
11. Maniglia AJ (1991) Fatal and other major complications of endoscopic sinus surgery. Laryngoscope 101:349–354
12. Matsuzaki H, Kunita M, Kawai K (1982) Optic nerve damage in head trauma: clinical and experimental studies. Jpn J Ophthalmol 26:447–461
13. Mattox DE (1990) External ethmoidectomy. In: Johns A, Price M, Miller J (eds) Atlas of head and neck surgery. Decker, Philadelphia, pp 218–222
14. Mattox DE, Kennedy DW (1990) Endoscopic management of cerebrospinal fluid leaks and cephaloceles. Laryngoscope 100:857–862
15. Neuhaus RW (1990) Orbital complications secondary to endoscopic sinus surgery. Ophthalmology 97:1512–1518
16. Niho S, Yasuda K, Sato T, et al. (1961) Decompression of the optic canal by transethmoid route. Am J Ophthalmol 51:659–665
17. Osguthorpe JD (1988) Optic nerve decompression. Otolaryngol Clin North Am 21:155–168
18. Panje W, Gross C, Anderson R (1981) Sudden blindness following facial trauma. Otolaryngol Head Neck Surg 89:941–948
19. Ramsay JH (1979) Optic nerve injury in fracture of the canal. Br J Ophthalmol 63:607–610
20. Rontal E, Rontal M, Guilford FT (1979) Surgical anatomy of the orbit. Ann Otol Rhinol Laryngol 88:382–386
21. Seiff S (1990) High-dose corticosteroids for treatment of vision loss due to indirect injury to the optic nerve. Ophthalmic Surg 21:389–395
22. Sofferman RA (1981) Sphenoethmoid approach to the optic nerve. Laryngoscope 91:184–196
23. Sofferman RA (1991) Transnasal approach to optic nerve decompression. Operative Tech Otolaryngol Head Neck Surg 2:150–156

24. Som PM (1985) CT of the paranasal sinuses. Neuroradiology 27:189
25. Spoor T, McHenry J (1996) Management of TON. J Craniomaxillofac Trauma 2:14–26
26. Spoor T, Hartel W, Lensink D, Wilkinson M (1990) Treatment of TON with corticosteroids. Am J Ophthalmol 110:665–669
27. Stammberger HR, Kennedy DW (1995) Paranasal sinuses: anatomic terminology and nomenclature. The Anatomic Terminology Group. Ann Otol Rhinol Laryngol Suppl 104:7–16
28. Stankiewicz JA (1987) Complications of endoscopic intranasal ethmoidectomy. Laryngoscope 97:1270–1273
29. Stankiewicz JA (1988) Blindness and intranasal endoscopic ethmoidectomy: prevention and management. Otolaryngol Head Neck Surg 101:320–329
30. Stankiewicz JA (1989) Complications of endoscopic surgery. Otolaryngol Clin North Am 22:749–768
31. Takahashi M, Itoh M, Ishii J, Yoshida A (1989) Microscopic endonasal decompression of the optic nerve. Arch Otorhinolaryngol Head Neck Surg 246:113–116
32. Wigand ME (1990) Endoscopic surgery of the paranasal sinuses and anterior skull base. Thieme, New York
33. Levine, HL, May M (1993) Endoscopic sinus surgery. Thieme, New York.

Cerebrospinal Fluid Rhinorrhea – Transnasal Micro-endoscopic Surgery

ALDO CASSOL STAMM and LUIZ A. SILVA FREIRE

Introduction

"The initial surgical treatment for CSF rhinorrhea must be a rhinological extracranial approach, unless a precise indication for neurosurgical exploration is present" [29]. CSF rhinorrhea is an abnormal communication between the anterior or middle skull base or temporal bone and the nose; it causes CSF to be present in the nasal cavity. The appropriate treatment requires prompt diagnosis (including determination of the site of the fistula) and surgery to correct the defect.

The first description of CSF rhinorrhea after skull-base fracture was published in 1745, according to Thompson [38]. In 1826, Miller reported the first case of idiopathic CSF rhinorrhea [39]. In 1913, after the introduction of X-rays, Luckett [16] published the first diagnosis of a pneumoencephalocele after a skull-base fracture.

Dandy first successfully carried out the surgical treatment of CSF rhinorrhea in 1926 [6], repairing a dural laceration by suturing a graft of fascia lata via a frontal craniotomy; in 1927, Cushing [5] employed a similar technique. Since those first descriptions, there has been continued interest in surgical repair, and many authors have recommended various techniques.

In 1944, German [9] successfully repaired fistulas of the cribriform plate in five patients by covering the region of the crista galli with dural flaps. In 1948, Dohlman [7] repaired a defect of the cribriform plate through an external ethmoidectomy using a mucosal flap from the nasal septum. In 1952, Hirsch [11] proposed treating CSF fistulas of the sphenoid sinus using a septal mucoperichondrial flap via an endonasal approach. In 1964, Vrabec-Hallberg [40] treated CSF rhinorrhea via an intranasal approach, using a pedicled mucosal flap of the middle turbinate to close cribriform-plate fistulas.

Montgomery [22], in 1966, recommended an external ethmoidectomy approach to repair CSF fistulas of the ethmoid and sphenoid sinuses and, in 1973, he proposed middle turbinectomy followed by nasal mucoperiosteal-flap rotation from the septum to cover the dural defect [23]. Mattox and Kennedy [20] proposed a transnasal endoscopic technique to repair fistulas at the roof of the ethmoid and sphenoid sinuses and at the cribriform plate. They used a contralateral mucoperiosteal free graft (from the septum and/or ipsilateral pedicled septal flap) and fibrin glue.

J. Heermann [10] used an endonasal surgical microscope and temporalis fascia and cartilage grafts to repair CSF rhinorrhea. Using a surgical microscope via an intranasal approach, Leher and Deustch [14] reported the treatment of two patients with fistula.

Stamm et al. [33, 35] and Amedee et al. [2] utilized transnasal microsurgery to repair idiopathic, traumatic and iatrogenic CSF fistulas of the ethmoid sinus, cribriform plate and sphenoid sinus. Wigand [41] recommends an endoscopic endonasal surgical approach and advocates the use of a mucoperiosteal free graft from the inferior turbinate to repair the fistulas. The graft is positioned at the site of the fistula and is fixed with fibrin glue. Levine and May [15] used a middle-turbinate pedicled flap to repair dural fistulas via an endoscopic approach. In the absence of the middle turbinate, they used fascia lata, fat and muscle.

Stankiewicz [37] has reported the results of his surgical treatment of iatrogenic CSF rhinorrhea using endonasal rigid endoscopes. In order to close the fistula, he used temporalis fascia, muscle and Gelfoam. For fistulas in the sphenoid sinus, he used Gelfoam and fibrin glue.

Diagnosis

CSF rhinorrhea is diagnosed using the clinical history, neurologic and otorhinolaryngologic examinations, laboratory tests and imaging studies. The clinical history includes questions about the presence of watery rhinorrhea, which is usually unilateral, of variable amounts and may persist when sleeping. These characteristics differentiate CSF rhinorrhea from the rhinorrhea of patients with allergic, vascular or infectious rhinopathy because, in the latter cases, rhinorrhea is bilateral, is generally preceded by sneezing and is either mucus (identified by its ability to precipitate when in contact with acetic acid) or is purulent in character.

In some patients without the above symptoms, there may be a history of recurrent meningitis or headache, suggesting the presence of a fistula. Patients are questioned about their history of craniofacial trauma, previous surgery and idiopathic drainage.

In addition to evaluating the rhinorrhea, other steps are necessary during the general physical examination, including a Valsalva maneuver with anterior flexion of the head to see if the rise in intracranial pressure increases the flow of liquid to the nasal cavity. The otorhinolaryngologic examination is carried out with 0°, 30° and 70° endoscopes to document the presence of (and hopefully to locate) the CSF fistulas. This examination is also useful for surgical planning, as it can help determinate whether septal deviation and/or hypertrophy of the turbinate are present.

Laboratory examinations may be helpful. A glucose content of greater than 30 mg/100 ml suggests the presence of CSF [3]. Oberascher and Arrer [26] suggested the use of intrathecal fluorescein combined with nasal endoscopy to locate the fistula; this is a quick, highly sensitive, low-risk method. Oberascher and Arrer [26] also used an immunologic method of identification; they looked for β-2 transferrin, which is characteristic of CSF.

One of the main problems in diagnosing patients with CSF rhinorrhea is determining the exact location of the lesion. Various techniques have been developed to locate the fistula. Axial and (especially) coronal computerized tomography (CT) scans with high resolution are very important. Axial CT better demonstrates the frontal sinus, while the coronal projections are better for the areas of the ethmoid sinus, cribriform plate and sphenoid sinus.

The presence of air inside the cranial cavity (pneumocephalus) indicates the existence of a dural defect, which is usually associated with a CSF fistula. In 1977, Drayer et al. [8] and Manelfe et al. [18] used a non-ionic, hydrosoluble contrast substance known as metrizamide [24] in combination with CT to locate the exact sites of the fistulas; this test is called CT cisternography. In 1982, Manelfe et al. [19] emphasized the use of metrizamide with computerized tomography for the diagnosis and location of CSF rhinorrhea. Using CT with metrizamide, Chow et al. [4] had a success rate of 76% in locating the fistulas in his series of 17 patients.

Using CT cisternography, the success rate varies between 22% and 100% [14, 21]. This wide variation may represent different radiological criteria or the possible absence of an active CSF rhinorrhea while the examinations were carried out. Shaeffer [31] felt that the use of high-resolution CT together with cisternography employing metrizamide or iopamidol was the best technique for locating active CSF fistulas. McCormack et al. [21] also recommend CT cisternography for the diagnosis and localization of fistulas; they prefer the coronal projection for depiction of the ethmoid–cribriform-plate complex and the sphenoid sinus and the axial projection for depiction of the frontal sinus.

In our department, diagnosis is established using clinical history and endoscopic examination. The site of the CSF fistula is established using CT cisternography with a hydrosoluble, non-ionic contrast substance (axial and coronal projections) [17]. For this examination, 6–8 ml of contrast material is injected via a lumbar puncture; the patient is then positioned with his head angled 45° downward for 5–10 min (Fig. 38.1) [4]. Coronal, axial and sagittal magnetic-resonance (MR) sections (mainly T2-weighted images) are also helpful for the identification of the site of a CSF fistula (Fig. 38.2). Recent MR studies based on T2-weighted sequences and thin sections allowed better visualization of CSF in the nose [34]. CSF has high T2-signal intensity (a myelographic effect). Three-dimensional reconstruction using T2-weighted MR images is also possible. We believe that MR combined with CT cisternography is the method of choice for the topodiagnosis of CSF fistulas.

Classification

A majority of authors classify CSF fistulas according to their traumatic or non-traumatic etiology. Traumatic fistulas may be iatrogenic or due to craniofacial trauma, in which case they are generally located in the region of the cribriform plate and at the roof of the ethmoid sinus, because the bone there is thin and the dura more adherent. This area is also more vulnerable to iatrogenic trauma during nasal, craniofacial and sinus surgeries. CSF rhinorrhea may be detected at the time of the trauma or weeks, months or even years later.

Lesions that generate intracranial hypertension, primary or metastatic intracranial tumors, extracranial tumors, cranial osteomyelitis or brain cysts can cause non-traumatic fistulas [3]. The

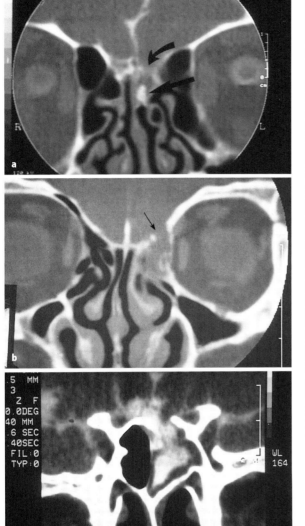

Fig. 38.1 a–c. Computed-tomography cisternography (coronal projection). **a** Idiopathic CSF fistula located at the left cribriform plate. **b** Defect of the roof of the ethmoid sinus (*arrow*); there is soft-tissue density (probably CSF) inside the sinus (post-traumatic fistula). **c** Idiopathic CSF fistula of the sphenoid sinus

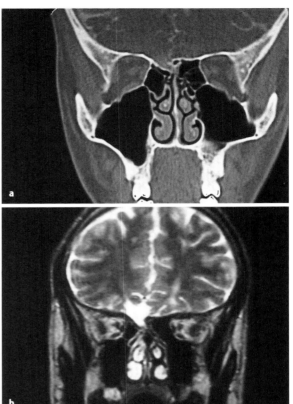

Fig. 38.2 a,b. A 16-year-old patient with a traumatic CSF fistula at the roof of the ethmoid sinus. **a** Coronal computed tomography, showing the fistula. **b** T2-weighted coronal magnetic-resonance image, showing the same fistula

cause of most non-traumatic fistulas is not known. They may be due to primary CSF rhinorrhea, frequently coming from congenital defects of the skull base at the level of the cribriform plate, ethmoid roof or sphenoid sinus. In patients with small meningoceles, there is a risk of rupture and CSF leakage even with normal intracranial pressure [3, 25]. It is extremely important to distinguish the cause of the fistula, as the fistulas differ considerably regarding their presentations, prognostic and treatment recommendations.

Park et al. [29] report that 88% of their cases were traumatic and 12% were non-traumatic, the same percentages reported by Yessenow and McCabe [42]. Settipane [30] reported that 80% of the fistulas were due to craniofacial trauma; 16% were iatrogenic, and only 3–4% were idiopathic. Jones and Moore [12] report that approximately 90% of their cases were traumatic and 10% were non-traumatic. McCormack et al. [21], in a series of 37 patients, observed that the etiology was iatrogenic in 60% of the cases, craniofacial trauma in 16% and idiopathic in 24%. In our series of 66 patients (30 female and 36 male), the ages of the patients ranged between 4 years and 73 years, and the fistulas were classified according to their location and etiology.

The fistulas were traumatic (craniofacial trauma and iatrogenic) in 73% of the cases and were non-traumatic fistulas (idiopathic or secondary to intracranial lesions) in 27%. Twenty-six patients (39%) had fistulas at the roof of the ethmoid sinus,

15 (23%) at the cribriform plate, 14 (21%) at the sphenoid sinus, nine (14%) at the ethmoid roof and cribriform plate and two (3%) at the roofs of both the ethmoid and sphenoid sinuses (Table 38.1).

Medical Treatment

CSF rhinorrhea can be treated non-operatively. The measures used to reduce CSF pressure and production include positioning of the patient in a posture that minimizes the leak [such resting in bed in a supine position, with the head slightly elevated (~30–45°)] and avoiding any physical effort or maneuver (such as a Valsalva maneuver or nose blowing) that can raise the intracranial pressure. These measures are also important in preventing complications, especially pneumocephalus and meningitis. Although still controversial, prophylactic antibiotics are given to prevent meningitis, and oral administration of carbonic-anhydrase inhibitors (acetazolamide, 250–750 ml/day) are recommended in order to decrease CSF production; this seems to promote dural healing. These conservative ways of treatment can be continued for 2 weeks or 3 weeks and can sometimes allow healing of the fistula, depending on the cause, site and size of the defect.

Serial lumbar punctures or continuous drainage of the subarachnoid space can also achieve reduction of CSF pressure. In a majority of cases, a silicone catheter is inserted via the spine (at the L2–L3 interspace) and is affixed to a sterile collecting bag through a closed external system that allows regulation of the amount of drainage. The amount of fluid to be evacuated must be strictly controlled by measuring its volume and checking for any clinical symptom of CSF hypotension, pneumocephalus or meningitis (headaches, vomiting, lethargy). The rate of drainage in adults should be approximately 30–50 ml every 8–10 h, and a decrease of the compression of the subarachnoid space can be accomplished in 4–5 days.

Surgical Treatment

Various techniques have been described for the repair of CSF fistulas located in the cribriform plate and the ethmoid and sphenoid sinuses. Intracranial approaches are frequently used by neurosurgeons and have the advantage of direct visualization of the dural defect and the simultaneous treatment of any associated lesions, as reported by Ommaya in 1976 [27, 28]. A craniotomy should be carried out in patients with brain hemorrhage or open wounds that communicate directly with brain tissue [1]. There are disadvantages to the intracranial approaches. These include:

- Difficult surgical access to fistulas in the sphenoid sinus [11]
- The need for brain retraction (which may lead to cerebral edema, or intracerebral hemorrhage) in order to achieve exposure of the skull base
- Frequent postoperative anosmia.

Speltzler [32] reported that one third of the patients treated by the craniotomy approach had persistence of CSF rhinorrhea.

Extracranial approaches may be either intranasal or extranasal. Dohlman [7] repaired fistulas at the cribriform plate via an extracranial, external approach; he used ethmoidectomy via a naso-orbital incision. The disadvantage of this approach is the external scar. The endonasal surgical approach first described by Hirsch in 1952 [11] can be used for the correction of all CSF fistulas of the sinuses except those localized in the frontal sinus. The presence of infection in the nasal cavity or the paranasal sinuses is a contraindication for surgery through an endonasal approach. The disadvantage of the extracranial approach is the inability to visualize some brain lesions or treat associated intracranial diseases.

Indications for the transnasal micro-endoscopic surgical approach to CSF rhinorrhea depend on previous identification of the fistula site via imaging

Table 38.1. CSF rhinorrhea fistulas according to the anatomical site and etiology ($n=66$)

Location	Etiology Traumatic	Iatrogenic	Non-traumatic	Total
Ethmoid sinus	15	7	4	26 (39%)
Cribriform plate	4	2	9	15 (23%)
Sphenoid sinus	4	6	4	14 (21%)
Ethmoid sinus/cribriform plate	3	5	1	9 (14%)
Ethmoid and sphenoid sinuses	1	1	–	2 (3%)
Total	27 (41%)	21 (32%)	18 (27%)	66 (100%)

[CT cisternography and MR images (high T2 signal)]. This surgical technique is usually indicated when CSF rhinorrhea (1) is persistent and intermittent and conservative treatment has failed or (2) is associated with a pneumocephalus and previous bouts of bacterial meningitis.

According to Park [29], the surgical treatment of CSF rhinorrhea should be performed via an extracranial rhinologic approach unless a positive indication for neurosurgical exploration is present. The success rate of micro-endoscopic repairs of CSF fistulas has been reported to range from 76% to 100% [20, 33–35, 37].

Transnasal Micro-Endoscopic Surgery

Transnasal micro-endoscopic surgery is carried out under general anesthesia with controlled hypotension. The surgical technique is performed with the operating microscope or with endoscopes.

An important consideration in choosing the surgical technique is whether there is simple rupture of the dura with CSF leakage or rupture of the dura with brain tissue protruding through the dura. If a traumatic or congenital encephalocele is present, treatment must include bipolar fulguration of the herniated brain tissue before closing the bony skull-base defect [13].

Different surgical techniques have been used for the transnasal surgical approach to CSF fistulas [34]. They are named A–E, according to the location of the fistulas: ethmoid sinus (A), cribriform plate (B), cribriform plate and fovea ethmoidalis (C), sphenoid sinus (D) or ethmoid plus sphenoid sinuses (E). These, in turn, are subdivided into A-1, A-2, B-1, B-2, C, D-1, D-2 and E techniques, according to the location, size and anatomical features of the fistula and the amount of fluid leaked (Table 38.2). The surgical techniques follow similar principles. A mucoperiosteal graft of the middle or inferior turbinate (secured by fibrin glue) is used in fistulas less than 5 mm in diameter. In cases where the fistula is more than 5 mm in diameter, two grafts are used: the first is a fascia-lata graft, the second is a mucoperiosteal graft. Both are secured by fibrin glue. A mucoperiosteal flap from the nasal septum can also be used. The nose is packed for 5–7 days, and antibiotics are used. In patients with very large defects, a septal bone or cartilage graft is also used to close the defect.

Micro-Endoscopic Surgical Techniques

Fistulas of the Fovea Ethmoidalis

The surgical approach to fistulas in the roof of the ethmoid sinus is through the middle meatus. When surgery is done with the microscope, the inferior turbinate is fractured, and the middle turbinate is gently displaced medially. With a microsurgical self-retaining nasal speculum, the structures of the middle meatus are exposed. When 4-mm endoscopes (0° and 30°) are used, the middle turbinate is gently displaced against the nasal septum. In both techniques, the uncinate process and the ethmoid bulla are removed to gain access to the ethmoid cells (which are removed to a variable extent depending on the location of the fistula) and the ethmoid skull base, where the defect can be identified (Fig. 38.3).

Precise identification of the site of the CSF fistula is crucial to successful repair. The adjacent mucosa is removed, leaving the bone exposed.

The surgical technique varies according to the size and amount of leak [36]. For fistulas smaller than 0.5 cm, a type A-1 technique is used. In larger fistulas, we use an A-2 technique [33, 34]. For high-flow fistulas we also rotate a mucoperiosteal flap from the middle turbinate, achieving a double-layer closure.

Technique A-1

Technique A-1 is our most commonly used technique. It is indicated for fistulas smaller than 0.5 cm. After adequate exposure of the fistula, the repair is accomplished with a mucoperiosteal free graft from the inferior turbinate or the floor of the nasal cavity. A graft, larger than the opening of the fistula, is placed with its raw surface toward the site of the fistula and is fixed by fibrin glue and Gelfoam pledges applied over the graft to keep it in position. The nasal cavity and the ethmoid sinus are packed with rayon gauze or Merocel (saturated with antibiotic ointment) for 5–7 days (Fig. 38.4).

Table 38.2. Surgical techniques, categorized according to the anatomical site and size of the defect

Location	Defect size	
	<0.5 mm	>0.5 mm
Ethmoid sinus	A-1	A-2
Cribriform plate	B-1	B-2
Cribriform plate/fovea ethmoidalis	C	
Sphenoid sinus	D-1	D-2
Ethmoid and sphenoid sinuses	E	

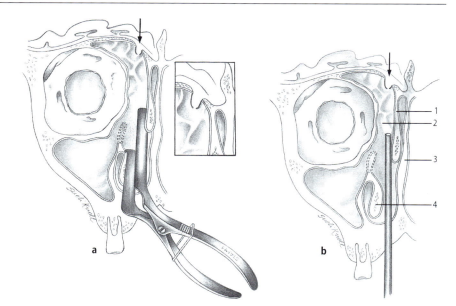

Fig. 38.3. Schematic drawing of a coronal view, showing a fistula at the roof of the ethmoid sinus (*arrow*) after an ethmoidectomy. The structures shown are: *1*, the middle turbinate; *2*, ethmoid sinus; *3*, nasal septum; *4*, inferior turbinate. **a** View through the nose, using a self-retaining speculum. **b** View available to the endoscope, **c** Coronal computed tomography showing a defect of the roof of the ethmoid sinus (post-traumatic fistula)

Technique A-2

Technique A-2 is used for the management of CSF fistulas of the fovea ethmoidalis that are larger than 0.5 cm, whether or not they are associated with an encephalocele. If an encephalocele is present, a bipolar fulguration of herniated brain tissue should be performed before identifying and closing the skull-base defect. Once identified, the fistula is repaired with a fascia-lata graft, which is tucked into the cranial cavity beneath the edges of the bony defect of the ethmoid roof. In large defects, it is advisable to hold the brain tissue inside the cranial cavity with a septal cartilage or bone graft in addition to with the fascia-lata graft. The graft is held in place with fibrin glue. The mucoperiosteum graft from the inferior turbinate or the floor of the nose covers the fascia-lata graft. Another layer of fibrin glue is then applied, and Gelfoam pledges are used to hold the grafts in position. The ethmoid sinus and the nasal cavity are packed with gauze (rayon or Merocel) for 5–7 days (Fig. 38.5).

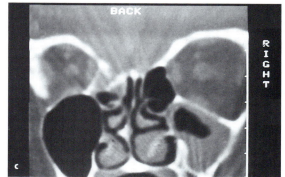

Fistulas of the Cribriform Plate

CSF fistulas located at the cribriform plate can be repaired using mucoperiosteal free grafts from the middle or inferior turbinate, ipsilateral septal pedicled flaps or a combination of both (Fig. 38.6). The surgical technique begins with the mobilization of the nasal septum (through a septoplasty, when necessary). This allows easier placement of the self-retaining speculum when a microscope or endoscope is used. The final goal is better exposure of the surgi-

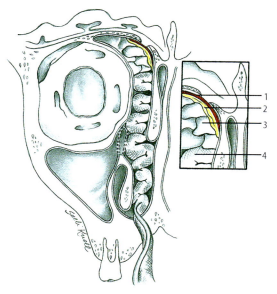

Fig. 38.4. Repair of a fistula at the roof of the ethmoid sinus, and the graft used for closure. *1*, Mucoperosteal free graft; *2*, fibrin glue; *3*, Gelfoam; *4*, nasal pack

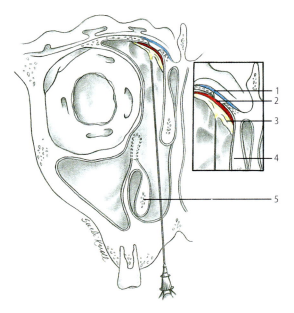

Fig. 38.5. Repair of a CSF fistula larger than 0.5 cm located at the roof of the ethmoid sinus. *1*, Fascia-lata graft; *2*, mucoperostial graft; *3*, fibrin glue; *4*, middle turbinate; *5*, inferior turbinate

cal field. The first surgical step is the total resection of the middle turbinate. The middle-turbinate bone is removed, and the mucoperiosteum is separated and preserved. Once the site of the fistula is identified, the mucoperiosteal graft is used to close the fistula (Fig. 38.7).

Technique B-1

Technique B-1 is indicated in small defects of the cribriform plate. If the defect is small, one layer should suffice. In such a case, a mucoperiosteal free graft from the middle or inferior turbinate is used. The graft is positioned with its periostial side toward the fistula and is fixed with fibrin glue. Gelfoam pledges are applied on the graft to keep it in place. The nasal cavity is packed as previously described (Fig. 38.8).

Technique B-2

Technique B-2 is used for larger defects of the cribriform plate. In such a case, a double-layered closure is preferred. This includes a mucoperiosteal free graft reinforced by a pedicled flap that can be rotated from the septum (Fig. 38.9).

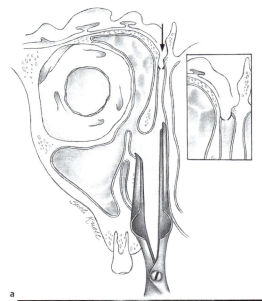

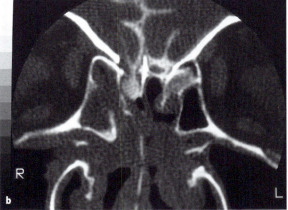

Fig. 38.6. a A CSF fistula at the right cribriform plate. b Computed-tomography cisternography (coronal projection) shows a CSF fistula of the right cribriform plate

Fistula of the Cribriform Plate and Fovea Ethmoidalis

Fistula of the cribriform plate and fovea ethmoidalis (technique C) is used for traumatic or idiopathic fistulas that involve the cribriform plate and the ethmoid-sinus roof (Fig. 38.10a). A septoplasty is done at the beginning of the procedure, if necessary. The defect is completely exposed by removing the middle turbinate, the adjacent olfactory mucosa and scar tissue until reaching the ethmoidal region (Fig. 38.10b). Successful control of larger defects involves sealing the defect by tucking the fascia lata into the anterior skull base, over which a mucoperiosteal free graft from

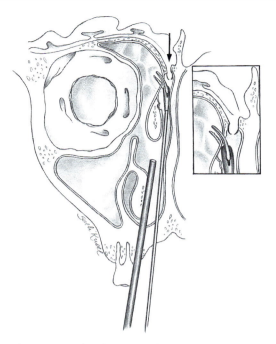

Fig. 38.7. Fistula of the cribriform plate (*arrow*) and endoscopic resection of the middle turbinate

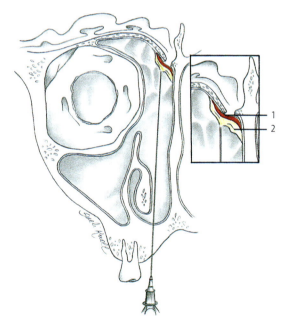

Fig. 38.8. Repair of a cribriform-plate CSF fistula. *1*, Mucoperiosteal graft from the middle turbinate; *2*, fibrin glue

the middle or inferior turbinate is placed; this is then sealed in place with fibrin glue and Gelfoam. The nasal cavity is packed as described previously (Fig. 38.10c).

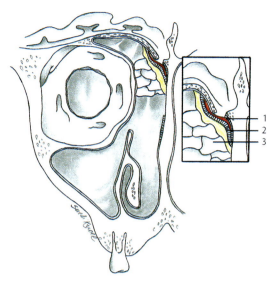

Fig. 38.9. Repair of a large cribriform-plate CSF fistula via: *1*, a mucoperiosteal free graft; *2*, a septal mucopericondreal flap; *3*, Gelfoam

Fistulas of the Sphenoid Sinus

Fistulas of the sphenoid sinus may be idiopathic, secondary to surgical procedures or due to craniofacial trauma. Surgery may be carried out via three basic approaches:
- A transethmoid approach when the posterior ethmoid sinus is involved. After opening the ethmoid sinus through the middle meatus, the anterior wall of the sphenoid sinus is identified by the 90° angle of its junction with the ethmoid roof.
- A trans-septal approach when both sphenoid sinuses are involved with large defects. The sphenoid sinus is opened at the midline after removing the rostrum.
- A direct transnasal approach to the sphenoid sinus. When approaching the sphenoid sinus transnasally, the important anatomical landmarks are the superior edge of the choana and the caudal ends of the middle and superior turbinates. The sphenoid ostium is approximately 1 cm above the superior choanal edge between the septum and the superior turbinate (Fig. 38.11). Coronal and axial CT are useful for surgical planning, because they provide precise identification of other important anatomical landmarks (Fig. 38.12). Septal deviations must be addressed at the beginning of the procedure.

A D-1 technique is used for the closure of fistulas smaller than 0.5 cm in diameter, and the D-2 technique is used for larger defects.

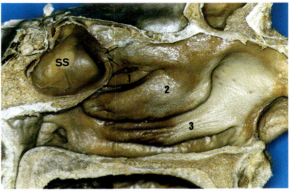

Fig. 38.11. Lateral nasal wall. *SS*, sphenoid sinus; *1*, Superior turbinate; *2*, middle turbinate; *3*, inferior turbinate; The *arrow* indicates the sphenoid-sinus ostium

Technique D-1

Small, unilateral fistulas of the sphenoid sinus (Fig. 38.13a) are repaired through a direct transnasal approach using an operating microscope or endoscopes. The first step is the identification of the sphenoid-sinus ostium and its anterior wall. The caudal portion of the middle and superior turbinates is removed to gain better exposure of the sphenoid sinus. The anterior wall of the sphenoid sinus is removed, and the site of the fistula is identified. If the fistula is located laterally, in a well-pneumatized sphenoid sinus, 30° and 70° telescopes are used. The next step is removal of the surrounding mucosa in order to clearly define the dural defect. Closure is carried out with a mucoperiosteal free graft from the middle or inferior turbinate; the graft is placed with its raw surface toward the fistula and is fixed with fibrin glue and Gelfoam. Finally, the nasal cavity is packed (especially in its posterior–superior area) with rayon gauze or Merocel saturated with an antibiotic ointment. The packing is removed after 5–7 days (Fig. 38.13b).

Technique D-2

Technique D-2 is used for fistulas larger than 0.5 cm in diameter. The micro-endoscopic surgical approach

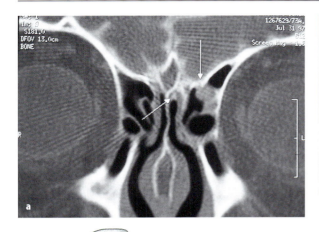

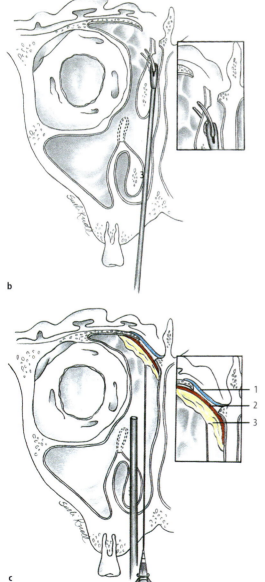

Fig. 38.10. **a** Computed-tomography cisternography (coronal projection) showing a CSF fistula (*arrows*) at the cribriform plate and ethmoid roof. **b** Resection of the middle turbinate with microscissors. **c** Repair of a CSF fistula (located at the cribriform plate and ethmoid sinus) using: *1*, fascia lata; *2*, a mucoperiosteal graft; *3*, fibrin glue

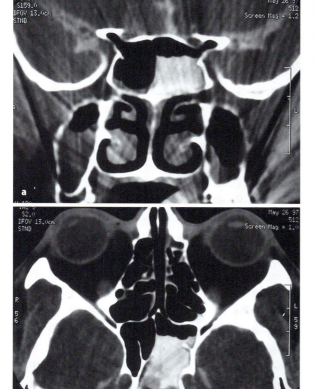

Fig. 38.12 a,b. Computed-tomography cisternography showing a CSF fistula at the sphenoid sinus. **a** Coronal projection. **b** Axial projection

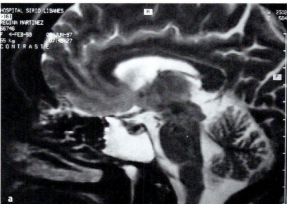

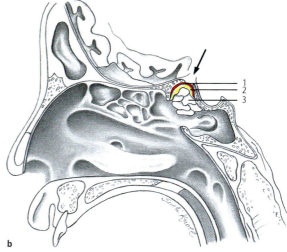

Fig. 38.13. a Sagittal magnetic resonance in a T2 sequence, showing the sphenoid sinus filled with CSF. **b** Repair of a small sphenoid-sinus fistula. *1*, Mucoperiosteal graft; *2*, fibrin glue; *3*, Gelfoam

may be transnasal or trans-septal, depending on the site and extension of the defect. A fascia-lata and/or septal-cartilage free graft is introduced inside the cranial cavity, underneath the edges of the bony defect. Additionally a mucoperiosteal free graft is placed over the defect and is sealed in place with fibrin glue and Gelfoam (Fig. 38.14). The nasal cavity is packed with rayon gauze or Merocel (saturated with antibiotic ointment) for 5–7 days.

Fistulas of the Ethmoid and Sphenoid Sinuses

The first surgical step in treating fistulas of the ethmoid and sphenoid sinuses (technique E) is the removal of the middle turbinate. This is followed by resection of the uncinate process and a complete sphenoethmoidectomy (until complete exposure of the bony defect and the dura is achieved). Surgicel, fascia lata or septal cartilage is placed intracranially through the defect. Fibrin glue is then applied to keep it in position. A mucoperiosteal free graft from the middle turbinate is glued to the first layer, keeping the raw surface in contact with the previous layer. Pieces of Gelfoam are then placed, and the ethmoid and sphenoid sinuses are packed with rayon or Merocel. Antibiotics are given for 7–10 days (Fig. 38.15).

Results

Our results were based on the postoperative presence or absence of CSF rhinorrhea, as described by the patient or seen using microscopic or endoscopic examinations. CT and MR were not routinely used postoperatively.

In our series, 60 of 66 patients with fistulas were successfully treated using the micro-endoscopic

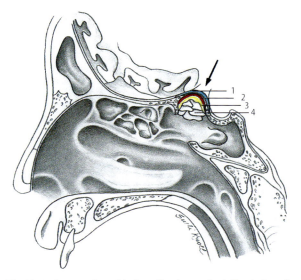

Fig. 38.14. Large sphenoid-sinus fistula repair. *1*, Fascia lata; *2*, mucoperiosteal graft; *3*, fibrin glue; *4*, Gelfoam

technique; an overall success rate of 91% was achieved. Six patients had recurrence of their CSF leak after surgery. Three were successfully repaired by the same transnasal approach. Two were re-operated by a frontal craniotomy, and one was lost on follow-up. The sixth patient had repeated endonasal surgery (which failed) and had a posterior frontal craniotomy, which also failed; this patient was lost on follow-up.

Complications

Minor complications, such as synechiae, crusts and bleeding and secondary sinusitis, were observed in 30% of our patients; these were successfully treated. Approximately 70% of our patients with cribriform-plate involvement developed some degree of hyposmia. Two patients experienced meningitis during the postoperative period and were treated successfully. One patient presented with a small posterior septal perforation.

Comments

- Unilateral rhinorrhea, headache and repeated bouts of meningitis usually indicate the presence of a CSF fistula.
- High-resolution CT (coronal projections, in association with an intrathecal, non-ionic, hydrosoluble contrast substance) and MR (mainly in T2-

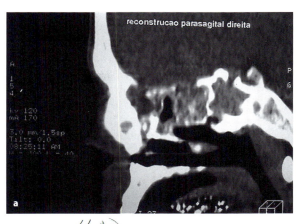

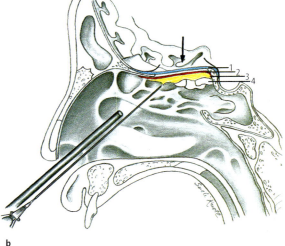

Fig. 38.15. a Sagittal computed tomography showing a large defect involving the ethmoid and sphenoid sinuses at the anterior skull base. b Endoscopic repair of an ethmoid- and sphenoid-sinus fistula using: *1*, a large piece of fascia lata; *2*, a mucoperiosteal graft; *3*, fibrin glue; *4*, Gelfoam

weighted sequences) are essential for precise location of a fistula located in the ethmoid or sphenoid sinus or cribriform plate.
- The majority of idiopathic fistulas and meningoceles presented in women 35 years and older. They were more commonly seen at the cribriform plate. Cases of traumatic fistulas were more frequent in men and were more frequently located at the fovea ethmoidalis. There was no statistical difference with regard to sex in cases of iatrogenic CSF fistulas.
- All the described operative techniques have the same surgical principles:
  1. Precise identification of the site of the fistula
  2. Mucoperiosteal free grafts from the middle or inferior turbinates, and fascia-lata grafts for larger defects (fistulas larger than 5 mm)

3. Fibrin glue
   4. Gelfoam
   5. Nasal packing
- Continuous, subarachnoid, closed lumbar drainage can also be used in cases of large or high-flow fistulas. The drain can be left in place for 3–5 days. Antibiotic therapy is used in all cases, for a minimum of 10 days. In cases of idiopathic fistulas, a permanent lumbo-peritoneal shunt is recommended.
- A solution of fluorescein (5%) is used intrathecally in cases where the site of the fistula could not be precisely identified during the preoperative assessment; this is a frequent problem with idiopathic fistulas. A mixture of 0.2 cm^3 of a 5% solution of sodium fluorescein diluted with 10 cm^3 of the patient's CSF is slowly injected through a cervical puncture 15–20 min before the operation.
- Due to its high success rate and low morbidity, transnasal micro-endoscopic surgery is the treatment of choice for CSF fistulas of the cribriform plate or the roof of the ethmoid and sphenoid sinuses. Transnasal micro-endoscopic surgery should be avoided in any case where there is a nasal or sinus infection.

References

1. Adson AW, Uihlein A (1949) Repair of defects in ethmoid and frontal sinuses resulting in cerebrospinal rhinorrhea. Arch Surg 58:623–634
2. Amedee RG, Mann WJ, Gilsbach J (1993) Microscopic endonasal surgery for CSF leaks. Am J Rhinol 7:1–5
3. Calcaterra TC (1985) Diagnosis and management of ethmoid cerebrospinal rhinorrhea. Otolaryngol Clin North Am 18:99–105
4. Chow JM, Goodman D, Mafee MF (1989) Evaluation of CSF rhinorrhea by computerized tomography with metrizamide. Otolaryngol Head Neck Surg 100:99–105
5. Cushing H (1927) Experiences with orbito-ethmoidal osteomata having intracranial complications, with the report of four cases. Surg Gynecol Obstet 44:721–742
6. Dandy WE (1926) Pneumocephalus (intracranial pneumatocele por aerocele). Arch Surg 12:949–982
7. Dohlman G (1948) Idiopathic cerebro-spinal rhinorrhea: case operated by rhinologic methods. Acta Otolaryngol 67:20–23
8. Drayer BP, Wilkins RH, Boehnke M, et al. (1977) Cerebrospinal fluid rhinorrhea demonstrated by metrizamide CT cisternography. AJR Am J Roentgenol 129:149–151
9. German MG (1944) Cerebro-spinal rhinorrhea surgical repair. J Neurosurg 1:60–66
10. Heermann J (1979) CSF leakage. Arch Otolaryngol Head Neck Surg 223:457–458
11. Hirsch O (1952) Successful closure of cerebrospinal fluid rhinorrhea by endonasal surgery. Arch Otolaryngol 56:1–12
12. Jones DT, Moore GF (1987) Evaluation and management of cerebrospinal spinal fistulas: a logical approach. American Academy of Otolaryngology, Head and Neck Surgery, Washington
13. Lanza DC, O'Brien DA, Kennedy DW (1996) Endoscopic repair of cerebrospinal fluid fistulas and encephaloceles. Laryngoscope 106:1119–1125
14. Lehrer J, Deutsch H (1970) Intranasal surgery for cerebrospinal fluid rhinorrhea. Mt Sinai J Med 37:133–138
15. Levine HL (1991) Endoscopic diagnosis and management of cerebrospinal fluid rhinorrhea. Operative Tech Otolaryngol Head Neck Surg 2:282–284
16. Luckett WH (1913) Air in the ventricle of the brain following fracture of the skull. Surg Gynecol Obstet 17:237
17. Mamo L, Cophingnon J, Rey A, Thurrel C (1982) A new radionuclide method for the diagnosis of post-traumatic cerebrospinal fistulas. J Neurosurg 57:92
18. Manelfe C, Guirand B, Tremoulet M (1977) Diagnosis of CSF rhinorrhea by computadorized cisternography using metrazamide. Lancet 2:1073
19. Manelfe C, Callerier P, Sobel D, Proust C, Bonafe A (1982) Cerebrospinal fluid rhinorrhea: evaluation with metrazamide cisternography. AJR Am J Roentgenol 138:471–476
20. Mattox DE, Kennedy DW (1990) Endoscopic management of cerebrospinal fluid leaks and cephaloceles. Laryngoscope 100:857–862
21. McCormack B, Cooper, Persky M, Rothstein S (1990) Extracranial repair of cerebrospinal fluid fistulas: technique and results in 37 patients. J Neurosurg 27:412–417
22. Montgomery WW (1966) Surgery of cerebrospinal fluid rhinorrhea and otorrhea. Arch Otolaryngol 84:92–104
23. Montgomery WW (1973) Cerebrospinal rhinorrhea. Otolaryngol Clin North Am 6:757–771
24. Naidich TP, MoramCJ (1980) Precise anatomic localization of cerebrospinal fluid rhinorrhea by metrizamide CT cisternography. J Neurosurg 53:222–228
25. Northfield JWC (1973) The surgery of the central nervous system. Blackwell Scientific, Oxford
26. Oberascher G, Arrer E (1986) Efficiency of various methods of identifying cerebrospinal fluid in oto- and rhinorrhea. ORL J Otorhinolaryngol Relat Spec 48:320–325
27. Ommaya AK (1976) Spinal fluid fistulas. Clin Neurosurg 23:363–392
28. Ommaya AK, Dichiro G, Baldwin M, Pennybacker JB (1968) Non-traumatic CSF rhinorrhea. J Neurol Neurosurg Psychiatry 31:214–225
29. Park JL, Strelzow VV, Friedman WH (1983) Current management of cerebrospinal fluid rhinorrhea. Laryngoscope 93:1294–1300
30. Settipane GA (1987) Systemic diseases associated with nasal symptoms. Am J Rhinol 1:33–44
31. Shaefer SD, Diehl JT, Brirggs WH (1980) The diagnosis of CSF rhinorrhea by metrizamide CT scanning. Laryngoscope 90:871–875
32. Speltzler RF, Wilson CB (1978) Management of recurrent CSF rhinorrhea of the middle and posterior fossa. J Neurosurg 49:393–397
33. Stamm A (1994) Microcirurgia transnasal no tratamento da rinoliquorréia etmoido-esfenoidal. Escola Paulista de Medicina, São Paulo
34. Stamm A, Pignatari SS (1998) Transnasal micro-endoscopic surgery for CSF rhinorrhea. In: Stammberger H, Wolf G (eds) Congress of the European Rhinologic Society (Vienna, 1998). Monduzzi, Bologna, pp 329–335

35. Stamm A, Neto LM, Braga FM, Souza HL (1992) Microcirurgia transnasal no tratamento da rinoliquorréia. Rev Bras ORL 58:176–184
36. Stammberger H (1991) Detection and treatment of cerebrospinal fluid leaks. In: Stammberger H, Hawke M (eds) Functional endoscopic sinus surgery. Decker, Philadelphia, pp 436–441
37. Stankiewicz JA (1991) Cerebrospinal fluid fistula and endoscopic sinus surgery. Laryngoscope 101:250
38. Thomson S (1899) The cerebro-spinal fluid: its idiopathic escape from the nose. Cassell, London, p 8
39. Thomson S, Negus VE (1947) Diseases of the nose and throat, 5th edn. Appleton-Century-Crafts, New York, p 104
40. Vrabec DP, Hallberg OE (1964) Cerebrospinal fluid rhinorrhea. Arch Otolaryngol Head Neck Surg 80:218–229
41. Wigand ME (1990) Endoscopic surgery of the paranasal sinuses and anterior skull base. Thieme, New York, p 128
42. Yessenow RS, McCabe BF (1989) The osteo-mucoperiosteal flap in repair of cerebrospinal rhinorrhea: a 20-year experience. Otolaryngol Head Neck Surg 101:555–558

Endoscopic Repair of Cerebrospinal Fluid Rhinorrhea

Alfredo Herrera and Emiro Caicedo

Introduction

CSF rhinorrhea can be defined as the presence of cerebrospinal fluid (CSF) in the nasal cavity; the CSF comes from a dural tear in the anterior cranial fossa. CSF rhinorrhea can also appear paradoxically, due to lesions in the middle or posterior cranial fossa, with fluid accumulation in the middle ear draining into the nasopharynx and nasal cavity via the Eustachian tube.

CSF rhinorrhea has been recognized for over a century and, to avoid devastating complications (such as meningitis and cerebral abscess), management has focused on diligent diagnosis and treatment. Treatment modalities include medical and surgical intervention. A variety of surgical techniques have been described for the management of CSF leaks and include intracranial and extracranial approaches that will be discussed thoroughly in this chapter.

Classification

The need for a universal classification system for CSF rhinorrhea has been recognized for decades [3, 35, 38]. We currently classify CSF rhinorrhea according to the anatomic site of the lesion and according to the etiology (Tables 39.1, 39.2). We feel that this classification accurately represents the clinical situations of our patients and has inherent therapeutic implications.

In our classification system, spontaneous CSF rhinorrhea refers to cases in which a traumatic event is probably responsible for the presence of the CSF fistula. Classification is not always an easy task. Some patients may overlook or forget minor traumatic events that may have etiologic significance. We often have to decide whether a particular minor craniofacial trauma is relevant or not as a predisposing factor for a CSF leak that may appear many years later; this can be quite challenging.

Table 39.1. Classification of cerebrospinal fluid (CSF) fistulas according to the anatomic site of the lesion

Frontal sinus
Ethmoid sinus
Lateral lamella of the cribriform plate
Anterior fovea ethmoidalis
Posterior fovea ethmoidalis
Cribriform plate
Anterior
Posterior
Sphenoid sinus
Lateral wall
Superior wall
Posterior wall
Paradoxical (through the Eustachian tube)
From the middle cranial fossa
From the posterior cranial fossa

Table 39.2. Classification of cerebrospinal fluid (CSF) fistulas according to etiology

Traumatic
After craniofacial trauma
Acute (appears in less than 14 days)
Delayed (appears in more than 14 days)
Associated with encephaloceles, intracranial hypertension, other conditions
Postsurgical (iatrogenic)
Acute (appears in less than 14 days)
Normal part of a surgical procedure (pituitary surgery, optic-nerve decompression, others)
Complication during surgery
Delayed (appears in less than 14 days)
Associated with encephaloceles, empty-sella syndrome, benign intracranial hypertension, other conditions
Non-traumatic (spontaneous)
Primary (pure)
Secondary [associated with encephaloceles, benign intracranial hypertension, hydrocephalus, space-occupying lesions (tumors), empty-sella syndrome, osteomyelitis, other conditions]

It is generally accepted that spontaneous CSF leaks may occur due to a previously occult congenital defect in the skull base. Hooper [5] studied 138 human sphenoid bones and found that remnants of the craniopharyngeal canal were present in 18% of the bones, with 5% having defects connecting the cranial vault with the sphenoid sinus.

An association between CSF rhinorrhea and empty sella has been reported in the literature [15] and should be considered. The appearance of these symptoms may occur spontaneously or following pituitary surgery or radiotherapy in this area.

It is also important to define some of the terminology frequently used to describe this pathological state, as the terminology can be confusing and, on occasions, misleading. For example, the term "CSF fistula" has been used interchangeably with terms such as "CSF leak" or "CSF rhinorrhea".

CSF fistula usually refers to a communication between the sterile intracranial cavity and the non-sterile nasosinus cavity. The terms CSF leak and CSF rhinorrhea imply the presence of CSF in the nose.

Diagnosis

Diagnosis of CSF rhinorrhea is primarily based on clinical history, a thorough otorhinolaringologic and neurologic exam and high-resolution radiological studies. Other diagnostic procedures include fluorescein endoscopy, the measurement of the glucose concentration and the presence of β2-transferrin in nasal secretions, all of which are useful in certain cases.

Clinical History

Patients with a CSF leak usually present with a unilateral, clear, watery rhinorrhea, which can appear abruptly or in association with certain positions of the head, such as extreme forward flexion. The quantity and frequency of the CSF leak varies from one patient to another and probably depends on a number of factors, which include the etiology and the size of the defect.

Diagnosis is usually easy when there is a recent history of trauma and a profuse leakage. Nevertheless, in the absence of trauma, one must always remember that a spontaneous leak may occur in an otherwise healthy person.

It is important to remember that patients with inactive CSF fistulas should have a thorough work-up. In our series, we had patients with inactive fistulas that were undiagnosed for many years. In one case, a 64-year-old patient with a 12-year history of intermittent CSF rhinorrhea cancelled surgery several times due to spontaneous remission. When the patient was referred to us after an episode of meningitis and a 9-month remission of his CSF rhinorrhea, magnetic resonance imaging (MRI) revealed a large encephalocele in the right sphenoid sinus (Fig. 39.1).

Patients frequently have other symptoms, such as headaches. In our series, two patients, both with a spontaneous CSF leak at the lateral wall of the right sphenoid sinus, experienced severe episodes of right facial pain in the infraorbital nerve area; these were exacerbated by walking or mild tapping of the heels and had been misinterpreted as sinusitis. The pain completely disappeared immediately after surgical repair of the CSF leak and has been absent ever since.

Meningitis can occur following traumatic CSF leaks; this usually occurs 1–2 weeks after the initial trauma, but can occur up to 20 years later [38]. Meningitis can also be a late complication in cases of non-traumatic CSF leaks.

Patients with recurrent episodes of meningitis or with an intracranial infection due to an upper respiratory tract pathogen, such as *Pneumococcus*, require a thorough work-up to search for an occult dural lesion with or without active CSF rhinorrhea [46]. In our series, one patient, who had a history of multiple episodes of meningitis beginning 11 years after suffering trauma to the frontal region, was found to

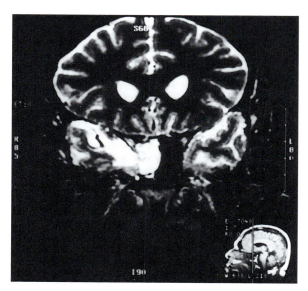

Fig. 39.1. Magnetic resonance. Coronal view in a T2-weighted sequence demonstrating a large encephalocele in the sphenoid sinus. Notice the temporal lobe protruding through a bony defect in the lateral wall of the right sphenoid sinus

have a meningoencephalocele in the right anterior fovea ethmoidalis. This meningoencephalocele was associated with an inactive CSF leak, which was successfully closed via a transnasal endoscopic approach.

However, two of our patients developed profuse unilateral CSF rhinorrhea 5 years and 7 years before they were referred to our institution, where a definite diagnosis of spontaneous CSF leak was made; despite this long period of time, neither had developed meningitis. Anosmia is another symptom that can appear; it is usually associated with defects of the cribriform plate. If a CSF leak occurs during sinus surgery and bleeding is not excessive, the washout sign, where clear fluid leaking through a dural defect washes away the blood from the surrounding structures, will be evident.

Diagnostic Laboratory Tests

Non-invasive diagnostic tests, such as glucose concentrations and β2-transferrin in nasal secretions, are useful in cases in which we suspect CSF rhinorrhea but are unable to make a definite diagnosis. The patient must have an active CSF leak so that enough fluid for a quantitative test may be collected.

When testing for glucose in nasal secretions, in order to avoid false positive results, it is important that the sample not be contaminated by blood. Using quantitative methods to measure the glucose concentrations in these secretions, the accuracy of the diagnosis of CSF rhinorrhea is very high when glucose concentrations are greater than 30 mg/100 ml. Glucose oxidase papers are unreliable and should be avoided, because reports have shown false positive results in up to 75% of cases) due to the presence of reducing substances in natural tears [39]). β2-Transferrin is a reliable, specific marker for CSF, and the mere presence of this protein in nasal secretions is diagnostic for CSF rhinorrhea [41, 42].

Radiologic Diagnosis

Radiologic diagnosis is essential for the confirmation of a CSF leak (in some cases), for preoperative localization of the exact site of the defect and CSF leak and as a surgical road map for evaluation of the individual sinus/skull-base anatomy (for surgical planning).

Radionuclide Cisternography

In 1964, Di Chiro [10, 11] described the method of radionuclide cisternography. Since then, many authors have reported the benefits of this test for the diagnosis of CSF rhinorrhea [27, 43]. Nevertheless, the true diagnostic accuracy for the test is controversial. Although the test confirms the presence of an active CSF leak, it has little or no value in localizing the exact site of the leak [4, 23, 33].

High-Resolution Computed Tomography Scans

High-resolution computed tomography (CT) scans are a non-invasive diagnostic procedure that is very useful, especially when a cisternography is contraindicated. This includes patients with a history of iodine allergy or a seizure disorder refractive to medical treatment and patients presenting with an acute traumatic CSF leak [8].

A bony dehiscence with a soft-tissue density in a paranasal sinus suggests a CSF leak at that site. Nevertheless, some cases may be challenging when trying to differentiate between a natural and a traumatic bony dehiscence in the skull base, particularly in the lateral lamella and cribriform-plate areas. CT scans do not clearly identify a soft-tissue density in a sinus as CSF rhinorrhea, blood or inflammatory disease.

Water-Soluble Contrast CT Cisternography

Water-soluble contrast CT cisternography is now considered to be the gold standard for the diagnosis of CSF rhinorrhea [12, 45, 54]. It is an excellent study for the localization of the exact site of the skull-base defect but should be performed only when the patient has an active CSF leak (Fig. 39.2). This can be a challenging situation in patients who have only occasional episodes of CSF rhinorrhea with low flow. Metrizamide has been replaced by iopamidol, which is a non-ionic, water-soluble contrast agent with low neurotoxicity effects.

Magnetic Resonance Imaging

MRI is an accurate, non-invasive exam that has proved to be very useful for the localization of inactive and spontaneous CSF leaks, especially in T2-

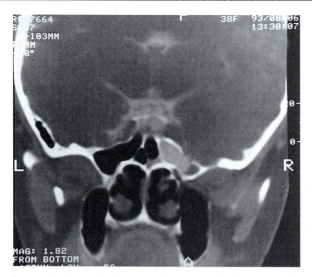

Fig. 39.2. Water soluble computed tomography cisternography using iopamidol. Demonstration of contrast material penetrating the sphenoid sinus through a bony defect in the lateral wall of the right sphenoid sinus

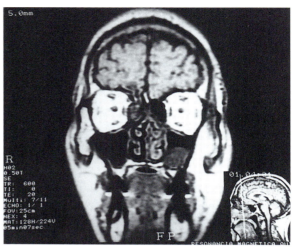

Fig. 39.3. Magnetic resonance. Coronal view in a T1-weighted image demonstrating a type-I encephalocele in the right anterior fovea ethmoidalis

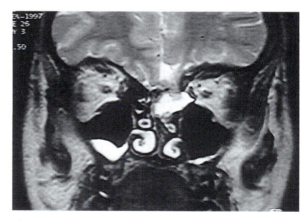

Fig. 39.4. Magnetic resonance. Coronal view in a T2-weighted image demonstrating a type-II encephalocele in the left anterior fovea ethmoidalis. Notice the hyperintense signal of the predominantly cystic portion of this encephalocele and the isointense signal of the relatively small brain herniation

weighted images [14, 31]. Remember that, on T2-weighted images, CSF has a very high signal, whereas bone has a very low signal. This high contrast between bone and CSF enables us to see the site of the lesion as a defect in the dark line representing the dura and cortical bone. A CSF signal in a paranasal sinus continuous with that within the basal cistern in T2 images is diagnostic of CSF rhinorrhea. If a CSF signal is seen in one of the paranasal sinuses and is not continuous with the basal cisterns, T1-weighted images before and after intravenous gadolinium diethylene triaminopentaacetic acid (Gd-DTPA) administration should be obtained to differentiate CSF from inflammatory disease. According to Eljamel [14], if the CSF signal does not enhance with Gd-DTPA on a T1-weighted sequence, a diagnosis of suspected CSF fistula should be made. In his study, the three patients with a MRI diagnosis of suspected CSF fistula were confirmed at surgery. This study also showed that MRI was 100% accurate in localizing the site of the defect in 21 patients with inactive CSF leaks, whereas CT scans missed 36% of these lesions. In this study, CT scans also yielded false positive results 9.5% of the time.

In cases in which CT shows a soft-tissue density adjacent to a bony dehiscence in the skull base, the presence of an encephalocele should be suspected. In such cases, MRI will determine if it is indeed an encephalocele and will give important information regarding how much of the herniated tissue is actually brain and how much is an accumulation of CSF.

We currently classify encephaloceles as type I or type II, based on information from MR studies in T1- and T2-weighted images (Fig. 39.3, 39.4).
- **Type I.** Predominantly solid (brain herniation)
- **Type II.** Predominantly cystic (CSF)

Type-I encephaloceles tend to be more challenging, and one must always remember that, in cases with large type-I encephaloceles, there is a chance that the herniated tissue may contain a large artery that could be inadvertently injured during surgery. In such a case, MRI angiography can be helpful. If a large vessel is seen within the herniated tissue, it is a relative contraindication for the transnasal approach.

due to the difficulty of controlling the bleeding from below. In such cases, an intracranial approach is preferred.

Intrathecal Fluorescein Endoscopy

The use of intrathecal fluorescein endoscopy (IFE) for the diagnosis of CSF rhinorrhea has been a controversial issue in the literature. Early studies using various intrathecal dyes were associated with a high, unacceptable morbidity rate due to chemical meningitis probably related to high concentrations of these dyes. More recent studies have discussed anecdotal cases of serious – but always transient – complications (including grand mal seizures) when performing this test at normal or slightly elevated concentrations [28, 30].

However, there are many studies that prove that IFE is a safe and useful diagnostic procedure when used with standard doses and concentrations [50]. There are no studies citing permanent complications associated with the use of intrathecal fluorescein.

We are currently using IFE as an intraoperative test (Fig. 39.5), but it can also be used preoperatively in selected cases. We use 0.2 cm^3 of a 5% solution of fluorescein mixed with 5–10 cm^3 of the patient's CSF. This solution is slowly injected intrathecally at the level of the L3–L4 intervertebral space, and the patient is put in a Trendelenburg position to enhance diffusion. Usually, after 20 min, the bright yellowish-green solution can be clearly seen in the nose and followed back to the exact site of the dural lesion at the skull base. With these concentrations, we do not need special light sources or filters in order to identify the fluorescein-stained CSF.

A recent study by Lanza et al. [26] shows that intraoperative IFE was useful in localizing the exact site of the skull-base defect in 64% of this series of 36 patients. In our series of 22 CSF fistulas, intraoperative IFE accurately localized the exact site of the CSF leak in 30% of our patients in which this technique was used (i.e., three of ten patients). In seven patients, an IFE was unable to identify the site of the leak despite the fact that a fracture line was clearly seen at the lateral lamella of both cribriform plates.

As a test for localizing the exact site of the CSF leak, IFE had little value in our series, due to the fact that the test was positive mainly in cases with large defects, where we would have found the site of the CSF leak without the help of the test. Unfortunately, in difficult cases in which there were small defects with intermittent or low-flow CSF rhinorrhea, the fluorescein did not appear and, therefore, was not helpful in identifying the site of the defect.

We conclude that the main benefit of this test is intraoperative evaluation of the efficacy of the CSF-leak surgical closure. If fluorescein-stained CSF is seen leaking around the graft, we can reposition it until we achieve a watertight seal.

Management

Management options should be selected based on the etiology, site and size of the defect and each surgeon's personal experience with the various surgical techniques available today, because these have proved to be key factors in successful closures [25].

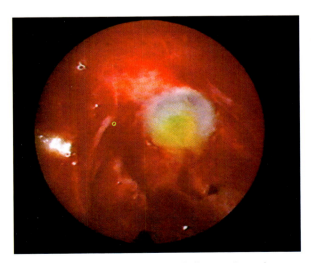

Fig. 39.5. Intraoperative intrathecal fluorescein endoscopy. Note the typical yellowish-green fluorescein, demonstrating an active CSF leak in the right sphenoid sinus

Management According to Etiology

Post-craniofacial Trauma CSF Fistulas

Most patients presenting with a CSF leak following craniofacial trauma respond well to conservative medical treatment alone [7]. A study by Shapiro and Skully [47] reports that 94% of their cases of post-traumatic CSF leaks were successfully closed nonsurgically, using a lumbar subarachnoid drainage procedure.

Initial conservative therapy for the CSF leak should not delay the surgical management of facial fractures, and realignment of displaced fractures seems to promote dural healing. If a CSF leak persists after 10 days of medical management, then a surgical closure of the dural defect is warranted. In patients with a CSF leak that begins several months or years after craniofacial trauma, early surgical intervention is favored.

Post-surgical CSF Fistulas

Post-surgical CSF rhinorrhea can occur as a complication of intracranial procedures, trans-septosphenoid hypophysectomies, or during paranasal-sinus surgery. CSF rhinorrhea has also been reported to occur following rhinoplasty, although this is certainly an infrequent complication of the procedure [18]. Charles et al. [6] report four cases of CSF leaks following 100 hypophysectomies. CSF rhinorrhea as a complication of endoscopic sinus surgery has been reported in between 0.34% and 1.3% of cases [51, 57].

Most authors agree that iatrogenic CSF leaks that are recognized intraoperatively should be closed during the course of the operation [17, 52]. If the CSF leak is recognized during the postoperative period, small defects will probably close with conservative management alone. In our institution, we have seen that 100% of the CSF leaks that appeared following trans-septosphenoid hypophysectomy closed successfully with medical management (which included a lumbar-drainage procedure). Nevertheless, clinical judgement should prevail in cases in which a profuse CSF rhinorrhea is detected postoperatively; in such a case, an immediate surgical exploration and closure of the defect is probably warranted.

Primary Non-traumatic (Spontaneous) CSF Fistulas

Due to the unlikelihood of a definite closure of the dural defect with medical therapy alone, management options for spontaneous CSF leaks favor early surgical intervention. A retrospective study from the Mayo Clinic [23] reviews a series of 28 patients diagnosed as having a spontaneous CSF leak; three patients were treated non-surgically, due to an abrupt remission of their symptoms. This study revealed that, during the following 24 months, all three patients had recurrence of their CSF leaks, which ultimately required surgical management.

Secondary Non-traumatic CSF Fistulas

The surgical options for non-traumatic CSF-fistula patients depend entirely on the associated intracranial pathology present in each specific case. Patients with a CSF leak associated with hydrocephalus will require some type of shunt procedure. Patients with a CSF leak associated with a space-occupying lesion will require a craniotomy for resection of the lesion, and simultaneous closure of the dural defect causing the CSF leak (via the same approach). CSF leaks that are associated with encephaloceles can be treated successfully with a transnasal endoscopic approach [2, 26, 29], as can CSF leaks associated with an empty sella.

Medical Management

The medical management of CSF rhinorrhea requires strict bed rest with a head elevation of 30°. Though controversial, spinal fluid drainage procedures (either through repeated lumbar punctures or by means of an indwelling lumbar subarachnoid catheter) can be beneficial in selected cases.

Spinal fluid drainage procedures should be done with a strict aseptic technique and with the patient awake to minimize trauma to the nerve roots and spine elements. Ideally, these procedures are performed with the patient in a sitting position, because the dural sac will be distended, making this procedure easier [32]. We typically drain between 30 cm^3 and 40 cm^3 daily.

If an indwelling lumbar subarachnoid catheter is used, the patient's head should not be elevated more than 10°, and the patient should not be allowed to sit up, because excessive drainage or a siphon effect with pneumocephalus can then occur. A siphon effect occurs when a negative-gradient CSF pressure permits air to enter the intracranial cavity through the open dural fistula [37]. We do not recommend the use of prophylactic antibiotics to prevent meningitis as part of the medical treatment of CSF leaks, because several studies have shown that they are not effective in preventing intracranial infections. In some cases, the use of antibiotics has been associated with severe complications due to infection by resistant organisms [24, 40].

Surgical Management

Surgical alternatives include intracranial and extracranial approaches. Each has specific indications, advantages and disadvantages that should be considered when planning surgery.

The Intracranial Approach

In 1926, Dandy [9] was the first to report the successful intracranial repair of a CSF leak through a bifrontal craniotomy. Since then, a variety of intracranial techniques using different types of grafts and flaps have been described [21, 34].

The main advantage of the intracranial approach is the possibility of treating associated problems, such as intracranial bleeding or tumors, and closing any associated dural defect. The main disadvantage of the intracranial approach is the loss of olfaction that occurs during mobilization of the anterior cranial base. Success rates for the intracranial closure of CSF leaks have been between 67% and 73% after the first procedure and up to 90% after multiple procedures [39].

The Extracranial Approach

The extracranial approach has been widely used since Dohlman described the first closure of a CSF leak via an external ethmoidectomy in 1948. Since then, a variety of extracranial procedures using different types of grafts and local flaps with good results have been described [22, 55, 58].

In 1981, Wigand [56] presented the first reported cases of CSF leaks treated via a transnasal endoscopic approach. Following this report, experienced micro-endoscopic surgeons have described a variety of transnasal microsurgical [49] and endoscopic [1, 2, 17, 19, 26, 29, 37, 52] techniques for the treatment of CSF fistulas, with a high rate of successful closures and a low morbidity rate.

Table 39.3 summarizes various reports of success rates for the closure of CSF leaks using the transnasal endoscopic approach. The following is our rationale for choosing the types of grafts, the graft-positioning techniques and the flaps for our transnasal endoscopic approaches.

Graft Selection

Mucosal Grafts

The most commonly used mucoperiostial grafts are those taken from the contralateral inferior turbinate or septum. Taking the graft from the opposite side of the CSF leak prevents blood from obscuring the surgical field of vision.

Fairly large grafts can be transferred (easily and with low morbidity) to the donor site, which will usually develop mucosa in 3–4 weeks [29]. The size of the graft should be 30% larger than the defect in order to compensate for postoperative graft shrinkage. Precise trimming and positioning of the graft is important in order to avoid tenting and folding of the graft, which would compromise the seal.

An experimental, histological animal study by Geode [17] showed that, 1 week after operation, the mucoperiostium tightly adhered to the surrounding bone and, after 3 weeks, it was completely transformed into fibrous connective tissue. Although mucosal grafts can be used as overlay or underlay grafts (see "Graft-Positioning Techniques"), we prefer to use them only as overlay grafts, because we are concerned that even a slight possibility of mucosal entrapment could compromise the seal. We frequently use these grafts as a single overlay free graft for the treatment of CSF leaks at the cribriform plate or the lateral lamella of the cribriform plate.

Connective-Tissue Grafts

The most frequently used autologous connective-tissue grafts are fascia-lata and/or muscle grafts, temporalis-fascia and/or muscle grafts, and cartilage or bone grafts taken from the nasal septum. We commonly use these grafts in both overlay and (especially) underlay techniques.

There is some controversy regarding the use of septal-cartilage or bone grafts as an additional support when dealing with large bony defects of the skull base. Nevertheless, because we have seen encephaloceles protrude through bony defects as small as 1.2 cm, we feel that bone grafts can be beneficial when dealing with skull-base defects larger than 1 cm. This is especially true if they can be placed between the dura and the intracranial surface of the skull base, above the defect.

Other Grafts

Other grafting materials include lyophilized dura and synthetic dura substitutes. Gates [16] reported on the use of a new alloplast known as biocompatible osteoconductive polymer solution/powder, which is a powder form of methyl methacrylate, which was used for the closure of a large defect that had failed to close by more conventional techniques.

The biocompatible glass–ionomer bone cement can also be an interesting alternative for difficult cases involving large bony defects, such as those seen after removal of large skull-base tumors. The value of these grafts compared with autologous grafts has yet to be demonstrated in large randomized studies, and their long-term results are not yet known.

Table 39.3. Closure of cerebrospinal fluid (CSF) fistulas. Success rates for the transnasal endoscopic approach

Authors	Total number of patients	Successful first attempts	Successful second attempts	Craniotomies
Papay et al.	4	4 (100%)	–	0
Mattox and Kennedy	7	6 (86%)	7 (100%)	0
Stankiewicz	6	6 (100%)	–	0
Gjuric et al.	33	32 (97%)	32 (97%)	1
Kelley et al.	8	7 (88%)	8 (100%)	0
Burns et al.	42	35 (83%)	38 (90%)	3
Lanza et al.	36	34 (94%)	35 (97%)	0

Graft-Positioning Techniques

Overlay Technique

With the overlay technique (Figs. 39.6, 39.7), once the skull-base defect has been identified, the surrounding mucosa is elevated and removed for at least 3–5 mm in all directions. This will prevent the mucosa from interfering with graft adhesion to the underlying bone.

If the bony defect is small, we place a free graft (mucosal or connective), which is glued over the defect and surrounding bone with fibrin glue, as described in the literature [13, 20, 36, 44, 53]. We will typically use this technique for small defects or defects that are localized to the cribriform plate or the lateral lamella of the cribriform plate. With this technique, we most often use a contralateral inferior-turbinate mucosal graft.

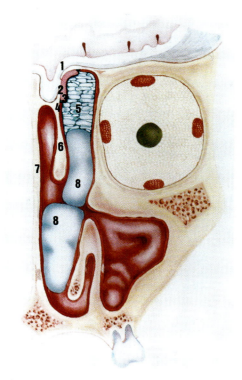

Fig. 39.6. The overlay technique. Schematic drawing representing the closure of a CSF leak at the lateral lamella of the left cribriform plate. *1*, Dura; *2*, lateral lamella of the cribriform plate; *3*, contralateral inferior-turbinate free graft; *4*, ethmoid-sinus mucoperiostium; *5*, Gelfoam; *6*, middle turbinate; *7*, septum; *8*, Merocel sponge

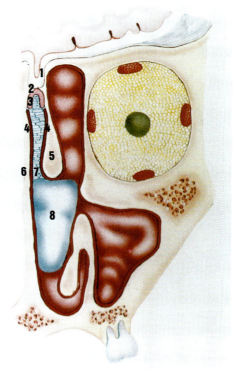

Fig. 39.7. The overlay technique. Representation of the closure of a CSF leak at the left cribriform plate. *1*, Dura; *2*, cribriform plate; *3*, contralateral inferior-turbinate free graft; *4*, mucosa; *5*, middle turbinate; *6*, septum; *7*, Gelfoam; *8*, Merocel sponge

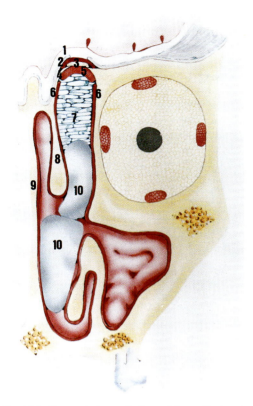

Fig. 39.8. The underlay technique. Schematic drawing representing the closure of a CSF leak at the left fovea ethmoidalis. *1*, Dura; *2*, fascia-lata underlay graft; *3*, septal-bone underlay graft; *4*, fovea ethmoidalis; *5*, contralateral inferior turbinate overlay graft; *6*, ethmoid-sinus mucoperiostium; *7*, Gelfoam; *8*, middle turbinate; *9*, septum; *10*, Merocel sponge

The Underlay Technique

After we have localized the exact site of the defect, we proceed to elevate the dura off the skull base surrounding the bony defect for 2–3 mm, thus creating an epidural space. We frequently use various neuro-otologic elevator instruments, and care is taken not to damage the dura surrounding the defect.

We then place a connective-tissue graft (fascia lata) in this epidural pocket above the bony defect. If the bony defect is larger than 1 cm, we will also place a septal bone graft in this pocket.

We then proceed to remove the mucosa surrounding the bony defect from the nasal side of the skull base, as we would in an overlay technique. We then reinforce our closure by placing an overlay inferior-turbinate mucosal graft. With this technique (Fig. 39.8), we achieve a triple-layered seal that separates the intracranial structures from the nasal cavity.

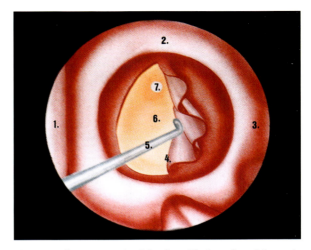

Fig. 39.9. The mucoperiostial sphenoid-sinus flap. Schematic drawing representing the closure of a CSF leak at the superior wall of the left sphenoid sinus. *1*, Septum; *2*, superior border of the sphenoidotomy; *3*, lateral nasal wall; *4*, mucoperiostial flap being elevated laterally; *5*, disector; *6*, exposed bone; *7*, site of the CSF leak

Local Flaps

A variety of local flaps (primarily taken from the nasal septum and middle turbinate) have been described for the closure of CSF leaks [22, 49, 55, 58]. In our series, we did not feel the need to rotate septal or middle turbinate flaps for the treatment of CSF leaks in the ethmoid-sinus/cribriform-plate areas. Nevertheless, we believe that a mucoperiostial flap can be beneficial in selected cases of CSF leaks in the sphenoid sinus. Since rotating a septal or turbinate flap into the sphenoid sinus is somewhat cumbersome, and normal postoperative retraction of this flap can mobilize the grafts and compromise the seal, we are now using what we call the mucoperiostial sphenoid-sinus flap.

The Mucoperiostial Sphenoid-Sinus Flap

Once we have created a large sphenoidotomy through the natural ostium, we proceed to elevate the mucoperiostium inside the sphenoid sinus. We begin elevating the flap at the superior and medial walls of the sinus and elevate at least 3–5 mm of the mucoperiostium surrounding the bony defect and the CSF leak. This flap is pedicled laterally and inferiorly. We then place connective-tissue grafts (fascia lata and/or bone), using either an underlay or overlay technique (Figs. 39.9, 39.10).

Once we have controlled the CSF leak via these grafts, we replace the vascularized mucoperiostial

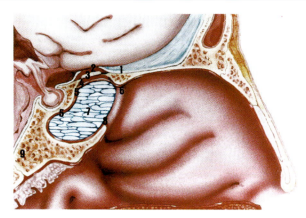

Fig. 39.10. Sagittal view of the mucoperiostial sphenoid-sinus flap with an underlay/overlay graft. *1*, Dura; *2*, fascia-lata underlay graft; *3*, septal-bone underlay graft; *4*, fascia lata overlay graft; *5*, mucoperiostial sphenoid sinus flap; *6*, superior border of the sphenoidotomy; *7*, Gelfoam; *8*, inferior border of the sphenoidotomy; *9*, clivus

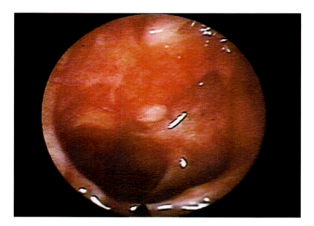

Fig. 39.11. Mucoperiostial sphenoid-sinus flap. Postoperative view 2 months after the closure of a post-traumatic CSF leak in the right sphenoid sinus

flap so that it covers the grafts when it returns to its original position. The fact that this flap is returned to its original position seems to minimize the postoperative retraction of the flap and the mobilization of the grafts.

At the end of the procedure, the sphenoid sinus is packed with Gelfoam, which serves as a temporary support while the mucoperiostial flap and the grafts adhere to the bony walls of the sinus. Once the Gelfoam reabsorbs, what remains is a functional sphenoid sinus that can easily be inspected endoscopically in the office (Fig. 39.11).

Other techniques include obliteration of the sphenoid sinus, which requires complete resection of all the sinus mucosa in order to avoid late complications, such as mucoceles. This can be quite difficult, especially in a sinus with small lateral recesses, and can also be dangerous if there is a dehiscent carotid artery or optic nerve. In 8% of normal sphenoid sinuses, there is a dehiscent carotid artery and, in 4%, there is a dehiscent optic nerve [48]. Inadvertent injury to either structure during mucosal elevation can have devastating consequences. The mucoperiostial sphenoid-sinus flap is an effective functional technique that avoids obliteration of the sphenoid sinus and the late complications associated with this technique.

The Transnasal Endoscopic Closure of CSF Leaks

The following is a summary of the technique we use for the closure of CSF leaks of the anterior skull base:
1. We usually prefer general anesthesia.
2. We always insert a spinal lumbar subarachnoid catheter while the patient is awake or in a sitting position. In certain situations, especially when closing large defects or encephaloceles, we may need to decrease the intracranial pressure. This can be achieved by removing 30–40 cm^3 of CSF through the catheter. In selected cases, especially when treating small defects that are not easily identified or if multiple sites of CSF leaks are suspected, we use intrathecal fluorescein (at the concentrations mentioned before) as an aid to exact localization of the defect.
3. The endoscopic approach depends on the suspected site of the lesion. CSF leaks at the ethmoid sinus or the lateral lamella of the cribriform plate will require a complete endoscopic anterior and posterior ethmoidectomy to gain wide exposure of the skull base. CSF leaks at the cribriform plate are approached directly through the olfactory groove. Lateral displacement of the middle turbinate using delicate maneuvers may be necessary to enhance visualization. Sphenoid-sinus CSF leaks can be approached endoscopically in many ways, including a trans-septosphenoid approach, a direct transnasal approach through the sphenoethmoid recess or a transethmoid approach. When we use a direct transnasal or a transethmoid approach to the sphenoid sinus, we usually resect the superior turbinate to gain a wide access to the anterior wall of the sphenoid sinus, and we perform a wide sphenoidotomy via the natural ostium with the aid of a sphenoid punch or drill.
4. Once the defect is identified, the surrounding mucosa is elevated and resected for at least

3–5 mm in all directions (except in selected cases in which we plan to perform a sphenoid-sinus mucoperiostial flap, as described previously).
5. At this point, we place the grafts according to the techniques and principles we have mentioned previously. In the majority of our cases, we use fibrin glue as a tissue adhesive and to promote watertight seals.
6. Once we have completed the closure of the defect and have our grafts in position, we always double check to be sure that the CSF leak has stopped. If we have used intrathecal fluorescein, even a mild, persistent CSF leak is enhanced. In such a case, we review our technique and probably reposition our grafts and again cover them with fibrin glue until we achieve a watertight seal.
7. We then place multiple layers of Gelfoam over our grafts and/or flap and place a Merocel sponge beneath them to secure adequate support. The Merocel will be kept in place for 5–7 days, and the Gelfoam is left to reabsorb by itself.
8. The patient stays in the hospital for 5–8 days to assure adequate postoperative treatment during this crucial period.
9. Postoperative management includes strict bed rest with a head elevation of 30°, stool softeners and diuretics. We use postoperative antibiotics to protect against bacterial infections induced by the nasal packing. We typically use antibiotics that protect against *Staphylococcus* but do not penetrate the blood–brain barrier, because we do not seek prophylaxis against meningitis (for the reasons mentioned previously). Early in our series, we utilized a spinal-drainage procedure for 2–3 days, though this is still a controversial issue in the literature and among various endoscopic sinus surgeons. In our last ten cases, this procedure was not performed, and no changes were seen in our results. Therefore, we do not routinely perform postoperative spinal drainage, although it may be helpful when rhinorrhea is evident and abundant during the postoperative period.

Clinical Cases

Table 39.4 summarizes our experiences in the treatment of CSF leaks and encephaloceles of the anterior skull base using the transnasal endoscopic technique. This series includes ten CSF leaks in nine patients who were treated endoscopically by the senior author (Herrera).

In our series, a successful repair was accomplished in all of the 22 CSF fistulas treated surgically, with a 100% closure rate after a single surgical procedure.

Nevertheless, one of our patients (patient 12) presented with a postoperative CSF leak at a different site than that of the previously treated leak and was not considered as a surgical failure. Interestingly, this patient was initially diagnosed as having a spontaneous CSF leak in the right sphenoid sinus but, 2 months after the operation, the patient presented with a CSF leak coming from the non-operated left ethmoid sinus. In this case, a postoperative lumbar puncture revealed intracranial pressure, which ultimately required a ventriculo-peritoneal shunt procedure.

Postoperative follow-up ranged from 3 months to 90 months (mean: 25 months). There were no significant complications, except for patient 2, who presented with meningitis without evidence of CSF rhinorrhea 3 months after we had performed an endoscopic closure of a CSF leak in her right sphenoid sinus. She was treated medically for the meningitis, which resolved with no sequelae. She has been asymptomatic since then for over 6 years.

Another patient (patient 11) presented with a frontal-lobe abscess without CSF rhinorrhea 5 months after endoscopic closure of a CSF fistula associated with a type-II encephalocele of the left fovea ethmoidalis. She was treated with antibiotics, and the abscess resolved without sequelae, and has been asymptomatic for the last 10 months.

Nine of our patients were initially diagnosed as having spontaneous CSF fistulas, and all of them had normal preoperative intracranial pressure measurements. Three of those patients had a benign intracranial hypertension that became evident after the third postoperative week, requiring a shunt procedure; the diagnoses were confirmed by repeated lumbar puncture measurements. Based on our results, we now routinely perform intracranial pressure measurements (preoperatively and 3–4 weeks after surgery) in patients with spontaneous CSF fistulas. This helps to more accurately identify cases of CSF fistulas associated with benign intracranial hypertension, which can be a cause of persistent headaches or recurrent CSF leaks.

Our casuistic includes five encephaloceles and one arachnoid cyst treated endoscopically with success:

- **Patient 2.** This 39-year-old female presented with a spontaneous CSF fistula associated with a small type-II encephalocele protruding through a large bony defect in the lateral wall of the right sphenoid sinus (Fig. 39.12). The encephalocele was reduced by draining 40 cm^3 of the patient's CSF through a spinal catheter, and closure of the defect was achieved using an underlay technique.
- **Patient 8.** This 26-year-old male presented with a late post-cranial facial-trauma CSF fistula associated with a type-I encephalocele in the right ante-

Table 39.4. Patients with cerebrospinal fluid (CSF) fistulas treated surgically by Dr. Alfredo Herrera between 1991 and 1997

Patient number	Age (years)	Gender	Etiology	Anatomical site	Technique
1	45	Female	Spon	R sphenoid sinus	TE (O)
2	39	Female	Spon (A-Encef)	R sphenoid sinus	TE (U)
3	51	Male	LP-Traum	R sphenoid sinus	TE (U/O)
4	40	Female	Iatro-Ref	B lamella cribriform	TE (O)
5	64	Female	Iatro-Ref	R cribriform plate	TE (O)
6	30	Male	Iatro-Pers (OND)	L sphenoid sinus	TE (O)
7	24	Male	LP-Traum (FP-FC)	R sphenoid sinus	TE (O/MSSF)
8	26	Male	LP-Traum (A-Encef)	R A fovea ethmoidalis	TE (U/O)
9	62	Female	Spon (A-ES and A-BIH)	R sphenoid sinus	TE (O)
10	37	Female	Spon	R cribriform plate	TE (O)
11	22	Female	LP-Traum (A-Encef)	L A fovea ethmoidalis	TE (U/O)
12	46	Female	Spon (A-ES and BIH)	R sphenoid sinus	TE (U/O)
13	15	Male	LP-Traum (FP-FC)	R cribriform plate	TE (O)
				L lateral lamella cribriform	TE (O)
14	28	Female	Iatro-Ref (TN Hyp)	ST sphenoid sinus	TE (U/O)
15	58	Female	Spon (A-BIH)	L cribriform plate	TE (O)
16	47	Female	Iatro-Ref (TN Hyp-R-ES)	ST sphenoid sinus	TE (U/O)
17	38	Female	Spon	L cribriform plate	TE (O)
18	31	Male	Iatro-Pers (A-Encef, FESS)	L P cribriform plate	TE (O)
19	52	Female	Spon (A-AC)	R sphenoid sinus	TE (U/O)
20	47	Female	Spon (A-Encef)	L cribriform plate	TE (O)

A-AC, associated with an arachnoid cyst; *A-BIH*, associated with benign intracranial hypertension; *A-Encef*, associated with an encephalocele; *A-ES*, associated with empty-sella syndrome; *FESS*, after functional endoscopic sinus surgery; *FP-FC*, failed previous frontal craniotomy; *Iatro-Pers*, iatrogenic, personal (i.e. not referred by another surgeon); *Iatro-Ref*, iatrogenic, referred by another surgeon; *LP-Traum*, late post-traumatic; *OND*, part of an optic-nerve decompression; *R*, after radiotherapy; *Spon*, spontaneous; *TN Hyp*, after transnasal hypophysectomy; *A*, anterior; *B*, bilateral; *L*, left; *P*, posterior; *R*, right; *ST*, sella turcica; *MSSF*, mucoperiostial sphenoid-sinus flap; *O*, overlay technique; *TE*, transnasal endoscopic approach; *U*, underlay technique

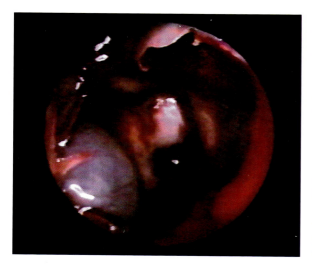

Fig. 39.12. Endoscopic view of an encephalocele protruding through a bony defect in the lateral wall of the right sphenoid sinus

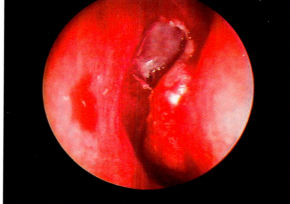

Fig. 39.13. Endoscopic view of an encephalocele protruding through a bony defect in the right anterior fovea ethmoidalis

rior fovea ethmoidalis (Fig. 39.13). The encephalocele was reduced by bipolar cautery in order to expose the bony skull base defect, which was closed using a combination underlay/overlay technique with a triple-layer seal that included an underlay septal-bone graft (Fig. 39.14), an underlay fascia-lata graft and a contralateral inferior-

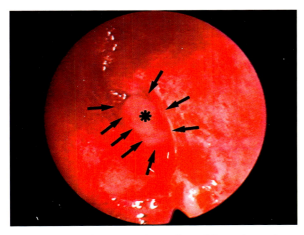

Fig. 39.14. Endoscopic view after fulguration of an encephalocele in the right anterior fovea ethmoidalis using bipolar cautery. Note the septal bone graft (*) placed as an underlay graft. The *arrows* delineate the bony defect in the skull base

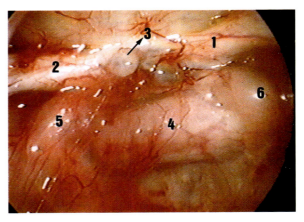

Fig. 39.15. Intraoperative endoscopic view of an arachnoid cyst adjacent to a dehiscent optic nerve in the right sphenoid sinus; the cyst is associated with a spontaneous CSF leak. *1*, Left optic nerve; *2*, right optic nerve; *3*, arachnoid cyst; *4*, Tuberculum sellae; *5* right internal carotid artery canal; *6*, left internal carotid artery canal

turbinate mucosal graft, which was placed as an overlay graft.
- **Patient 11.** This 22-year-old female also presented with a late post-cranial facial-trauma CSF fistula associated with a type-II encephalocele in the left anterior fovea ethmoidalis; she was treated in the same fashion as patient 8.
- **Patient 18.** This 31-year-old male had an iatrogenic CSF fistula associated with a type-I encephalocele in the posterior part of the left cribriform plate; the fistula occurred as an acute complication during a revision endoscopic-sinus procedure. The lesion occurred during use of a microdebrider in the upper medial posterior ethmoid sinus. An overlay technique was used to close the defect, and the patient has been completely asymptomatic for over 8 months. This case represents the only iatrogenic CSF fistula in over 1800 endoscopic procedures, and it is a good example of how cautious one must be while working in this particularly high-risk area, especially when using a microdebrider.
- **Patient 19.** This 52-year-old female presented with a spontaneous CSF leak associated with a small arachnoid cyst located in the right sphenoid sinus, adjacent to a dehiscent optic nerve (Fig. 39.15). After removing the arachnoid cyst, the defect was closed using an underlay/overlay technique without obliterating the sphenoid sinus. The functional sphenoid cavity can easily be inspected in the office. This patient has remained asymptomatic for over 5 months.
- **Patient 20.** This 47-year-old female presented with a spontaneous CSF leak associated with a small encephalocele in the left anterior cribriform plate. The patient was treated with an overlay technique and has been asymptomatic for over 5 months.

Two of our patients (7 and 13) had previous frontal craniotomies that failed to close the CSF fistulas. One of these patients also underwent a shunt procedure, which failed to close the leak. Both patients were treated with a transnasal endoscopic technique and have been asymptomatic ever since.

Summary

CSF fistula is a potentially fatal illness that requires precise diagnosis and diligent treatment. The transnasal endoscopic repair of CSF leaks has had a high success rate and low morbidity when performed by experienced endoscopic sinus surgeons. It should be considered a first choice for all surgical cases, except when intracranial pathology (such as an intracranial hemorrhage or tumor) is associated with the CSF leak. The importance of treating these patients with a multidisciplinary-team approach (which has proved to be beneficial for all the members of the team and is especially important for the patient and for the progress of science) should be vemphasized.

References

1. Anand VK, Liberatore LA (1996) Endoscopic cerebrospinal fluid repair. Operative Tech Otolaryngol Head Neck Surg 7:269–274
2. Burns JA, Dodson EE, Gross CW (1996) Transnasal endoscopic repair of cranionasal fistulae: a refined technique with long-term follow-up. Laryngoscope 106:1080–1083
3. Cairns H (1937) Injuries of the frontal and ethmoidal sinuses with special references to cerebrospinal rhinorrhea and aeroceles. J Laryngol Otol 52:589–623
4. Calcaterra TC (1980) Extracranial surgical repair of cerebrospinal rhinorrhea. Ann Otol Rhinol Laryngol 89:108–116
5. Calcaterra TC (1985) Diagnosis and management of ethmoid cerebrospinal rhinorrhea. Otolalaryngol Clin North Am 18:99–105
6. Charles DA, Snell D (1979) Cerebrospinal fluid rhinorrhea. Laryngoscope 89:822–826
7. Clemenza JW, Kaltman SI, Diamond DL (1995) Craniofacial trauma and cerebrospinal fluid leakage: a retrospective clinical study. J Oral Maxilofac Surg 53:1004–1007
8. Creamer MJ, Blendonohy P, Katz R, Russell E (1992) Coronal computerized tomography and cerebrospinal fluid rhinorrhea. Arch Phys Med Rehabil 73:599–602
9. Dandy WD (1926) Pneumocephalus (intracranial pneumocele or aeroscele). Arch Surg 12:949–982
10. Di Chiro G, Reames PH (1964) Isotopic localization of cranionasal cerebrospinal fluid leaks. J Nucl Med 5:376
11. Di Chiro G, Reames PH, Mathews WB (1964) RIHSA ventriculography and RIHSA cisternography. Neurology (Minneapolis) 14:185–191
12. Drayer BP, Wilkins RH, Boehnke M, Horton JA, Rosenbaum AE (1977) Cerebrospinal fluid rhinorrhea demonstrated by metrizamide CT cisternography. Am J Roentgenol 129:149–151
13. Dresdale A, Rose EA, Jeevanandam V, Reemtsma K, Bowman F, Malm JR (1985) Preparation of fibrin glue from single-donor fresh-frozen plasma. Surgery 97:750–754
14. Eljamel MS, Pidgeon CN (1995) Localization of inactive cerebrospinal fluid fistulas. J Neurosurg 83:795–798
15. Garcia-Uria J, Carrillo R, Serrano P, Bravo G (1979) Empty sella and rhinorrhea a report of eight treated cases. J Neurosurg 50:466–471
16. Gates GA, Sertl GO, Grubb RL, Wippold FJ (1994) Closure of clival cerebrospinal fluid fistula with biocompatible osteoconductive polymer. Arch Otolaryngol Head Neck Surg 120:459–461
17. Gjuric M, Goede U, Keimer H, Wigand ME (1996) Endonasal endoscopic closure of cerebrospinal fluid fistulas at the anterior cranial base. Ann Otol Rhinol Laryngol 105:620–623
18. Hallock GG, Trier WC (1983) Cerebrospinal fluid rhinorrhea following rhinoplasty. Plast Reconstr Surg 71:109–113
19. Hao SP (1996) Transnasal endoscopic repair of cerebrospinal fluid rhinorrhea: an interposition technique. Laryngoscope 106:501–503
20. Hartman AR, Galanakis DK, Honig MP, Seifert FC, Anagnostopoulos CE (1992) Autologous whole plasma fibrin gel. Arch Surg 127:357–359
21. Hasegawa M, Torii S, Fukuta K, Saito K (1995) Reconstruction of the anterior cranial base with the galeal frontalis myofascial flap and the vascularized outer table calvarial bone graft. Neurosurgery 36:725–731
22. Hirsch O (1952) Successful closure of cerebrospinal fluid rhinorrhea by endonasal surgery. Arch Otolaryngol Head Neck Surg 56:1–12
23. Hubbard JL, McDonald TJ, Pearson BW, Laws ER (1985) Spontaneous cerebrospinal fluid rhinorrhea: evolving concepts in diagnosis and surgical management based on the Mayo Clinic experience from 1970 through 1981. Neurosurgery 16:314–321
24. Ignelzi RJ, Van der Ark GD (1975) Analysis of the treatment of basilar skull fractures with and without antibiotics. J Neurosurg 43:721–726
25. Kelley TF, Stankiewicz JA, Chow JM, Origitano TC, Shea J (1996) Endoscopic closure of postsurgical anterior cranial fossa cerebrospinal fluid leaks. Neurosurgery 39:743–746
26. Lanza DC, O'Brien DA, Kennedy DW (1996) Endoscopic repair of cerebrospinal fluid fistulae and encephaloceles. Laryngoscope 106:1119–1125
27. Lewis DH, Rajendran J, Grady MS (1995) Radionuclide demonstration of cerebrospinal fluid rhinorrhea after a closed head injury. AJR Am J Roentgenol 165:958
28. Mahaley MS, Odom GL (1966) Complication following intrathecal injection of fluorescein. Case report. J Neurosurg 25:298–299
29. Mattox DE, Kennedy DW (1990) Endoscopic management of cerebrospinal fluid leaks and cephaloceles. Laryngoscope 100:857–862
30. Moseley JI, Carton CA, Stern WE (1978) Spectrum of complications in the use of intrathecal fluorescein. J Neurosurg 48:765–767
31. Murata Y, Yamada I, Isotani E, Suzuki S (1995) MRI in spontaneous cerebrospinal fluid rhinorrhea, Neuroradiology 37:453–455
32. Myers DL, Sataloff RT (1984) Spinal fluid leakage after skull base surgical procedures. Otolaryngol Clin North Am 17:601–612
33. Naidich TP, Moran CJ (1980) Precise anatomic localization of a traumatic sphenoethmoidal cerebrospinal fluid rhinorrhea by metrizamide CT cisternography, J Neurosurg 53:222–228
34. Normington EY, Papay FA, Yetman RJ (1996) Treatment of recurrent cerebrospinal fluid rhinorrhea with a free vascularized omental flap: a case report. Plast Reconstr Surg 98:514–519
35. Ommaya AK (1976) Spinal fluid fistulae. Clin Neurosurg 23:363–392
36. Oz MC, Jeevanandam V, Smith CR, Williams MR, Kaynar AM, Frank RA, Mosca R, Reiss RF, Rose EA (1992) Autologous fibrin glue from intraoperatively collected platelet-rich plasma. Ann Thorac Surg 53:530–531
37. Papay FA, Maggiano H, Dominquez S, Hassenbusch SJ, Levine HL, Lavertu P (1989) Rigid endoscopic repair of paranasal sinus cerebrospinal fluid fistulas. Laryngoscope 99:1195–1201
38. Park JI, Strelzow VV, Friedman WH (1983) Current management of cerebrospinal fluid rhinorrhea. Laryngoscope 93:1294–1300
39. Persky MS, Rothstein SG, Breda SD, Cohen NL, Cooper P, Ransohoff J (1991) Extracranial repair of cerebrospinal fluid otorhinorrhea. Laryngoscope 101:134–136
40. Picardi JL (1975) Analysis of the treatment of basilar skull fractures with and without antibiotics. J Neurosurg 43:721–726

41. Porter MJ, Brookes GB, Zeman ZJ, Keir G (1992) Use of protein electrophoresis in the diagnosis of cerebrospinal fluid rhinorrhea. J Laryngol Otol 106:504–506
42. Ryall RG, Peacock MK, Simpson DA (1992) Usefulness of β2-transferrin assay in the detection of cerebrospinal fluid leaks following head injury. J Neurosurg 77:737–739
43. Salar G, Carteri A, Zampieri P (1978) The diagnosis of CSF fistulas with rhinorrhea by isotope cisternography. Neuroradiology 15:185–187
44. Saltz R, Sierra D, Feldman D, Saltz MB, Dimick A, Vasconez LO (1991) Experimental and clinical applications of fibrin glue. Plast Reconstr Surg 88:1005–1015
45. Schaefer SD, Briggs WH (1980) The diagnosis of CSF rhinorrhea by metrizamide CT scanning. Laryngoscope 90:871–875
46. Schick B, Draf W, Kahle G, Weber R, Wallenfang T (1997) Occult malformations of the skull base. Arch Otolaryngol Head Neck Surg 123:77–80
47. Shapiro SA, Skully T (1992) Closed continuous drainage of cerebrospinal fluid via a lumbar subarachnoid catheter for treatment or prevention of cranial-spinal cerebrospinal fluid fistula. Neurosurgery 30:241
48. Sofferman RA (1995) The recovery potential of the optic nerve. Laryngoscope 105:1–38
49. Stamm AC (1995) Rinoliquorreia: microcirurgia transnasal. In: Stamm AC (ed) Microcirurgia naso-sinusal. Revinter, Rio de Janeiro, pp 265–277
50. Stammberger H (1991) Detection and treatment of cerebrospinal fluid leaks. In: Hawke M (ed) Functional endoscopic sinus surgery. Decker, Philadelphia, pp 436–441
51. Stankiewicz JA (1991) Cerebrospinal fluid fistula and endoscopic sinus surgery. Laryngoscope 101:250–256
52. Stankiewicz JA (1995) Cerebrospinal fluid fistula and endoscopic sinus surgery. In: Hurley R (ed) Advanced endoscopic sinus surgery, 1st edn. Mosby Year Book, St. Louis, pp 81–86
53. Stechison MT (1992) Rapid polymerizing fibrin glue from autologous or single donor blood: preparation and indications. J Neurosurg 76:626–628
54. Tolley NS, Lloyd GAS, Williams HOL (1991) Radiological study of primary spontaneous CSF rhinorrhea. J Laryngol Otol 105:274–277
55. Vrabec, DP, Hallberg OE (1964) Cerebrospinal fluid rhinorrhea. Arch Otolaryngol Head Neck Surg 80:218–229
56. Wigand ME (1981) Transnasal ethmoidectomy under endoscopic control. Rhinology 19:7–15
57. Wigand ME, Hosemann W (1990) Operations on the anterior base of the skull, including the roof of the ethmoids and the wall of the sphenoid sinus. In: Wigand ME (ed) Endoscopic surgery of the paranasal sinuses and anterior skull base, 1st edn. Thieme Medical, New York, pp 127–133
58. Yessenow RS, McCabe BF (1989) The osteo-mucoperiostial flap in repair of cerebrospinal fluid rhinorrhea: a 20-year experience. Otolaryngol Head Neck Surg 101:555–558

Endonasal Micro-endoscopic Surgery of Nasal and Paranasal Sinuses Tumors

Wolfgang Draf, Bernhard Schick, Rainer Weber, Rainer Keerl, and Anjali Saha

Introduction

While the frequency of benign tumors of the nose and paranasal sinuses is not known, malignant tumors in this region are said to make up approximately 0.1–1% of all malignancies and 3–5% of all head and neck carcinomas in the world [9, 23]. The history and symptoms are non-specific for long periods of time. Nasal obstruction, blood-stained nasal discharge or a vague feeling of pressure or headache may be the first symptoms. When these symptoms occur unilaterally, it may be the first suggestion of a tumor. Extension of symptoms beyond the limits of the nose or sinuses further suggests a tumor (Table 40.1).

Depending on the type and extent of the tumor, surgical radiotherapeutic or chemotherapeutic measures are appropriate [22]. In surgery, the important goal is complete tumor removal and successful management of any complications that may be present (for example, a subperiostal orbital abscess or meningitis from an osteoma occluding frontal-sinus drainage).

The choice of surgical approach is made more difficult if one also attempts to prevent secondary problems that can occur after surgery, such as:
- Late complications (like mucopyoceles)
- Major scarring
- Visible deformity
- Damage to nerves (with hypesthesia or pain)

The endonasal approach is the least traumatic, but the surgeon must be very familiar with the surgical technique and the prevention and management of complications (intraorbital hematoma, damage to the lacrimal system, CSF leaks, major bleeding). Since both the microscope and endoscope have advantages and disadvantages, the surgeon should be able to use both in order to optimize his work.

In most cases, inflammatory sinus disease can be successfully treated with endonasal surgery [5, 7, 10–12, 15, 21, 32]. Based on these positive experiences, we (and other authors) have extended the indications for endonasal surgery to include tumors of the paranasal sinuses and the anterior skull base [3, 21, 30, 32, 36]. Endonasal surgery, therefore, means an approach to the anterior skull base exclusively through the nasal openings, without any external incision, assisted by optical aids (such as the endoscope and microscope) [5]. The approach can be limited to the extracranial, extradural region, extended intracranially and extradurally or can include a combined extra- and intradural resection. This paper evaluates endonasal surgery in tumorous diseases of the nose and paranasal sinuses, based on experiences with 86 cases in the Fulda Hospital.

Operative Technique

The operative technique basically corresponds to that described in Chap. 20. In order to work as precisely as possible and to preserve the most mucosa, we make use of a microdebrider. This draws in pathological tissue and cuts it cleanly (Fig. 40.1) [8, 19]. We also avoid gross avulsion with blunt instruments but use sharp cutting forceps of various sizes and angles, as suggested by Moriyama (Fig. 40.2).

Table 40.1. Extension of symptoms beyond the limits of the nose or sinuses suggests a tumor

Extension	Suggestive symptoms
Orbit	Reduction in vision, double vision, dislocation of the bulb
Dura, brain	CSF leak, meningitis, hyposmia, visual disturbance
Cheek and forehead soft tissues, lacrimal system	External swelling, irritation of the first or second branch of the trigeminal nerve (hyperesthesia and pain), epiphora
Pterygopalatine fossa	Trismus, irritation of the second branch of the trigeminal nerve (pain, hyperesthesia)

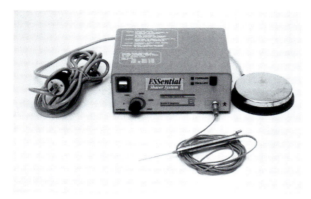

Fig. 40.1. Essential shaver system (Richards HNO, Smith and Nephew GmbH, Surgical Division, Schenefeld/Hamburg)

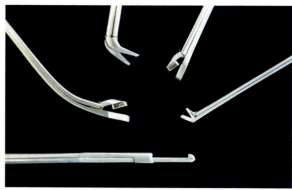

Fig. 40.2. Cutting instruments of varying sizes and angles recommended by Moriyama (Stuemer, Würzburg)

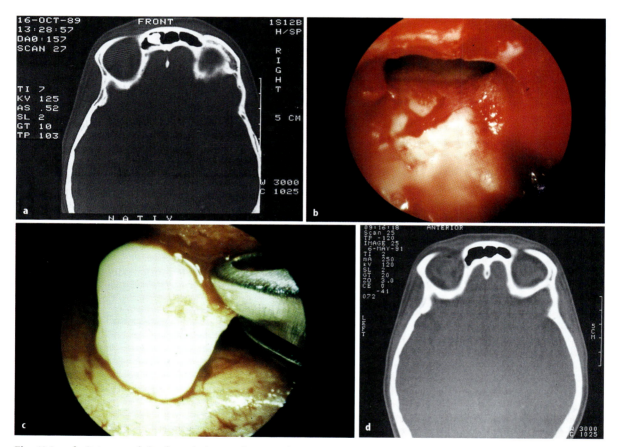

Fig. 40.3 a–d. Osteoma of the frontal sinus. **a** Preoperative computed tomography (CT; axial). **b** Endoscopic view after preparation of the tumor. **c** Endoscopic view after tumor removal. **d** CT 3 years after operation

Smaller tumors are dissected on all sides so that they can be removed *in toto* (Fig. 40.3). Stepwise reduction of tumor size is recommended for the removal of certain benign tumors, such as ossifying fibroma.

Exclusively endonasal resection of malignant tumors (Figs. 40.4, 40.5) requires extensive experience. In cases of doubt, wide exposure with temporary removal of surrounding bone is preferred [6, 17]. In many cases, resection of dura is then necessary. The dural defect must be sealed at the time of duraplasty (via the endonasal approach, if possible) [31].

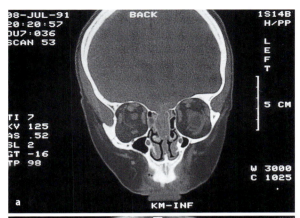

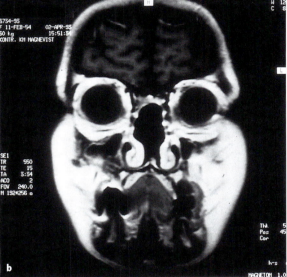

Fig. 40.4 a,b. Transitional-cell carcinoma of the skull base. **a** Preoperative coronal magnetic resonance image. **b** Coronal computed tomography 3 months after operation

For work on the skull base, we utilize a set of especially long instruments (Fig. 40.6). They allow precise dissection of mucosa, dura and tumor. As graft material, we usually use antogenous or allogenous fascia (Tutoplast; Biodynamics International, Erlangen). The following techniques may be used [31]:
1. The underlay technique, which may involve the ethmoid roof, the sphenoid sinus or the posterior basal wall of the frontal sinus. The intact dura is released from the edge of the bony defect in order to obtain adequate retraction for stable insertion of the graft. If possible, the graft is fitted so that it can be pushed in a few millimeters on all sides between the bony edge of the skull-base defect and the elevated intact dura. Additional fixation is provided with fibrin glue (Tisseel; Immuno AG, Vienna). Whenever possible, a neighboring mucosal flap (lateral nasal wall, middle turbinate, nasal septum) may be swung over the graft and similarly fixed with fibrin glue.
2. The onlay technique is used whenever there is a definite risk of injury to nerves or vessels while elevating intact dura from the surrounding bone or inserting the graft or when this is technically not possible (at the cribriform plate, sphenoid sinus or the attachment of the middle turbinate to the ethmoid roof). In these cases, bone removal for exposure of the dural defect is avoided. The graft is spread out widely, is glued over the region of the dural lesion and is covered with a neighboring mucosal flap, if possible.
3. Modified techniques in the sphenoid sinus:
   - Wedging pieces of fat into the bony slit before carrying out the onlay technique [1, 31]
   - The tobacco-pouch technique described by Kley [13] after stripping the whole mucosa

Although very serious and penetrating complications (Table 40.2) [4] requiring preoperative counseling can occur, they are quite rare [14, 25]. It must be realized that total tumor removal must include excision of involved meninges in some cases, especially for malignant tumors.

Patients and Results

Between 1980 and 1995, 77 benign and nine malignant tumors were treated endonasally in our department in Fulda. The tumor types and previous treatment results are compiled in Table 40.3. Five optic-nerve decompressions were performed within the framework of tumor therapy.

Except for two osteomas (frontal sinus) and three inverted papillomas (one frontal and two maxillary sinuses), all the benign tumors could be resected endonasally *in toto*, without recurrence. In these five cases, revision was performed via an osteoplastic

Table 40.2. Specific complications in endonasal sinus surgery

Postoperative sinusitis
Severe bleeding with blood transfusion (this occurs in less than 5% of cases)
Meningeal injury with cerebrospinal fluid leak, meningitis – a fluorescein test is necessary 6 weeks after operation in order to check the duraplasty seal
Hyposmia or anosmia
Visual disturbances or blindness
Periorbital hematoma with double vision
Nasolacrimal-duct injury with consequent stenosis

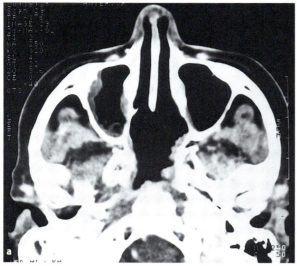
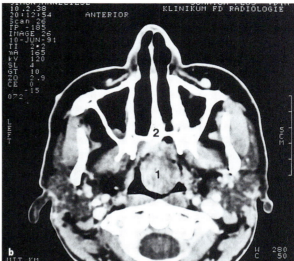

Fig. 40.5 a,b. Esthesioneuroblastoma (*1*) arising from the posterior part of the nasal septum (*2*). Axial computed tomography. **a** Preoperative. **b** One year after operation

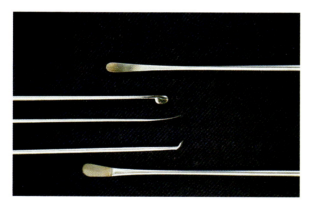

Fig. 40.6. Instruments for endonasal duraplasty (various elevators, round table knife and sickle knife)

frontal or maxillary operation. There were two malignant recurrences, which were similarly treated endonasally.

There were no intraoperative complications, as can be seen from Table 40.3. However, in two cases operated on for frontal-sinus osteoma, control computed tomography (CT) scans 1 year after operation showed early mucocele formation in the frontal sinus, without any specific symptoms. In one of the two cases, we carried out an endonasal revision with the formation of a type-III drainage. The second patient did not agree to a revision operation at that time.

Discussion

Based on our experience, we recommend the following groups of indications for endonasal surgery of neoplastic space-occupying lesions in the nose, sinuses or anterior skull base.

1. *Biopsy of suspicious tumorous lesions* (Fig. 40.7). This involves very little risk and has minimal effects on the patient, but it can establish an accurate diagnosis. The procedure does not influence or affect treatment (surgery via the endonasal or external approach, chemotherapy or radiotherapy).
2. *Decompression of the optic nerve* when adjacent space-occupying lesions, especially ossifying fibromas (Fig. 40.8) or fibrous dysplasia (Fig. 40.9), compromise the nerve (Fig. 40.8).
3. *Complete excision of localized benign tumors* of the sinuses and the bordering skull base (Fig. 40.10) [6, 26]. *Osteomas* form the largest group of benign neoplasms treated endonasally (28 of 77; 36.4%). In the head and neck region, the slow-growing osteomatous mass appears mainly in the frontal and ethmoid sinuses. Small asymptomatic osteomas that do not show much growth require no treatment. Operative indications for osteomas of the frontal and ethmoid sinuses are: (1) increase in size, (2) extension beyond the frontal sinus, (3) localization near the frontonasal duct, (4) coexisting chronic sinusitis, (5) involvement of the ethmoid sinuses, irrespective of size, and (6) headache without proof of any other etiology [18].

Osteomas of the ethmoid sinus can be removed endonasally without any problem. Additionally,

Table 40.3. Tumors endonasally treated micro-endoscopically (dates of operations: 1980–1995)

Tumor	Number	Follow-up (years)	Results	Complications
Benign tumors	73			
Osteoma	28	1–7	2 recs	Late (mucocele)
Ossifying fibroma	5	1–5	No rec	None
Angiofibroma	4	1–7	No rec	None
Fibrous dysplasia	4			
Total resection	2	3	No rec	
Subtotal resection	2	3	Optical decompression	None
Inverted papilloma	23	1–7	3 Recs (2 maxillary sinus, 1 frontal sinus)	None
Exophytic papilloma	6	1–2	No rec	None
Clivus chordoma	2	0.5–1	No rec	None
Hemangioendothelioma	1	1	No rec	None
Hypophyseal adenoma	1	3	Liquorrhea after hormone therapy, with tumor extension to the sphenoid sinus	None
Tornwaldt cyst	2	1–3	No rec	None
Malignant tumors	9			
Transitional-cell carcinoma	2	5	No rec, or rec treated elsewhere	None
Plasmocytoma	2	1	Revision because of rec 1 year and 4 years after operation; since then, the patient has been tumor free	None
		8	No rec	
Leukemic infiltration	1		No rec	None
Esthesioneuroblastoma		1.5 and 5	No rec	None
Adenoid cystic carcinoma	4	2	External, postoperative, localized rec; endonasal revision	
Metastatic breast carcinoma		2	No rec; death due to cardiac causes	None
Metastatic prostate cancer		2	No rec	None

rec, recurrence

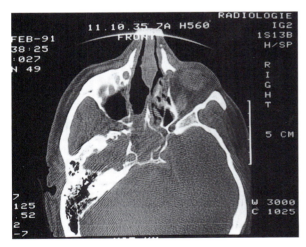

Fig. 40.7. Metastasis of prostatic cancer in the left sphenoid sinus (axial computed tomography)

one can dissect or resect adjacent structures, such as the lacrimal sac, nasal septum, periorbit and dura mater according to the individual need.

Reconstructions like dacryocystorhinostomy [27] or endonasal duraplasty [31] should be performed immediately.

If the anatomy of the nose and the approach to the frontal sinus is not too crowded, even osteomas in the region of the posterior frontal-sinus wall situated close to the infundibulum (Fig. 40.3) can be successfully removed endonasally. In this way, external scars, sensory disturbances in the forehead and other side effects of external incisions can be avoided. However, compared with the osteoplastic-flap-obliterating frontal sinus operation, postoperative care is intensive and requires daily cleaning of the nose by an otorhinolaryngologist at the beginning of treatment and douching by the patient later [4]. In our opinion, the minimally invasive endonasal micro-endoscopic surgical technique justifies removal of osteomas near the infundibulum during their early stages, before serious symptoms and complications arise and require a more extensive operative approach.

Because of our experience in two cases where the Draf type-II [2] frontal sinus-drainage procedure

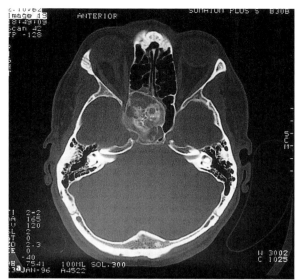

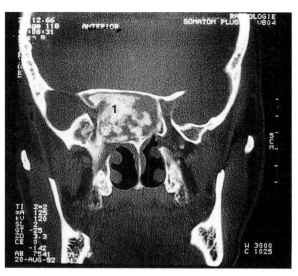

Fig. 40.9. Fibrous dysplasia of the ethmoid and sphenoid sinuses (*1*), with compression of both optic nerves. Coronal computed tomography

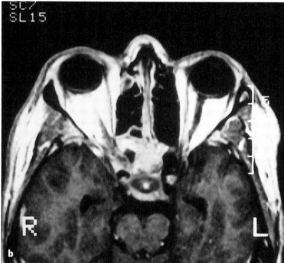

Fig. 40.8 a,b. Ossifying fibroma of the sphenoid sinus, with compression of the right optic nerve. **a** Preoperative computed tomography (axial). **b** Preoperative magnetic-resonance image (axial)

was inadequate and led to mucocele formation, we now suggest the primary formation of type-III drainage to minimize the possibility of delayed complications from stenosis. With very large tumors or in osteomas located laterally or on the anterior frontal-sinus wall, we prefer the osteoplastic approach [28–30]. Broad-based osteomas are removed stepwise with the cutting and diamond burrs. Osteomas with a narrow pedicle may be released with a chisel, gouge, flag, circular knife or some other instrument. In these cases too, the base must be smoothed with a diamond burr so as to protect the dura and remove the tumor *in toto*.

Sufficient tumor material must always be sent for histopathological examination.

In many cases, *papillomas* can also be successfully treated endonasally [24]. While fungiform (exophytic) papillomata are mostly localized to the nasal septum, inverted papillomas present on the lateral nasal wall and the paranasal sinuses. A 25% recurrence rate and an expected malignant transformation in 1.5–56% are described [15, 16, 20]. In our cases, we had a recurrence rate of 13% (3 out of 23) after endonasal treatment of inverted papillomas. Papillomas extending into the maxillary or frontal sinus should be operated on via an osteoplastic approach to ensure complete tumor removal.

4. *Resection of small malignant growths* by experienced surgeons. Primary malignant-tumor resection must be differentiated from surgery of metastases. With metastases, endonasal micro-endoscopic palliative surgery may also be indicated, not for the purpose of cure but as an attempt to achieve considerable improvement in the quality of life. Metastasis at the skull base may lead to persistent cerebrospinal-fluid leak and meningitis by destruction of the dura or (when situated in the sphenoid sinus between the optic nerves) to visual disturbances. Duraplasty and the removal of metastases cause little stress to the patient and provide considerable benefit for the rest of his (eventually limited) life. Endonasal surgery of malignant tumors requires frozen sections for control of free margins during surgery and requires both short-term and continuous follow-

found the osteoplastic-flap approach to be effective [28–30].

Conclusion

It is always essential to check whether the endonasal micro-endoscopic resection of tumorous, space-occupying lesions of the nose, sinuses and anterior skull base is possible. Individual anatomy and pathology and the personal experience of the surgeon ultimately lead to the choice of approach and operative treatment. The following preconditions should be realized for successful removal of the disease and consideration of functional and aesthetic aspects.
1. Extensive experience in endonasal surgery (sinusitis, traumatology)
2. Vast experience in tumor surgery
3. Optimized technical equipment (both microscope *and* endoscope)
4. Control of complications
5. Short-term and continuous follow-ups using endoscopy, CT and MRI

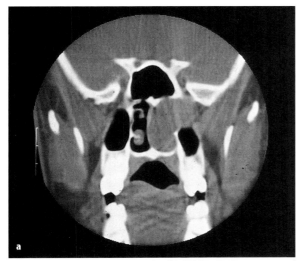
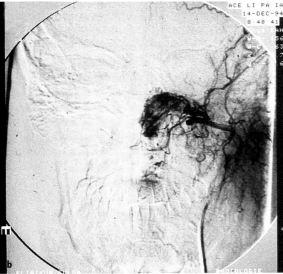

Fig. 40.10 a,b. Juvenile angiofibroma. **a** Preoperative computed tomography (coronal). **b** Preoperative angiography

up. One should not perform endonasal removal of malignant tumors without adequate follow-up. The follow-up includes endoscopy, CT and MRI in order to detect residual or recurrent tumors and to allow early revision surgery, if necessary.
5. *Mucopyoceles of the sinuses* may have the appearance of tumorous, space-occupying lesions. In our experience, mucoceles of the ethmoid and sphenoid sinuses may be operated on conservatively through the nose. By resecting the wall separating them from the nasal cavity, the mucoceles can be changed into an adjoining inlet. Similarly, it is often possible to marsupialize medially situated frontal-sinus mucoceles. Laterally located mucoceles must be operated on externally; we have

References

1. Draf W (1983) Fronto-basal injuries – principles in diagnosis and treatment. In: Samii M, Brihaye J (eds) Traumatology of the skull base. Springer, Berlin Heidelberg New York, pp 61–69
2. Draf W (1991) Endonasal micro-endoscopic frontal sinus surgery, the Fulda concept. Operative Tech Otolaryngol Head Neck Surg 2:234–240
3. Draf W, Berghaus A (1993) Tumoren und Pseudotumoren der frontalen Schädelbasis – Rhinochirurgisches Referat. Eur Arch Otorhinolaryngol Suppl 1:105–186
4. Draf W, Weber R (1992) Endonasale Chirurgie der Nasennebenhöhlen – das Fuldaer mikro-endoskopische Konzept. In: Ganz H, Schätzle W (eds) HNO Praxis Heute, vol 12. Springer, Berlin Heidelberg New York, pp 59–80
5. Draf W, Weber R (1993) Endonasal micro-endoscopic pansinusoperation in chronic sinusitis. I. Indications and operation technique. Am J Otolaryngol 14:394–398
6. Draf W, Weber R, Keerl R (1995) Endonasale Chirurgie von Tumoren der Nasennebenhöhlen und Rhinobasis. Med Bild 2:13–17
7. Draf W, Weber R, Keerl R, Constantinidis J (1995) Aspects of frontal sinus surgery. I. Endonasal frontal sinus drainage for inflammatory sinus disease. HNO 43:352–357
8. Grevers G (1995) Ein neues Operationssystem für die endoskopische Nasennebenhöhlenchirurgie. Laryngorhinootologie 74:266–268
9. Hamaker RC, Singer MI (1986) Cancer of the paranasal sinuses. Clin Oncol 5:549–556
10. Hosemann W, Göde U (1994) Epidemiology, pathophysiology of nasal polyposis, and spectrum of endonasal sinus surgery. Am J Otolaryngol 15:85–98

11. Keerl R, Weber R, Draf W (1996) Multimedia in ENT surgery. The endonasal pansinus operation. Ullstein Mosby, Wiesbaden
12. Kennedy DW (1992) Prognostic factors, outcomes and staging in ethmoid sinus surgery. Laryngoscope 102[suppl]:1–18
13. Kley W (1967) Diagnostik und operative Versorgung von Keilbeinhöhlenfrakturen. Laryngorhinootologie 46: 469–475
14. Lawson W (1994) The intranasal ethmoidectomy: evolution and an assessment of the procedure. Laryngoscope 104[suppl]:1–49
15. Lawson W, Le Benger J, Som P, Bernard PJ, Biller HF (1989) Inverted papilloma: an analysis of 87 cases. Laryngoscope 99:1117–1124
16. Myers EN, Fernau JL, Johnson JT, Tabet JC, Barnes EL (1990) Management of inverted papilloma. Laryngoscope 100:481–490
17. Samii M, Draf W (1989) Surgery of the skull base. An interdisciplinary approach. Springer, Berlin Heidelberg New York
18. Savic D, Djeric D, Jasovic A (1989) Oncocytome du nez et de sinus ethmoidaux et sphenoidaux. Oncocytoma of the nose and ethmoidal and sphenoidal sinuses. Rev Laryngol Otol Rhinol (Bord) 110:481–483
19. Setliff RC, Parsons DS (1994) The "Hummer" – new instrumentation for functional endoscopic sinus surgery. Am J Rhinol 8:275–278
20. Siivonen L, Virolainen E (1989) Transitional papilloma of the nasal cavity and paranasal sinuses. ORL J Otorhinolaryngol Relat Spec 51:262–267
21. Stammberger H (1991) Functional endoscopic sinus surgery. Decker, Philadelphia
22. Thiel H-J, Rettinger G (1986) Malignant tumors of the nasal cavity and paranasal sinuses: present situation of therapy and results. HNO 34:96–107
23. Waitz G, Wigand ME (1992) Results of endoscopic sinus surgery for the treatment of inverted papilloma. Laryngoscope 102:917–922
24. Weber R, Draf W (1992) Complications of endonasal micro-endoscopic ethmoidectomy. HNO 40:170–175
25. Weber R, Draf W (1992) Endonasale mikro-endoskopische Pansinusoperation bei chronischer Sinusitis. II. Ergebnisse und Komplikationen. Otorhinolaryngol Nova 2:63–69
26. Weber R, Draf W (1994) Reconstruction of lacrimal drainage after trauma or tumor surgery. Am J Otolaryngol 15:329–335
27. Weber R, Draf W, Keerl R (1995) Aspects of frontal sinus surgery. II. External frontal sinus surgery: osteoplastic procedure. HNO 43:358–363
28. Weber R, Draf W, Constantinidis J (1995) Aspects of frontal sinus surgery. III. Indications and results of osteoplastic frontal sinus surgery. HNO 43:414–420
29. Weber R, Draf W, Constantinidis J, Keerl R (1995) Aspects of frontal sinus surgery. IV. Management of frontal sinus osteomas. HNO 43:482–486
30. Weber R, Keerl R, Draf W, Schick B, Mosler P, Saha A (1996) Management of dural lesions occuring endonasal sinus surgery. Arch Otolaryngol Head Neck Surg 122:732–736
31. Wigand ME (1990) Endoscopic surgery of the paranasal sinuses and anterior skull base. Thieme, New York

Micro-endoscopic Surgery of Benign Sino-nasal Tumors

Aldo Cassol Stamm, Cleonice Hirata Watashi, Paulo Fernando Malheiros,
Lee Alan Harker, and Shirley Shizue Nagata Pignatari

Introduction

Sino-nasal tumors represent a large collection of diverse lesions. Histopathologically, they are divided into tumors and pseudotumors. Although there is no one predominant type, approximately half of the neoplasms are malignant, and one fourth originate from osseous structures [35]. According to Willis [68], the tumors are classified depending on their histogenesis and origin; they are classified as epithelial, mesenchymal or neural (Table 41.1). The distribution of our cases can be seen in Table 41.2.

The signs and symptoms of the nasal and paranasal-sinus tumors depend mostly on their location and size. The most frequent signs include nasal obstruction, rhinorrhea (mucus, purulent or serosanguineous), epistaxis, diplopia, hyposmia or anosmia, headache, cerebrospinal-fluid leakage and craniofacial deformities.

The diagnostic investigation should always include endoscopy, and imaging studies, such as computed tomography (CT) and magnetic resonance (MR). They are important for evaluation of the location and extent of the lesion, surgical planning and postoperative follow-up. The vascularity of the lesions can be evaluated with angioresonance or digital subtraction angiography.

In this chapter, the authors emphasize the clinical aspects of the most frequent benign tumors and some primitive and relatively rare tumors of the sino-nasal region, such as hemangiopericytoma, aneurysmal bone cyst, esthesioneurocytoma, nasal schwannoma, meningioma, fibrous dysplasia and nasal glioma. The surgical treatment of sino-nasal tumors is accomplished through classic approaches and some new surgical techniques that often employ the operating microscope and endoscopes.

Sino-Nasal Polyposis

Nasal polyposis is a chronic inflammatory disease of the mucosa of the nasal cavity and paranasal sinuses. Polyps are prolapses of the respiratory epithelium and exhibit mucosal edema, goblet cell hyperplasia, basement-membrane thickening and numerous leukocytes (predominantly eosinophils) [37].

The polyps usually originate in the middle-meatal area (middle turbinate, ethmoid bulla, uncinate

Table 41.1. Anatomic–pathologic classification of benign sino-nasal tumors

Pseudotumors
Polyposis
Polyp
Mucous-retention cyst
Aneurysmal bone cyst
Mucocele
Meningoencephalocele
Tumors
Epithelial
Papilloma
Inverted papilloma
Adenoma
Mesenchymal
Hamartoma
Hemangioma
Hemangiopericytoma
Paraganglioma
Angiofibroma
Fibrous dysplasia
Fibroma
Fibromyxoma
Osteoma
Neural
Nasal glioma
Neurilemmoma (schwannoma)
Esthesioneurocytoma

Table 41.2. Frequency of pseudotumors and benign tumors of the nasal and paranasal sinus during the 1983–1999 period ($n=837$)

Type	Number
Nasal polyposis	632
Antrochoanal polyp	17
Aneurysmal bone cyst	1
Mucocele	34
Encephalocele	4
Papilloma	4
Inverted papilloma	32
Adenoma	1
Hamartoma	1
Hemangioma	3
Hemangiopericytoma	4
Paraganglioma	1
Angiofibroma	64
Fibromyxoma	2
Osteoma	21
Fibrous dysplasia	5
Meningioma	4
Nasal glioma	4
Neurilemmoma (schwannoma)	2
Esthesioneurocytoma	1
Total	837

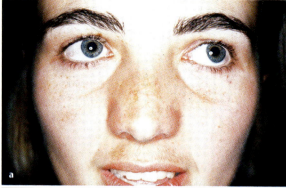
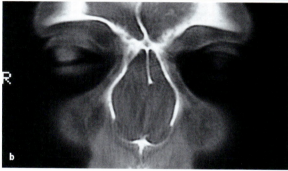

Fig. 41.1. a A 20-year-old female patient with deforming polyposis (Woake's disease). **b** Coronal computed tomography of the same patient, showing enlargement of the piriform aperture

process and sphenoethmoidal recess) [64], but nasal polyposis is a diffuse process that is usually bilateral, progressive and eventually involves all the paranasal sinuses and the nasal cavity itself.

The signs and symptoms depend on the extent of the process and frequently include nasal obstruction, mucus or purulent rhinorrhea, symptoms of sinusitis, localized headache and hyposmia or anosmia. Approximately one third of patients with polyposis have a concomitant disease of the lower respiratory tract, such as bronchitis, bronchiectasis or recurrent pneumonia. Asthma coexists with nasal polyposis in 20–40% of cases [22].

It has been suggested that irritating agents, metabolic diseases and psychogenic factors play a role in the etiopathogenesis of sino-nasal polyposis. Alterations of the sinus mucosa are related to the onset of the disease and to etiologic factors (bacterial, fungal or allergy).

The clinical nasal exam can reveal polyps of different sizes and consistencies. In the initial phase, there are few polyps, and they are often restricted to the middle-meatal and sphenoethmoidal-recess areas; however, in extensive cases, they can cause deformity of the middle third of the face, widening the base of the nasal pyramid (Woakes' disease; Fig. 41.1).

Examination of the nasal cavity is best performed with endoscopes, which facilitate identification of small polyps in the early stages of the disease. CT in axial, coronal and sagittal reconstruction projections is the method of choice for evaluation of the extent of the disease (Fig. 41.2). The combination of endoscopy and CT permits a surgeon to stage the nasal polyposis for the purpose of deciding on surgical treatment (if appropriate). The combination of CT and endoscopy is also essential in the following-up of the patients.

Treatment

Polyposis can be treated medically or surgically. Medical treatment consists of a combination of topical and systemic steroids and is recommended in early cases or as an adjunct to surgery to facilitate the surgical procedure, diminishing the risk of bleeding and increasing the chances of complete removal of the diseased tissue. Postoperatively, topical steroids act in direct contact with the mucosa, facilitating the healing process.

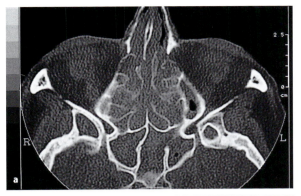

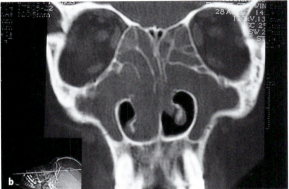

Fig. 41.2. Axial (a) and coronal (b) computed tomography of a stage-IV polyposis patient

Fig. 41.3. Microdebrider blades for polyposis surgery (Xomed, Jacksonville)

Surgery is usually indicated in extensive cases that are refractory to medical treatment or when the patient presents intolerance [64]. However, some authors recommend surgical treatment regardless the extent of the disease, varying the surgical technique according to the severity of the case [57]. Surgical treatment attempts to completely remove the diseased tissue, and incomplete procedures lead to limited patient benefit.

Surgical Technique

The intranasal surgical procedure can be performed with an operating-microscope objective lens of 250–300 mm and/or with 4-mm, 0° or 45° endoscopes. In cases of stage-I polyposis (limited to the middle meatus or sphenoethmoid recess), surgery is restricted to the removal of polyps, uncinectomy and/or turbinoplasty. In stage-II polyposis (involving the middle meatus and ethmoid sinus), surgical treatment consists of partial ethmoidectomy and removal of the polyps. Stage-III polyposis (stage II plus one sinus) usually requires partial ethmoidectomy and opening of the involved sinus. In cases of stage-IV polyposis (stage II plus two sinuses), the surgical procedure is similar, with opening of an additional sinus; in stage-V polyposis (involvement of all sinuses), the surgical procedure is performed in all sinuses [59]. In cases of massive polyposis, it is very important to identify anatomical landmarks (middle turbinate, uncinate process and ethmoid bulla). This step is usually performed with the help of a microdebrider (Fig. 41.3).

To approach the maxillary sinus, a middle-meatal antrostomy is performed; the medial wall of the maxillary sinus is usually thinner than normal, due to the pressure of the polyps. In very severe and extensive cases, it is not possible to identify the maxillary-sinus ostium, and a seeker probe or curved aspirator is introduced into the sinus cavity as a guide. The posterior fontanelle is removed with biting forceps, and the anterior portion is removed with backbiting forceps. All the diseased tissue is removed from this region, leaving a wide passage connecting the nasal cavity and the sinus through the middle meatus. In order to completely remove the intrasinus disease, an antrum punch is recommended. Removal of polyps of the roof of the nasal cavity between the nasal septum and the middle turbinate (i.e., on the cribriform plate) must be done very carefully due to the risk of causing CSF leakage.

The anterior ethmoid cells cannot be directly visualized with an operating microscope, and their removal is better accomplished with a 4-mm, 45° endoscope and angled forceps. The posterior ethmoid cells are removed after opening the basal lamella, and the surgery progresses to the anterior wall of the sphenoid sinus which, in severe cases, may already be opened. Surgical removal should be complete, involving the entire ethmoid complex. Surgical opening of the sphenoid sinus is usually accomplished via the sphenoethmoidal recess after identification of the superior turbinate and the sphenoid-sinus ostium.

At the end of the surgery, the ethmoid complex and the sphenoid sinus are marsupialized, becoming a single cavity. The operating cavity is usually packed with Merocel or rayon gauze, which is removed after 24–48 h.

Antrochoanal Polyp

Antrochoanal polyps have also been referred to as Killian polyps, nasopharyngeal benign polyps, retronasal polyps and recurrent polyps. The antrochoanal polyp is usually a unilateral lesion originating from the posteriolateral wall of the maxillary sinus; it reaches the nasal cavity through the accessory foramen of the maxillary sinus, which is enlarged in these situations (Fig. 41.4). Inside the sinus cavity, the antrochoanal polyp is cystic, while the nasal portion is usually solid and polypoid. It is most common in teenagers and young adults and represents 3.5–6.5% of all nasal polyps [6].

It is visible at the choana, completely filling the nasopharynx. It occasionally originates from other regions, such as the ethmoid or sphenoid sinus. Symptoms most often consist of unilateral or bilateral nasal obstruction and mucous or purulent rhinorrhea. Clinical examination often shows a single polyp leaving the middle meatus and projecting into the nasal cavity and choana, sometimes partially or completely filling the nasopharynx. When the polyp is large, it can be seen in the oropharynx or by posterior rhinoscopy. In general, there are no associated lesions.

The polyp's mucosal surface, especially the portion that extends to the nasopharynx, is metaplasic respiratory epithelium. CT usually shows homogeneous opacification of the maxillary sinus and of the ipsilateral nasal cavity (Fig. 41.5). Differential diagnosis, especially in this age group, includes mucous-retention cysts, mucocele, tumors of the maxillary sinus, angiofibroma, meningoencephalocele and chronic maxillary sinusitis.

The treatment necessitates surgical removal of the polyp, and several routes of surgical access have been proposed. Particularly in children, the most used surgical technique is micro-endoscopic. Although one can relieve the nasal obstruction temporarily by resecting the nasal portion of the polyp transnasally, the definitive removal of this polyp includes resection of its implantation in the sinus cavity. This can be done micro-endoscopically through the middle meatus or through a mini-aperture in the anterior wall of the maxillary sinus, wide enough to permit simultaneous passage of the endoscope and a surgical instrument. Extensive openings of the anterior wall of the maxillary sinus should be avoided in small children because of the presence of roots of the deciduous teeth.

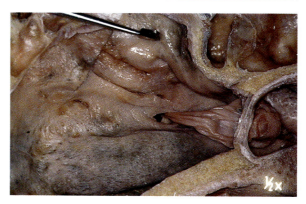

Fig. 41.4. Anatomical view of the lateral wall of the nasal cavity, showing an antrochoanal polyp protruding into the middle meatus through the accessory ostium of the maxillary sinus

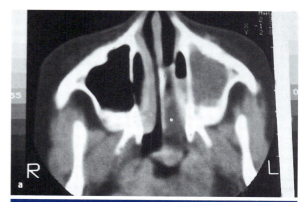

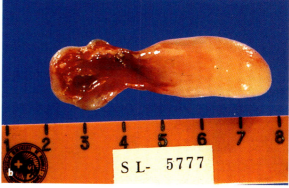

Fig. 41.5. a Axial computed tomography shows a typical image of an antrochoanal polyp (unilateral opacification of the maxillary-sinus and nasal cavities). **b** Antrochoanal polyp after surgical removal

Mucous-Retention Cysts

Mucous-retention cysts are most common in the maxillary sinus and result from inflammatory obstruction of a seromucinous gland within the serum mucosal lining [54]. It may also reflect respiratory allergy or apical dental disease. The incidence is approximately 1.9–9.6% in the general population [7].

A majority of the cysts are asymptomatic, and approximately 10% are accidentally found on plain X-ray films. Diagnosis can be made by a conventional X-ray or CT (Fig. 41.6).

Retention cysts and mucoceles are similar histopathologically. The mucosal wall contains fibrous tissue and congested and dilated vessels but no cellular atypia.

Treatment is indicated if the cyst is large enough to cause local pain or chronic infection from obstruction of the ostium. They can be removed endoscopically via an inferior antrostomy or a mini fenestration at the anterior wall of the maxillary sinus, but middle-meatal antrostomy is usually not recommended unless disease of the middle meatus is also present.

Aneurysmal Bone Cyst

The aneurysmal bone cyst is a non-neoplastic vascular lesion, usually single, that grows rapidly, expands and destroys the involved bone. It was first described by Jaffe and Liechtenstein [39] in 1942, who established a particular diagnostic criterion to separate this lesion from other isolated cystic lesions. This lesion may appear in any part of the skeleton but predominates in long bones and vertebral column. It is rare in the head and neck regions (2.5–6%); however, it has been described in the maxilla, mandible, orbital roof, temporal bone, cranium and maxillary and sphenoid sinuses [1]. It is more frequent in young females under the age of 20 years [39].

The etiopathology of the aneurysmal bone cyst is still controversial. Liechtenstein [44] believed that the lesion could be a result of a local hemodynamic anomaly, with increased venous pressure subsequently transforming the area into an engorged and dilated vascular bed. Other suggested local alterations include thrombosis of large veins and anomalous arteriovenous communications. According to some authors [8], the lesion is the unorganized or recanalized part of the cyst. Recent findings have given a new dimension to the pathophysiology of the aneurysmal bone cyst, and it is believed that this lesion can be secondary to vascular alterations initiated by a primary, pre-existing bony lesion [10, 13].

Occasionally the aneurysmal bone cyst is associated with other lesions, such as fibrous dysplasia, solitary bone cysts, osteoclastoma, chondromyxoid fibroma, non-ossifying fibroma, spontaneous fractures of the long bones or cervical fusion. Although not considered a real neoplasm, the aneurysmal cyst grows rapidly, developing in phases. In the first stage, osteolysis and elevation of the periosteum occurs, followed by bony destruction. Later, the process stabilizes and progressively ossifies.

The clinical picture always reflects the speed of growth and the involvement of the surrounding structures. When a palpable mass is present, it is usually tender.

Although plain X-rays may show areas of calcification and pathological bony destruction, CT and MR better define the magnitude of the process (Fig. 41.7). Angiography of an aneurysmal bone cyst shows abnormal vascularity and, when it is present in other body parts, arteriovenous shunts can be seen.

The preferred treatment is surgical, and the approach depends on the location and extent of the tumor. Profuse bony bleeding during tumor removal is a significant problem that can be controlled by bone wax or packing.

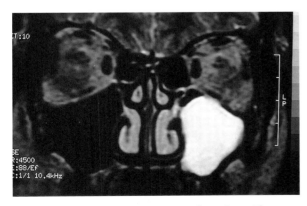

Fig. 41.6. Coronal magnetic resonance of a patient with a maxillary-sinus mucous-retention cyst

Mucocele

A mucocele is a chronic, expansive, benign cystic lesion limited by the mucosa of the paranasal sinus, with thick, translucent, sterile mucous secretions; it features a pseudostratified or low-columnar epithelial lining containing occasional goblet cells [6]. In 1896, Rollet introduced the term mucocele and, in

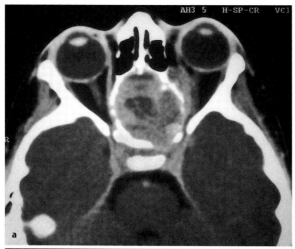

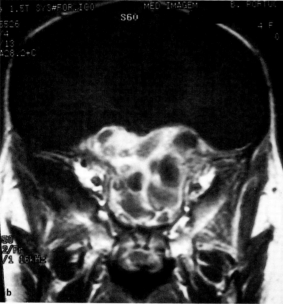

Fig. 41.7. Axial computed tomography (a) and magnetic resonance (b) of a 4-year-old girl with proptosis, presenting with a sphenoidal aneurysmal bone cyst

The etiopathogeny of the mucocele is not completely understood. It is believed that this disease is secondary to obstruction of sinus drainage, leading to stagnation of the secretion within the cavity. Fractures, mucosal edema, polyps, tumors, surgical trauma and chronic sinusitis can be predisposing factors.

Mucoceles are classified according to the sinus of origin. The frontal sinus is the most common site, followed by the ethmoid, maxillary and sphenoid sinuses (Fig. 41.8).

Diagnosis

The diagnosis of the mucocele is made via a clinical history and an otolaryngologic examination complemented by image studies (CT, MR). The signs and symptoms of the mucocele are related to the location of the lesion and reflect the distension of the sinus and its effect on the neighboring structures. Because of its slow progression and development, the mucocele is asymptomatic and sub-clinical for a long time, and symptoms can develop slowly during a range of time lasting between a few days and many years.

Although plain films may demonstrate opacification, bone erosion and expansion caused by the mucocele, CT is necessary to precisely determine the anatomical alteration caused by the mucocele's expansion, particularly the limits of the paranasal sinuses and the involvement of the surrounding structures [69]. MR may be useful to differentiate the mucocele from tumors originating in this region, such as meningiomas, meningoceles, craniopharyngiomas, epidermoid tumors, neurofibromatosis, aneurysms, abscess, cysts, polyps, optic-nerve gliomas and carcinomas.

Frontal- and/or Frontoethmoidal-Sinus Mucoceles

The most common clinical features of the fronto and frontoethmoidal mucocele are summarized as follows. Edema, proptosis and inferior lateral ocular displacement are the cardinal manifestations. Diplopia, pain and frontal swelling due to frontal-sinus anterior-wall erosion are less common. The specific symptoms are: periorbital edema (in 80–90% of patients), proptosis (70–80%), displacement of the ocular globe (70–80%), diplopia (40–50%), frontal, periorbital, retro-orbital headaches (40–50%) and frontal tumefaction (5–10%) [40, 48].

1901, Onodi first described its histologic characteristics [14, 25]. In 1949, Schuknecht and Lindsay [52] referred to the mucocele as an accumulation of products originating from secretion, desquamation and chronic inflammatory processes within a paranasal sinus, accompanied by distension of the sinus wall.

Although considered a benign lesion, the expansive character of the mucocele promotes slow erosion of the adjacent bone via compression and consequent bone absorption. The distension of the sinus and involvement of the surrounding structures are responsible for the signs and symptoms.

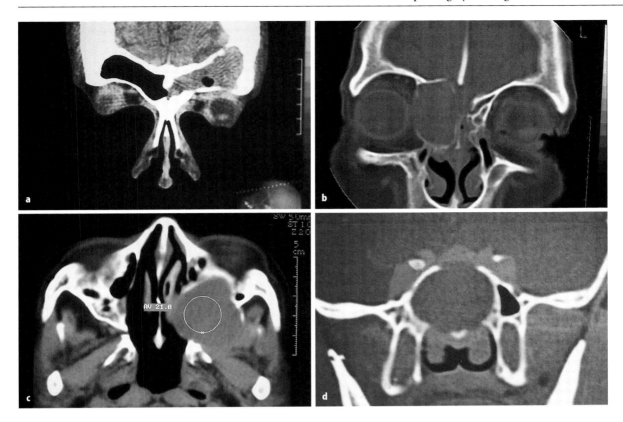

Fig. 41.8 a–d. Computed-tomography-scan images of mucoceles. a Frontoethmoid region. b Ethmoid sinus. c Maxillary sinus. d Sphenoid sinus

Maxillary- and/or Maxilloethmoidal-Sinus Mucoceles

Maxillary- and maxilloethmoidal-sinus mucoceles rarely achieve sufficient size to provoke bone erosion, and symptoms usually occur only when the ostium of the maxillary sinus is obstructed by the mucocele [40]. Signs and symptoms include pain in the periorbital, malar and nasal regions, tumefaction of the malar region (from erosion of the anterior maxillary wall), edema of the periorbital and malar regions and displacement of the ocular globe to the opposite side of the mucocele's expansion, with or without associated diplopia.

Sphenoidal-Sinus Mucoceles

The patients may experience headaches and visual disturbance, sometimes occurring for long periods of time. In the cases described in the literature, the period of time between the beginning of the symptomatology and the diagnosis varied between 4 weeks and 20 years [41]. Similar to the symptoms seen in the other sinuses, sphenoidal-mucocele symptoms are mostly related to the pressure and expansion directed at the neighboring structures: the cranial nerves (I, III, IV, V and VI), dura, internal carotid artery, cavernous sinus and pituitary gland.

Headaches occur in 80–90% of cases and is probably secondary to the pressure of the sphenoid wall on the dura. It is generally frontal and retro-orbital; less often, it is occipital. The most frequent ophthalmologic symptoms are ophthalmoplegia, reduced visual acuity and proptosis. Ophthalmoplegia occurs in 50% of the cases and usually progresses from involvement of the III, IV and VI cranial nerves. The reduction of visual acuity is usually unilateral due to compression, vascular thrombosis and ischemia of the optic nerve. Other signs and symptoms include nasal obstruction (in 40% of patients), anosmia (10–15%) and hypopituitarism (5%).

Treatment

Treatment of the mucocele is surgical, and several techniques specific for each sinus have been described. Among these techniques, some are more radical, removing the mucoperiosteum of the entire involved sinus and collapsing or obliterating the cavity [38, 50], such osteoplastic surgery of the frontal sinus [46].

External frontoethmoidectomy (Lynch) and Caldwell-Luc access to the maxillary sinus are currently restricted to special situations. Most surgeons today prefer more functional and conservative approaches, such as transnasal micro-endoscopic surgery techniques.

In 1991, Draf [21] recommended marsupialization of frontal-sinus mucoceles by removing the floor of the sinus via a transnasal micro-endoscopic approach. This surgical access to the frontal sinus by use of an endoscope is performed via the frontal recess after uncinectomy, preserving the ethmoid bulla as much as possible (Fig. 41.9). Appropriate instrumentation and angled endoscopes enormously facilitate the surgical procedure.

To approach the sphenoid sinus using an endoscope, any of the three approaches previously described (direct endonasal, trans-septal and transethmoidal) can be utilized. The direct approach is preferable if the disease is isolated to the sphenoid sinus (Fig. 41.10).

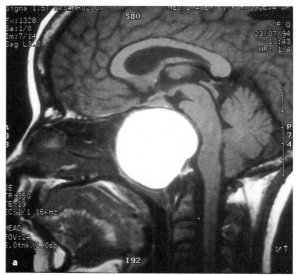

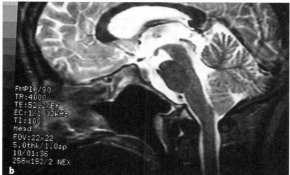

Fig. 41.10. Preoperative (a) and postoperative (b) sagittal magnetic resonance of a patient with a sphenoidal mucocele

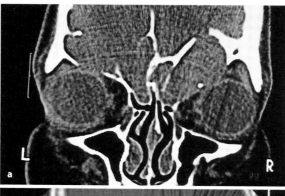

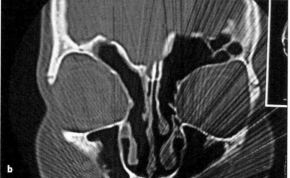

Fig. 41.9. Preoperative (a) and postoperative (b) coronal computed tomography of a patient with a frontoethmoidal mucocele

Ethmoid-sinus mucoceles are treated surgically with a transnasal micro-endoscopic approach, with marsupialization of the mucocele in the middle-meatus region (Fig. 41.11). Maxillary-sinus mucoceles can also be approached through the middle-meatus by performing a middle-meatus antrostomy.

The surgical treatment of a mucocele usually consists of wide marsupialization and drainage, preserving the mucosa. It is important to attempt to identify any causal obstructive factor in order to avoid or reduce the chances of recurrence. Postoperative care is extremely important to obtain a good result; one must avoid scar bands, infection and polyps in the region of the marsupialization.

Meningoencephalocele

Meningoencephaloceles are non-neoplastic, benign lesions that consist of ectopic neural or meningeal

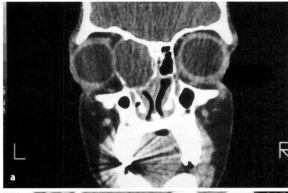
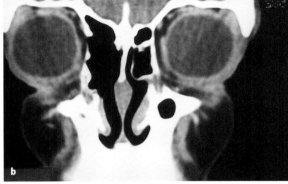

Fig. 41.11. Preoperative (a) and postoperative (b) coronal computed tomography of a patient with an ethmoidal mucocele

tissue. They usually appear at the midline of the cranium base or vertex and may be traumatic or congenital. Two theories explain its congenital origin: (1) incomplete closure of the cranium, through which meningeal and neural tissue insinuates and (2) excessive growth of the neural tube, impeding cranium closure. Neuroglial elements and meninges can be identified as part of the herniation, which can be composed of meninges alone (meningocele) or meninges and brain tissue (meningoencephalocele).

Its incidence is estimated to be one in 4000 live births; there is no difference of incidence between the genders, nor is there any evidence of familial predisposition [60]. A majority of patients present with signs and symptoms during the first year of life; in a small percentage of patients, symptoms occur between the ages of 5 years and 10 years.

Occurrence after childhood can be explained by diagnostic failure or by traumatic or acquired bone defects. Anatomically, the lesions can be classified into three groups according to their location [11]:
1. Occipital (75% of cases)
2. Sincipital (15% of cases)
3. Basal (10% of cases)

The first two forms are visible externally. The basal form is internal, occurring in the nasal and pharyngeal spaces. The sincipital and basal forms occur in otolaryngologic, head and neck areas.

The sincipital (or extranasal) lesion usually appears in the nasofrontoethmoidal region as a soft, non-painful, reducible mass that expands during crying or on compression of the jugular vein (Furstemberg's sign). It is subdivided into three types:
1. Nasofrontal, characterized by protrusion between the nasal and frontal bones
2. Nasoethmoidal, characterized by protrusion through the foramen cecum, which is separated from the nasal cavity by the ethmoid process
3. Naso-orbital, characterized by protrusion through the medial wall of the orbit; this involves the frontal, ethmoidal and lachrymal bones

The basal or intranasal meningoencephalocele is usually misinterpreted as a polyp, although it is more firm and brilliant and less translucent. It frequently causes nasal obstruction, epistaxis and (sometimes) cerebrospinal rhinorrhea, which predisposes the patient to recurrent bouts of meningitis.

The basal lesion is subdivided into four types:
1. Transethmoidal, characterized by protrusion to the superior meatus through a defect of the cribriform plate
2. Sphenoethmoidal, characterized by protrusion into the nasopharynx through a defect between the sphenoid and posterior ethmoid cells
3. Trans-sphenoidal, characterized by protrusion into the nasopharynx through the craniopharyngeal canal
4. Spheno-orbital, characterized by protrusion (as a mass) to the medial branch of the mandibular nerve through the supra-orbital fissure

When both extranasal and intranasal components are present, there is communication between them. There may also be an intracranial link, usually through the cribriform plate. Image evaluation by CT and/or MR is always necessary to define the correct location and extent of the link (Fig. 41.12). The isotopic transit may confirm the existence of connections to the subarachnoid space and can identify cerebrospinal rhinorrhea. If a CSF leak is already known to be present, this exam is indispensable. Usually, it is not urgent to resect the lesion, because its development is slow, mirroring the growth of the patient, and it neither invades surrounding structures nor affects normal development. However, this is not true of the intranasal forms when cerebrospinal rhinorrhea is present, when there is a previous history of meningitis or even when a functional disturbance of the respiratory nasal function is present [29].

Many different surgical approaches have been described, including frontal craniotomy with opening of the dura (sometimes including ligation of the sagittal sinus in order to remove the extradural cerebral tissue and close the bony defect). We prefer an endonasal micro-endoscopic approach.

Inverted Papilloma

In the medical literature, there are more than 20 different designations for inverted papilloma; these include inverted papilloma, Schneiderian papilloma, Ewing papilloma, papillary sinusitis, true papilloma, cylindric-cell papilloma, soft papilloma and transitional papilloma. A viral etiology is suggested by several authors, especially for those cases that coexist with vocal-cord papillomas, for which a viral etiology has already been established. Studies in molecular biology have shown the presence of human papilloma virus in these tumors [12].

Inverted papillomas are unilateral in most cases and represent approximately 2–4% of all nasal and paranasal tumors. They are three to five times more frequent in males. Patients with these tumors are usually in the fifth to seventh decades of life, though tumors have been documented in all age groups [20]. Nasal obstruction usually prompts referral; epistaxis and purulent rhinorrhea are also common.

Intranasally, inverted papilloma usually appears as a pedicled or sessile exophytic tumor that is histologically pseudoglandular, with various degrees of nuclear atypia. Although histologically benign, inverted papillomas should be viewed as locally malignant. They have been described in all regions of the nasal cavity, but the primary origin is the lateral wall, frequently at the angle formed by the maxillary and ethmoid sinuses. The lesions can grow in any direction, at times inferiorly, presenting in the middle meatus or extending through the sinus ostia and growing into the maxillary, ethmoid or sphenoid sinuses. When manipulated, they bleed easily. Because of the longstanding pressure exerted by the tumor, the bony walls of the sphenoid sinus can be thin or may even be destroyed, and it is not uncommon to find a large dural exposure, which permits identification of the contours of the frontal lobe. The extent of the disease is evaluated via CT- and MR-image studies (Fig. 41.13).

Inverted papilloma usually grows slowly, needing several years to cause symptoms, but it has a high tendency to recur and to become malignant. According to Cody [17], incomplete surgical removal results in recurrence in approximately 43% of the cases. The incidence of malignancy is approximately 10%, which justifies a histopathological examination of every papilloma removed from the nasal cavity and paranasal sinuses [17].

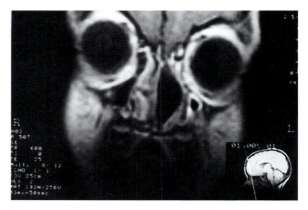

Fig. 41.12. Coronal magnetic resonance of a 4-year-old boy, with an intranasal meningoencephalocele

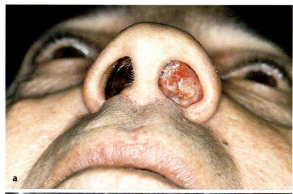

Fig. 41.13. a Inverted papilloma completely occupying the nasal cavity of a 54-year-old male patient. **b** Coronal computed tomography of the same patient, showing a grade-II inverted papilloma

Treatment

Treatment of inverted papilloma consists of surgical removal. Although several surgical techniques to remove this type of tumor have been proposed, the choice should be based on the extent and location of the lesion [58].

The most common surgical approaches are endonasal micro-endoscopic approaches and external approaches, such as the external ethmoidectomy, Caldwell-Luc, lateral rhinotomy and midfacial degloving. The endonasal micro-endoscopic approach is used for localized lesions (at the middle meatus, ethmoid and sphenoid sinuses, medial wall of the maxillary sinus). In large tumors with involvement of the lateral wall of the maxillary sinus, the midfacial degloving approach combined with an operating microscope/endoscope and bipolar coagulation is recommended. The lateral nasal wall is then removed "en bloc" until healthy tissue can be identified. In order to completely eradicate the tumor, the ethmoid-sinus cells must also be removed, as must the lateral portion of the middle turbinate (Fig. 41.14).

According to Harrison [35], lateral rhinotomy also allows adequate removal of the papilloma, minimizing the risk of leaving malignant tissue. Radiotherapy is not indicated and may induce malignancy.

Hamartoma

Hamartomas in the upper respiratory tract are unusual and are extremely rare in the paranasal sinuses. Although the larynx, trachea and lungs may harbor cartilaginous hamartomas, glandular hamartomas are uncommon in the upper airway.

Hamartomas are considered to be malformations derived from the spontaneous growth of components of local tissue, although they often present clinical features of neoplasia. Their growth is self-limited and usually ceases when a large number of cells reach maturity. In many instances, the resulting lesions seem to represent a simple exaggeration of a normal physiological process with this more general definition. Hemangiomas and lymphangiomas could be considered hamartomatous lesions, as could congenital lipomas, tuberous sclerosis and its congeners, multiple enchondroses, multiple exostoses, neurofibromas and melanotic nevi [6].

We have encountered one case of paranasal-sinus hamartoma in our series: a 72-year-old male presenting with nasal obstruction and bilateral hyposmia. The tumor was surgically treated through a

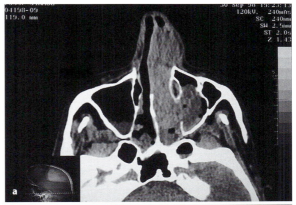

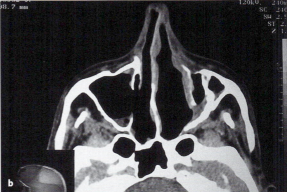

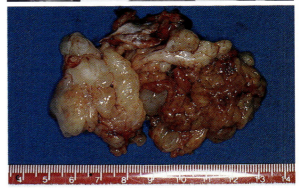

Fig. 41.14. Preoperative (**a**) and postoperative (**b**) computed tomography of an inverted-papilloma case. **c** Post-surgical aspect of the same inverted papilloma after *en bloc* removal

transnasal endoscopic approach. Figure 41.15 shows the pre- and postoperative CT appearances.

Adenoma

Mixed tumors (pleomorphic adenomas) are very frequent in the major salivary glands, but they are rarely present in the upper respiratory tract. In the paranasal-sinus region, the nasal cavity is the

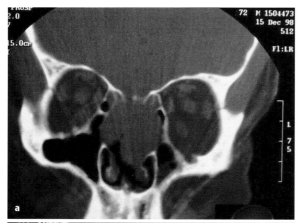
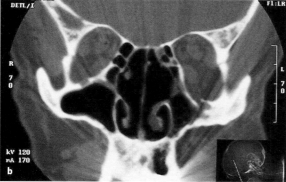

Fig. 41.15. a Adenomatous hamartoma originating from the roof of the nasal cavitiy. b Postoperative coronal computed tomography

favored site of origin, followed distantly by the maxillary sinus and nasopharynx. In the nasal cavity, these lesions most often arise from the bony or cartilaginous parts of the nasal septum; one fifth of cases arise from the lateral wall and usually involve a turbinate. They usually have a benign behavior, extending into the adjacent sinuses infrequently.

There is no gender predominance, and the tumor can occur at any age. The initial clinical picture is nonspecific, consisting of nasal obstruction and (less frequently) epistaxis, which commonly delays the diagnosis. Typically, the tumor appears as a homogeneous, lobular mass, occasionally bosselated or cystic, but more often polypoid and translucent.

Treatment consists of local or total surgical excision. Recurrences occur in 10% of patients, usually when the lesion extends into the paranasal sinuses. Follow-up should last beyond the traditional 5-year period [6].

We had one adenoma patient, a 52-year-old woman with an adenoma originating in the middle-meatus region. Her main complaint was left nasal obstruction. CT findings included a circumscribed lesion that extended into the left nasal cavity in the middle-meatus region (Fig. 41.16). The lesion was removed through a transnasal endoscopic access.

Angiofibroma

Angiofibroma is a histologically benign tumor composed of connective tissue intertwined with blood vessels. Although considered a benign neoplasm, it can be very invasive locally.

It is the most frequent benign neoplasm of the nasopharynx and is responsible for approximately 0.5% of head and neck tumors [3, 9, 33, 55]. According to some authors [28], the frequency of angiofibroma may be underestimated, because many small, asymptomatic tumors may disappear spontaneously. Because of its strong predilection for young males, an endocrinologic etiology is suspected. Because of its clinical and anatomical characteristics, the angiofibroma is frequently referred to as juvenile and nasopharyngeal in the medical literature. There are a few reports of angiofibroma occurring in adults, and we treated one adult in our series of patients [56].

Angiofibroma usually arises from the posteriolateral wall of the nasal cavity, where the sphenoid process of the palatine bone meets the sagittal wing of the vomer and the pterygoid process of the sphenoid bone [34]. More specifically, it originates at the sphenopalatine foramen and the posterior end of the middle turbinate, from which the angiofibroma can extend into the nasal cavity and nasopharynx, the paranasal sinuses, the pterygomaxillary, zygomatic and infratemporal spaces and the cranial base. The maxillary artery is the main blood supplier, but the angiofibroma may receive blood from secondary branches of the internal carotid artery. Other sources of blood include the pharyngeal, palatine and recurrent meningeal arteries. When the angiofibroma extends into the infratemporal fossa, there may be additional vascularization from the temporal and facial arteries [15, 55].

Differential diagnosis includes other nasopharyngeal lesions, such as hemangioma, craniopharyngioma, inflammatory polyps, antrochoanal polyps and malignant tumors (such as hemangiopericytoma and nasopharyngeal carcinoma). Histologically, angiofibroma consists of collagen-connecting angiomatous tissue. The muscular layer of the dilated vessels is incomplete, and the vessel walls may contain a single layer of endothelial cells. The connective tissue is wide and the nucleus:cytoplasm ratio of the endothelial and fibroblast cells is normal [60].

Diagnosis

The diagnosis of angiofibroma is essentially clinical and requires weighing of the clinical findings (Fig. 41.17) and the CT and MR imaging results (Fig. 41.18). The major signs and symptoms observed in our series of 64 patients were nasal obstruction (in 60% of patients) and epistaxis (in 42%). Rhinorrhea, a bulge in the soft palate, proptosis and facial deformity, headache, diplopia, rhinolalia or the presence of a tumor in the nasal cavity or nasopharynx (among others signs and symptoms) were also observed. The initial symptoms are occasionally caused by extranasal extension of the tumor. Clinical examination is best accomplished by endoscopes, and the lesion usually appears as a smooth, reddish-blue, semi-firm tumor with surrounding fibrinous secretions. In some cases, there are ulcerated areas and signs of recent bleeding. Coronal and axial CT with sagittal reconstruction is the method of choice in evaluating the location and extent of the angiofibroma, and MR is useful in evaluating previously operated cases and patients with suspected intracranial invasion. Although the follow-up is mostly accomplished with endoscopic clinical evaluation, CT is used postoperatively to detect residual lesions or verify "cure". Angiography can be used to identify the arterial supply of the lesion (which may be bilateral) and has been recommended in cases of large tumors with intracranial invasion and in cases where a preoperative embolization is planned.

Treatment

Although the definitive treatment of angiofibroma is surgical, preoperative adjuvant use of estrogen-hormone therapy to decrease the size of the tumor and diminish its vascularity has been advocated. However, other authors have been unable to demonstrate any arteriographic changes in the vascular patterns of angiofibromas before or after estrogen administration [60]. Some authors have used megavoltage radiotherapy to treat angiofibromas (primarily or as an adjunct to surgery) [19, 26]. This is not considered an appropriate treatment option in young patients

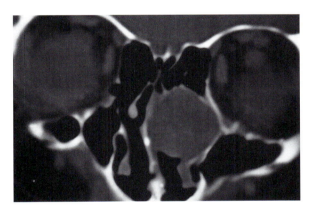

Fig. 41.16. Coronal computed tomography showing a pleomorphic adenoma originating in the left middle-meatus region

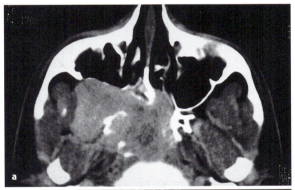

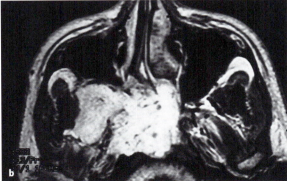

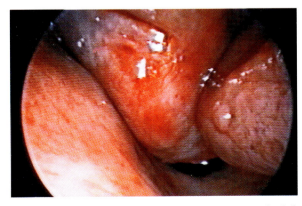

Fig. 41.17. Endoscopic view of an angiofibroma in the left nasal cavity (*)

Fig. 41.18. Axial computed tomography (a) and magnetic resonance (b) of a class-IIIa angiofibroma in a 17-year-old male

and is mainly indicated in invasive and recurrent tumors.

Chemotherapy has been advocated for treatment of extensive angiofibromas considered to be inoperable using protocols similar to those for malignant diseases [31]. Embolization is a useful preoperative adjunct in patients with significant vascular anomalies and in cases where the tumor involves the cavernous sinus, internal carotid-artery regions or the intradural space. Embolization can diminish the vascularity of the tumor considerably but must be performed no more than 2–3 days before the surgery.

Several surgical techniques for the removal of angiofibroma are described in the literature [2, 36], depending on the extent and stage of the tumor (as determined by accurate clinical and image examinations). The authors have used the staging system proposed by Fisch [27], as seen in Table 41.3. Different staging classes of angiofibromas can be seen in Fig. 41.19.

The techniques and surgical approaches used to treat angiofibroma include transnasal, transpalatal, unilateral or bilateral transantral, transmandibular, craniofacial, transtemporal/infratemporal, transzygomatic and suprahyoid approaches, lateral rhinotomy, maxillectomy and midfacial degloving. When the tumor is restricted to the nasal cavity and/or the nasopharynx, surgical access can be made through a natural orifice (the nose) using an operating microscope or endoscopes. Resection of the tumor under endoscopic control is usually indicated in limited lesions [42, 60, 65]. In 1997, Stammberger reported several cases of angiofibroma operated on endoscopically and preceded by superselective embolization [61].

The transpalatine approach may also be utilized to resect small tumors restricted to the posterior region of the nasal cavity and nasopharynx. This approach can lead to growth anomalies and functional disturbances of the palate years later, especially when performed on very young patients. The transantral approach (with an operating microscope) is recommended by Terzian-Naconecy [63] to remove angiofibromas, regardless of their sizes and locations.

The transpharyngeal–suprahyoid surgical approach can be used to remove small tumors limited to the nasopharynx. This surgical approach always leaves an external scar in the anterior aspect of the neck and does not provide accesses adequate for the removal of tumors that extend beyond the epipharynx.

When large lesions occupy the nasal cavity, nasopharynx, sphenoid sinus, ethmoid complex or the infratemporal or middle cranial fossae, the best surgical approaches are the transnasal retromaxillary approach (via a sublabial incision), lateral rhinotomy or the midfacial-degloving approach. These techniques are all performed with an operating microscope.

Ten to twenty percent of angiofibromas exhibit intracranial extension without invasion of the intradural space, merely elevating the dura [60]. Only rarely does the tumor penetrate the intradural space. When that occurs, surgical access is usually neurootolaryngologic. Another approach for large tumors is the pre- [49] or post-auricular [27] infratemporal-fossa approach. This technique also allows removal of large tumors with intracranial intradural extensions. However, with this approach, patients can have postoperative conductive hearing loss, facial-nerve paresis or paralysis, temporomandibular-joint dysfunction and an extensive skin scar.

We consider the midfacial *degloving* approach utilizing an operating microscope and bipolar cautery to be the preferred approach in removing class-II, -IIIa and -IIIb tumors. For larger class-IVa and -IVb tumors, it can be combined with neurosurgical techniques. The degloving approach provides excellent exposure of the middle third of the face, paranasal sinuses, skull base and vasculonervous structures (including the internal carotid artery and the cavernous sinus) [56]. The surgical field provided by this access is superior to the field provided by the previously discussed approaches; it also provides visualization of the tumor and causes minimal functional and cosmetic sequelae (Figs. 41.20, 41.21).

The main risk in angiofibroma resection is operative bleeding, which can be minimized by early ligation of the maxillary artery and complete tumor removal. Incomplete removal of the tumor can result in trans-section of its feeding vessels, sometimes precipitating profuse bleeding.

Between 1983 and 1999, we evaluated 64 patients with clinical and histopathological diagnoses of angiofibroma. The majority of patients were 11–20 years of age. There was one 51-year-old patient. According to the Fisch [27] staging system, eight patients were class I, 35 were class II, 17 were considered class IIIa or IIIb and four were class IVa or IVb; all the patients were treated surgically (Table 41.4).

Paraganglioma

Paragangliomas rarely originate in the paranasal-sinus region and more commonly represent extensions from nearby glomus jugulare, glomus vagale or orbital paraganglioma. Less frequently, they can also arise primarily from the mucous membranes of the nasal cavity, paranasal sinuses or nasopharynx. A

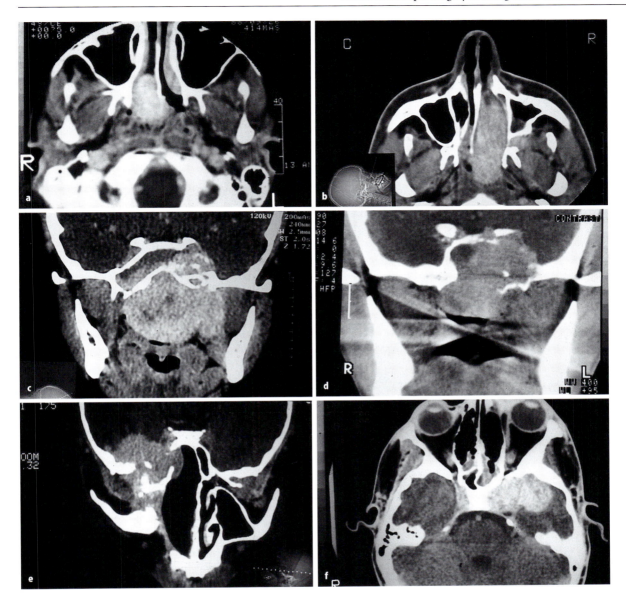

Fig. 41.19 a–f. Angiofibroma staging class (from image studies), according to Fisch [27]. **a** Class I. **b** Class II. **c** Class IIIa. **d** Class IIIb. **e** Class IVa. **f** Class IVb

third (very unusual) form consists of tumors that have apparently metastasized to the nasal cavities from a remote paraganglioma. Although the origin of nasopharyngeal paragangliomas is not clear, there is some evidence that paraganglioma tissue is normally present at birth, surrounding the terminal part of the maxillary artery in the pterygopalatine fossa ("glomus nasopharyngis").

There is no specific surgical procedure consistently used for these tumors [6]. In our series of benign paranasal tumors, we had one case of paraganglioma of the sphenoid sinus in a 56-year-old male patient. The MR appearance can be seen in Fig. 41.22. The tumor was removed surgically via a transnasal micro-endoscopic approach, and the final diagnosis was made by histopathology.

Hemangiomas

Vascular soft-tissue and mucosal tumors of the head and neck represent a frustrating group of lesions, because it is more difficult to characterize them clinically than anatomically or pathologically. Distin-

Table 41.3. Staging of nasopharyngeal angiofibroma [27]

Stage	Characteristics
Class I	Tumor limited to the sphenopalatine foramen, nasopharynx and nasal cavity (no bony destruction)
Class II	Tumor invading the paranasal sinuses and pterygopalatine fossa (with bony destruction)
Class IIIa	Tumor invading the orbital region and infratemporal fossa (no intracranial extension)
Class IIIb	Tumor invading the orbital region and infratemporal fossa (with extradural intracranial extension)
Class IVa	Tumor invading the intracranial and intradural spaces (no infiltration of the cavernous sinus, pituitary fossa or optic chiasma)
Class IVb	Tumor invading the intracranial and intradural spaces (with infiltration of the cavernous sinus, pituitary fossa and optic chiasma)

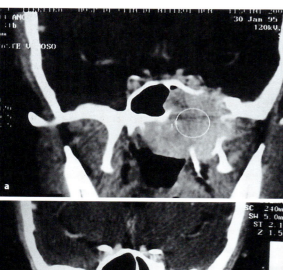
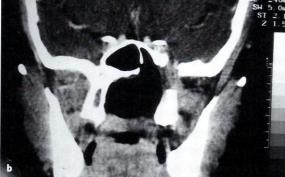

Fig. 41.20. Preoperative (a) and postoperative (b) coronal computed-tomography image of a 15-year-old patient operated on because of a class-IIIa angiofibroma

guishing whether they are malformations or true tumors with vascular components and separating vascular tumor tissue from granulation tissue is sometimes impossible [6].

Hemangiomas are common congenital lesions that occur predominantly in the head and neck regions. Together with lymphangiomas, they constitute approximately 30% of all oral tumors in children. The incidence in adults is considerably lower [24].

Whether they are considered malformations or neoplasms, they are always classified as capillary, cavernous, mixed, hypertrophic or juvenile hemangiomas, according to their predominant histological architectural pattern. There is no clear distinction among them (especially between the capillary and cavernous types). Zones of transition suggest that the hypertrophic form is an immature capillary hemangioma and that the cavernous type is a result of the maturation of a capillary variety [6].

Mucosal hemangiomas can develop in the oropharynx or any upper-respiratory-tract location. The oral cavity and the nasal cavity are the most frequent sites [6]. Granulation tissue and other inflammatory pseudotumors can present clinical aspects similar to those of hemangiomas.

The diagnosis is based on the patient's history and clinical picture. The clinical picture depends on the location and size of the lesion. Hemangioma of the nasal cavity ("bleeding polyp") is a polypoid, red-wine-colored lesion or a lobulated, sessile mass that progressively enlarges, producing nasal obstruction and bleeding after trauma.

Hemangiomas situated at the anterior septum are usually of the capillary type [6, 24]. Unlike the capillary variety, cavernous hemangiomas are usually located subcutaneously, and their depth is often poorly defined. Clinically, the tumors are soft, poorly defined and compressible [6, 24].

Ultrasonography, CT, MR and angiography are useful in the diagnostic work-up, depending on the information needed for treatment (Fig. 41.23). The treatment depends on the age of the patient, the size and localization of the lesion and the stage of development. Generally, observation alone is adequate management for small tumors that don't interfere with function, because the majority of congenital hemangiomas involute spontaneously [6, 23, 24, 32].

In the past, cryosurgery, radiation, electrocauterization and sclerosing agents were used to treat hemangiomas, with variable results. These treatment modalities carry a high rate of complications (especially scarring, in the treatment of hemangiomas of the face). With radiation, there is also the possibility of inducing malignancy and retarding growth, which makes that modality inappropriate in children. There are still reports of local and systemic steroids for use in treating hemangiomas (mostly for rapidly growing hemangiomas) [24].

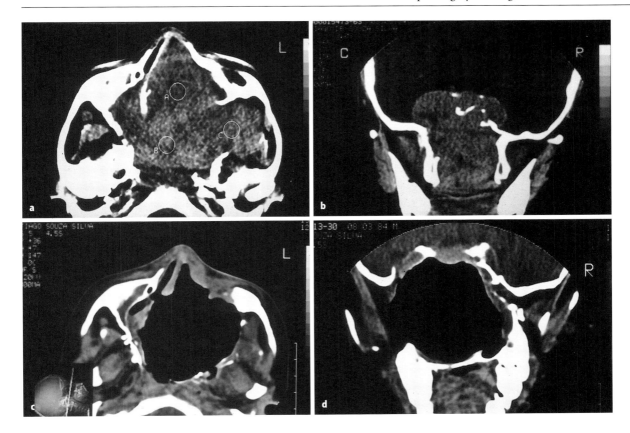

Fig. 41.21. Preoperative axial (**a**) and coronal (**b**) and postoperative axial (**c**) and coronal computed tomography (**d**) of a 11-year-old male operated on because of a class-IIIb angiofibroma

In many centers, the laser is used to remove these lesions; this technique results in good hemostasis and healing. Although embolization without surgery has been recommended in the treatment of large tumors, the affected region may become too large, and ischemia and infarction can occur, increasing morbidity.

Hemangiopericytoma

Hemangiopericytoma is a vascular tumor derived from the capillary pericytes that surround smooth-muscle cells. Hemangiopericytoma presents as a network of vascular structures that exhibit a large variation in the caliber of the endothelial lining and have a very fine wall (regardless of their diameter). Fifteen to twenty-five percent of hemangiopericytomas occur in the head and neck region, most commonly in the orbit and nasopharynx [62]. Malignancy is established only by the presence of metastases (not by histopathology alone), which occur less frequently in nasal, paranasal-sinus and nasopharyngeal hemangiopericytomas [30, 66]. Forty percent of Walike's [66] cases of head and neck hemangiopericytoma recurred locally, and 10% metastasized to distant regions.

Most paranasal-sinus hemangiopericytomas actually originate in extrasinusal soft tissues. Those that do arise from inside the paranasal sinuses are usually seen in the sphenoid and ethmoid sinuses. Compagno et al. [18] did a retrospective study of 23 patients with hemangiopericytoma at the Army Force Institute in Washington and observed that the sino-nasal hemangiopericytomas mimicked nasal tumors, with nasal obstruction, nasal secretion and epistaxis.

Hemangiopericytoma occurs most commonly in adults during the sixth and seventh decades of life, with no gender predominance. We treated two female patients aged 62 years and 82 years, and two males aged 56 years and 52 years. In the nose, hemangiopericytomas appear as friable, gray-white or translucent polypoid lesions. The differential diagnosis should include:
1. Malignant schwannoma
2. Leiomyosarcoma

Table 41.4. Distribution of angiofibroma patients seen between 1983 and 1999, categorized by stage (Fisch) [27], operative technique and recurrences (n=64)

Staging	Number	Operative technique	Recurrences	Re-operation technique	Number lost on follow-up
I	8	Transnasal micro-endoscopic	1	Midfacial degloving	1
II	35	Midfacial degloving	2	Midfacial degloving	6
IIIa	12	Midfacial degloving	1	Midfacial degloving	–
IIIb	5	Midfacial degloving	1	Pre-auricular, infratemporal	–
IVa	2	Pre-auricular, infratemporal	–	–	–
IVb	2	Pre-auricular, infratemporal	1	Stereotaxic radiotherapy	–

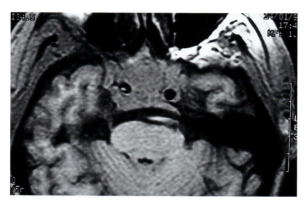

Fig. 41.22. Axial magnetic resonance of a sphenoid sinus paraganglioma in a 56-year-old male

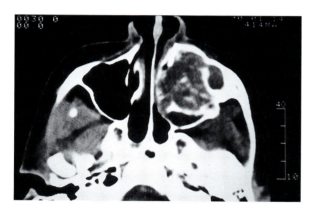

Fig. 41.23. Axial computed tomography of a cavernous hemangioma of the maxillary sinus in a 10-year-old girl

3. Cellular hemangioma
4. Olfactory neural tumors
5. Angiofibroma
6. Embryonal sarcoma
7. Fibromas
8. Malignant angiomatous tumors

Evaluation consists of endoscopic examination and CT- and MR-image studies. CT usually shows a lesion occupying the upper part of the nasal cavity, extending into or originating from the paranasal sinuses. Sometimes it is difficult to define the exact source of the tumor.

Treatment

Treatment is surgical; the technique depends on the location and extent of the tumor. The surgical approaches can include external ethmoidectomy, intranasal sphenoidectomy (with or without turbinectomy), the Caldwell-Luc technique, midfacial degloving and intranasal micro-endoscopic approaches. Our four patients were treated surgically. Two were operated on via a transnasal, microscopic approach; one was operated on via a midfacial degloving approach, and one was operated on via external frontoethmoidectomy (Fig. 41.24).

Meningioma

Meningiomas constitute approximately 15% of intracranial tumors, and extracranial or extraspinal meningiomas are uncommon. The few that do occur are found in the orbital-cavity bones, temporal bone or the skin. Meningiomas of the nasal cavity and paranasal sinuses are rare and, by 1980, only 19 cases had been described in the medical literature [45]. Intracranial meningiomas extending to the nasal cavity and paranasal sinuses are also rare. The epidemiology of primary meningiomas of nasal cavity and paranasal sinuses is quite different from the epidemiologies of meningiomas that extend from an intracranial origin. The average age of patients with intracranial meningiomas is 45 years, and 60% are females; sino-nasal meningiomas predominate in younger patients (44% occur in patients under 20 years of age). It is more common in males [60]. Meningiomas usually originate due to the prolifera-

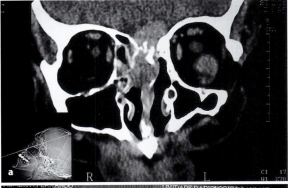

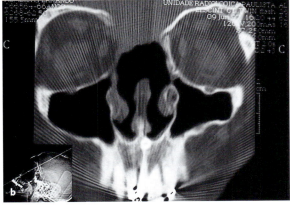

Fig. 41.24. Preoperative (a) and postoperative (b) coronal computed tomography of a 65-year-old male patient with a hemangiopericytoma of the roof of the ethmoid sinuses, extending to the intradural space. The tumor was operated on via an external frontoethmoidectomy

psammomatous (containing a hyalin mass in the center) and fibroblastic (similar to fibromas, with meningothelial components). The mixed form is composed of two or more types together.

The most common histologic types in the nasal cavity and paranasal sinuses are the meningothelial and psammomatous types. Regardless of the histological type, the tumor behavior is usually the same. Anaplasia, mitosis, giant cells and necrotic areas characterize the malignant meningioma, which is extremely rare.

Meningioma of the nasal cavity and paranasal sinuses behaves in a manner similar to any benign tumor of this region. Usually, the clinical picture includes nasal obstruction, exophthalmos, facial deformity, epistaxis and visual impairment.

Anterior rhinoscopy and endoscopic examination may show a tumor located in the upper region of the nasal cavity, which is sometimes also visible on posterior rhinoscopy. CT- and MR-image studies better evaluate the extent of the meningioma and help the surgeon during surgical planning and follow-up.

Treatment

Primary meningiomas of the nasal cavity and paranasal sinuses are treated surgically. Although the surgical technique may vary according to the location and extent of the tumors, the goal of the surgery is complete removal of the tumor. However, if the meningioma affects the orbit or the anterior or middle cranial fossae, the surgical objective may be limited to decompression with partial tumor removal, because the growth rate is usually slow.

Surgical access can combine intracranial and sublabial approaches. If the tumor is restricted to the nasal cavity, the recommended approach is transnasal and micro-endoscopic. When the meningioma expands into the paranasal sinuses, the midfacial degloving approach provides excellent exposure of the middle third of the face, and the lesion can be removed more easily. If surgical removal is complete, patients with primary meningioma of the nasal cavity and paranasal sinuses have an excellent prognosis, and recurrence of tumor is unusual (Fig. 41.25).

Fibrous Dysplasia

Fibrous dysplasia is a benign, slowly progressive chronic bone disorder of unknown etiology, in which normal bone is replaced by a variable amount

tion of meningocytes (arachnoid or meningothelial cells), which cover the villus projections of the arachnoid.

There exist small numbers of meningocytes outside of the arachnoid; they occur in the cranium, the spinal nerves, vertebrae, cranium periosteum and in the sheaths of the cranial nerves. They are rare outside the brain or meninges but may occasionally be seen along the midline of the head, neck and trunk (due to migration of the meningocytes during embryonic development). All the ectopic locations of the arachnoid cells can give rise to extracranial meningiomas. Intranasal and paranasal-sinus tumors arise because of their contiguity with the structures of the prosencephalon during the embryonic period.

Meningioma is a solitary, encapsulated, slow-growing tumor that has intratumor calcified areas and tends to cause an osteoblastic reactions in adjacent bones. Malignant transformation occurs but is rare. Histologically, three types of meningiomas are described: meningothelial (meningothelial cells),

of fibrous tissue and woven bone. Fibrous dysplasia has a broad, diverse spectrum of clinical and histological findings that reflect the various affected sites. When fibrous dysplasia is associated with abnormal pigmentation of the skin, retardation of growth and early puberty, it is called Albright's syndrome.

Fibrous dysplasia is classified into three groups:
1. Monostotic (one or more lesions involving a single bone)
2. Polyostotic (multiple lesions in more then one bone, with a tendency to be monomelic and to have unilateral presentation)
3. Disseminated (polyostotic lesions with extraskeletal manifestations)

Some of the monostotic lesions that present a large amount of fibrous tissue are designated ossifying fibroma or fibrous osteoma and are considered by some authors to be variations of fibrous dysplasia.

Fibrous dysplasia occurs most often in the femur, tibia, ribs and facial bones and, in the polyostotic form, skull lesions are seen in over 50% of patients [47]. Histologically, there are three different types, each of them presenting variability in clinical behavior. The first type is the active form, exhibiting a sparsely cellular matrix, many mitoses, intercellular collagen and a well-defined osseous matrix. It is more common in young patients. The second type is the potentially active form, presenting more mature connective tissue with few mitoses and a predominant osseous component. The third (very uncommon) form is an inactive form characterized by degeneration of the connective tissue, with a rare bony component. Regardless the origin or type, the bony tissue is absorbed and replaced by fibrous tissue with some poorly developed bony trabeculae. Usually, the lesion is surrounded by intact tissue [51]. Sarcomatous degeneration occurs in less than 1% of cases (usually following irradiation).

Clinically, signs and symptoms of fibrous dysplasia of the nasal cavity and paranasal sinuses appear at approximately the second decade of life, although the lesion is already present in childhood. It is a tumor of slow growth and is usually asymptomatic. During puberty, the rate of growth may increase, giving rise to the initial signs and symptoms, which are usually those already described for other benign tumors of this region, such as exophthalmos and facial deformity. Pain, nasal obstruction and an increase in the growth rate are suggestive of malignant degeneration.

Radiographically, three types of fibrous dysplasia are described: sclerotic (which follows the bone contours), lytic (which is more common and is seen as

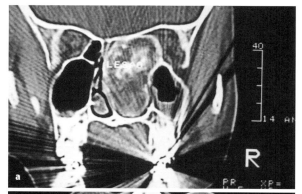

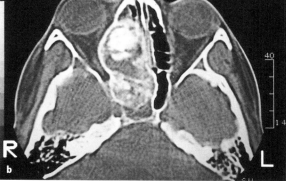

Fig. 41.25. Coronal (a) and axial (b) computed tomography of a meningioma involving the nasal cavity and ethmoid and sphenoid sinuses in a 14-year-old girl

an expansion of the bone cortex) and unilocular. Pre- and postoperative CT or MR are important in evaluating the extent of the tumor and for the follow-up control (Fig. 41.26). The presence of craniofacial fibrous dysplasia does not necessitate immediate treatment. Treatment should only be considered if clinical manifestations occur. Surgical approaches and techniques are chosen based on the size and degree of suspicious of malignancy of the tumor. If malignancy is suspected, an intraoperative histopathological exam can determine any need for a wider resection (including a secure margin).

In benign cases, the surgical resection can be more sparing, because a total removal can result in an extensive cosmetic defect. The surgical approach should always consider the age of the patient in addition to the extent of the disease. Resections of malignant cases should include neck dissection, because metastases occur via hematogenous routes (especially those leading to the lungs). Radiotherapy is contraindicated because of the risk of sarcomatous degeneration [46]. If the lesion is completely removed, prognosis is good, with a low rate of recurrence.

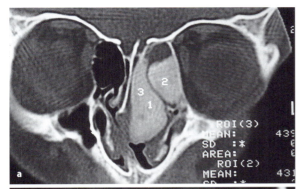
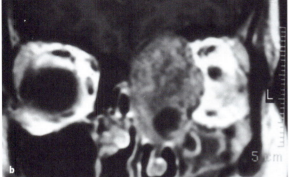
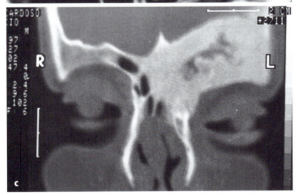

Fig. 41.26 a–c. Fibrous dysplasia involving the nasal cavity and paranasal sinuses, with different clinical manifestations. a Nasal obstruction. b Proptosis. c Cosmetic defect. The operations were performed via distinct approaches

Osteoma

Osteoma is a slowly growing, osteogenic tumor that appears in cranial and facial bones. Only occasionally does it cause symptoms, which are related to its location and anatomical relationships with the surrounding structures.

The several etiologic theories do not explain all the varieties. The three classic theories are:
1. Embryonic (based on the initial growth potential of embryonic cells trapped between areas of endochondral and osteomembraneous tissues)
2. Infectious (which postulates that osteoblasts stimulated by infection produce local inflammation followed by calcium-salt deposition and ossification)
3. Traumatic (which suggests that bone trauma can lead to cyst formation, which is followed by fibrous osteitis and osteoma formation)

Histologically, osteomas are classified into three groups:
- Spongy osteoma occurs mostly next to the bone margin, compact and lobulated with radiated septae. Active osteoblasts are usually present, especially in small tumors.
- Compact osteoma appears as a compact lammelar structure with numerous Haversian canals and osteocytes.
- Mixed osteoma contains elements of both spongy and compact osteomas. It is the most frequent in the paranasal sinuses.

Osteomas occur throughout between the second and sixth decades and, in most series, male patients predominate 2:1 [67].

Childrey [16] reviewed 3510 plain X-rays of the paranasal sinuses and calculated the incidence of osteoma to be 0.43%. This is an underestimate, because most osteomas are not detected unless symptoms promote investigation or complications occur. They are frequently detected when plain X-rays of the cranium or paranasal sinuses are made for other reasons. They are most frequent in the frontal, ethmoid and maxillary sinuses and are rare in the sphenoid sinus [4].

Diagnosis

The clinical picture has minimal or no symptoms and is mostly related to the affected surrounding structures. Small osteomas are nearly always asymptomatic and are rarely diagnosed. Osteomas of the ethmoid sinus usually produce earlier symptoms: frontomaxillary headaches, nasal secretions, periorbital edema and sometimes proptosis. Signs and symptoms from intracranial involvement are rare. If the osteoma is located near and obliterates a drainage ostium, it can lead to sinus infection. The diagnosis, location and extent of the lesion are usually determined with plain X-rays and CT (Fig. 41.27).

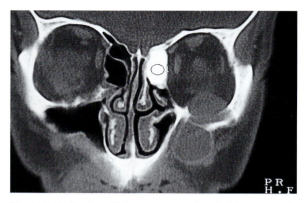

Fig. 41.27. Left ethmoidal osteoma associated with a mucocele of the maxillary sinus

Treatment

Most authors agree that surgical intervention is only indicated when symptoms are present, and asymptomatic patients should be followed-up with periodic X-rays. When removal is contemplated, several surgical techniques have been proposed depending on the tumor size and location. For the frontal-sinus osteomas, an osteoplastic surgical approach with bitemporal or supraorbital incisions is recommended [46]. Osteomas in the ethmoid sinus are usually removed via the "Lynch" technique and, if there is extension to the nasal cavity, a lateral rhinotomy or degloving approach can be used (Fig. 41.28).

Small ethmoid tumors can be removed via transnasal micro-endoscopic surgery. Either the Caldwell Luc or midfacial degloving approach can be used to remove osteomas of the maxillary sinuses. Tumors extending into the anterior cranial fossa require the addition of craniotomy to these techniques.

Nasal Glioma

Nasal glioma is not considered a true neoplasm but is instead a form of encephalocele. Sino-nasal presentations of glioma are often misinterpreted as nasal polyps, though gliomas are more firm and gray and are less translucent. Approximately 60% of gliomas are extranasal; 30% are intranasal and 10% have both components [11]. Usually, they are located in the roof of the nasal cavity and cause nasal obstruction. The majority of the signs and symptoms appear early in infancy and childhood. There is no familial tendency or gender predominance. Usually, these tumors exhibit slow growth.

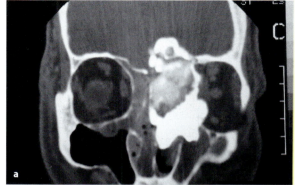

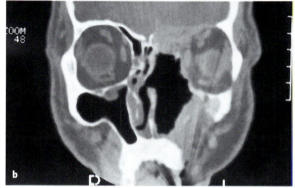

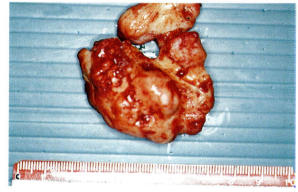

Fig. 41.28. Preoperative (a) and postoperative (b) computed tomography of a patient operated on because of an osteoma involving the nasal cavity, ethmoid sinus and orbit, with intracranial involvement. c Appearance of the tumor (osteoma) after surgical removal

Diagnosis is based on clinical and image studies and CT and MR findings (Fig. 41.29). Once the diagnosis is established, treatment consists of complete surgical removal of the tumor, employing a variety of approaches (including lateral rhinotomy, midfacial degloving and coronal flaps). Intranasal gliomas are preferably treated via a transnasal micro-endoscopic approach. Recurrences are common if removal is incomplete.

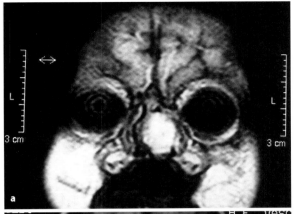

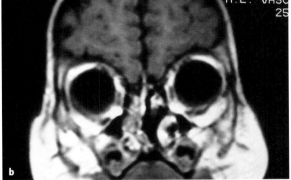

Fig. 41.29 a,b. A newborn with a nasal glioma operated on using a transnasal endoscopic technique. Preoperative (a) and postoperative (b) coronal magnetic resonance

The tumors frequently involve the nasal cavity and ethmoid sinus, but they can occur in the maxillary and sphenoid sinuses. In general, the tumors are solitary and manifest themselves between the ages of 25 years and 55 years [43].

The location of the tumor dictates the signs and symptoms. Epistaxis occurs when the nasal cavity and ethmoid complex are involved. Pain, exacerbated by local pressure, is experienced when the maxillary sinus is affected [6].

Exam of the nasal cavity reveals a tumor similar in appearance to an angiofibroma or a fibrotic nasal polyp. Like angiofibroma, it bleeds easily, and severe hemorrhage can result from biopsy, due to the high vascularization of these tumors. CT is necessary to evaluate the extent and exact location of the lesion and to provide the optimum information for surgical planning.

Treatment consists of complete tumor removal via a surgical approach adequate for the extent and region of growth. The transnasal micro-endoscopic and midfacial-degloving surgical approaches are the most likely to be applicable (Fig. 41.30).

Schwannomas

They are benign tumors that originate from cells of the Schwann sheath and usually exhibit very slow growth. Histologically, schwannomas consist of a proliferation of Schwann cells in two distinct patterns.
1. *Antoni A*: cells form a palisade arrangement in irregular fascicles
2. *Antoni B*: cells present myxoid degeneration of the stroma

Schwannomas are extremely rare in the nasal cavity, although there are reports of these tumors in other head and neck regions, such as the pharynx, parapharyngeal spaces, tongue, soft palate, larynx, external ear, trachea and, of course, the internal auditory canal [43].

The nerves from which the intranasal tumors originate are branches of the ophthalmic and maxillary division of the trigeminal nerve and branches of the autonomous nervous system. The olfactory nerve does not give rise to schwannomas, because it has no Schwann-cell sheath.

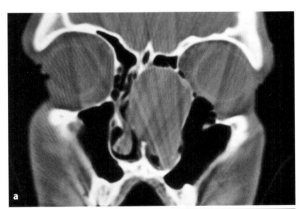

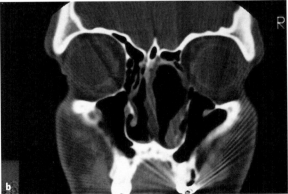

Fig. 41.30 a,b. A 20-year-old male with a schwannoma occupying the nasal cavity and ethmoid sinus. Preoperative (a) and postoperative (b) coronal computed tomography

Esthesioneurocytoma

Esthesioneurocytoma is a tumor of the olfactory plate and can be considered a type of esthesioneuroepithelioma of neural tube origin. The embryonically derived esthesioneuromas are histologically classified as esthesioneuroepithelioma, esthesioneuroblastoma and esthesioneurocytoma, which constitutes 40% of the total [6]. It occurs slightly more often in males (60%) and is most common between 10 years and 40 years of age [5].

The clinical picture may be subtle but includes nasal obstruction, mucosanguineous nasal secretions, anosmia and headache. Epistaxis is a late manifestation. Symptoms are only slowly progressive and may exist for several years before medical consultation is sought. Examination of the nasal cavity reveals a tumor of variable consistency that can be as firm as cartilage and exhibits a blue–red color. CT and MR are essential to define the exact location and extent of invasion (Fig. 41.31).

The final diagnosis is established by histological examination of a biopsy and can be difficult for the pathologist if the sample is too small. The tumor is composed of neurocytes and neuroblasts, with a predominance of one type.

Treatment is essentially surgical after staging the lesion. The most frequently recommended surgical approaches are midfacial degloving, paralateral rhinotomy and craniofacial approaches, either isolated or combined.

In cases of small localized tumors, removal can be performed via a transnasal, micro-endoscopic surgical approach. Radiotherapy is reserved for cases presenting postoperative persistence or recurrence or when metastases are present [53].

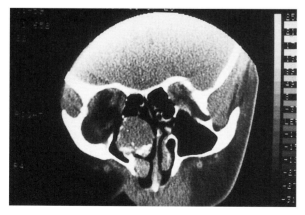

Fig. 41.31. A 40-year-old female with an esthesioneurocytoma of the middle meatus region

References

1. Ameli NO, Abbassion K, Azod A, Saleh H (1984) Aneurysmal bone cyst of the skull. Can J Neurol Sci 11:466–471
2. Andrews JC, Fisch U, Valvanis A, Aeppli U, Makek MS (1989) The surgical management of extensive nasopharyngeal angiofibromas with infratemporal fossa approach. Laryngoscope 99:429–437
3. Antonelli AR, Cappielo J, Dilorenzo D (1987) Diagnosis, staging and treatment of juvenile nasopharyngeal angiofibroma. Laryngoscope 97:1319–1325
4. Atallah N, Jay MM (1981) Osteomas of the paranasal sinuses. J Laryngol Otol 95:291–304
5. Baitey BJ, Barton S (1985) Olfactory neuroblastoma management and prognosis. Arch Otolaryngol Head Neck Surg 101:1
6. Batsakis JG (1979) Tumors of the head and neck, 2nd edn. Williams and Wilkins, Baltimore
7. Batsakis, JG (1980) The pathology of head and neck tumours: nasal cavity and paranasal sinuses. Part 5. Head Neck 2:410
8. Bernier JL, Bhasker SN (1958) Aneurysmal bone cyst of mandibula. Oral Surg 11:1018
9. Bremem JW, Neel HB, DeSanto LW, Jones G (1986) Angiofibroma: treatment trends in 150 patients during 40 years. Laryngoscope 96:321–329
10. Bresecker JL, Marcone RC, Huvos SG, Mike V (1970) Aneurysmal bone cysts: a clinicopathologic study of 66 cases. Cancer 26:615
11. Brown K, Brown OE (1998) Congenital malformations of the nose. In: Cummings CW, Fredricson JM, Harker LA, et al. (eds) Otolaryngology, Head and Neck Surgery. Mosby, St. Louis, pp 92–103
12. Buchwald C, Franzmann MB, Tos M (1995) Sinonasal papillomas: a report of 82 cases in Copenhagen county, including a longitudinal epidemiological and clinical study. Laryngoscope 105:72–79
13. Buraczewski J, Dabska M (1971) Pathogenesis of aneurysmal bone cyst: relationship between the aneurysmal bone cyst and fibrosis dysplasia of bone. Cancer 28:597
14. Canalis RF (1982) Frontal mucoceles. In: English GM (ed) Otolaryngology. Harper and Row, Philadelphia, pp 1–11
15. Chandler JR, Goulding R, Moskowitz L (1984) Nasopharyngeal angiofibroma: staging and management. Ann Otol Rhinol Laryngol 93:322–329
16. Childrey JH (1939) Osteoma of sinuses, the frontal and sphenoid bone: report of fifteen cases. Arch Otolaryngol Head Neck Surg 30:63–72
17. Cody DT II, DeSanto LW (1998) Neoplasms of the nasal cavity. In: Cummings CW, Fredricson JM, Harker LA, et al. (eds) Otolaryngology, Head and Neck Surgery. Mosby, St. Louis, pp 883–901
18. Compagno J, Hyams VJ, Caprain MC (1976) Hemangiopericytoma like intranasal tumors. Am J Clin Pathol 66:677–683
19. Cummings BJR, Keane T (1980) Primary radiation therapy for juvenile nasopharyngeal angiofibroma. Head Neck Surg 3:21–26
20. Dolgin SR, Zaveri VD, Cassiano RR, Maniglia AJ (1992) Different options for treatment of inverting papilloma of the nose and paranasal sinuses: a report of 41 cases. Laryngoscope 102:231–236

21. Draf W (1991) Endonasal micro-endoscopic frontal sinus surgery. The Fulda concept. Operative Tech Otolaryngol Head Neck Surg 2:234–240
22. Drake-Lee A (1997) The pathogenesis of nasal polyps. In: Settipane GA, Lund VJ, Bernstein JM, Tos M (eds) Nasal polyps: epidemiology, pathogenesis and treatment. Oceanside, Providence, pp 57–64
23. Easterley NM, Soloman LM (1972) Neonatal dermatology. II. Pigmentary lesions and hemangiomas. J Pediatr 81:1003
24. Edgerton MT (1976) The treatment of the hemangiomas with special reference to the role of steroid therapy. Am Surg 183:517
25. Evans C (1981) Aetiology and treatment of fronto-ethmoidal mucocele. J Laryngol Otol 95:361–375
26. Fields JN, Haverson KJ, Devineni VR, Oimpeou JR, Perez CA (1990) Juvenile nasopharyngeal angiofibroma: efficacy of radiation therapy. Radiology 176:263–265
27. Fisch U (1989) Surgical management of extensive nasopharyngeal angiofibromas with the infratemporal fossa approach. Laryngoscope 99:429
28. Fitzpatrick PJ, Briant TDR, Berman JM (1980) The nasopharingeal angiofibroma. Arch Otolaryngol 106:234
29. Fleury P, Narcy PH, Basset JM, Bobin S (1983) Tumeurs benignes du nez et des sinus. Med Chir Otorhinolaryngol 29:400–410
30. Fu YS, Perzin KH (1974) Non-epithelial tumors of the nasal cavity, paranasal sinuses, and nasopharynx: a clinicopathologic study. I. General features and vascular tumors. Cancer 33:1275
31. Geopfert H, Cangir A, Lee YY (1985) Chemotherapy for agressive juvenile nasopharyngeal angiofibroma. Arch Otolaryngol 111:285–289
32. Grabb WC, Dingman RO, Oneal R (1980) Facial hamartomas in children: neurofibroma, lymphangioma and hemangioma. Plast Reconstr Surg 66:509
33. Grybauskas V, Parker J, Friedman M (1986) Juvenile nasopharyngeal angiofibroma. Otolaryngol Clin North Am 19:647–657
34. Harrison DFN (1976) Juvenile postnasal angiofibroma an evaluation. Clin Otolaryngol 1:187
35. Harrison DFN (1979) Tumors of the nose and sinuses. In: Maran AGD, Stell PM (eds) Clinical otolaryngology. Blackwell Scientific, Oxford, pp 424–432
36. Haughey BH, Wilson JS, Barber CS (1988) Massive angiofibroma: a surgical approach and adjuntive therapy. Otolaryngol Head Neck Surg 98:618–623
37. Hellquist HB (1997) Histopathology. In: Settipane GA, Lund VJ, Bernstein JM, Tos M (eds) Nasal polyps: epidemiology, pathogenesis and treatment. Oceanside, Providence, pp 31–39
38. Howarth WG (1921) Mucocele and pyocele of the nasal accessory sinuses. Lancet 2:744–746
39. Jaffe HL, Lichtenstein L (1942) Solitary unicameral bone cyst with emphasis on roentgen, picture, pathologic appearance and the pathogenesis. Arch Surg 44:1004–1025
40. Johnson JT, Ferguson BJ (1998) Infection. In: Cummings CW, Fredricson JM, Harker LA, et al. (eds) Otolaryngology, Head and Neck Surgery. Mosby, St. Louis, pp 1107–1118
41. Josepphson JS, Herrera A (1995) Mucocele of the paranasal sinuses: endoscopic diagnosis and treatment. In: Stankiewicz JA (ed) Advanced endoscopic sinus surgery. Mosby, St. Louis pp 51–59
42. Kamel RH (1992) Transnasal endoscopic surgery in juvenile nasopharyngeal angiofibroma. J Laryngol Otol 110:962–968
43. Kaufman SM, Conrad LP (1976) Schwannoma presenting as nasal polyp. Laryngoscope 86:595–597
44. Lichtenstein L (1950) Aneurysmal cyst. A pathological entity commonly and osteogenic sarcoma. Cancer 3:279–289
45. Ho KL (1980) Primary meningioma of the nasal cavity and paranasal sinuses. Cancer 46:1442–1447
46. Montgomery WW (1979) Surgery of the upper respiratory system, 2nd edn. Lea and Febiger, Philadelphia, pp 563
47. Nadol JB Jr (1974) Positive "fistula sign" with an intact tympanic membrane. Arch Otolaryngol Head Neck Surg 100:273
48. Nativig K, Larsen TE (1978) Mucocele of the paranasal sinuses. A retrospective clinical and histological study. J Laryngol Otol 92:1075
49. Panje WR, Gross C (1987) Surgery of the nasopharynx. In: Thawley SE, Panje WR (eds) Comprehensive management of head and neck tumors. Saunders, Philadelphia, pp 662–682
50. Riedel R (1978) The paranasal sinuses: surgery and technique. In: Ritter FN (ed) Mosby, St. Louis, pp 136–145
51. Robbins SL, Cotran RS (1983) Patologia estrutural e funcional, 2nd edn. Interamericana, Rio de Janeiro, pp 1241
52. Schuknecht HF, Lindsay JR (1949) Benign cysts of the paranasal sinuses. Arch Otolaryngol Head Neck Surg 49:604
53. Skolnik EM, Masari FS (1966) Olfactory neuroephitelioma, review of world literature and a presentation of two cases. Arch Otolaryngol Head Neck Surg 84:644
54. Som P (1991) Sinonasal cavity. In: Som P, Bergeron T (eds) Head and neck imaging. Mosby, St. Louis
55. Spector JG (1988) Management of juvenile angiofibroma. Laryngoscope 98:1016–1026
56. Stamm A (1991) Abordaje quirurgico mediofacial "degloving" de la base de craneo. An Otorrinolaringol (Uruguay) 57:75–84
57. Stamm A (1992) A surgical staging system for sinonasal polyposis. 23rd Pan-American Congress of ENT, Head and Neck Surgery. Grune and Stratton, Orlando, p 115
58. Stamm A (1997) Surgical "grading" system for inverting papilloma. In: McCafferty G, Coman W, Carroll R (eds) 16th World Congress of Otolaryngology, Head and Neck Surgery (Sydney). Monduzzi, Bologna, pp 1423–1427
59. Stamm A (1999) Cirurgía microendoscópica. Conceptos básicos. An ORL (Peru) 6:27–36
60. Stamm A Burnier M Jr (1989) Tumores benignos nasosinusais. In: Brandão LG, Ferraz AR (eds) Cirurgia de cabeça e pescoço, vol 1. Roca, São Paulo, pp 465–491
61. Stammberger H (1997) Endoscopic diagnosis in surgery of the paranasal sinuses and anterior skull base. Karl Storz, Tuttlingen
62. Steanhouse D, Mason DK (1968) Oral hemangiopericytoma: a case report. Br J Oral Maxillofac Surg 6:114
63. Terzian AE, Naconecy C (1985) Juvenile nasopharyngeal angiofibroma. Microsurgical approach in 25 cases as unique treatment. In: Meyers EN (ed) New dimensions in otorhinolaryngology, head and neck surgery. Elsevier, New York, pp 505–506
64. Tos M (1997) Early stages of polyp formation. In: Settipane GA, Lund VJ, Bernstein JM, Tos M (eds) Nasal polyps: epidemiology, pathogenesis and treatment. Oceanside, Providence, pp 65–72

65. Tzeng HZ, Chao HY (1997) Transnasal endoscopic approach for juvenile nasopharyngeal angiofibroma. Am J Otolaryngol 18:151–154
66. Walike JW, Bailey BJ (1971) Head and neck hemangiopericytomas. Arch Otolaryngol Head Neck Surg 93:345
67. Weymuller EA (1998) Neoplasms. In: Cummings CW, Fredricson JM, Harker LA, et al. (eds) Otolaryngology, Head and Neck Surgery. Mosby, St. Louis, pp 1119–1134
68. Willis RA (1948) Pathology of tumors. Butterworth, London, p 992
69. Zinreich SJ, Kennedy DW, Rosenbaum AE, et al. (1987) Paranasal sinuses: CT imaging requirements for endoscopic surgery. Radiology 163:769–775

Juvenile Nasopharyngeal Angiofibroma – Transantral Microsurgical Approach

Alejandro E. Terzian

Introduction

Juvenile nasopharyngeal angiofibroma (JNA) is a highly vascularized, benign neoplasm that almost exclusively affects peripubescent male patients. Its most remarkable features are the site of origin and implantation, its vascular feeding vessels, its extension to surrounding structures, the difficulties in finding an appropriate surgical technique and its high rate of recurrence.

Epidemiology

Incidence

The incidence of JNA has been estimated at 1:5000 to 1:6000 otolaryngological admissions per year and approximately 1:150,000 young males per year in the general population [3, 40]. The tumor is estimated to account for 0.5% of all neoplasms of the head and neck (Table 42.1).

Age

Forty-two percent of our 149 cases were between 14 years and 16 years old. This age coincides with high levels of testosterone in serum. The rest of the cases ranged from ages 6–28 years.

Gender

Cases in females are extremely unusual. We could find fewer than ten cases in the literature, and we had none in our series. In 1968, Rominger [46] reported an unusual case of angiofibroma in an elderly black woman. Batsakis [4] reported two cases in females, while Lasjaunias described the case of a 17-year-old in her last months of pregnancy. In 1989, Palmer [41] reported the last female case. No more cases were found in the literature until 1999.

Hormonal Influence

Hays Martin [35] and Maurice Schiff [49, 50] reported hormonal influences in the development of the tumor and described its treatment with sex hormones. In 1987, Farag [15] demonstrated specific androgen receptors with a higher affinity for dihydrotestosterone than for testosterone in JNA. Alterations in the serum levels of sex hormones have not been reported in patients with JNA, nor have abnormalities in the maturity of their sexual organs.

Pathogenesis

Many different pathogenic theories have been offered, but none have been scientifically proved. In 1959, Schiff suggested that the oncogenetic focus could be found within the nasopharyngeal periostium. In 1987, Harrison [23] suggested that unilateral hamartomatous ectopic tissue nests in the region of the sphenopalatine foramen and at the base of the pterygoid plate could respond to endogenous testosterone.

Site of Origin and Implantation

Nelaton first suggested that the tumor could originate from the periosteum of the vault of the nasopharynx in 1853 and, in 1878, Tillaux stated that the tumor almost invariably arose from the basilar fibrocartilaginous tissue extending from the atlas to

Table 42.1. Incidence rates for different authors

Author	Year	ENT external consultations	Period (years)	Cases	Annual incidence
Batsakis [3, 4]	1979	1:15000	NA	NA	NA
Ungkanont [60]	1997	NA	38	43	1.1
Witt [67]	1983	NA	31	41	1.3
Spector [52]	1988	1.6:16000	17	27	1.6
Harrison [23]	1987	NA	20	44	2.2
Economou [14]	1985	NA	25	83	2.7
Krause [28]	1982	NA	4	14	4.6
Terzián [57] (1)	1999	1:7500	20	149	7.4 (2)

ENT, ear, nose and throat; *NA*, no available data (*1*) National Referential Center (*2*) Updated data

the lower surface of the sphenoidal bone. However, in 1992, Schiff argued that the areas most frequently involved by the tumor were the basiocciput and the medial pterygoid lamina, from which the anlage migrates forward to form the inferior turbinate. Our experience suggests that the implantation area is located unilaterally, deep in the basisphenoid (the embryonic sphenoidal bone separated from the occipital bone's basilar apophysis by a synchondrosis that later becomes the anterior clivus in adult life; Fig. 42.1).

Vascular Pedicle

The most important arteries consistently nourishing the tumor arise from the terminal branches of the third portion of the internal maxillary artery (IMA) linking the tumor to the pterygomaxillary fossa. The two most important are the sphenopalatine and the descending palatine arteries. Occasionally, the ascending pharyngeal artery also plays an important role via its internal superior branch [31, 32].

The internal carotid artery can supply the tumor via the foramen rotundum artery or the anterior and posterior ethmoidal branches of the ophthalmic artery in a few cases. Another source of blood supply can arise from multiple small arteries from the dura mater of the cavernous sinus.

Signs and Symptoms

- *Epistaxis* is the most frequent presenting symptom and can be mild, moderate or severe. When patients present with mild or moderate epistaxis, JNA may not be suspected; it is only in cases where the epistaxis is persistent or severe that there is a thorough search for the cause before the diagnosis is made.
- *Nasal obstruction* was the first complaint in 20% of our patients with JNA who had no previous epistaxis (Fig. 42.2).
- *Facial deformity* occurs when the tumor expands out of the pterygomaxillary fossa and follows the path of least resistance. It pushes forward, displacing and then penetrating the maxillary sinus; it then extends outward, invading the zygomatic fossa. Ultimately, it extends inferiorly and anteriorly, surrounding the tuberosity of the maxillary bone and eventually appearing in the cheek (Fig. 42.3).
- *Orbital and ocular symptoms* can also occur when the orbit and its contents are compromised by tumor extension. In Stern's revision series of 218 cases [55], 14% presented with exophthalmos, 5% with decreased visual acuity and 2% with partial ophthalmoplegia.
- A *nasopharyngeal mass* is present in all cases to some degree.
- A *bulging soft palate and a gingivobuccal mass* occur with anterior extension from the nasopharynx.
- A *unilateral anterior nasal mass* is uncommon but occurs when the tumor mass fills the nasal cavity and extends to the nostril (Fig. 42.4).

Intracranial Invasion

There have been many reports of intracranial extension of JNA, including direct invasion of the brain. However, in our experience with 29 cases exhibiting intracranial tumor extension, there was no evidence of transdural invasion of the cerebral parenchyma or invasion of the cavernous sinus. The tumor always remained extradural (Fig. 42.5).

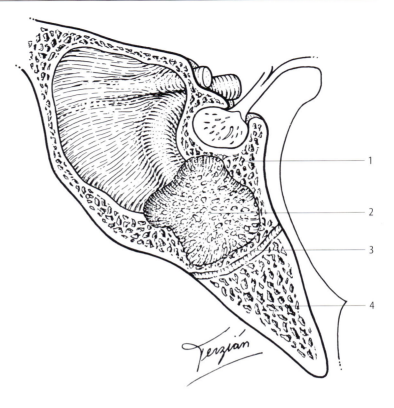

Fig. 42.1. Implantation of the tumor in the basisphenoidal region. *1*, Basisphenoid; *2*, insertion bed; *3*, sphenobasilar synchondrosis; *4*, occipital basilar process

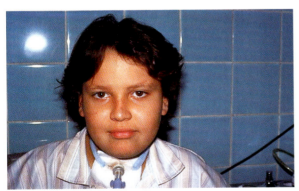

Fig. 42.2. Uncommon juvenile nasopharyngeal angiofibroma case presenting bilateral and total nasal obstruction but no epistaxis. Tracheostomy was performed in the preoperative period because of severe episodes of sleep apnea refractory to usual treatments

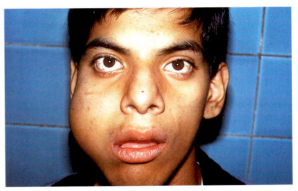

Fig. 42.3. Hemifacial and cheek deformities caused by juvenile nasopharyngeal angiofibroma

Imaging Studies

It is conceptually useful to divide the cranium into neurocranial and viscerocranial portions. While the neurocranium is mainly occupied by soft tissue and is best studied by magnetic-resonance imaging (MRI). The viscerocranium is composed principally of bone and air-filled spaces limited by bone, making computed tomography (CT) the ideal method for its study.

- CT demonstrates normal and distorted bone and areas of tumor implantation and its extension (Fig. 42.6). The addition of contrast is very helpful in establishing the diagnosis of the tumor by virtue of its particular pattern of vascular supply. Only hemangioma, hemangiopericytoma, plasmocytoma and olfactory neuroblastoma have patterns that are similar (Fig. 42.7). Erosion of the base of the pterygoid process and its medial lam-

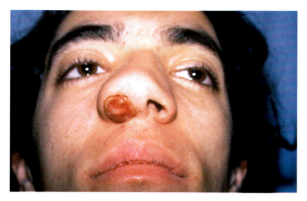

Fig. 42.4. Exteriorization of a juvenile nasopharyngeal angiofibroma from the nose

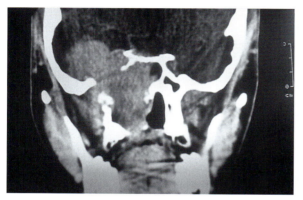

Fig. 42.5. Coronal computed tomography. Extradural intracranial extension of juvenile nasopharyngeal angiofibroma

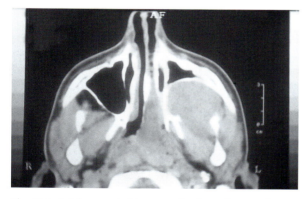

Fig. 42.6. Axial computed tomography. Juvenile nasopharyngeal angiofibroma involving the pterygomaxillary fossa, which displace the posterior wall of the maxillary sinus forward (antral sign) and displace the pterygoid muscles backward

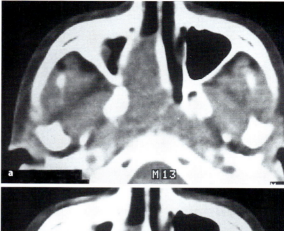

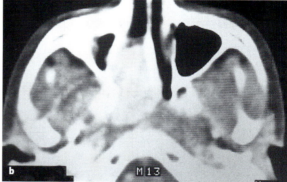

Fig. 42.7. Preoperative axial computed tomography without (**a**) and with (**b**) contrast. Notice the enhancement of the tumor, which is clearly differentiated from the surrounding tissues

ina and vaginal process can be seen together with extension of the tumor into the basisphenoid.

- Some authors believe that MRI is the ideal modality for investigation of the tumor. Gadolinium contrast helps to enhance and delineate the lesion and its extensions, especially in the neurocranial cavity [63]. Sagittal-plane MRI scans are very helpful in determining the degree of erosion of the sphenoidal body (basisphenoid). The enhancement of the fatty marrow must be taken into consideration (Fig. 42.8) [2].
- Superselective angiography (Fig. 42.9) has been used extensively to show the vascularization of JNA, making differentiation from angiomatous polyps [24], hemangioma and esthesioneuroblastoma much easier. Because of the risk of complications, angiography is not universally recommended [43]. It is advocated when it is necessary to know the specific vascular supply before embolization is performed.

Histopathological Findings

JNA is characterized by a characteristic histologic pattern of angiomatous and fibrous structures. The angiomatous tissues can be loose, edematous with star-shaped fibroblasts and numerous mast cells or dense, acellular and highly collagenized tissue. The tumor vessels are always without elastic fibers. The propensity for significant hemorrhage from these lesions can be explained by the irregularity of the tumor, the lack of an elastic layer in the vascular wall or the lack of elastic stromal fibers [56]. In older, more mature lesions, there is a prevalence of fibrous tissue; the capillary-fibroblastic cambium zone disappears, and areas of hyalinization are increased [5].

Differential Diagnosis

Sophisticated imaging studies, such as CT, MRI and superselective angiography, and a more complete understanding of the biologic behavior of JNA have facilitated better diagnostic precision. Among the benign tumors in the differential diagnosis are angiomatous polyps [24], hemangiomas, craniopharyngiomas, Thornwaldt's cyst, inflammatory polyp, pyogenic granuloma and antrochoanal polyps [6]. Malignant tumors to include in the differential diagnosis are rhabdomyosarcoma, nasopharyngeal carcinoma, hemangiopericytoma, extramedullary plasmocytoma and esthesioneuroblastoma; these also have rich vascular supplies.

Diagnosis

In almost all cases, the medical history (gender, age and symptoms) and imaging studies (without and with contrast) are sufficient to establish the JNA diagnosis and to exclude other pathologies (Fig. 42.7). Biopsy is to be avoided. If absolutely necessary, we recommend that biopsy be performed in an operative setting because of the risk of severe bleeding (Fig. 42.10).

Staging

The purpose of tumor staging is to:
1. Standardize the different tumor presentations by size and extent
2. Determine surgical strategies for the different stages
3. Predict the prognosis
4. Predict which cases are likely to recur

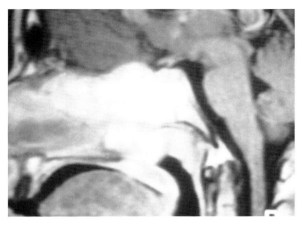

Fig. 42.8. Sagittal magnetic-resonance imaging. Clear differentiation of the basilar process and sphenopalatine synchondrosis. The tumor invades the basisphenoid above the posterior face

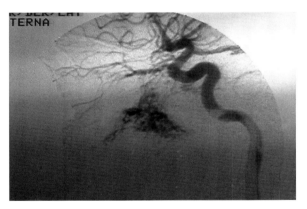

Fig. 42.9. Angiography of the internal carotid artery. The superior portion of the tumor is irrigated by the ophthalmic artery via its ethmoidal branches

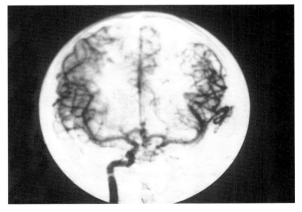

Fig. 42.10. Ligature of the common carotid artery due to severe hemorrhage after biopsy. Courtesy of Pedro Elizalde Children's Hospital, Buenos Aires

Standardization is probably the most important of these goals and is the reason we adopted Chandler's [9] classification (Tables 42.2, 42.3) to clearly define the stage based on different degrees of tumor extension from the site of origin.

Treatment

Non-Surgical Treatment

Radiotherapy

The use of radiotherapy in primary cases is controversial. Preoperative radiotherapy has been used in the past to reduce tumor size and intraoperative bleeding [61]. The usual dosages range from 4000 rad to 4500 rad [37]. Lower doses seemed to be less effective, and there are reports that show that treatment with 3200 rad was followed by recurrence in 80% of cases.

Nowadays, however, radiotherapy is rarely justified, at least in primary cases. Radiotherapy is sometimes used in healthy, young patients during their growing period or when there is tumor compromise of vital anatomic elements, such as the pituitary gland, optic nerves or brain stem. It is more widely accepted in cases of invasive intracranial tumors or recurrences that cannot be re-operated [16].

Osteoradionecrosis [11] and induced malignant tumor transformation into malignant fibrous histiocytoma [51], fibrous sarcoma [10] or sarcomas [4, 12] have been reported to result from radiotherapy. Spector [52] described two cases of maxillary-sinus sarcoma more than 40 years after radiotherapy. We observed a patient who developed an extensive maxillary-sinus carcinoma as a result of radiation 38 years before (Fig. 42.11).

Chemotherapy

Systemic chemotherapy has been used in the treatment of recurrent JNA with intracranial extension that was considered inoperable. Adriamycin and decarbazine have been used in combination, although adriamycin can cause significant bone-marrow and cardiac toxicity [19, 20]. However, for most authors, chemotherapy is justified only if the tumor is unresectable [48]. We have never used this treatment modality.

Hormonal Therapy

Because sex hormones have an effect on the tumor, both estrogen and testosterone have been used for its treatment. In 1954, Martin employed testosterone. Schiff [49, 50] and Walike [62] had success with stilbestrol, noting a reduction in both the size and vascularity of the tumors. However, neither Krause

Table 42.2. Chandler's juvenile-angiofibroma classification

Stage	Extension
I	Nasopharynx
II	Stage I plus nasal fossa and/or sphenoidal sinus
III	Stage II plus maxillary and/or ethmoidal sinus
	Orbit and/or cheek
	Subtemporal fossa
	Pterygomaxillary fossa
IV	Stage III plus intracranial invasion

Table 42.3. Stages in our 149 juvenile-nasopharyngeal-angiofibroma cases (1999)

Stage	Number	Percentage
I	0	0
II	27	18
III	92	82
IV	30	20

Fig. 42.11. Patient presenting with an extensive maxillary-sinus carcinoma 38 years after irradiation of the juvenile nasopharyngeal angiofibroma. Courtesy of Dr. G. Haedo, M.D.

[28] nor Weidenbecher [64] observed significant reduction in size or vascularity on estrogen treatment. In 1992, Gates [18] reported that he obtained reduction of JNA size, as measured on pre- and post-treatment CT scans using anti-androgens (flutamide).

The side effects of this type of therapy must be taken into account, especially considering the young age of the patients and the benign nature of the disease. Discussing these therapies in 1987, Harrison asked whether "their use can be considered ethically defensible". There has been no conclusive demonstration of the efficacy of this therapy in reducing tumor size or vascularity, either in clinical or experimental studies.

Embolization

The value of presurgical embolization for JNA treatment is also a controversial matter. Those in favor of embolization argue that it is a well-established technique that reduces tumor blood supply and facilitates tumor removal. The timing between embolization and surgery is important, and surgery must be performed between 24 and 72 hs after the embolization. Those against embolization point out that an 11% complication rate is unacceptably high for a benign tumor [1]. Neurological complications from embolization have included hemiplegia, facial palsy, blindness and cerebral infarction, among others.

Some authors reported a relationship between preoperative embolization and the frequency of tumor recurrence; they suspected that embolization shrinks the tumor and allows small parts of it to remain undetected [36]. In our opinion, this procedure is advisable when the surgical strategy does not include ligation and division of the branches of the IMA in the pterygomaxillary fossa, which is usually an important and early step in our surgery. In our experience, one of the principal drawbacks of embolization is an increased difficulty in dissecting the tumor from the surrounding tissues because of peritumor inflammation induced by the procedure itself.

Intratumoral Embolization

In order to prevent the potential side effects of embolization, direct intratumoral embolization was proposed by French authors following the technique described by Casasco for craniofacial tumors. The materials used were cyanoacrylate, lipiodal and tungsten powder. According to the authors, this technique induces a marked devascularization and necrosis of the tumor, thus reducing its volume and facilitating its surgical removal [58]. However, there is the possibility of side effects due to cyanoacrylate cytotoxicity and to uncontrolled diffusion of it.

Surgical Treatment

Historical Remarks

Six different ways to access the tumor have been described:
1. An anterior transcutaneous route via an external rhinotomy (incision of paranasal skin), as described by Michaux in 1848 and Moure in 1902 for ethmoidalis cells resection [38]. Since 1954, Handousa [21] has used this approach for resection of JNA. Krespi [30], Janecka [25] and others have described other anterior transcutaneous approaches.
2. A lateral transcutaneous route, e.g., Kremen's transmandibular approach [22], Samii's [47] transzygomatic approach, Fisch's [16] infratemporal approach.
3. The transpalatal approach used in North America by Walker, Brown, Heck and Hemley [22]. In 1957, Wilson [66] refined this technique, which has been adopted more frequently since then.
4. Bocca's [7] transpharyngeal–suprahyoid approach.
5. A combined transcranial approach [29, 54].
6. Sublabial approaches [8].

Evolution of the Sublabial Approaches and Deep Transantral Microsurgery

The practice of performing sublabial antrostomies started in 1802 with Desault, who performed a simple antrostomy. Caldwell (in 1893) and Luc (in 1897) added a counter-opening in the inferior meatus [38]. In 1911, Denker enlarged the antrostomy by removing the bony strip that separates it from the nasal fossa. Since then, many variations have been performed, such as the techniques of Albrecht (1931) and Ardourin (1955) [22]. However, it was not until the publications of Krause [28], Chandler [9], Terzián [57] and Price [44] that the concept of *deep transantral microsurgery*, which involves access to the pterigomaxillary fossa and the skull base structures, was developed. The advantages of this technique are direct access to the IMA (for ligation) and to the basisphenoid implantation area; this allows the region to be better exposed.

All tumoral extensions of the angiofibroma can be removed via this approach. In addition to the excellent cosmetic and functional results, this technique allows easy postoperative examination via anterior rhinoscopy and/or endoscopic examination of the resultant large rhinosinual cavity.

Deep Transantral Microsurgery Technique

Despite the apparent ease of the approach, the proposed technique has some technical difficulties. However, the advantages are so numerous that it is worthwhile to dedicate time and effort to mastering it.

The technique requires (1) extensive experience in the use of the surgical microscope, (2) technical skill and (3) anatomical knowledge of the way to approach the pterigomaxillary fossa. Adequate instrumentation and a versatile surgical microscope with a 300-mm objective are necessary (Fig. 42.12).

Main Steps

- Opening of the maxillary sinus. An incision is made in the gingival sulcus, as in a conventional Caldwell-Luc procedure. A wide resection of the anterior wall of the antrum is performed, leaving the infraorbital opening and its contents in place. It is important to remove all the mucous lining of the opened sinus, because it bleeds and diminishes the visibility of the operative area. The cheek flap is reflected upward and laterally by two self-retaining retractors (Fig. 42.13). The surgical microscope is then brought into the field and used throughout the rest of the operation.
- Opening of the posterior wall of the maxillary sinus. When there is tumoral pterygomaxillary extension, the thickness of the bone may vary, depending mainly on the size and consistency of the adjacent tumor. In many cases, the wall has the consistency of an eggshell and, in other cases, it may even disappear.
- Ligation and section of the pedicle. Once the aponeurosis of the pterygomaxillary fossa has been exposed, it must be approached carefully. The third portion of the IMA is found in the anterior superficial vascular plane, along with its branches, all of which are embedded in Bichat's fat. These arteries are exposed with the aid of a vessel hook, are dissected to the point where they enter the tumor, and are ligated and sectioned. Yasargil's ligature guide is a very helpful tool for these maneuvers. As a result, three arteries ligated and sectioned: the terminal portion of the IMA and its two branches, the sphenopalatine and the descending palatine, which are the arteries that irrigate the tumor (Fig. 42.14, step 1). In cases where the tumor occupies the pterygomaxillary fossa, it is possible to push it (the tumor) aside (toward the midline) until the Bichat's fat becomes visible. The appearance of this fat marks the lateral limit of the tumor's extension, where entry of the aforementioned arteries into the tumor is seen.
- Opening of the antronasal wall and nasal mucosa. The entire medial wall is removed using delicate, curved chisels. The mucoperiosteum of the external nasal wall can then be seen. An incision is

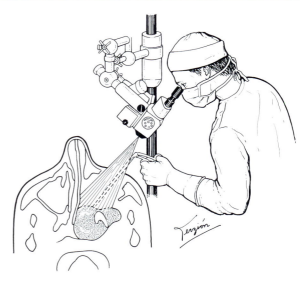

Fig. 42.12. Use of the operating microscope in juvenile nasopharyngeal angiofibroma surgery

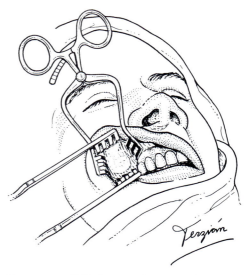

Fig. 42.13. Sublabial approach. Two self-retaining retractors expose the surgical field

made on the mucosa at the level of the inferior meatus. On lifting the mucosa, the inferior turbinate is encountered.
- Posterior de-insertion of the turbinates and transnarial elevation ("curtain maneuver"). The inferior and middle turbinate are de-inserted from their posterior attachments. Two tractor stitches are placed in the distal ends of the turbinates and are carried through the nostril until all the visible portions of the turbinates are lifted. Therefore, just as the raising of a theater curtains allows us to see the stage, lifting the posterior portion of the turbinate allows us to see the tumor in the nasochoanal and nasopharyngeal regions (Fig. 42.15). To facilitate this maneuver, the bony component of the inferior turbinate can be resected via the same route.
- Opening of the ethmoid sinus. The ethmoidal cells are resected via the supero-internal corner of the maxillary sinus. All the bones and mucosa of the various cells must be removed. The last cell, which is usually larger (Onodi's cell), is also resected, allowing immediate exposure of the sphenoid sinus. If the tumor occupies the ethmoid sinus, it can be carefully dissected and displaced toward the midline.
- Opening of the sphenoid sinus. The anterior wall of the sphenoid sinus can be completely or partially penetrated by the tumor. In the latter situation, the tumor penetrates in the shape of an hourglass. Therefore, this wall must be carefully resected using Kerrison's rongeurs. The implantation of the superior and posterior border of the vomer and its articular alae must be resected along with the rostrum of the sphenoid bone to remove the anterior wall of the sinus. As a result, the base of the sphenoidal sinus is properly exposed, making visible the superior border of the sphenoidal body, or basisphenoid bone. It is precisely in this area, directly in the bone's trabeculae, that penetration by the tumor is seen. We observed that this is the area of implantation of the tumor in all our cases.
- De-insertion of the tumor. The tumor is held and mobilized in different directions using forceps with sharp edges (like Foerster's forceps) to determine its extent and adhesions. These adhesions may involve:
  1. The mucosa of the posterior wall of the rhinopharynx
  2. The mucosa of the posterior third of the nasal septum
  3. The aponeurosis of the orbicularis oris
  4. The dura mater of the cavernous sinus (the dura mater of the anterior and middle fossa seldom adhere to the tumor)

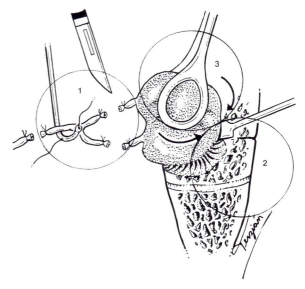

Fig. 42.14. The three most important steps of juvenile-nasopharyngeal-angiofibroma surgery. *1*, Ligation and sectioning of the vascular feeding vessels; *2*, basisphenoid de-insertion; *3*, complete rotation and removal of the tumor

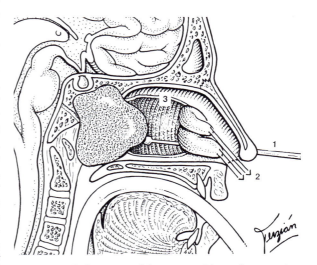

Fig. 42.15. Displacement of the tumor with an elevator, via a maxillary sublabial antrostomy (*1*). The posterior two thirds of the middle and inferior turbinates are de-inserted and secured with stitches near the nares (the "curtain maneuver", *2*). The medial wall of the maxillary sinus is removed and, together with the curtain maneuver, this allows wider vision through the antrostomy (*3*)

  5. The superior maxillary nerve (often embedded in the tumor)
  6. The basisphenoidal-implantation area
- Basisphenoidal de-insertion. The tumor is firmly entrenched in the trabeculae of the basisphenoid. The surgeon must employ all his skill in order to

de-bulk and extract these tumoral invaginations. Occasionally, the tumor penetrates so deeply that the pre-pontine dura mater can be visualized. The most severe hemorrhage of the surgery may occur at this time and will not cease until the invaginated, millimeter-sized tumor remnants are removed. We believe that these tumoral remnants are the most frequent cause of recurrence (Fig. 42.14, step 2).
- Mobilization and rotation. The maneuvers of de-invagination, detachment and de-insertion must be performed until the tumor is completely loose. This occurs when the tumor can rotate freely around its axis and in different directions. These maneuvers must be performed with the aid of cottonoids and dull dissection (Fig. 42.14, step 3).
- Removal. The tumor is removed in one piece via the path previously made. There are two situations in which this might be a difficult maneuver: when the tumor is large and hard, or when the patient has a small maxillary sinus and a large tumor. This situation was found in only five of our 149 cases. Pushing the tumor toward the nasopharynx or oropharynx can allow it to be extracted via the mouth.
- Final look. Once the tumoral mass has been removed, a great cavity is exposed, and all its crannies and nooks must be searched for hemorrhagic spots although, in most cases, no blood is seen. If hemorrhagic spots are found, it is very possible that some tumor tissue remains, especially in the basisphenoidal area. All such remnants must be removed, always using the surgical microscope.
- Final steps and closure. After the entire region has been explored and there is absolute certainty of complete hemostasis, the incision can be closed. The posterior portions of both turbinates must be loosely sutured to the soft tissue of the anterior side of the maxillary sinus to avoid intranasal adhesions. The function of both turbinates remains unaltered. The sublabial suture is performed using inverted stitches. There is usually no need for packs or drains.
- Postoperative care. Postoperative care is no different than that needed after routine antrostomy. The patient is discharged from the hospital between the third and fifth day after surgery. Imaging diagnostic studies must be avoided during the first 6 months, if they are not indispensable, because the regenerating tissues may produce false images. These studies must be done every 6 months during the first two postoperative years, then annually for three more years.

Endoscopic Surgery

Transoral and/or transnasal endoscopic resection of JNA has been described for limited lesions. Those in favor argue that the advantages of this technique include less bleeding and a shorter procedure time, but experienced surgeons consider it a valid alternative only for very specific cases [26, 59]. In 1996, Stammberger [53] presented several cases with previous embolization operated on using this procedure.

Recurrences

Recurrence, or *regrowth*, is the reappearance of the tumor, with or without symptoms, after the primary resection (Table 42.4). A *residual* or *persistent* tumor is any part of the tumor that remains after the surgical procedure.

Recurrence is the regrowth of residual tumor tissue. After a retrospective analysis of the different possible causes of recurrence in our cases, we found that the origin of the regrowths was minimal tumoral tissue left in the basisphenoidal area.

Table 42.4. Juvenile nasopharyngeal angiofibroma recurrence rates for different authors

Author	Year	Cases	Recurrences	Recurrence rate (%)
Spector [52]	1988	28	2	7.1
Economou [14]	1988	68	5	7.3
Terzián [57]*	1999	149	18	12
Da Costa [13]	1992	24	5	21
Radkowski [45]	1996	23	5	21.7
Malik [34]	1991	27	5	24
García-Cervignon [17]	1988	58	11	35
Achouche [1]	1992	34	13	38.2
Kosokovic [27]	1987	28	13	46
McCombe [36]	1990	33	17	50

* These results are updated data

In accordance with the above-mentioned concepts, a study conducted in France [1] found that erosion of the clivus and displacement of the cavernous sinus seemed to be two factors significantly associated with the development of recurrent lesions. In London, McCombe [36] suggested that "the stronger predictor of recurrence was preoperative embolization, with early and multiple recurrences".

Tumoral Involution

Theoretically, since the tumor is hormone dependent, its activity and size can decrease when intense hormone activity diminishes after puberty. In 1988, Stiller found that, in older lesions, fibrous tissue prevails, the capillary fibroblastic cambium zone disappears and areas of hyalinization are enlarged. In 1992, Phelps [33, 42] described spontaneous regression in a recurrence case after many surgeries; this recurrence was controlled by serial CT.

In 1991, Weprin [65] reported a well-documented case of an 11-year-old boy who was diagnosed by biopsy but not treated. After a 12-year follow-up, total involution of the tumor was documented by CT.

In our experience, an asymptomatic persistence of the tumor was seen in a 58-year-old patient operated on and irradiated 42 years ago in another city. The tumoral persistence was detected on a CT examination performed for an unrelated pathology.

In conclusion, three situations are possible: total disappearance, symptomatic persistence or asymptomatic persistence of the tumor. The last situation can be explained by fibroblastic changes and increased hyalinization of the tumor with time, thus making the angiomatous component less important. Subsequently, the tumor is reduced in size, and the symptoms become minimal or are unnoticed.

The possibility of spontaneous tumor regression does not justify expectant behavior. Surgery at the time of the diagnosis, when possible without risk, is the treatment of choice. Involution of the tumor is an extremely infrequent and improbable outcome and, if it occurs, it does so after many years.

Mortality

Despite our extensive bibliographic research, we have found no reports of death or very serious complications during the natural evolution of the tumor. Unfortunately, there have been many published reports of death and very serious complications due to surgical intent and overaggressive treatment (Fig. 42.16; Table 42.5).

Treatment Results

There are four possible results of the treatment:
1. Total cure: no symptoms and negative imaging studies 12 months after initial treatment
2. Symptomatic recurrence (which, after re-treatment, can end in total cure)
3. Asymptomatic persistence of the tumor (which may be re-treated until it is completely cured or until either the physician or the patient declines re-treatment)
4. Symptomatic persistence of the tumor that is not re-treatable (either because of patient refusal or because of medical decision)

Follow-Up

A patient can be considered completely cured if endoscopic, imaging and clinical controls are negative 12 months after treatment. Control CT scans should be performed 6, 12, 18, 24, 36, 48 and

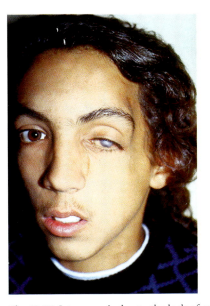

Fig. 42.16. Iatrogenesis due to the lack of appropriate surgical treatment (transfacial approach). Notice the external-carotid-artery ligature in the neck. The patient had left amaurosis due to pre- and postoperative radiotherapy. Moreover, the tumor could not be extracted. Courtesy of Pedro Elizalde Children's Hospital, Buenos Aires

Table 42.5. Juvenile nasopharyngeal angiofibroma mortality rates for different authors

Author	Year	Cases	Mortalities	Mortality rate (%)
Handousa [21]	1954	61	2	3
Piquet [43]	1985	29	1	3
Figi [22]	1950	50	2	4
Malik [34]	1991	25	1	4
Härmä [22]	1958	48	3	6.25
Neel [39]	1973	56	5	9
Martin [35]	1948	29	2	9
Kobylinski [22]	1908	NA	NA	17
Stöppler [22]	1952	13	5	38

NA, no available data

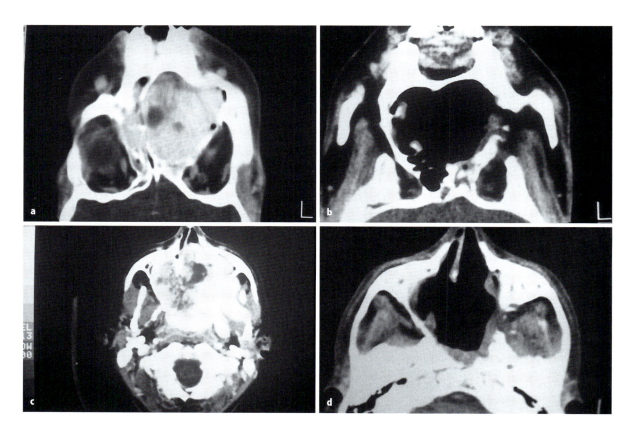

Fig. 42.17 a–d. Coronal computed tomography (CT) of a juvenile-nasopharyngeal-angiofibroma patient: preoperative (a) and at a 2-year postoperative follow-up (b). Axial CT of the same patient: preoperative (c) and at a 2-year postoperative follow-up (d)

60 months after surgery (Fig. 42.17). Possible outcomes of treatment are shown schematically in Fig. 42.18.

Acknowledgements. The author gratefully acknowledges Isabel Kaimen-Maciel, M.D., from Buenos Aires, Argentina, and Carlos Naconecy, M.D., from Blumenau, Brazil, for their assistance in researching the references and in the preparation of the manuscript. Special thanks to my secretary, Ms. Maggie Mizsey.

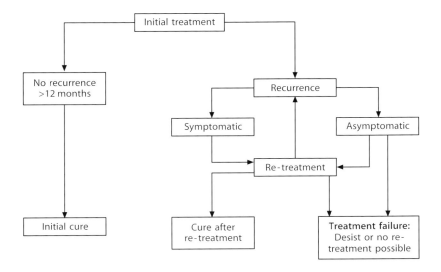

Fig. 42.18. Possible outcomes of treatment

References

1. Achouche J, Laccoureye D, De Gaudemar I, et al. (1992) Intra and extracranial nasopharyngeal fibroma. Contribution of imaging and study of local failure. Report on 34 patients. Ann Otol Chir Cervicofac 109:223–230
2. Applegate GR, Hirsch WL, Applegate LJ, Curtin HD (1992) Variability in the enhancement of the normal central skull base in children. Neuroradiology 34:217–221
3. Batsakis JG (1979) Tumors of the head and neck, 2nd edn. Williams and Wilkins, Baltimore, pp 291–312
4. Batsakis JG, Kloop CT, Newman W (1956) Fibrosarcoma arising in a juvenile nasopharyngeal angiofibroma following extensive radiation therapy. Am Surg 21:786–793
5. Beham A, Fletcher CD, Kainz J, et al. (1993) Nasopharyngeal angiofibroma: an immunohistochemical study of 32 cases. Virchows Arch 423:281–285
6. Bernat Gili A, Garcia Garcia B (1991) Craniopharyngioma of the nasopharynx. Apropos of a case. Acta Otorrinolaringol Esp 42:269–272
7. Bocca E (1971) Transpharyngeal approach to nasopharyngeal fibroma. Ann Otol Rhinol Laryngol 80:171–176
8. Brunner H (1942) Nasopharyngeal angiofibroma. Ann Otol Rhinol Laryngol 51:29–63
9. Chandler JR, Goulding R, Moskowitz L, Quencer RM (1984) Nasopharyngeal angiofibromas: staging and management. Ann Otol Rhinol Laryngol 93:322–329
10. Chen KT, Bauer FW (1982) Sarcomatous transformation of nasopharyngeal angiofibroma. Cancer 49:369–371
11. Conley J, Healy WV, Blaugrund SM, et al. (1968) Nasopharyngeal angiofibroma in the juvenile. Surg Gynecol Obstet 126:825–837
12. Cummings BJ, Blend R, et al. (1984) Primary radiation therapy for juvenile nasopharyngeal angiofibroma. Laryngoscope 94:1599–1605
13. Da Costa DM, Franche GL, Gessinger RP, Strachan D (1992) Surgical experience with juvenile nasopharyngeal angiofibroma. Ann Otolaryngol Chir Cervicofac 109:231–234
14. Economou TS, Abemayor E, Ward PH (1988) Juvenile nasopharyngeal angiofibroma: an update of the UCLA experience, 1960–1985. Laryngoscope 98:170–175
15. Farag MM, Ghanimah SE, Ragaie A, Saleem TH (1987) Hormonal receptors in juvenile nasopharyngeal angiofibroma. Laryngoscope 87:208–211
16. Fisch U (1983) The infratemporal fossa approach for nasopharyngeal tumors. Laryngoscope 93:36–44
17. Garcia-Cervignon E, Bien S, Rufenacht D, Reizine D (1988) Preoperative embolization of nasopharyngeal angiofibroma. Report of 58 cases. Neuroradiology 30:556–560
18. Gates GA, Rice DH, Koopmann CF, Schuller DE (1992) Flutamide-induced regression of angiofibroma. Laryngoscope 102:641–644
19. Göepfert H, Cangir A, Ayala AG, Eftekhari F (1982) Chemotherapy of locally aggressive head and neck tumors in the pediatric age group. Desmoid fibromatosis and nasopharyngeal angiofibroma. Am J Surg 144:437–444
20. Göepfert H, Cangir A, Lee YY (1985) Chemotherapy for aggressive juvenile nasopharyngeal fibroma. Arch Otolaryngol Head Neck Surg 111:285–289
21. Handousa A, Farid H, Elwi AM (1954) Nasopharyngeal fibroma: clinicopathological study of 70 cases. J Laryngol Otol 68:647–66
22. Härmä RA (1958) Nasopharyngeal angiofibroma. Acta Otolaryngol Suppl (Stockh) 146:1–76
23. Harrison DFN (1987) The natural history, pathogenesis and treatment of juvenile angiofibroma. Arch Otolaryngol Head Neck Surg 113:936–942
24. Irnberger T (1985) Computed tomographic diagnosis and differential diagnosis of juvenile angiofibroma and angiomatous polyps. Rofo Fortschr Geb Rontgenstr Neuen Bildgeb Verfahr 142:391–394
25. Janecka IP, Sen C, Sekhar LN, Nuss DW (1993) Facial translocation approach to nasopharynx, clivus, and infratemporal fossa. In: Sekhar LN, Janecka IP (eds) Surgery of cranial-base tumors. Raven, New York, pp 245–259
26. Kamel RH (1996) Transnasal endoscopic surgery in juvenile nasopharyngeal angiofibroma. J Laryngol Otol 110:962–968
27. Kosokovic F, Danic D (1987). Juvenile angiofibroma. Laryngorhinootologie 66:494–497
28. Krause CJ, Baker SR (1982) Extended transantral approach to pterygomaxillary tumors. Ann Otol Rhinol Laryngol 91:395–398

29. Krekorian EA, Kato RH (1977) Surgical management of nasopharyngeal angiofibroma with intracranial extension. Laryngoscope 87:154–164
30. Krespi Y, Har-El G (1993) The transmandibular–transcervical approach to the skull base. In: Sekhar LN, Janecka IP (eds) Surgery of cranial-base tumors. Raven, New York, pp 261–265
31. Lasjaunias PL (1980) Nasopharyngeal angiofibromas: hazards of embolization. Radiology 136:119–123
32. Lasjaunias PL (1983) Craniofacial and upper cervical arteries. Williams and Wilkins, London, pp 65
33. Lloyd GA, Phelps PD (1986) Juvenile angiofibroma: imaging by magnetic resonance, CT and conventional techniques. Clin Otolaryngol 11:247–259
34. Malik MK, Kumar A, Bhatia BP (1991) Juvenile nasopharyngeal angiofibroma. Indian J Med Sci 45:336–342
35. Martin H, Ehrlich HE, Abels JC (1948) Juvenile nasopharyngeal angiofibroma. Am Surg 127:513
36. McCombe A, Lund VJ, Howard DJ (1990) Recurrence in juvenile angiofibroma. Rhinology 28:97–102
37. McGahan RA, Durrance FY, Parke RB, Easley JD, Chou JL (1989) The treatment of advanced juvenile nasopharyngeal angiofibroma. Int J Radiat Oncol Biol Phys 17:1067–1072
38. Myers EN (1987) Lateral rhinotomy. In: Goldman JL (ed) The principles and practice of rhinology. Wiley, New York, pp 475–498
39. Neel HB III, Whicker JH, Devine KD, et al. (1973) Juvenile angiofibroma: review of 120 cases. Am J Surg 126:547–556
40. Nemes Z, Szucs J, Racz T (1993) Retrospective clinicopathological study of juvenile nasopharyngeal angiofibroma in a 20-year case load. Orv Hetil 134:1695–1698
41. Palmer FJ (1989) Pre-operative embolization in the management of juvenile nasopharyngeal angiofibroma. Australas Radiol 33:348–350
42. Phelps PD, Lloyd GAS, Cheesman AD (1992) Juvenile angiofibroma: the natural history and imaging assessment. Abstracts of the First International Skull-Base Congress, Hannover. National Institutes of Health, Bethesda, p 32
43. Piquet JJ, Vaneecloo FM, Moreau P, et al. (1985) Nasoryngeal fibroma. Apropos of 29 cases. Acta Otorhinolaryngol Belg 39:994–1000
44. Price JC, Holliday MJ, et al. (1988) The versatile midface degloving approach. Laryngoscope 98:291–295
45. Radkowski D, McGill T, Healy GB, Ohlms L, Jones DT (1996) Angiofibroma. Changes in staging and treatment. Arch Otolaryngol Head Neck Surg 122:122–129
46. Rominger CJ, Santore FJ (1968) Juvenile nasopharyngeal angiofibroma in female adult. Arch Otolaryngol 88:85–87
47. Samii M, Draf W (1989) Surgery of the skull base. Springer, Berlin Heidelberg New York, p 201
48. Schick B, Kahle G, Hassler R, Oraf W (1996) Chemotherapy of juvenile angiofibroma – an alternative? HNO 44:148–152
49. Schiff M (1959) Juvenile nasopharyngeal angiofibroma: a theory of pathogenesis. Laryngoscope 69:981–1016
50. Schiff M, Gonzalez AM, et al. (1992) Juvenile nasopharyngeal angiofibroma contains an angiogenic growth factor: basic FGF. Laryngoscope 102:940–945
51. Spagnolo DV, Papadimitrou JM, Archer M (1984) Postirradiation malignant fibrous histiocytoma arising in juvenile nasopharyngeal angiofibroma and producing α1-antitrypsin. Histopathology 8:339–352
52. Spector JG (1998) Management of juvenile angiofibromata. Laryngoscope 98:1016–1026
53. Stammberger H (ed) (1996) Abstracts of the Second International Skull-Base Congress, San Diego. National Institutes of Health, Bethesda
54. Standefer J, Holt RG, Brown WE, Gates GA (1983) Combined intracranial and extracranial excision of nasopharyngeal angiofibroma. Laryngoscope 93:772–779
55. Stern RM, Beauchamp GR, Berlin AJ (1986) Ocular findings in juvenile nasopharyngeal angiofibroma. Ophthalmic Surg Lasers 17:560–564
56. Stiller D, Kuttner K (1988) Growth patterns of juvenile nasopharyngeal fibromas. A histological analysis on the basis of 40 cases. Zentralbl Allg Pathol 134:409–422
57. Terzián AE, Naconecy C (1985) Juvenile nasopharyngeal angiofibroma. Microsurgical approach in 25 cases as unique treatment. In: Meyers EN (ed) New dimensions in otorhinolaryngology head and neck surgery. Elsevier, New York, pp 505–506
58. Tranbahuy P, Borsik M, Herman P, Wasseff M, Casasco A (1994) Direct intratumoral embolization of juvenile angiofibroma. Am J Otolaryngol 15:429–435
59. Tseng HZ, Chao WY (1997) Transnasal endoscopic approach for juvenile nasopharyngeal angiofibroma. Am J Otolaryngol 18:151–154
60. Ungkanont K, Byers RM, et al. (1996) Juvenile nasopharyngeal angiofibroma: an update of therapeutic management. Head Neck 18:60–66
61. Vadivel SP, Bosch A, Jose B (1980) Juvenile nasopharyngeal angiofibroma. J Surg Oncol 15:323–326
62. Walike JW, Mackay B (1970) Nasopharyngeal angiofibroma: light and electron microscope changes after stilbestrol therapy. Laryngoscope 80:1109–1121
63. Weber AL (1992) Computed tomography and magnetic resonance imaging of the nasopharynx. Isr J Med Sci 28:161–168
64. Weidenbecher M (1984) Juvenile nasopharyngeal fibroma. Report of experience. Laryngolrhinootologie 63:184–188
65. Weprin LS, Siemers PT (1991) Spontaneous regression of juvenile nasopharyngeal angiofibroma. Arch Otolaryngol Head Neck Surg 117:796–799
66. Wilson CP (1957) Observations on the surgery of the nasopharynx. Ann Otol Rhinol Laryngol 66:5–40
67. Witt TR, Shah JP, Sternberg SS (1983) Juvenile nasopharyngeal angiofibroma. A 30-year clinical review. Am J Surg 146:521–525

Endoscopic Approach to Lesions of the Anterior Skull Base and Orbit

Dharambir S. Sethi

Introduction

During the past decade, the use of endoscopes has revolutionized the diagnosis and treatment of sinusitis [15, 24]. Initially used only for the diagnosis of sinusitis, endoscopes were subsequently utilized to perform limited surgery within the ostiomeatal complex (aiming to reverse the sinus pathology), and the term "functional endoscopic sinus surgery" was used. Endoscopes provided excellent illumination and a better understanding of the pathophysiology of sinusitis. It soon became apparent that endoscopes could be used for a range of other sino-nasal conditions, such as nasal polyps, mucoceles and choanal atresia (CA). With increased understanding of the endoscopic anatomy of the paranasal sinus and familiarity with endoscopic instrumentation, the use of endoscopes to perform surgery on adjacent areas of the sinuses (such as the orbit and the skull base) was natural. Transnasal endoscopic surgery obviated the need for more extensive external approaches, thus reducing morbidity while offering equivalent (if not superior) results. In the past 3 years, in collaboration with neurosurgeons and ophthalmologists, technically challenging and interesting surgeries of the skull base have been performed at the Singapore General Hospital. The use of endoscopes for surgery of pituitary tumors has been particularly successful. Endoscopic surgery for orbital decompressions, orbital apex biopsies, and drainage of orbital abscesses are increasing. The use of endoscopes in otolaryngology has been expanded beyond surgery for chronic sinusitis to include several other sino-nasal conditions, such as mucoceles, CA, control of epistaxis, closure of CSF fistulas and surgery of benign tumors (such as inverted papillomas). In this chapter, our experiences with endoscopic approaches to skull-base lesions and the orbit are discussed.

Endoscopic Skull-Base Surgery

Repair of CSF Leaks

CSF may leak into the nasal cavity if there is a breach in the tissue layers that separate the CSF from the nasal cavity, the arachnoid, dura, bone and mucosa. The bone tightly adheres to the dura at the anterior skull base; therefore, it is at greater risk of developing a fistula. The abnormal communication between the intracranial and extracranial cavities increases the risk of meningitis and brain abscesses. In fact, the risk of meningitis during the first 3 weeks of a CSF leak is between 3% and 10% [10]. The risk of meningitis, which may develop even after a leak appears to close spontaneously, may be as high as 10%. The overall risk of meningitis in untreated cases of CSF leak is up to 25% [48].

A variety of intracranial, extracranial and transnasal approaches have been utilized for surgical repair of CSF fistulas [3, 4, 6, 13]. These include bifrontal craniotomy [4], a naso-orbital approach [6], an external ethmoidectomy approach and transnasal approaches using a combination of headlight illumination and the operating microscope [3, 13]. More recently, endoscopes have been used to repair CSF leaks in the anterior skull base [23, 29, 43, 51]. An endoscopic approach to the anterior skull base has significant advantages over craniotomy. Endoscopes permit precise localization of the area of dehiscence and enable direct repair with minimal morbidity (Figs. 43.1–43.4). The use of intrathecal fluorescein may be helpful in localizing a small dehiscence that may not be demonstrated by computed tomography (CT) or magnetic-resonance imaging (MRI). The literature cites numerous materials recommended for closure of such defects, and all are probably equally effective. In general, autologous materials are preferable to artificial materials. There is no risk of infectious transmission or rejection. Bone grafts are not necessary unless there is a large bony defect. The precise positioning of the graft material over the

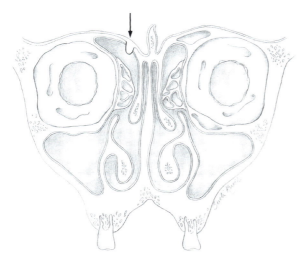

Fig. 43.1. Graphic illustration of a patient with a right encephalocele

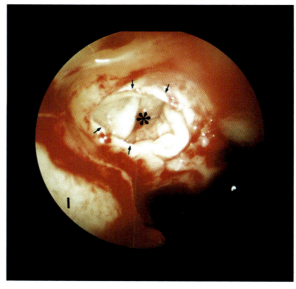

Fig. 43.3. Free mucoperiosteal graft (asterisk) from the middle turbinate used to repair the defect (viewed with a 30°, 4-mm endoscope). *l*, Lamina papyracea

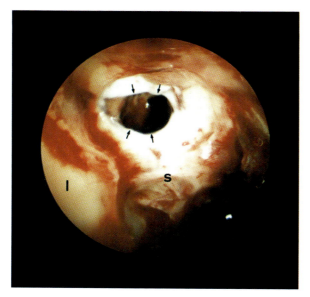

Fig. 43.2. Defect in the skull base following removal of the encephalocele (viewed with a 30°, 4-mm endoscope). *Small arrows* point to the margins of the defect that have been denuded of the mucosa. *l*, Lamina papyracea; *s*, skull base

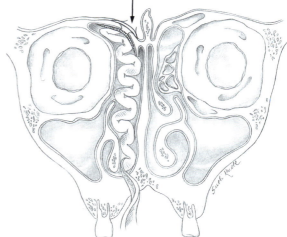

Fig. 43.4. Graphic illustration of layers of the repair

defect is more important. Our preference is a free mucoperiosteal graft from the middle turbinate, held in place with Gelfoam and fibrin glue. The postoperative use of lumbar drains is required in long-standing, profuse leaks. During the past 3 years, five CSF leaks have been successfully repaired in our institution using minimally invasive techniques [39]. The endoscopic method is not recommended for leaks associated with high-pressure conditions, brain injury, frontal sinus leaks, CSF leaks with large bony defects or sphenoid leaks in which there is extensive sphenoid pneumatization.

Pituitary Surgery

The concept of trans-sphenoidal pituitary surgery is well established. The approach was described as early as 1907 [34] but fell into disfavor because of a high incidence of complications. The headlamp commonly used as a light source during that period provided poor illumination of the deep, narrow, trans-septal, trans-sphenoidal tunnel that was created for exposure of the sella. The introduction of the operating microscope a few decades later improved illumination and provided magnification. The introduction of radiofluroscopy to intraoperatively visualize the depth and position of the surgical instruments improved the safety of the procedure [9]. The operating microscope, combined with radiofluroscopy, became the standard approach for a majority of sellar lesions [11].

Several modifications of this approach to the sphenoid sinus were subsequently described [19, 49, 52]. The sub-labial, trans-septal, trans-sphenoidal approach was favored over other approaches. The midline approach was safer and provided equal access to both sides of the sphenoid sinus and sella. We studied the endoscopic anatomy of the sphenoid sinus and the sella in 30 cadavers to assess the feasibility of using the endoscopes for pituitary surgery; subsequently, endoscopes were used to perform surgery on pituitary macro- and microadenomas [40]. Pituitary surgery has been routinely performed with endoscopes at the Singapore General Hospital since that time [37, 38] and has been used for more than a 100 pituitary tumors (Fig. 43.5) [36]. Since most pituitary tumors are in the midline, a trans-septal, trans-sphenoidal approach provides direct access to the sella (Figs. 43.6–43.10). The approach is fairly straightforward, utilizing only a hemitransfixion incision. The technique is minimally invasive and obviates the need for a sub-labial incision combined with a transfixion incision, which might result in dental injury or numbness of the lips. In addition to the excellent illumination, magnification and a panoramic perspective of the operative field, angled endoscopes provide the opportunity to examine the

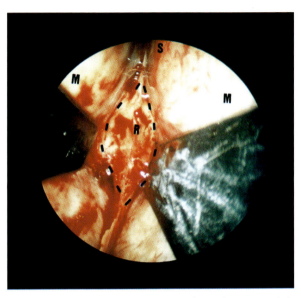

Fig. 43.6. Endoscopic view of the sphenoid rostrum (*R*), viewed with a 0°, 4-mm endoscope. The mucoperichondrial flaps (*M*) are retracted laterally by a transnasally placed self-retaining retractor. The bony nasal septum has been removed. *Dotted lines* indicate the extent of the sphenoidotomy

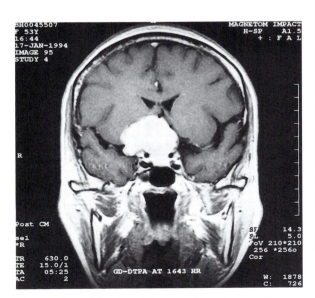

Fig. 43.5. T2-weighted, gadolinium-enhanced magnetic resonance imaging scan of a patient with a pituitary tumor

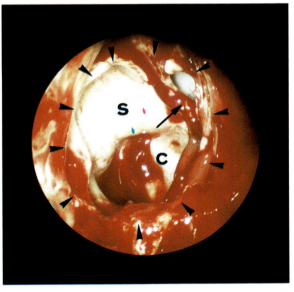

Fig. 43.7. A sphenoidotomy has been performed (viewed with a 0°, 4-mm endoscope). *Arrowheads* indicate the extent of the sphenoidotomy and point to the intersinus septum terminating at the left internal carotid artery. *c*, Clivus; *s*, anterior wall of the sella

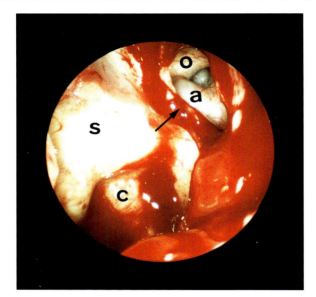

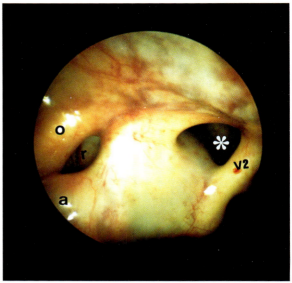

Fig. 43.8. Intraoperative view of the posterior and left lateral walls of the sphenoid sinus (viewed with a 30°, 4-mm endoscope). *a*, Left internal carotid artery; *c*, clivus; *o*, left optic nerve; *s*, anterior wall of the sella. The intersinus septum terminates at the left internal carotid artery and optic nerve

Fig. 43.10. Endoscopic view of the left lateral wall using a 70°, 4-mm endoscope. *a*, Internal carotid artery; *o*, optic nerve; *r*, optic–carotid recess; *V2*, second branch of trigeminal nerve as it courses through the lateral aspect of the sphenoid sinus, which is well pneumatized in this subject (*white asterisk*)

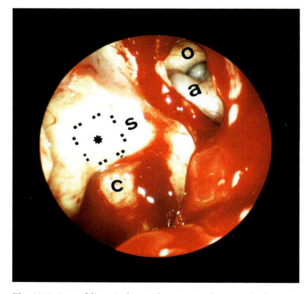

Fig. 43.9. *Dotted lines* indicate the extent of removal of anterior sellar wall (*asterisk*)

sella for any residual tumor after the tumor is removed. A more accurate and complete tumor removal is therefore possible. Bleeding is not a major problem and can be minimized by taking certain precautions [36, 38]. The surgery has been particularly successful for microadenomas. This minimally invasive technique has resulted in decreased morbidity and shorter hospital stays for patients, reducing the overall cost of the surgery.

Surgical Technique

The surgery is performed under general anesthesia. Oral endotracheal intubation is used, and a pharyngeal pack is placed in the pharynx.

A spinal catheter may be placed in the subarachnoid space, depending on individual assessment of the patient. [A Foley urinary catheter is routinely inserted to monitor urinary output intraoperatively and postoperatively].

The patient is positioned supine on the operating table, with the head elevated to 40°. An antiseptic solution (such as povidone) is applied to the nose and mouth, and the area is draped with sterile towels and Steridrapes. The lower abdomen is prepared and draped to obtain adipose tissue for later insertion into the sella and sphenoid sinus. The nasal cavity is decongested with topical application of 4% cocaine and local infiltration of 1% lidocaine with 1:80,000 epinephrine. Particular attention is given to the sphenoethmoid recess and the anterior wall of the sphenoid sinus.

The surgical team is comprised of an ENT surgeon and a neurosurgeon. A three-chip video camera is attached to the eyepiece of the endoscope, and the entire procedure is monitored on a 14-in., high-resolution color video monitor. The entire surgical procedure is recorded on a super-VHS videotape for documentation. An image intensifier and fluoroscope are kept on stand-by and are used to determine the intrasellar location of the endoscope and the instruments during the procedure.

A transnasal, trans-septal approach is routinely employed. A left hemitransfixion incision is made and extended onto the floor of the nose. A left mucoperichondrial flap is elevated from the septal cartilage and extended posteriorly to reveal the bony septum. The mucoperiostium is elevated from the floor of the nose. Both tunnels are connected by dividing the bony, cartilaginous intersection. A 4-mm, 0° endoscope with a video camera is introduced into the tunnel. Under endoscopic visualization using a high-resolution, 14-in. video monitor, the quadrilateral cartilage is disarticulated from the maxillary crest to allow the cartilage to be displaced. The bony, cartilaginous junction with the perpendicular plate of the ethmoid bone is disarticulated next, and bilateral posterior mucoperiosteal flaps are elevated to the sphenoid rostrum. At this point, a small, self-retaining sphenoid retractor is inserted. The bony septum consisting of the vomer and the perpendicular plate of the ethmoid bone is then removed. This bone may be needed later to aid in the closure of the sphenoid rostrum. Mucosa is then elevated from the face of the sphenoid (Fig. 43.6). A sphenoidotomy is made close to the rostrum and is laterally enlarged as far as the sphenoid ostia using a Kerrison sphenoid punch. A sphenoidotomy sufficiently large to allow passage of the 4-mm endoscope and instruments is performed (Fig. 43.7)

The sphenoid cavity is examined with 0°, 30° and 70° endoscopes. The location of the anatomical landmarks in the sphenoid sinus are noted. The internal carotid arteries, the optic nerves, optic chiasma, intersinus septa and the sella turcica are of particular importance (Figs. 43.8, 48.9). Bony septa may have to be removed at this point; however, extreme caution should be exercised, as these often terminate on the carotid canal. A midline intersinus septum that extends posteriorly onto the sella, if present, will have to be removed for access to the sella. The intersinus septum is removed by elevating the mucosa from the septum and carefully removing the thin bone with a true-cutting forceps. In situations where the midline septum is absent, accessory septae are present and the approach to the sella is unobstructed, it is not necessary to remove the accessory septa. The accessory septa provide important landmarks of the lateral limit of the surgical field. However, in situations where the intersinus septa have to be removed for access to the sella, the termination of the intersinus septa should be located on the CT scan and with angled endoscopes prior to removing it. Extreme caution should be exercised in removing it if it terminates at the cavernous carotid artery or optic nerve. Caution is also exercised to avoid stripping the sphenoid mucosa, as this may result in considerable bleeding. Once a panoramic view of the entire sphenoid cavity and the surgical landmarks is obtained and access to the sella turcica is complete, a Freer elevator is used to fracture the anterior sellar wall. If the sellar wall is thick, light mallet taps may be used to fracture it. A small, backbiting rongeur or a 2-mm Kerrison punch is used to delicately remove the bone overlying the dura. Bipolar diathermy is used for hemostasis of the dura before incising it. The pituitary tumor is removed using a combination of blunt ring curettes and pituitary forceps.

At this point, a 30°, 4-mm or 2.7-mm endoscope is used to examine the lateral and superior reaches of the sella turcica for any remnant tumor, "tumor-capsule defect", CSF leak or dural defect. It is the surgeon's intent to leave behind an intact tumor capsule. A small tumor-capsule defect without any CSF leak is managed by plugging the sella with Surgicel and fibrin glue. In the event a CSF leak is recognized intraoperatively, the sphenoid cavity is also plugged with abdominal fat and fibrin glue, and the sphenoid-rostrum defect is reconstructed with a small piece of bone from the vomer to prevent the fat from herniating into the mucosal layers.

Closing involves reattaching the caudal end of the septal cartilage to the maxillary spine. The mucosal layers are returned to their original positions, and the incision is closed with interrupted sutures. Silicone intranasal splints are placed lateral to the nasal septum, and the nasal cavity is packed with Merocel in a latex finger stall. The Merocel is removed after 24–48 h, and the intranasal splints are removed on the fifth postoperative day.

Tumors of the Anterior Skull Base

Craniofacial resection is the standard approach for benign or malignant tumors involving the paranasal sinuses and extending to the skull base. In recent years, endoscopes have been successfully used in the management of benign lesions of the paranasal sinus that extend to the skull base. One such lesion is the inverted papilloma (Fig. 43.11), variously known as Schneiderian papilloma, Ewing's papilloma and transitional-cell papilloma. The aggressive and recurrent nature of inverted papilloma is well known [35], as is the approximately 10% risk of the development of squamous cell carcinoma. Traditional surgical management involves lateral rhinotomy and ipsi-

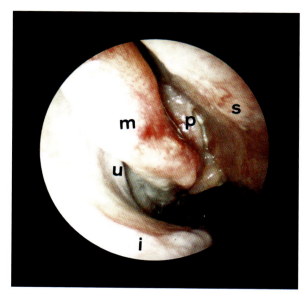

Fig. 43.11. Inverting papilloma arising from the skull base, presenting as a polyp in the right sphenoethmoid recess (viewed with a 30°, 4-mm endoscope). *i*, Inferior turbinate; *m*, middle turbinate; *p*, inverted papilloma; *s*, nasal septum; *u*, uncinate process

lateral medial maxillectomy. The recurrence rate with this approach is 7% and has been reported to be as high as 58% when a transnasal approach is used [31, 32].

More recently, endoscopic removal of inverted papilloma has become the treatment of choice in situations in which the lesion is small and can be easily removed with a margin of normal tissue. With skull-base involvement in the area of the ethmoid or sphenoid sinuses, the endoscopic approach allows the tumor to be resected as completely as any approach except the more massive craniofacial operation (Fig. 43.12), which most surgeons feel is not required for this benign lesion. The medial wall of the maxillary sinus can be removed with down- and backbiting forceps to create an opening that extends inferiorly to the floor of the nose, anteriorly to the nasolacrimal duct and posteriorly to the pterygoid plate. Attachment of the tumor within the frontal sinus is a contraindication for the endoscopic approach.

Endoscopic Orbital Surgery

Orbital Decompression

Graves' ophthalmopathy, or dysthyroid orbitopathy, is the most common cause of unilateral or bilateral

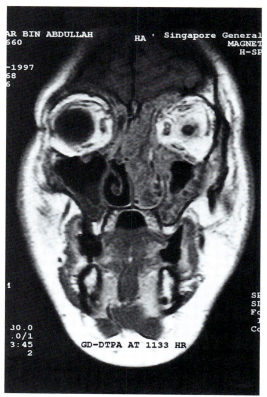

Fig. 43.12. T2-weighted, gadolinium-enhanced magnetic resonance imaging scan of a patient with inverting papilloma, showing the frontal sinus and skull-base involvement

proptosis in adults. Persistent proptosis may lead to exposure keratitis, and visual acuity can be compromised by compression of the optic nerves at the orbital apex (Fig. 43.13) [32, 42]. It is usually seen in patients with thyrotoxicosis, diffuse toxic goiter and, rarely, Hashimoto's thyroiditis (Fig. 43.14), but it can also be seen in hypothyroid and euthyroid patients [46]. An autoimmune mechanism is thought to be responsible for the development of orbitopathy in Graves' disease. Endomysial fibroblasts produce mucopolysaccharides, which contribute to the inflammatory process, ultimately leading to the degeneration of the extraocular muscles and their replacement by fat [14]. As the muscles are replaced by fat and fibrosis, they cause an increase in orbital volume. This mass effect leads to restricted ocular mobility and proptosis and crowding at the orbital apex. As the eye continues to expand, visual loss occurs, with loss of color vision happening first [33].

The current treatment of Graves' ophthalmopathy includes steroids, immunosuppressive agents, radiation therapy and surgical decompression of orbital walls. Surgery is the mainstay of therapy for treat-

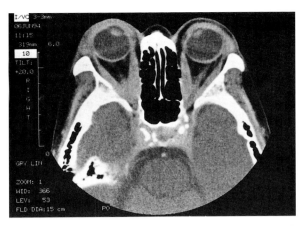

Fig. 43.13. Axial computed tomography scan of a patient with Graves' disease, showing optic-nerve compression at the apex

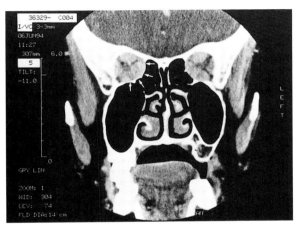

Fig. 43.14. Coronal view computed-tomography scan showing crowding at the orbital apex in a patient with Graves' disease. *White arrows* indicate the extent of removal of the orbital walls

ment of acute sight-threatening Graves' disease. It is the most direct and effective way to treat orbital-apex syndrome. The goals of surgery are to expand the orbital confines, thereby reducing intraorbital pressure, relieving optic-nerve compression and eliminating corneal exposure and cosmetic disfigurement associated with the condition [50].

Several external approaches for decompression of the lateral, superior medial and inferior walls of the orbit have been described [8, 12, 28, 41]. Decompression of the medial orbital wall and the orbital floor using an external ethmoidectomy incision is the most commonly used method. This approach provides limited access to the orbital apex. The goals of surgical orbital decompression are achieved best by use of the endoscopic surgical technique. Use of endoscopes allows excellent visualization of the landmarks and full decompression of the medial orbital wall, which may be extended as far as the optic canal [16]. The thicker sphenoidal bone overlying the optic nerve may also be removed using a drill designed for this purpose. The inferior wall can be decompressed up to the infraorbital nerve via a wide middle-meatal antrostomy. The access may be technically more difficult than the conventional external approach in this respect, but it results in a similar degree of decompression. The endoscopic approach avoids a scar and carries a much smaller risk for the nasolacrimal system and infra-orbital nerve.

Orbital-Apex Lesions

Orbital-apex lesions often present a diagnostic dilemma. Though clinical findings, CT and MRI provide some information regarding the lesion, a tissue histological diagnosis is often necessary [5]. Several approaches to the orbital apex have been described [22]. These are invasive surgical procedures associated with considerable morbidity and mortality. Whereas resectable lesions affecting the orbital apex may be approached via these invasive routes, a less invasive technique for obtaining a biopsy of the lesion is needed. A CT-guided fine-needle aspiration biopsy technique has been used in some institutions [17]. There are several pitfalls to this technique. Exact diagnosis cannot always be made with cytological analysis. Furthermore, posterior apex lesions are often difficult to aspirate, and there is always risk of injury to the optic nerve and contents of the superior orbital fissure. With endoscopes, it is possible to approach the orbital apex transnasally for tissue biopsy or decompression. The transnasal approach to the orbital apex is direct but has not been well utilized because of poor illumination. Endoscopes provide brilliant illumination and excellent visualization of this anatomical region (Figs. 43.15, 43.16). The endoscopic technique obviates major surgical approaches. Its advantage over fine-needle aspiration biopsy lies in the ability to obtain a larger tissue biopsy under direct vision. This minimizes the possibility of an inconclusive diagnosis, which often results from fine-needle aspiration biopsy. The technique has the additional advantage of permitting surgical decompression of the orbital apex and the optic nerve when necessary. The ability to biopsy lateral or superior lesions of the orbital apex is limited, and medial lesions are more suitable for this technique (Figs. 43.17, 43.18).

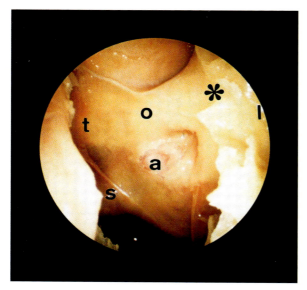

Fig. 43.15. Endoscopic view of left orbital apex (*asterisk*), viewed with a 30°, 4-mm endoscope. *a*, Internal carotid artery; *l*, lamina papyracea; *o*, optic nerve; *s*, sella turcica; *t*, tuberculum sella

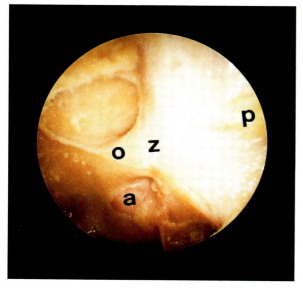

Fig. 43.16. Endoscopic view of left orbital apex. The overlying bone has been drilled to expose the annulus of Zinn (*z*). *a*, Internal carotid artery; *o*, optic nerve; *p*, periorbita

Operative Technique for Orbital Apex Decompression and Biopsy

Surgery is performed under general anesthesia. Patient preparation, preoperative procedures and nasal preparation are as described earlier for pituitary surgery. A complete sphenoethmoidectomy

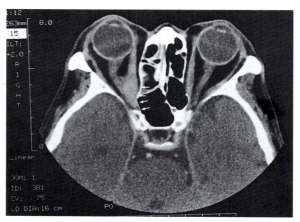

Fig. 43.17. Axial computed-tomography scan showing right optic-nerve compression from a lesion of the medial rectus. *White arrows* indicate the extent of endoscopic removal of the lamina papyracea and orbital apex

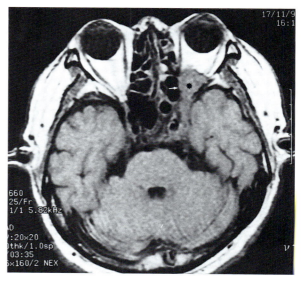

Fig. 43.18. T2-weighted axial magnetic resonance imaging scan of a patient with a left orbital-apex lesion (*asterisk*). An endoscopic biopsy taken from the apex (*white arrow*) revealed the lesion to be metastatic nasopharyngeal carcinoma

with a wide middle-meatal antrostomy is performed as detailed by Kennedy [16]. Care is taken to identify and skeletonize the medial orbital wall and the skull base. A very generous middle-meatal antrostomy is performed, creating an opening that extends anteriorly to the nasolacrimal duct, inferiorly to the inferior turbinate and posteriorly to the posterior limit of the sinus. Superiorly, the orbital floor is identified. A wide sphenoidotomy is also essential for exposure of the orbital apex and the lateral wall of the sphenoid sinus. Care is taken to identify the anterior sphenoid wall before performing a sphenoidotomy. The posterior ethmoidal neurovascular bundle and the sphenoid ostium are useful landmarks. A sufficiently large sphenoidotomy is necessary to allow the passage of the 4-mm endoscope and the instruments into the sphenoid sinus. The orbital apex is then identified. The medial orbital wall is left intact to avoid prolapse of the orbital fat, which obscures the view of the orbital apex. The medial wall of the orbital apex is drilled with a diamond burr to expose the annulus of Zinn. The annulus is incised to gain access to the medial confines of the orbital apex. A tissue biopsy is obtained from the area of interest and is sent for frozen-section examination and tissue diagnosis. Tissue is then sent for paraffin sectioning and definitive tissue diagnosis.

If orbital decompression is deemed necessary, the medial orbital wall is removed and the periorbita is incised. Merocel in latex finger stalls is used for nasal packing after completion of the procedure. The packing is removed 24 h after operation.

Dacrocystorhinostomy

Dacrocystorhinostomy (DCR) is a procedure performed to drain the lacrimal sac in instances of nasolacrimal-duct obstruction. The nasolacrimal duct may be approached by an external incision or endonasally. The concept of intranasal DCR is not new. The approach was introduced by Caldwell [2] almost 100 years ago but failed to gain wide acceptance because of difficult visualization within the narrow confines of the superior nasal vault. Therefore, traditionally, ophthalmologists perform this procedure using an external incision. Recently, there has been renewed interest in intranasal DCR due to the advent of endoscopic instrumentation for sinus surgery. Endoscopes provide the surgeon with excellent intranasal visualization and access to the lacrimal sac. Endoscopic DCR allows drainage of an obstructed lacrimal sac without the need for a skin incision. The exact position of the sac can be readily determined by a fiberoptic light fiber passed through one of the canaliculi. The medial wall of the canal is removed by a drill, curette or laser [27]. The use of stents may be unnecessary when marsupialization of the sac is wide.

Optic-Nerve Decompression

Traumatic optic neuropathy is perhaps the most common indication for this procedure. Surgical decompression of the optic nerve for the treatment of loss of visual acuity has been described for orbital pseudotumor, Graves' orbitopathy and fibro-osseous lesions of the optic canal. The precise surgical technique and indications remain controversial [47].

Sinonasal Conditions

Mucoceles

Mucoceles are epithelium-lined, mucus-containing sacs commonly occurring in the frontoethmoid region; they also occur in the sphenoid and maxillary sinuses (Figs. 43.19–43.25). Whether mucoceles result from trauma, long-term sino-nasal disease or sinus surgery, the "final common pathway" involves sinus obstruction with gradual build-up of mucus until the limits of the cavity are reached. Slow continued expansion and growth result in bone erosion. Pressure-induced osteolysis is probably one significant factor leading to bone erosion and extra-sinus extension; however, some investigators hypothesize that mucocele expansion is secondary to locally increased levels of the cytokines interleukin-1 (IL-1) and IL-6 in a chronic inflammatory environment [21]. Ultimately, intracranial and/or intraorbital extension sometimes cause proptosis, diplopia and pressure effects on cranial nerves. Temporary or permanent focal neurological deficits (including blindness) may occur, necessitating surgical intervention.

Traditional surgery for mucoceles has depended on the particular sinus involved. Ethmoid-sinus involvement usually required external ethmoidectomy; frontal-sinus involvement necessitated a Lynch procedure or an osteoplastic flap. Sphenoid-sinus

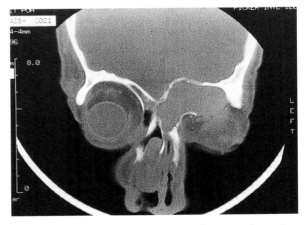

Fig. 43.19. Coronal computed-tomography scan of a patient with nasal polyposis and left frontal mucocele. Note the anterior and downward displacement of the left eye

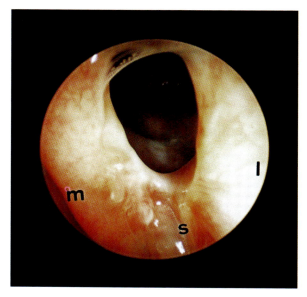

Fig. 43.20. View of the left frontal opening in the patient in Fig. 43.19 3 weeks after operation (viewed with a 30°, 4-mm endoscope). *l*, Lateral; *m*, medial; *s*, skull base

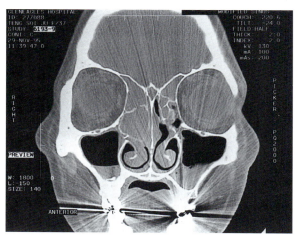

Fig. 43.22. Coronal computed tomography scan of the patient in Fig. 43.21, showing a mucocele in the concha bullosa and extending into the frontal sinus

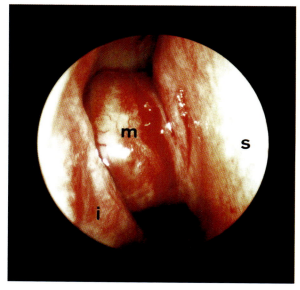

Fig. 43.21. Endoscopic view of a right mucocele (*m*), viewed with a 30°, 4-mm endoscope. *i*, Inferior turbinate; *s*, nasal septum

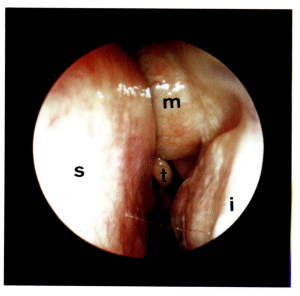

Fig. 43.23. Endoscopic view of a left ethmoid mucocele viewed with a 30°, 4-mm endoscope. *i*, Inferior turbinate; *m*, mucocele; *s*, nasal septum; *t*, middle turbinate

mucoceles necessitated an intranasal sphenoidotomy or trans-septal sphenoidotomy, and maxillary-sinus involvement necessitated a Caldwell-Luc procedure. The surgical aim was always to marsupialize the lesion into the nose as widely as possible while simultaneously ensuring that as much of the osteitic bone was removed as possible and that the opening created communicated with the natural ostium of the sinus.

Frontoethmoid, ethmoid sphenoid and maxillary mucoceles may now be marsupialized endoscopically using minimally invasive techniques. The likely success of the approach can be determined on coronal CT scans, which will also show whether more than one locus is present and demonstrate any contributory pathology. In the frontal region, significant distortion of the anatomy (for example due to midfacial trauma, abnormally thick bone or a laterally

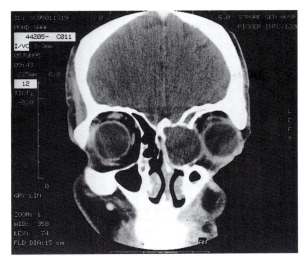

Fig. 43.24. Coronal computed tomography scan of the patient in Fig. 43.23, showing an ethmoid mucocele. Note the destruction of the medial orbital wall

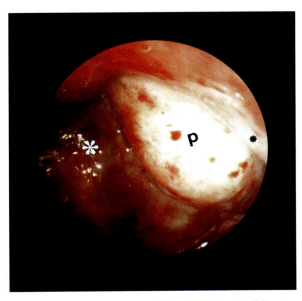

Fig. 43.25. Intraoperative view of the ethmoid cavity following endoscopic drainage of the mucocele in Fig. 43.23. *Small asterisk* indicates the dehiscent nasolacrimal duct. *p*, Periorbital bulging into the ethmoid cavity (*white asterisk*) on applying orbital pressure

placed lesion) may preclude an endoscopic approach. However, even in these circumstances, the external and endoscopic approaches can be usefully combined, and stenting can be avoided. In our experience, most frontal mucoceles are amenable to exclusively endoscopic drainage, and ethmoidal and sphenoidal mucoceles are almost always accessible to endoscopic drainage alone.

Sphenoid Lesions

In the past, the sphenoid sinus was considered to be the sinus least accessible to the surgeon. Interest in this "neglected" sinus has been revived with current imaging and endoscopic techniques. Isolated sphenoid pathology, once regarded as uncommon because of lower diagnostic capability, is now being increasingly diagnosed. During the past 2 years, we have managed 18 isolated sphenoid lesions, including sphenoiditis, mucoceles, inverted papillomas and spheno-choanal polyps. Traditionally, surgical access to the sphenoid sinus has been via a transnasal, transethmoid or trans-septal approach using a headlight or operating microscope. The sphenoid sinus can be accessed similarly accessed with an endoscope [25, 26, 44, 51]. Different transethmoid approaches have been described by Wigand and Messerklinger [25, 51]. The trans-septal approach is routinely used in our institution for endoscopic trans-sphenoid access to the sella turcica [36–38]. The transnasal approach is direct and, in our opinion, is ideal for sphenoid mucoceles (Figs. 43.26, 43.27).

Choanal Atresia

CA, if bilateral, typically presents during the neonatal period with airway obstruction. Unilateral CA is more common and may present later in life with unilateral rhinorrhea and nasal obstruction. The presumptive diagnosis, usually made by a pediatrician when a nasal catheter cannot be passed beyond 3.5 cm, should be confirmed by flexible endoscopic examination and axial CT. The obstruction may be membranous or bony, and the atresia plate may be thin or fairly thick. Many surgical approaches, including transnasal, sub-labial, transpalatal and external rhinoplasty, have been proposed for CA [18]. Several investigators have recently described CA repair with rigid endoscopes [7, 20, 45]. The atresia-plate area, which is usually fibrous and bony, can be precisely removed endoscopically. Strong bone punches and an otologic drill with cutting burrs are usually necessary. Some authors have described the use of a protected spinning blade within a non-rotating sheath and continuous suction to repair CA [1, 28, 30]. Often, excision of the posterior nasal septum is necessary to obtain a wide opening. As the nasal passage is often too narrow to accommodate both the drill and the endoscope, a 70° endoscope may be introduced through the oral cavity, which is held open with a Boyle-Davies gag. The atresia is visualized on the video monitor and drilled with a cutting

Fig. 43.26. Computed tomography scan demonstrating a sphenoid mucocele

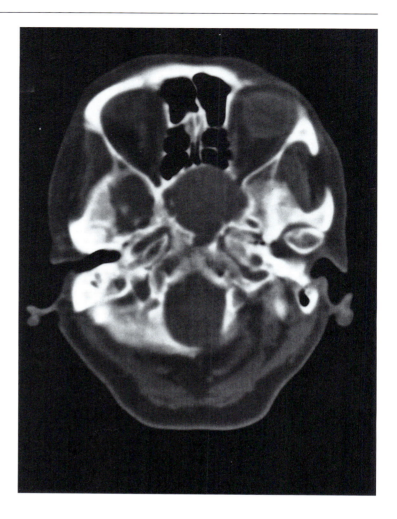

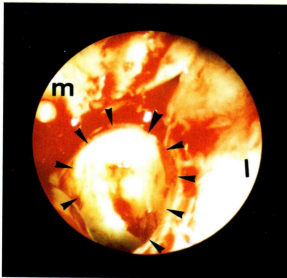

Fig. 43.27. Endoscopic view of transnasal endoscopic sphenoidotomy in the patient in Fig. 43.26. *Arrowheads* indicate the extent of the sphenoidotomy. *l*, Lateral; *m*, medial

burr. Stenting of the choana is routinely done, though the ideal stent and duration of stenting remain controversial.

Summary

Extended application of endoscopic techniques to the skull base and orbit and to non-inflammatory sino-nasal conditions is safe and effective. Endoscopic approaches provide improved visualization and are a superior alternative to open surgical approaches in a majority of cases.

References

1. April MM, Ward RF (1996) Channel atresia repair: use of powered instrumentation. Operative Tech Otolaryngol Head Neck Surg 7:248–251 1996

2. Caldwell GW (1893) Two new operations for obstruction of the nasal duct with preservation of the canaliculi and an incidental description of a new lacrimal probe. NY Med J 57:581 1893
3. Chandler JR (1983) Traumatic cerebrospinal fluid leakage. Otolaryngol Clin North Am 16:623–632
4. Dandy WD (1926) Pneumocephalus (intracranial pneumocele or aerocele). Arch Surg 12:949–982
5. Daniels DL, Yu S, Peck P, et al. (1987) Computed tomography and magnetic resonance imaging of orbital apex. Radiol Clin North Am 24:803–817
6. Dohlman G (1948) Spontaneous cerebrospinal rhinorrhoea. Acta Otolaryngol (Stockh) 67:20–23
7. El-Guindy A, El-Sherief S, Hagrass M, Gamea A (1992) Endoscopic endonasal surgery of posterior choanal atresia. J Laryngol Otol 106:528–529
8. Golding-Wood PH (1969) Trans-antral ethmoid decompression in malignant exophthalmos. J Laryngol Otol 83:683–694
9. Guiot G (1973) Trans-sphenoidal approach in surgical treatment of pituitary adenomas: general principles and indications in non-functioning adenomas. In: Kohler PO, Ross GT (eds) Diagnosis and treatment of pituitary tumors. Excerpta Medica, Amsterdam, pp 159–178
10. Handley GH, Goodson LA, Real TH (1993) Transnasal endoscopic closure of anterior cranial fossa cerebrospinal fluid fistula. South Med J 86:217–219
11. Hardy J (1969) Microneurosurgery of the hypophysis: a sub-nasal trans-sphenoidal approach with television magnification and televised radiofluroscopic control. In: Rand RW (ed) Microneurosurgery. Mosby, St Louis, pp 87–103
12. Hirsch O (1950) Surgical decompression of malignant exophthalmus. Arch Otolaryngol Head Neck Surg 51:325–334
13. Hirsch O (1952) Successful closure of cerebrospinal fluid rhinorrhea by endonasal surgery. Arch Otolaryngol Head Neck Surg 56:1–13
14. Jakobiec FA, Font RL (1986) Orbit. In: Spencer W (ed) Ophthalmic pathology. Saunders, Philadelphia, pp 2765–2777
15. Kennedy DW (1985) Functional endoscopic sinus surgery: technique. Arch Otolaryngol Head Neck Surg 111:643–649
16. Kennedy DW, Goodstein ML, Miller NR, et al. (1990) Endoscopic transnasal orbital decompression. Arch Otolaryngol Head Neck Surg 116:275–282
17. Kennerdell JS, Dubois PJ, Dekker A, et al. (1980) CT-guided fine needle aspiration biopsy of orbital optic nerve tumors. Ophthalmology 87:491–496
18. Koltai PJ (1994) Surgical management of choanal atresia. Operative Tech Otolaryngol Head Neck Surg 5:5–11
19. Koltai PJ, Goldstein JC, Parnes SM, et al. (1985) External rhinoplasty approach to trans-sphenoidal hypophysectomy. Arch Otolaryngol Head Neck Surg 111:456–458
20. Lazar RH, Younis RT (1995) Transnasal repair of choanal atresia using telescopes. Arch Otolaryngol Head Neck Surg 121:517–520
21. Lund VJ, Henderson B, Song Y (1993) Involvement of cytokines and vascular adhesion receptors in the pathology of fronto-ethmoidal mucoceles. Acta Otolaryngol (Stockh) 113:540–546
22. Maroon JC, Kennerdell JS (1984) Surgical approaches to the orbit. Indication and techniques. J Neurosurg 60:1226–1235
23. Mattox DE, Kennedy DW (1990) Endoscopic management of cerebrospinal fluid fistulas and cephaloceles. Laryngoscope 100:857–862
24. Messerklinger W (1978) Endoscopy of the nose. Urban and Schwarzenberg, Munich
25. Messerklinger W (1978) Uber die menschlichen nasennebenhohlen unter nor malen und pathologischen Bendingungen. In: Messerklinger W (ed) Ohrenhelkunde und Laryngo-rhinologie. Urban and Schwarzenberg, Vienna, pp 55–56
26. Metson R, Gliklich RE (1996) Endoscopic treatment of sphenoid sinusitis. Otolaryngol Head Neck Surg 114:736–744
27. Metson R, Woog JJ, Puliafito CA (1994) Endoscopic laser dacrocystorhinostomy. Laryngoscope 104:269–274
28. Nafzigger HC (1931) Progressive exophthalmus following thyroidectomy, its pathology and treatment. Ann Surg 94:582–586
29. Papay FA, Maggiano H, Dominquez S, Hassenbusch SJ, Levine HL, Lavertu P (1989) Rigid endoscopic repair of paranasal sinus cerebrospinal fluid fistulas. Laryngoscope 99:1195–1201
30. Parsons DS (1996) Rhinologic uses of powered instrumentation in children beyond sinus surgery. Otolaryngol Clin North Am 29:105–114
31. Pelausa EO, Fortier MA (1992) Schneiderian papilloma of the nose and paranasal sinuses: the University of Ottawa experience. J Otolaryngol 21:9–15
32. Phillips PP, Gustafson RO, Facer GW (1990) The clinical behavior of inverted papilloma of the nose and paranasal sinus: report of 112 cases and review of literature. Laryngoscope 100:463–469
33. Sacks EH, Anand VJ, Lisman RD (1993) Orbital decompression: endoscopic perspective. In: Anand VJ, Panje W (eds) Practical endoscopic sinus surgery. McGraw Hill, New York, pp 138–159
34. Schloffer H (1907) Erfolgreiche operation eines hypophysen-tumors auf nasalem Wege. Wien Klin Wochenschr 20:621–624
35. Seshul MJ, Eby TL, Crowe DR, Peter GE (1995) Nasal inverted papilloma with involvement of middle ear and mastoid. Arch Otolaryngol Head Neck Surg 121:1045–1048
36. Sethi DS, Pillay PK (1996) Endoscopic surgery for pituitary tumors. Operative Tech Otolaryngol Head Neck Surg 7:264–268
37. Sethi DS, Pillay PK (1995) Endoscopic management of lesions of sella turcica. J Laryngol Otol 109:956–962
38. Sethi DS, Pillay PK (1996) Endoscopic pituitary surgery. A minimally invasive technique. Am J Rhinol 10:141–147
39. Sethi DS, Chan C, Pillay PK (1996) Endoscopic management of cerebrospinal fluid fistula and traumatic encephalocele. Ann Acad Med Singapore 25:724–727
40. Sethi DS, Stanley RE, Pillay PK (1995) Endoscopic anatomy of sphenoid sinus and sella turcica. J Laryngol Otol 109:951–956
41. Sewall EC (1936) Operative control of progressive exopthalmus. Arch Otolaryngol Head Neck Surg 24:621–624
42. Sisler HA, Jakobiec FA, Trokel SL (1986) Ocular abnormalities and orbital changes of Graves' disease. In: Duane TD, Jaeger EA (eds) Clinical ophthalmology, vol 2. Harper and Row, Philadelphia, pp 19–86
43. Stankiewicz JA (1989) Complications in endoscopic intranasal ethmoidectomy: an update. Laryngoscope 99:686–690
44. Stankiewicz JA (1989) The endoscopic approach to the sphenoid sinus. Laryngoscope 99:218–221

45. Stankiewicz JA (1990) Endoscopic repair of choanal atresia. Otolaryngol Head Neck Surg 104:931–937
46. Tami H, Nakagawa N, Oshako N, et al. (1980) Changes in thyroid function in patients with euthyroid Graves' disease. J Clin Endocrinol Metab 50:108–112
47. Thaler ER, Lanza DC, Kennedy DW (1996) Endoscopic optic nerve decompression for traumatic optic neuropathy. Operative Tech Otolaryngol Head Neck Surg 7:293–296
48. Tolley NS, Brooks GB (1992) Surgical management of cerebrospina, fluid rhinorrhea. J R Coll Surg Edinb 37:112–115
49. Tucker HM, Hahn JF (1982) Transnasal, trans-septal approach to hypophysectomy. Laryngoscope 92:55–57
50. Walsh TE, Ogura JH (1957) Transantral orbital decompression for malignant exophthalmos. Laryngoscope 67:544–569
51. Wigand ME (1981) Transnasal ethmoidectomy under endoscopic control. Rhinology 19:7–15
52. Wilson WR, Khan A, Laws ER Jr (1990) Trans-septal approaches for pituitary surgery. Laryngoscope 100:817–819

Chapter 44

Endonasal Microscopic Transseptalsphenoidal Approach to Sellar and Parasellar Lesions

Ricardo Sergio Cohen, Aldo Cassol Stamm, and André Bordasch

Introduction

The sella turcica can be surgically approached either transcranially or transsphenoidally through the ethmoid bone, maxillary sinus, nasal septum and the nasal passages (Tables 44.1–44.5). Operative exposure of the sellar region for the treatment of disorders of the pituitary gland has been a challenge for both neurosurgeons and otolaryngologists since the last part of the nineteenth century.

Table 44.1. Transseptalsphenoidal approach to the sellar and parasellar regions during the period 1980–1997 ($n=82$: 47 female, 35 male).

Surgery type	Number
Submucosal resection	2
Rhinoplasty	3
Septorhinoplasty	4
Septoplasty	7

Previous rhinologic-surgery types ($n=16$). Ages ranged from 12 years to 82 years. Mean age: 40 years

Table 44.2. Transseptalsphenoidal approach to the sellar and parasellar regions during the period 1980–1997: etiologies

Etiology	Number of patients		
	Total	Female	Male
Somatotroph adenoma	26	3	23
Prolactinoma	36	31	5
Empty-sella syndrome	3	2	1
Breast-cancer metastasis	1	1	0
Pituitary carcinoma	1	1	0
Non secreting adenoma	3	1	2
Craniopharyngioma	3	2	1
ICA aneurysm	1	1	0
Sphenoid sinus mucocele	5	3	2
CSF leak (sphenoid sinus)	3	2	1
Total	82	47	35

ICA, internal carotid artery

Table 44.4. Transseptalsphenoidal approach to the sellar and parasellar regions during the period 1980–1997. Postoperative complications ($n=82$)

Complication	Number
Transient diabetes insipidus	10
Transient CSF leak	8
Pneumocephalus	1
Total	19

CSF, cerebrospinal fluid

Table 44.5. Transseptalsphenoidal approach to the sellar and parasellar regions during the period 1980–1997: postoperative sequelae ($n=82$)

Sequelae	Number
Cosmetic deformity of the labial–septal–columellar complex; transient sensitive disturbances of the anterior teeth; cicatricial changes of the gingival–labial sulcus	3
Intranasal crusting (lasting more than 3 months)	5
Anosmia	1
Stenosis of the nasal vestibule (slight)	1
Sinequiae	1

Table 44.3. Transseptalsphenoidal approach to the sellar and parasellar regions during the period 1980–1997: intraoperative complications ($n=82$)

Complication	Number
Massive hemorrhage of the ICA	1
Mucosal tearing	6
Intraoperative bleeding	3
Total	10

ICA, internal carotid artery

Neurosurgery and otolaryngology have contributed significantly to the development of safe and effective means of surgical management of sellar and parasellar diseases. Both disciplines agree that – except in unusual circumstances – the operation of choice is the direct midline transseptal, transsphenoidal approach. The otolaryngologist, being well equipped to perform this type of surgery, should play an integral role as a member of the surgical team. In recent years, advances in microsurgery of the nose and paranasal sinuses and modern micro-surgical instrumentation (plus the latest refinements in the use of nasal endoscopes, which can now reach areas beyond the sinuses, including the hypophysis) demonstrate the soundness of the participation of the rhinologist in a team approach to the surgery of the sellar and parasellar regions.

Historical Aspects

In 1897, Giordano [34] approached the hypophysis transsphenoidally for the first time. In 1907, Schloffer [86] was the first surgeon to perform a hypophysectomy in a living patient, using a very radical, direct, external, facial approach. This approach was simplified by Hochenegg [92], Exner [82] and Von Eiselberg [95–97]. In 1911, Cushing [21] reported that a simplified external facial approach was the method that he used in his first experience with the trans-sphenoidal technique in 1909 [20]. Further modifications of Von Eiselberg's technique were suggested by Moskowicz [75]; these tried to avoid the high incidence of meningitis that followed these extensive procedures. These surgical modifications were also soon abandoned because of their adverse effects on the nose and the severe facial deformity that resulted. Proust (in 1908) [80] and Kocher, (in 1909) [3] also made contributions to the development of a better transsphenoidal approach.

In 1909, Kanavel [54] devised a U-shaped infranasal incision. His approach was further modified by Knavel and Grinker [59] and by Mixter and Quakenboss [72]. Also in 1909, at the Johns Hopkins Hospital, Cushing collaborated with Crowe and Homans [19] to perform more than 100 canine pituitary procedures, experimenting with various approaches. In 1909, Hirsch [48–50, 59] described an endonasal approach to the sella; this approach avoided the significant cosmetic deformities of the earlier approaches. At the same time, West [98], a laryngologist at the Johns Hopkins Hospital, published a two-stage approach that appears to be the first proposal of a combined otolaryngological–neurosurgical approach to the sella.

In 1909, Segura [3] also developed an endonasal approach (via the septum) to the sella. A very skillful rhinologist, he conceived a surgical method that preserved the anatomic structures of the nose; he designed surgical instruments that are very similar to the modern instruments used for the transseptal approach. The contributions by Killian (submucosal resection of the septum) and by Hajek (his approach to the sphenoid sinus) were introduced to pituitary surgery by Hirsch and Segura, thus offering a new philosophy for the management of septal and sinus structures.

Halstead [40] of Chicago modified the inferior-rhinotomy approach of Kanavel by placing the infranasal incision at the gingivo-labial sulcus. This improved the operative field and left no cosmetic defects. In 1912, Cushing [22] described his surgical approach, combining the sublabial incision of Halstead and the sub-mucosal resection used by Hirsch.

Cushing performed 74 pituitary operations during his work at the Johns Hopkins Hospital; these were performed primarily via the transsphenoidal route, with low mortality. He also performed 338 hypophysectomies in Boston, and he was able to reduce surgical mortality by more than 50% in his later transsphenoidal series. A detailed review of Cushing's series was published by Henderson [47] in 1939.

During the last years of his career, Cushing reverted almost completely to the transfrontal approach (described by Krause [62] in 1905 and developed and improved by Dandy in 1938 [23, 24]). By 1940, the trans-septal approach went out of fashion in North America, although Hirsch [51] and Segura [3] continued to practice it in their respective countries. Cushing's pupil, Dott [26], carried this technique to Edinburgh and then to Paris, where he introduced the operation to Guiot [35–38], who performed it in France in association with Bouche [6], a rhinologist. A quick but comprehensive review of the history of transsphenoidal surgery must mention the names of Chiari [10] and Nager [76] in connection with the transethmoidal route. Angell-James [2] and Bateman [4] combined the transethmoidal approach with the transnasal or trans-septal technique.

In 1900, König [61] was the first to attempt to reach the pituitary via the sphenoid sinus; he recommended a transpalatal approach. Lowe [69], Fein [32], Preysing [79] and Tiefenthal [93] made further modifications to the transpalatal technique, but these have fallen into disuse [39].

In 1947, Lautenschlager [63] described the technique by West and Claus, a modification of Hirsch's approach. He also described the bucconasal approach proposed by Fein in 1910 and carried out by Denker in 1921 [25]. Finally, Lautenschlager described his own approach to the pituitary (via the maxillary sinus).

The introduction of the operating microscope and special microinstruments caused a revival of the discussion among neurosurgeons and otorhinolaryngologists regarding the most suitable route by which to gain better surgical access to the sella [67, 84, 94]. Among the neurosurgeons, Hardy [43–45] must be credited with the repopularization of the transsphenoidal method in North America, and the contributions of Fahlbusch [31], Guiot [37] and Nicola [77], together with Basso and Pardal [3] in Argentina, must be remembered. Among the otorhinolaryngologists, Bateman [4], Burian [7–9], Escher [30], Hamburger [41], Hamburger and Hammer [42] and Nager [76] took an active role in the development of the microsurgery of pituitary tumors. Nager's techniques were adopted by Angell-James [1, 2], Montgomery [73, 74] and Richards et al. [81].

Cottle's application of anatomic and physiologic concepts to the management of the septal structures [18], in conjunction with the historical role played by rhinologists in the evolution of the transsphenoidal approach, led to the participation of otolaryngologists in the team approach to operations on sellar and parasellar lesions via the sphenoid sinus. Several otolaryngologists enriched the rhinologic literature on this subject [5, 11, 12–14, 17, 33, 55, 57, 58, 60, 66, 70, 101].

With the work of Messerklinger [71], a new chapter in the history of pituitary surgery began. The use of the nasal endoscope for the treatment of sinonasal diseases was developed by several authors, such as Stammberger [89, 90], Wigand [99], Kennedy [56] and several others [27, 28, 68, 91]. Draf [27, 28] and Rudert [83] combined the use of the microscope and endoscopes in sino-nasal surgery. In 1992, Jankowski et al. [53] described an endonasal endoscopic technique for gaining access to the sphenoid sinus and the pituitary gland. The same Jankowski [52] recently published his transseptal endoscopic approach and compared this approach to the endonasal endoscopic approach and the microscopic approach.

The Endonasal, Transeptalsphenoidal Approach to Sellar and Parasellar Lesions

The endonasal, transseptalsphenoidal technique is basically an extracranial, extradural approach to sellar and perisellar lesions. It is accomplished entirely through the nose, almost as a standard septum procedure.

The endonasal, transseptalsphenoidal approach (ETA) is a quick, safe, relatively simple approach to the pituitary. It provides midline access to the sphenoid sinus and the sella, excellent visualization through the microscope and has proven to offer excellent operative sterility. The ETA also offers the advantages of a very low incidence of complications and mortality and considerable potential for the preservation of nasal shape and function.

Several transsphenoidal techniques have been described [12, 14, 84]. The sublabial, transseptalsphenoidal approach (STA) and the ETA are the transsphenoidal methods most widely used for gaining access to the sella (Table 44.6). Comparison of the relative advantages and disadvantages of these two widespread transseptalsphenoidal methods leads to the conclusion that the ETA is more convenient for both the surgeon and the patient [14].

Advantages of the ETA

1. It is familiar to the rhinologist.
2. It allows excellent management and preservation of septal structures.
3. There is no oral contamination of the wound.
4. There are no oral incisions (a benefit for denture wearers).
5. The speculum is in the midline.
6. The speculum easily stays at a good angle to the pituitary gland.
7. There is a very low incidence of functional and/or cosmetic complications.
8. The size of nose is irrelevant if the proper speculum and lateral alotomy are used.
9. There is better accomplishment of functional corrections (if needed).
10. The postoperative evolution is similar to that of a standard septum procedure.
11. There are no external scars except when a lateral alotomy is performed.

Table 44.6. Surgical techniques in patients with no previous rhinologic surgery ($n=66$)

Sub-labial transseptalsphenoidal approach	16
With removal of nasal spine and piriform crest	3
With preservation of nasal spine and piriform crest	7
Prior endonasal septal dissection (Kern)	6
Endonasal transseptal sphenoidal approach	26
With no lateral alotomy	21
With lateral alotomy	5
Microscopic endonasal transseptalsphenoidal approach	20
Micro-endoscopic endonasal transseptalsphenoidal approach	4

12. Use of the microscope from the beginning of the operation (microsurgical ETA; META) has these further advantages:
    – It provides optical magnification.
    – It provides good illumination.
    – Like other microsurgical techniques, it is gentle and precise.
    – It is useful for teaching and documentation purposes.

Advantages of STA

1. It is relatively easy for the rhinologist.
2. Good management and preservation of septal structures is possible.
3. The speculum is in the midline.
4. The speculum stays at a good angle relative to the pituitary gland.
5. There is a low incidence of functional and/or cosmetic complications.
6. The size of the nose is almost completely irrelevant.
7. Functional corrections are possible.
8. The postoperative evolution is good.
9. There are no external scars.
10. Use of the microscope from the beginning of the operation is very difficult.

Disadvantages of ETA

1. The nasal size affects the choice of proper retractors and lateral alotomy.
2. External scars, when present, are hardly noticeable.
3. Disadvantages of the microsurgical technique include:
    – Mandatory microsurgical training in nasal surgery
    – Relatively less mobility for the surgeon
    – A relatively restricted surgical field

Disadvantages of STA

1. Oral contamination of the wound is possible.
2. There are potential problems for denture wearers.
3. There are potentially sensitive dental complications.
4. There is a possibility of cosmetic deformity of the nasal tip or naso-labial angle when the spine is resected.

5. Potential intranasal complications include:
    – Mucosal tears and scars
    – Intraoperative and/or postoperative bleeding
6. Dissection of the nasal floor and septo-vomerine junction is difficult.
7. The speculum is sometimes difficult to introduce.
8. The speculum is pushed down by the upper lip and is sometimes not at the best angle relative to the sella.
9. The upper lip is sometimes interposed in the upper part of the field.

In 1957, Heerman [46] introduced the microscope in surgery of the nose and paranasal sinuses. Prades [78], of Spain, must be credited for the introduction of the use of the microscope in nasal surgery in Spanish-speaking countries (especially in South America, where nasal microsurgery achieved outstanding development, as evidenced by authors like Stamm [88], from Brazil, and others from other Latin American countries).

The META [11, 12–14, 17] for the sphenoid sinus and the sella is defined as an ETA procedure carried out with an operating microscope throughout the operation. The META allows the rhinologist trained in microsurgery to be more precise and gentle in dealing with the surgical anatomy of the nose, the sphenoid sinus and the sellar and perisellar regions.

The nasal endoscope has provided increased visualization for the surgeon performing sinus surgery, and the use of this technology has been proposed as an alternative to the use of a microscope for pituitary surgery. The main advantage of the use of the endoscope is the panoramic view it gives inside the sphenoid sinus and inside the pituitary fossa. Despite this advantage, it cannot replace the binocular vision and the two-handed instrumentation allowed by the microscope via the wide field provided by the META. Theoretically, it is useful to think not in terms of endoscopic surgery *versus* microsurgery but in terms of microsurgery *and* endoscopic surgery. From this point of view, the practical advantages of microsurgery are strongly enhanced by the use of the endoscope, thus allowing the surgeon a better inspection of the critical anatomic landmarks of the sphenoid sinus, and an accurate control of the complete removal of a sellar tumor. The micro-endoscopic technique, featuring use of both the microscope and the endoscope, may be a new trend for the future of the transseptal sphenoidal surgery of sellar and parasellar lesions.

Choice of the Surgical Approach

The choice of the surgical approach to the sellar region depends on the results of neuroradiologic studies [12]. Magnetic-resonance imaging (MRI) of the sellar region and the sphenoid sinus is the most helpful exam in the neuroradiologic work-up of patients with sellar and parasellar tumors. MRI depicts the tumor's size, extent and characteristics, such as hemorrhagic and cystic changes. This exam also helps to delineate the tumor from the surrounding anatomical structures, and the position of the carotid artery is, in most instances, well depicted by coronal MRI (Fig. 44.1) [29, 31, 64]. However, the relationship of the tumor to the osseous structures is best demonstrated by computed tomography (CT) [15].

Basic Criteria for Patient Selection

Two basic criteria for patient selection for trans-septal sphenoidal surgery must be considered [12]:
1. All intrasellar pituitary tumors (and those with symmetrical, moderate suprasellar extension) can be operated via this approach. Initially conceived for the treatment of intrasellar tumors, the transsphenoidal approach can be used in the surgery of some perisellar lesions (chordoma, craniopharyngioma) or in the debulking of lesions with suprasellar extension [31, 64, 65].
2. The sella turcica must be surgically accessible via an adequately pneumatized sphenoid sinus. There are three described anatomical varieties of sphenoid sinus: conchal (non-pneumatized), presellar and sellar. The sellar type is the most frequent and suitable for transsphenoidal surgery (Fig. 44.2).

Indications for the ETA to the Sella

1. The excision of pituitary tumors
2. The excision or debulking of non-pituitary skull-base tumors
3. Biopsy and culture of sphenoid sinus diseases
4. Marsupialization of a sphenoid sinus mucocele
5. Repair of CSF leakage into the sphenoid sinus [12, 17]

Contraindications for the ETA

1. Absolute contraindications
   - Active sino-nasal infection
   - Excessive tumoral extension
   - Diffuse tumoral invasion by non-capsulated tumors
2. Relative contraindication
   - A conchal or presellar type of sphenoid sinus [12, 17]

The Micro-Endoscopic Surgical Technique

Four important remarks should be recalled before proceeding with the description of the surgical technique in a step-by-step fashion:
1. Close teamwork between the neurosurgeon and the otorhinolaryngologist is advised.
2. Surgery is routinely carried out under general anesthesia and oral endotracheal intubation.
3. Septal structures are handled following the principles of Cottle's "maxilla–premaxilla" approach to the septum, with very few modifications.
4. A surgical microscope is used throughout the operation.

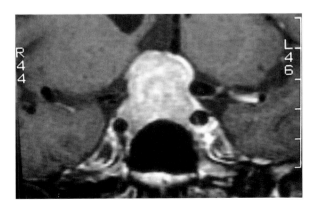

Fig. 44.1. Coronal magnetic resonance showing a macroadenoma of the pituitary gland and its relationship to the internal carotid arteries

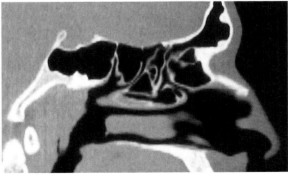

Fig. 44.2. Sagittal reconstructed computed tomography. Sphenoid sinus (sellar type)

The Surgical Technique Step by Step

1. Once anesthetized, the patient is placed in a semi-recumbent position, the head reclined to the left and the face turned to the right, allowing a right-handed surgeon to stand in a sagittal plane without leaning over the patient's torso.
2. The surgeon or his assistant removes autograft muscle from the thigh or the upper lateral abdomen. Some fascia lata and fat can also be removed. The fascia can be used for reconstruction in the event of injury to the diaphragma sellae.
3. Draping is carried out after cleansing the skin with an antiseptic solution and protecting the eyes. The upper lips, nose, eyes and cheeks are left exposed. Some authors prefer to cover the face and leave uncovered only the nose.
4. Topical vasoconstriction of the nose with gauze strips soaked in saline solution (with 1 drop of 1:1000 adrenaline per ml of solution) is carried out.
5. The columella, floor of the nose and the caudal end of the septum are infiltrated with approximately 5 ml of 1% lidocaine containing epinephrine (1:100,000).
6. The caudal end of the septum is exposed with a nasal speculum, a Cottle clamp or a blunt, double-ended hook.
7. A right hemitransfixion incision is made with a no. 15 blade and is made 2 mm behind the free border of the septal cartilage. It reaches from the floor of the nose to the septum and extends upward, parallel to the caudal margin, to a point near its junction with the major alar cartilage. The shape of the incision is that of an inverted "L" with its horizontal line pointing to the left of the surgeon (right of the patient; Fig. 44.3). The soft tissues surrounding the incision, especially on the right side of the septum, are detached from the underlying surfaces to avoid unnecessary tension.
8. In women with small noses, the nostril may be enlarged by an incision at the base of the ala, which is later connected to the incision in the skin of the floor of the nasal cavity (lateral alotomy).
9. The superior lip is undermined for exposure of the maxilla and nasal spine.
10. The anterior nasal spine is dissected on both sides through the lower part of the hemitransfixion incision. The piriform crest of the nasal cavity is then dissected on both sides.
11. A right sub-mucoperichondrial dissection extending to the posterior area of the perpendicular plate of the ethmoid bone is performed. This dissection is carried out posteriorly using progressively longer nasal specula.
12. A left superior tunnel is created by dissecting the mucoperichondrium with scissors and a suction elevator. We do not preserve the connection between the mucoperichondrium and the septal cartilage on one side, because we feel that doing so may lead to asymmetrical nutrition, asymmetrical growth of the cartilage and septal deviation.
13. A left inferior tunnel is created.
14. With sharp dissection along the chondro-osseous junction of the septal cartilage with the vomer and premaxilla, the mucosa is progressively elevated from the posterior portion of the septum. Bony spines and crests are dissected carefully and removed, thus avoiding tearing of the mucosa and consequent bleeding and posterior scarring.
15. The cartilaginous septum is mobilized by luxating its lower part from its osseous groove on the nasal spine, premaxilla and vomer. This posterior attachment to the perpendicular plate of the ethmoid bone is separated. The premaxilla and the vomer may be shortened with a chisel, if necessary (Fig. 44.4).
16. The posterior portion of septal bone, which obstructs access to the sphenoid rostrum, is resected with a sharp 4-mm chisel and septal forceps. The mucosa is laterally elevated from the external surface of the sphenoid sinus.
17. A self-retaining speculum (Stamm no. 3) is introduced between the two separated layers of the septal mucosa, with the septal cartilage luxated to any side, according to the surgeon's preference. The speculum is pushed firmly but gently

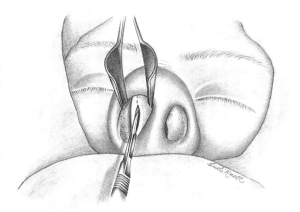

Fig. 44.3. Hemitransfixion incision behind the free border of the cartilaginous nasal septum

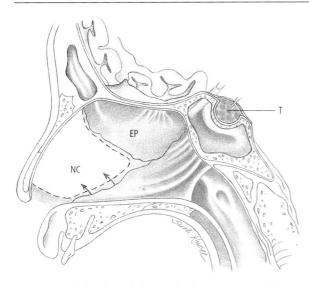

Fig. 44.4. Mobilization of the cartilaginous septum via luxation of its lower part from the pre-maxilla and vomer. *EP,* ethmoid plate; *NC,* nasal cartilage; *T,* tumor

against the anterior wall of the sphenoid sinus. An imaginary line along the superior edge of the speculum blades must point to the tuberculum sellae if checked radiologically.

18. The rostrum of the sphenoid bone and sphenoid ostia are identified.
19. The anterior wall of the sphenoid sinus (which proceeds from the superior ostial-orifice level and is enlarged with Kerrison forceps, this foramina being the lateral landmark for the opening; Fig. 44.5) is resected. This is done with flat chisels (4–5 mm) and a variety of punches. The septum of the sphenoid sinus is excised, and an adequate amount of sinus mucosa is removed. The sphenoid sinus is carefully inspected with 0° and 30° nasal telescopes, and some prominences can be noted, such as the carotid prominence, the bulge of the optic canal and the prominence formed by the maxillary nerve on the lateral wall of the sinus.
20. The frontobasal face of the sella is opened with a small, flat chisel or a micro-gauge and is removed in a piecemeal fashion. The dura is carefully separated and exposed. The opening is made as large as possible, reaching the cavernous sinus on both sides and ending 1–2 mm below the inferior surface of the tuberculum sellae. The inferior edge of the opening lies on the horizontal part of the sellar floor. This step of the operation can also be controlled using televised fluoroscopy.
21. The exposed dura bulges out of the opening in the bony sellar floor in most cases. Careful hemostasis is done via coagulation with a bipolar forceps.
22. After a preliminary needle aspiration, the dura is opened by means of a cruciate incision made with a micro-sickle knife, and the flaps are elevated with right-angle hooks. The soft, gray–purple tumor tissue will then extrude through the opening if the tumor is a prolactinoma. Somatotroph adenomas are whitish in color, resembling cerebral white matter.
23. The adenoma tissue is removed with appropriate forceps, malleable ring curettes and spoons (Fig. 44.6). Adequate suction is mandatory. An ultrasonic dissector is also used for the removal of the tumor. The adenoma tissue and a layer of normal pituitary tissue are both resected to ensure complete tumor extirpation. Tumor biopsy specimens are sent for histology, inmunocytochemistry and ultrastructure examination. The complete removal of the sellar tumor is controlled by means of 0° and 30° nasal endoscopes.
24. Bleeding from the tumor bed usually stops after complete extirpation of the adenoma has been achieved. Some remaining oozing can be controlled by a 5-min application of cottonoids soaked in saline solution and monopolar coagulation of any remaining small vessels. Bleeding from the anterior intercavernous sinus is stopped by coagulation with bipolar, right-angled forceps and/or Surgicel.
25. Small pieces of muscle or fat tissue removed at the beginning of the operation from the thigh or abdominal fat soaked in gentamicin solution are placed in the tumor cavity to secure hemostasis and avoid CSF leakage. Overpacking must be avoided, because it may lead to optic-chiasma compression.
26. A piece of septal cartilage or bone is inserted between the margins of the bony sellar floor and the basal dura after packing the sella with muscle. The entire opening is sealed with a generous application of fibrin glue. The sphenoid sinus can be filled completely if the diaphragma sellae has been damaged during the operation. The sphenoid sinus can be left with no filling at all. The speculum is removed.
27. The nasal structures are repositioned. The septal cartilage is returned to its correct midline position. Corrections of any remaining septal deformity that could impair the respiratory function of the nose are performed. Autologous cartilage can be reimplanted in the places from which septal cartilage has been resected, restoring the anatomy and consistency of this nasal structure, with all its functional implications. The mucosal incision in the right nasal antrum is closed with interrupted 3-gauge chromic catgut sutures. The

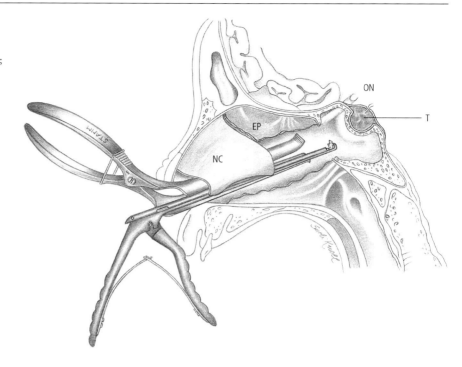

Fig. 44.5. Resection of the anterior wall of the sphenoid sinus, enlarged with a Kerrison forceps. *EP,* ethmoid plate; *NC,* nasal cartilage; *ON,* optic nerve; *T,* tumor

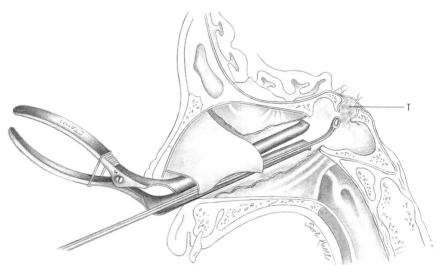

Fig. 44.6. Removal of the pituitary adenoma via a ring curette. *T,* tumor

ala, if incised basally, is closed with interrupted 5-gauge prolene or silk sutures.
28. Two plastic stents are placed on each side of the septum and fixed with 4-gauge prolene or silk sutures or a mattress suture.
29. The nasal cavities are packed with gauze strips soaked with antibiotic ointment, following the usual technique that begins on the floor of the nasal cavity and reaches the nasal cavity's roof.
30. The packing is removed on the second postoperative day. A nasal ointment is used for some days after retiring the packing, to prevent crusting.
31. The plastic stents are removed between the fifth day and the seventh day, depending on the amount of septal work performed.

The Transseptalsphenoidal Approach in Previously Operated Noses

Previous pituitary surgery represents a special challenge for the rhinologist. The iatrogenic absence of various anatomical landmarks is noted in these cases.

A careful midline opening of the anterior soft tissues of the sphenoid is advised if there are no remnants of septal structures. The use of fluoroscopy is helpful in this step of the operation and can be combined with the use of endoscopes before and after the midline incision. An endoscope can be introduced in the sphenoid sinus through a small opening, thus giving the surgeon the opportunity for direct inspection of the status of this cavity before widening the approach. With the widespread use of transseptalsphenoidal approaches for pituitary surgery, a need has developed for a transseptal approach for reoperation or for noses that have undergone previous rhinologic surgery [16].

Prior surgical manipulation of the septum usually causes areas of missing bone or cartilage. In these areas, the septal mucosa of one side directly adheres to mucosa on the opposite side, with little intervening fibrous tissue or with atrophic changes in both mucosal layers. To avoid septal perforations and to obtain a good trans-septal approach to the sphenoid sinus in these circumstances, different surgical strategies have been found to be useful [12, 16, 85, 100].

The technique of the septal displacement, first described by Sofferman [87] as an alternative to lateral rhinotomy and used successfully by different authors on previously operated noses, is the method of choice in these cases [16]. The septal-displacement approach to the sphenoid sinus provides a method that allows the surgeon to avoid re-dissection of the nasal septum after transseptal pituitary surgery or septoplasty or in patients with septal perforations. In these situations, time-consuming and meticulous dissection of the adherent mucosal layers and surgical techniques that involve extensive mobilization of nasal or facial structures are alternatives. However, they do not always achieve satisfactory results and are uncomfortable for the patient (Fig. 44.7).

In the septal-displacement technique, a complete transfixing incision is made on the right side of the nose and is then extended laterally around the piriform margin. An incision paralleling the septum is made laterally in the mucosa of the floor of the right nasal cavity, near the inferior turbinate. Both incisions are connected, allowing the septum and attached mucosal flaps of floor of the nose to be dissected and displaced laterally and to the left. Dissection progresses posteriorly until residual septal bone (vomer) is found, providing valuable surgical orientation. The mucosa of the opposite nasal floor is also elevated.

A self-retaining speculum (Stamm no. 3) is carefully inserted in the right nasal fossa to a depth just anterior to the solid septal-bone remnant, displacing the mobile septum and its attached nasal floor mucosal flap to the left. A mucosal incision is made at the leading edge of the bone remnant, cutting until reaching the septal bone. Mucoperiostial envelopes are developed on either side of the bony septum. Dissection then progresses posteriorly until the anterior aspect of the sphenoid sinus is found. The use of the microscope is not advised until safe anatomical landmarks are reached.

At the end of the operation, the posterior septal incision and the floor of the nose incision are not sutured after the flaps are replaced in proper position. The transfixing incision is closed with interrupted absorbable sutures. The nose is carefully packed. The septal-displacement technique, carried out in the way explained above, is a quick approach to the sphenoid sinus in patients who have undergone previous rhinologic surgery (Table 44.7).

Table 44.7. Surgical techniques in patients with previous rhinologic surgery (n=16)

Technique	Number of patients
Septal dissection	4
Degloving	1
External rhinoplasty	2
Septal displacement	9
Total	16

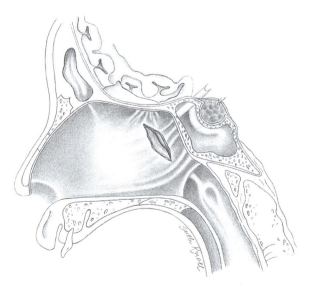

Fig. 44.7. Access to the sphenoid sinus in a previously operated nose. An incision is made in the posterior region of the nasal septum

Parasellar Lesions of Non-Pituitary Origin

Treatment of sphenoidsinus mucoceles is described in detail in another chapter, as is treatment of CSF leaks. The transseptalsphenoidal approach also allows resection of chordomas restricted to the clivus (Fig. 44.8).

Once the sphenoid sinus is exposed, its posterior wall is opened, beginning at a midline level, posterior and inferior to the dorsum sellae. This opening is initially created using a diamond burr 4 mm or 5 mm in diameter and is then widened using a micro-Kerrison punch until both carotid arteries and the cavernous sinus are reached laterally. Posteriorly and inferiorly, the osseous matrix of the clivus is removed with a long, angled forceps or a diamond burr. The lesion, in the case of chordomas, can usually be removed with suction, and the resection is then completed with angled curettes. After total removal of the tumor, the nasal structures are repositioned as described for the transseptalsphenoidal approach to the sella.

Conclusion

The history of transsphenoidal hypophysectomy demonstrates the importance of the concept of approaching the sellar structures in a midline fashion via the nose. The transseptal, transsphenoidal approach, refined by recent advances in microsurgical techniques and instrumentation and by the technology of the nasal endoscope, is a simple, reliable, and quick technique that provides excellent extradural access to the sphenoid sinus and the pituitary gland. Because the transseptalsphenoidal route involves the nose, whose function and shape must be preserved, close teamwork between the neurosurgeon and the rhinologist is highly recommended. The otolaryngologist's experience will prove to be of the utmost importance in transnasally approaching the sphenoid sinus in patients with previous pituitary or septal surgery and in taking advantage of the technical possibilities of nasal endoscopes in sellar surgery.

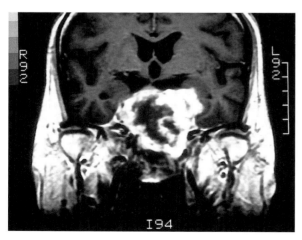

Fig. 44.8. Coronal magnetic resonance demonstrating a large lesion (chordoma) involving the sellar and parasellar regions

References

1. Angell-James J (1967) The hypophysis. The Semon lecture, 1965. J Laryngol Otol 81:1283–1307
2. Angell-James J (1969) Trans-sphenoidal hypophysectomy. In: Hamburger CA, Wersall J (ed) Disorders of the skull base region. Almquist and Wiksell, Stockolm, pp 163–169
3. Basso A, Pardal E (1980) Estado actual de la cirugia de la hipofisis. Rev Assoc Med Argent 1980:89–98
4. Bateman GH (1962) Trans-sphenoidal hypophysectomy. J Laryngol 76:442
5. Blaugrund SM (1987) Rhinologic approach to the hypophysis. In: Goldman JL (ed) The principles and practice of rhinology. Churchill-Livingstone, New York, pp 581–591
6. Bouche J (1973) The technique of trans-sphenoidal hypophysectomy. Rhinology 11:153–159
7. Burian K (1969) Die Hypophysekmtomie aus des sicht der Rhinologen. HNO 17:193
8. Burian K, Pendl G, Salah S (1961) Technik, Indikation und Erfahrungen bei partieller Hypophysen: Ausschaltung. Laryngorhinolootologie 40:186
9. Burian K, Pendl G, Salah S (1970) Uber die rezidivhaufigkeit von hypophysen adenomen nach transfrontaler, trans-sphenoidal oder zweizeitig kombinierter operation. Wien Med Wochenschr 120:833–836
10. Chiari O (1912) Uber eine Modifikation der Schlofferschen Operation von Tumoren der Hypophyse. Klin Wochenschr Wien 25:5
11. Cohen RS (1994) Cirugia transesfenoidal de la region selar. Fol Med (Bras) 109:25–29
12. Cohen RS (1994) Microcirurgia transnasal: via transeptoesfenoidal a hipofise e estruturas selares. In: Stamm AC (ed) Microcirurgia naso-sinusal. Revinter, Rio de Janeiro, pp 321–331
13. Cohen RS (1995) La participacion del rinologo en la cirugia de la region selar y periselar. An Otorrinolaringol Peru 4:35–36
14. Cohen RS (1995) Abordajes transeptales en cirugia pituitaria. nuestra experiencia. An Otorrinolaringol Peru 4:37–43
15. Cohen RS, Pacios AR (1994) Multiplanar computerized tomography in the surgical decisions of the pathology of the sellar region: tumors that can be approached by the trans-septal sphenoidal. Fol Med (Bras) 108:165–167
16. Cohen RS, Pacios AR (1994) The trans-septosphenoidal approach to the hypophysis: our technique in previously operated noses. Fol Med (Bras) 109:111–113

17. Cohen RS, Pacios AR (1995) Trans-septal sphenoidal surgery of the sellar and perisellar region. Comments on 57 cases. An Otorrinolaring Mex 40:13–16
18. Cottle MH (1958) The "maxilla–premaxilla" approach to extensive nasal septum surgery. Arch Otolaryngol Head Neck Surg 68:301–313
19. Crowe SJ, Homans J (1909) Hypophysectomy. Bull Johns Hopkins Hosp 20:102
20. Cushing H (1909) Partial hypophysectomy for acromegaly. Ann Surg 50:1002
21. Cushing H (1911) The pituitary body and its disorders: clinical states produced by disorders of the hypophysis cerebri. Lippincott, Philadelphia, p 296
22. Cushing H (1914) Surgical experiences with pituitary disorders. JAMA 63:1515–1525
23. Dandy WE (1938) Hirnchirurgie. Barth, Leipzig
24. Dandy WE (1966) The brain. Prior, Hagerstown
25. Denker A (1921) Drei Falle (von Hypophysentumoren) Operiert nach Transmaxillarer Methode. Int Zentralbl Laryngol 37:225
26. Dott NM, Bailey PA (1925) Consideration of the hypophyseal adenomata. Br J Surg 13:314–366
27. Draf W (1978) Endoskopie der Nasennebenhohlen. Springer, Berlin Heidelberg New York
28. Draf W (1982) Die Chirurgische Behandlung entzundlicher Erkrankungen der Nasennebenhohlen. Arch Otorhinolaryngol 235:133–305
29. Elster AD (1993) Modern imaging of the pituitary. Radiology 187:1–14
30. Escher F (1965) Resultate und Erfahrungen mit der Hypophysen ausschaltung. Fortschr Hals Nasen Ohrenheilkd 12:129–160
31. Fahlbusch R, Buchfelder M (1993) The trans-sphenoidal approach to invasive sellar and clivus lesions. In: Sekhar LN, Janecka IP (eds) Surgery of cranial base tumors. Raven, New York, pp 337–349
32. Fein J (1910) Zur Operation der Hypophyse. Wien Klin Wochenschr 23:1305
33. Gammert C (1990) Rhinosurgical experience with the trans-septal trans-sphenoidal hypophysectomy: technique and long-term results. Laryngoscope 100:286–289
34. Giordano A (1987) Compendio di chirurgia operatoria italiana, vol 2. Monduzzi, Bologna, p 100
35. Guiot G (1958) Adenomes hypophysaires. Masson, Paris
36. Guiot G (1973) Trans-sphenoidal approach in surgical treatment of pituitary adenomas: general principles and indications in non-functioning adenomas. Excerpta Med, Amsterdam, pp 159–178
37. Guiot G, Derome P (1976) Surgical problems of pituitary adenomas. (Advances and technical standards in neurosurgery, vol 3) Springer, Berlin Heidelberg New York
38. Guiot G, Bouche J, Oppotu A (1967) Les indications de l'abord trans-sphenoidal des adenomes hypophysaires. Presse Med 75:1563
39. Guleke N (1950) Die Eingriffe am Gehirnschadel, Gehirn, au der Werbelsaule und am Ruckenmark. In: Guleke N, Zenker R (eds) Allgemeine und spezielle chirurgische operations. Springer, Berlin Heidelberg New York
40. Halstead AE (1910) Remarks on the operative treatment of tumors of the hypophysis: with the report of two cases operated on by oro-nasal method. Surg Gynecol Obstet 10:494–502
41. Hamburger CA (1961) Trans-sphenoidal hypophysectomy. Arch Otolaryngol Head Neck Surg 74:2–8
42. Hamburger CA, Hammer G (1964) Der Transnasale Weg der Hypophysektomie. In: Berendes J, Link R, Zollner F (eds) Hals Nasen Ohren Heilkunde. Thieme, Stuttgart, pp 795–818
43. Hardy J (1969) Trans-sphenoidal microsurgery of the normal and pathologic pituitary. Clin Neurosurg 16:185
44. Hardy J (1973) Trans-sphenoidal surgery of hypersecreting pituitary tumors. In: Kohler PO, Ross GT (eds) Diagnosis and treatment of pituitary tumors. Excerpta Medica. Elsevier, New York
45. Hardy J (1971) Trans-sphenoidal hypophysectomy. J Neurosurg 34:582–594
46. Heermann H (1958) Endonasal surgery with the use of the binocular Zeiss operating microscope. Arch Klin Exp Ohren Nasen Kehlkopfheilkd 171:295–297
47. Henderson WR (1939) The pituitary adenomata: a follow up study of the surgical results in 338 cases (Dr. Cushing's series). Br J Surg 26:811–921
48. Hirsch O (1909) Eine Neue Method der Endonasalen Operation von Hypophysen Tumoren. Wien Med Wochenschr 22:636
49. Hirsch O (1910) Endonasal method of removal of hypophyseal tumors. JAMA 55:772–774
50. Hirsch O (1911) Uber Endonasale Operations. Method bei Hypophysis-Tumoren mit Bericht uber 12 operierte falle. Klin Wochenschr (Berl) 48:1933
51. Hirsch O (1959) Life-long cures and improvements after trans-sphenoidal operation of pituitary tumors. Arch Otolaryngol Head Neck Surg 55:5–60
52. Jankowski R (1995) Endoscopic pituitary surgery. In: Stankiewicz JA (ed) advanced endoscopic sinus surgery. Mosby, St. Louis
53. Jankowski R, Auque J, Simon C, Marchal JC, Hepner H, Wayoff M (1992) Endoscopic pituitary tumor surgery. Laryngoscope 102:198–202
54. Kanavel AB (1909) Removal of tumors of the pituitary body by an infranasal route. JAMA 53:1704–1707
55. Kenan PD (1979) The rhinologist and the management of pituitary disease. Laryngoscope 89[suppl 14]:1–26
56. Kennedy DW (1992) Prognostic factors, outcomes and staging in ethmoid sinus surgery. Laryngoscope 102:[suppl 12]:1–18
57. Kennedy DW, Cohn ES, Papel ID, Holliday MJ (1984) Trans-sphenoidal approach to the sella: the Johns Hopkins experience. Laryngoscope 94:1066–1074
58. Kern EB, Pearson BW, McDonald TJ, Laws ER Jr (1979) The trans-septal approach to lesions of the pituitary and parasellar regions. Laryngoscope 89[suppl 15]:1–34
59. Knavel AB, Grinker J (1910) Removal of tumors of the pituitary body. Surg Gynecol Obstet 10:414–418
60. Koltai P, Goldstein JC, Parnes SM, Price JC (1985) External rhinoplasty approach to trans-sphenoidal hypophysectomy. Arch Otolaryngol Head Neck Surg 111:456–458
61. Konig F (1900) Die Eingrifte am Gehirnschade, Gehirn, an der Wirbersaule und am Ruckenmark. In: Guleke N, Denker R (eds) Allgemeine und spezielle chirurgische Operationeslehre, vol 2, 2nd edn. Springer, Berlin Heidelberg New York, p 335
62. Krause T (1905) Hirnchirurgie (Freilegung der Hypophyse). Dtsch Klin 144:1004–1012
63. Lautenschlager A (1947) Operaciones en el oido, nariz, garganta y vias respiratorias. In: Kirschner M (ed) Tratado de tecnica operatoria general y especial. vol 3. Labor, Barcelona, pp 165–168

64. Laws ER (1987) Pituitary surgery. In: Molitch M (ed) Pituitary tumors: diagnosis and management. (Clinics of North America, vol 16) Saunders, New York, pp 647–665
65. Laws ER (1993) Clivus chordomas. In: Sekhar LN, Janecka IP (eds) Surgery of cranial base tumors. Raven, New York, pp 679–685
66. Lee KJ (1978) The sub-labial trans-sphenoidal approach to the hypophysis. Laryngoscope 88[suppl 10]:1–65
67. Lenhardt E (1979) Septum – Schleimhautplastik Zur Daver drainage der Keilbeinhohle. Arch Otolaryngol Head Neck Surg 222:43–46
68. Levine HL, May M (1993) Endoscopic sinus surgery. Thieme, New York
69. Lowe L (1905) Die Eingriffe am Gehinschadel Gehirn, und der Wirbelsaule und am Ruckemmark. In: Guleke N, Denker R (eds) Allgemeine und spezielle chirurgische Operationeslehre, vol 2, 2nd edn. Springer, Berlin Heidelberg New York, p 335
70. MacBeth R, Hall M (1962) Hypophysectomy as a rhinological procedure. Arch Otolaryngol Head Neck Surg 75:440–450
71. Messerklinger M (1978) Endoscopy of the nose. Urban and Schwarzenberg, Baltimore
72. Mixter SJ, Quakenboss A (1910) Tumors of the hypophysis (with infantilism). Trans Am Surg Assoc 27:94–109
73. Montgomery WW (1963) Transethmoidosphenoidal hypophysectomy with septal mucosal flap. Arch Otolaryngol Head Neck Surg 78:68–77
74. Montgomery WW (1971) Surgery of the upper respiratory tract, vol 1. Lea and Febiger, Philadelphia, pp 71–93
75. Moszkowicz L (1907) Zur technik der operationen on der hypophyse. Wien Klin Wochenschr 20:792–795
76. Nager FR (1940) The paranasal approach to intrasellar tumors. Semon lecture, 1939. J Laryngol 55:361
77. Nicola G (1972) Trans-sphenoidal surgery for pituitary adenomas with extrasellar extension. Prog Neurol Surg 6:19–72
78. Prades J (1977) Microcirugia endonasal. Gersi, Madrid
79. Preysing A (1913) Beitrage zur operation der Hypophyse. Ver Dtsch Laryngol 20:51
80. Proust R (1908) La chirurgie de l' hypophyse. J Chir 1:665–680
81. Richards SM et al. (1974) Transethmoidal hypophysectomy for pituitary tumors. Proc R Soc Med 67:889–892
82. Ruckgang A (1909) Der akromegalischen orscheimingen nach operation eines hypophysentumors. Wien Klin Wochenschr 23:108–109
83. Rudert H (1988) Mikroskop-und endoskopgestützte Chirurgie der entzündlichen Nasennebenhöhlenerkrankungen (Der Stellenwert der Infundibulotomie nach Messerklinger). HNO 36:475–482
84. Samii M, Draf W (1989) Surgery of the skull base. Springer, Berlin Heidelberg New York, pp 273–296
85. Sawyer R (1991) Nasal approach to the sphenoid sinus after prior septal surgery. Laryngoscope 101:89–91
86. Schloffer H (1906) Zur frage der operationen an der hypophyse. Beitr Klin Chir 1:767–817
87. Sofferman RA (1988) The septal translocation procedures: an alternative to lateral rhinotomy. Otolaryngol Head Neck Surg 98:18–25
88. Stamm AC (ed) (1994) Microcirurgia naso-sinusal. Revinter, Rio de Janeiro
89. Stammberger H (1989) Functional endoscopic nasal and paranasal sinus surgery – the messerklinger technique. Decker, Toronto
90. Stammberger H, Hawke M (1993) Essentials of functional endoscopic sinus surgery. Mosby, St. Louis
91. Stankiewicz JA (ed) (1995) Advanced endoscopic sinus surgery. Mosby, St. Louis
92. Stumme E (1908) Akromegalie und hypophyse. Arch Klin Chir 87:437–466
93. Tiefenthal A (1920) Technik der Hypophysen Operation. Munch Med Wochenschr 67:794
94. Uffenorde W (1923) Orbitale Stirnhohlenoperation. Z Hals Nas Ohrenheilkd 6:117
95. Von Eiselberg A (1907) Über Operative Freilegung der Tumoren in der Hypophysengegend. Neurol Zentralbl 26:994
96. Von Eiselberg A (1910) My experience about operations upon the hypophysis. Trans Am Surg Assoc 28:55
97. Von Eiselberg A (1912) Zur Operation der Hypophygeschwülste Langenbecks. Arch Klin Chir 100:8
98. West JM (1910) A combined otolaryngologic-neurosurgical approach to the sella. JAMA 54:1132–1134
99. Wigand M (1991) Endoscopic surgery of the paranasal sinuses and anterior skull base. Thieme, New York
100. Wilson WR, Laws ER Jr (1992) Transnasal septal displacement approach for secondary trans-sphenoidal pituitary surgery. Laryngoscope 102:951–953
101. Wilson WR, Khan A, Laws ER Jr (1990) Trans-septal approaches for pituitary surgery. Laryngoscope 100:817–819

Transnasal Endoscopic Surgery of the Sella and Parasellar Regions

ALDO CASSOL STAMM, ANDRÉ BORDASCH, EDUARDO VELLUTINI and FELIX PAHL

Introduction

Interest in pituitary diseases has increased since Pierre Marrie first described acromegaly in 1886 in Paris; a neurosurgical odyssey ensued, pursuing surgical approaches to the sella and parasellar region while attempting to lower morbidity and mortality rates [33, 37]. After several cadaver studies in 1907, Schloffer performed the first sellar exploration, mobilizing the entire nose on a pedicle [28]. For the next several years, surgical indications were mainly visual defects and headache, which surgeons attempted to relieve by decompressing the tumor. Although this was often successful, many patients died during surgery or subsequently, from meningitis [4].

Kanavel [16] and Halstead [8] first devised an intranasal approach (and later a sub-labial approach) to the sellar region. In 1910, Cushing [5, 11, 26] was the first neurosurgeon to use the sub-labial approach; he reported using that approach on over 200 cases, with a mortality rate of 5.6%. However, patients with suprasellar tumors were not surgically accessible sub-labially and, in 1929, he changed to the intracranial approach.

In the 1950s, the introduction of cortisone and antibiotics made surgery of the pituitary gland safer and more successful. In 1965, Hardy [9, 10] first used the operating microscope (OM) and intraoperative radiofluoroscopy for total hypophysectomy in breast cancer patients and for selective anterior hypophysectomy in the treatment of diabetic retinopathy. He used the midline oronasal route and used a microscope and specially designed instruments; he reported no deaths or serious complications in his first 50 patients. Considerable time elapsed between the first nasal endoscopy using a cystoscope (by Hirschmann, in 1902) and the development of an optical system good enough to facilitate improved understanding of the anatomy and physiology of the nose and paranasal sinuses.

Between 1951 and 1956, Professor H. Hopkins was responsible for many contributions to the development and improvement of endoscopes, but it was only after the 1969 publications of Messerklinger and subsequent reports of Wigand, Draf, Stammberger and Kennedy that naso-sinus endoscopy became accepted worldwide. Today, it is also extensively used in the surgical treatment of lesions of the anterior skull base and the sella and parasellar regions [6, 17, 18, 23, 35, 39].

The combination of the OM and endoscopes has largely been the result of the work of Draf and Rudert [6, 27]. While the OM provides binocular, three-dimensional vision and allows bimanual work, the endoscope gives better freedom of mobility, a panoramic view, angulation of vision and improved image resolution for documentation, learning and teaching, making paranasal-sinus operations safer and more dynamic. Endoscopic assistance represents an enormous help, especially in surgery of the sphenoid sinus, which is deeper and is in direct contact with important lateral structures (such the optic nerve and the internal carotid artery), where an angled view is essential.

Jankowski et al. first described the successful endonasal endoscopic resection of pituitary adenomas in three patients in 1992 [12, 13]. In those cases, the middle turbinates were removed to gain access to the anterior wall of the sphenoid sinus.

In 1994, Gamea et al. reported ten cases of pituitary tumors removed via a sub-labial, trans-septal, trans-sphenoidal pituitary approach using a combined technique including the OM and an endoscope. They concluded that the endoscope facilitated the dissection of the tumor from the normal pituitary gland, resulting in maximum preservation of pituitary function [7].

Many authors believe that the endoscope has replaced the OM for surgery of pituitary adenomas and other sellar lesions [3, 12–15, 29–31]. In 1995, Sethi et al. described the application of the transnasal endoscopic technique for pituitary adenomas and craniopharyngiomas in 40 successfully treated patients [29–31].

In 1996 and 1997, Jho et al. published their experience with an endoscopic, direct, transnasal approach in 50 patients, using a holder attached to the endoscope after completion of the anterior sphenoidotomy. They concluded that the holder provided a steady video image in addition to freeing both of the surgeon's hands to maneuver surgical instruments simultaneously [14, 15].

Endoscopic Anatomy of the Sphenoid Sinus and the Sellar Region

The sphenoid sinus has been classified into three types (conchal, presellar and sellar), depending on the extent of the pneumatization of the sphenoid bone (Fig. 45.1). In the conchal type, the area below the sella is a solid block of bone without an air cavity. In the presellar type, the posterior limit of the air cavity is a perpendicular plane projected on the sellar wall. The sellar type is the most common, occurring in 86% of individuals; in it, the air cavity extends into the body of the sphenoid below the sella and may extend posteriorly as far as the clivus [19]. The conchal type is very common in children under the age of 12 years, after which pneumatization begins within the sphenoid sinus [22].

The ostium of the sphenoid sinus most often lies in the "superior quadrant", a few millimeters of the cribriform plate [19, 24]. Less commonly, it lies close to the floor of the sinus or opens into a posterior ethmoid cell. In most cases, the sphenoid ostium can be identified using the superior edge of the choana and the superior turbinate as landmarks, without removal of the middle turbinate. The ostium is located in the sphenoethmoidal recess and varies in size from 1 mm to 4 mm.

The location of the intersinus septum is highly variable. While it often traverses the sphenoid sinus and inserts into the anterior surface of the sella, it occasionally terminates on the bulge of the internal carotid artery or, rarely, on the optic canal (Fig. 45.2A) [29–31].

The internal carotid artery is the most medial structure in the cavernous sinus. It rests directly against the lateral surface of the body of the sphenoid sinus, creating a prominence there, and its course is marked by a groove in the bone (the carotid sulcus), which defines the course of the intracavernous portion of the carotid artery (Fig. 45.2B). This prominence is greatest in those with maximal sphenoid pneumatization. In very rare situations, the carotid artery may be dehiscent inside the sinus.

The optic canal protrudes bilaterally into the superolateral part of the sphenoid sinus. The surgeon must be cognizant of anatomic variations of the relationship between the optic canal and the Onodi cell.

The pterygoid canal has an average length of 16 mm and can run inferior to, at the same level as or (sometimes) within the sphenoid sinus itself. The bone of the canal's roof is dehiscent in 10% of the cases [19]. A segment of the maxillary division of the trigeminal nerve frequently produces a bulge in the lateral sinus wall below the sella, especially when the sinus is well pneumatized.

The anterior wall of the sella is recognized by its midline bulge, inferior to the tuberculum sella. The sellar wall is approximately 0.5–1 mm thick and bulges markedly into the sphenoid sinus Fig. 45.3 [19, 24]. After the removal of the anterior sellar wall, the dura is identified.

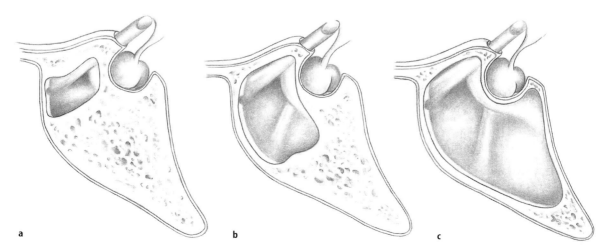

Fig. 45.1 a–c. Three types of sphenoid sinus, based on the pneumatization. **a** Conchal. **b** Presellar. **c** Sellar

Transnasal Endoscopic Surgery of the Sella and Parasellar Regions   557

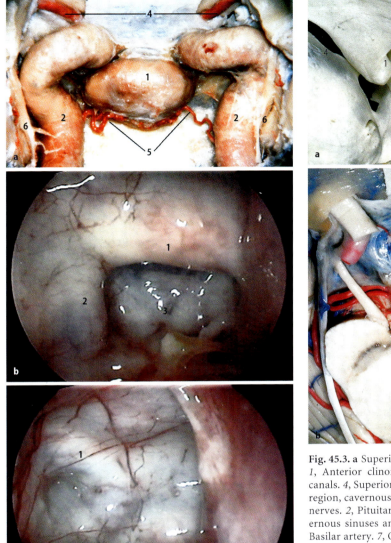

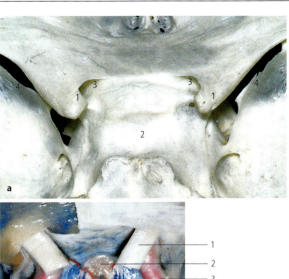

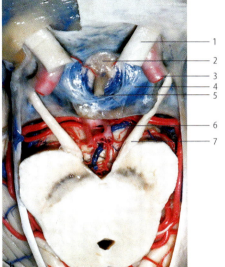

Fig. 45.3. a Superior view of the sellar and parasellar regions. *1*, Anterior clinoid process. *2*, Tuberculum sellae. *3*, Optic canals. *4*, Superior orbital fissure. **b** Superior view of the sellar region, cavernous sinuses and posterior cranial fossa. *1*, Optic nerves. *2*, Pituitary gland. *3*, Internal carotid arteries. *4*, Cavernous sinuses and their venous plexes. *5*, Pituitary stalk. *6*, Basilar artery. *7*, Oculomotor nerves

Fig. 45.2. a Anterior view into the sphenoid sinus. The bony walls have been removed, as has the dura that covers the pituitary gland (anterior and posterior lobes), internal carotid arteries, optic nerves, ophthalmic arteries, inferior hypophyseal arteries and sympathetic nerves (courtesy of T. Inoue). *1*, Pituitary gland. *2*, Internal carotid arteries. *3*, Optic nerves. *4*, Ophthalmic arteries. *5*, Inferior hypophyseal arteries. *6*, Sympathetic nerves. **b** Endoscopic view inside the sphenoid sinus. *1*, Anterior wall of sella. *2*, Internal carotid-artery canal. *3*, Clivus. **c** Same patient, showing a more superior–lateral region. *1*, Optic-nerve canal. *2*, Anterior wall of the sella. *3*, Internal carotid-artery canal

Pre- and Perioperative Evaluation

Patients with pituitary-gland disease usually present endocrinological and visual symptoms. A complete preoperative evaluation should be done by endocrinologists and ophthalmologists before selecting patients for pituitary surgery.

The visual disturbances are mostly related to compression of the optic chiasma and oculomotor dysfunction, usually due to involvement of the cavernous sinus. Pituitary adenomas are either non-secreting or secreting tumors. Non-secreting tumors should be removed only if progressive growth or compression of the optic nerve or optic chiasma occurs. The majority of the pituitary-gland tumors that produce prolactin and cause significant visual symptoms respond to medical treatment with bromocriptine, even if they are large. Surgical treatment is usually indicated for patients who cannot tolerate medical therapy or who have cystic prolactinomas. The tumors that produce growth hormone and adrenocorticotropic hormone should be surgically removed in order to achieve endocrinological cure.

Several factors must be considered before selecting the surgical approach. Preoperative knowledge of the anatomy of the sellar region (based on diagnostic imaging studies) is one such factor. The data information obtained by the computed tomography (Fig. 45.4) and magnetic resonance (MR) are complementary. The MR image in Fig. 45.5 shows a pituitary lesion and the relationship between normal pituitary gland, the optic chiasm and the carotid artery.

Although the size and pneumatization of the sphenoid sinus are sometimes limiting factors, more than 95% of the pituitary adenomas are operated on via a trans-sphenoidal route. With large and extensive tumors, especially if there is parasellar or suprasellar involvement, craniotomy may be considered.

Preoperative endoscopic examination and treatment of nasal and paranasal sinus infections until there is complete return to normal are essential. Perioperative prophylactic antibiotics are routinely used. Corticosteroids are useful in edematous lesions.

Some diseases can mimic pituitary adenomas. When an aneurysm is suspected, angiography should be considered. Differential diagnosis should also be made with meningioma, craniopharyngioma and empty-sella syndrome.

An endocrinologist should assist with the care of the patient during the entire perioperative period. If transient diabetes insipidus occurs, postoperative desmopressin can be effective. Approximately 1 week after operation, baseline pituitary-hormone functions are again studied, and appropriate replacement is initiated.

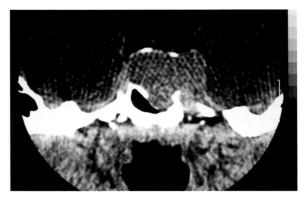

Fig. 45.4. Coronal computed tomography showing a large pituitary tumor extending into the sphenoid sinus and to the suprasellar region

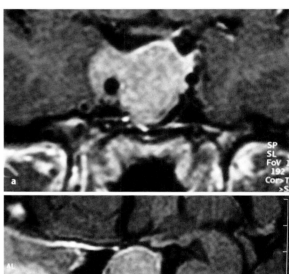

Fig. 45.5. a Coronal magnetic resonance (MR) demonstrating a macroadenoma involving both internal carotid arteries and touching the optic chiasma. b Sagittal MR, showing a pituitary tumor protruding into the sphenoid sinus

Surgical Technique

The patient is prepared in the usual fashion for transnasal endoscopic surgery. The patient lies in a supine position on the operating table, with the head elevated 30° and turned toward the surgeon. No head fixation is used.

Surgery is performed under general anesthesia with oral endotracheal intubation. To maintain a low systolic blood pressure, intravenous propofol is given prior to intubation.

The lower abdomen is prepared as the donor site for a free adipose graft. The video-camera equipment is then aseptically draped. The patient eyes are kept uncovered during the entire procedure. The nasal cavity is decongested for 5–10 min with topical cottonoids moistened with naphazoline hydrochloride and saline solution. To reduce perioperative bleeding, the mucous membrane of the septum (including the posterior part), the superior turbinate and the anterior wall of the sphenoid sinus are infiltrated with a 2% lidocaine solution containing a 1:100,000 dilution of epinephrine.

A video camera is attached to the eyepiece of the endoscope; the entire procedure is miniaturized by a high-resolution video monitor and is recorded on a super-VHS videotape. The surgery is performed using 4-mm sino-nasal rigid endoscopes with 0°, 30°, 45° and sometimes 70° lenses. A lens-cleansing irrigation–suction system is attached to the endoscope and controlled with a pedal by the surgeon. This system eliminates the need to remove the endoscope from the surgical field to clean or de-fog the lens.

The surgeon may operate from the right or left side of the patient, depending on which is the surgeon's dominant hand. The endoscope is held in the surgeon's non-dominant hand, and the instruments are held in the dominant hand. Because endoscopic surgery is a very dynamic procedure, normally no holder is attached to the endoscopes.

At the beginning of the procedure, a systematic nasal endoscopy exam is performed, initially using a 0° degree, 4-mm rigid endoscope. Three approaches are used to access the sphenoid sinus and the sellar region:
1. An endoscopic, transnasal, direct approach
2. An endoscopic, transnasal/trans-septal approach
3. An endoscopic, transethmoid approach

In our series, only the transnasal, direct approach and the transnasal/trans-septal approach were used [34].

Endoscopic Transnasal Direct Approach

The same positioning and preparation are used for all three approaches. The side of the nose to be used for access is based on the endoscopic examination and radiological findings. The decision depends on the width of the nasal cavity and the laterality of the tumor. The operation can be carried out through one nostril or through both nostrils if the nasal cavity is very narrow and limits the passage of the endoscope and operating instruments [21]. Septoplasty is done first, if necessary.

The direct approach to the sphenoid sinus is the fastest of the three types. Identification of the choanal arch and the end of the middle turbinate is performed with a 0°, 4-mm endoscope. Looking superiorly the superior edge of the choana, the superior turbinate and the anterior wall of the sphenoid sinus at the sphenoethmoid recess are identified (Fig. 45.6A). When the middle turbinate obstructs access to the sphenoethmoid recess, careful partial resection of the middle turbinate with curved scissors allows a better view of this narrow space. The ostium seeker may be necessary when the superior turbinate covers the sphenoid ostium.

After identification of the ostium, it is enlarged with a micro-Kerrison punch to a size that is sufficiently large to simultaneously admit passage of the endoscopes and surgical instruments (Fig. 45.6B). The posterior third of the septum is completely resected after adequate elevation of the mucoperiostium, followed by the removal of the sphenoid

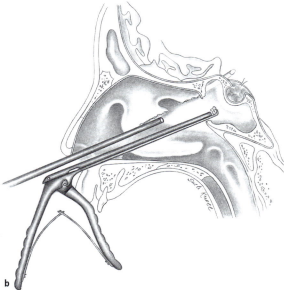

Fig. 45.6. a Lateral nasal wall, showing the relationship between the superior, middle and inferior turbinates and the sphenoid sinus. **b** Transnasal, direct approach to the sellar region. The posterior portion of the middle turbinate is resected as is the anterior wall of the sphenoid sinus

rostrum (Fig. 45.7). Next, using a micro-Kerrison punch, the anterior wall of the sphenoid sinus is totally removed, and careful resection of the intersinus septum is begun (Fig. 45.8). Because the endonasal approach is located a few degrees away from the midline, a laterally positioned lesion is best approached from the opposite side of the nose.

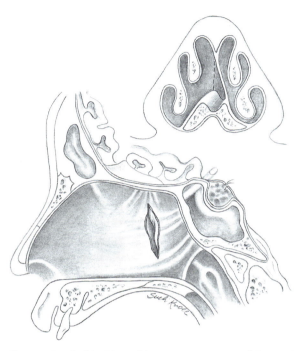

Fig. 45.7. Surgical incision of the mucoperiosteum of the nasal septum just anterior to the sphenoid rostrum

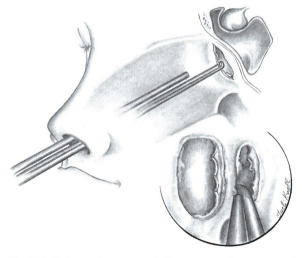

Fig. 45.8. Endoscopic, transnasal, direct approach. Aperture of the anterior wall of the sphenoid sinus with a micro-Kerrison punch

Endoscopic Transnasal/Trans-Septal Approach

After infiltration of the mucoperichondrial and mucoperiostial layers of the septum, the right nostril is incised (in an inverted "L" fashion with a no. 15 blade) approximately 2 mm behind the free border of the septal cartilage, extending to the floor of the nose in a hemitransfixion fashion. The mucoperichondrial flap of the septal cartilage and the mucoperiostium of the bony septum are then elevated, and the bony–cartilaginous junction is separated. Using an elevator, the lower, dissected part of the cartilaginous septum is luxated out of its pre-maxillary, osseous groove. The posterior attachment of the septum to the perpendicular lamina of the ethmoid is fractured. The posterior part of the septal bone, which obstructs access to the sphenoid rostrum, is resected with a Jansen-Middleton forceps. Any large fragment of cartilage or of the perpendicular plate of the ethmoid bone is set aside, because it may be needed later for closure of the anterior wall of the sella. The mucoperiostium of the anterior wall of the sphenoid sinus is elevated until the sinus ostia on both sides are visualized.

At this point, a self-retaining speculum is introduced between the two separated layers of the septal mucosa, and the anterior wall of the sphenoid sinus is entirely exposed (Fig. 45.9). The anterior wall of the sphenoid sinus is then opened with a bayonet-shaped chisel (Fig. 45.10) and is enlarged with a micro-Kerrison punch. The opening extends to the two ostia of the sphenoid sinus and is enlarged inferiorly (until reaching the floor) and laterally. The sphenoidotomy is made large enough to allow easy passage of the 4-mm endoscope and the instruments. At this time, the surgeon must keep in mind the possibility of anatom-

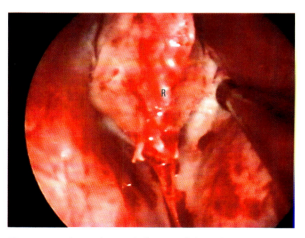

Fig. 45.9. Endoscopic surgical view of the sphenoid rostrum (*R*) after resection of the posterior nasal septum

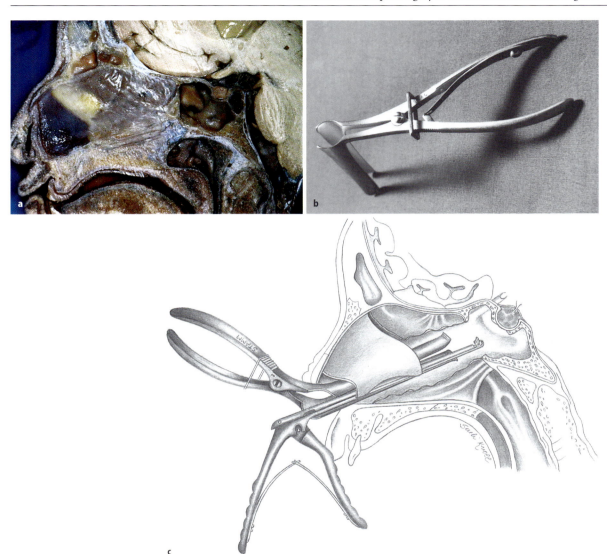

Fig. 45.10. a The relationship between the nasal septum, sphenoid sinus and pituitary gland in a sagittal view. **b** A self-retaining speculum modified by the senior author. **c** The anterior wall of the sphenoid sinus is opened with a micro-Kerrison punch

ical variations in the lateral sphenoidal wall. Superior enlargement of the ostium is not recommended because of the proximity of the cribriform plate and the risk of cerebrospinal-fluid leak.

Once the sphenoid sinus is completely opened, the intersinus septum is very cautiously resected, because it can insert into the carotid canal, and damage can precipitate catastrophic injury to the internal carotid artery. Care must be taken not to strip the sphenoid mucosa, avoiding any chance of bleeding.

At this point, the anatomical landmarks within the sphenoid sinus are noted, and the access to the sella turcica is completed.

Opening and Closure of the Sellar Region

The opening of the sella turcica is done by gently fracturing the bony lamina with a bayonet-shaped chisel or a Freer elevator; the opening is then enlarged with a micro-Kerrison punch (Fig. 45.11). This step can be also done with a microdrill, depending on the thickness of the anterior sella wall (Fig. 45.12). The dura mater is widely exposed and is opened in an inverted "U" shape using a bayonet-shaped scalpel handle or micro-cautery with a low intensity (Fig. 45.13). Many authors prefer a cruciate incision [3, 9, 10, 14, 15]. The pituitary tumor is resected using a combination of blunt ring curettes and pituitary forceps (Fig. 45.14). A microdebrider with a special atraumatic tip developed by Stamm (Xomed) is also very useful for removal of the tumor (Fig. 45.15) [32].

If intrasellar bleeding occurs during the resection of the tumor, an instrument with a double function, such as the microdebrider (which has the capability of simultaneous resection and suction), can be helpful. Angled endoscopes are particularly useful at this stage to visualize every step of tumor removal and to look for any residual tumor.

If no CSF leakage has occurred, Surgicel and fibrin glue are used to securely seal the sella (Fig. 45.16). When bleeding or CSF leakage are present, vastus lateralis muscle or fat with fibrin glue may be necessary. If an endoscopic, transnasal, direct approach has been chosen and it has been necessary to partially resect the middle turbinate, the mucoperiostium of the middle turbinate is carefully removed and used to repair the anterior wall of the sella. At the end of the entire procedure, Gelfoam and Merocel are placed. Sometimes, nasal splints are very important to avoid synechia and allow optimal postoperative follow-up. Packing is removed on the second postoperative day and, to prevent pneumocephalus, the patient is instructed not to blow the nose for at least 1 week after surgery. The first postoperative evaluation is performed 7–10 days after surgery with endoscopic control.

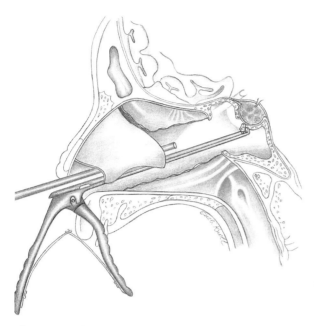

Fig. 45.11. Enlargement of the aperture of the sella turcica with a micro-Kerrison punch during a trans-septal approach

Results and Complications

Transnasal micro-endoscopic surgery for pituitary tumors is a procedure with low morbidity, though it

Fig. 45.12. Opening the anterior wall of the sella turcica with a microdrill

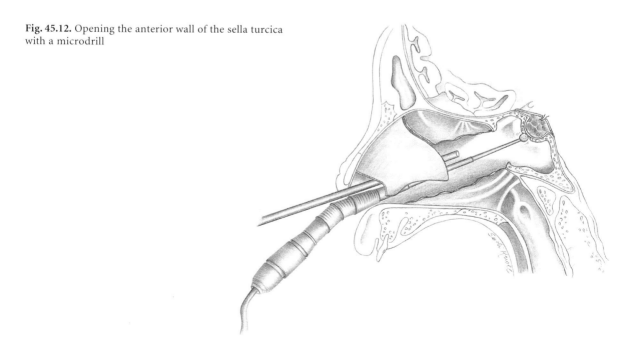

Fig. 45.13. Opening the dura in an inverted "U" shape via microcautery

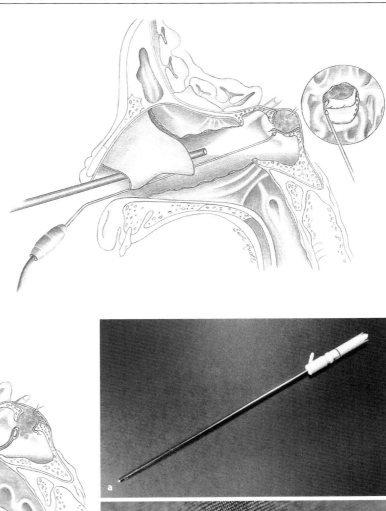

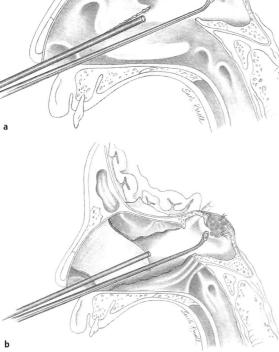

Fig. 45.14. a Resection of the tumor with ring curettes in a transnasal, direct approach. **b** Resection of the tumor with ring curettes in a trans-septal approach

Fig. 45.15 a, b. A special atraumatic cutting blade used in the shaver system (Xomed, Jacksonville)

carries the risk of serious consequences, even when performed by experienced surgeons. The surgeon must be able to recognize and effectively manage any operative incidents in order to avoid complications.

The most common operative complication is CSF rhinorrhea. According to the literature, the incidence

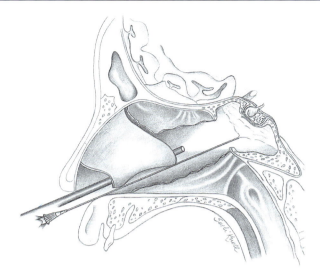

Fig. 45.16. Closure of the sella with Surgicel and fibrin glue

of leakage ranges between 1.5% and 4.4% [38]. The usual cause of CSF rhinorrhea is curettage against the diaphragm of the sella or the use of instruments that remove tissue from the diaphragm. Such maneuvers must be very gentle to avoid damage to this delicate structure. This can occur in patients with microadenomas and in those with large tumors. If CSF leakage is identified, the sella should be packed with abdominal fat and fibrin glue, and continuous lumbar drainage for 5 days should be begun. Black et al. [2] believe that fat retains its viability and is not reabsorbed (as muscle can be). However, many authors utilize muscle [20]. Sometimes, rhinorrhea persists even after lumbar drainage, and re-operation is necessary to seal the leakage. A second complication is pneumocephalus, which can be even more serious than CSF leakage and is the reason patients must be told not to blow the nose after surgery.

Although meningitis is not a common complication after trans-sphenoidal operation, it does occur and can be fatal. The microorganisms include *Staphylococcus aureus*, *Streptococcus pneumoniae* and enteric organisms. Broad-spectrum antibiotics should be used as the management of choice, together with attempts to identify the etiologic agents via CSF cultures.

The incidence of *diabetes insipidus* as a complication of trans-sphenoidal operation ranges from 0.4% to 17% [2, 20]. It can be transient or permanent and, after identification, it should be treated with intranasal administration of desmopressin acetate (1 cm^3 twice per day). Permanent diabetes insipidus probably represents pituitary-stalk injury during the surgical procedure, and transient diabetes insipidus may be a result of excessive manipulation.

Transient or permanent worsening of vision or oculomotor palsy may be present postoperatively. Visual-system complications occur most often as a result of hematoma or direct damage to optic, oculomotor, trochlear or abducens nerves.

Profuse bleeding that is difficult to control should alert one to the possibility of intracranial vascular trauma. Bleeding from the carotid artery should be suspected if the surgeon is working laterally and deeper than 7 cm from the nasal spine in adults. The nose and sinus cavities should be promptly packed, the carotid artery compressed in the neck and hypotensive anesthesia and blood transfusions initiated. If needed, neurosurgical consultation should be obtained, and electroencephalographic monitored arteriography with balloon test occlusion performed [36].

Cavernous sinus bleeding should be suspected when venous bleeding fills the surgical field; it can be repaired with Surgicel, muscle and fibrin glue. Injury of the III, IV or VI cranial nerves can occur when the cavernous sinus is damaged [34]. Direct injury to the hypothalamus is the major cause of surgical death and occurs most often in patients with large tumors characterized by significant intratumoral bleeding [2]. Special care must also be taken to avoid minor postoperative problems that are very frequent in transnasal operations, such as epistaxis, adhesions, sinusitis and septal perforation.

In our department, 62 patients with pituitary-gland lesions diagnosed by laboratory exams and histopathologic examination (including immunohistochemistry) were treated via a transnasal micro-endoscopic surgical technique (Table 45.1). The results showed a rate of success of 93.5% (58 patients; Figs. 45.17–45.19). Residual lesions were observed in four patients, and they underwent a second surgical procedure using the same technique.

Table 45.1. Distribution of pituitary lesions treated via the micro-endoscopic technique ($n=62$)

Lesion	Number
HNA	25
GH	13
Prolactinoma	8
FSH	6
Cushing's disease	5
Chronic inflammatory process	2
LH	2
Rathke's cyst	1

FSH, follicle-stimulating hormone; *GH*, growth hormone; *HNA*, hormone non-functioning adenoma; *LH*, leutenizing hormone

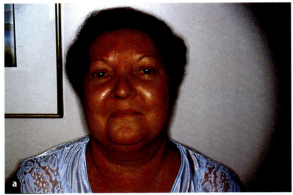

Fig. 45.17 a,b. A 52-year-old female with Cushing's disease. **a** Preoperative appearance. **b** Postoperative appearance 8 months after complete tumor removal

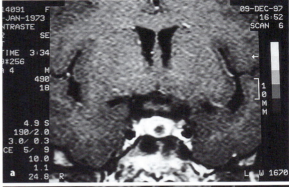
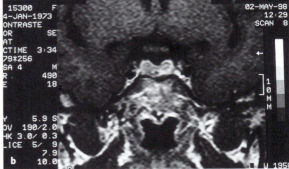

Fig. 45.18 a,b. Coronal magnetic resonance of a 26-year-old female patient, showing a hormone-functioning microadenoma (lutenizing hormone). **a** Preoperative. **b** Postoperative

Four patients (6.4%) experienced CSF leak and were treated intraoperatively immediately; three of them had been submitted to a previous craniotomy. These four patients were treated successfully with external subarachnoid lumbar drainage. Endocrinological disturbances, such as diabetes insipidus and hypopituitarism, occurred in two patients.

One of the patients operated via the direct, transnasal approach presented with profuse nose bleeding 1 week after surgery. The cause of this hemorrhage was an injury of the septal artery, a terminal nasal branch of the maxillary artery located along the anterior wall of the sphenoid sinus, which was diagnosed and coagulated in the operating room, under endoscopic vision.

Final Considerations

Our experience in transnasal micro-endoscopic surgery for pituitary tumors has some advantages relative to the conventional trans-sphenoidal route to the hypophysis [25]. There is a very low morbidity rate associated with tumor removal. Avoidance of a sub-labial incision has resulted in fewer septal perforations and less labial and dental numbness [3, 14, 15, 37]. One additional disadvantage of the standard sub-labial approach is the tendency to orient the tips of the speculum slightly higher on the face of the sphenoid, increasing the risk of CSF leak, creating an optic chiasm lesion and decreasing the chances of total tumor removal.

In our opinion, the advantages of endonasal endoscopic pituitary surgery are manifold when compared with those of microscopic surgery. Some of the optical properties of the endoscope are superior to those of the OM. The endoscopic system provides a panoramic view of the sphenoid sinus, in contrast to the limited visualization of the anterior wall of the sella afforded by microscopic exposure. In addition, the angled endoscopes enable identification of the structures of the lateral sphenoid wall (such as optic protuberances, carotid prominence and the clival region), minimizing the chance of complications.

After removal of a tumor from the sella region, the 30° or 45° endoscope provides excellent visualization of the diaphragm sella, arachnoid membrane and the internal carotid-artery pulsation, facilitating complete resection of the tumor. Pituitary surgery has benefited from the refinements provided by endo-

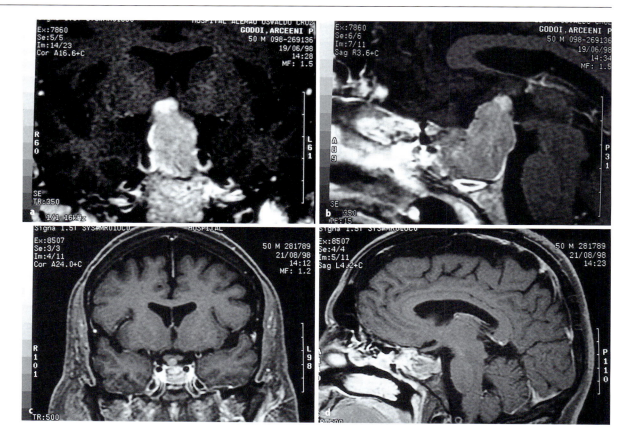

Fig. 45.19 a–d. Magnetic resonance (MR) of a 50-year-old male patient, showing a pituitary macroadenoma. **a** Preoperative coronal MR. **b** Preoperative sagittal MR. **c** Postoperative coronal MR. **d** Postoperative sagittal MR

scopes, computer-guided technology and high-definition television monitors. We believe that endoscopic endonasal pituitary surgery may become the method of choice for treatment of trans-sphenoidal pituitary lesions [1].

References

1. Allen MB Jr, Gammal TE, Nathan MD (1981) Trans-sphenoidal surgery on the pituitary. Am Surg 47:291–306
2. Black PM, Zervas NT, Candia GL (1987) Incidence and management of complications of trans-sphenoidal operation for pituitary adenomas. Neurosurgery 20:920–924
3. Carrau RL, Jho HD, Ko Y (1996) Transnasal trans-sphenoidal endoscopic surgery of the pituitary gland. Laryngoscope 106:914–918
4. Cope VZ (1916) The pituitary fossa, and the methods of surgical approach there to. Br J Surg 4:107–144
5. Cushing H (1912) The pituitary body and its disorders, clinical states produced by disorders of the hypophysis cerebri. Lippincott, Philadelphia, pp 296–305
6. Draf W (1991) Endonasal micro-endoscopic frontal sinus surgery: the Fulda concept. Operative Tech Otolaryngol Head Neck Surg 2:234–240
7. Gamea A, Fathi M, El-Guindy A (1994) The use of the rigid endoscope in Trans-sphenoidal pituitary surgery. J Laryngol Otol 108:19–22
8. Halstead AE (1910) Remarks on the operative treatment of tumors of the hypophysys. With the report of two cases operated on by an oronasal method. Trans Am Surg Assoc 28:73–93
9. Hardy J (1962) L'exérèse des adénomes hypophysaires par voie trans-sphénoidale. Union Med Can 91:933–945
10. Hardy J (1971) Trans-sphenoidal hypophysectomy. J Neurosurg 34:582–594
11. Hirsch O (1910) Demonstration eines nach einer neuen methode operiten hypophysentumors. Verch Dtsch Ges Chir 39:51–56
12. Jankowski R (1995) Endoscopic pituitary surgery. In Stankiewicz JA (ed) Advanced endoscopic sinus surgery. Mosby, Saint Louis, pp 95–102
13. Jankowski R, Auque J, Simon C, Marchal JM, Hepner H, Wayoff M (1992) Endoscopic pituitary tumor surgery. Laryngoscope 102:198–202
14. Jho HD, Carrau RL (1996) Endoscopic endonasal trans-sphenoidal surgery: experience with 50 patients. Neurosurg Focus 1:1–6
15. Jho HD, Carrau RL, Ko Y, Daly MA (1997) Endoscopic pituitary surgery: An early experience. Surg Neurol 47:213–223
16. Kanavel AB (1909) The removal of tumors of the pituitary body by an infranasal route. JAMA 53:1704–1707

17. Kennedy DW (1985) Functional endoscopic sinus surgery. Theory technique, and patency. Arch Otolaryngol 111:576
18. Kennedy DW, Cohn ES, Papel ID, Holliday MJ (1984) Transsphenoidal approach to the sella: the Johns Hopkins experience. Laryngoscope 94:1066–1074
19. Lang J (1989) Clinical anatomy of the nose, nasal cavity and paranasal sinuses. Thieme, New York
20. Laws ER Jr, Kern EB (1976) Complications of trans-sphenoidal surgery. Clin Neurosurg 23:401–416
21. Lekas MD (1988) The nasal gateway for pituitary surgery. Am J Rhinol 2:193–200
22. Mattox DE, Carson BS (1991) Transpalatal trans-sphenoidal approach to the sella in children. Skull Base Surg 1:177–182
23. Messerklinger W (1978) Endoscopy of the nose. Urban and Schwartzenberg, Baltimore
24. Navarro JAC (1997) Seio esfenoidal. In: Navarro JAC (ed) Cavidade do Nariz e Seios Paranasais. AllDent, Bauru, pp 109–116
25. Parnes SM, Koltai PJ (1990) External rhinoplasty approach to trans-sphenoidal hypophysectomy. Ear Nose Throat J 70:438–440
26. Rosegay H (1981) Cushing's legacy to trans-sphenoidal surgery. J Neurosurg 54:448–454
27. Rudert H (1988) Mikroskop und enedoskopgestützte chirurgie der entzündlichen nasennebenhohlenerkrankungen. HNO 36:475–482
28. Schloffer H (1906) Zur frage der operation an der hypophyse. Bruns' Beitr Klin Chir 50:767–817
29. Sethi DS, Pillay PK (1995) Endoscopic management of lesions of the sella turcica. J Laryngol Otol 109:956–962
30. Sethi DS, Pillay PK (1996) Endoscopic pituitary surgery: a minimally invasive technique. Am J Rhinol 10:141–147
31. Sethi DS, Stanley RE, Pillay PK (1995) Endoscopic anatomy of the sphenoid sinus and sella turcica. J Laryngol Otol 109:951–955 1995
32. Shikani AH, Kelly JH (1993) Endoscopic debulking of a pituitary tumor. Am J Otolaryngol 14:254–256
33. Skinner DW, Richards SH (1988) Acromegaly – the mucosal changes within the nose and paranasal sinuses. J Laryngol Otol 102:1107–1110
34. Stamm A, Bordasch A, Vellutini E, Pahl F (1998) Transnasal micro-endoscopic surgery of the pituitary gland. In: Stammberger H, Wolf G (eds) Congress of the European Rhinologic Society (Vienna, 1998). Monduzzi, Bologna, pp 341–347
35. Stammberger H (1991) Functional endoscopic sinus surgery. Decker, Philadelphia
36. Terrel JE (1998) Primary sinus surgery. In: Cummings CW, Fredrickson JM, Harker LA, et al. (eds) Otolaryngology, head and neck surgery. Mosby, St. Louis, pp 1145–1172
37. Tucker HM, Hahn JF (1982) Transnasal, trans-septal sphenoidal approach to hypophysectomy. Laryngoscope 92:55–57
38. Welbourn RB (1986) The evolution of trans-sphenoidal pituitary microsurgery. Surgery 100:1185–1190
39. Wigand EM (1990) Endoscopic surgery of the paranasal sinuses and anterior skull base. Thieme, New York

Midfacial Degloving – Microsurgical Approach

Aldo Cassol Stamm, Moacir Pozzobon, and Oswaldo L. Mendonça Cruz

Introduction

The success of a surgical approach for the management of lesions of the middle third of the face, nasal cavity, paranasal sinuses, nasopharynx, skull base and especially the clivus requires the best possible exposure of the surgical field. This exposure should allow optimal direct visualization of the lesions, which are usually deeply seated.

Among the several surgical technical options, the midfacial degloving approach is very important, because it provides excellent exposure of the middle third of the face, the anterior skull base and most of the clivus. The midfacial degloving approach utilizes a sub-labial incision combined with the intercartilaginous and transfixion incisions frequently used in rhinoplasty procedures [5]. The soft tissues of the face and dorsum of the nose are elevated, exposing the anterior wall of both maxillae without an external incision, thus avoiding scars.

This approach has proved to be useful for the management of benign tumors, such as angiofibroma and inverted papilloma, and for the treatment of maxillo-orbitozygomatic fractures and some malignant tumors. In cases of larger lesions presenting orbital and intracranial involvement, this technique can be combined with various external approaches, such as coronal, frontal, temporal and palatine incisions, providing wider surgical access.

By the end of the last century, Caldwell (in 1893) [8] and Luc in (1897) [16] independently developed an access to the maxillary sinus via a sub-labial incision [4]. Denker and Khaler [11] extended the approach to the nasal cavity, enlarging the Caldwell-Luc procedure and resecting the lateral nasal wall, which improved the local access but was still limited by the soft tissues of the lip, nose and face. The Weber-Fergusson incision (lateral rhinotomy) solved part of this problem by lateralizing the soft tissues of the face. However, this method also had limitations, such as unilateral exposure and the inevitable external facial scar [13]. In 1927, Portmann and Retrouvey [19] carried out a partial maxillectomy via sub-labial access. In 1951, Wilson [28] described transpalatine access for the treatment of nasopharyngeal lesions with a satisfactory exposure. In 1967, Butler et al. [7] described a sub-labial approach combined with the transpalatine access for the removal of angiofibromas with large lateral extensions. In 1974, Casson et al. [9] were the first to report the midfacial "degloving" access for management of midfacial fractures and for reconstructive procedures on the middle third of the face. Conley and Price [10] published the use of this approach to treat 26 tumors, extending its indication and showing its versatility. Sachs et al. [24] described the use of this approach for the removal of 48 cases of inverted papilloma. Allen and Siegel [1] treated a patient with telangiectasia by this approach and removed both lateral nasal walls.

Terzian and Naconecy [27] combined the degloving approach with microsurgical techniques to remove angiofibromas in 25 patients. Maniglia [17] reported 30 cases of radical maxillectomy with or without orbital exenteration through a midfacial access. The approach can be an alternative for central skull base lesions, as described by Price [20], who also reported 29 cases in which he combined midfacial access with frontal and temporal craniotomies for craniofacial resections [20, 21]. Doley et al. [12] describe combining the sublabial incision with modified lateral rhinotomy for cases of tumors extending into the maxillary, ethmoid and sphenoid sinuses. Berghaus and Jovanovic [2, 3] recommend the degloving approach for removing tumors like inverted papilloma and angiofibroma. In 1992, Lenarz and Keiner [15] reported a series of 40 patients operated on via a midfacial access to treat a variety of tumors (juvenile angiofibroma, inverted papilloma, clivus chordoma, esthesioneuroblastoma, squamous-cell carcinoma). After using the approach in 36 patients with different types of lesions, Howard and Lund [14] found it to have significant advantages, including excellent surgical exposure, the possibility of easy modification and extension and the absence of significant cosmetic sequelae if the integrity of the cartilaginous septum is preserved. Rabadan and Conesa [23] described modifications of the classic technique in four areas: antromaxillary, nasal, sphenoidal and clival. The wider access of this approach is achieved mainly via osteotomy of the frontal process of the maxilla, which transforms the nasal cavity and antrum into a single cavity while preserving the functional anatomy of the nose. Nishigawa et al. [18] also presented the degloving technique (in association with other incisions) for "*en bloc*" craniofacial resections. Buchwald et al. [6] proposed a technical modification in which they transected the nasal spine together with the anterior and superior portion of the nasal septum and the superior and inferior lateral cartilage.

Author's Experience

Between 1985 and 1999, 114 patients with various types of tumors and traumatic lesions of the middle third of the face and anterior skull base were treated surgically in our department. Of the 114 patients in this study, 92 were males and 22 were females. The ages ranged from 8 years to 80 years (Table 46.1). The most frequent signs and symptoms were unilateral or bilateral nasal obstruction, epistaxis, proptosis, the presence of a retropharyngeal or nasal mass, anosmia and diplopia. The diagnostic evaluation included clinical history and otolaryngologic physical examination, including surgical microscopy, flexible and rigid endoscopy, conventional X-rays, coronal and axial computed tomography (CT, with and without contrast), magnetic resonance imaging (MRI, with and without paramagnetic contrast, i.e., gadolinium) and digital subtraction arteriography.

CT was performed pre- and postoperatively in most of the patients. MRI and digital arteriography were performed in special cases in which there was suspicion of an intracranial or intradural invasion and in patients who had undergone previous surgery.

Table 46.2 lists the lesions by frequency. Most of the male preponderance was accounted for by angiofibroma, inverted papilloma and trauma. Of the 78 patients with those conditions, only six were women. Less than 5% of the cases had malignancies, although many of the "benign" conditions are very locally aggressive. The surgical procedures following the midfacial degloving approach were partial maxillectomy (in 67 cases), combined procedures (18 cases), trauma repair (eight cases), total maxillectomy (five cases) and clivus resection (four cases; Table 46.3).

Table 46.1. Midfacial degloving approach: ages and genders of patients (n=114)

Age	8–80 years
Gender	
Male	92 (81%)
Female	22 (19%)

Table 46.2. Midfacial degloving approach: diagnoses (1986–1999; n=114)

Lesion	Number	Gender Male	Female
Angiofibroma	48	48	–
Inverted papilloma	22	18	4
Craniofacial fracture	8	6	2
Chordoma	4	2	2
Nasal polyposis	3	2	1
Fibrous dysplasia	3	1	2
Meningioma	3	2	1
Hemangiopericytoma	3	1	2
Esthesioneuroblastoma	3	1	2
Neurolemmoma (schwannoma)	3	2	1
Basilar impression (platybasia)	2	2	–
Osteoma	2	2	–
Odontogenic cyst	2	2	–
Fibromyxoma	1	–	1
Cavernous hemangioma	1	–	1
Malignant conditions	6	3	3
Total	114	92	22

Table 46.3. Midfacial degloving approach: procedures (n=114)

Procedures	Number
Partial maxillectomy	79
Trauma repair	8
Total maxillectomy	5
Clivus resection	4
Combined procedures	18

Surgical Technique

The midfacial degloving approach is performed under general anesthesia. The patient is kept in a supine position, and the head is elevated 20°–30°. A tarsorrhaphy is performed with a 5-gauge or 6-gauge mononylon suture.

First, the nasal cavity is packed with cotton soaked in lidocaine (2%) and epinephrine. The anterior portion of the nasal septum, vestibule, buccogingival sulcus and canine fossa are infiltrated with a (2%) lidocaine solution with 1:100,000 epinephrine. Surgery begins with a complete transfixion incision of the membranous septum (Fig. 46.1). The incision is extended around the piriform aperture to the space between the superior and inferior lateral cartilages. At this stage, a complete circumvestibular incision, which could induce postoperative vestibular stenosis, is avoided (Fig. 46.2). The soft tissues of the nasal dorsum are then elevated in a subperichondreal and subperiosteal plane by using an elevator and a Metzenbaum scissors (Fig. 46.3).

A sub-labial incision extending from the midline to one or both maxillary tuberosities, depending on whether the lesion is unilateral or bilateral, is created (Fig. 46.4). In patients with preserved dentition, the incision can be made at the gingivodental border. The remaining connections between the columella and the anterior nasal spine are dissected transnasally, joining the nasal cavity to the sublabial incision (Fig. 46.5). In the next step, the periosteum is elevated, exposing the anterior maxillary wall, the ascending branch of the maxilla and the piriform fossa (Fig. 46.6). The degloving approach is then completed by elevating the soft tissues of the upper lip, nasal dorsum and superior maxillary region, thus exposing the bony structures of the middle third of the face (up to the infraorbital foramen and the infraorbital rim; Figs. 46.7, 46.8).

Once the bony structures of the middle third of the face are exposed, a variety of surgical procedures can be performed, depending on the diagnosis and extent of the lesion. Some of the techniques include partial or total maxillectomy, trauma repair, clivus resection and combined procedures.

In all cases, the next step is removal of the anterior wall of the maxillary sinus (thus exposing the sinus cavity) and resection of the lateral nasal wall. This resection varies according to the extent of the lesion. The anterior maxillary wall (extending to the

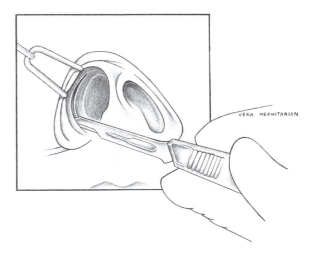

Fig. 46.2. Circumvestibular incision

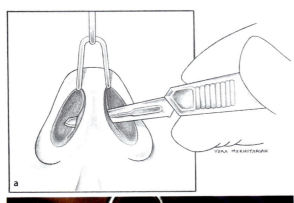

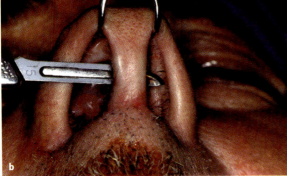

Fig. 46.1. **a** Transfixion incision at the level of the membranous septum of the nose using a scalpel with a no. 15 blade. **b** Surgical view of **a**

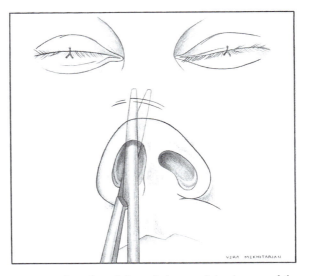

Fig. 46.3. Dissection of the soft tissues of the dorsum of the nose using Metzenbaum scissors

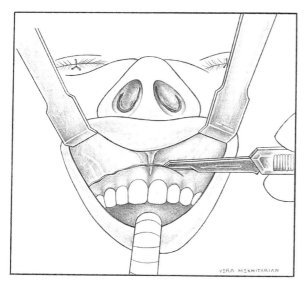

Fig. 46.4. Sub-labial incision using a scalpel with a no. 15 blade

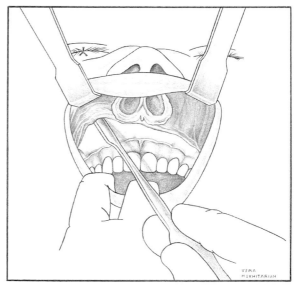

Fig. 46.6. Subperiosteal elevation of the soft tissues of the middle third of the face

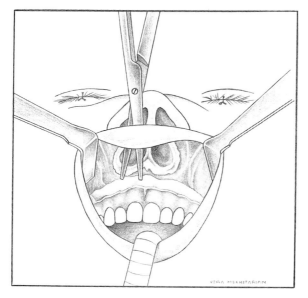

Fig. 46.5. Communication of the nasal and sub-labial incisions using Metzenbaum scissors

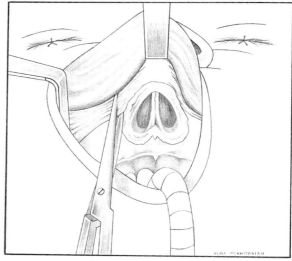

Fig. 46.7. Dissection of the soft tissues of the middle third of the face using a Metzenbaum scissors

inferior orbital rim superiorly, the coronoid process of the maxilla laterally, the dental roots or hard palate inferiorly and ascending branch of the maxilla and the nasal bones medially) is usually completely removed.

The bone can be resected with osteotomes, electric saws or Kerrison forceps. An osteoplastic flap can also be performed, using precise bone incisions in order to preserve a bone flap that can be replaced at the end of the procedure, secured by nylon sutures or miniplates (Fig. 46.9). The next step is to free the lateral nasal wall by detaching the mucoperiostium (degloving) of the piriform aperture and floor of the nasal cavity. After an anterior and medial maxillectomy, the mucoperiostial flap, including the inferior and middle turbinates (which are always preserved), is elevated (Fig. 46.10) [25, 26]. If the lesion extends to the frontal-recess or cribriform-plate areas, a medial, lateral and transverse osteotomy should be performed to gain a wider exposure.

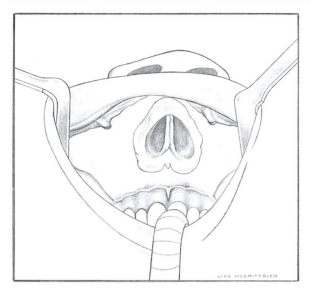

Fig. 46.8. Exposure of the anterior wall of the maxillary sinus, with visualization of the infraorbital nerve and frontal maxillary process

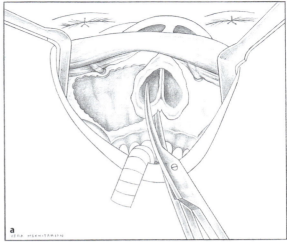

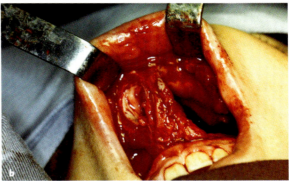

Fig. 46.10. a Incision and dissection of the lateral nasal wall using scissors after performing an anterior and medial maxillectomy. **b** Surgical view after connecting the maxillary sinus with the nasal cavity

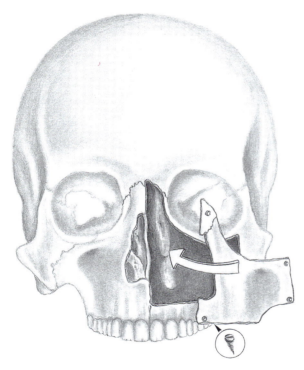

Fig. 46.9. Repositioning of the anterior wall of the maxillary sinus and fixation using miniplates

From this point on, an operating microscope and bipolar coagulation are usually used. In order to avoid trauma to the nasolacrimal duct and the artery of the inferior turbinate, the bone dissection must be as precise and atraumatic as possible during the medial maxillectomy.

After the removal of the anterior and medial walls of the maxillary sinus, the mucosa of the posterior maxillary wall is cauterized circumferentially, and the posterior bony wall is removed with a hammer and chisel, an osteotome or diamond burrs. A Kerrison punch can be used to enlarge the opening, but the periosteum of the posterior wall is preserved to protect the vascular and nervous contents of the pterygomaxillary and zygomatic fossae and avoid fat protrusion into the maxillary cavity. The microscope is used when the periosteal incision is made with a microsurgical blade and for any dissection and ligature or cauterization of vessels [25, 26].

Removal of the vertical plate of the palatine bone helps to identify the sphenopalatine foramen and the terminal branches of the maxillary artery on their way to the nasal cavity. These vessels can be electrocoagulated, which is especially useful in cases of vascular tumors, such as angiofibromas.

The pterygoid process and muscles are identified, and the pterygopalatine and zygomatic fossae are dissected (if necessary) until reaching the infratemporal fossa posteriorly and the floor of the middle cranial fossa superiorly. This approach provides excellent exposure for ethmoidectomy and sphenoidotomy also gives wide access to the nasopharynx.

The main limits of this access are the internal carotid artery and the optic nerve, posterolateral to the sphenoid sinus. In very large midline lesions with significant lateral extension, a bilateral degloving approach may be required (Fig. 46.11). In these cases, the nasal septum must be displaced laterally from the midline, and the separation must be maintained with a self-retaining retractor. At the end of the procedure, the nasal septum should be sutured to the anterior nasal spine with a 4-gauge mononylon suture through a small drill hole on each side of the spine. At the end of the operation, the surgical cavity is inspected with both the operating microscope and angled endoscopes. Rigid, angled endoscopes are used to explore "blind areas" that cannot be seen directly with the microscope and to search for residual disease and minor bleeding. Hemostasis is achieved with bipolar coagulation or pieces of Surgicel and, when the bleeding is from small vessels in the bone, it can be stopped either with bone wax or by drilling with a diamond burr.

When the degloving approach is bilateral, with mobilization of the nasal septum, reconstruction of the cavity begins with fixation of the nasal septum at the nasal spine, as previously mentioned. The middle and inferior turbinates are sutured to the periosteum of the inferior orbital border (Fig. 46.12). The sub-labial incision is closed with an absorbable suture (catgut or vicryl, 3-gauge; Fig. 46.13), starting at the labial frenulum. The transfixion incision of the nasal septum is also closed with an absorbable suture (Fig. 46.14). The circumvestibular incisions are closed with fine, absorbable sutures (4- or 5-gauge).

The surgical cavity is carefully packed and, in some cases, a Foley catheter can be placed into the nasopharynx in order to prevent posterior displacement of the nasal pack. A bandage is placed on the nasal dorsum to promote soft-tissue adhesion to the nasal skeleton, preventing postoperative hematoma.

The Foley catheter is left in place for 24 h, and nasal packing is removed 2–4 days after operation. The patient is instructed to irrigate the nasal cavity with saline solution three or four times per day. Postoperative removal of blood and fibrin clots from the cavity is performed under endoscopic control (with topical anesthesia) to prevent adhesions and stenosis.

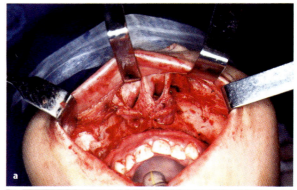
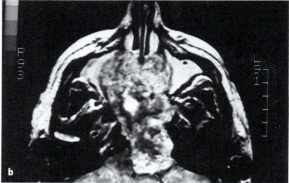

Fig. 46.11. a Surgical view of the bilateral degloving approach. **b** Axial magnetic resonance image showing an extensive lesion (clivus chordoma) occupying the sphenoid and ethmoid sinuses and clivus, with compression of the anterior brain stem. **C** Sagittal magnetic resonance of the same lesion (clivus chordoma)

The midfacial degloving approach can be modified, depending on the size and extent of the lesion. If warranted, the sub-labial incision may be unilateral or bilateral and may be extended laterally to the level of the third molar (or beyond this point). The midfacial degloving approach can be used in conjunction with neurosurgical procedures for the treatment of craniofacial tumors (for example, with a frontobasal incision to access the anterior fossa or with an orbitozygomatic approach for middle-fossa

Fig. 46.12. a Fixing the middle and inferior turbinates to the periosteum of the inferior orbital rim. **b** Surgical view of the fixation of the nasal turbinates to the infraorbital periosteum. MT, middle turbinate; IT, inferior turbinate. **c** Axial computed tomography showing a postoperative view of the nasal turbinates in position

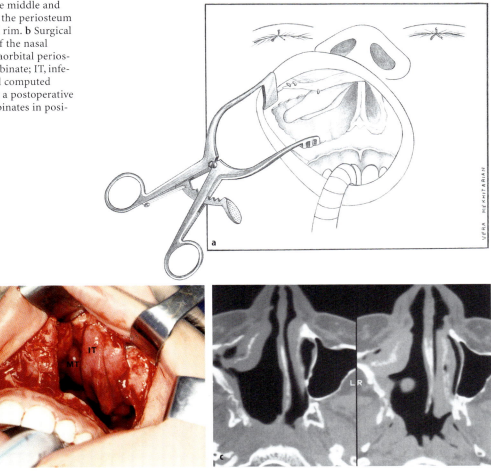

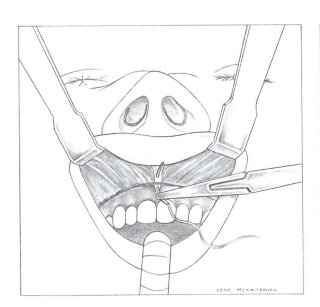

Fig. 46.13. Suture of the sub-labial incision at the labial frenulum with vicryl sutures (3-gauge)

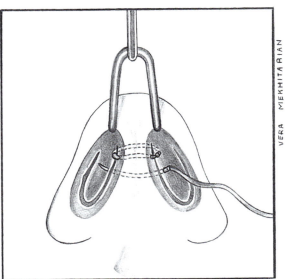

Fig. 46.14. Suture of the transfixion and vestibular incisions with absorbable sutures (4- or 5-gauge)

tumors extending into the infratemporal or nasopharyngeal areas).

Case Reports

The following four clinical cases serve to exemplify some indications for the midfacial degloving approach and to illustrate the results.

Case 1

A 15-year-old boy with a 6-month history of nasal obstruction and diplopia was referred. The otolaryngologic exam revealed a vascular, smooth, violet-colored tumor occupying the right nasal cavity and displacing the nasal septum toward the left. The coronal and axial CT showed a large naso-sinus meningioma (Fig. 46.15a). The histopathologic exam confirmed the diagnosis. The postoperative coronal CT showed complete resection of the lesion and no recurrence 5 years later (Fig. 46.15b).

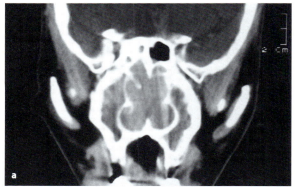

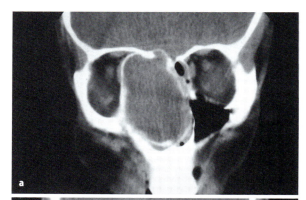

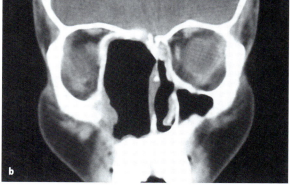

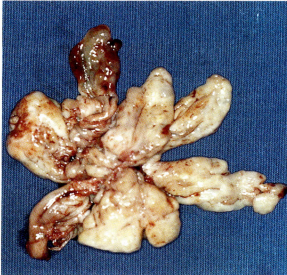

Fig. 46.16. a Coronal computed tomography (CT) demonstrating involvement of the sphenoid and ethmoid sinuses bilaterally, with partial destruction of the nasal septum. **b** Surgical specimen (inverted papilloma). **c** Postoperative coronal CT showing the preservation of the right inferior turbinate and part of the posterior bony septum

Fig. 46.15 a,b. Coronal computed tomography showing the pre- and postoperative appearances of a naso-sinus meningioma removed via a midfacial degloving approach

Case 2

A 56-year-old man with a 1-year history of nasal obstruction and bleeding was referred. The otorhinolaryngologic exam revealed a firm polypoid mass occupying the left nasal cavity. Coronal CT showed an extensive lesion occupying the nasal cavity and paranasal sinus with destruction of the nasal septum (Fig. 46.16a). A histopathologic exam revealed an inverted papilloma (Fig. 46.16b). The postoperative coronal CT showed resection of the middle turbinate and no residual lesion (Fig. 46.16c).

Case 3

A 15-year-old boy presented with a 1-year with history of left nasal obstruction and epistaxis. The otolaryngologic exam revealed a firm, smooth, vascular tumor occupying the left nasal cavity and nasopharynx. Coronal and axial CT showed a large tumor of the sphenopalatine and nasopharyngeal regions (Fig. 46.17a,b). The histopathologic diagnosis was angiofibroma (Fig. 46.18). Postoperative coronal and axial CT after tumor removal is shown in Fig. 46.17c,d.

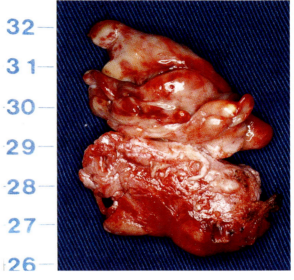

Fig. 46.18. Surgical specimen (angiofibroma)

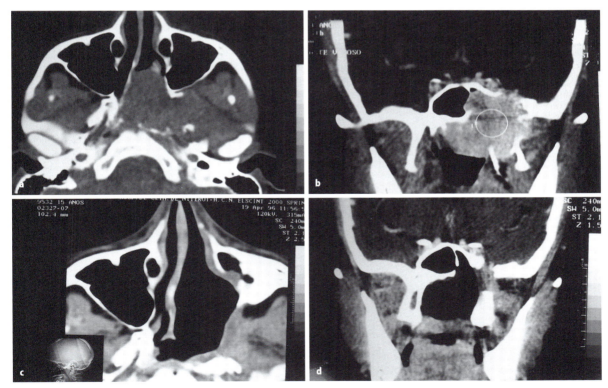

Fig. 46.17. a Axial computed tomography (CT) demonstrating a tumor (angiofibroma) occupying the nasal cavity, nasopharynx, pterygopalatine and zygomatic fossae, with widening of the sphenopalatine foramen. **b** Coronal CT showing the lateral extension of the angiofibroma in relation to the cavernous sinus and ipsilateral internal carotid artery. **c,d** Postoperative axial and coronal CT demonstrating complete removal of the angiofibroma using the midfacial degloving approach

Case 4

A 54-year-old man with a 5-year history of bilateral nasal obstruction and diplopia was seen. The otolaryngologic exam revealed left ocular proptosis, facial asymmetry and a polypoid mass occupying both nasal cavities, as shown by coronal and axial CT (Fig. 46.19a,b). The surgical treatment combined the midfacial degloving approach with a frontal craniotomy. Histopathologic examination revealed nasosinus osteoma and inflammatory polyps. Postoperative coronal and axial CT showed complete resection of the lesion (Fig. 46.19c,d).

Complications of the Midfacial Degloving Approach

One of our patients died due to severe hemorrhage. This patient had a very large angiofibroma infiltrating the internal carotid artery and cavernous sinus.

Significant bleeding occurred in four patients; three of them required blood transfusion, with a mean of 1500 ml/person. The most frequent complications (according to the patients) are nasal crusting and infraorbital or dental paresthesias, which usually improve within 2–8 months after operation.

Vestibular stenosis that did not require surgical correction occurred in one patient. One patient developed a columellar abscess, and another experienced epiphora. Septal perforations, hyposmia and anosmia were occasionally seen (Table 46.4).

Table 46.4. Midfacial degloving approach: sequelae and complications

Paresthesia (infraorbital; dental numbness)
Nasal crusting
Sinonasal secretion
Subcutaneous hematoma
Facial edema
Bleeding (trans- or postoperative)
Vestibular stenosis
Epiphora
Septal perforation
Hyposmia and anosmia

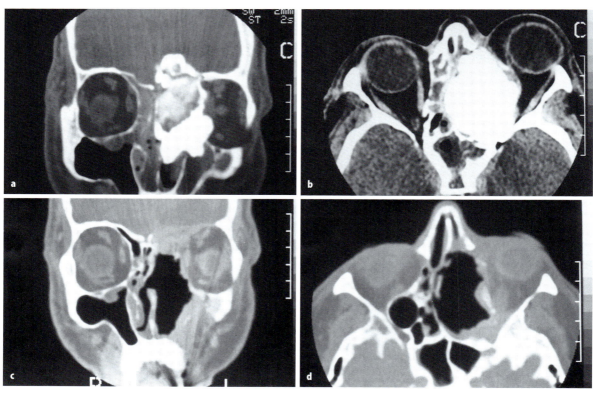

Fig. 46.19 a,b. Coronal and axial computed tomography (CT) showing an osteoma occupying the ethmoid sinus and part of the maxilla, with erosion of the medial orbital wall, proptosis and invasion of the anterior skull base, nasal cavity, ethmoid sinus and left maxillary sinus (with polyps). **c,d** Coronal and axial CT showing the postoperative surgical cavity

Conclusions

We believe that the midfacial degloving approach:
1. Is an excellent approach to several anatomical regions, including the nasal cavity, paranasal sinuses, nasopharynx and skull base
2. Can be used in combination with other surgical approaches (transpalatine and lateral infratemporal craniotomies)
3. Allows bilateral access to the naso-facial structures, without leaving a facial scar
4. Is the preferable approach for treatment of angiofibroma and some cases of inverted papilloma when performed with the aid of the operating microscope and a bipolar coagulation system

References

1. Allen GW, Siegel GJ (1981) The sublabial approach for extensive nasal and sinus resection. Laryngoscope 91:635–640
2. Berghaus A (1990) Midfacial degloving. HNO 38:7–11
3. Berghaus A, Jovanovic S (1991) Technique and indications of extended sublabial rhinotomy. Rhinology 29:105–110
4. Bernstein L (1971) The Caldwell-Luc operation. Otolaryngol Clin North Am 4:69
5. Bingham BJG, Griffiths MV (1989) Sublabial rhinotomy septal transfixion as on approach to the nasal fossa, paranasal sinuses and nasopharynx. J Laryngol Otol 103:661–663
6. Buchwald C, Bonding P, Kirkby B, Fallentim E (1995) Modified midfacial degloving. A practical approach to extensive bilateral benign tumors of the nasal cavity and paranasal sinuses. Rhinology 33:39–42
7. Butler RM, Nahum AM, Hanafee W (1967) New surgical approach to nasopharingeal angiofibroma. Trans Am Ophthalmol Soc 71:92
8. Caldwell GW (1893) Discares of the accessory sinuses of the nose and improved method of treatment for suppuration of the maxillary antrum. NY Med J 4:18–93
9. Casson PR, Bonnano PC, Converse JM (1974) The midfacial degloving procedure. Plast Reconstr Surg 53:102–113
10. Conley J; Price JC (1979) Sublabial approach to the nasal and nasopharyngeal cavities. Am J Surg 138:615–618
11. Denker A, Khaler O (1926) Handbuch der Hals-Nasen-Ohren-Heikunde. Springer, Berlin Heidelberg New York
12. Doley PJ, Riding K, Kahn K (1987) Management of nasopharyngeal angiofibroma. J Otolaryngol 6:224
13. Fergunson W (1845) A system of practical surgery. Operations of the upper jaw, 2nd edn. Lea and Blanchard, Philadelphia
14. Howard DJ, Lund V (1992) The midfacial degloving approach to sinonasal disease. J Laryngol Otol 106:1059–1062
15. Lenarz TH, Keiner S (1992) Midfacial Degloving: ein alternativer zugangs weg zur frontobasis, der Nasenhaupt-und den Nasennebenhöhlen. Laryngorhinootologie 71:381–387
16. Luc H (1897) Une nouvelle methode operatoire pour la cure radicale de l'empyeme chronique du sinus maxillaire. Arch Laryngol 6:275
17. Maniglia AJ (1986) Indications and midfacial degloving: a 15-year experience. Otolaryngol Head Neck Surg 112:750–752
18. Nishikawa K, Nishioko S, Aoji K, Koike S, Nameki H (1993) Skull base surgery using the degloving technique – an approach without facial scarring. Nippon Jibiinkoka Gakkai Kaiho 96:1447–1456
19. Portman G, Retrouvey H (1927) Le Cancer du Nez. Doin, Paris
20. Price JC (1986) The midfacial degloving approach to the central skull base. Ear Nose Throat J 65:176–180
21. Price JC (1987) Midfacial degloving approach. In: Goldman JL (ed) The principles and practice of rhinology. Wiley, New York, pp 615–638
22. Price JC, Holliday MJ, Kennedy DW (1988) The versatile midface degloving approach. Laryngoscope 98:291–295
23. Rabadan A, Conesa H (1992) Transmaxillary–transnasal approach to the anterior clivus: a microsurgical anatomical model. Neurosurgery 30:473–482
24. Sachs ME, Conley J (1984) Degloving approach for total excision of inverted papilloma. Laryngoscope 94:1595
25. Stamm A (1991) Abordaje quirurgico mediofacial "degloving" de la base de craneo. An Otorrinolaringol (Uruguay) 57:75–84
26. Stamm A, Mekhitarian L (1995) Acesso Microcirúrgico Mediofacial-Degloving. In: Stamm A (ed) Microcirurgia Naso-Sinusal. Revinter, Rio de Janeiro, pp 331–342
27. Terzian AE, Naconecy C (1985) Juvenile nasopharyngeal angiofibroma; microsurgical approach in 25 cases as unique treatment. In: Meyer EN (ed) New dimensions in otorhinolaryngology head and neck surgery. Elsevier, New York, pp 505–506
28. Wilson CP (1951) The approach to the nasopharynx. J R Soc Med 44:353

Complications of Micro-endoscopic Sinus Surgery

Aldo Cassol Stamm

"Theoretically the paranasal sinuses operation is easy, however, it has proved to be one of the easiest ways to kill a patient" [13].

Introduction

The paranasal sinuses present a complex anatomy surrounded by important structures, such as the orbit and its contents, the anterior and middle cranial fossae, cranial nerves I–VI, vascular structures (including the internal carotid arteries, the anterior and posterior ethmoid arteries, the maxillary arteries and their branches) and the cavernous sinus.

Because of the strategic position of the paranasal sinuses, the risk of damaging the surrounding structures during surgery must be considered. Severe headache, ophthalmoplegia, proptosis, ecchymosis, alteration of pupil size, severe bleeding, loss of vision and changes in behavior are signs and symptoms of complications that necessitate immediate imaging studies or consultation with ophthalmologists or neurologists. Complications of micro-endoscopic surgery of the paranasal sinuses can be classified according to severity as *minor* or *major* and according to the time of appearance as *immediate* or *delayed*.

Minor complications are those that present little morbidity and do not compromise the life of the patient. They occur in between 2% and 21% of patients who undergo endoscopic surgery of the paranasal sinuses [3, 19, 20]. Minor complications include synechias, crusts, minor bleeding, nasal-septum perforation, headache, facial pain, alteration of dental sensitivity, edema, local infection, periorbital ecchymosis, palpebral edema, subcutaneous emphysema, stenosis of sinus ostia, hyposmia, epiphora, exacerbation of bronchial asthma and postoperative sinusitis.

Major complications present significant morbidity and a possibility of mortality. The principal major complications are:
1. Orbital: injury to the optic nerve or the extraocular muscles, resulting in orbital hematoma, diplopia, proptosis, decrease of visual acuity or blindness
2. Intracranial: cerebrospinal fluid (CSF) leakage, intracranial hemorrhage and hematoma, damage to the brain itself, meningitis, cerebral abscess, damage to the olfactory nerve, injury to cranial nerves III, IV, V or VI, pneumocephalus or stroke
3. Bleeding: injury to the arteries (internal carotid, ethmoidal and maxillary arteries, including their branches), damage to the cavernous sinus or intracranial bleeding

Studies comparing the various techniques of intranasal sphenoethmoidectomy reveal similar incidences of major complications, which vary from 1% to 3% [9]. Other authors, such as Wigand [25], Stammberger [17], Vleming et al. [24], May et al. [12] and Kinsella et al. [8], report an incidence between 0.75% and 8%.

Table 47.1 displays the complications associated with micro-endoscopic surgery of the paranasal sinuses via different surgical techniques, as reported by different authors. Since the major complications are similar in the different techniques, we should consider how radical each technique is in addition to considering its surgical results.

The most frequent *immediate* complications are: CSF leakage, intraoperative bleeding, orbital hematoma and injury to the brain. *Delayed* complications include progressive loss of vision or smell, meningitis, bleeding, synechia and infection. Table 47.2 lists the major complications that we observed in 632 consecutive patients operated on between 1985 and 1999 due to naso-sinusal polyposis.

Preoperative Evaluation

Prevention of the complications of micro-endoscopic sinus surgery begins with an adequate preoperative evaluation of the patient. The diagnostic methods most useful for this purpose are the clinical his-

Table 47.1. Complications of micro-endoscopic sinus surgery, as reported by different authors

Authors	Year	Surgery type	Complications Major (%)	Minor (%)
Heermann and Neues [6]	1986	Intranasal microsurgery	0.4	NR
Stevens and Blair [21]	1988	Traditional endonasal	3.0	4.3
Friedman and Katsantonis [5]	1990	Traditional endonasal	0.94	2.06
Friedman and Katsantonis [5]	1990	Traditional transantral	1.45	5.8
Stammberger and Posawetz [18]	1990	Endoscopic surgery	0.15	NR
Wigand [25]	1990	Endoscopic surgery	2.2	NR
Lawson [9]	1991	Traditional endonasal surgery	1.1	NR
Levine and May [10]	1993	Endoscopic surgery	0.85	6.9
Draf [4]	1993	Intranasal micro-endoscopic surgery	2.3	NR
Levine and May [10]	1993	Traditional surgery	1.3	2.8
Levine and May [10]	1993	Endoscopic surgery	1.1	5.4
Stamm [15]	1995	Transnasal micro-endoscopic surgery	0.94	NR

NR, not reported

Table 47.2. Complications of micro-endoscopic sinus surgery, from the author's series (nasal polyposis; n=632)

	Number	Percentage
CSF Leak	6	0.94
Hemorrhage (major)	6	0.94
Meningitis	1	0.15
Tension pneumocephalus	1	0.15
Smell disturbance	3	0.47
Facial pain	3	0.47
Asthma attack (postoperative)	3	0.47
Periorbital ecchymosis	5	0.79
Periorbital emphysema	1	0.15

tory (use of medications, previous operations, etc.), sino-nasal endoscopy and imaging studies – principally high resolution computed tomography (CT) in coronal and axial projections, with sagittal reconstruction. Magnetic resonance (MR), should always be considered if an intracranial complication is suspected.

Tomography provides the best information for the planning surgery and provides medico-legal documentation. Anatomic variations (especially of the sphenoid and ethmoid sinuses) are common and can be reviewed and studied on the images. CT studies are mandatory in paranasal-sinus surgery. Sinus endoscopy permits the surgeon to evaluate the nasosinusal mucosa and the potentially critical areas that will be approached during the operation, especially those that have the greatest possibility of causing major bleeding.

Major Complications

Orbital Complications

An orbital complication has occurred whenever one can detect (either intra- or postoperatively) ophthalmoplegia, proptosis, ecchymosis, alterations of pupil diameter or loss of vision. In these cases, immediate ophthalmological evaluation and imaging studies are necessary [22–24].

Orbital complications usually result from removal of a portion of the orbital wall of the ethmoid sinus, the lateral wall of the sphenoid sinus or the roof of the maxillary sinus. Orbital complications range from orbital edema to sectioning of the optic nerve.

Accidental fracture or removal of the orbital wall of the ethmoid sinus (lamina papyracea) can occur without causing problems if the periosteum remains intact. Damage to the orbit can occasionally occur during middle-meatal antrostomy, usually when the maxillary sinus is small and the inferior turbinate has a high bony insertion.

The greatest danger of injury to the orbital cavity occurs during ethmoidectomy, especially in previously operated patients. This occurs because of the extensive and intimate anatomic relationship between the ethmoid sinus and the orbit and because of the loss of surgical landmarks in previously operated patients. During ethmoid-sinus surgery for massive polyposis in which the periosteum has been disrupted, there is a danger that the surgeon may not be able to differentiate orbital fat from polyps. Inadvertent removal of fat can result in damage to the medial rectus muscle, resulting in diplopia or direct or indirect damage to the cranial nerves (which may require between 6 months and 1 year to recover). Lesions of the extraocular muscles and the cranial nerves that fail to recover are indications for ophthalmologic surgery. Other complications include the disruption of blood vessels and their retraction into the orbit, producing hematoma with acute proptosis in the orbital cavity. In this case, it is necessary to remove the nasal packing and perform external ethmoidectomy to alleviate the intraocular pressure. Lateral canthotomy is indicated in emergency situations when the patient experiences acute postoperative hemorrhage with loss of vision [21]. If the surgeon continues to remove orbital fat, damage to the optic nerve (with consequent loss of vision) can occur; this is usually irreversible if the nerve is damaged directly. To avoid major complications during surgery, it is important to keep the patient's eyes uncovered and to delicately palpate the globe. In the case of a defect of the orbital wall, with disruption of the periorbita, the orbital contents protrude into the ethmoid sinus (Fig. 47.1). However, unduly prolonged compression of the globe can result in bradycardia (oculovagal reflex).

Blindness

Loss of vision resulting from naso-sinus surgery can be *temporary* or *permanent*. When it is due to orbital hematoma, the blindness can be reversible. Direct or indirect damage to the optic nerve usually occurs from the use of forceps or electrocautery on the superior–lateral sphenoid sinus wall or in the posterior ethmoid cells, especially the Onodi cell (Fig. 47.2). In these situations, loss of vision can be partial or total. Another cause of blindness during ethmoidectomy occurs when a surgical instrument penetrates the orbital cavity and produces damage or section of the optic nerve (Fig. 47.3). When there is no direct injury to the optic nerve, vision can recover; nasal packing is removed, mega-doses of steroids are administered, and the optic nerve is decompressed.

Orbital Hematoma

Inadvertent entry into the orbital cavity by removal of the lamina papyracea with opening of the periosteum and removal of fat can also produce damage to vascular structures, resulting in hemorrhage and the development of a retrobulbar hematoma. Another type of orbital hemorrhage occurs when the anterior ethmoid artery is damaged and retracts into the interior of the orbit. In this case, the hematoma can develop immediately or several days after the trauma (Fig. 47.4). This complication can be manifested by alterations of the conjunctiva, pupillary dilatation, proptosis or progressive loss of vision. For these reasons, the patient must be continuously observed.

Treatment of these complications must be undertaken as quickly as possible, between 1 h and 2 h

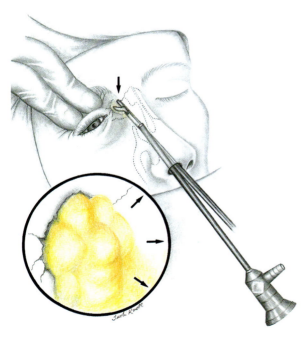

Fig. 47.1. When the orbital periosteum is injured, gentle palpation of the globe demonstrates orbital fat in the ethmoid sinus

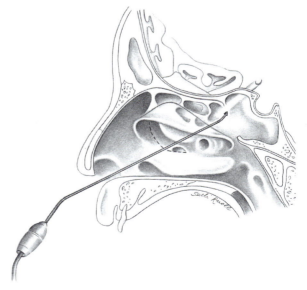

Fig. 47.2. Working with a monopolar cautery system too close to the optic nerve canal is not advisable

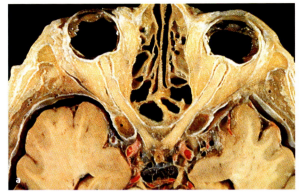

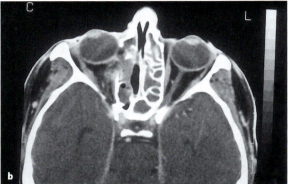

Fig. 47.3. a Anatomic transverse section through the ethmoid sinus (seen from above), showing the relationship of the optic nerves to the paranasal sinuses. **b** Axial computed tomography of a patient with nasal polyposis after sectioning of the right optic nerve during ethmoidectomy

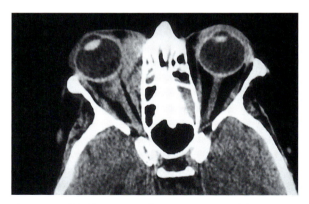

Fig. 47.4. Orbital hematoma due to retraction of the anterior ethmoid artery into the orbit after its violation

that the ocular pressure will diminish 15–30 min after the start of the infusion. When attempts to reduce orbital pressure are unsuccessful and acute loss of vision occurs, lateral canthotomy and cantholysis of the lower lid are indicated to temporarily relieve intraorbital pressure. External ethmoidectomy with orbital decompression and ligation of the anterior and posterior ethmoid arteries is indicated when these methods are unsuccessful.

Diplopia

Diplopia occurs primarily as a result of damage to the extraocular muscles or the III, IV or VI cranial nerves. The muscle most commonly affected is the medial rectus, and the superior oblique is involved least often. Extraocular-muscle damage can be direct or indirect (through the nervous or vascular supply). Indirect lesions are most frequently produced by heat transmitted via monopolar electrocautery or the use of a laser. The region where direct and indirect injuries most commonly occur is at the level of the mid-portion of the orbital wall of the ethmoid sinus (lamina papyracea). To investigate the possibility of reparative surgery, persistent diplopia needs to be evaluated by CT (Fig. 47.5) and by an ophthalmologist.

Intracranial Lesions

Damage to the anterior cranial base can produce leakage of CSF, meningitis, tension pneumocephalus, brain hemorrhage, neurologic deficit, brain abscess or death [1, 2, 11, 15, 16, 18–20].

CSF Leakage

One of the most frequent neurologic complications that occurs during intranasal sinus surgery is CSF leakage. The usual location is at the level of the roof of the ethmoid sinus (fovea ethmoidalis) and the cribriform plate; less commonly, it occurs in the sphenoid sinus. The majority of CSF leaks occur during ethmoidectomies during which the dura mater is disrupted with forceps or by the inappropriate use of electrocautery along the roof of the ethmoid sinus.

The CSF leaks that arise from the cribriform plate can be produced the same way those of the roof of the ethmoid sinus are produced. To avoid them, the surgeon must have the insertion of the middle turbinate as a reference, recognizing that the cribriform plate is located between the nasal septum and

after the onset of the hematoma, in order to avoid secondary damage to the optic nerve [22, 23]. Removal of nasal packing is the first step, followed by consultation with an ophthalmologist. If the loss of vision progresses, we recommend the use of mannitol (0.5–1.0 mg/kg) and corticosteroids. It's hoped

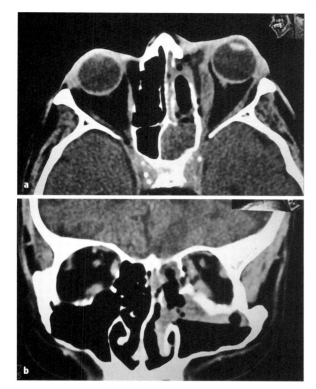

Fig. 47.5 a,b. Damage of the medial rectus muscle following sphenoethmoidectomy. **a** Axial computed tomography (CT) view. **b** Coronal CT section

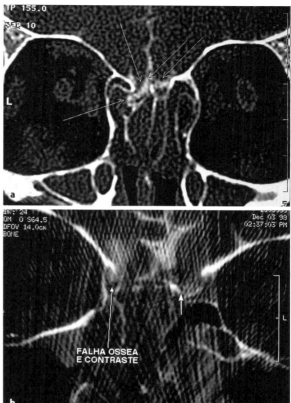

the septal face of the middle turbinate. The most frequent sites of iatrogenic CSF leakage are (1) in the area of the insertion of the middle turbinate into the cranial base, (2) in the region of the anterior ethmoid artery and (3) near the lateral lamella of the olfactory fossae (Fig. 47.6).

When a CSF leak is identified, it must be repaired during the same operation. Regardless of its location, correction of an iatrogenic leak follows the same principles. Fistulas smaller than 5 mm are repaired using a mucoperiosteal graft (from the middle or inferior turbinate) and fibrin glue. Defects greater than 5 mm in diameter are corrected with fascia lata and a mucoperiosteal graft. In both cases, nasal packing is mandatory. In large fistulas with a high flow of CSF, a lumbar catheter is used for 2–3 days to provide drainage and to stop the flow through the defect [15, 16].

Meningitis

All patients who develop CSF leaks can potentially develop meningitis once the natural barriers of the dura and arachnoid are opened. Accordingly, it is

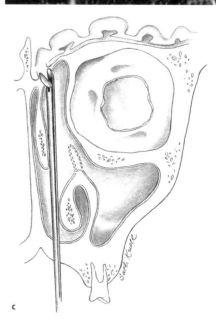

Fig. 47.6. a Coronal computed tomography cisternography, showing a CSF fistula at the fovea ethmoidalis after an ethmoidectomy for treatment of polyposis. **b** CSF fistulas in both fovea ethmoidalises following surgery for nasal polyposis. **C** The most common site of penetration is at the junction of the anterior ethmoid artery and the lateral lamella of the cribriform plate

advisable to give prophylactic antibiotics until the leak is totally closed. If a CSF leak develops during surgery on infected paranasal sinuses, it is advisable to administer antibiotics in doses equivalent to those for treating meningitis.

Tension Pneumocephalus

The presence of air in the intradural space indicates that disruption of the dura mater has occurred, whether or not the patient has evidence of a CSF leak. Determination of the extent of the pneumocephalus can be accomplished using a simple anterior–posterior projection X-ray, though CT and MR imaging are more accurate techniques for this determination. If the pneumocephalus increases, the defect is still open. Pneumocephalus drainage is advisable when there is brain compression due to hypertensive pneumocephalus (Fig. 47.7) [1].

Brain Abscess

A brain abscess can be secondary to direct injury of cerebral tissue or can be a consequence of venous extension of infection in the sinuses. The abscess can be treated medically or surgically. The choice of medical treatment depends on its location, extent and clinical evolution. When the response has not been satisfactory, with persistence of fever, headache and neurologic manifestations despite medical treatment, surgery is indicated (Fig. 47.8).

Cranial-Nerve Injuries

The most frequent injury to a cranial nerve is damage to the terminal branches of the olfactory nerves, usually accompanied by a CSF fistula. Involvement of the III, IV and VI nerves is secondary to an injury of the cavernous sinus. The infraorbital branch of the V nerve may be traumatized in the lateral wall of the sphenoid sinus or as a result of surgical manipulation at the roof of the maxillary sinus during surgery via a maxillary antrostomy. To minimize this risk, these manipulations should only be performed under direct endoscopic vision.

Brain Injury

Inadvertant intradural penetration can cause direct damage to cerebral tissue and is a potentially fatal complication [11]. The effect of simple penetration with a forceps, without removal of brain tissue, may only be cerebral edema; however, damage of small

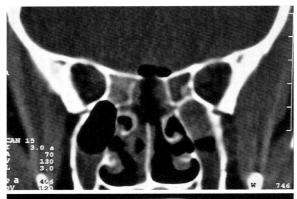

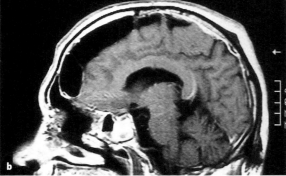

Fig. 47.7. a Small amount of pneumocephalus treated conservatively following septoplasty and ethmoidectomy. **b** Sagittal magnetic resonance, showing tension pneumocephalus provoked by vigorous nose blowing after sphenoethmoidectomy. Frontal trephination was necessary to drain the pneumocephalus

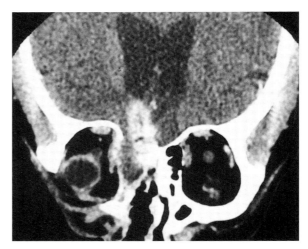

Fig. 47.8. Coronal computed tomography of a patient with a brain abscess secondary to frontal-sinus surgery and ethmoidectomy

blood vessels can result in sub-arachnoid hemorrhage, the most important symptom of which is headache. In these cases, neurologic consultation and imaging studies are mandatory. In the patient shown in Fig. 47.9, the surgical instrument entered the brain and caused sub-arachnoid hemorrhage. The patient had intense, persistent headaches and successfully responded to medical treatment.

When brain tissue is damaged or removed, the patient presents a clinical picture of transient or permanent neurologic deficit or even death. Figure 47.10 shows pre- and postoperative CT of a patient who experienced frontal-lobe damage associated with pneumocephalus, CSF leakage and meningitis. This patient was operated on to repair the CSF leak; however, he continued to have permanent neurologic deficits.

Hemorrhage

Intraoperative or postoperative bleeding resulting from operations on the paranasal sinuses originates from the internal-carotid-artery system via the anterior or posterior ethmoid arteries, which are branches of the ophthalmic artery. Bleeding can also originate from the external carotid artery via the maxillary artery and its terminal branches.

There is an increased risk for bleeding in patients who have undergone previous surgery, due to the formation of new vessels in the operated area. Due to retraction of the ethmoid artery into its canal, bleeding from the artery can be difficult to identify when it occurs at the point where it leaves the canal. Control of this type of bleeding is undertaken with electrocautery under microscopic or endoscopic vision, using great care not to introduce the points of the electrocautery forceps into the arterial canal, which can result in a CSF leak. When cautery is performed at the level of the posterior ethmoid artery, propagation of heat can damage the optic nerve. If it is impossible for bleeding originating from the ethmoid arteries to be controlled via the intranasal route, it is advisable to utilize an external approach, electrocoagulating the anterior and posterior ethmoid arteries with bipolar cautery and the operating microscope. In operations involving the nasal turbinates, it is important to use preoperative meas-

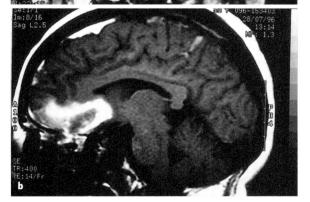

Fig. 47.9 a,b. Subarachnoid hemorrhage following ethmoidectomy. **a** Coronal magnetic resonance (MR). **b** Sagittal MR

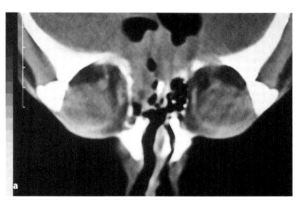

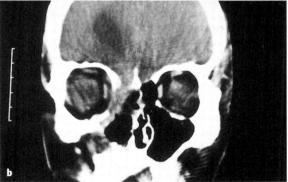

Fig. 47.10. **a** An injury of the frontal lobe, associated with pneumocephalus, CSF leak and meningitis in a patient after surgery of the maxillary and ethmoid sinuses. **b** Coronal computed tomography after operative repair of the dural defect, with subsequent frontal-lobe gliosis

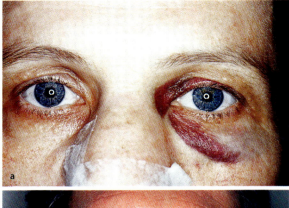
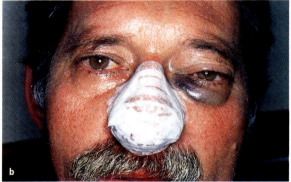

Fig. 47.12 a,b. Periorbital ecchymosis in two patients following re-operations for massive polyposis

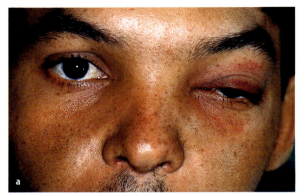
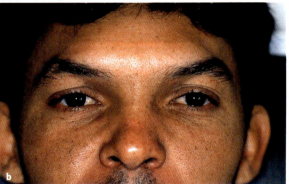

Fig. 47.13. a A 21-year-old patient with upper-eyelid edema after left-ethmoid-sinus surgery. b The same patient, following conservative medical treatment

head of the inferior turbinate is left intact. Injuries of the nasolacrimal system that result in epiphora can be treated by canalization of the duct or by transnasal dacryocystorhinostomy [14].

Epistaxis (Minor)

Minor epistaxis during the postoperative period is common and may become significant after removal of crusts. Regenerating mucosa forms small areas of granulation tissue that can result in small, spontaneous episodes of bleeding for 4–6 weeks. More severe bleeding requires packing, surgery or even blood transfusion.

Facial or Dental Pain

Facial or dental pain occurs most commonly in the malar region and the anterior wall of the maxillary sinus; it results from injuries to the infraorbital nerve and its alveolar branches, from excessive dissection of the interior of the maxillary sinus or, occasionally, from puncture of the canine fossa.

Disturbance of Smell

Disturbance of smell can occur after surgical procedures in the olfactory region, such as removal of polyps, tumors or mucous membrane between the septal surface of the medial turbinate and the roof of the nose. The diminution of olfaction can be temporary or permanent. In the case of simple edema with obstruction and reduction of air flow in the olfactory region, the use of steroids can restore the sense of smell [10, 12, 17].

Risk Factors

- Previous sinus surgery
- Absence of anatomic landmarks
- Bleeding
- Infection
- Extensive disease
- Inexperience of the surgeon

Complications of micro-endoscopic naso-sinus surgery can occur during the operation or the immedi-

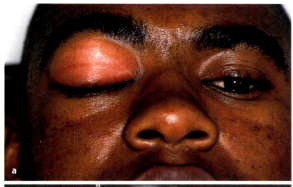

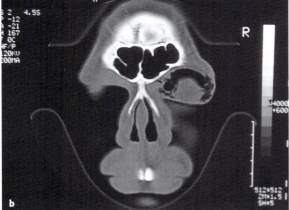

Fig. 47.14. a A patient exhibiting right-eyelid emphysema provoked by vigorous nose blowing after ethmoidectomy. There was spontaneous resolution (without surgery). b Periorbital emphysema of the right superior eyelid, as seen by coronal computed tomography

ate postoperative period or can be delayed in their appearance. To decrease the incidence of surgical complications, pre- intra- and postoperative precautions should be observed.

Recommendations to Reduce Surgical Complications

Preoperative

- Preoperative naso-sinus evaluation with endoscopy and high-resolution CT in axial, coronal and sagittal planes permits establishment of the extent of the disease and verification of existing anatomic variations.
- Patients should not take anticoagulants, drugs with anti-platelet activity, anti-inflamatory agents, aspirin or fibrinolytics because of the resultant increased risk of intra- and postoperative bleeding.
- Topical and/or systemic steroids are important to diminish edema and mucosal inflammation in the nasal cavity, especially in naso-sinusal polyposis.
- Preoperative antibiotic therapy should be used in suspected or documented concomitant infection.
- In very extensive surgery (for recurrent polyposis or tumors) with a risk of significant bleeding autotransfusion should be considered.
- Sufficient familiarity with the anatomy and the surgical techniques is needed to develop an appropriate surgical plan.
- Training in microscopic and endoscopic techniques in the paranasal sinuses is necessary.
- All the possible complications should be discussed with the patient.

Operative

- Keeping the patient's eyes uncovered in the surgical field facilitates the surgeon's orientation in relation to the orbit, possibly allowing earlier identification of any injuries to the orbital cavity.
- When the surgeon detects the presence of orbital fat in the surgical field, he should avoid its removal and watch cautiously for the development of ecchymosis, edema or proptosis. Delicate palpation of the ocular globe during surgery (ethmoidectomy) can identify any defect in the orbital wall by causing medial motion of the periorbita or prolapse of fat into the interior of the ethmoid sinus.
- The surgeon must carefully identify the site of origin of any CSF leak. If the surgeon has just begun the procedure, it is recommended that he terminate the operation and repair the injured area as the final step.
- The surgeon should always have surgical instrumentation appropriate for either children or adults available.
- The surgeon should take measures, such as (1) controlled hypotensive anesthesia, (2) the use of topical and injected vasoconstrictors and (3) delicate manipulation of surgical instruments, to minimize bleeding.
- If necessary, surgery of the nasal septum or the turbinates should be performed to enlarge the surgical field.
- The surgeon should choose the surgical technique that best suits his experience and that of the entire operating-room team, including both nursing and anesthesia personnel.
- In re-operations and cases of very extensive disease, where there is loss of surgical landmarks, a

Fig. 47.15. a Sagittal section through the lateral nasal wall. Protuberance of the nasolacrimal duct. **b** Endoscopic view, illustrating damage to the nasolacrimal duct when the surgeon works endoscopically on a middle-meatus antrostomy

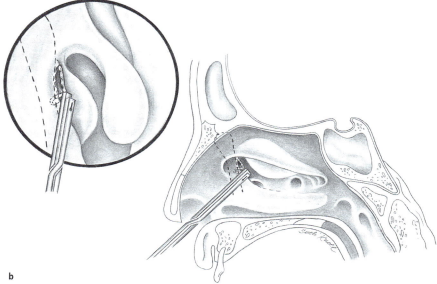

retrograde technique should be used; in other words, the surgery should begin at the sphenoid sinus and proceed in a posterior-to-anterior direction.
- A nasal splint should be used to avoid scar-tissue formation and to maintain the septum in the proper position.
- An unstable middle turbinate should be partially removed or fixed to the nasal septum.
- Computer-guided surgical assistance can be very useful, especially in revision cases.

Postoperative

- The patient should be instructed not to blow the nose during the immediate postoperative period because of the risk of producing bleeding, pneumocephalus, ascending infection and subcutaneous emphysema.
- The sinus cavities should be inspected and cleaned (with simultaneous endoscopic visualization) during each postoperative visit until complete healing has occurred.
- Crusts should not be prematurely removed, because bleeding and scar-tissue formation may be induced.
- Vigorous irrigation with physiologic solutions and antibiotics diminishes the chance of postoperative infection.

Final Considerations

- Complications of intranasal surgery of the paranasal sinuses are related to multiple factors, principally their complex anatomy, the type and extent of the disease, the history of previous sinus surgery and, above all, the experience of the surgeon.

- Complications have decreased, mainly because of diagnostic advances in imaging (CT and MR), use of the surgical microscope and endoscopes, modern anesthesia techniques and improved use of instruments, clinical resources and therapies.
- Reduction of complications begins with preventive measures that include meticulous preoperative evaluation with endoscopes and careful scrutiny of the imaging studies. Reduction of complications also results from the methods used to reduce intraoperative bleeding.
- The surgeon must always remember that the surgical procedure should not be more aggressive than the normal evolution of the disease.

References

1. Clevens R, Bradford C, Wolf G (1994) Tension pneumocephalus after endoscopic sinus surgery. Ann Otol Rhinol Laryngol 103:235
2. Corey JP, Bumsted R, Panje W, Namon A (1993) Orbital complications in functional endoscopic sinus surgery. Otolaryngol Head Neck Surg 109:814
3. Davis WE, Templer JW, Lamear WR, Davis WE Jr, Craig SB (1991) Middle meatus antrostomy: patency rates and risk factors. Otolaryngol Head Neck Surg 104:467
4. Draf W (1993) Personal communication
5. Friedman WH, Katsantonis GA (1990) Intranasal and transantral ethmoidectomy. A 20-year experience. Laryngoscope 100:343–348
6. Heermann J, Neues D (1986) Intranasal microsurgery of all paranasal sinuses, the septum, and the lacrimal sac with hypotensive anesthesia. 25-year experience. Ann Otol Rhinol Laryngol 95:631–637
7. Hollis LJ, Walsh RM, Bowdlere DA (1994) Radiology in focus: massive epistaxis following sphenoid sinus exploration. J Laryngol Otol 108:171 1994
8. Kinsella JB, Calhoun KH, Bradfield JJ, Hokanson JA, Bailey BJ (1995) Complications of endoscopic sinus surgery in a residency training program. Laryngoscope 105:1029
9. Lawson W (1991) The intranasal ethmoidectomy: an experience with 1077 procedures. Laryngoscope 101:367–371
10. Levine HL, May M (1993) Endoscopic sinus surgery. Thieme, New York, p 272
11. Maniglia AJ (1991) Fatal and other major complications of endoscopic sinus surgery. Laryngoscope 101:349
12. May M, Schaitkin B (1994) Complications of endoscopic sinus surgery: analysis of 2108 patients – incidence and prevention. Laryngoscope 104:1080
13. Mosher HP (1929) The surgical anatomy of the ethmoid labyrinth. Trans Am Acad Ophtalmol Otolaryngol 31:376–410
14. Serdahl C, Beris, Chole R (1990) Nasal lacrimal duct obstruction after endoscopic sinus surgery. Arch Ophthalmol 108:391–372
15. Stamm A (1995) Complicações da cirurgia microendoscópica naso-sinusal. In: Stamm A (ed) Microcirurgia naso-sinusal. Revinter, Rio de Janeiro, pp 255–263
16. Stamm A, Freire LAS (1997) Complicações da cirurgia microendoscópica naso-sinusal. Fed Med (Bras) 115: 127–132
17. Stammberger H (1991) Functional endoscopic sinus surgery. Mosby, St. Louis
18. Stammberger H, Posawetz W (1990) Functional endoscopic sinus surgery. Eur Arch Otorhinolaryngol 247:63–76
19. Stankiewicz JA (1989) Complications in endoscopic sinus surgery. Otolaryngol Clin North Am 22:749
20. Stankiewicz JA (1991) Avoiding orbital and lacrimal complications of sinus surgery. Operative Tech Otolaryngol Head Neck Surg 2:285–288
21. Stevens HE, Blair NJ (1988) Intranasal sphenoethmoidectomy: 10-year experience and literature review. J Otolaryngol 17:254–259
22. Terrel JE (1998) Primary sinus surgery. In: Cummings CW, Fredrickson JM, Harker LA, et al. (eds) Otolaryngology, head and neck surgery. Mosby, St. Louis pp 1145–1172
23. Thompson RF, Glukman JL, Kulwin D (1990) Orbital hemorrhage during ethmoid sinus surgery. Otolaryngol Head Neck Surg 102:45
24. Vleming M, Middelweerd RJ, De Vries N (1992) Complications of endoscopic sinus surgery. Arch Otolaryngol Head Neck Surg 118:617–623
25. Wigand ME (1990) Endoscopic surgery of the paranasal sinus and anterior skull base. Thieme, New York, pp 134–41

Subject Index

A
abdominal fat   564
abscess
– brain   11, 586
– dental   133
– epidural   54
– intramucosal   300
– lacrimal duct   429
– orbital tissue   352
– periorbital   185
– subdural   54
– subperiostal   54, 270, 352, 354, 381, 481
acetazolamide   454
acetylcholine   87
acromegaly   555
activating signal   329
acuity, visual   447
adenocarcinoma   120, 418
adenoid facies   89
adenoid hyperplasia   128
adenoid tissue   36, 411
adenoidectomy   258, 382, 383
adenoiditis   128
adenoma   549
– macro- / microadenoma   531, 532
– pituitary   557
– pleomorphic   120, 499, 500
adenoma theory, nasal polyp   111, 112
adenosine monophosphate, cyclic (cAMP)   85
adenotonsillar hypertrophy   212
adenotonsillectomy   347, 353
adenoviruses   98
adhesion   145, 167, 168, 175, 400, 588
– fibrous   413
aditus, nasal   17
adrenal suppression   433
adrenocorticotropic hormone   558
adriamycin   520
age, nasal polyposis   289
agger nasi   5, 19, 28, 29, 37, 63, 191, 205, 207, 224, 227, 230, 240, 260
– enlarged cells   73
– pneumatisation   144
AIDS (aquired immunodeficiency syndrome)   129, 313
air condition   164
Albright's syndrome   508
albumin   97, 98
algorithm   53
"alio loco"   9

allergy   4, 83–94, 128, 141, 147, 290, 395
– challenge phase   85, 86
– complementary diagnostic protocol   89
– food   94
– "in vivo" tests   89, 90
– late phase   86, 87, 91
– nasal   88, 89
– – treatment   91–94
– sensitization phase   84, 85
alotomy, lateral   548
alternaria   100
amphotericin   312
– liposomal   313
amplifier, rhinomanometry   47
anatomical factors, nasal polyposis   290, 291
anatomy, three-dimensional   155
androgen   515
anemia   395
anesthesia   217, 238, 350, 408
– hypotensive   2, 186, 201, 212, 317, 426, 588, 591
– local   167, 185, 253
– neuroleptanalgesia   259
aneurysm   494, 558
angiofibroma   365, 395, 485, 492, 500–502, 511, 569, 573, 578, 579
– juvenile   120, 385, 569
– – diagnosis   519
– – embolization   10, 502, 521
– – epidemiology   515
– – follow-up   525, 526
– – histopathological findings   518, 519
– – imaging studies   517, 518
– – intracranial invasion   516, 517
– – involution   525
– – mortality   525
– – origin / implantation   515, 516
– – pathogenesis   515
– – recurrence   524, 525
– – signs / symptoms   516
– – staging   519, 520
– – treatment   520–525
– – vascular pedicle   516
angiogenesis   329
angiography   393, 493, 501, 504
– digital subtraction   396, 570
– electroencephalographic monitored   564
– superselective   518
angioleiomyoma   120
annulus, *Zinn*   435, 444, 446, 447, 537

anosmia 5, 400, 467, 570
- recurrence 305
anterior-to-posterior dissection 186–188, 228
anticholinergics 92, 166
anticoagulants 591
antifungal agent, oral 313
antihistamines 91, 92, 166
anti-hypertensives 208
anti-inflammatory drugs 133
anti-leukotrienes 92
anti-thyreoglobulin 433
Antoni A / B 511
antroscopy 131
antrostomy 218, 262, 300, 524
- anterior 214
- combined 214
- endonasal 249
- inferior-meatus 214, 219–221, 251, 254, 255, 347
- inframeatal 353
- intranasal 182
- maxillary 190
- middle-meatus 214, 218, 219, 251, 253, 254, 435, 536
- - revision 372, 373
- radical 56
- sub-labial 521
antrum-grasping punch 216
apex 17
apnea, sleep- 395
appendectomy 267
approach
- combined 222
- craniofacial 385, 512
- endonasal 375
- endoscopic transnasal direct / transseptal 559–561
- ETA (endonasal, trans-septal, sphenoidal) 545–552
- external 301
- extracranial 471
- fronto-orbital 340
- intracranial 470, 471
- maxilla-premaxilla 547
- midfacial degloving 365
- naso-orbital 529
- orbitozygomatic 574
- osteoplastic 147, 486
- osteoplastic-flap 384, 487
- STA (sub-labial, trans-septal, sphenoidal) 545–552
- sub-labial 147, 359, 531, 555
- transantral 183, 347
- transethmoid 231, 458
- transmandibular 521
- transnasal direct 214, 230, 231, 359, 458, 460, 474, 539
- transpalatal 186
- transpharyngeal suprahyoid 521
- trans-septal 385, 458, 460, 531, 539, 544
- trans-septosphenoidal 366, 550, 551
- trans-sphenoidal 531, 539, 544
- transzygomatic 521
arachidonic acid 85
arachnoid 442
- embryogenic cells 365
- cyst 477
arch, anterior 218, 224, 230
artery
- basilar 68

- carotid, internal 30, 68, 73, 155, 157, 187, 198, 199, 204, 209, 211, 231, 258, 374, 393, 434, 443, 533, 556, 574, 578, 588
- - ligation 397
- central retinal 443
- cerebral, anterior 588
- ethmoidal
- - anterior / posterior 4, 23, 136, 153, 155, 196, 197, 202, 204, 209, 226, 228, 243, 261, 318, 443
- - - cauterization 5, 6, 393
- - - electrocoagulation 398–400
- - - ligation 397, 400
- - middle 23
- facial 162, 394
- infraorbital 33
- maxillary 22, 26, 32, 33, 162, 171, 516
- - electrocoagulation 398
- - embolization 397
- - ligation 397
- nasal, posterior lateral 23, 162, 394, 398
- nasopalatine 23
- ophthalmic 23, 26, 198, 442, 443
- palatine 394
- septal 23, 162, 393, 394, 398, 446, 588
- sphenopalatine 261, 515
ASA triad 188
aspergillosis, invasive 77
aspergillus 100, 309, 310, 312
asphyxia 405, 408
aspiration 395
aspirators, atraumatic 233
aspirin 591
- idiosyncrasy 4, 290
- intolerance 129, 141, 181, 264, 289
astemizole 92
asthenic 295
asthma 4, 5, 83, 103, 141, 142, 181, 212, 264, 287, 289, 290, 353
atmospheric-pressure change, sudden 395
atopy 83, 84
atrophy, optical-disk 444
autoimmune disorder 433
autonomic nervous system 87, 163, 288
axial plane 53
azelastin 92

B

backbiting forceps 216, 436, 533
bacteroides 99
ball point explorer 254
ballottement, orbital 316
barotrauma 97, 128, 264
basalioma 429
basophils 87, 107
beak, nasofrontal 380
Beck's burr-hole 9
beclomethasone dipropionate 93, 121, 166, 292, 297, 342
Benfield maneuver 170, 409
Bichat's fat 522
biopsy 55, 484
- orbital apex 529, 536, 537
black spots 312
Blakesley's forceps 216, 374, 377
bleeding 175, 258, 564, 581
blind areas 574

blindness   258, 447, 521, 583
– temporary   10, 429
blockade theory, nasal polyp   113
blood transfusion   265
blue eye   242
blue line   243
Boeck's canal (pterygopalatine fossa)   31–33, 201, 213, 393
bone forceps   255
bone graft   456, 471, 529
bone-cutting burrs   380
Bochdalek's valve   416
Bowman dilator   421
Bowman probe   416, 427, 428
brain
– abscess   11, 586
– cyst   452
– herniation   468
– injury   448, 586, 587
bromocriptin   558
bronchiectasis   143
bronchitis   353
bronchodilator   317
bronchopulmonary disturbance   89
budesonide   93, 123, 166, 332, 341, 342
built-in irrigation   379
bulb-pressing test   262
bulbus pressure maneuver   242
bulla
– ethmoid   20, 28, 36, 61, 191, 206, 228, 242, 246
– frontal   72
– lamella   284
bullectomy   252, 363
bullotomy   214
Bangerter probe   425
burr
– bone-cutting   380
– mastoid   281
– nasal   383, 384
buttress   199

C

Ca2+, intracellular   85
cadaver dissection   154, 159, 179, 390, 423
Caldwell view   56, 208, 281
Caldwell-Luc surgery   1, 3, 99, 148, 182, 249, 332, 353, 434, 437, 496, 499, 506, 510, 538
– mini-   301
canal
– carotid   561
– craniopharyngeal   466
– infraorbital   26, 348
– lacrimal-nasal   254
– optic   66, 435, 556
– pterygoid   213, 556
– vidian   69
canaliculitis   418
candida albicans   100
canine fossa puncture   222
canthal tendon, internal   416
canthotomy, lateral   318, 583, 584
carbon dioxide   412
carbonic-anhydrase inhibitor   454
carcinogens, environmental   147

carcinoma   40
– adenocarcinoma   120
– squamous cell carcinoma   120
cardiac defects   405
cardioarrhythmia   91
cardiopulmonary reserves   395
cardiorespiratory complications   185
cartilage graft   471
cartilaginous septum, anterior   3
cataract   433
cationic protein, eosinophilic   88
caucasian skulls   284, 405
cauterization   6, 7, 169
– chemical   393, 395
– electro   305, 393, 395
– submucosal   167, 168
cavernous erectile tissue   163
cavity, nasal   17
CD4+ cells   84, 98
CD8+ cells   84
cell adhesion molecule (CAMs)   84
cellulitis, orbital   352
cephalgia, midfacial   318–320
cetirizine   92
cetoconazol   91
cGMP   87
challenge test, nasal   90
Charcot-Leyden crystals   311
CHARGE   405
chemotactic signal   329
chemotherapy   481, 502, 520
chewing gum   310
choana   3, 18, 209, 261
choanal atresia   167, 212, 341, 358, 359, 383, 529, 539
– diagnosis   405–407
– treatment   382, 407–412
cholecystectomy, laparoscopic   387
choloboma   405
chondrosarcoma   311
chordoma   547
– clivus   485, 552
chromosome   83
Churg-Strauss-syndrome   92, 103, 111
cicatrization   326, 375
ciliary dysfunction syndrome   103, 129, 135, 142, 289, 349, 361
ciliary dyskinesia   101, 381
cisternography   452, 455
– pneumocisternography   365
– radionuclide cisternography   467
cleaning, atraumatic   341
clearance, mucociliary   180, 181
clinoid process, anterior   69, 70
clivus   525, 569
– chordoma   485, 552
– resection   570
clogging   378
clots   234
coagulation cascade   329
coagulopathy   367
coaxial illumination   1
cocaine   129
cold   339
cold light   2

– extracorporal   347
collagen slurry, microfibrillar   318
collagen synthesis   330
collagen, typ III   330
collapse, valvular   51
colonizers, normal   99
columella   162, 548
combined approach   222
complement   87
complementary diagnostic protocol   89
complications
– micro-endoscopic sinus surgery   581–593
– microscope / endoscope   10–14
computed tomography (CT)   4, 53, 130, 141, 155, 179, 239, 258, 281, 332, 347, 350, 378, 406, 434, 582
– axial   445
– biplanar   180
– contrast media   54, 467
– coronal   445
– helical   209
– high-resolution   467
– three-dimensional   387
concha
– bullosa   5, 21, 37, 59, 119, 128, 144, 161, 165, 291
– nasomaxillary   20
conchotomy   262
condrovomeral junction   62
congenital defect, nasolacrimal system   418
conjunctival sac   418
connective-tissue graft   471
contrast substance, non-ionic hydrosoluble   452
conventional technique, sinus surgery   182
corneal epithelium   416
coronal images   53
coronal incision, bitemporal   265, 275
coronavirus   98
corticosteroids   92, 165, 166
– high-dose   445
– intravenous   318
corynebacteria   99
Cottle's sign   50
cough   353, 361
– persistent   129
cranialization   268
craniofacial approach   385, 512
craniofacial resection   148, 150, 312
– "en bloc"   570
craniofacial trauma   445, 458
craniopharyngioma   120, 365, 494, 500, 519, 547, 558
craniotomy   366, 454
– bifrontal   471
– frontal   7, 451, 529
– lateral   579
– transpalatine   579
Crawford catheter   420–422
crest
– lacrimal   29
– maxillary   174, 417
cribriform plate   3, 5, 19, 58, 78, 183, 195, 198, 204, 454, 483
– fistula   457, 458, 461
Crigler dilator   422
crista galli   18, 204
cromolyn sodium   92, 166
crooked nose   173

cross-links   330, 340
crowded housing   128
cruciate incision   562
crusting   145, 150, 234, 413, 461, 588
– blood   333
cryosurgery   167, 397
cryotherapy   167, 169
CSF (cerebrospinal fluid) leakage   78–80, 150, 175, 196, 204, 213, 228, 274, 318, 320, 354, 365, 366, 398, 434, 584, 585
– iatrogenic   429
– repair   529, 530
– spontaneous   7
"cul-de-sac"   415
cup-type forceps   325, 377
cure   501
curtain maneuver   523
cutting forceps   325
cyanoacrylate   521
cyanosis   405
cycle, nasal   44
cyclic nucleotides   87
cyclo-oxygenase   85
cyst   74
– aneurysmal bone   493
– arachnoid   477
– brain   452
– cholesterol   300
– dermoid   120, 360
– mucous-retention   364, 492, 493
– nasolacrimal duct   120, 360, 361
– submucous   252
– *Thornwaldt*   485, 519
cystic fibrosis (mucoviscidosis)   101, 103, 111, 129, 135, 142, 189, 264, 289, 290, 297, 349, 353, 361, 362
cytokines   84, 87, 117, 288, 311, 537
cytology, nasal   90
cytometry, nasal   130
cytotoxic effect   128

D

dacryocystitis   150, 361, 415
– suppurative   418, 423
dacryocystography   418
dacryocystorhinostomy   441, 485, 537
– endonasal   237, 415
– endoscopic   384
– external   415
– transnasal   590
dacryolith   427
dacryostenosis   417, 423, 428, 431
debridement   292, 390
– microdebriders   186, 215, 217, 325, 351, 380, 408, 409, 419, 477, 481, 491
decarbazine   520
decompression
– orbital   4, 318, 387, 529, 534, 535
– – non-surgical treatment   433
– – surgical treatment   433–438
– optic nerve   237, 245, 384, 484, 537
– – traumatic optic neuropathy   441–448
decongestants   92, 165
– nasal   133
deep transantral microsurgery   521–524

Subject Index

degloving, midfacial 147, 499, 502, 506, 510, 510
- microsurgical approach 569–579
Denny's marks 88
dental pain 590
dermoid cyst 120, 360
dermoplasty, septal 397
detachment, orbital 444
diabetes insipidus 564, 565
diabetes mellitus 143, 166, 208, 313
diaphanoscopy 130
diaphragm, sella 564, 565
dihydrotestosterone 515
dilator 420
- *Bowman* 421, 422
- *Crigler* 422
diplopia 150, 296, 434, 570, 584
- postoperative 438, 447
discharge
- nasal 144, 349, 353
- postnasal 319, 320
diseases (*see* syndromes)
diuretics 475
dorsal flexion, head 253
double curve 62
double vision 198, 447
douche, cold / cool 341
drainage, venous 24
drill 220, 390
- bone-cutting 378, 379, 380
- diamond 409
- *Stammberger* 419
drilling of nasofrontal beak 279, 282
- history 279–281
drugs, chronic use 212
duct
- frontonasal 267
- naso-frontal 1, 2, 374
- naso-lacrimal 27, 35, 40, 71, 137, 219, 428, 436
- – cyst 120, 360, 361
- *Stensen's* duct 98
dura 442
- allogenous 483
- defect 459
- lyophilized 471
- lesion 153
duraplasty 267–270, 482, 486
- endonasal 485
dysautonomia 290

E

ears, small 405
ebastine 92
ecchymosis 219, 448, 582, 588, 589
eczema 83
edema 88, 145, 287, 374, 411
- facial 413
- gray 337
- lid 150, 258
- mucosal 349
- periorbital 494, 588, 589
- polypoid 296, 306
education, how to improve 154–159
electrocauterization 305, 393

electrocoagulation 5, 397–400, 403
electron microscopy, scanning (SEM) 323
electronystagmography 43
ELISA (enzyme-linked immunosorbent assay) 91
embolization 7
- angiofibroma 10, 502, 521
- selective 393, 395
emotional disturbance 166
emphysema
- periorbital 589
- subcutaneous 592
empyema 4, 185, 213
ems brine 341
encephalocele 120, 359, 360, 456, 466, 468–477, 510
- traumatic / congenital 455
encephalomeningocele 359
enchondroses, multiple 499
endocrinologist 434, 557
endonasal approach 375
endonasal sinus surgery, microscopic (MESS) 189–192
endoscope 2, 315
endoscopic diagnosis, nasosinus disorders 35–41
endoscopic surgical technique, combined functional
    rhinologic 316–320
endoscopy
- bimeatal 222
- flexible 119, 233, 349
- nasosinus 133, 179, 180
- rhinosinusal 131
- rigid 141
- virtual 80
endoscrub 377, 383
endosheat 377
enteric feeding 359
enterobacteriacea 100
eosinophil chemotactic factor 290
eosinophilia syndrome 88, 311
eosinophilia, mucosal 263
eosinophilic polyps 4
eosinophilic rhinitis syndrom, nonallergic (NARES) 90, 288, 290
eosinophils 87, 88, 287, 288
ephedrine, pseudo- 92
epidermoid tumors 494
epinastine 92
epinephrine 174, 253
epiphora 150, 361, 415, 417, 428
epistaxis 4, 150, 365, 367, 500, 511, 512, 516, 590
- cauterization (*see there*)
- classification 395
- micro-endoscopic surgical technique 393–403
- treatment 395–398
epithelial tumor 120
epithelial-rupture theory, nasal polyp 114, 115
epithelium
- corneal 416
- pseudostratified 104, 106, 163, 493
- stratified squamous, non-keratinized 104, 106
- transitional 104, 106
esthesioneuroblastoma 5, 368, 518, 519, 569
esthesioneurocytoma 512
estrogen-hormone therapy 501
ETA (endonasal, trans-septal, sphenoidal approach)
    545–552

– advantages 545, 556
– disadvantages 546
– indications / contraindications 547
ethmoid
– anterior area 2
– bone 161
– – perpendicular plate 18
– – power instrumentation 382
– cell
– – anterior 429
– – prebullar 26
– – supra-orbital 187, 202
– compounded disease 302
– forceps 242
– infundibulum 206
– labyrinth 135
– mucosa 290
– surgery 1, 4
ethmoid sinus 28–30, 315
– anatomy 202–204, 357
– anterior 347
– conventional operation 183, 184
– fistula 460
– roof 327
– surgery 214, 228–230
– – combined microscopic and endoscopic surgery 237–246
ethmoid sinus-middle meatus complex 201
ethmoidal cleft, anterior 118
ethmoidectomy 123, 190, 212, 213, 228, 362, 363, 399, 454
– anterior 300
– antero-posterior 201
– classic 292
– endoscopically guided 316
– external 148, 184, 201, 347, 434, 437, 451, 471, 499, 529
– intranasal 183, 347
– partial 214, 491
– posterior 300, 301, 474
– radical 214, 230
– revision 373, 374
– total 214, 228
– transmaxillary 332
ethmoiditis, purulent 11
ethmomaxillary plate 65
Eustachian tube 35, 348
– catheter 188
– dysfunction 88
– ipsilateral orifice, deformity 405
euthyreoidism 433
evoked-response testing, visually 444
Ewing's papilloma (inverted papilloma) 64, 103, 119, 120, 147, 270, 385, 483, 485, 498, 499, 533, 539, 569, 577, 579
excretory duct, cystic dilatation 112
exocytosis 85
exophthalmometer, *Hertel* 434, 437
exophthalmos 4, 10, 365, 507, 508
exostoses, multiple 499
expansion, soft-tissue 350
external approach 301
external nose 17
extracellular matrix 288
extracranial approach 471
extraocular muscle 447
extravasation 335

F
facial deformity 365, 516
facial pain 144, 397, 466, 590
facial paralysis 358, 397, 405, 521
fascia lata 451, 548
– graft 455, 458, 460, 471, 473
fenestration 4
– superior / inferior 12
ferromagnetic elements 311
FESS (functional endoscopic sinus surgery) 184, 185, 347, 377, 529
fetid breath 129
fexofenadine 92
fiberoptic light fiber 537
fibrin 330
– exsudation 340
fibrin glue 267, 274, 411, 483
fibrinolytics 591
fibroblastic cambium zone, capillary 525
fibroblasts 288, 330, 337, 519
– endomysial 534
fibroma 120
– chondromyxoid 493
– non-ossifying 493
– ossifying 120, 482, 485, 508
– osteofibroma 264, 270
fibroma theory, nasal polyp 111, 112
fibronectin matrix 337
fibrous dysplasia 264, 270, 485, 493, 507, 508
fibrous tissue 327
final common pathway 537
fistula
– carotid-cavernous 445, 448
– cribriform plate 457, 458
– CSF (cerebrospinal fluid) 7, 78–80, 150, 175, 196, 204, 213, 228, 274, 318, 320, 354, 365, 366, 398, 429, 434, 469, 470, 584, 585
– – iatrogenic 477
– – repair 529, 530
– – spontaneous 470
– ethmoid sinus 460
– fovea ethmoidalis 455–458, 477
– idiopathic 461
– oroantral 3, 182
– sphenoid sinus 458–461
flap
– local 473
– *Lopez* Infante flap 5
– middle-turbinate pedicled 451
– mucoperichondrial 451, 560
– mucoperiosteal 414, 451, 455, 456, 459, 473, 474, 475
– osteoplastic 183
flare-up, inflammatory 306
flunisolide 342
fluorescein
– intrathecal 452, 462, 474
– – endoscopy 469
– surgery 466
– test 418
fluoroscopy 551
fluticasone 166, 342
Foerster's forceps 523
Foley catheter 574
fontanelle, anterior / posterior 20, 219, 323, 491

food allergy 94
foramen
- cecum 359
- ethmoid, anterior 29
- palatine 434
- pterygopalatine 32
- rotundum 24, 30, 31
- sphenopalatine 20, 31, 32, 65, 262, 398, 500, 515, 573
foreign body 39, 56, 127, 208, 213, 367, 368
fossa
- anterior cranial 28
- canine 218
- lacrimal 29, 427
- nasal 300
- olfactory 243
- pituitary 30
- pterygomaxillary 250, 522, 573
- pterygopalatine 31–33, 201, 213, 393, 398
- zygomatic 213, 573
foul odor 170
fovea ethmoidalis 28, 58, 73, 78, 183, 202, 228, 584
- fistula 455, 456, 461, 467, 475
fracture 56, 208, 268–270
- frontal sinus 271
- lateral out-fracture 168
- maxillo-orbitozygomatic 569
Frankfurt plane 417
Friedman staging 141
frontobase, dorsal 245
frontal nails 298, 299
frontal pain 303
frontal sinus 10, 25, 26, 136, 137, 144, 494
- anatomy 205
- conventional operation 182, 183
- extension 72, 73
- fracture 271
- mucocele 494
- supernumerary 58
- surgery 213, 214, 223–227
- – endonasal and external micro-endoscopic 257–276
- – mini-anterior and combined sinusotomy (MAFS) 279–285
- – power instrumentation 384
frontal-sinus-drainage type 243, 260, 261, 273
frontoethmoidectomy, external 147, 150, 224, 496
frontonasal drainage surgery 279
fronto-orbital approach 340
frontotomy, external 301
frozen sections 486
fulguration, bipolar 455
fungal infection 4, 77, 131, 350, 418
fungus ball 223, 232, 371
Furstenberg's sign 360, 497
fusarium 100

G
gadolinium diethylene triaminopentaacetic acid 468, 518
ganglion
- pterygopalatine 24
- stellate, block 10, 431
- superior cervical 163
gastric feeding 359
Giraldes' ostia 26, 250, 254

glabella region 359
gland
- intraepithelial 106
- lacrimal 415
- mucous 107
- tuboalveolar 114
glandular cyst theory, nasal polyp 112
glandular hyperplasia theory, nasal polyp 114
glas-ionomer cement 276, 471
Gliklich and Metson staging 142
glioma 120, 494
- nasal 360, 510
goblet cell 104, 105
glomus
- jugulare 502
- nasopharyngis 503
- vagale 502
glucocorticoid steroids 121
glucose content 452
glucose oxidase 467
glycoproteins 97
grading
- inflammatory sinus disease 141–145
- inverting papilloma 147–151
graft
- bone 529
- cartilage 471
- connective-tissue 471
- fascia lata 455, 458, 460, 471, 473
- free septal-cartilage 460
- mucosal 471
- muscle 471
- overlay 471, 472
- underlay 471, 473, 483
graft-positioning technique 471
granulation, exuberant 330
granulation tissue 234, 333
granulocyte-macrophage colony-stimulating factor (GM-CSF) 84, 85, 288
granuloma 167, 418
- paraffin 340
- pyogenic 519
Graves' orbitopathy 433, 434, 534
green phase 154
Grocott stain 311
growth hormone 558
growth retardation 405
Guedel cannula 359, 407
Gussenbauer's incision 10

H
haemophilus influenzae 99, 100, 131
Haller's cell 63, 64, 144, 155, 206–208, 218, 219
hamartoma 499
hardware 159
Hashimoto's thyreoiditis 534
Hasner's valve 220, 262, 416, 417
headache 144, 296, 353, 361, 363, 452, 454, 461, 466, 494, 587
- intense 130, 133
- postoperative frontal 138
- retro-orbital 303
hearing loss 405
heart defect 358

Heermann concept 12
helical contiguous images 53
helix 265
hemangioma 120, 360, 364, 418, 500, 503–505, 517, 519
hemangiopericytoma 500, 505, 506, 517, 519
hematologic malignancy 77
hematoma 296
- orbital 185, 198, 258, 444, 481, 581, 583, 584
- periorbital 274
- postoperative 3
- subdural 445
hematopoetic disorder 367
hemiplegia 521
hemitransfixion incision 174
hemodynamic anomaly 493
hemodynamic change 395
hemophilia 395
hemorrhage 296, 320, 442, 587, 588
- intracranial 175, 477
- intraorbital 198
- vitreous 444
hemostasis 524
hemostatic plug 329
heredity, nasal polyposis 289
herniation, brain 468
Hertel exophtalmometer 434, 437
hiatus semilunaris 21, 29, 136, 181, 206, 207, 240, 250, 419
high-risk area, endoscopic sinus surgery 195–199
Hirtz-position 406
histamine 115, 290
histiocytoma, malignant fibrous 520
Hochstetter membrane 405
Hodgkin's disease 395
hollow chisel 427
hollow punch 255
Hopkin's small-diameter 347, 415
Hopkin's telescope 237, 254, 256, 349
hormones 87
- therapy 520
Horner's muscle 416
hummer 260
hyalinization 525
hydrocephalus 470
hyperactivity, nasal (NHR) 89
hypereosinophilia 290
hyper-reactivity, inflammatory 287
hypersecretion 353
hypersensitivity 83, 276
hypertension 166, 208, 212
- intracranial 452, 475
hypesthesia 481
- transient 175
hyphema 444
hypogenitalism 405
hypophysectomy 544
- total 555
- trans-sphenoidal 552
hypophysis 155, 238
hypopituitarism 495, 565
hyposmia 5, 361, 400, 461
hypotensive anesthesia 2, 186, 201, 212, 317, 426, 588, 591
hypothalamic-pituitary-adrenal axis 93
hypothalamus 564

hypothyreoidism 433
hypoxemia, serious 395

I

iatrogenic problems 304
ICAM-1 (intercellular cell-adhesion molecule 1) 87
image investigation 53–80
- examination technique 53, 54
- objectives 54–56
immune deficiency syndrome 349, 353
- acquired (AIDS) 129, 313
- congenital 289
immunocompetence 309, 312
immunoglobulin (Ig) 88
- IgA 83, 98
- IgE 83, 84, 288
- – total serum level 90
- IgG 98
- IgM 83
immunology 83–94
immunopathology, nose 83
immunotherapy 93, 166
"in vitro" laboratory test 90
"in vivo" tests 89, 90
infarction, cerebral 521
infection 290, 347, 413, 418
- fungal 4
infernal trio 99
inflammation 87, 88
- neurogenic 290
- polypoid / granulomatous 312
inflammatory complications 347
inflammatory phase 329
influenza virus 98
infundibulotomy 242
infundibulum 21, 136, 186, 206, 250, 485
- frontal recess 136
- frontal-sinus 260
injection, pterygopalatine-fossa 393, 397
inspiratory peak flow 119
Instratrak surgical navigation system 387
integrin 88
interferon 87, 99
interfrontal septum 10
interleukin 84, 85, 87, 117, 311, 537
- interleukin-2 receptor 98
intermediate images 53
intersinus septum 268
intestinal peptide, vasoactive 115
intracranial approach 470, 471
intracranial complications 581
intubation 350
iopamidol 452, 467
ipatropium bromide 92
isoprenaline 289
isthmi, postoperative, development / prevention 138
isthmus surgery 135
itraconazole 312

J

Jansen-Middleton rongeur 169, 560
Jansen-Ritter operation 264, 268, 270, 340

JNA   518
Jones procedure   418
Jorgensen, scoring system   142

K

Kartagener's syndrome   103, 111, 189, 264, 290, 349, 353
keel   417
Kennedy staging   141
keratopathy, exposure   433
Kerrison punch, micro-   228, 230, 231, 398, 533, 549, 559, 560, 572
ketoconazole   312
keystone   162
Killian's polyp   37, 75, 364, 492, 500, 519
Killian procedure   183, 265
Kiesselbach's plexus   162, 367, 393, 395
Kofler punch   237, 240, 245
Krause's valve   416
Kuhn curettes   187
Kuhn-Bolger giraffe forceps   191

L

laceration, corneal / scleral   444
lacrimal bone   19, 161, 305
lacrimal canaliculi   416
lacrimal duct   27, 35, 40, 71, 137, 219
– aplasia, congenital   428, 429, 431
– cyst   120, 360, 361
– prosthesis   428
– stenosis   417, 423, 428, 431
lacrimal gland   415
lacrimal pore   418
lacrimal sac   4, 19, 240, 262, 284, 415, 420
– endonasal surgery   427
– fornix   427
lacrimal semiology   40
lactoferrin   97, 98
lamella, basal   20, 186, 373, 388
lamina
– cribrosa   54, 197, 241
– elastic   330
– papyracea   21, 59, 155, 180, 183, 195, 204, 224, 228, 262, 320, 327, 352, 388, 435, 582, 583
– perpendicularis   62
– propria   114, 115, 330
– pterygoid, medial   516
laser   5, 297, 423
– vaporization   167
lavage, antral   347
learning curve   153, 154, 188
leiomyoma   120
leiomyosarcoma   505
Lemoyne's nail   298, 299
lethargy   454
leukemia   395
leukocytes
– multinucleated   326
– polymorphonuclear   329
leukoprotease inhibitors   97
leukotriene   85, 115
levator palpebrae superioris muscle   433
Levine and May staging   141

lid edema   150, 258
lidocaine   174, 253
ligation, vessel   393, 397, 400
limen, nasal   17
lipid layer, external   416
lipid-vacuole-rich tumor   429
lipiodal   521
lipoma   499
liquid stratum, trilaminar   416
liquorrhea   241
Little's area   163, 367, 393, 395
liver disorders   395
L-nasal support   174
local anesthesia   167, 185, 253
local flap   473
Lopez Infante flap   5
loratadine   92
loss of tip projection   173
Lothrop procedure, modified transnasal   224, 284
L-shape   172
lumbar puncture   454
luminous probe   423
Lund-Mackay scoring system   142–145
lymph, cervical   163
lymph nodes   163
lymphangioma   360, 504
lymphatic system, nasal septum   163
lymphocytes   87, 98
lymphoma   74, 368, 395, 418, 429
Lynch procedure   183, 264, 496, 510
lysozyme   97

M

macrolides   91
macrophage   329
MAFS (mini-anterior frontal sinusotomy)   279–285
– type A   281, 282
– type B   282
magnetic resonance imaging   4, 131, 210, 258, 350, 467–469, 517, 582
malformation, congenital   358
mannitol   318
Marcus-Gunn pupil   444
marsupialization   298, 496
mast cell   87, 107, 519
– nonimmune degranulation   87
– stabilizers   92
MAST-CLA (multiallergen specific test-chemiluminescent assay)   91
mastoid surgery   378
maxilla   161
– frontal process   12
maxilla-premaxilla approach   547
maxillary line   419
maxillary palatine process   19
maxillary sinus   1, 137, 184, 327
– anatomy   205, 249, 250, 357
– conventional operation   182
– endoscopic diagnosis   39, 40
– extension   71, 72
– malformation   405
– mucocele   495
– surgery   3, 4, 213, 214, 218, 249–256

maxillary window, anterior 222
maxillectomy 502
– medial 572, 573
– partial 570, 571
– radical 569–571
maxillofacial trauma 128
McGovern nipple 407
meatal route, inferior 316
meatotomy
– inferior 218
– middle 292, 300
– wide 305
meatus, nasal
– common 21, 427
– inferior / middle 20, 35, 218
mediators, chemical 115–117
melanotic nevi 499
membrane
– buconasal 405
– bucopharyngeal 405
– phospholipids 288
meningioma 120, 365, 494, 506, 507, 558
– naso-sinus 576
meningitis 11, 429, 431, 454, 461, 466, 481, 564, 581, 585, 586
– bacterial 455
– chemical 469
– recurrent 452
meningocele 9, 453, 461, 494
meningocytes 507
meningoencephalocele 359, 360, 467, 492, 496–498
– intranasal 497
menopause 289
mequitazine 92
merocel 396
MESS (microscopic endonasal sinus surgery) 189–192
Messerklinger's concepts 180, 181
β-methasone 121
methylmethacrylate 270, 276, 471
metrizamide 452, 467
Metzenbaum scissors 571
microcephaly 405
microdebriders 186, 215, 217, 325, 351, 380, 408, 409, 419, 477, 481, 491
micrognathia 405
microosteotome 220
microtia 358
midfacial degloving approach 365
minimal access surgery 181
mini-trephine 226
miscellaneous 418
mitose 330
mongoloid skulls 284
monitoring, nerve 391
moraxella catarrhalis 131
morphing 331
MRSA (methicillin-resistant staphylococcus aureus) 390
mucocele 4, 10, 59, 72, 76, 77, 150, 171, 185, 213, 268, 363, 364, 371, 438, 474, 484, 486, 492, 493–496, 529, 537–539
– ethmoidal 423
mucociliary function 127, 323
mucolytic drugs 133
mucoperichondrial flap 451, 560
mucoperiosteal flap 414, 451, 455, 456, 459, 473, 474, 475
mucopolysaccharides 534

mucoproteins 98
mucopyoceles 270, 271, 481, 487
mucor 313
mucosa
– eosinophilia 263
– ethmoid 290
– hypertrophic 324
– olfactory 22, 458
– pituitary 22
– respiratory 127
– sinus, wound healing 323–327
mucosal exudate theory, nasal polyp 112
mucosal graft 471
mucosal hypertrophy 165
mucosal strips 340
mucositits, chronic respiratory 135
mucous blanket 181
mucoviscidosis (cystic fibrosis) 101, 103, 111, 129, 135, 142, 189, 264, 289, 290, 297, 349, 353, 361, 362
multimedia system 155
multiple myeloma 143
muscle graft 471
mycetoma 100, 309, 310
mycoplasma pneumoniae 98
myofibroblast 340

N

nasal blood supply, anatomy 393–395
nasal bone 161
nasal mucosa, wound healing 330, 332
nasal septum 172–177, 412, 414
– anatomy 161–164
– deviation 127, , 128, 161, 291, 316, 405
– microsurgery 3
– physiology 164
– surgery 172, 383
nasal speculum, self-holding 237, 259
nasal surgery, flanking 135
nasal valve angle 162, 164
nasal wall, lateral 36, 37
nasalization 292, 298, 301, 305, 306
nasal-tip surgery 10
nasal-valve collapse 173
nasoantral window 182
nasolacrimal system 415
– anatomy 415–417
– diagnosis 418
– pathology 417, 418
– surgical technique 419–423
– – endonasal 425–431
nasolacrimal-duct injury 589, 590
naso-orbital approach 529
nasopharyngeal mass 516
nasopharynx 35, 36
natural-killer cells 326
navigation sinus surgery, image guided 188
– instrumentation, technology 387–391
necrotizing ethmoiditis theory, nasal polyp 112
neglected sinus 539
Nelaton catheter 191
neo-choana 411
neoplasm, benign 364, 365
nephritis, chronic 395

Subject Index

nerve
- alveolar 249
- cranial 30, 205
- - injury 586
- facial 163
- infraorbital 249, 250, 437, 535
- intermedius 33
- maxillary 24, 33, 73, 443
- olfactory 22
- optic 28, 68, 155, 157, 183, 187, 198, 204, 209, 211, 245, 318, 434, 441, 533, 574
- petrosal, greater superficial 163
- pterygoid 205
- sphenopalatine 186
- trigeminal 27, 33, 481
- vidian 73, 163
- zygomatic orbital 24
neuroblastoma, olfactory 517
neurofibroma 120, 499
neurofibromatosis 494
neuroleptanalgesia 259
neurosurgery 390, 543
neurotoxin, eosinophil 88
neurotransmitters 87
neurovascular bundle, supraorbital 268
neutropenia 77
neutrophil chemotactic factor, high molecular weight- 115
neutrophils 130
Newman, grading system 142
nipple
- *McGovern* 407
- orthodontic 359
norepinephrine 87
nose
- artificial 47
- surgical anatomy 17–24, 179

O

obstruction
- nasal 144, 306, 318, 349, 353, 361, 365, 516
- - mixed type 51
- - mucous / functional 51
- - structural / anatomic 51
- tubal 395
- vessel 112
occipito-frontal view 258
occlusion, balloon 448, 564
ocular muscle 155
oculovagal reflex 583
odontogenetic causes, sinusitis 129
olfaction 164
olfactory bulb 197
olfactory mucosa 22, 458
olfactory nerve 22
onlay technique 483
Onodi's cell 66, 68, 198, 204, 228, 232, 239, 244, 523, 556
operating microscope 1
ophthalmologist 390, 429, 434, 557
ophthalmology, american academy 423
ophthalmoplegia 495, 582
optic canal 66, 435, 556
optic chiasma 533, 565

optic nerve 28, 68, 155, 157, 183, 187, 198, 204, 209, 211, 245, 318, 434, 533, 574
- decompression 237, 245, 384, 441–448, 484, 537
optic neuropathy
- ischemic 433
- traumatic
- - anatomy 442–444
- - diagnosis 444, 445
- - pathophysiology 441, 442
- - therapy 445–448
optic protuberance 565
orbit 4, 28
- lesions, endoscopic approach 529–540
orbital apex 535
- biopsy 529, 536, 537
- lesions 535–537
orbital complications 219, 581–584
orbital fat, extraconal 59
orbital fissure, superior 24
orbital roof, pneumatization 58
orbitopathy
- dysthyroid 433, 534
- *Graves'* 433, 434, 534
orbitotomy, lateral 433, 438
orbitozygomatic approach 574
orthopedics 390
oscillating saw 267
osteitis 227
osteocartilaginous skeleton 17
osteoclast 330
osteoclastoma 493
osteofibroma 264, 270
osteoma 9, 264, 268, 270, 332, 483–485, 508–510, 578
osteomyelitis 54, 185, 353
- cranial 452
osteoplastic approach 147, 486
osteoplastic flap 183
- approach 384, 487
osteoplastic obliteration 264
osteoplastic surgery 201, 224, 496
osteoporosis 433
osteoradionecrosis 520
osteotome 572
osteotome, micro 409
ostia, accessory 26
ostiomeatal complex 43, 128, 130, 171, 180, 315, 335
- anatomy 205
- anatomic defects 128
- evaluation 56
ostioplasty 218f
- frontal-sinus duct 202
- maxillary sinus 214
ostium
- frontal 300, 340, 375
- maxillary-sinus 187, 241, 242, 491
- sinus 127
- sphenoid-sinus 36
ostrom forceps 190
othodontic nipple 359
otitis media 88
otitis, secretory 395
otolaryngologist 434, 543
otosurgery, microscopic 317
out fracture, lateral 168

overlay graft 471, 472
oxymetazoline 350

P

packing, nasal 255, 393, 395
pain
- dental 590
- facial 144, 397, 466, 590
- frontal 303
palatine bone 305
palatine process 19
pallor 444
palpator 216
palpebral ligament, medial 429
pansinusitis 361
papilloma 55, 74, 120, 486
- basal cell 429
- columnar 147
- exophytic 147, 485
- inverted (transitional cell) 64, 103, 119, 120, 147, 270, 385, 483, 485, 498, 499, 533, 539, 569, 577, 579
- - surgical grading 147–151
- squamous 364
papilloma virus, human 147
paradoxical curvature 416
paraganglioma 502, 503
parainfluenza virus 98, 99
paralysis, facial 358, 397, 405, 521
paranasal sinuses and anterior skull base, compact disk 154
parasellar region
- lesions, approach 543–552
- transnasal endoscopic surgery 555–566
parasympathetic autonomic fibers 163
paresthesia, infraorbital / dental 578
pearly white color 88
pediatric endoscopic sinus surgery 347–354
peptic ulcer 166
peptococcus 99
perforation, septal 395, 551
periorbital fat 381
periosteum, nasopharyngeal 515
periphlebitis / perilymphangitis theory, nasal polyp 114
peroxidase, eosinophilic 88
petrous pyramid 213
pharyngitis, recurrent acute 129
photocoagulation 297, 305
photodocumentation 2
phycomycetes 313
pia mater 442
pituitary 384, 385
- mucosa 22
- surgery 441, 466, 530–533
- tumor 562
pituitary-gland surgery 4
planum sphenoidale 73
plasma proteins 97
plasmapheresis 433
plasmocytoma 517
- extramedullary 519
platelets 329
platelet activating factor (PAF) 115
platelet-derived growth factor (PDGF) 288
pneumatization 58, 59, 423

pneumatosinus 270
pneumocephalus 175, 454, 470, 564, 587
- tension 586
pneumocephalus 452
pneumocisternography 365
pneumococci 99, 131, 466
pneumoencephalocele 451
pneumotachymeter 47
pneumatization 171
polycytemia 395
polyethylene 415
polyp 74, 138, 171, 212
- angiomatous 518, 519
- antrochoanal (*Killian's* polyp) 37, 75, 364, 492, 500, 519
- inflammatory 500, 519
- nasal 37, 38, 127, 185, 332, 529
- - autopsy material 117–119
- - chronic inflammatory 110
- - definition 103
- - diagnosis 119, 120
- - edematous 110
- - eosinophilic (allergic) 4, 110
- - epithelium 104
- - etiology 111
- - histological classification 110, 111
- - incidence 103, 104
- - morphology 104–110
- - origin 117
- - pathogenesis 111–117
- - stroma 107
- - treatment 120–123
- - power instrumentation 381, 382
- sphenochoanal 75, 539
polypectomy
- laser 297
- medical 295, 297, 305
- simple 121, 122
- surgical 297
polypoid reaction 310
polyposis 4, 5, 64, 88, 181, 210, 212, 261, 371
- recurrent 172, 353
- - follow-up 305, 306
- - localized 296
- - risk factors 287–295
- - treatment 295–305
- sino-nasal 362, 489–492
positioning, patient 238, 239
- *Hirtz*-position 406
- semi-*Fowler's* position 2, 425, 426
- semi-recumbent position 398
- semi-sitting position 215
- *Trendelenburg* position 434, 469
posterior-to anterior dissection 188, 201
potassium titanyl phosphate 316, 412
Pott's puffy tumor 353
power instrumentation 377–380
- clinical applications 380–385
prednisolone 121
prednisone 292
pregnancy 166, 289
premaxillae 22
premedication 2
preoperative preparation 2
prick test 90

prime movers 295
PRIST (paper-radioimmunosorbent test) 91
prolactin 558
prolactinoma, cystic 558
proliferative phase 329
proliferative signal 329
propionibacterium 99, 100
propofol 426
proptosis 363, 433, 447, 448, 582
– bilateral 313
prostaglandin 85, 87, 115, 117
proteic hormones 87
pseudomonas aeruginosa 129, 390
pseudotumors 489
psychopathy 395
pterygoid / pharyngeal plexus 163
pterygoid plexus 24
pterygoid process 69, 574
– vertical 18
pterygoid process, medial plate 20
pulsation, internal carotid artery 565
puncture
– canine fossa 222
– maxillary sinus 133
pupillary function 444
purpura 395
pyocele 268, 276
pyramid, nasal 17, 316
piriform aperture, stenosis 359

R

radiofluoroscopy 531, 555
radiographs, plain 56, 130, 350
radiosurgery 466
radiotherapeutics 481
radiotherapy 422, 520
– megavoltage 501
RANTES (regulated on activation, normal T-cell expressed and secreted) 85
raspberry appearance 50
RAST (radio allergosorbent test) 90
ratchet mechanism 329
recess
– alveolar 71, 72
– carotid-optic 204
– frontal 26, 28, 36, 118, 184, 187, 197, 203, 207, 208, 224, 246, 375, 389
– – dilation 243, 244
– – stenosis 303
– – surgery 213, 223–227
– frontoethmoidal 25
– infraorbital 71
– infundibular 26
– palatine 71
– prelacrimal 71
– piriform 17
– retrobulbar 206
– sphenoethmoidal 21, 25, 36, 204, 244, 378, 490, 492, 532
– suprabulbar 206
– supraorbital 58
– terminalis 206, 223
– zygomatic 71, 72
recirculation phenomena 223

rectus muscle, medial 446, 582
recurrences, diffuse 302, 305
red cell concentrate 265, 275
red phase 154
regeneration, glandular 330
Riedel operation 183
remodeling 330
– mesenchymal 333
renal disease 166
resection, submucosal 169, 372
resistance, nasal 43
respiratory distress 361
respiratory mucosa 127
– wound healing 330
respiratory reflex 358
respiratory syncytial virus 98
respiratory tract disease, upper 89
retardation, developmental 358
retarded growth 405
retractor, self-retaining 400
revision operation 214
rhabdomyosarcoma 120, 360, 368, 519
rhinitis
– acute viral 128
– allergic 48, 83, 88, 94, 97, 212
– atrophic 167, 171
– dry 167
– hyperplastic 50
– hypertrophic 170
– nonallergic eosinophilic rhinitis syndrom (NARES) 90, 288, 290
– obstructive 290
– vasomotor 167
rhinogram 48
rhinologist 306
rhinomanometry
– acoustic 43, 119
– active 44
– anterior 44, 45, 119
– binasal 44
– calibration 47
– cases 48–51
– computerized 46
– history 43, 44
– hygiene 47
– with a mask 44
– methods 46, 47
– with olives 44
– passive 44, 45
– posterior 44–46
– recording 48
– results 48
– techniques 44–46
– terminology / definitions 44
– uninasal 44
rhinoplasty 408
– aesthetic, microsurgery 10
rhinorrhea 349, 361, 501
– cerebrospinal fluid (*see also* fistula) 441, 448
– – classification 452–454, 465, 466
– – clinical history 466, 467
– – diagnosis 451, 452, 467–469
– – treatment 454–462, 469–477
– clear 296

– little   133
rhinoscleroma   418
rhinoscopy
– anterior   361
– posterior   208
rhinosinobronchial syndrome   44
rhinosinusitis   361
– bacterial infection   99, 100
– etiology   97
– fungal infection   100
– viral infection   98, 99
rhinotomy
– external   521
– lateral   147, 148, 499, 510
– paralateral   512
rhinovirus   98
rhizopus   100, 313
rod / lens system   347
root   17
Rosenmüller's valve   416
rostrum   533, 549
rubber finger packs   337

S

saddling, dorsal   173, 175
saline solution   233, 341
salivary gland tumor   55
sarcoidosis   120, 143, 418
scarring / scar tissue   145, 327, 330, 429
Schneiderian papilloma (inverted papilloma)   64, 103, 119, 120, 147, 270, 385, 483, 485, 498, 499, 533, 539, 569, 577, 579
schwannoma   55, 120, 505, 511
sclerosing, invasive   313
scout view   53
seagull incision   265
secretion, rhinosinus   97
seeker   216, 390
E-selectin / P-selectin   87, 88
sella turcica   79, 533
– anatomy   556, 557
– empty sella   466, 470, 558
– lesions, approach   543–552
– opening / closure   561, 562
– transnasal endoscopic surgery   555–566
SEM (scanning electron microscopy)   323
semi-Fowler's position   2, 425, 426
sensitivity, disturbed   397
septal deviation   50, 62, 212
septal-cartilage free graft   460
septal-displacement technique   551
septoplasty   135, 172, 366, 367, 393, 457
– basic   174, 175
– localized   214, 216
– complications   175
– limited   175
septorhinoplasty   51, 423
septum
– cartilaginous, anterior   3
– interfrontal   10
– intersinus septum   268
– nasal (see there)
sequela, iatrogenic   294
sex, nasal polyposis   289

sharp cutting instruments   337
shaver   238, 259, 337
– soft tissue shaver   184, 378–380
silastic tube   412
silicon tube   268
silicone catheter   454
sinobronchial-syndrome   317
sinonasal surgery   94
sinoscopy   4
sinus mucosa, wound healing   323–327
sinus ostium   127
sinus surgery
– children   357–368
– complications   581–593
– endonasal   122, 123, 258
– external frontal   258
– FESS (functional endoscopic sinus surgery)   184, 185, 347, 377, 529
– high-risk area   195–199
– navigation, image-guided   188, 387–391
– pediatric   347–354
– power instrumentation   377–385
– radical   312
– revision   371–375
sinus system, developing   348
sinus
– anatomy   179
– brain   25
– cavernous   54, 205, 209, 312, 517, 523, 578, 588
– dental   25
– ethmoid (see there)
– ethmoidofrontal   25
– ethmoidomaxillary   25
– ethmoidosphenoidal   25
– frontal (see there)
– maxillary (see there)
– occipital venous   30
– paranasal   24, 25
– – anatomic development   357
– – micro-endoscopic surgery   201–234
– skull   25
– sphenoid (see there)
sinuscopy   39, 131
sinusitis   138, 438
– acute, complications   352, 353
– bacteriology   131
– chronic inflammatory, staging / grading / post-operative follow-up   141–145
– chronic paranasal, biomechanical effect of the ethmoidal isthmi   135–139
– chronic polypoid   271
– chronic recurrent / persistent
– – pathophysiology   349
– – signs and symptoms   348, 349
– conchal   171
– diagnosis   130, 131
– etiology   128, 129
– examination, otorhinolaryngologic   129, 130
– fungal   39, 100, 185, 213
– – allergic   103, 310–312
– – classification   309
– – invasive   312, 313
– – non-invasive   309, 310
– history   129

– hyperplastic 222
– maxillary 251
– medical management 132, 133
– postoperative frontal 437
– purulent 264
– rhinogenic 239
– rhinosinusitis (*see there*)
– secondary 461
– surgery 123
– – overview of technique 179–192
sinusotomy
– frontal 136
– inferior transnasal frontal 279
siphon 470
skiagram, sinus 258
skin cancer 418
skin test 90, 288
skull base 180, 261, 388, 569
– lesions
– – endoscopic approach 529–540
– – power instrumentation 385
– tumors 441
slit-lamp examination 434
small-hole surgery 218
– power instrumentation 382
smell, sense of 144, 310
– disturbance 590
snapshot 331
sneezing 144
Snellen chart 444
snoring 119
soft tissue shaver 184, 378–380
spasm, microvascular 442
speculum, self-retaining 215
speech resonation 97, 164
sphenoethmoidectomy 367, 460, 536
– endoscopic 434, 437, 446
– radical 123, 186
– total 445
sphenoid bone, inferior face 18
sphenoid lesions 539
sphenoid sinus 137, 454, 483
– anatomy 30, 204, 205
– conventional operation 184
– extensions 69–71
– fistula 458–461
– opening 244
– surgery 4, 214, 215, 230, 363
sphenoidectomy, intranasal 506
sphenoiditis 539
sphenoidotomy 213, 374, 384, 560
– intranasal 538
– pediatric 382
– trans-septal 538
– wide 474
sphenomaxillary plate 65
spina nasalis anterior 243
spine, nasal 17, 18, 282
splint, nasal 175, 562
splitscreen video technique 332
squamous cell carcinoma 120, 270, 368, 569
STA (sub-labial, trans-septal, sphenoidal) 545–552
– advantages 546
– disadvantages 546

Stammberger-drill 419
Stamm-retractor, self-retaining 400
Stankiewicz maneuver 316
stapes surgery 1
staphylococci 99, 131, 475
– MRSA (methicillin-resistant staphylococcus aureus) 390
stellate ganglion block 10
stenosis 377
– choanal atresia 407
– dacryostenosis 417, 423, 428, 431
– frontal recess 303–305
– ostial 188, 291
– piriform aperture 359
– tracheal 341
– vestibular 578
Stensen's duct 98
stent 227, 375, 411
– intranasal 414
– silicon 341
stenting, mid-meatal 315–320
stereoscopic view, three dimensional 12, 318
stereotactic surgical navigation 389
steroids 287, 317, 447
– corticosteroids 92, 165, 166, 318
– glucocorticoid steroids 121
– systemic 92, 93, 121, 232, 433, 591
– topical 93, 121, 312
Stevens tenotomy scissors 168
stones, lacrimal system 427
stool softener 475
streptococci, α-hemolytic 99
stress 289, 290
stroma removal, intra-turbinal 168
subarachnoid 462
sub-labial approach 147, 359, 531, 555
subluxation, corneal 444
submucous cyst 252
subperiostal resection 149
substance P 87, 115, 290
suction forceps 377
supraorbital cell 26
supraorbital ridge 267, 299
surgical planning 185
surgicel 460
swallowing problems 395
swelling
– edematous 333, 335
– facial 182
– periorbital 447
swimming 127
swinging door technique 172
sympathetic fibers 163
sympathetic plexus, pericarotid 24
syndromes / diseases (names only)
– *Albright's* syndrome 508
– *Churg-Strauss*-syndrome 92, 103, 111
– *Hodgkin's* disease 395
– *Kartagener's* syndrome 103, 111, 189, 264, 290, 349, 353
– *Young's* syndrome 103, 111, 142
– *Widal's* disease 297
– *Woake's* syndrome 264
synechia 317, 320, 333, 336, 461, 588
– middle meatus 305
– postoperative 5, 188, 223, 232, 291, 298, 302, 413

– scarred 337
– septo-turbinal 305
syphilis 418
systemic disease, management 181, 212

T

Taillefer's valve 416
Takahashi forceps 169
tamponade 262, 333, 340
teaching software 159
tear pump 429
teeth, deciduous 492
telangiectasia 167, 395
telescope, fiberoptic / rigid 179
template 267
temporalis fascia 7, 431
temporomandibular-joint surgery 260
tension
– nasal 51
– ocular 444
terfenadine 91
testosterone 520
– dihydrotestosterone 515
tetracycline-labeling studies 382
Thornwaldt cyst 485, 519
thrombosis 54, 442
– large veins 493
thrombosis, cavernous sinus 352
thromboxane 85
through-cutting 381
– forceps 324
thyreoiditis, *Hashimoto* 534
tics, facial 129
T-lymphocytes 87, 98
tobacco 147, 349
tonometry 434
Toti technique 428, 429
transantral approach 183, 347
transductor, rhinomanometry 47
transethmoid approach 231, 458
β2-transferrin 466, 467
transfixion
– complete 571
– suture 415
transforming growth factor (TGF) 115, 288
transition space 382
translucence 510
transmandibular approach 521
transnasal direct approach 214, 230, 231, 359, 458, 460, 474, 539
transnasal microsurgery 201
transpalatal approach 186
transpharyngeal suprahyoid approach 521
trans-septal approach 385, 458, 460, 531, 539, 544
trans-septosphenoidal approach 366, 550, 551
trans-sphenoidal approach 531, 539, 544
transzygomatic approach 521
trauma
– barotrauma 97, 128, 264
– craniofacial 445, 458, 469, 470
– CSF fistula (*see there*)
– epistaxis 395
– maxillofacial trauma 128

– nasolacrimal system 418
Trendelenburg position 434, 469
trephination 182, 191, 224, 226, 253, 257, 279, 282, 311
– minimal technique 374
triamcinolone 166
trigeminal nerve 27, 481
– branches 33
trismus 397
true-cutting 382
tubercel, optic 435
tubercle, optical 446
tuberculosis 166, 313, 429
tuberosities, maxillary 571
tuberous sclerosis 499
tumefaction
– frontal 494
– malar region 495
tumor necrosis factor α (TNFα) 84, 85, 115, 311
tumor-capsule defect 533
tumors 4, 270
– epithelial tumor 120
– *Pott's* puffy tumor 353
– salivary gland tumor 55
– sino-nasal, malignant 368
– nasal / paranasal, endonasal micro-endoscopic surgery 481–487
– sino-nasal, benign, micro-endoscopic surgery 489–512
tungsten powder 521
turbinate 523
– anatomy 161–164
– bullous 416
– diseases, treatment 164–172
– displacement 168
– hypertrophy 49, 76, 316
– inferior 19, 161, 220, 417
– lower 3
– micro-endoscopic surgery 161–177
– middle 300, 327, 417
– – basal lamella 204
– – lateralization 227
– – meatal surface 207
– – paradoxical 61, 144, 165
– – preservation 5
– – resection 318
– – treatment 241
– mulberry-like posterior 171
– nasal, pneumatization 59
– physiology 164
– polypoid 165
– superior 21
– supreme 22
– surgery 366, 367
turbinectomy
– anterior 170
– inferior 292
– partial inferior 169, 170, 382
– posterior 170
– total inferior 170, 171, 187
turbinoplasty 94, 491
– inferior 168, 169
two-one-one-half rule 400

U

U sign 50
ultrasound 130, 504
uncinate process 4, 28, 36, 62, 191, 202, 214, 217, 240, 257, 372, 382, 419
– anatomy 206
– deformed 128
– everted 144
– lateralization 64
uncinectomy 186, 214, 218, 219, 362, 491
underlay graft 471, 473, 483
unguis 416
URI (viral upper respiratory infection) 127
urticaria 83

V

Valsalva maneuver 452
valve, nasal 43
vasoconstrictor 47
vein, facial 163
venous congestion, orbital 88
Venturi effect 291
vertex 265
vessel
– ligation 393, 397, 400
– obstruction 112
vestibules, nasal 17
video
– animation 155
– documentation 420
video monitor, high-resolution 559
video-endoscopic technique 187
– continous 331
viral upper respiratory infection (URI) 127
visual acuity, loss of 365

visual loss 434, 444
voice, resonance 97, 164
vomer 3, 18, 22, 62, 548, 551
vomiting 454

W

Wagner forceps 255
Wagner terminals 254
washout sign 467
Waters view 208
Wegener granulomatosis 120
Widal's disease 297
Wigand technique 188, 201
Woake's syndrome 264
wound healing 138
– cutaneous 329, 330
– nasal mucosa 330, 332
– physiology / pathophysiology 337
– respiratory mucosa 330
– sinus mucosa, preservation 323–327

X

xylometazolin 238

Y

Yarsargil's ligature guide 522
yellow phase 154
Young's syndrome 103, 111, 142

Z

Zinn, annulus 435, 444, 446, 447, 537